Analysis and Design of Transimpedance Amplifiers for Optical Receivers

Analysis and Design of Transimpedance Amplifiers for Optical Receivers

Eduard Säckinger

WILEY

This edition first published 2018

Registered Offices
John Wiley & Sons, Inc., 111 River Street, Hoboken, NJ 07030, USA

Editorial Office
111 River Street, Hoboken, NJ 07030, USA

For details of our global editorial offices, customer services, and more information about Wiley products visit us at www.wiley.com.

Wiley also publishes its books in a variety of electronic formats and by print-on-demand. Some content that appears in standard print versions of this book may not be available in other formats.

Library of Congress Cataloging-in-Publication Data Applied For:

ISBN: 9781119263753

Cover design by Wiley
Cover image: © shuoshu/Gettyimages

Set in 10/12pt WarnockPro by SPi Global, Chennai, India

10 9 8 7 6 5 4 3 2 1

To Elizabeth and Marianne

Contents

Preface *xi*

1 **Introduction** *1*
1.1 Optical Transceivers *1*
1.2 Modulation Formats *5*
1.3 Transmission Modes *12*
References *19*

2 **Optical Fibers** *23*
2.1 Loss and Bandwidth *23*
2.2 Dispersion *26*
2.3 Nonlinearities *32*
2.4 Pulse Spreading due to Chromatic Dispersion *35*
2.5 Summary *38*
Problems *39*
References *40*

3 **Photodetectors** *43*
3.1 p–i–n Photodetector *43*
3.2 Avalanche Photodetector *58*
3.3 p–i–n Detector with Optical Preamplifier *65*
3.4 Integrated Photodetectors *77*
3.5 Detectors for Phase-Modulated Optical Signals *86*
3.6 Summary *94*
Problems *96*
References *97*

4 **Receiver Fundamentals** *107*
4.1 Receiver Model *107*
4.2 Noise and Bit-Error Rate *109*
4.3 Signal-to-Noise Ratio *117*

4.4 Sensitivity *120*
4.5 Noise Bandwidths and Personick Integrals *135*
4.6 Optical Signal-to-Noise Ratio *139*
4.7 Power Penalty *148*
4.8 Intersymbol Interference and Bandwidth *153*
4.9 Frequency Response *165*
4.10 Summary *170*
Problems *172*
References *174*

5 Transimpedance Amplifier Specifications *181*
5.1 Transimpedance *181*
5.2 Input Overload Current *186*
5.3 Maximum Input Current for Linear Operation *187*
5.4 Bandwidth *188*
5.5 Phase Linearity and Group-Delay Variation *189*
5.6 Timing Jitter *190*
5.7 Input-Referred Noise Current *191*
5.8 Crosstalk *197*
5.9 Product Examples *199*
5.10 Summary *201*
Problems *202*
References *202*

6 Basic Transimpedance Amplifier Design *207*
6.1 Low- and High-Impedance Front-Ends *207*
6.2 Shunt-Feedback TIA *210*
6.3 Noise Analysis *230*
6.4 Noise Optimization *242*
6.5 Noise Matching *254*
6.6 Summary *267*
Problems *269*
References *272*

7 Advanced Transimpedance Amplifier Design I *279*
7.1 TIA with Post Amplifier *279*
7.2 TIA with Differential Inputs and Outputs *284*
7.3 TIA with DC Input Current Control *289*
7.4 TIA with Adaptive Transimpedance *292*
7.5 Common-Base and Common-Gate TIAs *301*
7.6 Regulated-Cascode TIA *312*
7.7 TIA with Inductive Broadbanding *320*
7.8 Distributed-Amplifier TIA *325*

7.9 Summary *331*
Problems *333*
References *334*

8 Advanced Transimpedance Amplifier Design II *343*
8.1 TIA with Nonresistive Feedback *343*
8.2 Current-Mode TIA *349*
8.3 TIA with Bootstrapped Photodetector *351*
8.4 Burst-Mode TIA *352*
8.5 Analog Receiver TIA *360*
8.6 Summary *364*
Problems *365*
References *365*

9 Transimpedance Amplifier Circuit Examples *373*
9.1 BJT, HBT, and BiCMOS Circuits *373*
9.2 CMOS Circuits *380*
9.3 MESFET and HFET Circuits *387*
9.4 Summary *389*
References *392*

A Communication Signals *397*
A.1 Non-Return-to-Zero Signal *397*
A.2 Return-to-Zero Signal *401*
A.3 Pulse Amplitude-Modulated Signal *403*
A.4 Analog Television Signal *406*
A.5 Digital Television Signal *408*
References *411*

B Eye Diagrams *413*
References *420*

C Timing Jitter *421*
C.1 Data Jitter *421*
C.2 Clock Jitter *431*
C.3 Jitter, Phase Noise, and Bit-Error Rate *435*
Problems *438*
References *438*

D Nonlinearity *441*
D.1 Gain Compression *442*
D.2 Harmonic Distortions *443*
D.3 Intermodulation Distortions *444*

D.4 Composite Distortions *446*
Problems *449*
References *449*

E Adaptive Equalizers *451*
E.1 Feedforward and Decision-Feedback Equalizers *452*
E.2 Adaptation Algorithms *456*
E.3 Hardware Implementations *460*
Problems *463*
References *463*

F Decision-Point Control *469*
Problems *473*
References *473*

G Forward Error Correction *475*
Problems *481*
References *481*

H Second-Order Low-Pass Transfer Functions *483*
References *495*

I Answers to the Problems *497*
References *533*

J Notation *537*

K Symbols *539*

L Acronyms *549*

Index *557*

Preface

Transimpedance amplifiers (TIA) are used at the front end of optical receivers. They can also be found at the front end of read circuits for optical storage systems and laser RADAR systems for distance measurement. But TIAs are not limited only to optical applications; particle/radiation detector chips, vision sensor chips, biological sensor chips, motion sensors in microelectromechanical systems, and wideband radio receivers also make use of TIAs.

This broad range of applications is not surprising. The TIA is essentially a sensitive and fast current measurement device: A weak current signal, typically originating from a sensor such as a photodetector, a particle/radiation detector, a biological sensor electrode, a MEMS electrostatic transducer, or a radio receiver antenna, is amplified and converted into a voltage signal. The term *transimpedance* derives from the older term *transfer impedance*, which indicates that an input current at one port is producing an output voltage at another port.

The term transimpedance amplifier may evoke the image of a voltage amplifier with a shunt-feedback resistor. However, this is just one particular implementation. Several other topologies exist and novel TIA circuits are still being invented today. Each circuit presents a different trade-off between sensitivity (noise), speed (bandwidth), power, area, and other performance measures. With each application having its own set of requirements, different applications benefit from different circuit designs.

Book Outline. Chapters 1–4 provide background information on optical communication. This part of the book establishes useful context for the later chapters on TIA design. Readers who are not interested in optical applications may skip over much of this material.

Chapter 1 describes the components that make up conventional and digital coherent optical receivers and transmitters. Common modulation formats (NRZ, RZ, 4-PAM, QPSK, SCM, etc.), modulation codes, transmission modes (continuous mode and burst mode), and standards are introduced.

Chapter 2 is about the communication channel presented by the optical fiber. Its loss, bandwidth, various forms of dispersion, and nonlinearities are described. The compensation of loss and dispersion and the mitigation of nonlinear effects are discussed briefly.

Chapter 3 covers the relevant photodetectors. The responsivity, bandwidth, and noise properties of the p–i–n photodetector, the avalanche photodetector (APD), and the optically preamplified p–i–n detector are examined. Then, integrated detectors including detectors for silicon photonics are covered. Finally, detectors for phase-modulated signals (QPSK, DQPSK, etc.) including the coherent detector with phase and polarization diversity are discussed.

Chapter 4 deals with the receiver at the system level. An analysis of how noise in the receiver causes bit errors leads to the definition of the receiver sensitivity in unamplified transmission systems and the required optical signal-to-noise ratio (required OSNR) in amplified transmission systems. Power penalties due to receiver impairments, such as intersymbol interference (ISI), are discussed. An analysis of the trade-off between noise and ISI leads to recommendations for the receiver's bandwidth and frequency response.

The remainder of the book focuses on the analysis and design of TIAs.

Chapter 5 introduces the main specifications, such as the transimpedance, bandwidth, phase linearity, group-delay variation, jitter, input-referred noise current, maximum input current, and crosstalk. The measurement of some key parameters is discussed. Example values from recent product data sheets are given to illustrate the specifications.

Chapter 6 covers the popular shunt-feedback TIA in detail. The transimpedance, input impedance, and output impedance are calculated. The stability and the transimpedance limit of single and multistage implementations are analyzed. The noise performance of TIAs with FET and BJT front ends are derived. Ogawa's noise factor and its relationship to induced gate noise is explained. Then, the noise optimization of TIAs with FET and BJT front ends by means of device sizing and biasing is discussed. The impact of constraints, such as a constant gain-bandwidth product, on the noise optimum is examined. Finally, noise matching networks and their properties are investigated.

Chapter 7 extends the basic shunt-feedback TIA with practical features such as a postamplifier, differential inputs and outputs, DC input current control, and adaptive transimpedance. Then, the chapter turns to alternative TIA topologies such as the common-base TIA, common-gate TIA, the regulated-cascode TIA, and the distributed-amplifier TIA.

Chapter 8 examines additional TIA circuit techniques such as capacitive feedback, optical feedback, active feedback, current mode, and photodetector bootstrapping. Then, the chapter turns to TIAs for specialized applications, namely burst-mode TIAs (e.g., for passive optical networks) and analog-receiver TIAs (e.g., for hybrid fiber-coax networks or microwave photonic links).

Chapter 9 discusses published circuit examples in a variety of technologies (BJT, HBT, BiCMOS, CMOS, MESFET, and HFET) illustrating and solidifying the concepts covered in the earlier chapters. The chapter concludes with a list of recent TIA publications.

A number of appendices cover subjects related to, but not limited to, the design of optical receivers and TIAs.

Appendix A reviews the power spectral density, bandwidth, and signal-to-noise requirements of some common communication signals (NRZ, RZ, 4-PAM, CATV).

Appendix B discusses eye diagrams, eye openings, and eye margins, including their measurement and simulation.

Appendix C deals with data and clock jitter. The terminology and the measurement of jitter is discussed and the relationship between jitter, phase noise, and bit-error rate is explained.

Appendix D reviews nonlinearity and the resulting signal distortions, which are important in systems that perform linear signal processing (equalization, data conversion, etc.) and in applications that use higher-order or multicarrier modulation.

Appendix E provides an introduction to adaptive equalization. The basics of the feedforward equalizers (FFEs) and decision-feedback equalizers (DFEs) are covered.

Appendix F briefly discusses adaptive control of the decision threshold and sampling instant.

Appendix G provides an introduction to forward error correction (FEC).

Appendix H discusses second-order low-pass transfer functions, which are important for the analysis of TIAs. The frequency response, bandwidth, noise bandwidth, phase linearity, group-delay variation, overshoot, and jitter are covered.

Appendix I provides answers to all the end-of-chapter problems. This appendix also serves as a repository for additional material, such as derivations and generalizations, that would be too distracting to present in the main text.

Audience. It is assumed that the reader is familiar with basic analog IC design as presented, for example, in *Analysis and Design of Analog Integrated Circuits* by Gray et al. [1] or a similar book [2–4].

The book is written from the perspective of an electrical engineer. For example, whenever possible we use voltages and currents rather than abstract variables, we use one-sided power spectral densities as they would appear on a spectrum analyzer, we prefer the use of noise bandwidths over Personick integrals, and so forth. Examples are given frequently to make the material more concrete. Many problems, together with their answers, are provided for readers who want to practice and deepen their understanding of the learned material.

I hope this book will be useful to upper-level undergraduates and graduate-level students in integrated circuit design and optical communication. Professionals in the IC and optical industry may find this book to be a valuable reference as well.

This book grew out of an effort to make a second edition of my earlier book *Broadband Circuits for Optical Fiber Communication* [5]. As I was reworking chapter by chapter, covering new developments, treating subjects in more depth, and so forth, the length of each chapter doubled or tripled. For this reason, it became impractical to cover all the subjects of the original book in a single book. The present book covers the material from Chapters 1 to 5 of the original book.

Acknowledgments. I would like to thank all my colleagues at the Bell Laboratories (first of AT&T and then of Lucent Technologies), Agere Systems, Conexant, and Ikanos Communications from whom I have learned so much.

I am deeply indebted to the reviewers who have given freely of their time to read through the book, in part or in full. In particular, I am most grateful to Dr. Ricardo Aroca, Acacia Communications; Mr. Henry M. Daghighian, Finisar Corporation; Dr. Christopher Doerr, Acacia Communications; Dr. Yuriy M. Greshishchev, Ciena Corporation; Prof. Dan Li, Jiaotong University, Xi'an; Dr. Sunderarajan Mohan, Synopsys Inc.; Prof. Sung-Min Park, Ewha Women's University, Seoul; and Prof. Sorin Voinigescu, University of Toronto.

Despite the efforts made, there are undoubtedly some mistakes left in this book. If you have any corrections or suggestions, please e-mail them to edi@ieee.org. Thank you!

Rumson, NJ *E. Säckinger*
August 2016

References

1 P. R. Gray, P. J. Hurst, S. H. Lewis, and R. G. Meyer. *Analysis and Design of Analog Integrated Circuits*. John Wiley & Sons, Inc., New York, 5th edition, 2009.
2 P. E. Allen and D. R. Holberg. *CMOS Analog Circuit Design*. Oxford University Press, 2nd edition, 2002.
3 D. Johns and K. Martin. *Analog Integrated Circuit Design*. John Wiley & Sons, Inc., New York, 1997.
4 B. Razavi. *Design of Analog CMOS Integrated Circuits*. McGraw-Hill Education, 2000.
5 E. Säckinger. *Broadband Circuits for Optical Fiber Communication*. John Wiley & Sons, Inc., Hoboken, NJ, 2005.

1

Introduction

After a brief description of the components that make up an optical receiver and transmitter, we discuss how digital and analog information is modulated on a lightwave. We explain the difference between continuous-mode and burst-mode transmission and summarize applications and standards for both transmission modes.

1.1 Optical Transceivers

Figure 1.1 shows the block diagram of a conventional optical receiver and transmitter. On the transmitter side, a coder and/or scrambler preprocesses the parallel input data. Optionally, the coder adds redundancy to permit error detection and correction at the receiver end. These coding steps condition the data for the subsequent serial transmission through a band-limited and noisy channel. Next, a *multiplexer* (MUX) serializes the n-bit wide parallel data into a single high-speed bit stream. A *clock multiplication unit* (CMU) synthesizes the necessary bit-rate (or half bit-rate) clock from the n times slower word clock (or another convenient reference clock). After that, a *transmit equalizer* (TXEQ) may be used to shape (predistort) the serial high-speed signal in preparation of the band-limited channel. Finally, a *laser driver* or *modulator driver* drives the corresponding optoelectronic device. The laser driver modulates the current of a *laser diode* (LD), whereas the modulator driver modulates the voltage of a *modulator*, which in turn modulates the light from a *continuous wave* (CW) laser. Some laser/modulator drivers also retime the data to reduce jitter and thus require a clock signal from the CMU (dashed line in Fig. 1.1).

On the receiver side, the same process happens in reverse order. A *photodetector* (PD) receives the optical signal from the fiber and produces a small current in response to the optical signal. A *transimpedance amplifier* (TIA or TZA) amplifies and converts this current into a voltage. A *limiting amplifier* (LA) or an *automatic gain control amplifier* (AGC amplifier) further amplifies

Analysis and Design of Transimpedance Amplifiers for Optical Receivers, First Edition. Eduard Säckinger.

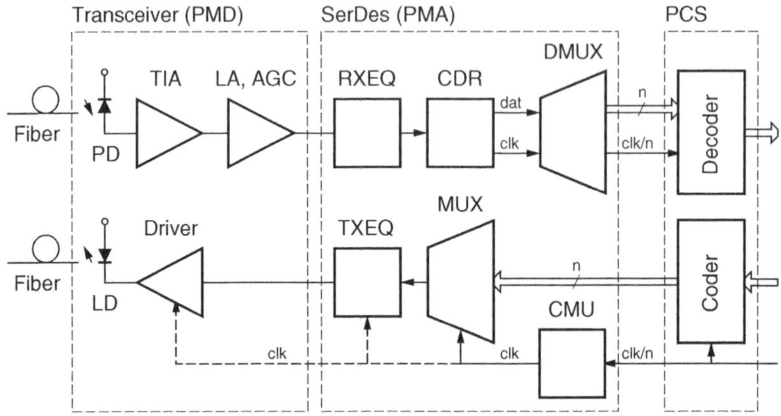

Figure 1.1 Block diagram of a conventional optical receiver (top) and transmitter (bottom).

this voltage signal. The LA and AGC amplifier are collectively known as *main amplifiers* (MAs) or *post amplifiers*. Next, a *receive equalizer* (RXEQ) may be used to undo some of the distortions accrued along the way. After that, a *clock and data recovery circuit* (CDR) extracts the clock signal and retimes the data signal. Finally, a *demultiplexer* (DMUX) converts the fast serial bit stream into *n* parallel lower-speed data streams that are processed by the subsequent decoder and/or descrambler. Optionally, the decoder performs error checks and error corrections. Subsequent digital blocks extract the payload data from the framing information, synchronize the received data to another clock domain, and so forth.

In practice, the blocks may not be as neatly delineated as shown in Fig. 1.1. For example, the MUX in the transmitter may be merged with the driver into one block [1]. On the other hand, if the driver is located far from the MUX (e.g., in a separate package), a CDR may be interposed to clean up the data (reduce the jitter) without the need for a clock signal from the CMU. Furthermore, the RXEQ and the CDR may be merged into a single block, especially when a decision feedback equalizer, which needs feedback from the data output of the CDR, is used. Finally, in parallel sampling architectures, the CDR performs some or all of the DMUX operation [2].

This book covers the optical fiber (Chapter 2), the photodetector (Chapter 3), and the TIA (Chapters 5–9) and provides an introduction to equalization (Appendix E) and forward error correction (Appendix G).

Modules and Subassemblies. A module containing a PD, TIA, MA, laser driver, and LD, that is, all the blocks shown inside the dashed box on the left of Fig. 1.1, is referred to as a *transceiver*. (The term *transceiver* is a contraction of the words "*trans*mitter" and "re*ceiver*.") Figure 1.2 shows a photograph of so-called XFP transceiver modules. The transceiver is often built around a *receiver optical*

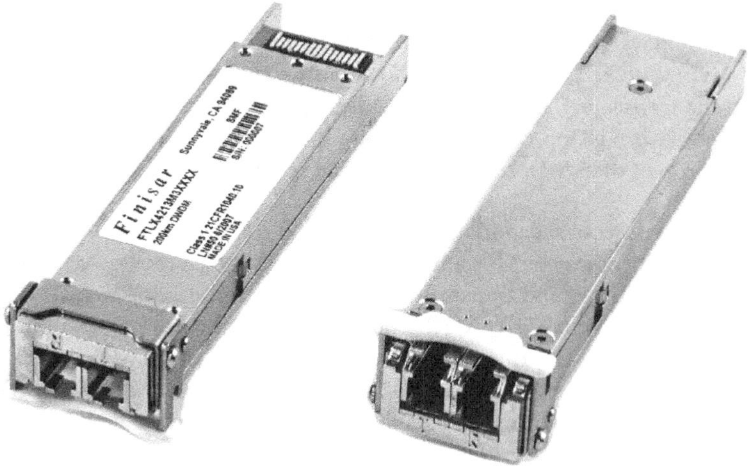

Figure 1.2 Two 10-Gb/s transceivers in small form factor packages following the XFP specification (7.8 cm × 2.2 cm × 1.3 cm). Two fibers are plugged in from the front of the package (LC connectors). *Source*: Reprinted by permission from Finisar Corporation.

Figure 1.3 A 10-Gb/s receiver optical subassembly (ROSA) with a GaAs p–i–n photodetector and a TIA (Ø 0.6 cm × 1.2 cm). *Source*: Reprinted by permission from Finisar Corporation.

subassembly (ROSA) and a *transmitter optical subassembly* (TOSA). The ROSA is a small package that contains the PD, in most cases the TIA, and optical components, such as a lens and means for optical fiber alignment (see Fig. 1.3). The TOSA is a small package that contains the LD, in some cases the driver, and optical components, such as an optical isolator, a lens, and means for optical fiber alignment. The blocks in the dashed box in the middle, namely the RXEQ, CDR, DMUX, CMU, MUX, and TXEQ form the *serializer/deserializer* or SerDes for short. A module that contains the functionality of the transceiver and the SerDes is frequently called a *transponder*.

OSI Layers. The functionality shown in Fig. 1.1 can be identified with the bottom *layer* of the OSI communication system model. This layer is known as the *physical layer* (PHY) and has three sublayers: the *physical medium dependent* (PMD) sublayer, the *physical medium attachment* (PMA) sublayer, and the *physical coding sublayer* (PCS). The PMD sublayer, at the very bottom, corresponds to the transceiver. In our case, the physical medium is the optical fiber. The PMA sublayer, on top of the PMD, corresponds to the SerDes. The PCS sublayer, on top of the PMD, corresponds to the coder and decoder.

DSP-Based Coherent Receiver and Transmitter. Around 2007, *digital signal processors* (DSP) and data converters became sufficiently fast to enable a new architecture for optical receivers and transmitters [3] (see Fig. 1.4). In this architecture only the front-end blocks, TIA, AGC amplifier, and modulator driver, remain in the analog domain. The functionality of the RXEQ, CDR, DMUX, MUX, and TXEQ blocks are implemented with a DSP. This approach permits the use of sophisticated algorithms for equalization, clock recovery, and so on but requires high-speed *analog-to-digital converters* (ADC) and *digital-to-analog converters* (DAC) at the interface between the analog transceiver section and the DSP (cf. Appendix E).

This DSP-based architecture is the preferred approach for 100- Gb/s transceivers with phase- and polarization-diverse coherent detection [4, 5]. In these transceivers, an optical-to-electrical (O/E) converter outputs four electrical signals corresponding to the in-phase and quadrature components and the x-polarized and y-polarized components of the incoming optical signal. (We discuss this type of detector in Section 3.5.) Similarly, an electrical-to-optical (E/O) converter, driven by four electrical signals, controls the phase and amplitude of the transmitted optical signal for both polarizations. As a result, four TIAs, four AGC amplifiers, and four ADCs are needed on the receiver side, and four DACs and four modulator drivers are needed on the transmitter side (see Fig. 1.4). The coherent approach permits the use of advanced modulation formats such as DP-QPSK and enables the effective compensation of large amounts of fiber dispersion in the electrical domain [5, 6].

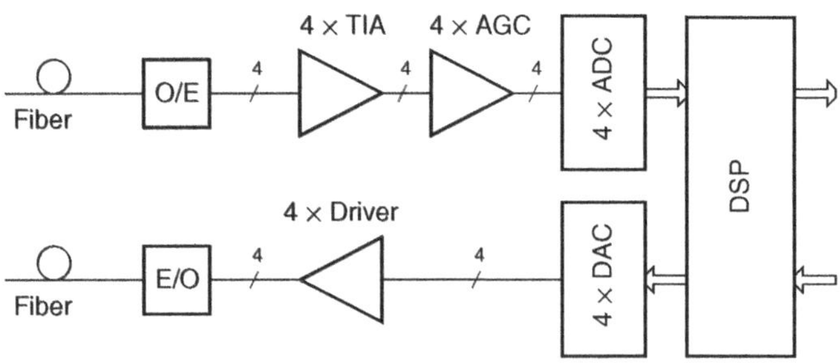

Figure 1.4 Block diagram of a DSP-based optical transceiver with phase- and polarization-diverse coherent detection.

1.2 Modulation Formats

Basic Modulation Formats. The most commonly used modulation format in optical communication is the *non-return-to-zero* (NRZ) format shown in Fig. 1.5(a). Despite the forbidding name, this modulation format is as simple as it gets. The laser light is turned on to transmit a one bit, and it is turned off to transmit a zero bit. When the light is on, it stays on for the entire bit period. The latter feature explains why this format is called *non-return-to-zero.* When transmitting the periodic bit pattern "010101010..." in the NRZ format at 10 Gb/s, a 5-GHz square wave with 50% duty cycle is produced. The NRZ format is used, for example, in SONET/SDH telecommunication systems as well as in Ethernet data communication systems (see Tables 1.1 and 1.2). Some standards, such as Fast Ethernet and FDDI, call for the *non-return-to-zero change-on-ones* (NRZI or NRZ1) format. The waveform for this format is the same as for NRZ, but the bit stream is preprocessed by a differential encoder that changes its (binary) output value when the input bit is a one and leaves the output value unchanged when the input bit is a zero.

The *return-to-zero* (RZ) format, shown in Fig. 1.5(b), shortens the pulses, which represent the one bits, to only a fraction of the bit period. The figure shows 50%-RZ pulses, but other fractions, such as 33% or 67%, are also used. In many situations, the RZ signal can be detected at a lower signal-to-noise ratio than the NRZ signal [7]. This can be understood intuitively by recognizing that, for the same average signal power, the narrower RZ pulses exhibit a larger signal swing, which can better overcome the noise. The RZ format also can tolerate more pulse distortion and spreading without disturbing the adjacent bits. On the downside, faster, more expensive transceiver components (laser/modulator, photodetector, front-end electronics, etc.) are required to handle the shorter pulses. Furthermore, because of its shorter pulses, the RZ signal occupies a wider bandwidth than the NRZ signal (for a given bit rate), making it less tolerant to chromatic fiber dispersion (cf. Section 2.2). The RZ format has been used extensively in undersea (submarine) lightwave systems [8]. In contrast to terrestrial systems, these custom-built lightwave systems tend to be very

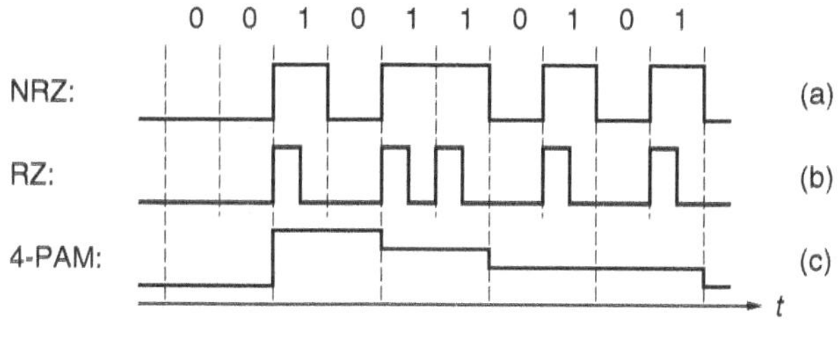

Figure 1.5 Modulation formats: (a) NRZ, (b) RZ, and (c) 4-PAM.

long (e.g., connecting two continents) and benefit from the robustness of the RZ format.

The four-level *pulse amplitude modulation* (4-PAM) format, shown in Fig. 1.5(c), extends the on/off concept and controls the brightness of the light source in four discrete steps: off, one-third on, two-thirds on, and fully on. By using a four-level signal, two bits can be transmitted in every signaling period: bit pair "00" selects the first (lowest) level, "01" selects the second level, "11" selects the third level, and "10" selects the fourth (highest) level. The coding is chosen such that an accidental confusion of two adjacent levels results in only a single-bit error (Gray code). Because of its capability to encode more than one bit per symbol, 4-PAM is known as a *higher-order modulation format*. Compared with the NRZ signal, the 4-PAM signal occupies only half the bandwidth for a given bit rate, making it a *bandwidth-efficient modulation* format. The reduced bandwidth helps to mitigate the effects of chromatic fiber dispersion, permitting an increased reach, and relaxes the speed requirements for the laser and the photodetector. However, the 4-PAM format requires a higher signal-to-noise ratio than the NRZ format for reliable detection. In other words, because the receiver has to discriminate between four levels, it is more affected by noise, leading to a substantially lower sensitivity. The 4-PAM format also requires a more complex multilevel transmitter and receiver. The upcoming 400-Gigabit Ethernet standard is expected to use the 4-PAM format (see Table 1.2). Other higher-order and bandwidth-efficient modulation formats have been studied as well [9–11].

An analytical comparison of the NRZ, 50%-RZ, and 4-PAM signals with respect to the required bandwidths and signal-to-noise ratios can be found in Appendix A.

Advanced Modulation Formats. Let us examine the modulation process in more detail. Modulation is the mapping of data bits (zeros and ones) into signal waveforms [12]. As illustrated in Fig. 1.6, modulation in an optical transmitter occurs in two steps. First, the bits are mapped to an electrical current or voltage waveform. Second, the electrical waveform is mapped to an optical field (electromagnetic field) waveform. The optical field oscillates at around 200 THz, which means that there are about 20,000 oscillations in a bit period of a 10-Gb/s signal, but for clarity, only two oscillations are shown in the figure. The NRZ, RZ, and 4-PAM signals, shown in Fig. 1.5, are outputs of the

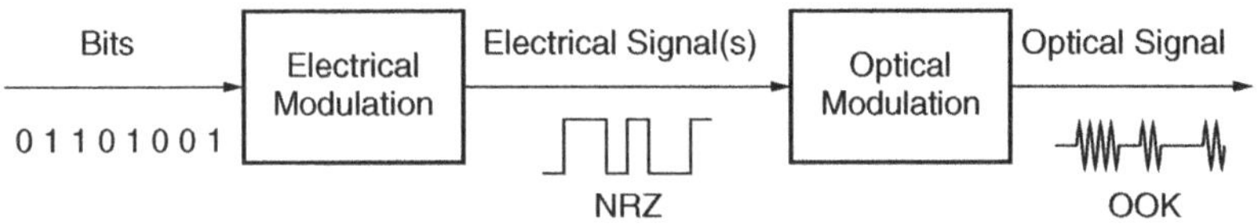

Figure 1.6 Modulation in an optical transmitter.

electrical modulation step. They are so-called *baseband signals* because their spectra extend down to DC. Modulating the intensity of the optical signal with an NRZ or RZ signal is known as *on-off keying* (OOK) and what we called NRZ and RZ formats earlier, more precisely should have been called NRZ-OOK and RZ-OOK formats. Similarly, modulating the intensity of the optical signal with a 4-PAM signal is known as four-level *amplitude-shift keying* (4-ASK). (Note that intensity modulation implies an amplitude modulation of the optical field.) The optical field signals are so-called *passband signals* because their spectra are concentrated around the optical carrier frequency. In practice, the electrical and optical modulation steps often are not clearly distinguished and by saying that an optical system uses NRZ modulation, NRZ-OOK is implied.

Now that we understand optical modulation as a two step process, it becomes clear that the second step does not necessarily have to be intensity or amplitude modulation alone but could include phase and frequency modulation. This possibility is exploited in advanced optical modulation formats. Some examples are illustrated in Fig. 1.7.

- The *optical duobinary* format combines amplitude and phase modulation [13]. The amplitude of the optical carrier is modulated with an NRZ signal, as in the OOK case, but additionally, the optical phase is shifted by 180° for one bits that are separated by an odd number of zero bits (see Fig. 1.7(a)).
- The *chirped return-to-zero* (CRZ) format modulates the amplitude of the optical carrier with an RZ signal, but additionally, some frequency modulation (chirp) is applied [14].
- The *carrier-suppressed return-to-zero* (CS-RZ) format also modulates the amplitude of the optical carrier with an RZ signal, but additionally, the optical phase is shifted by 180° for every bit (no matter if the bit is zero or one) with the result that the carrier becomes suppressed in the optical spectrum [15] (see Fig. 1.7(b)).

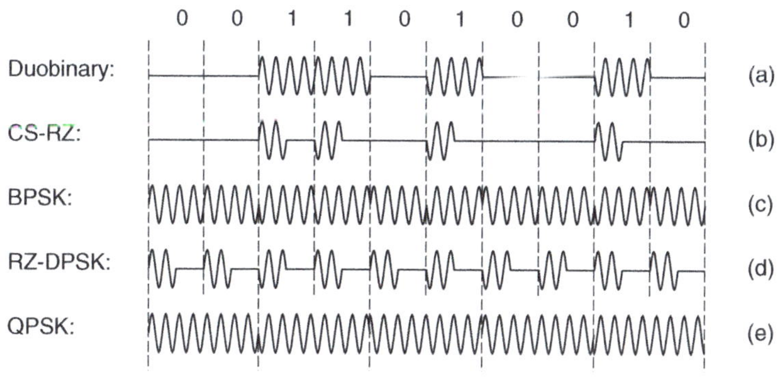

Figure 1.7 Advanced modulation formats (optical field): (a) optical duobinary, (b) CS-RZ, (c) BPSK, (d) RZ-DPSK, and (e) QPSK.

- The *binary phase-shift keying* (BPSK) format encodes the data bits in the phase of the optical signal while the amplitude remains constant. A one bit is encoded with a phase that is 180° shifted compared to the phase of a zero bit (see Fig. 1.7(c)).
- The *return-to-zero differential phase-shift keying* (RZ-DPSK) format modulates the amplitude of the optical carrier with an all-one RZ signal and encodes the data bits in the phase *difference* of two adjacent optical pulses [16]. A one bit is encoded with a 180° phase shift; a zero bit is encoded with a 0° phase shift (see Fig. 1.7(d)).

The purpose of these advanced modulation formats is to mitigate the detrimental effects of the optical communication channel (fiber dispersion, fiber nonlinearity, and crosstalk) in an attempt to increase the reach and capacity of the link at the lowest possible system cost [17].

Phase modulation can also be used to encode multiple bits per symbol. For example, the *quadrature phase-shift keying* (QPSK) format encodes pairs of bits with four different phase shifts relative to a reference phase: bit pair "00" selects a 0° phase shift, "01" selects 90°, "11" selects 180°, and "10" selects 270° (see Fig. 1.7(e)) [3]. Like for the 4-PAM format, the coding is chosen such that an accidental confusion of two adjacent phases results in only a single-bit error. The *differential quadrature phase-shift keying* (DQPSK) and the *return-to-zero differential quadrature phase-shift keying* (RZ-DQPSK) formats similarly encode pairs of bits with four different phase shifts, but this time the shift is relative to the phase of the *previous* symbol [18]. The advantage of these higher-order modulation formats is their bandwidth efficiency: a two-bit per symbol format requires only half the bandwidth of the corresponding one-bit per symbol format given the same bit rate. Conversely, twice the bit rate can be transmitted in the same optical bandwidth. The bandwidth efficiency can be further increased by using constellations with more phase/amplitude values (e.g., 16-QAM) or by using both optical polarizations to transmit information (polarization division multiplexing). The dual-polarization QPSK format (DP-QPSK) is used in commercial 100-Gb/s systems (see Table 1.2).

Direct detection receivers are insensitive to phase and frequency modulation. Thus, special phase-sensitive receivers are required to detect formats such as DPSK, DQPSK, BPSK, and QPSK (cf. Section 3.5). However, a direct detection receiver designed for NRZ can also receive an optical duobinary signal and a direct-detection receiver designed for RZ can also receive CRZ and CS-RZ signals.

Multicarrier Modulation Formats. Since the late 1980s, the TV signals in *community-antenna television* (CATV) systems are often transported first *optically* from the distribution center to the neighborhood before they are distributed to the individual homes on conventional coaxial cable. This combination, called *hybrid fiber-coax* (HFC), has the advantage over an

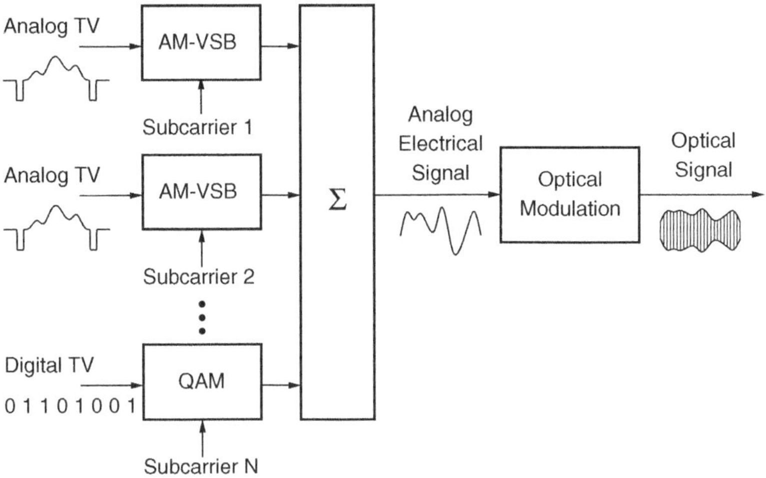

Figure 1.8 Subcarrier multiplexing (SCM) in a CATV/HFC system.

all-coax system of saving many electronic amplifiers (the loss of a fiber is much lower than the loss of a coax cable) and providing better signal quality (lower noise and distortions) [19].

The modulation process used in CATV/HFC systems is illustrated in Fig. 1.8. In North America, analog TV channels use *amplitude modulation with vestigial sideband* (AM-VSB) and digital TV channels use *quadrature amplitude modulation* (QAM). In either case, the TV signal is modulated on a radio frequency (RF) carrier. Many of these modulated carriers, each one at a different frequency, are combined (added) into a single analog broadband signal. To avoid interference between adjacent channels, a small guard band is left between the channels. Then, this analog broadband signal linearly modulates the intensity of a laser to produce the optical signal that is transmitted over the fiber from the distribution center (a.k.a. head end) to the remote node in the neighborhood. At the remote node, the analog broadband signal is recovered from the optical signal and distributed over coaxial cable to the homes. This method of modulation and aggregation is known as *subcarrier multiplexing* (SCM) [20]. The optical carrier can be regarded as the main carrier and the electrical RF carriers of the individual TV channels as the subcarriers. For a discussion of analog and digital TV signals, see Appendix A.

In contrast to NRZ and RZ modulations, which produce two-level digital signals, the AM-VSB and QAM modulation used in CATV applications produce *analog* signals. The latter signals are more easily corrupted by noise and, especially if many TV channels are multiplexed together, are very sensitive to nonlinear distortions (cf. Appendix D). For this reason, special analog optical receivers and transmitters featuring low noise and high linearity are required for CATV/HFC applications.

Although we have introduced SCM in the context of CATV systems, this is not the only application. SCM can also be used to transport digital data. The use of N subcarriers to transmit a total bit rate B, results in a bit rate of only B/N per subcarrier. The corresponding long bit (or symbol) interval makes the SCM signal insensitive to channel impairments (such as fiber dispersion). Moreover, SCM with high-order QAM-modulated subcarriers achieves a high spectral efficiency. Finally, the baseband section of an SCM system runs at only a fraction of the full speed (B/N), possibly simplifying the implementation and reducing the cost. For example, the SCM system in [21] uses 16 subcarriers, each one modulated with a 16-QAM signal running at a symbol rate of 666 Mb/s to transmit a total of 40 Gb/s in a bandwidth of only 14 GHz.

Even higher spectral efficiencies can be obtained with *orthogonal frequency division multiplexing* (OFDM). The guard bands in between the channels of an SCM signal are eliminated by making the carriers orthogonal. The orthogonality condition is satisfied when the carrier spacing is a multiple of the QAM symbol rate [22, 23]. The generation and demodulation of OFDM signals rely on DSPs and data converters.

To overcome the speed limitations of the DSP and the data converters, OFDM can be combined with SCM, resulting in *multiband OFDM*. For example, the system in [24] uses 8 bands, each one containing an OFDM signal consisting of 520 subcarriers, each one modulated with an 8-QAM signal at a symbol rate of 9.6 Mb/s to transmit a total of 100 Gb/s in a bandwidth of only 23 GHz.

Preprocessing. As we know from Fig. 1.1, the raw input data is first scrambled or coded (or both) before it is transmitted. Figure 1.9 shows conceptually how the *information bits* are transformed into *channel bits* before they are passed on to the modulator. The purpose of this preprocessing step is to shape the spectrum of the modulated electrical signal and give it the following desirable properties: DC balance, short run lengths, and a high transition density.

A *DC-balanced signal* has an average value (DC component) that is centered halfway between its minimum and maximum values. This property often permits the use of coupling capacitors (AC coupling) between circuit blocks. To obtain a DC-balanced NRZ signal, the transmitted bit sequence must contain on average the same number of zeros and ones. Equivalently, the average

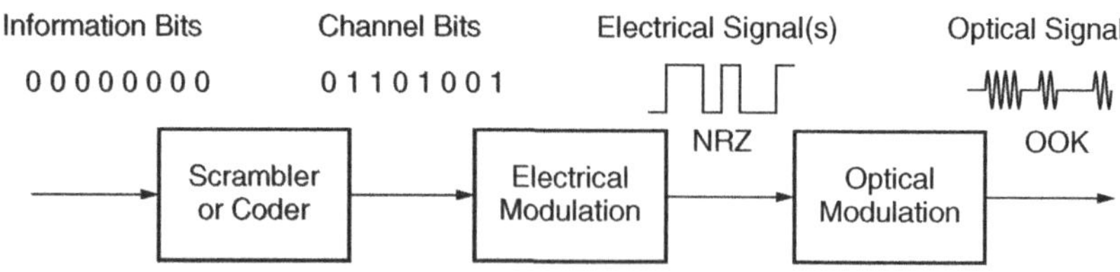

Figure 1.9 Scrambling, coding, and modulation.

mark density, defined as the number of one bits divided by the total number of bits, must be 50%.

The *run length* is the number of successive zeros or ones in a bit sequence. Keeping the run length short reduces the low-frequency content of the modulated NRZ signal (cf. Fig. A.2 on p. 399) and limits the associated *baseline wander* (a.k.a. *DC wander*) when AC coupling is used. Short runs also imply a high *transition density*, which aids the clock recovery process.

Scrambler. Figure 1.10 shows an example of a *scrambler*. It consists of a *pseudorandom bit sequence* (PRBS) generator, implemented with a feedback shift register, and an XOR gate that combines the PRBS with the information bit stream to form the scrambled channel bit stream.

Because two inversions restore the original bit value, the channel bits can be descrambled with the same arrangement, provided the descrambling PRBS generator is synchronized with the scrambling PRBS generator. Scrambling provides DC balance without adding overhead bits to the bit stream, thus keeping the bit rate unchanged. However, the maximum run length is not strictly limited, that is, there is a small chance for very long runs of zeros or ones, which can be hazardous. Equipment designed to process scrambled bit streams is usually tested with runs up to 72 bits [25]. The scrambling method is used, for example, in SONET and SDH telecommunication systems.

Modulation Codes. A *modulation code* replaces a contiguous group of information bits (a block) by another slightly larger group of channel bits such that the average mark density becomes 50% and DC balance is established. Modulation codes (a.k.a. *line codes*) are named after the block length before and after the encoding. Typical examples are the 4B/5B, 8B/10B, and 64B/66B codes:

- 4B/5B code. This code, as the name suggests, replaces 4-bit blocks with 5-bit patterns based on a look up table. The 4B/5B code is simple to implement

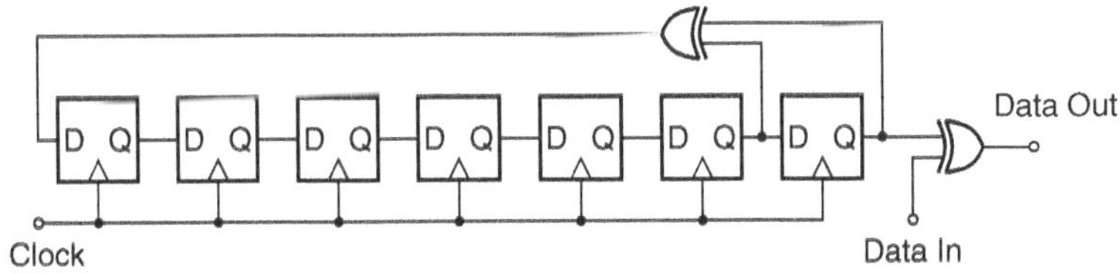

(a)

Data In: ... 00000000 00000011 00000101 00000001 ...

PRBS: ... 01101001 01110111 00110010 10101111 ...

Data Out: ... 01101001 01110100 00110111 10101110 ...

(b)

Figure 1.10 SONET scrambler: (a) implementation and (b) example bit patterns.

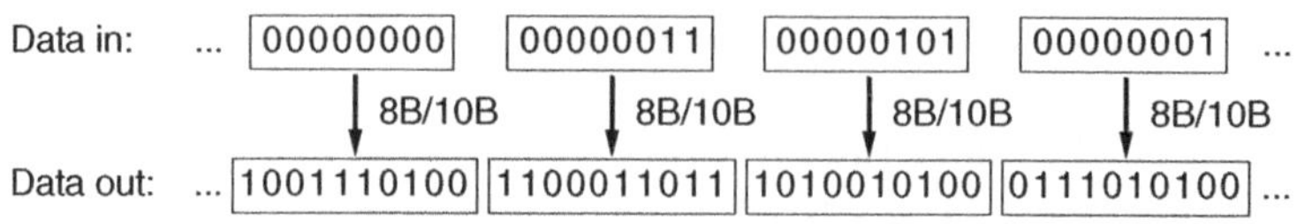

Figure 1.11 Example of 8B/10B encoding.

but increases the bit rate by 25% and does not achieve perfect DC balance; the worst-case unbalance is ±10% [26]. It is used, for example, in the Fast Ethernet and FDDI data communication systems.

- 8B/10B code. This code replaces 8-bit blocks with 10-bit patterns, as illustrated in Fig. 1.11 [27]. Again, the mapping is defined by look-up tables. Besides the 256 data codes, the 8B/10B code maps 12 control codes into the 10-bit code space. The 8B/10B code also increases the bit rate by 25%, but unlike the 4B/5B code, it does achieves exact DC balance. The maximum run length is strictly limited to five zeros or ones. The 8B/10B code is used, for example, in the Gigabit Ethernet (GbE) and Fibre Channel data communication systems.
- 64B/66B code. This modulation code is somewhat different from the 4B/5B and 8B/10B codes in that it partly relies on scrambling rather than a look-up table. The 64B/66B code takes a block of 64 information bits and appends the bit pattern "01" to the beginning. The resulting 66-bit block is scrambled (excluding the two-bit preamble) with a PRBS generated with a 58-bit feedback shift register, producing 66 DC-balanced channel bits. The 64B/66B code also permits the transmission of control information by appending the preamble "10" to a block of 64 control and data bits. The remaining two preambles, "00" and "11," are not used. The 64B/66B code strictly limits the run length to 66 bits by virtue of its two-bit preamble and increases the bit rate by only about 3%. This code is used in many high-speed (10 Gb/s and above) communication systems.

1.3 Transmission Modes

In the following, we explain the difference between continuous-mode and burst-mode transmission and discuss how these two transmission modes are used in optical point-to-point and point-to-multipoint networks.

Continuous Mode versus Burst Mode. Figure 1.12 schematically shows the difference between a *continuous-mode* signal and a *burst-mode* signal.

In continuous-mode transmission, a continuous, uninterrupted stream of bits is transmitted, as shown in Fig. 1.12(a). The transmitted signal usually is DC balanced by means of scrambling or coding.

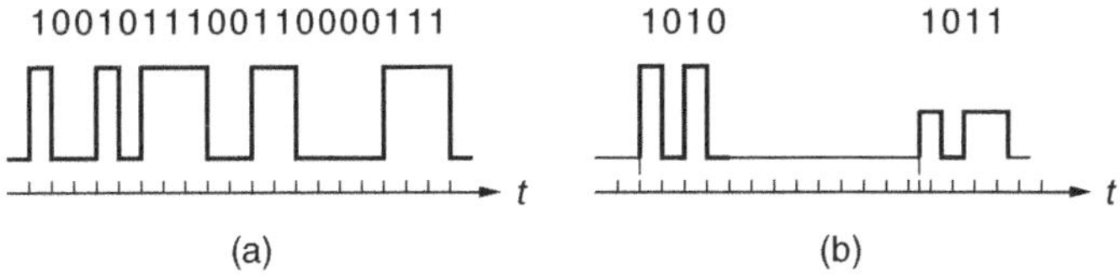

Figure 1.12 (a) Continuous-mode versus (b) burst-mode signals (schematically).

In burst-mode transmission, bits are transmitted in *bursts*, with the transmitter remaining idle (laser off) in between bursts, as shown in Fig. 1.12(b). Bursts typically are longer than 400 bits, but for clarity only 4 bits are shown in the figure. The average value (DC component) of a burst-mode signal varies with time, depending on the burst activity. If the activity is high, it may be close to the halfway point between the zero and one levels, as in a continuous-mode system; if the activity is low, the average drifts arbitrarily close to the zero level. This means that the (overall) burst-mode signal is *not* DC balanced and AC coupling generally cannot be used. (The signal within each burst, however, may be DC balanced.)

Bursts can have a fixed or variable length. For example, bursts that transport (asynchronous transfer mode) cells have a *fixed length*, whereas bursts that transport Ethernet frames have a *variable length*. In either case, the bursts consist of an overhead section followed by a framing structure, such as an ATM cell or an Ethernet frame. Figure 1.13 compares a fixed-length ATM burst with a variable-length Ethernet burst. The burst-mode receiver uses the preamble, which is part of the overhead section, to establish the appropriate gain and decision threshold and to synchronize the sampling clock with the incoming data. In passive optical networks, which we discuss shortly, bursts arrive *asynchronously* and with widely *varying power levels*. Therefore, the gain, the decision threshold, and the clock phase must be acquired for each individual burst.

The lack of DC balance and the fact that bursts may arrive with varying amplitudes necessitate specialized amplifier circuits for burst-mode applications.

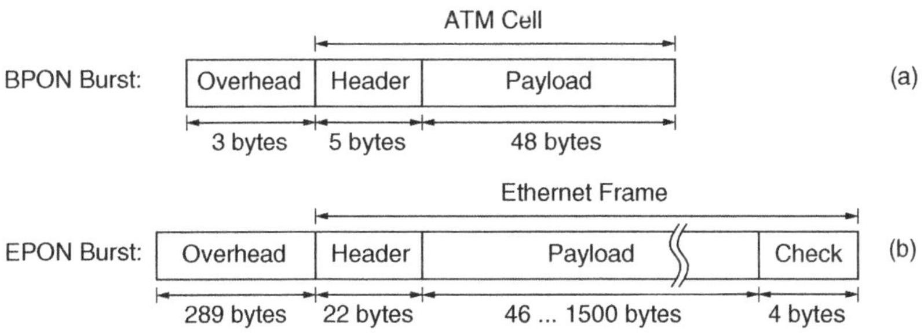

Figure 1.13 (a) Fixed length versus (b) variable length bursts.

Furthermore, the asynchronous arrival of the bursts requires fast-locking CDRs. The design of burst-mode circuits is particularly challenging for bursts with short preambles.

Optical Point-to-Point Connection. An optical *point-to-point connection* between two *central offices* (CO) is illustrated schematically in Fig. 1.14(a). The length of such a connection can range from a few kilometers to more than 10,000 km for the longest undersea lightwave systems. The operating speed per wavelength typically is in the range of 10 Gb/s to 100 Gb/s. In the United States, many point-to-point telecommunication links are based on the SONET (synchronous optical network) standard [28–30]. In Europe, Japan, and other countries the almost identical SDH (synchronous digital hierarchy) standard [25, 31, 32] is used. For example, a SONET OC-192 long reach link has a length of about 80 km and operates at 10 Gb/s (9.953 28 Gb/s to be precise), a bit rate that can carry about 130,000 voice calls. Newer telecommunication standards, such as OTN (optical transport network) [33], respond to the shift from voice traffic to predominantly data traffic as a result of the growing Internet. Besides a framing procedure for voice and data, OTN also supports forward error correction (FEC). See Table 1.1 for more information about the SONET, SDH, and OTN standards.

Point-to-point links are also used in data communication links, that is, in connections between computers. The best-known standard for electrical

Table 1.1 Point-to-point optical telecommunication standards.

Standard	Line speed (Mb/s)	Modulation code	Modulation format
SONET OC-1	51.84	Scrambling	NRZ
SONET OC-3 or SDH STM-1	155.52	Scrambling	NRZ
SONET OC-12 or SDH STM-4	622.08	Scrambling	NRZ
SONET OC-48 or SDH STM-16	2,488.32	Scrambling	NRZ
SONET OC-192 or SDH STM-64	9,953.28	Scrambling	NRZ
SONET OC-192 or SDH STM-64 + FEC (G.975/G.709)	10,664.23	Scrambling	NRZ
SONET OC-768 or SDH STM-256	39,813.12	Scrambling	NRZ
SONET OC-768 or SDH STM-256 + FEC (G.975/G.709)	42,656.91	Scrambling	NRZ
OTN OTU-1 (SONET OC-48 + FEC)	2,666.06	Scrambling	NRZ
OTN OTU-2 (SONET OC-192 + FEC)	10,709.23	Scrambling	NRZ
OTN OTU-3 (SONET OC-768 + FEC)	43,018.41	Scrambling	NRZ
OTN OTU-4 (100GBase + FEC)	111,809.97	64B/66B	NRZ

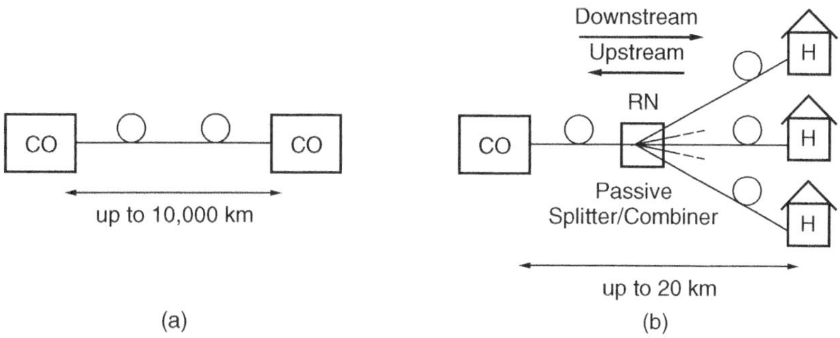

Figure 1.14 Example of (a) a point-to-point link and (b) a point-to-multipoint network.

and optical data communication is *Ethernet* [34]. Ethernet comes at different speed grades, currently ranging from 10 Mb/s to 100 Gb/s and soon 400 Gb/s, and typically resides within the extent of a campus or a building, making it a so-called *local area network* (LAN). For example, Gigabit Ethernet (GbE) operates at a line rate of 1.25 Gb/s providing a usable bit rate of 1.00 Gb/s after 8B/10B decoding. It comes in three main flavors: 1000Base-T, an electrical version operating over a cable with four twisted-pair wires, 1000Base-SX, an optical version using a short-wavelength laser (850 nm), and 1000Base-LX, an optical version using a long-wavelength laser (1,310 nm). Although the original electrical Ethernet was *not* a point-to-point network (but a bus network), all the optical Ethernet flavors are based on point-to-point connections. Other standards for data communication over optical fiber are the *Fiber Distributed Data Interface* (FDDI) [26, 35–37] and the *Fibre Channel* [38, 39]. See Table 1.2 for more information about data communication standards.

Point-to-point connections can be assembled into more complex structures such as *ring networks* and *active star networks* (see Fig. 1.15). Examples for ring networks are provided by SONET/SDH rings and FDDI token rings. An active star is formed, for example, by Gigabit Ethernet links converging into a hub. Each individual optical connection of the active star has a transceiver on both ends and therefore forms an optical point-to-point link. This contrasts with a *passive star network* or an optical point-to-multipoint network, where multiple optical fibers are coupled with a passive optical device (cf. Fig. 1.14(b)). We discuss the latter network type shortly.

Continuous-mode transmission is used on almost all point-to-point connections. One exception occurs in half-duplex systems, in which bidirectional communication is implemented by periodically reversing the direction of traffic following a ping-pong pattern. This method is known as *time compression multiplexing* (TCM; a.k.a. *time division duplexing*) and requires burst-mode transmitters and receivers. However, for bandwidth efficiency reasons, TCM systems are limited to relatively short links such as home networking applications. In all other cases of bidirectional transmission (i.e., with

Table 1.2 Point-to-point optical data communication standards.

Standard	Line speed (Mb/s)	Modulation code	Modulation format
Ethernet (10Base-F)	12.50	4B/5B	NRZI
Fast Ethernet (100Base-FX)	125.00	4B/5B	NRZI
Gigabit Ethernet (1000Base-SX/LX)	1,250.00	8B/10B	NRZ
10-Gigabit Ethernet (10GBase-LX4)	4 × 3,125.00	8B/10B	NRZ
10-Gigabit Ethernet (10GBase-SR/LR/ER/LRM)	10,312.50	64B/66B	NRZ
40-Gigabit Ethernet (40GBase-SR4/LR4)	4 × 10,312.50	64B/66B	NRZ
40-Gigabit Ethernet (40GBase-FR)	41,250.00	64B/66B	NRZ
100-Gigabit Ethernet (100GBase-SR10)	10 × 10,312.50	64B/66B	NRZ
100-Gigabit Ethernet (100GBase-LR4/ER4)	4 × 25,781.25	64B/66B	NRZ
100-Gigabit Ethernet (100GBase-ZR, non-IEEE)	120,579.00	64B/66B	DP-QPSK
400-Gigabit Ethernet (400GBase-FR8/LR8, proposal)	8 × 53,125.00	64B/66B	4-PAM
400-Gigabit Ethernet (400GBase-DR4, proposal)	4 × 106,250.00	64B/66B	4-PAM
Fiber Distributed Data Interface	100.00	4B/5B	NRZI
Fibre Channel (1GFC)	1,062.50	8B/10B	NRZ
Fibre Channel (2GFC)	2,125.00	8B/10B	NRZ
Fibre Channel (4GFC)	4,250.00	8B/10B	NRZ
Fibre Channel (8GFC)	8,500.00	8B/10B	NRZ
Fibre Channel (16GFC)	14,025.00	64B/66B	NRZ

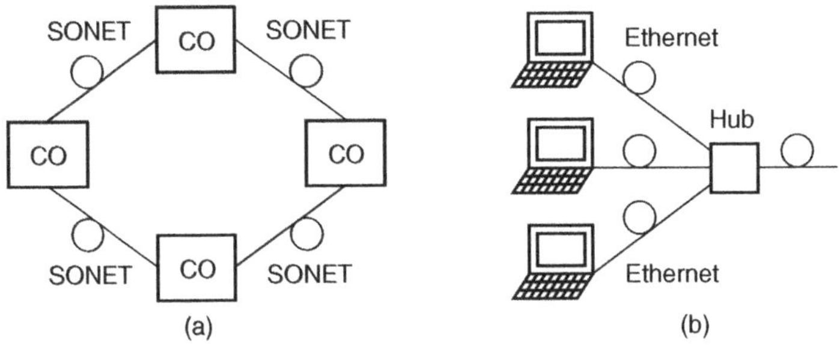

Figure 1.15 (a) SONET ring and (b) Ethernet active star.

two fibers, so-called *space division multiplexing* [SDM], or two wavelengths, so-called *wavelength division multiplexing* [WDM]), continuous-mode transmission is used.

Optical Point-to-Multipoint Network (Passive Optical Network). A *passive optical network* (PON) is illustrated schematically in Fig. 1.14(b). A feeder fiber from the central office (CO) runs to a *remote node* (RN), which houses a passive optical power splitter/combiner. From there, around 32 fibers branch out to the subscribers. If these fibers extend all the way to the *homes* (H), as shown in Fig. 1.14(b), this system is known as a *fiber-to-the-home* (FTTH) system. Alternatively, if the fibers terminate at the curb, the system is known as a *fiber-to-the-curb* (FTTC) system. The final distribution from the curb to the homes is accomplished, for example, by twisted-pair copper wires or radio. All systems that bring the fiber relatively close to the subscriber are collectively known as FTTx systems.

In a traditional telephony access network, the electrical or optical connection from the CO to the remote node is digital. The final distribution from the remote node to the subscribers, however, is accomplished with analog signals over twisted-pair copper wires. Thus, the remote node must be *active*; that is, it needs to be powered to perform the conversion from the high-speed digital signals to the analog signals. In contrast, a PON system is all optical and *passive*. Because a PON does not require outside power supplies, it is low in cost, reliable, and easy to maintain.

A PON is a point-to-multipoint network because the transmission medium is *shared* among the subscribers. Information transmitted downstream, from the CO to the subscriber, is received by all subscribers, and information transmitted upstream, from the subscribers to the CO, is superimposed at the passive combiner before it is received at the CO (see Fig. 1.16). To avoid data collisions in the *upstream direction*, the subscriber must buffer its data and transmit it in short bursts. The CO coordinates which subscriber sends a burst at which point in time. This method is known as *time division multiple access* (TDMA) and requires *burst-mode transmission*. The *downstream direction* is

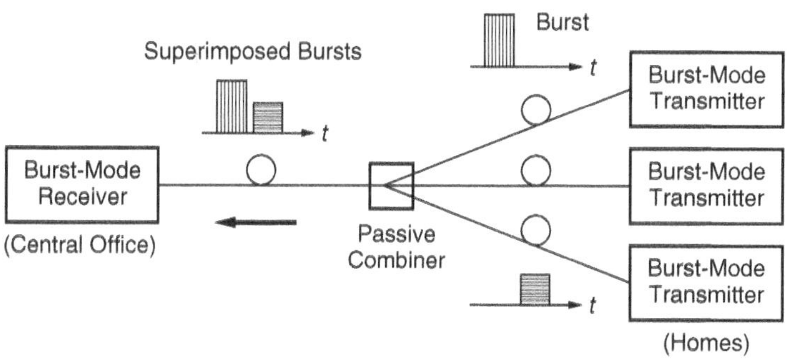

Figure 1.16 Burst-mode transmission in a PON (upstream direction).

more straightforward: the CO tags the data with addresses and broadcasts it to all subscribers in sequential order. Each subscriber simply selects the information with the appropriate address tag. This method is known as *time division multiplexing* (TDM), and conventional continuous-mode transmission can be used. Upstream and downstream transmissions usually are separated by means of two different wavelengths (WDM bidirectional transmission).

One of the first standardized PON systems with significant deployment was BPON (broadband passive optical network) [40, 41]. This system uses ATM cells to transport the data and hence is also known as ATM-PON or APON (cf. Fig. 1.13(a)). In a typical BPON FTTH scenario, 16 to 32 homes located within 20 km from the CO share a downstream bit rate of 622 Mb/s and an upstream bit rate of 155 Mb/s, giving each subscriber an average downstream speed of 20 to 40 Mb/s. This is sufficient for fast Internet access, telephone service, and video on demand. Sometimes, an all-optical CATV service is provided over the PON infrastructure by means of a third wavelength. A typical wavelength plan allocates 1,310 nm for PON upstream, 1,490 nm for PON downstream, and 1,550 nm for the CATV overlay service.

Higher speed PON systems have been standardized since then. EPON (Ethernet passive optical network) transports the data in Ethernet frames, as the name implies (cf. Fig. 1.13(b)), and operates at a line rate of 1.25 Gb/s or 10.3125 Gb/s [42, 43]. GPON (Gigabit-capable passive optical network) operates at bit rates up to 2.5 Gb/s and transports ATM or Ethernet traffic with a high bandwidth utilization [44]. See Table 1.3 for more information about PON standards.

Besides the TDM/TDMA approach outlined earlier, there are several other types of PON systems [45–47]. For example, WDM-PON systems, which use multiple wavelengths, have been studied extensively. In these systems, data collisions are avoided by assigning different wavelengths to different subscribers, thus making burst-mode transmission unnecessary. However, the optical WDM components required for such a system are expensive. Finally, hybrid PON systems combining the TDM and the WDM approach are also being studied.

Note About Numerical Examples. In the following chapters, we make extensive use of numerical examples. When we introduce a new quantity or relationship, we frequently illustrate it with *typical values*. In my own learning experience, this approach is most helpful: it makes the subject more concrete and promotes a feeling for the numerical values. However, specialists tend to be quite critical about such examples because the values are never quite right. Typical values may change over time as the field advances or they may depend on several conditions that may or may not be met in a particular case. It is therefore important to take the subsequent numerical examples only as an illustration and not as the basis for your next design project!

Table 1.3 Point-to-multipoint optical communication standards.

Standard	Downstream speed (Mb/s)	Upstream speed (Mb/s)	Layer 2 protocol	Modulation code	Modulation format
ATM-PON/BPON (option 1)	155.52	155.52	ATM	Scrambling	NRZ
ATM-PON/BPON (option 2)	622.08	155.52	ATM	Scrambling	NRZ
GPON	1,244.16	155.52	ATM, Ethernet, etc.	Scrambling	NRZ
GPON	1,244.16	622.08	ATM, Ethernet, etc.	Scrambling	NRZ
GPON	1,244.16	1,244.16	ATM, Ethernet, etc.	Scrambling	NRZ
GPON	2,488.32	155.52	ATM, Ethernet, etc.	Scrambling	NRZ
GPON	2,488.32	622.08	ATM, Ethernet, etc.	Scrambling	NRZ
GPON	2,488.32	1,244.16	ATM, Ethernet, etc.	Scrambling	NRZ
GPON	2,488.32	2,488.32	ATM, Ethernet, etc.	Scrambling	NRZ
EPON (1000Base-PX)	1,250.00	1,250.00	Ethernet	8B/10B	NRZ
10/10G-EPON	10,312.50	10,312.50	Ethernet	64B/66B	NRZ
10/1G-EPON	10,312.50	1,250.00	Ethernet	64B/66B, 8B/10B	NRZ

References

1 M. Möller, T. F. Meister, R. Schmid, J. Rupeter, M. Rest, A. Schöpflin, and H.-M. Rein. SiGe retiming high-gain power MUX for direct driving an EAM up to 50 Gb/s. *Electron. Lett.*, 34(18):1782–1784, 1998.

2 T. H. Hu and P. R. Gray. A monolithic 480 Mb/s parallel AGC/decision/clock-recovery circuit in 1.2-μm CMOS. *IEEE J. Solid-State Circuits*, SC-28(12):1314–1320, 1993.

3 K. Kikuchi. Coherent optical communications: historical perspectives and future directions. In M. Nakazawa, K. Kikuchi, and T. Miyazaki, editors, *High Spectral Density Optical Communication Technologies*, pages 11–49. Springer, Berlin Heidelberg, 2010.

4 B. Mikkelsen. Challenges and key technologies for coherent metro 100G transceivers. *Lightwave*, 29(6), 2012.

5 K. Roberts, S. H. Foo, M. Moyer, M. Hubbard, A. Sinclair, J. Gaudette, and C. Laperle. High capacity transport – 100G and beyond. *J. Lightwave Technol.*, LT-33(3):563–578, 2015.

6 S. J. Savory, G. Gavioli, R. I. Killey, and P. Bayvel. Electronic compensation of chromatic dispersion using a digital coherent receiver. *OSA Opt. Express*, 15(5):2120–2126, 2007.

7 R. Ludwig, U. Feiste, E. Dietrich, H. G. Weber, D. Breuer, M. Martin, and F. Küppers. Experimental comparison of 40 Gbit/s RZ and NRZ transmission over standard singlemode fibre. *Electron. Lett.*, 35(25):2216–2218, 1999.

8 G. Charlet and S. Bigo. Upgrading WDM submarine systems to 40-Gbit/s channel bitrate. *Proc. IEEE*, 94(5):935–951, 2006.

9 C. Cole, I. Lyubomirsky, A. Ghiasi, and V. Telang. Higher-order modulation for client optics. *IEEE Commun. Mag.*, 51(3):50–57, 2013.

10 J. Conradi. Bandwidth-efficient modulation formats for digital fiber transmission systems. In I. P. Kaminow and T. Li, editors, *Optical Fiber Telecommunications IVB*, pages 862–901. Academic Press, San Diego, CA, 2002.

11 S. Walklin and J. Conradi. Multilevel signaling for increasing the reach of 10 Gb/s lightwave systems. *J. Lightwave Technol.*, LT-17(11):2235–2248, 1999.

12 J. G. Proakis and M. Salehi. *Digital Communications*. McGraw Hill, New York, 5th edition, 2008.

13 K. Yonenaga, S. Kuwano, S. Norimatsu, and N. Shibata. Optical duobinary transmission system with no receiver sensitivity degradation. *Electron. Lett.*, 31(4):302–304, 1995.

14 B. Bakhshi, M. Vaa, E. A. Golovchenko, W. W. Patterson, R. L. Maybach, and N. S. Bergano. Comparison of CRZ, RZ, and NRZ modulation formats in a 64 × 12.3 Gb/s WDM transmission experiment over 9000 km. In *Optical Fiber Communication Conference (OFC), Anaheim, CA*, 2001.

15 A. Hirano, Y. Miyamoto, K. Yonenaga, A. Sano, and H. Toba. 40 Gbit/s L-band transmission experiment using SPM-tolerant carrier-suppressed RZ format. *Electron. Lett.*, 35(25):2213–2215, 1999.

16 W. A. Atia and R. S. Bondurant. Demonstration of return-to-zero signaling in both OOK and DPSK formats to improve receiver sensitivity in an optically preamplified receiver. In *Lasers and Electro-Optics Society (LEOS) 12th Annual Meeting, San Francisco, CA*, November 1999.

17 P. J. Winzer and R.-J. Essiambre. Advanced optical modulation formats. *Proc. IEEE*, 94(5):952–984, 2006.

18 R. A. Griffin and A. C. Carter. Optical differential quadrature phase-shift key (oDQPSK) for high capacity optical transmission. In *Optical Fiber Communication Conference (OFC), Anaheim, CA*, 2002.

19 W. Ciciora, J. Farmer, D. Large, and M. Adams. *Modern Cable Television Technology: Video, Voice, and Data Communications*. Morgan Kaufmann, San Francisco, CA, 2nd edition, 2004.

20 T. E. Darcie. Subcarrier multiplexing for multiple-access lightwave networks. *J. Lightwave Technol.*, LT-5(8):1103–1110, 1987.
21 J. Chen, F. Saibi, E. Säckinger, J. Othmer, M.-L. (Mark) Yu, F. Yang, J. Lin, T. Huang, T.-P. Liu, and K. Azadet. A multi-carrier QAM transceiver for ultra-wideband optical communication. *IEEE J. Solid-State Circuits*, SC-41(8):1876–1893, 2006.
22 G. P. Agrawal. *Fiber-Optic Communication Systems*. John Wiley & Sons, Inc. Hoboken, NJ, 4th edition, 2010.
23 X. Chen, A. A. Amin, A. Li, and W. Shieh. Multicarrier optical transmission. In I. P. Kaminow, T. Li, and A. E. Willner, editors, *Optical Fiber Telecommunications VIB*, pages 337–380. Academic Press, 2013.
24 S. L. Jansen, I. Morita, T. C. W. Schenk, and H. Tanaka. 121.9-Gb/s PDM-OFDM transmission with 2-b/s/Hz spectral efficiency over 1000 km of SSMF. *J. Lightwave Technol.*, LT-27(3):177–188, 2009.
25 ITU-T. Digital line systems based on the synchronous digital hierarchy for use on optical fibre cables, recommendation G.958. International Telecommunication Union, Geneva, Switzerland, November 1994.
26 W. E. Burr. The FDDI optical data link. *IEEE Commun. Mag.*, 24(5):18–23, 1986.
27 A. X. Widmer and P. A. Franaszek. A DC-balanced, partitioned-block, 8B/10B transmission code. *IBM J. Res. Dev.*, 27(5):440–451, 1983.
28 ANSI. Telecommunications – synchronous optical network (SONET) – basic description including multiplex structures, rates, and formats – ANSI T1.105-1995. American National Standards Institute, Washington, DC, October 1995.
29 Telcordia Technologies. SONET OC-192 transport system generic criteria, GR-1377-CORE, Issue 5. Telcordia Technologies (formerly Bellcore), Piscataway, NJ, December 1998.
30 Telcordia Technologies. Synchronous optical network (SONET) transport systems: common generic criteria, GR-253-CORE, Issue 3. Telcordia Technologies (formerly Bellcore), Piscataway, NJ, September 2000.
31 ITU-T. Optical interfaces for equipments and systems relating to the synchronous digital hierarchy (SDH), recommendation G.957. International Telecommunication Union, Geneva, Switzerland, June 1999.
32 ITU-T. Network node interface for the synchronous digital hierarchy (SDH), recommendation G.707. International Telecommunication Union, Geneva, Switzerland, October 2000.
33 ITU-T. Interfaces for the optical transport network (OTN), recommendation G.709/Y.1331. International Telecommunication Union, Geneva, Switzerland, February 2001.
34 IEEE. 802.3. Part 3: Carrier sense multiple access with collision detection (CSMA/CD) access method and physical layer specifications. IEEE Computer Society, New York, NY, 2005, http://www.ieee802.org/3/.

35 ANSI. Fiber Distributed Data Interface (FDDI) physical layer, medium dependent (PMD) – ANSI X3.166-1990. American National Standards Institute, Washington, DC, 1988. Also see ANSI X3.184 (SMF-PMD) and ISO 9314-3:1990.
36 ANSI. Fiber Distributed Data Interface (FDDI) physical layer (PHY) – ANSI X3.148-1988. American National Standards Institute, Washington, DC, 1988. Also see ANSI X3.231 (PHY-2) and ISO 9314-1:1989.
37 F. E. Ross. FDDI – a tutorial. *IEEE Commun. Mag.*, 24(5):10–17, 1986.
38 ANSI. Fibre Channel physical and signaling interface (FC-PH) – ANSI X3.230-1994. American National Standards Institute, Washington, DC, 1994. Also see ANSI X3.297 (FC-PH-2) and ANSI X3.303 (FC-PH-3).
39 M. W. Sachs and A. Varma. Fibre Channel and related standards. *IEEE Commun. Mag.*, 34(8):40–50, 1996.
40 FSAN. Full service access network. http://www.fsan.org.
41 ITU-T. Broadband optical access systems based on passive optical networks (PON), recommendation G.983.1. International Telecommunication Union, Geneva, Switzerland, October 1998.
42 IEEE. 802.3ah. Part 3: Carrier sense multiple access with collision detection (CSMA/CD) access method and physical layer specifications. Amendmend: Media access control parameters, physical layers, and management parameters for subscriber access networks. IEEE Computer Society, New York, NY, September 2004, http://www.ieee802.org/3/efm/.
43 G. Kramer. *Ethernet Passive Optical Networks*. McGraw-Hill, New York, 2005.
44 ITU-T. Gigabit-capable passive optical networks (GPON): General characteristics, recommendation G.984.1. International Telecommunication Union, Geneva, Switzerland, March 2003.
45 F. Effenberger. PONs: state of the art and standardized. In I. P. Kaminow, T. Li, and A. E. Willner, editors, *Optical Fiber Telecommunications VIB*, pages 891–926. Academic Press, 2013.
46 E. Harstead and P. H. van Heyningen. Optical access networks. In I. P. Kaminow and T. Li, editors, *Optical Fiber Telecommunications IVB*, pages 438–513. Academic Press, San Diego, CA, 2002.
47 U. Killat, editor. *Access to B-ISDN via PONs: ATM Communication in Practice*. John Wiley and B. G. Teubner, Chichester, England, 1996.

2
Optical Fibers

In the following, we introduce the communication channel presented by the optical fiber. We briefly describe the loss, various forms of dispersion, and nonlinearities of the fiber. For each effect, we discuss what can be done to counteract it in the optical as well as the electrical domain.

2.1 Loss and Bandwidth

Loss. As the optical signal propagates through a long stretch of fiber, it becomes attenuated because of scattering, absorption by material impurities, and other effects. The attenuation is measured in dBs ($10 \cdot \log$ of power ratio) and is proportional to the length of the fiber. *Fiber attenuation* or *fiber loss* therefore is specified in dB/km.

As shown in Fig. 2.1, silica glass has two low-loss windows, one around the wavelength $\lambda = 1.3$ μm and one around $\lambda = 1.55$ μm, both of which are used for optical fiber communication. The popular single-mode fiber has a loss of 0.18 to 0.25 dB/km at the 1.55-μm wavelength and 0.32 to 0.4 dB/km at the 1.3-μm wavelength. Because the loss is lower at 1.55 μm, this wavelength is preferred for long-haul communication. The two low-loss windows are separated by the so-called *water peak* at about 1.4 μm. In modern fiber, this peak is reduced significantly, essentially merging the two windows into one (dashed line). A third wavelength window around $\lambda = 0.85$ μm, where the loss is 1.8 to 2.5 dB/km, is used for short-reach (data) communication applications, mostly because low-cost optical sources and detectors are available for this wavelength.[1]

The loss of modern silica-glass fiber is phenomenally low compared with that of an RF coax cable at high frequencies. A high-performance RF coax cable operating at 10 GHz has an attenuation of about 500 dB/km. Compare

1 All wavelengths are defined in the *vacuum*. In a glass fiber, which has a refractive index of about 1.5, an optical signal with $\lambda = 1.55$ μm has a wavelength of about 1 μm.

Analysis and Design of Transimpedance Amplifiers for Optical Receivers, First Edition. Eduard Säckinger.

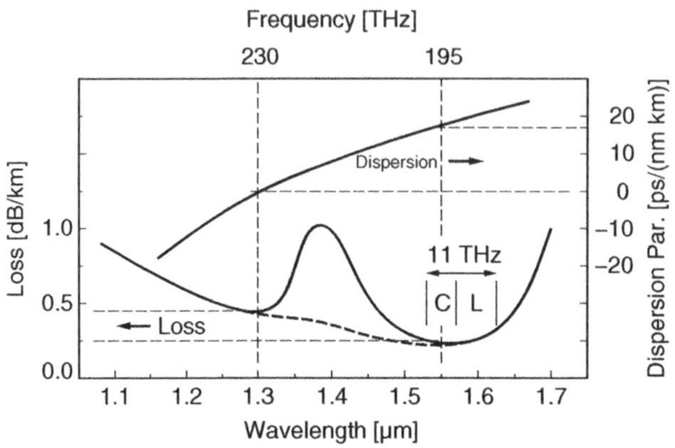

Figure 2.1 Loss and dispersion parameter D of a standard single-mode fiber.

this with 0.2 dB/km for a fiber! Interestingly, the low-loss fiber was not easy to produce [1]. In 1965, the best glass fiber had a loss of around 1,000 dB/km. It was estimated that for a fiber to be useful for optical communication, its loss must be reduced to 20 dB/km or less, that is, an improvement by *98 orders of magnitude* was required! It is therefore understandable that in 1965 most researchers thought that using glass fiber for optical communication was a hopelessly crazy idea. They spent their time working on "reasonable" approaches to optical communication such as metal pipes that contained periodically spaced lenses (so-called *confocal waveguides*) or pipes heated in such a way that the air in them formed *gas lenses*. Nevertheless, in 1970, a research team at the *Corning Glass Works* managed to reduce the fiber loss below 20 dB/km by using ultrapure *silica glass* rather than the ordinary compound glass. So, next time your circuit parameters are off by 98 orders of magnitude, don't give up …

In comparison with silica-glass fiber, *plastic optical fiber* (POF) is very cheap to manufacture and also permits the use of low-cost connectors because of its large core size of almost 1 mm. However, it has a huge loss of about 180 dB/km, even when operated in its "low-loss" window at 0.65 μm (visible red) [2]. POF is therefore restricted to very-short reach applications such as automatic control equipment in manufacturing, information and multimedia transmission in automobiles, and high-quality digital audio transmission [3].

Loss and Amplification. Although the loss of silica-glass fiber is very low, after about 100 km the optical signal becomes so weak that it must be regenerated. A straightforward solution is to detect the weak signal, electronically regenerate it, and retransmit it at a higher power level. This arrangement is shown in Fig. 2.2(a) and is known as an *optoelectronic repeater*. Unfortunately, these repeaters are expensive, complex, and typically work for only one particular bit

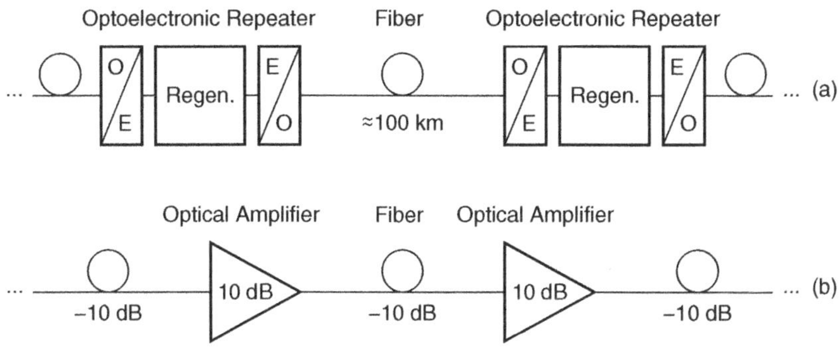

Figure 2.2 Fiber link with (a) optoelectronic repeaters and (b) optical amplifiers.

rate, modulation format, and wavelength, necessitating a replacement when the link capacity is upgraded.

An alternative strategy is to insert periodically spaced *optical in-line amplifiers*, as shown in Fig. 2.2(b). These amplifiers boost the signal, effectively compensating for the optical loss. The most common optical amplifiers are (i) the *erbium-doped fiber amplifier* (EDFA), which provides gain in the important 1.55-μm band [4, 5], and (ii) the *Raman amplifier*, which can provide distributed gain in the transmission fiber itself at a selectable wavelength (13 THz below the pump frequency) [6, 7]. A third type of amplifier is in the research stage: the *fiber optical parametric amplifier* (FOPA), which promises gain over a very wide bandwidth at a selectable wavelength (surrounding the pump frequency) and very low noise [8]. In Section 3.3, we have a closer look at some of these amplifiers. With the advent of reliable optical in-line amplifiers in the 1990s, ultra-long-haul communication (> 3,000 km) without optoelectronic repeaters has become possible [9] (cf. Section 4.6).

Bandwidth. In addition to the very low loss, silica optical fiber also has a huge bandwidth. With *bandwidth*, we mean the range of optical frequencies or wavelengths for which the fiber is highly transparent. For example, the C and L bands located in the low-loss window around the 1.55-μm wavelength (C stands for "conventional" and L stands for "long-wavelength") together provide a bandwidth of 95 nm, corresponding to about 11 THz (see Fig. 2.1). This means that with the simple non-return-to-zero (NRZ) modulation format, which achieves a spectral efficiency[2] of about 0.4 b/s/Hz, we can transmit approximately 4 Tb/s over a single fiber. (With a bandwidth-efficient modulation format, such as DP-QPSK, and using both low-loss windows, we could

2 The spectral efficiency of the NRZ format is not well defined because the NRZ spectrum is not strictly band limited (cf. Appendix A). Based on the null-to-null bandwidth, the efficiency is 0.5 b/s/Hz; here we use the more conservative value of 0.4 b/s/Hz [10], realized, for example, by a DWDM system with 40 Gb/s channels on a 100 GHz grid.

transmit at an even higher rate.) Thus, a backbone connection consisting of 100 parallel fibers could transport 400 Tb/s of information, or a fiber-to-the-home (FTTH) system based on a passive optical network (PON) in which each feeder fiber serves 100 subscribers could be upgraded to 40 Gb/s per user, if the demand for so much access bandwidth should arise! This is why FTTH advocates tout their system as *future proof*. [→ Problem 2.1.]

Bandwidth and Dispersion. Would it be possible to transmit a single lightwave modulated with a 4-Tb/s NRZ signal through the fiber? Besides the fact that the electronic circuits and the optoelectronic devices (laser, modulator, detector) would be too slow, the received signal would be totally distorted already after a very short distance because of *dispersive effects* in the fiber. The transmitted optical signal in our hypothetical system has a very large spectral width, filling all of the C and L bands. Although each spectral component is in the low-loss window and arrives intact at the other end of the fiber, each component is *delayed* by a different amount, and the superposition of all components, the received signal, is severely distorted. The dependence of delay on wavelength is known as *chromatic dispersion*, which we discuss in the next section.

It is important to distinguish between two types of *fiber bandwidths*: the *bandwidth of the low-loss window*, which is very high (> 10 THz), and the *bandwidth for the modulation signal*, which is limited by dispersion and is much, much smaller. For example, the modulation-signal bandwidth of 1 km of standard single-mode fiber is just a few 10 GHz, given a 1.55-μm source with a 1-nm linewidth. We discuss this bandwidth and its dependence on linewidth in Section 2.4. For 1 km of graded-index multimode fiber, the modulation-signal bandwidth is only 300 MHz to 3 GHz, and for a step-index multimode fiber, it is even lower, ranging from 6 to 50 MHz [11].

How can we take advantage of the huge fiber bandwidth? We can use multiple optical carriers at different wavelengths and modulate each one at a reasonable bit rate. Instead of one wavelength modulated at 4 Tb/s, we could use, for example, 100 wavelengths, each one modulated at 40 Gb/s or 400 wavelengths, each one modulated at 10 Gb/s. This approach is known as DWDM. In addition, we can use optical or electrical dispersion compensation.

2.2 Dispersion

Modal Dispersion. An optical fiber consists of a core surrounded by a cladding that has a slightly lower refractive index than the core such that the light beam is guided by total internal reflection, as shown in Fig. 2.3. In principle, air, which has a lower refractive index than glass, could act as the cladding. However, the fiber surface then would be extremely sensitive to dirt and scratches, and two fibers touching each other would leak light. The invention of the *clad fiber* was a major breakthrough on the way to a practical optical fiber [1].

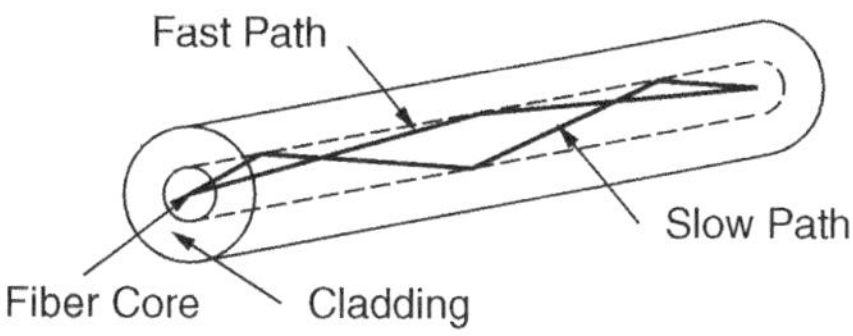

Figure 2.3 Modal dispersion in a multimode fiber.

Depending on the size of the fiber core, there are only a number of pathways (so-called modes) for the light beam to propagate through the fiber. The core of a *multimode fiber* (MMF) is large enough (50 to 100 μm) for the light to take multiple pathways from the transmitter to the receiver, as shown schematically in Fig. 2.3 (typically, several hundred modes exist). Each path has a slightly different propagation delay, thus producing a distorted (spread out) pulse at the receiver end. This effect is known as *modal dispersion*. The time difference between the longest and shortest path ΔT for a so-called *graded-index multimode fiber* (GRIN-MMF) can be approximated by [12]

$$\Delta T = \frac{(n_{\text{cor}} - n_{\text{clad}})^2}{8c \cdot n_{\text{cor}}} \cdot L, \tag{2.1}$$

where L is the fiber length, c is the speed of light in vacuum, and n_{cor} and n_{clad} are the refractive indices of the core and cladding, respectively. With the typical values $n_{\text{cor}} = 1.48$ and $n_{\text{cor}} - n_{\text{clad}} = 0.02$ and a fiber length of 1 km, we find a propagation-delay variation of about 113 ps. Thus, modal dispersion is significant even for short fiber links.

The core of a *single-mode fiber* (SMF) is much smaller (8 to 10 μm) and permits only one pathway (a single mode) of light propagation from the transmitter to the receiver, and thus distortions due to modal dispersion are suppressed.[3] Note that the word "mode" in single-*mode* fiber, multi*mode* fiber, and *modal* dispersion refers to *pathway* modes only. We see later that the single pathway mode in an SMF can be decomposed into two *polarization* modes, both of which propagate through the fiber and may cause polarization-mode dispersion.

SMF is preferred in telecommunication applications where distance matters (from access networks to ultra-long-haul transmission). MMF is mostly used within buildings for data communication (computer interconnects) and in consumer electronics. Because the MMF has a larger core size, alignment of the fiber with another fiber or a laser chip is easier. A transverse alignment error between a laser and an SMF of just 0.5 μm causes a power penalty of about

3 Why is a 8- to 10-μm core small enough to ensure single-mode propagation of light with a wavelength of 1.3 to 1.55 μm? It turns out that the condition for single-mode propagation is that the core diameter must be $d < \lambda \cdot 0.766/\sqrt{n_{\text{cor}}^2 - n_{\text{clad}}^2}$ [12]. Because the difference between n_{cor} and n_{clad} is small (less than 1% for an SMF), the core can be made quite a bit larger than the wavelength λ, simplifying the light coupling into the fiber. Another advantage of the clad fiber!

1 dB, whereas the laser-to-MMF alignment is about 5× less critical [13]. Thus, components interfacing to MMF generally are lower in cost. (However, MMF itself is more expensive than SMF.)

Modal dispersion in MMF can be compensated to some degree with receiver equalization [14]. This method is known as *electronic dispersion compensation* (EDC). The 10-Gb/s Ethernet standard 10GBASE-LRM makes use of this technique, enabling a reach of 220 m over legacy (FDDI-grade) MMF.

Chromatic Dispersion. *Chromatic dispersion,* also called *group-velocity dispersion* (GVD), is another source of signal distortions resulting from different wavelengths (colors) traveling at different speeds through the fiber. Figure 2.4 illustrates how the group delay varies with wavelength for 1 km of standard SMF (cf. Eq. (I.1) on p. 497). We recognize that the change in group delay is large around 1.55 μm, whereas it is nearly zero at 1.3 μm. In practice, chromatic dispersion is specified by the *change* in group delay per nm wavelength and km length:

$$D = \frac{1}{L} \cdot \frac{\partial \tau}{\partial \lambda}, \tag{2.2}$$

where D is known as the *dispersion parameter,* L is the fiber length, τ is the group delay, and λ is the wavelength. A standard SMF operated at 1.55 μm has $D = 17$ ps/(nm · km), which means that a change in wavelength of 1 nm will change the group delay by 17 ps in a 1-km stretch of fiber (cf. Fig. 2.4). The dependence of D on wavelength is plotted together with the fiber loss in Fig. 2.1. [→ Problem 2.2.]

How much pulse distortion results from a given amount of chromatic dispersion ($|D|L$) depends on the spectral linewidth of the transmitter. If the transmitter operates at precisely a single wavelength, which implies a zero-linewidth laser without modulation (continuous-wave operation), chromatic dispersion has no impact. However, if the transmitter operates over a range of wavelengths, as a real, modulated laser does, chromatic dispersion causes pulse distortions.

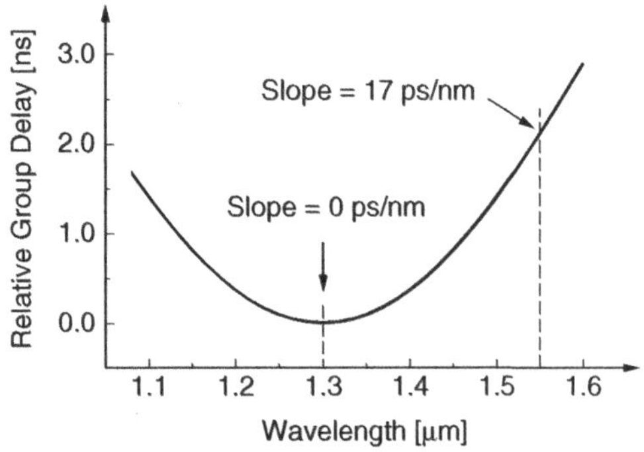

Figure 2.4 Relative group delay as a function of wavelength for 1 km of standard SMF.

The propagation time difference between the slowest and fastest wavelength for a transmitter that emits light over the range $\Delta\lambda$ can be derived from Eq. (2.2) as

$$\Delta T = |D|L \cdot \Delta\lambda. \tag{2.3}$$

We continue the discussion of pulse spreading due to chromatic dispersion in Section 2.4.

Chromatic Dispersion Compensation. What can we do to reduce chromatic dispersion? We can transmit at the 1.3-μm wavelength, where the dispersion parameter D of a standard SMF is much smaller than at 1.55 μm (see Fig. 2.1). But as we know, the fiber loss is higher at the 1.3-μm wavelength. To resolve this dilemma, fiber manufacturers have come up with the so-called *dispersion-shifted fiber* (DSF), which has a value of D close to zero at the 1.55-μm wavelength while preserving the low loss of an SMF. This fiber, however, suffers from other problems when used in WDM systems that we discuss in Section 2.3.

Because dispersion is a linear phenomenon, it can be reversed by applying an equal amount of *negative* dispersion. This method is known as *dispersion compensation,* and so-called *dispersion compensating fiber* (DCF) with a large negative value of D such as -100 ps/(nm · km) is available for this purpose. For example, to compensate for the positive dispersion of 100 km of standard SMF, we can append 17 km of DCF, as shown in Fig. 2.5(a). The result is an overall dispersion of zero: $17\ \text{ps/(nm}\cdot\text{km)} \cdot 100\ \text{km} + (-100\ \text{ps/(nm}\cdot\text{km))} \cdot 17\ \text{km} = 0$. In such a system, each bit spreads out over many adjacent bit slots, resulting in a total mess, until the DCF pulls all the bits back together again, producing a crisp optical signal (assuming nonlinear effects are negligible). In practice, perfect dispersion compensation can be achieved only at a single wavelength. Thus, in a DWDM system only the middle channel can be compensated perfectly, whereas the outer channels retain some residual dispersion.

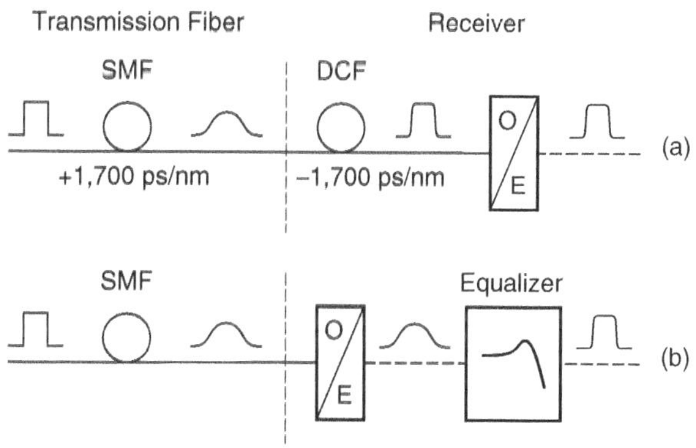

Figure 2.5 Chromatic dispersion compensation in (a) the optical and (b) the electrical domain.

Chromatic dispersion also can be compensated in the electrical domain with an equalizer, as shown in Fig. 2.5(b). Electronic dispersion compensation is more compact, more flexible, and lower in cost than optical dispersion compensation. Moreover, it avoids the loss of the DCF. The effectiveness of EDC depends significantly on whether a conventional direct-detection receiver, which loses the optical phase information in the detection process, or a coherent receiver, which preserves the phase information, is used. For comparison, the direct-detection receiver with EDC in [15] can compensate about 1,600 ps/nm of chromatic dispersion (about 100 km of SMF), whereas the coherent system with EDC in [16] can compensate about 300,000 ps/nm! This is sufficient for a transpacific application over non-dispersion compensated fiber. The introduction of coherent transmission systems with DSP-based EDC around 2007 revolutionized the field of dispersion compensation.

Besides dispersion compensation, the effects of chromatic dispersion can be reduced by using spectrally narrow modulation formats, such as 4-PAM, optical duobinary, or QPSK [17].

Polarization-Mode Dispersion. Another source of distortions is *polarization-mode dispersion* (PMD), which is caused by different polarization modes traveling at different speeds. This effect occurs in fibers with a slightly elliptic core or asymmetrical mechanical stress. Figure 2.6 shows a short fiber segment that suffers from PMD. Horizontally and vertically polarized light propagate at slightly different speeds. The difference in arrival time, ΔT, which grows with the fiber length, is known as the *differential group delay* (DGD). Because, the transmitter generally excites both polarization modes (horizontal and vertical), the receiver sees two time-shifted copies of the transmitted sequence superimposed on top of each other. The strength of each sequence depends on the alignment of the transmitter's (linear) state of polarization with the axes of the elliptic core.

In a long stretch of fiber, the situation is complicated by the fact that the fiber's effect on the polarization state (birefringence) changes randomly along its length. As a result, we have the following modifications. (i) The input polarization states that correspond to the fast and slow propagation are no longer

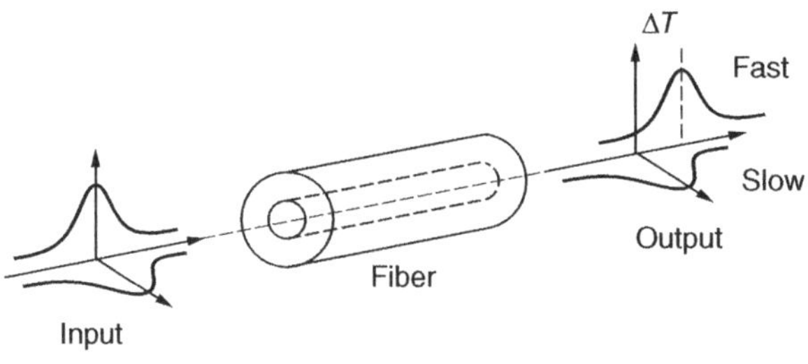

Figure 2.6 Polarization-mode dispersion in a short fiber segment.

linear states (i.e., horizontal and vertical) but general elliptic polarization states, which differ from fiber to fiber. These states are known as the input *principle states of polarization* (PSP). (ii) The differential group delay, averaged over many fibers, is proportional to the *square root* of the fiber length L (rather than the length itself) and can be written

$$\overline{\Delta T} = D_{PMD}\sqrt{L}, \tag{2.4}$$

where D_{PMD} is the *polarization-mode dispersion parameter.* Most annoyingly, PMD (PSP and DGD) also varies slowly (< 30 kHz) and randomly with *time.* Thus, ΔT becomes a function of time described by a probability distribution. As a rule of thumb, we must keep the average $\overline{\Delta T}$ below 10% of the bit interval ($0.1/B$) to keep the power penalty due to PMD almost permanently below 1 dB.[4] This unpredictable aspect of PMD has been likened to having "weather" in the fiber [19]: most of the time it is sunny and clear, occasionally there is fog, and rarely a storm hits! Another complication in real systems is the wavelength dependence of both the input PSPs and the DGD, which makes the PMD distortions look more complex than just a simple superposition of time-shifted copies of the transmitted signal (so-called *higher-order PMD* distortions).

Fiber installed after about 1998 has a very low PMD parameter of $D_{PMD} = 0.1\ \text{ps}/\sqrt{\text{km}}$ or less [20]. Thus, a 10,000-km stretch of fiber has an average DGD of 10 ps. Whereas PMD in these fibers is no problem at 10 Gb/s, it limits 40-Gb/s links to about 600 km. Older fiber, which is widely deployed and has a slightly elliptic cross section of the fiber core as a result of manufacturing tolerances, has a much larger PMD parameter up to $D_{PMD} = 2\ \text{ps}/\sqrt{\text{km}}$ [21]. In this case, PMD severely limits the length of the transmission system to < 25 km at 10 Gb/s and < 1.6 km at 40 Gb/s.

PMD Compensation. What can we do to counteract PMD? We can use optical or electronic PMD mitigation techniques. A simple optical PMD compensator is shown schematically in Fig. 2.7(a) [22]. A short fiber with strong intentional PMD, a so-called *polarization maintaining fiber* (PMF), is placed in front of the photodetector to undo the PMD accumulated during transmission (DGD = 50 ps in the figure). A polarization controller (PC) must be inserted between the transmission fiber and the compensation fiber to make sure that the fast output PSP is fed into the slow input PSP of the compensation fiber and vice versa. Because PMD is time varying, the polarization controller must be adjusted continuously in response to a feedback signal.

Figure 2.7(b) shows an electronic PMD compensator. After the distorted optical signal is converted to the electrical domain, an adaptive equalizer removes as much distortion as possible. Because of the time-varying nature

---

4 More precisely, to ensure an outage probability of $< 10^{-7}$, that is, < 3 seconds/year during which the power penalty exceeds 1 dB, we need $\overline{\Delta T} < 0.1/B$ for RZ modulation and $\overline{\Delta T} < 0.07/B$ for NRZ modulation [18].

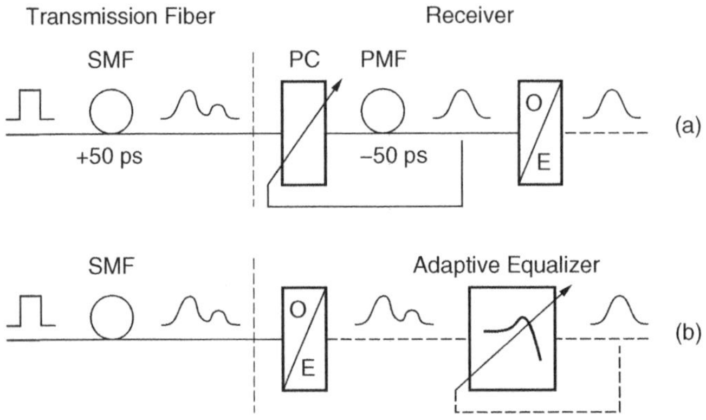

Figure 2.7 PMD compensation in (a) the optical and (b) the electrical domain.

of PMD, the equalizer must adapt continuously to the incoming signal. Electronic PMD compensation is more compact, more agile, and lower in cost than optical PMD compensation. The effectiveness of electronic PMD compensation depends significantly on whether a conventional direct-detection receiver, which loses the polarization information in the detection process, or a coherent receiver with polarization diversity (cf. Fig. 3.35), which preserves the polarization information, is used. For comparison, the direct-detection receiver with EDC in [15] can compensate about 60 ps of DGD, whereas the coherent system with polarization diversity and EDC in [16] can compensate more than 400 ps, making PMD a non-issue, even in older fiber links.

Besides PMD compensation, the effects of PMD can be reduced through the choice of the modulation format. In general, RZ formats can tolerate a larger amount of PMD than NRZ formats [17].

2.3 Nonlinearities

Attenuation and dispersion are *linear* fiber effects because they can be described by a linear relationship between the optical fields at the input and output of the fiber. Apart from these linear effects, the fiber suffers from a number of *nonlinear* effects that may distort and attenuate the optical signal or may produce crosstalk between optical channels. As shown in Fig. 2.8, these effects can be classified into two groups: one that is caused by the dependence of the refractive index on the light intensity (Kerr effect) and one that is caused by the (inelastic) scattering of light by material vibrations (phonons). The first group includes *self-phase modulation* (SPM), *cross-phase modulation* (XPM or CPM), and *four-wave mixing* (FWM); the second group includes *stimulated Raman scattering* (SRS) and *stimulated Brillouin scattering* (SBS)

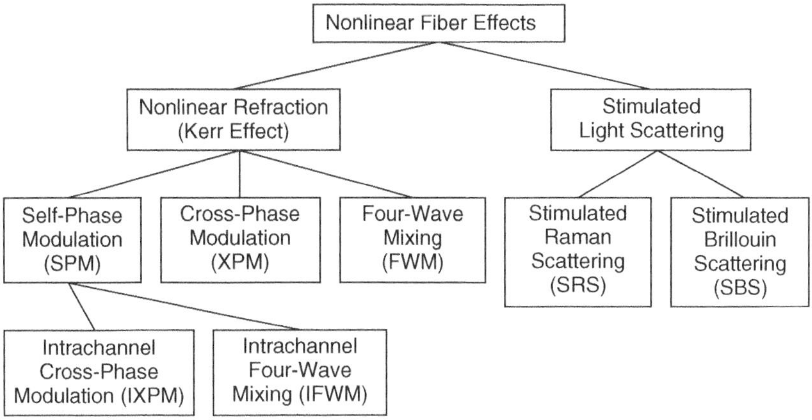

Figure 2.8 Nomenclature of nonlinear fiber effects.

[12, 21, 23, 24]. With the introduction of 40-Gb/s transmission systems, two subtypes of SPM became important and received their own names: *intrachannel cross-phase modulation* (IXPM) and *intrachannel four-wave mixing* (IFWM) [17, 23, 25, 26].

In single-wavelength systems, SPM, IXPM, IFWM, SRS, and SBS can cause nonlinear pulse distortion and nonlinear attenuation. In addition to that, XPM, FWM, and SRS can cause nonlinear crosstalk between optical channels in WDM systems. Whereas distortion within a channel is known as an *intra-channel* impairment, crosstalk between channels is known as an *interchannel* impairment. As you may have guessed from all this nomenclature, nonlinearity in optical fiber is a rather difficult subject. In the following we provide just two examples for illustration.

As an example for an intrachannel nonlinear impairment, we pick IFWM. We know that short optical pulses propagating through a dispersive fiber broaden rapidly, resulting in overlapping pulses. The overlap can span many bit periods, making the original signal shape unrecognizable [26], as sketched in the middle of Fig. 2.9. In the absence of fiber nonlinearities, a dispersion compensating fiber can exactly undo the broadening and restore the original signal. However, in the presence of IFWM the overlapping pulses exchange energy and the dispersion compensated signal is not perfect, as sketched on the right-hand side of Fig. 2.9. The nonlinear interaction that occurs during the overlap phase manifests itself as power fluctuations and shadow (or ghost) pulses in the dispersion compensated signal [23, 26].

As an example for an interchannel nonlinear impairment, let us look at the effect of SRS in a two-channel system. (Although FWM is often the dominant interchannel nonlinear impairment [24], we choose SRS as an example because it is easier to explain.) Figure 2.10 shows how SRS transfers optical energy

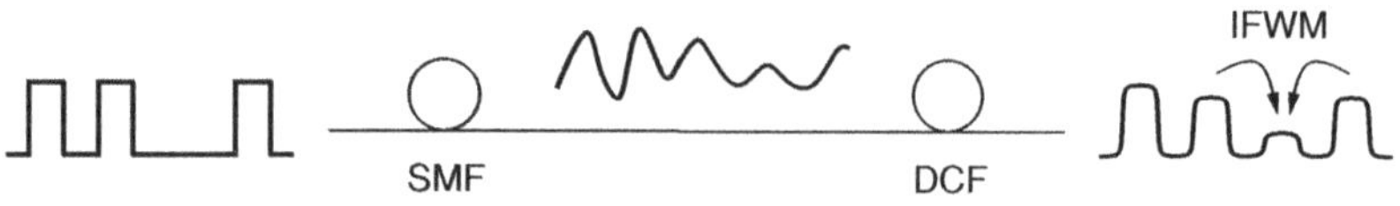

Figure 2.9 Power fluctuations and ghost pulses caused by IFWM.

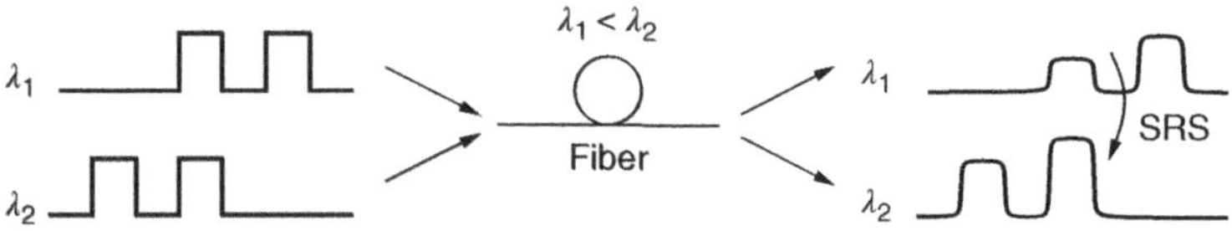

Figure 2.10 Crosstalk and power fluctuations caused by SRS in a two-channel system.

from the channel with the shorter wavelength (λ_1) to the channel with the longer wavelength (λ_2) [24]. This crosstalk results in power fluctuations in both channels. We can explain the SRS effect as follows. Stimulated by the photons in the long-wavelength channel, some of the photons in the short-wavelength channel "decay" into a lower-energy photon and a molecular vibration (a phonon). The lower-energy photon has exactly the same direction, wavelength, phase, and polarization as those in the long-wavelength channel, adding coherently to the signal in this channel. Thus signal energy is transferred from the short-wavelength channel to the long-wavelength channel.

Nonlinearity Mitigation. What can we do to mitigate these nonlinear effects? An obvious remedy is to reduce the optical power launched into the fiber. For very small optical powers, the fiber behaves linearly. With increasing power, nonlinear effects become significant one by one as the power crosses the associated thresholds. A particular power threshold depends on the type of nonlinearity, the transmission fiber (effective length, effective core area, dispersion), the bit rate, the modulation format, and the number of WDM channels [21]. Nonlinear effects thus can be suppressed by keeping the transmitted optical power below the relevant thresholds. To lower the transmitted power without impacting the bit-error rate performance, we can employ one (or several) of the following techniques: shorten the span length between optical amplifiers (cf. Section 4.6), use distributed Raman amplifiers (cf. Section 3.3), or use forward error correction (cf. Appendix G).

Another approach to combat nonlinear effects is to exploit the interplay between fiber nonlinearity and chromatic dispersion. As we have seen, the bits in different channels of a DWDM system interact with each other through nonlinear effects such as SRS, resulting in crosstalk and power fluctuations. The longer the interacting bits stay together, the stronger the nonlinear crosstalk distortions become. For this reason, it helps if the signals in the different wavelength channels propagate at somewhat different speeds, that is, if there is some chromatic dispersion. A special fiber, called *nonzero dispersion-shifted fiber* (NZ-DSF), has been developed that has a small value for $|D|$ around 1 to 6 ps/(nm · km) [21], large enough to create a "walk off" between the channel bits that reduces nonlinear interactions, but small enough to limit the needed dispersion compensation (or to avoid it). In coherent transmission systems with powerful EDC, a regular SMF or a large-core fiber with a higher nonlinear power threshold can be used to create the walk off [16].

Finally, the effects of fiber nonlinearity can be reduced through the choice of the modulation format. For example, at 10 Gb/s and above, the RZ format is more resistant to nonlinearity than the NRZ format [17] permitting a higher optical power [27] and, at 40 Gb/s, CS-RZ and RZ-DPSK show superior resistance to IFWM [17].

2.4 Pulse Spreading due to Chromatic Dispersion

Modal dispersion can be eliminated with an SMF, PMD is only a concern in long-haul and high-speed systems, but almost every optical fiber link is affected by chromatic dispersion. In the following, we investigate the pulse distortions resulting from chromatic dispersion in more detail.

Nonlinear Character of Intensity Modulation and Direct Detection. Conventional optical communication systems transmit a signal by modulating the *intensity* of a light source and receive the signal by detecting the *intensity* of the incoming light. This method is known as *intensity modulation* and *direct detection* (IM/DD). As we know, the alternative to direct detection is coherent detection, but for the following discussion we assume a DD receiver.

The complete communication channel can be modeled as shown in Fig. 2.11. The electrical signal is converted to light with a proportional intensity; the light is described by an optical field (an electromagnetic field at optical frequencies) with a magnitude that is proportional to the square root of the intensity; the optical field propagates through the fiber and disperses linearly; the field at the end of the fiber is characterized by an intensity that is proportional to the square of the optical field's magnitude; finally, the light intensity is detected and an electrical signal proportional to it is generated. This is a *nonlinear* system, despite fiber dispersion being a *linear* effect! Note that the nonlinearity here occurs in the photodetector, not the fiber. It is due to the square law relating the intensity and the field and has nothing to do with the nonlinearity in fibers discussed in the previous section. (This same nonlinearity will bother us again when discussing optical and electrical dBs, and then again when we talk about optical noise.)

Now we understand that we cannot simply apply linear system theory to analyze the pulse distortions caused by fiber dispersion. Fortunately, however,

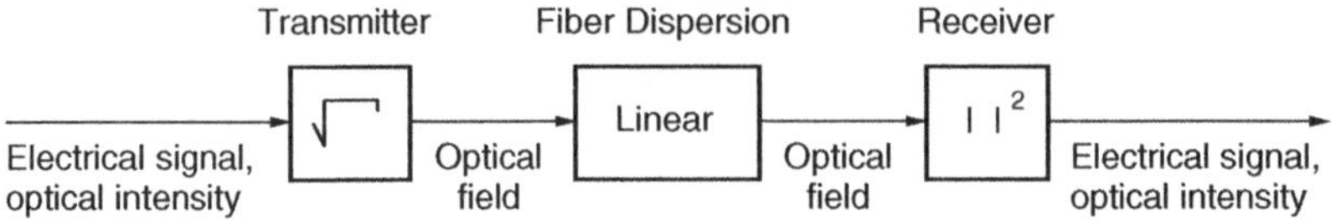

Figure 2.11 Communication channel with intensity modulation, fiber dispersion, and intensity detection.

there is a linear approximation that we can use under certain conditions. If we use a light source with a bandwidth much larger than the signal bandwidth, we can approximately describe the channel with a linear response [28]. With such a wide-linewidth source, the transmitter linewidth $\Delta\lambda$ is approximately equal to the source linewidth $\Delta\lambda_S$, that is, the effects of the modulation on $\Delta\lambda$ can be neglected. In practice, this is the case for transmitters with a light-emitting diode (LED) or a Fabry–Perot (FP) laser source. If we further assume that the source spectrum is Gaussian and we are operating at a wavelength far from zero dispersion, the impulse response of the channel in Fig. 2.11 (ignoring the propagation delay) turns out to be

$$h(t) = h(0) \cdot \exp\left(-\frac{t^2}{2(\Delta T/2)^2}\right), \tag{2.5}$$

where

$$\Delta T = |D| L \cdot \Delta\lambda \tag{2.6}$$

and $\Delta\lambda$ is the 2σ-linewidth of the transmitter (or source). In other words, a Dirac pulse injected into a fiber spreads out into a Gaussian pulse with a 2σ-width, ΔT, that grows linearly with distance traveled. The dispersion parameter D tells us how rapidly the pulse is spreading with distance. For example, for a standard SMF it spreads out to 17 ps after 1 km given a 1.55-μm source with a linewidth of 1 nm.

Time-Domain Analysis. Now that we have a linear model, we are on familiar territory and we can calculate how a regular (non-Dirac) data pulse spreads out. The math is easiest if we assume that the transmitted pulse is Gaussian. The convolution of the (Gaussian) input pulse with the (Gaussian) impulse response produces a Gaussian output pulse (cf. Eq. (I.2)). We find the relationship between the 2σ-width of the input pulse, T_{in}, and the 2σ-width of the output pulse T_{out} to be

$$T_{out} = \sqrt{T_{in}^2 + \Delta T^2}. \tag{2.7}$$

This situation is illustrated in Fig. 2.12. For example, given a source linewidth of 1 nm at 1.55 μm, a 100-ps pulse will broaden to $\sqrt{(100\text{ ps})^2 + (17\text{ ps})^2} =$ 101.4 ps after 1 km of standard SMF. [→ Problem 2.3.]

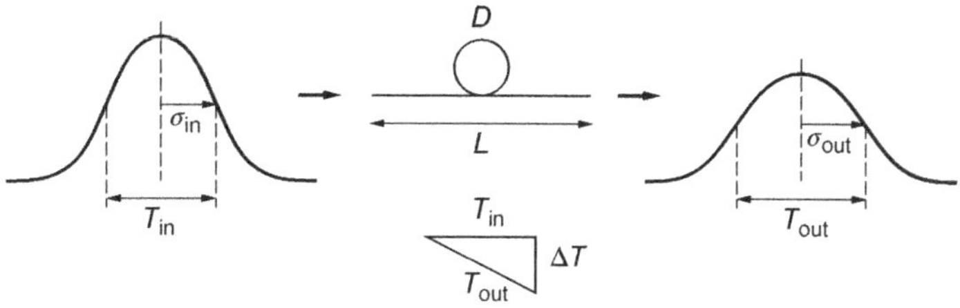

Figure 2.12 Pulse spreading due to chromatic dispersion.

Interestingly, Eq. (2.7) holds even for non-Gaussian pulse shapes, if we interpret $T_{in}/2$, $T_{out}/2$, and $\Delta T/2$ as the *rms pulse widths* of the (arbitrary) input pulse, output pulse, and impulse response, respectively [29].

The maximum amount of spreading, ΔT, that normally can be tolerated in an NRZ-modulated system (without equalizer) is equal to half a bit period [30][5]:

$$\Delta T \leq \frac{1}{2B}. \tag{2.8}$$

This amount of spreading increases the pulse width by $\sqrt{1^2 + 0.5^2} = 1.12$ or about 12% and causes a power penalty of approximately 1 dB [30].

This simple calculation helps us to understand under which circumstances chromatic dispersion becomes important. We can also see from Eqs. (2.6) and (2.8) that the linewidth of the source is of critical importance in determining the amount of pulse spreading in a dispersive fiber.

Frequency-Domain Analysis. Given the expression for the impulse response of a dispersive fiber, we can easily transform it into the frequency domain and discuss the corresponding bandwidth. Transforming the Gaussian impulse response in Eq. (2.5) yields the Gaussian frequency response (cf. Eq. (I.3))

$$H(f) = H(0) \cdot \exp\left(-\frac{(2\pi f)^2(\Delta T/2)^2}{2}\right). \tag{2.9}$$

The 3-dB bandwidth of this optical power response can be found by setting the equation equal to $\frac{1}{2}H(0)$ and solving for f. Together with Eq. (2.6), we find [12][6]

$$BW_{3dB} = \frac{0.375}{\Delta T} = \frac{0.375}{|D|L \cdot \Delta\lambda}. \tag{2.10}$$

This is the modulation-signal bandwidth due to chromatic dispersion first introduced in Section 2.1. Its value decreases as the fiber length L, the linewidth $\Delta\lambda$, or the dispersion parameter D increases. For example, 1 km of

---

5 In terms of the rms impulse spread $\sigma_T = \Delta T/2$, this limit is $\sigma_T \leq 1/(4B)$.
6 In the electrical domain, this bandwidth is the 6-dB bandwidth, because three optical dBs convert to six electrical dBs (cf. Section 3.1).

standard SMF with a source linewidth of 1 nm at 1.55 μm has a bandwidth of just 22 GHz. If we replace the SMF with an NZ-DSF that has the lower dispersion $D = 2$ ps/(nm · km), the bandwidth increases to 188 GHz. [→ Problem 2.4.]

Finally, what is the interpretation of the spreading limit, Eq. (2.8), in the frequency domain? Inserting Eq. (2.8) into Eq. (2.10), we find

$$BW_{3\text{dB}} \geq 0.75B. \tag{2.11}$$

This means that the fiber bandwidth must be made larger than 75% of the bit rate to avoid excessive distortions. [→ Problem 2.5.]

Narrow-Linewidth Source. What happens if we use a light source with a bandwidth much smaller than the signal bandwidth? In practice, this situation occurs for transmitters with a distributed feedback (DFB) laser. Under these circumstances, the source linewidth $\Delta\lambda_S \ll \Delta\lambda$ can be neglected and the transmitter linewidth $\Delta\lambda$ is determined by the modulation format, the bit rate, and possibly spurious frequency modulation of the optical carrier, known as *chirp*.

As we said earlier, for a narrow-linewidth source and a direct-detection receiver, the communication channel becomes nonlinear and cannot be described by linear system theory. In particular, the concept of modulation-signal bandwidth cannot be applied strictly. We can still interpret Eqs. (2.5) and (2.6) loosely to mean that the pulse spreading, ΔT, tends to increase with increasing transmitter linewidth, $\Delta\lambda$, that is, with higher bit rates and stronger chirp. However, there are important exceptions to this rule, such as for optical pulses with *negative* chirp and for so-called *solitons*. Pulses with negative chirp are characterized by a temporary decrease in optical frequency (red shift) during the leading edge and an increase in frequency (blue shift) during the trailing edge. Although negative as well as positive chirp broadens the transmitter linewidth, pulses with negative chirp become *compressed* up to a certain distance in a dispersive medium with $D > 0$ [12]. Solitons are short (≈10 ps) and powerful optical pulses of a particular shape (hyperbolic secant). Although the shortness of these pulses implies a wide transmitter linewidth, they do not broaden at all. The reason is that chromatic dispersion is counterbalanced by self-phase modulation, one of the nonlinear fiber effects [12]. [→ Problem 2.6.]

2.5 Summary

Optical silica-glass fiber is characterized by a very low loss of about 0.2 dB/km (at the 1.55-μm wavelength) and a huge bandwidth of more than 10 THz (in the C and L bands).

Various types of dispersion cause the optical pulses to spread out in time and to interfere with each other. The following types of dispersion can be distinguished:

- Modal dispersion, which only occurs in multimode fibers.
- Chromatic dispersion, which is small at 1.3 μm, but presents a significant impairment at the 1.55-μm wavelength in standard single-mode fibers.
- PMD, which is significant in high-speed and long-haul transmission, especially when using older fiber types. PMD is slowly and randomly varying with time.

Modal dispersion can be suppressed in the optical domain by using single-mode fibers. Chromatic and polarization-mode dispersion can be compensated in the optical domain by using special fibers. All types of dispersion can be compensated, to a varying degree, in the electrical domain using equalizers. Electronic dispersion compensation (EDC) is particularly effective when used with a coherent receiver with phase and polarization diversity.

A spectrally narrow transmitter keeps the effects of chromatic dispersion small.

At elevated power levels, nonlinear fiber effects cause undesirable attenuation and pulse distortions. In dense wavelength division multiplexing (DWDM) systems, nonlinear effects can cause crosstalk. Keeping the optical power low and the chromatic dispersion nonzero helps to suppress nonlinear effects.

Problems

2.1 **Wavelength and Frequency.** (a) At what frequency oscillates the optical field of a 1.55-μm lightwave? (b) An optical filter has a bandwidth of 0.1 nm at 1.55 μm. What is its bandwidth in Hertz?

2.2 **Group-Delay Variation.** Assume that the dispersion parameter D depends linearly on the wavelength. It has a value of zero at 1.3 μm and 17 ps/(nm · km) at 1.55 μm. Calculate the dependence of the group delay on wavelength.

2.3 **Pulse Spreading.** Derive the pulse-spreading rule for Gaussian pulses given in Eq. (2.7) from the impulse response given in Eq. (2.5).

2.4 **Fiber Response.** Derive the frequency response given in Eq. (2.9) from the impulse response given in Eq. (2.5).

2.5 **1-dB Dispersion Penalty.** What is the highest bit rate, B, at which we can transmit an NRZ signal while incurring a power penalty of less than

1 dB due to chromatic dispersion? Use Eq. (2.9) to estimate the penalty (attenuation) and assume that most of the data signal's energy is located at the frequency $B/2$ (as for the "101010···" sequence).

2.6 **Pulse Compression.** Explain qualitatively why a pulse with negative chirp initially becomes compressed for a fiber with $D > 0$.

2.7 **Transmission System at 1,310 nm.** A 1.31-μm transmitter with a 2-nm linewidth launches a 10-Gb/s NRZ signal with 1 mW into a standard SMF. (a) How long can we make the fiber before the power is attenuated to −21.5 dBm? (b) How long can we make the fiber before chromatic dispersion causes too much pulse spreading? Assume $D = 0.3$ ps/(nm · km).

2.8 **Transmission System at 1,550 nm.** Now we use a 1.55-μm transmitter with the same linewidth, bit rate, and launch power as in Problem 2.7. How does the situation change?

2.9 **Transmitter Linewidth.** (a) In which system, Problem 2.7 or 2.8, would it make sense to use a narrow-linewidth transmitter? (b) How far could we go, if we reduce the linewidth to 0.02 nm?

2.10 **Fiber PMD.** We are using fiber with $D_{PMD} = 0.1$ ps/$\sqrt{\text{km}}$. Do we have to be concerned about PMD in any of the transmission systems analyzed in Problems 2.7–2.9?

References

1 J. Hecht. *City of Light: The Story of Fiber Optics.* Oxford University Press, New York, NY, 1999.
2 H. J. R. Dutton. *Understanding Optical Communications.* Prentice Hall PTR, Upper Saddle River, NJ, 1998.
3 Y. Dong and K. W. Martin. Gigabit communications over plastic optical fiber. *IEEE Solid-State Circuits Mag.*, 3(1):60–69, 2011.
4 A. K. Srivastave and Y. Sun. Advances in erbium-doped fiber amplifiers. In I. P. Kaminow and T. Li, editors, *Optical Fiber Telecommunications IVA*, pages 174–212. Academic Press, San Diego, CA, 2002.
5 J. L. Zyskind, J. A. Nagel, and H. D. Kidorf. Erbium-doped fiber amplifiers for optical communications. In I. P. Kaminow and T. L. Koch, editors, *Optical Fiber Telecommunications IIIB*, pages 13–68. Academic Press, San Diego, CA, 1997.

6 S. Namiki, K. Seo, N. Tsukiji, and S. Shikii. Challenges of Raman amplification. *Proc. IEEE*, 94(5):1024–1035, 2006.

7 K. Rottwitt and A. J. Stentz. Raman amplification in lightwave communication systems. In I. P. Kaminow and T. Li, editors, *Optical Fiber Telecommunications IVA*, pages 213–257. Academic Press, San Diego, CA, 2002.

8 J. Hansryd, P. A. Andrekson, M. Westlund, J. Li, and P.-O. Hedekvist. Fiber-based optical parametric amplifiers and their applications. *IEEE J. Sel. Top. Quantum Electron.*, 8(3):506–520, 2002.

9 W. C. Barnett, H. Takahira, J. C. Baroni, and Y. Ogi. The TPC-5 cable network. *IEEE Commun. Mag.*, 34(2):36–40, 1996.

10 J. Conradi. Bandwidth-efficient modulation formats for digital fiber transmission systems. In I. P. Kaminow and T. Li, editors, *Optical Fiber Telecommunications IVB*, pages 862–901. Academic Press, San Diego, CA, 2002.

11 J. M. Senior. *Optical Fiber Communications: Principles and Practice.* Prentice Hall, Harlow, England, 3rd edition, 2009.

12 G. P. Agrawal. *Fiber-Optic Communication Systems.* John Wiley & Sons, Inc., Hoboken, NJ, 4th edition, 2010.

13 P. W. Shumate. Lightwave transmitters. In S. E. Miller and I. P. Kaminow, editors, *Optical Fiber Telecommunications II*, pages 723–757. Academic Press, San Diego, CA, 1988.

14 O. E. Agazzi, M. R. Hueda, D. E. Crivelli, H. S. Carrer, A. Nazemi, G. Luna, F. Ramos, R. López, C. Grace, B. Kobeissy, C. Abidin, M. Kazemi, M. Kargar, C. Marquez, S. Ramprasad, F. Bollo, V. Posse, S. Wang, G. Asmanis, G. Eaton, N. Swenson, T. Lindsay, and P. Voois. A 90 nm CMOS DSP MLSD transceiver with integrated AFE for electronic dispersion compensation of multimode optical fibers at 10 Gb/s. *IEEE J. Solid-State Circuits*, SC-43(12):2939–2957, 2008.

15 H.-M. Bae, J. B. Ashbrook, J. Park, N. R. Shanbhag, A. C. Singer, and S. Chopra. An MLSE receiver for electronic dispersion compensation of OC-192 fiber links. *IEEE J. Solid-State Circuits*, SC-41(11):2541–2554, 2006.

16 K. Roberts, S. H. Foo, M. Moyer, M. Hubbard, A. Sinclair, J. Gaudette, and C. Laperle. High capacity transport – 100G and beyond. *J. Lightwave Technol.*, LT-33(3):563–578, 2015.

17 P. J. Winzer and R.-J. Essiambre. Advanced optical modulation formats. *Proc. IEEE*, 94(5):952–984, 2006.

18 H. Kogelnik, R. M. Jopson, and L. E. Nelson. Polarization-mode dispersion. In I. P. Kaminow and T. Li, editors, *Optical Fiber Telecommunications IVB*, pages 725–861. Academic Press, San Diego, CA, 2002.

19 H. Yaffe. Taking the guesswork out of PMD tolerance testing. *Lightwave*, 24(12)2007.

20 A. Gladisch, R.-P. Braun, D. Breuer, A. Ehrhardt, H.-M. Foisel, M. Jaeger, R. Leppla, M. Schneiders, S. Vorbeck, W. Weiershausen, and F.-J. Westphal.

Evolution of terrestrial optical system and core network architecture. *Proc. IEEE*, 94(5):869–891, 2006.

21 R. Ramaswami, K. N. Sivarajan, and G. Sasaki. *Optical Networks: A Practical Perspective*. Morgan Kaufmann Publishers, San Francisco, CA, 3rd edition, 2010.

22 F. Roy, C. Francia, F. Bruyère, and D. Penninckx. A simple dynamic polarization mode dispersion compensator. In *Optical Fiber Communication Conference (OFC)*, San Diego, CA, February 1999.

23 P. Bayvel and R. Killey. Nonlinear optical effects in WDM transmission. In I. P. Kaminow and T. Li, editors, *Optical Fiber Telecommunications IVB*, pages 611–641. Academic Press, San Diego, CA, 2002.

24 F. Forghieri, R. W. Tkach, and A. R. Chraplyvy. Fiber nonlinearities and their impact on transmission systems. In I. P. Kaminow and T. L. Koch, editors, *Optical Fiber Telecommunications IIIA*, pages 196–264. Academic Press, San Diego, CA, 1997.

25 G. Charlet and S. Bigo. Upgrading WDM submarine systems to 40-Gbit/s channel bitrate. *Proc. IEEE*, 94(5):935–951, 2006.

26 R.-J. Essiambre, G. Raybon, and B. Mikkelsen. Pseudo-linear transmission of high-speed TDM signals: 40 and 160 Gb/s. In I. P. Kaminow and T. Li, editors, *Optical Fiber Telecommunications IVB*, pages 232–304. Academic Press, San Diego, CA, 2002.

27 R. Ludwig, U. Feiste, E. Dietrich, H. G. Weber, D. Breuer, M. Martin, and F. Küppers. Experimental comparison of 40 Gbit/s RZ and NRZ transmission over standard singlemode fibre. *Electron. Lett.*, 35(25):2216–2218, 1999.

28 M. J. Bennett. Dispersion characteristics of monomode optical-fiber systems. *IEE Proceedings, Pt. H*, 130(5):309–314, 1983.

29 R. G. Smith and S. D. Personick. Receiver design for optical fiber communication systems. In H. Kressel, editor, *Topics in Applied Physics: Semiconductor Devices for Optical Communication*, volume 39. Springer-Verlag, Berlin, Germany, 1982.

30 P. S. Henry, R. A. Linke, and A. H. Gnauck. Introduction to lightwave systems. In S. E. Miller and I. P. Kaminow, editors, *Optical Fiber Telecommunications II*, pages 781–831. Academic Press, San Diego, CA, 1988.

3

Photodetectors

The first element of an optical receiver is the *photodetector*. The characteristics of this device have a significant impact on the receiver's performance. To achieve a good receiver sensitivity, the photodetector must have a large response to the received optical signal, have a bandwidth that is sufficient for the incoming signal, and generate as little noise as possible.

We start with the three most common photodetectors: the p–i–n photodetector, the avalanche photodetector (APD), and the optically preamplified p–i–n detector, discussing their responsivity, bandwidth, and noise characteristics. Then, we turn our attention to photodetectors that are suitable for integration in a circuit technology, in particular, detectors compatible with CMOS technology (silicon-photonics detectors). Finally, we explore detectors for phase-modulated optical signals, such as QPSK and DQPSK, including the coherent detector with phase and polarization diversity.

3.1 p–i–n Photodetector

The *p–i–n photodetector* (or *p–i–n photodiode*) shown schematically in Fig. 3.1(a) and (b) is one of the simplest detectors. It consists of a p–n junction with a layer of intrinsic (undoped or lightly doped) semiconductor material sandwiched in between the p- and the n-doped material. The junction is reverse biased with $-V_{PIN}$ to create a strong electric field in the intrinsic layer. The light enters through a hole in the top electrode (anode), passes through the p-doped material, and reaches the i-layer, which is also known as the *absorption layer*. The photons incident on the absorption layer knock electrons from the valence band to the conduction band creating electron–hole pairs. These pairs become separated by the strong electric drift field with the holes traveling to the negative terminal and the electrons traveling to the positive terminal, as indicated in Fig. 3.1(a). As a result, the photocurrent i_{PIN} appears at the diode terminals. Figure 3.1(c) shows the circuit symbol for the photodiode.

Analysis and Design of Transimpedance Amplifiers for Optical Receivers, First Edition. Eduard Säckinger.

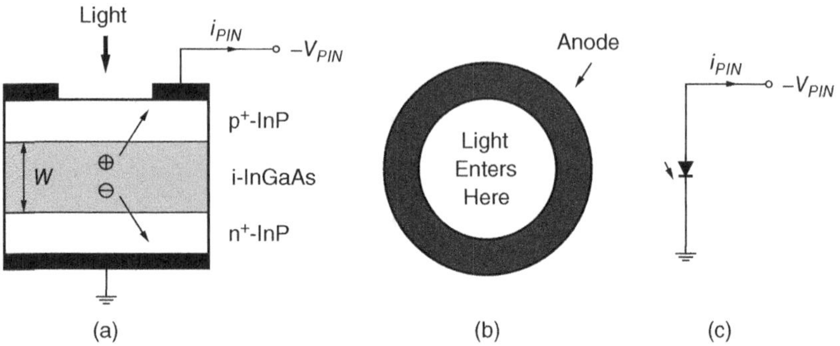

Figure 3.1 Vertically illuminated p–i–n photodetector: (a) cross-sectional view, (b) top view, and (c) circuit symbol.

Quantum Efficiency. The fraction of incident photons that results in electron–hole pairs contributing to the photocurrent is an important performance parameter known as the *quantum efficiency* η. An ideal photodetector has a 100% quantum efficiency, $\eta = 1$.

In a vertically illuminated photodetector, as the one in Fig. 3.1, the quantum efficiency depends on the width W of the absorption layer. The wider W is made, the better the chances that a photon is absorbed in this layer become. More specifically, the photon absorption efficiency is $1 - e^{-\alpha W}$, where $1/\alpha$ is the *absorption length* (a.k.a. *penetration depth*). The InGaAs material, for example, has an absorption length of about 1 μm when illuminated at the 1.3 to 1.5-μm wavelength [1, 2]. If the absorption layer width is made equal to the absorption length ($\alpha W = 1$), the photon absorption efficiency is around 63%; for widths much larger than the absorption length, the photon absorption efficiency asymptotically approaches 100%; for widths much smaller than the absorption length, the photon absorption efficiency is approximately proportional to W. A technique for improving the photon absorption efficiency is to make the bottom electrode reflective thus sending the not-yet-absorbed photons back up into the absorption layer giving them another chance to make a useful contribution to the photocurrent (double-pass scheme) [3].

The quantum efficiency also depends on how much light is coupled from the fiber into the detector. To that end, the sensitive area of the photodetector should be made large enough to completely cover the light spot from the fiber and the detector's surface should be covered with an antireflection coating to maximize the light entering the detector. Another factor affecting the quantum efficiency is the fraction of electron–hole pairs that is collected by the electrodes and contributes to the photocurrent as opposed to the fraction that is lost to recombination.

Overall, the quantum efficiency can be understood as the product of three factors: fiber-to-detector coupling efficiency, photon absorption efficiency, and electron–hole pair collection efficiency. Sometimes the term *external quantum efficiency* is used for this overall quantum efficiency, whereas the term *internal quantum efficiency* refers to an internal aspects of the detection process, such as the photon absorption efficiency [3] or the electron–hole pair collection efficiency [4]. (Caution: Not all authors use these two terms in the same way.)

Spectral Response. Most semiconductor materials are transparent at the 1.3- and 1.55-μm wavelengths commonly used in telecommunication applications, that is, they do *not* absorb photons at these wavelengths. For example, silicon absorbs photons for $\lambda < 1.09$ μm only, gallium arsenide (GaAs) for $\lambda < 0.87$ μm only, and indium phosphide (InP) for $\lambda < 0.92$ μm only [2]. For a semiconductor to absorb photons, its bandgap energy E_g must be smaller than the photon energy: $E_g < hc/\lambda$, where h is the Planck constant. Only then do the photons have enough punch to knock electrons from the valence band into the conduction band. Therefore, the absorption layer in a photodetector must be made of a semiconductor compound with a sufficiently narrow bandgap. Nevertheless, the bandgap should not be made too narrow either to avoid an excessive thermally generated dark current.

The quantum efficiency of a photodetector degrades toward the long-wavelength *and* the short-wavelength ends of the spectrum. The long-wavelength cutoff results from a lack of absorption when the photon energy drops below the bandgap energy, as discussed earlier. Interestingly, for high-energy photons (short wavelengths) the absorption length becomes so short that most photons are absorbed near the surface where many of the generated electron–hole pairs recombine before they reach the electrodes [5]. In other words, at long wavelengths the detector is limited by a low photon absorption efficiency and at short wavelengths the detector is limited by a low electron–hole pair collection efficiency.

For photodetectors that are sensitive at the 1.3- and 1.55-μm wavelengths, a common choice for the absorption-layer material is indium gallium arsenide (InGaAs or more precisely $In_{0.53}Ga_{0.47}As$), which has the important property of being lattice matched to the InP substrate (cf. Fig. 3.1(a)). InGaAs has a bandgap of 0.75 eV making the detector sensitive to wavelengths with $\lambda < 1.65$ μm [2]. Choosing InP for the p- and n-layers has the advantage that they are transparent at the wavelengths of interest, permitting a top or bottom illumination of the InGaAs absorption layer. Another absorption-layer material suitable for the 1.3- and 1.55-μm wavelengths is germanium (Ge). It has a bandgap of 0.67 eV and can be grown on a silicon substrate by epitaxy (4% lattice mismatch), making it of particular interest for silicon photonics.

Detectors for the 0.85-μm wavelength (commonly used in data-communication applications) are typically based on silicon or GaAs. Whereas silicon is lower in cost, GaAs offers a higher speed. Silicon has a longer absorption length than GaAs, because its indirect bandgap (at 0.85 μm) requires the participation of a phonon to conserve momentum as well as energy. This low absorption rate must be compensated with a wider absorption layer, which makes the silicon detector slower [1, 2].

Bandwidth. The speed of a p–i–n photodetector depends mainly on the following factors: the width of the absorption layer W, the reverse bias voltage V_{PIN}, the presence of slow diffusion currents, the photodiode capacitance, and packaging parasitics. We briefly discuss these factors in this order.

The width of the absorption layer W determines the time it takes for the electrons and holes to traverse it. To obtain a fast response, this *transit time* must be kept short. For example, whereas $W = 2.7$ μm is fine for a 10-GHz InGaAs photodetector [6], the width must be reduced to 0.4 μm for a 40-GHz detector [7] or even 0.2 μm for a 100-GHz detector [3]. The problem with reducing W is that it also reduces the quantum efficiency. Whereas a 10-GHz detector still has a good quantum efficiency (its absorption layer width is more than twice the absorption length), at 40 GHz and above the quantum efficiency quickly becomes unacceptable, prompting an alternative photodetector design. The solution is to replace the vertically illuminated p–i–n detector from Fig. 3.1(a) with a so-called edge-coupled photodetector, which we discuss shortly.

The transit time not only depends on the width W but also on the strength of the electric drift field in the absorption layer. With increasing field strength E the carrier velocity increases and (after a possible velocity overshoot) saturates at $\upsilon_{c,\mathrm{sat}}$. For holes in InGaAs $\upsilon_{c,\mathrm{sat}} \approx 50$ μm/ns (for electrons $\upsilon_{c,\mathrm{sat}} \approx 70$ μm/ns) and is reached for $E > 5$ V/ μm [6]. Thus, to obtain the minimum transit time $W/\upsilon_{c,\mathrm{sat}}$, the bias voltage, V_{PIN}, must be high enough such that velocity saturation is reached. For a 10-GHz photodetector the bandwidth saturates at around 4 to 10 V [6], for a 40-GHz detector at around 2 to 3 V [7], and for a 100-GHz detector at around 1.5 to 2 V [3]. As the width of the absorption layer is reduced for higher speed devices, less voltage is needed to reach the field at which the velocity saturates. On the high end, the bias voltage is limited by the onset of avalanche breakdown. At this point, the reverse current increases rapidly, as illustrated in Fig. 3.2. (The I/V characteristic of a dark p–i–n photodetector is identical to that of a regular p–n junction; when illuminated, it shifts down along the current axis by the amount of the photocurrent I_{PIN}.) Power dissipation ($V_{PIN} \times I_{PIN}$) is another consideration limiting the bias voltage, especially when the photocurrent is large as, for example, in a coherent receiver.

Photons absorbed outside of the drift field create slowly diffusing carriers, which when eventually stumbling into the drift field make a delayed contribution to the photocurrent. For example, photons absorbed in the (neutral) n-layer of a silicon p–i–n photodetector create electron–hole pairs. The holes, which are the minority carriers, take about 1.6 ns to diffuse through 2 μm of silicon [2] (and four times as long for twice this distance). As a result, the desired current pulse corresponding to the optical signal is followed by a

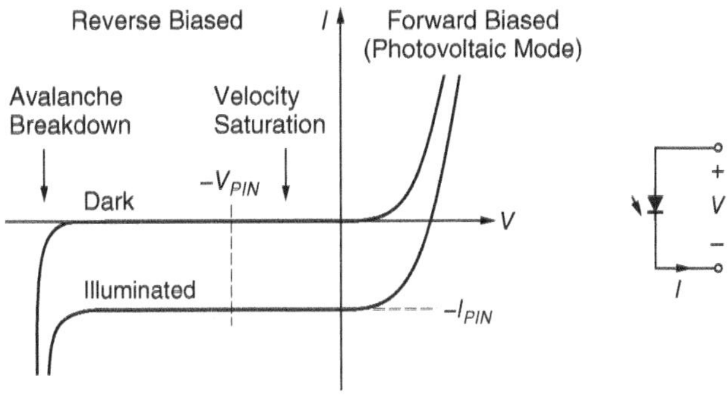

Figure 3.2 I/V characteristics of a dark and an illuminated p–i–n photodetector.

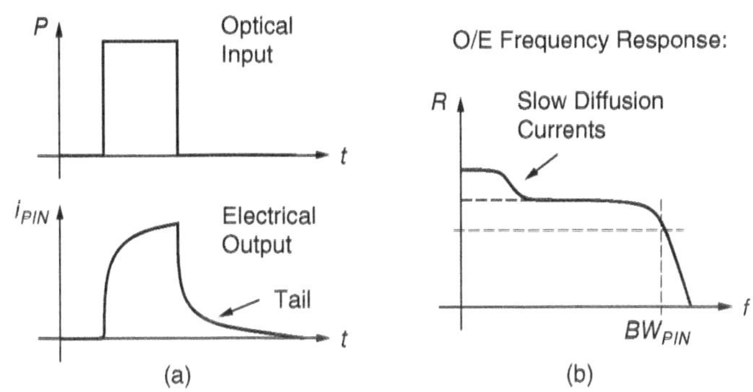

Figure 3.3 Impact of diffusion currents in (a) the time domain and (b) the frequency domain.

spurious current tail, as shown in Fig. 3.3(a) [1, 8]. In the frequency response, the diffusion currents manifest themselves as a hump at low frequencies, as shown in Fig. 3.3(b) [8]. Diffusion currents can be minimized by using either transparent materials for the p- and n-layers or by making the layers very thin and aligning the fiber precisely to the active part of the absorption layer. Diffusion currents are particularly bothersome in burst-mode receivers, where the tail of a very strong burst may mask the subsequent (weak) burst [9].

The capacitance of the p–i–n photodetector C_{PD} together with the contact and load resistance present another speed limitation. Figure 3.4(a) shows an equivalent AC circuit for a bare p–i–n photodetector (without packaging parasitics). The current source i_{PIN} represents the photocurrent generated in the p–i–n structure. Besides the photodiode junction capacitance C_{PD}, the combination of contact and spreading resistance is modeled by R_{PD}. Given the load resistance R_L, often assumed to be 50 Ω, the time constant of this RC network is $(R_{PD} + R_L)C_{PD}$.

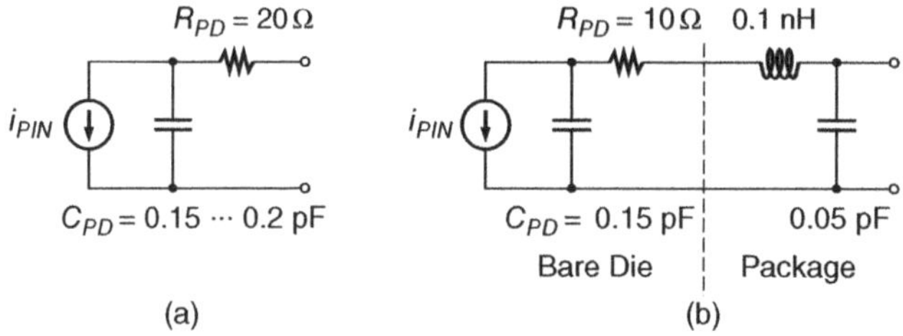

Figure 3.4 Equivalent AC circuits for 10-Gb/s p–i–n photodetectors: (a) bare photodiode [10] and (b) photodiode with packaging parasitics [11].

The bandwidth due to this RC network alone, that is, the RC-limited bandwidth, follows easily as $1/[2\pi(R_{PD} + R_L)C_{PD}]$. The bandwidth due to the transit time alone, that is, the transit-time-limited bandwidth, can be approximated as $3.5v_{c,\text{sat}}/(2\pi W)$. (The numerical factor is given variously as 2.4 [5], 2.8 [12], 2.4 to 3.4 [4], and 3.5 [3].) Combining these two bandwidths results in the following bandwidth estimate for the bare p–i–n photodiode [3]:

$$BW_{PIN} = \frac{1}{2\pi} \cdot \frac{1}{\sqrt{\left(\frac{W}{3.5v_{c,\text{sat}}}\right)^2 + \left[(R_{PD} + R_L)\, C_{PD}\right]^2}}. \tag{3.1}$$

As we make the absorption layer thinner and thinner to reduce the transit time, unfortunately, the diode capacitance gets larger and larger, possibly making the RC time constant in Eq. (3.1) the dominant contribution. One solution is to reduce the area of the photodetector (which, however, may also reduce the coupling efficiency), another solution is to replace the lumped photodiode capacitance with a distributed one, leading to the traveling-wave photodetector, which we discuss shortly.

In addition to C_{PD} and R_{PD}, the packaged photodetector has LC parasitics caused by wire bonds, lead frames, and so forth, as shown in Fig. 3.4(b). In high-speed photodetectors, these parasitics can significantly impact the overall bandwidth and close attention must be payed to them [3, 6].

The equivalent AC circuits in Fig. 3.4 can be extended to model the transit-time effect by replacing the current source i_{PIN} with a voltage-controlled current source, connected to the output of a noiseless RC low-pass filter with time constant $W/(3.5v_{c,\text{sat}})$ [13].

p–i–n Photodetectors for 40 Gb/s and Faster. As we have seen, the vertically illuminated photodetector suffers from a rapidly diminishing quantum efficiency at speeds of 40 Gb/s and above. The bandwidth-efficiency product ($BW_{PIN} \times \eta$) of vertically illuminated p–i–n detectors tops out at about 20 to 35 GHz [14]. This issue can be resolved by illuminating the photodetector from the side rather than from the top, as shown in Fig. 3.5. This configuration is known as an

Figure 3.5 Waveguide p–i–n photodetector.

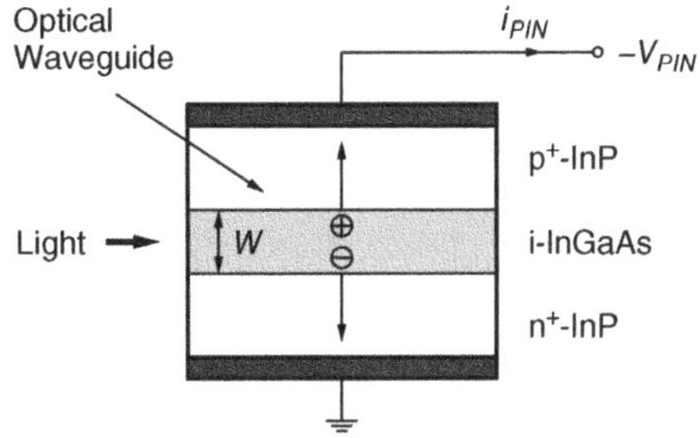

edge-coupled photodetector or a *waveguide photodetector*. Now, the quantum efficiency is controlled by the horizontal dimension, which can be made large, whereas the transit time is controlled by the vertical dimension W, which can be made small.

However, this is easier said than done. The main difficulty is to efficiently couple the light from the fiber with a core of 8 to 10 μm into the absorption layer with a submicrometer width. For comparison, vertically illuminated photodetectors have a diameter of 20 μm or more. Even when focused by a lens, the light spot is still too large for the thin absorption layer. One solution, the so-called *double-core waveguide photodetector*, is to embed the thin absorption layer into a larger optical multimode waveguide that couples more efficiently to the external fiber [3]. Another solution, the so-called *evanescently coupled waveguide photodetector*, is to place an optical waveguide designed for good coupling with the external fiber in parallel to (but outside of) the absorption layer and take advantage of the evanescent field (near field), which extends outside of the optical waveguide, to do the coupling [3].

For example, the 40-Gb/s InGaAs evanescently coupled waveguide photodetector reported in [7] achieves a 47-GHz bandwidth and a 65% quantum efficiency, the InGaAs double-core waveguide photodetector in [3] achieves a 110-GHz bandwidth and a 50% quantum efficiency, and the GaAs (short wavelength) waveguide photodetector in [15] achieves a 118-GHz bandwidth and a 49% quantum efficiency. For a packaged waveguide p–i–n photodetector, see Fig. 3.7.

Even after edge coupling, the photodiode junction capacitance and its associated RC time constant is still a limiting factor, especially for high-speed detectors. The solution to this problem is to replace the photodiode contact pad by a terminated transmission line. The transmission line still has a large capacitance, but now it is distributed in between inductive elements that make the overall transmission line impedance real valued. Figure 3.6 shows a so-called *traveling-wave photodetector* terminated by the resistor R_T, which matches the characteristic impedance of the transmission line. The light pulse enters from the left and gets weaker and weaker as it travels through the absorption layer. The photogenerated carriers get collected by the waveguide

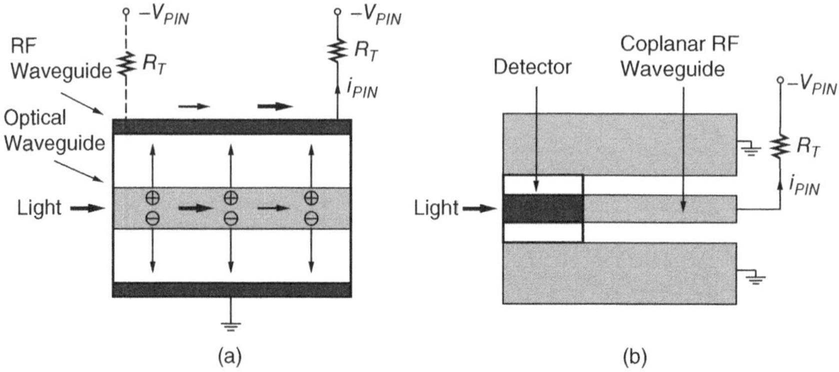

Figure 3.6 Traveling-wave p–i–n photodetector: (a) cross-sectional view and (b) top view (different scale).

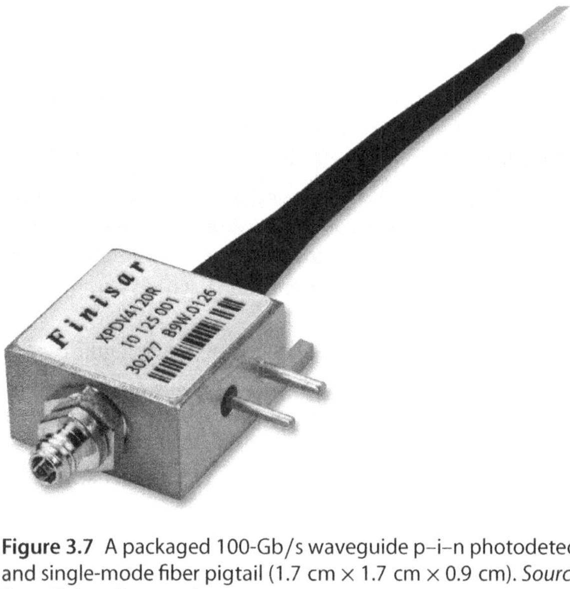

Figure 3.7 A packaged 100-Gb/s waveguide p–i–n photodetector with biasing network and single-mode fiber pigtail (1.7 cm × 1.7 cm × 0.9 cm). *Source*: Reprinted by permission from Finisar Corporation.

at the top producing a stronger and stronger electrical pulse as it travels to the right. The top view in Fig. 3.6(b) shows how the photodiode electrode is made part of a coplanar waveguide. The idea behind the traveling-wave photodetector is the same as that behind the distributed amplifier (cf. Section 7.8), except that the electrical input is replaced by an optical input.

In principle, the bandwidth of the traveling-wave photodetector is independent of the detector's length. However, in practice the bandwidth is limited by the velocity mismatch between the optical traveling wave and the electrical traveling wave, which is not easy to keep small [3]. Another issue is the backward-traveling electrical wave, which can be terminated into another resistor (R_T with dashed lines in Fig. 3.6(a)) or can be left open. In the first case, the efficiency is cut in half as a result of the current lost in the back termination; in the second case, the reflected backward-traveling wave reduces the bandwidth of the photodetector, especially when the photodetector is long [3].

For example, the GaAs traveling-wave photodetector reported in [15] achieves a 172-GHz bandwidth and a 42% quantum efficiency, demonstrating the very high bandwidth-efficiency product of 72 GHz.

Responsivity. Let us calculate the current i_{PIN} produced by a p–i–n photodetector that is illuminated with the optical power P. Each photon has the energy hc/λ. Given the incident optical power P, the photons must arrive at the average rate $P/(hc/\lambda)$. Of all those photons, the fraction η creates electron–hole pairs that contribute to the photocurrent. Thus the average electron rate becomes $\eta P/(hc/\lambda)$. Multiplying this rate by the electron charge q gives us the "charge rate," which is nothing else but the photocurrent:

$$i_{PIN} = \eta \frac{\lambda q}{hc} P. \tag{3.2}$$

The factor relating i_{PIN} to P is known as the *responsivity* of the photodetector and is designated by the symbol $\mathcal{R}$:

$$i_{PIN} = \mathcal{R}P \quad \text{with} \quad \mathcal{R} = \eta \frac{\lambda q}{hc}. \tag{3.3}$$

For example, for the commonly used wavelength $\lambda = 1.55$ μm and the quantum efficiency $\eta = 0.64$, we obtain the responsivity $\mathcal{R} = 0.8$ A/W. This means that for every milliwatt of optical power incident onto the photodetector, we obtain 0.8 mA of current. The responsivity of a typical InGaAs p–i–n photodetector is in the range 0.6 to 0.9 A/W [1].

A Two-for-One Special. The relationship in Eq. (3.3) has an interesting property: If we double the light power, the photodiode current doubles as well. Now this is very odd! Usually, power is related to the *square* of the current rather than the current directly. For example, if we double the RF power radiated at a wireless receiver, the antenna current increases by a factor $\sqrt{2}$. Or, if we double the current flowing through a resistor, the power dissipated into heat increases by 4×. This square-law relationship between power and current is the reason why we use "10 log" to calculate power dBs and "20 log" to calculate current or voltage dBs. When using this convention, a 3-dB increase in RF power translates into a 3-dB increase in antenna current, or a 6-dB increase in current results in a 6-dB increase in power dissipation in the resistor. For a photodetector, however, a

3-dB increase in optical power translates into a 6-dB increase in current. What a bargain! [→ Problems 3.1 and 3.2.]

Wireless Receiver with a Photodetector? Unlike optical receivers, wireless receivers use antennas to detect electromagnetic waves. The rms current that is produced by an antenna under matched conditions is [16]

$$i_{\mathrm{ANT}}^{rms} = \sqrt{\frac{P}{R_{\mathrm{ANT}}}}, \tag{3.4}$$

where P is the received power (more precisely, the power incident on the effective aperture of the antenna) and R_{ANT} is the antenna resistance. For example, for a −50 dBm signal (10^{-8} W), we obtain approximately 14 μA rms from an antenna with $R_{\mathrm{ANT}} = 50\ \Omega$.

What if we replace our antenna with a hypothetical hyperinfrared photodetector that can detect a 1-GHz RF signal? Let us assume we succeeded in making a photodetector with a very small bandgap that is sensitive to low-energy RF photons (4 μeV) and suppressing thermally generated dark currents by cooling the detector to a millikelvin or so. We can then calculate the responsivity of this detector with Eq. (3.3) to be an impressive 120 kA/W assuming that $c/\lambda =$ 1 GHz and $\eta = 0.5$. So, for the same received power level of −50 dBm, we obtain a current of 1.2 mA, that is, almost 100× more than with the old-fashioned antenna! The reason for this, of course, is that the photodetector produces a current proportional to the *square* of the electromagnetic field, whereas the antenna produces a current directly proportional to the field.

But do not launch your start-up company to market this idea just yet! What happens if we reduce the received power? After all, it is for weak signals where the detector's responsivity matters the most. The signal from the photodetector decreases *linearly*, whereas the signal from the antenna decreases more slowly following a *square-root* law. Once we are down to −90 dBm (10^{-12} W), we obtain approximately 140 nA from the antenna and 120 nA from the photodetector (see Fig. 3.8), which is about the same!

The aforementioned comparison is meant to illustrate signal detection laws and for simplicity disregards detector noise. To be fair, we should compare the

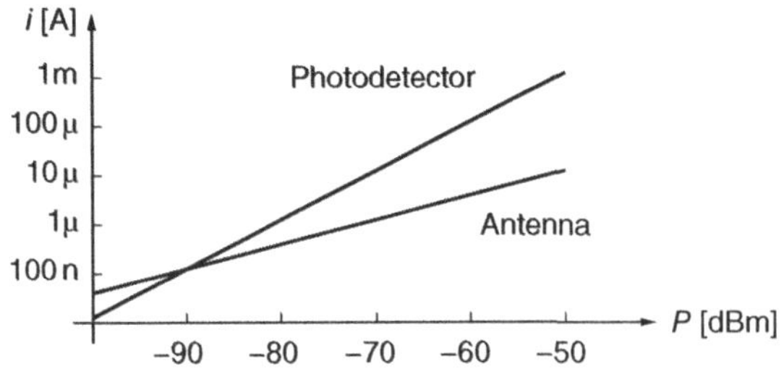

Figure 3.8 Responsivity of an antenna and a photodetector at 1 GHz.

detector *sensitivities* defined as the (optical or RF) input power required to make the (electrical) output signal power equal to the (electrical) output noise power. Even then, it turns out that the regular antenna already reaches the fundamental sensitivity limit [17], leaving no hope for the photodetector to beat it. [→ Problem 3.3.]

Optical Receiver with an Antenna? What is the fundamental reason why photodetectors respond to the intensity rather than to the optical field? The processes within the photodetector (carrier transport and relaxation processes) are too slow to track the rapid field variations that occur at around 195 THz [18].

Could we help the detector by converting the optical frequencies down to RF frequencies using a mixer and a local oscillator just like in a superheterodyne radio receiver? Yes, this is possible with the *heterodyne receiver*[1] setup shown in Fig. 3.9. The incoming optical signal is combined with (added to) the beam of a continuous-wave laser operating at a frequency that is offset by, say, 1 GHz from the signal frequency. The latter laser source is known as the *local oscillator* (LO). The square-law photodetector acts as the mixer nonlinearity producing a spectral component at the 1-GHz intermediate frequency (IF).

The rms current that is produced by the optical heterodyne receiver is [1, 2]

$$i_{PIN,\text{het}}^{rms} = \mathcal{R}\sqrt{2P_{LO}P}, \tag{3.5}$$

where P is the received power, P_{LO} is the power of the LO, and $P_{LO} \gg P$ is assumed. Lo and behold, the current is now proportional to the square root of the power, just like for an antenna! We have converted a square-law detector into a linear one. Besides the detection law, the optical heterodyne receiver shares several other properties with the antenna of an RF receiver [19].

Unlike our hypothetical hyperinfrared photodiode, the optical heterodyne receiver is a practical invention used in many commercial products. It and other coherent receivers have been studied thoroughly [1, 2]. The heterodyne

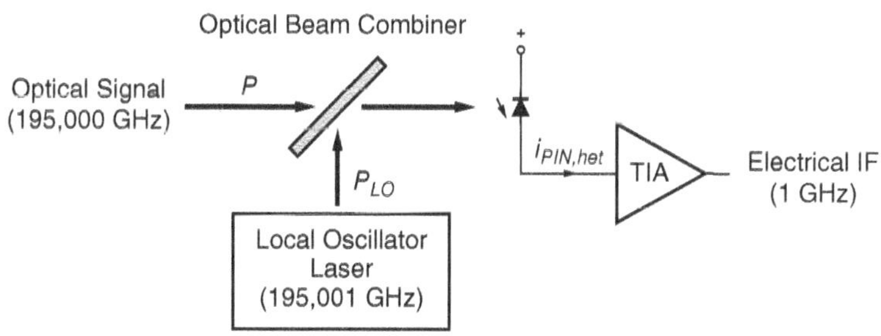

Figure 3.9 Optical heterodyne receiver.

1 The name *heterodyne* stands for "different" (=*hetero*) "power" (=*dyn*), that is, the *power* of the local oscillator in this receiver is at a frequency that *differs* from that of the received signal.

receiver is sensitive to the phase of the incoming signal, permitting the reception of phase-modulated optical signals. We continue the discussion of coherent receivers in Section 3.5.

Shot Noise. A p–i–n photodetector illuminated by a noise-free (coherent) continuous-wave source not only produces the DC current I_{PIN} but also a noise current known as *shot noise*. This fundamental noise appears because the photocurrent is composed of a large number of short pulses that are distributed randomly in time. Each pulse is caused by an electron–hole pair, which in turn was created by an absorbed photon. The area under each pulse (its integral over time) equals the electron charge q. If we approximate these pulses with Dirac delta functions, we obtain the instantaneous current shown in Fig. 3.10(a). In practice, the bandwidth of the photodetector is finite causing the individual pulses to smear out and overlap. To analyze the band-limited shot noise, we make use of the conceptually simple rectangular filter, which outputs the moving average over the time window T (cf. Section 4.8). Filtered in this way, the band-limited current can be written as $i_{PIN}(t) = n(t)q/T$, where $n(t)$ is the number of pulses falling into the window starting at time $t - T$ and ending at time t. The band-limited current, illustrated in Fig. 3.10(b), can be thought of as a superposition of the average photocurrent I_{PIN} and the shot-noise fluctuations. The average current is $I_{PIN} = Mq/T$, where M is the average number of pulses falling into the window T.

For example, a received optical power of 1 μW generates an average current of 0.8 μA, assuming $\mathcal{R} = 0.8$ A/W. From $I_{PIN} = Mq/T$, we can calculate that the electrons in this current move at an average rate of five electrons per picosecond ($M = 5$ and $T = 1$ ps). If the electrons were marching through the detector like little soldiers, with exactly five passing every picosecond, then the band-limited photocurrent would be noise free. However, in reality the

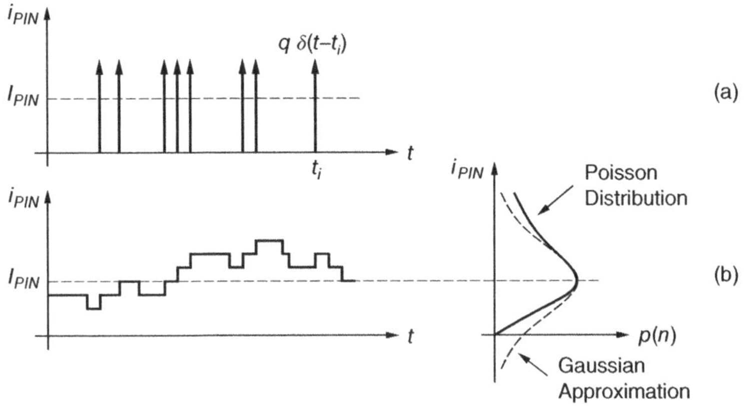

Figure 3.10 Fluctuations in the (a) wide-band and (b) band-limited photocurrent (ps scale).

electrons are moving randomly and shot noise is produced. For a coherent optical source, the number of electrons passing through the detector during the time interval T follows a *Poisson distribution*:

$$\text{Poisson}(n) = e^{-M} \cdot \frac{M^n}{n!}, \tag{3.6}$$

where $n = i_{PIN}T/q$ and $M = I_{PIN}T/q$. This distribution is shown on the far right of Fig. 3.10(b). Note its asymmetric shape: whereas the current never becomes negative, there is a chance that the current exceeds $2I_{PIN}$. For this reason, the *Gaussian distribution* (shown with a dashed line in Fig. 3.10(b)) cannot accurately describe the distribution. However, the larger the average M becomes, the better the Gaussian approximation fits. For small photocurrents, I_{PIN}, the Poisson distribution must be used, but for large currents, the Gaussian distribution may be a more convenient choice.

We are now ready to derive the mean-square value of the shot noise [17, 20]. The standard deviation of a Poisson distribution with average M is $\sqrt{M}$. Thus the rms noise current is $i^{rms}_{n,PIN} = \sqrt{M}q/T$ and the mean-square noise current is $\overline{i^2_{n,PIN}} = M(q/T)^2$. Inserting $M = I_{PIN}T/q$ for the average number of electrons, we find $\overline{i^2_{n,PIN}} = qI_{PIN}/T$. Finally, using the fact that the noise bandwidth of the rectangular filter is $BW_n = 1/(2T)$ (cf. Section 4.8), the mean-square value of the shot noise measured in the bandwidth BW_n becomes

$$\overline{i^2_{n,PIN}} = 2qI_{PIN} \cdot BW_n. \tag{3.7}$$

Thus, the 0.8-μA current from our earlier example produces a shot-noise current of about 51 nA rms in a 10-GHz bandwidth. The signal-to-noise ratio of this DC current can be calculated as $10\log(0.8\ \mu\text{A}/51\ \text{nA})^2 = 24$ dB. [→ Problem 3.4.]

The power spectral density (PSD) of the wide-band photocurrent in Fig. 3.10(a) is shown in Fig. 3.11(a). It has a Dirac delta function at DC, which corresponds to the average signal current I_{PIN} and a white noise component with the (one-sided) PSD $I^2_{n,PIN} = 2qI_{PIN}$, which corresponds to the shot noise.[2] The PSD of the band-limited photocurrent, shown in Fig. 3.11(b), is shaped by the low-pass response of the rectangular filter. The shot-noise PSD becomes $I^2_{n,PIN} = 2qI_{PIN}|H(f)|^2$, where $|H(f)|$ is the frequency response of the rectangular filter.

It is clear from the aforementioned considerations and Eq. (3.7) that the shot-noise current is *signal dependent*, that is, it is a function of I_{PIN}. If the received optical power increases, the noise increases, too. But fortunately, the

2 We follow the notation in [21] and write the power spectral density of the noise current i_n as I^2_n and the root spectral density as I_n, rather than the more conventional $\overline{i^2_n}/\Delta f$ and $\sqrt{\overline{i^2_n}/\Delta f}$, respectively. See Appendix J for more on the notation used in this book.

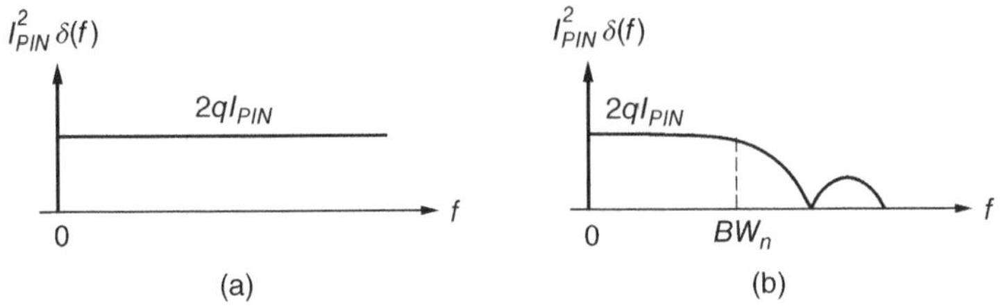

Figure 3.11 Power spectral density of the (a) wide-band and (b) band-limited photocurrent.

rms noise grows only with the *square root* of the signal amplitude, so we still gain in terms of signal-to-noise ratio. If we double the optical power in our previous example from 1 to 2 μW, we obtain an average current of 1.6 μA and a shot-noise current of 72 nA; thus, the signal-to-noise ratio improved by 3 dB from 24 to 27 dB. Conversely, if the received optical power is reduced, the noise reduces, too. For example, if we reduce the optical power by 3 dB, the signal current reduces by 6 dB, but the signal-to-noise ratio degrades by only 3 dB.

If we receive a (noise free) non-return-to-zero (NRZ) signal with a p–i–n photodetector, the electrical noise on the one bits is much larger than that on the zero bits. In fact, if the transmitter light source turns off completely during the transmission of zeros (infinite extinction ratio) and the photodetector is free of dark current (to be discussed shortly), then there is no current and therefore no shot noise. Let us suppose that the received optical signal is DC balanced (same numbers of zeros and ones), has a high extinction ratio, and has the average power $\overline{P}$. Then, the optical power for the ones is $P_1 = 2\overline{P}$ and that for the zeros is $P_0 \approx 0$. Thus with Eq. (3.7), we find the noise currents for zeros and ones to be

$$\overline{i^2_{n,PIN,0}} \approx 0 \quad \text{and} \tag{3.8}$$

$$\overline{i^2_{n,PIN,1}} = 4qR\overline{P} \cdot BW_n. \tag{3.9}$$

If incomplete extinction and the dark current are taken into account, $\overline{i^2_{n,PIN,0}} = 2q(RP_0 + I_{DK}) \cdot BW_n$. Figure 3.12 illustrates the signal and noise currents produced by a p–i–n photodetector in response to an optical NRZ signal with DC balance and high extinction. Signal and noise magnitudes are expressed in terms of the average received power $\overline{P}$.

Dark Current. The p–i–n photodetector produces a small amount of current even when it is in total darkness. This so-called *dark current*, I_{DK}, depends on the diode material, temperature, reverse bias, junction area, and processing. For a high-speed InGaAs photodetector at room temperature and a reverse bias of 2 V, it is usually less than 2 nA. Photodiodes made from materials with a

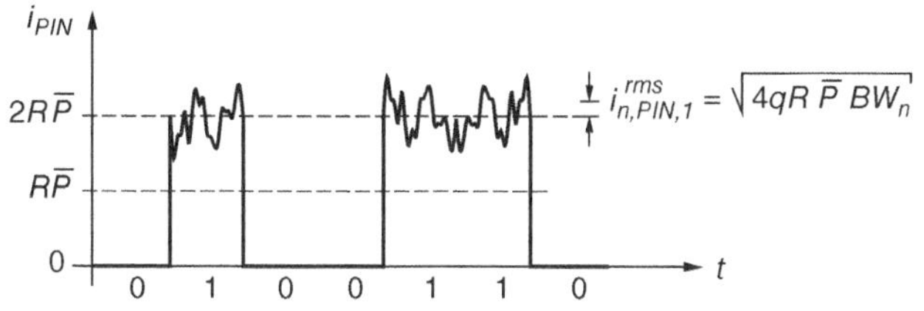

Figure 3.12 Signal and noise currents from a p–i–n photodetector.

smaller bandgap (such as germanium) suffer from a larger dark current because thermally generated electron–hole pairs become more numerous at a given temperature. Similarly, the dark current increases with temperature because the electrons become more energetic and thus are more likely to jump across a given bandgap.

The dark current and its associated shot-noise current interfere with the received signal. Fortunately, in high-speed p–i–n receivers (> 1 Gb/s), this effect usually is negligible. Let us calculate the optical power for which the worst-case dark current amounts to 10% of the signal current. As long as our received optical power is much larger than this, we are fine:

$$\overline{P} \gg 10 \cdot \frac{I_{DK}(\text{max})}{\mathcal{R}}. \tag{3.10}$$

With the values $\mathcal{R} = 0.8$ A/W and $I_{DK}(\text{max}) = 2$ nA, we find $\overline{P} \gg -46$ dBm. In Section 4.4, we see that high-speed p–i–n receivers require much more signal power than this to work at an acceptable bit-error rate (to overcome the TIA noise), and therefore we do not need to worry about dark current in such receivers. However, in low-speed p–i–n receivers or APD receivers, the dark current can be an important limitation. In Section 4.7, we formulate the impact of dark current on the receiver sensitivity in a more precise way.

Saturation Current. Whereas the shot noise and the dark current determine the lower end of the p–i–n detector's dynamic range, the *saturation current* defines the upper end of this range. At very high optical power levels, a correspondingly high density of electron–hole pairs is produced, which results in a space charge that counteracts the bias-voltage induced drift field. The consequences are a decreased responsivity (gain compression) and a reduced bandwidth. Moreover, the power dissipated in the photodiode ($V_{PIN} \times I_{PIN}$) causes heating, which results in high dark currents or even the destruction of the device.

For a photodiode preceded by an optical amplifier, such as an erbium-doped fiber amplifier (EDFA), its quantum efficiency and responsivity are of secondary importance (we discuss photodetectors with optical preamplifiers in Section 3.3). Low values of these parameters can be compensated for with a higher optical gain. In contrast, the bandwidth and the saturation current are of primary importance. The saturation current, in particular, limits the voltage swing that can be obtained by driving the photocurrent directly into a 50 Ω resistor. For example, the 80-Gb/s detector reviewed in [22] is capable of producing a 0.8-V swing, which is sufficient to directly drive a decision circuit.

Photodiodes for analog applications, such as CATV/HFC, must be highly linear to minimize distortions in the sensitive analog signal (cf. Appendix D) and therefore must be operated well below their saturation current. A beam splitter, multiple photodetectors, and a power combiner can be used to increase the effective saturation current of the photodetector [23].

There are two approaches for increasing the saturation current of a photodiode. One way is to distribute the photogenerated carriers over a larger volume. In a vertically illuminated photodetector this can be done by overfilling the absorbing area such that 5 to 10% of the Gaussian beam is beyond the active region [14]. In an edge-coupled photodetector the carrier density can be reduced by making it longer, however, this measure may lower the bandwidth. The second way is to increase the carrier velocity by exploiting the fact that under certain conditions electrons (but not holes) can drift much faster (e.g., five times faster) than at their saturated velocity [3]. This fast drift velocity is known as the *overshoot velocity*. The latter approach led to the development of the *uni-traveling-carrier* (UTC) *photodiode*, which employs a modification of the p–i–n structure that eliminates the (slow) holes from participating in the photodetection process [22]. UTC photodiodes come in all the flavors known from p–i–n photodiodes: vertically illuminated, waveguide, and traveling wave. Some of the highest reported saturation currents for UTC photodiodes are in the 26 to 76 mA range, whereas those for p–i–n photodiodes are in the 10 to 32 mA range [14].

3.2 Avalanche Photodetector

The basic structure of the *avalanche photodetector* (APD) is shown in Fig. 3.13. Like the p–i–n detector, the avalanche photodetector is a reverse biased diode. However, in contrast to the p–i–n photodetector, it features an additional layer, the *multiplication region*. This layer provides internal gain through avalanche multiplication of the photogenerated carriers.

The vertically illuminated InGaAs/InP APD, shown in Fig. 3.13, is sensitive to the 1.3 and 1.55-μm wavelengths common in telecommunication systems. It operates as follows. The light enters through a hole in the top electrode and passes through the transparent InP layer to the InGaAs absorption layer. Just like in the p–i–n structure, electron–hole pairs are generated and separated by the electric field in the absorption layer. The holes move upward and enter the multiplication region. Accelerated by the strong electric field in this region the holes acquire sufficient energy to create secondary electron–hole

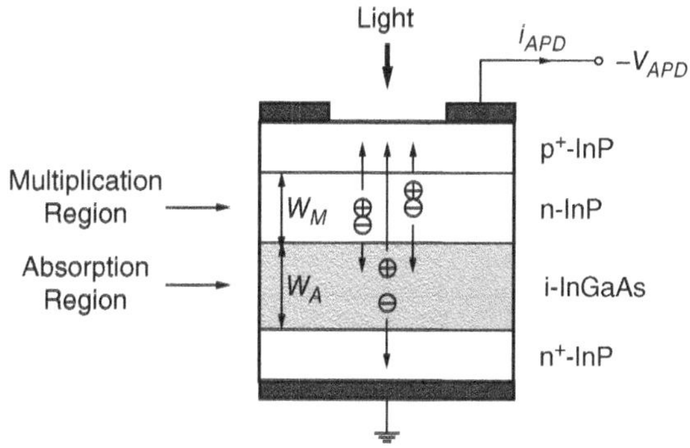

Figure 3.13 Avalanche photodetector (vertically illuminated structure).

pairs. This process is known as *impact ionization.* The figure shows one primary hole creating two secondary holes corresponding to an avalanche gain of three ($M = 3$). In InP, holes are more ionizing than electrons hence the multiplication region is placed on the side of the absorption region where the primary holes exit. (In silicon, the opposite is true and the multiplication layer is placed on the other side.) Like for the p–i–n photodiode, the width of the absorption layer W_A impacts the quantum efficiency. The width of the multiplication layer W_M together with the bias voltage V_{APD} determines the electric field in this layer. A smaller width leads to a stronger field. The wider bandgap InP material is chosen for the multiplication region because it can sustain a higher field than InGaAs and is transparent at the wavelengths of interest.

In practice, APD structures are more complex than the one sketched in Fig. 3.13. A practical APD may include a guard ring to suppress leakage currents and edge breakdown, a grading layer between the InP and InGaAs layers to suppress slow traps, a charge layer to control the field in the multiplication region, and so forth [24].

Responsivity. The gain of the APD is called *avalanche gain* or *multiplication factor* and is designated by the letter M. A typical value for an InGaAs/InP APD is $M = 10$. The optical power P is converted to the electrical current i_{APD} as

$$i_{APD} = M\mathcal{R}P, \tag{3.11}$$

where $\mathcal{R}$ is the responsivity of the APD *without* avalanche gain. The value of $\mathcal{R}$ is similar to that of a p–i–n photodetector. Assuming $\mathcal{R} = 0.8$ A/W and $M = 10$, the APD generates 8 A/W. We can also say that the APD has the total responsivity $\mathcal{R}_{APD} = M\mathcal{R}$, which is 8 A/W in our example, but we have to be careful to avoid confusion between $\mathcal{R}_{APD}$ and $\mathcal{R}$. The total responsivity, $\mathcal{R}_{APD}$, of a typical InGaAs/InP APD is in the range of 5 to 20 A/W [1].

For the avalanche multiplication process to set in, the APD must be operated at a reverse bias, V_{APD}, that is significantly higher than that of a p–i–n photodetector. For a typical 2.5 Gb/s InGaAs/InP APD, the reverse voltage

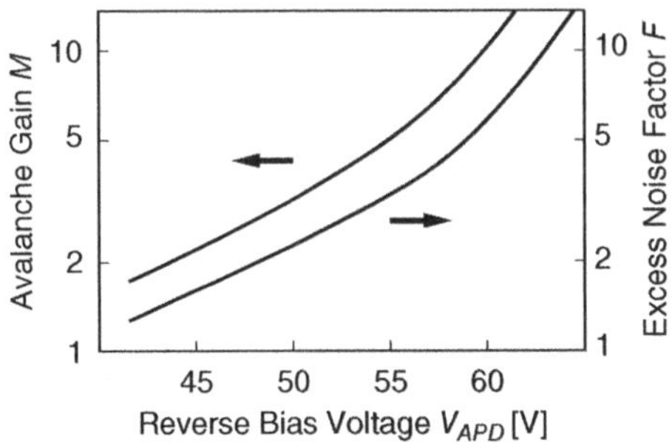

Figure 3.14 Avalanche gain and excess noise factor as a function of reverse voltage for a typical 2.5 Gb/s InGaAs/InP APD.

is about 40 to 60 V (cf. Fig. 3.14). For a 10 to 40 Gb/s InGaAs/InAlAs APD, the required voltage is in the 10 to 20 V range [24]. Faster devices require less voltage to reach the field necessary for avalanche multiplication because of their thinner layers.

Figure 3.14 shows how the avalanche gain M varies with the reverse bias voltage. For small voltages, it is close to one, like for a p–i–n photodiode, but when approaching the reverse-breakdown voltage it increases rapidly. Moreover, impact ionization and thus the avalanche gain also depend on the temperature. To keep the avalanche gain constant, the reverse bias voltage of an InGaAs/InP APD must be increased at a rate of about $0.2\%/°C$ to compensate for a decrease in the ionization rate. Finally, the necessary reverse voltage also varies from device to device.

APD Bias Circuits. A simple circuit for generating the APD bias voltage is shown in Fig. 3.15(a). A switch-mode power supply boosts the 5-V input voltage to the required APD bias voltage. Thermistor R_T measures the APD temperature and an analog control loop, consisting of resistors R_1 to R_3 and an op amp, adjusts the APD bias voltage to

$$V_{APD}(T) = \left(\frac{R_1}{R_2 + R_3 \parallel R_T(T)} + 1 \right) V_{\text{REF}}. \tag{3.12}$$

With the appropriate choice of R_1 to R_3, the desired APD voltage and an approximately linear temperature dependence of $0.2\%/°C$ can be achieved.

A more sophisticated APD bias circuit with digital control is shown in Fig. 3.15(b) [25, 26]. Here, an A/D converter digitizes the value of thermistor R_T and a digital controller determines the appropriate APD bias voltage with a look-up table. A scaled-down version of that voltage is converted back to the analog domain and subsequently boosted to its full value with the switch-mode power supply. The advantages of this approach are that the look-up table permits the bias voltage to be optimized for every temperature point and can correct for thermistor nonlinearities.

In some optical receivers, the dependence of the avalanche gain on the bias voltage is exploited to implement an *automatic gain control* (AGC) mechanism

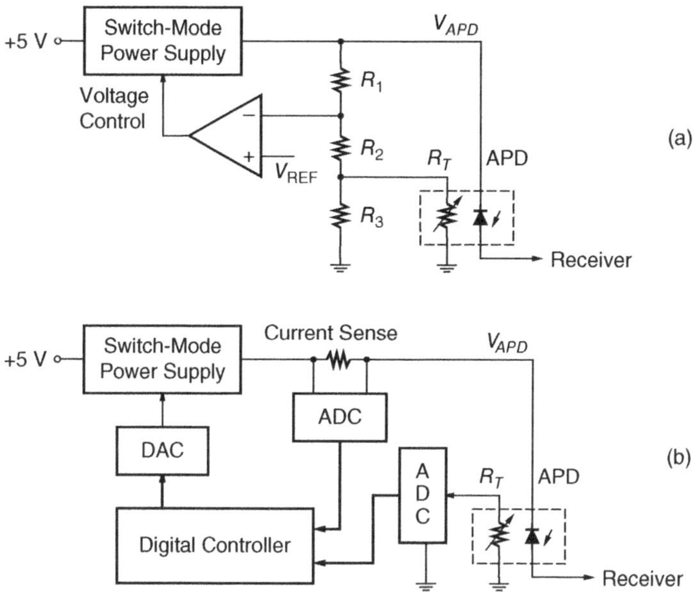

Figure 3.15 Temperature-compensated APD bias circuits with (a) analog and (b) digital control.

that acts right at the detector. Controlling the avalanche gain in response to the received signal strength with an AGC loop increases the dynamic range of the receiver (cf. Section 7.4). To determine the received signal strength, the average APD current can be sensed, as shown in Fig. 3.15(b).

To avoid sensitivity degradations, it is important that the bias voltage supplied to the APD contains as little noise and ripple as possible. To that end, the voltage from the switch-mode power supply must be passed through a filter (not shown in Fig. 3.15) before it is applied to the cathode of the APD. A typical filter is comprised of a series inductor (ferrite bead) with two capacitors on each side to ground. To avoid damaging the APD during optical power transients, which can excite the LC filter, it is recommended to put a resistor (about 500 Ω) in series with the APD [13].

Bandwidth. All the bandwidth limiting mechanisms that we discussed for the p–i–n photodetector also apply to the APD. To obtain a high speed, we must minimize the carrier transit time through the absorption layer, avoid slow diffusion currents, and keep the photodiode capacitance and package parasitics small. But in addition to those there is a new time constant associated with the avalanche region, known as the *avalanche build-up time.* Therefore, APDs are generally slower than p–i–n photodetectors. Often, the avalanche build-up time dominates the other time constants thus determining the APD's speed.

Without going into too much detail, we state here an approximate expression for the APD bandwidth assuming it is limited by the avalanche build-up time [8, 12, 24]

$$BW_{APD} \approx \frac{1}{2\pi} \cdot \frac{v_{c,\text{sat}}}{W_M} \cdot \frac{1}{Mk_A}, \quad \text{if} \quad Mk_A > 1, \tag{3.13}$$

where M is the avalanche gain at DC and k_A is the so-called *ionization-coefficient ratio.* If electrons and holes are equally ionizing, k_A reaches its maximum value of one. If one carrier type, say, the electrons, is much more ionizing than the other, k_A goes to zero. In the first case, electrons and holes participate equally in the avalanche process, whereas in the second case only one carrier type (electrons *or* holes) participates in the avalanche process. For example, in InP holes are more ionizing than electrons and $k_A = 0.4$ to 0.5 [24] (depending on the electric field); in silicon electrons are much more ionizing than holes and $k_A = 0.02$ to 0.05 [1].

Examining Eq. (3.13), we see that the second factor is the reciprocal value of the transit time through the multiplication layer. Not surprisingly, to obtain a fast APD, W_M must be made small. The third factor indicates that the bandwidth shrinks with increasing avalanche gain M, as illustrated in Fig. 3.16. For large gains ($M > 1/k_A$), the gain-bandwidth product is constant, reminiscent of an electronic single-stage amplifier. At low gains ($M < 1/k_A$), the bandwidth remains approximately constant. According to Eq. (3.13), a low k_A results in a high gain-bandwidth product.

Clearly silicon would be a better choice than InP for the multiplication layer. But getting a silicon multiplication layer to cleanly bond with an InGaAs absorption layer is a challenge [24]. Good results have been achieved by combining a silicon multiplication layer with a germanium absorption layer [27] (cf. Section 3.4). Another material that has been used successfully in high-speed APDs is indium aluminum arsenide (InAlAs or more precisely $In_{0.52}Al_{0.48}As$), which has $k_A = 0.3$ to 0.4 [24] and is lattice matched to both the InP substrate and the InGaAs absorption layer. Finally, a multiple quantum well (MQW) structure, which can be engineered to attain a low k_A value, can be used for the multiplication region [24]. Incidentally, the same measures

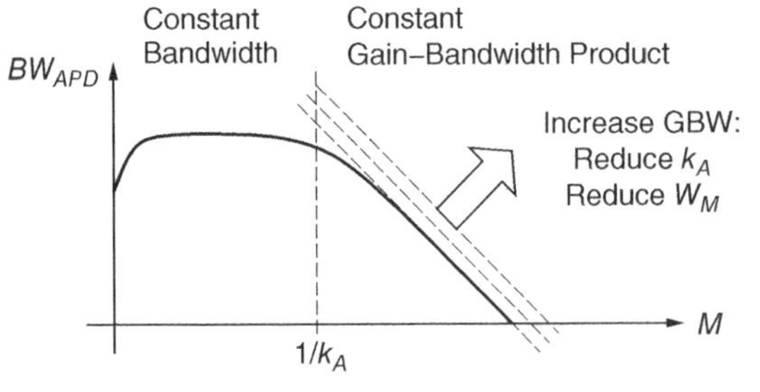

Figure 3.16 Dependence of the APD bandwidth on the DC avalanche gain.

that improve the APD speed (small W_M and small k_A) also improve its noise characteristics, which we discuss shortly.

Why does the participation of only one type of carrier in the avalanche process (small k_A) lead to a higher APD speed? Imagine a snow avalanche coming down from a mountain. On its way down, the amount of snow (= carriers) in the avalanche grows because the tumbling snow drags more snow with it. But once it's all down, the avalanche stops. Now, imagine a noisily rumbling snow avalanche that sends tremors (= second type of carriers) back up the mountain that in turn trigger more snow avalanches coming down. This new snow brings down more snow and rumbles enough to start another mini avalanche somewhere further up the mountain. And on and on it goes. In an analogous manner, an electron-only (or hole-only) avalanche comes to an end sooner than a mixed electron–hole avalanche [1, 12].

The equivalent AC circuit for an APD is similar to that for the p–i–n photodetector shown in Fig. 3.4, except that the current source i_{PIN} must be replaced by a current source i_{APD} that represents the *multiplied* photocurrent.

APDs for 10 Gb/s and Faster. Vertically illuminated APDs (cf. Fig. 3.13) are in widespread use for receivers up to and including 2.5 Gb/s. However, at 10 Gb/s and beyond, the necessary thin absorption layer results in a low quantum efficiency (as discussed for the p–i–n photodetector) and edge-coupled APDs become preferable. Besides very thin absorption and multiplication regions, high-speed APDs also employ low-k_A materials in their multiplication region.

For example, the 10-Gb/s InGaAs/InAlAs waveguide APD reported in [28] achieves a bandwidth of 11 GHz when biased for a DC avalanche gain of 10, thus achieving a gain-bandwidth product of 110 GHz. The InGaAs absorption layer and the InAlAs multiplication layer both are 0.2 μm thick. The Ge/Si APD reported in [27] achieves a bandwidth of 11.5 GHz and a gain-bandwidth product of 340 GHz using a 1-μm thick Ge absorption layer and a 0.5-μm thick Si multiplication layer.

The InGaAs/InAlAs waveguide APD reported in [29] achieves a bandwidth of 28 GHz when biased for a DC avalanche gain of 6 and a gain-bandwidth product of 320 GHz at around 20 GHz, making it marginally suitable for 40-Gb/s applications. The InGaAs absorption layer of this APD is 0.2 μm and the InAlAs multiplication layer is 0.15 μm thick. A similar device with a bandwidth of 30 to 34 GHz at a DC avalanche gain of 2 to 3 and a gain-bandwidth product of 130 to 140 GHz, was successfully incorporated into a high-sensitivity 40-Gb/s receiver [30].

With APDs running out of steam at about 40 Gb/s, optically preamplified p–i–n detectors, which we discuss in Section 3.3, take over. These detectors are more expensive than APDs but feature superior speed and noise performance.

Avalanche Noise. Unfortunately, the APD not only provides a stronger signal but also more noise than the p–i–n photodetector, in fact, *more* noise than simply the amplified shot noise that we are already familiar with. At a microscopic level, each primary carrier created by a photon is multiplied by a random gain factor: for example, the first photon ends up producing nine electron–hole pairs, the next one 13, and so on. The avalanche gain M, introduced earlier, is really just the *average* gain value.

The mean-square noise current of an APD illuminated by a noise-free (coherent) continuous-wave source can be written as [1, 12, 24]

$$\overline{i^2_{n,APD}} = FM^2 \cdot 2qI_{PIN} \cdot BW_n, \tag{3.14}$$

where F is the so-called *excess noise factor* and I_{PIN} is the primary photodetector current, that is, the current before avalanche multiplication ($I_{PIN} = I_{APD}/M$). Equivalently, I_{PIN} can be understood as the current produced in a p–i–n photodetector with a matching responsivity $\mathcal{R}$ that receives the same amount of light as the APD under discussion. In the ideal case, the excess noise factor is one ($F = 1$), which corresponds to the situation of deterministically amplified shot noise. For a practical InGaAs/InP APD, the excess noise factor is more typically around $F = 6$. [→ Problem 3.5.]

Just like the p–i–n photodetector noise, the APD noise is signal dependent, leading to unequal noise for the zeros and ones. The noise currents for a DC-balanced NRZ signal with average power $\overline{P}$ and high extinction can be found with Eq. (3.14):

$$\overline{i^2_{n,APD,0}} \approx 0 \quad \text{and} \tag{3.15}$$

$$\overline{i^2_{n,APD,1}} = FM^2 \cdot 4q\mathcal{R}\overline{P} \cdot BW_n. \tag{3.16}$$

If incomplete extinction and the primary dark current are taken into account, $\overline{i^2_{n,APD,0}} = FM^2 \cdot 2q(\mathcal{R}P_0 + I_{DK}) \cdot BW_n$.

As plotted in Fig. 3.14, the excess noise factor F increases with increasing reverse bias, roughly tracking the avalanche gain M. Under certain assumptions, such as a relatively thick multiplication layer, F and M are related as follows [1, 12, 24]:

$$F = k_A M + (1 - k_A)\left(2 - \frac{1}{M}\right), \tag{3.17}$$

where k_A is the same ionization-coefficient ratio that we encountered when discussing the bandwidth. For an InGaAs/InP APD, which has a relatively large k_A, the excess noise factor increases almost proportional to M, as illustrated in Fig. 3.14; for a silicon APD of Ge/Si APD, which has a very small k_A, the excess noise factor increases much more slowly with M. Not surprisingly, an orderly one-carrier-type avalanche is less noisy than a reverberating two-carrier-type one. For very thin multiplication layers we get some unexpected help. The avalanche multiplication process becomes less random resulting in an excess noise that is lower than predicted by Eq. (3.17) [24].

Because the avalanche gain of a given APD can be increased only at the expense of more detector noise (Eq. (3.17)), there is an optimum APD gain at which the receiver becomes most sensitive. As we'll see in Section 4.4, the value of this optimum gain depends, among other things, on the APD material (k_A).

The amplitude distribution of the avalanche noise, which is important for calculating the bit-error rate of APD receivers (cf. Section 4.2), is non-Gaussian. Like the shot noise from the p–i–n photodetector, the avalanche noise has an asymmetric distribution with a steep left tail (the detector current is always positive). Unfortunately, avalanche noise is much harder to analyze than shot noise. Not only does the primary current from the detection process have a random (Poisson) distribution, but the avalanche multiplication process also has a random distribution [31]. The latter distribution depends on the number of primary carriers, n, that initiate the avalanche as well as the material constant k_A and is strongly non-Gaussian, especially for small values of n and large values of k_A [32, 33]. The exact mathematical form of the gain distribution is very complex but several approximations have been found [8, 31].

Dark Current. Just like the p–i–n photodetector, the APD also suffers from a dark current. The *total dark current* appearing at the APD terminals can be separated into a multiplied and an unmultiplied component [8, 34]. The multiplied dark current, MI_{DK}, arises from the *primary dark current*, I_{DK}, which is multiplied like a signal current. The unmultiplied dark current, results from surface leakage and is not multiplied. The unmultiplied dark current is often negligible when compared with the multiplied dark current. Dark current also produces noise. Like a signal current, the primary dark current I_{DK} produces the avalanche noise $FM^2 \cdot 2qI_{DK} \cdot BW_n$.

As we know, thermally generated dark current gets worse with increasing temperature and decreasing bandgap energy. A typical high-speed InGaAs/InP APD has a primary dark current of less than 10 nA at room temperature, resulting in a total dark current of less than 100 nA for $M = 10$. At elevated temperatures, the total dark current can go up into the microamps.

We can again use Eq. (3.10) to judge if a given amount of dark current is harmful. With the values $\mathcal{R} = 0.8$ A/W and $I_{DK}(\text{max}) = 10$ nA, we find that we are fine as long as $\overline{P} \gg -39$ dBm. Most high-speed APD receivers require more signal power than this to work at an acceptable bit-error rate (cf. Section 4.4), and in this case dark current is not a concern.

3.3 p–i–n Detector with Optical Preamplifier

An attractive alternative to the APD in direct-detection receivers is the p–i–n detector with optical preamplifier or simply the *optically preamplified p–i–n detector*. Rather than amplifying the photogenerated electrons, we amplify the photons before they reach the detector. Optical amplifiers come in various types [1, 2, 12, 35, 36] and several of them have been used as preamplifiers. For example, the *semiconductor optical amplifier* (SOA) is small and can be integrated together with an edge-coupled p–i–n photodiode on the same substrate [37], making it an attractive candidate for an optical preamplifier. Alternatively, the *erbium-doped fiber amplifier* (EDFA) features high gain and low noise and operates in the important 1.55-μm band, making it a popular choice for high-performance telecommunication receivers [38, 39].

The EDFA was co-invented around 1987 by two teams, one at the University of Southampton in England and one at AT&T Bell Laboratories in New Jersey, and has revolutionized the field of optical communication. The EDFA provides high gain over a huge bandwidth, eliminating the gain-bandwidth trade-off known from APDs. The overall detector bandwidth is limited only by the p–i–n photodetector, which can be made very large. Furthermore, the EDFA-preamplified p–i–n detector has superior noise characteristics when compared with an APD. The downsides of the EDFA are mainly its large size and high cost. In the following, we discuss the EDFA-preamplified p–i–n detector in more detail.

Erbium-Doped Fiber Amplifier. Figure 3.17 shows the operating principle of an EDFA-preamplified p–i–n detector. A WDM coupler combines the received weak optical signal with the light from a strong continuous-wave laser, known as the *pump laser*. The pump laser typically provides a power of about 10 mW at either the 1.48-μm or 0.98-μm wavelength (the 0.98-μm wavelength is preferred for low-noise preamplifiers), whereas the received signal is at the 1.55-μm wavelength. The signal and the pump light are sent through an erbium-doped fiber, typically 10 to 20 m long, where the amplification takes place. An optical isolator prevents reflections of the optical signal from entering back into the amplifier, which could cause instability and extra noise. An optical filter with bandwidth BW_O reduces the optical noise of the amplified signal before it is converted to an electrical signal with a p–i–n photodetector. (We discuss the effect of the optical filter on the electrical noise shortly.)

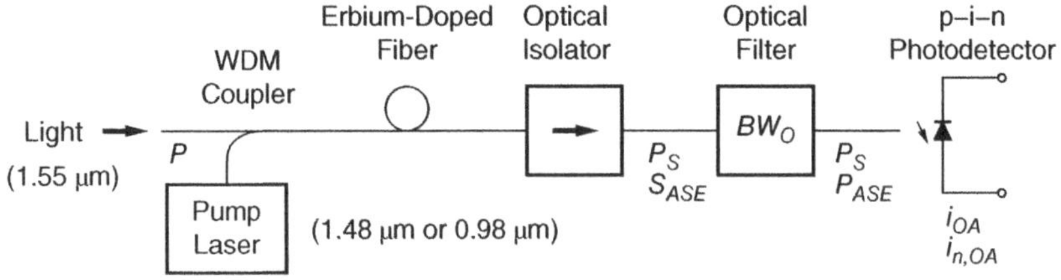

Figure 3.17 A p–i–n photodetector with erbium-doped fiber preamplifier.

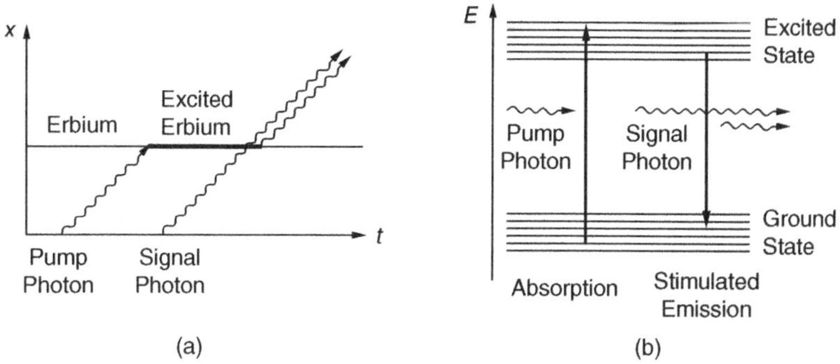

Figure 3.18 Pumping and stimulated emission: (a) space–time diagram and (b) energy diagram.

The light amplification in the erbium-doped fiber can be explained as follows (see Fig. 3.18). The pump light is absorbed by the erbium ions (Er^{3+}) elevating them into an excited state. In their excitement they *almost* radiate a photon with a wavelength around 1.55 μm, but they need a little encouragement from an incident photon. When that happens, the radiated photon, being a fashion-conscious boson, travels in the same direction, has the same wavelength, the same phase, the same polarization, and carries the same style hand bag as the incident photon. This process is known as *stimulated emission.* The radiated photon can stimulate the emission of further photons, and so forth, creating an avalanche of perfectly identical and coherent photons. At the end of the erbium-doped fiber, the optical signal is amplified, while much of the pump power is used up.

Occasionally, it happens that one of the excited erbium ions cannot wait and radiates a photon even without stimulation. The resulting photon has a random direction, random wavelength (within the amplifier's bandwidth), random phase, and random polarization. This process is known as *spontaneous emission.* Unfortunately, a small fraction of these errant photons is radiated in the same direction as the signal photons and thus gets amplified just like those. The result is an optical noise called *amplified spontaneous emission* (ASE) noise.

The basic EDFA preamplifier shown in Fig. 3.17 can be enhanced in a number of ways [39]. The WDM coupler can be moved to the point *after* the gain medium, injecting the pump light backward through the erbium-doped fiber. This variation results in a different trade-off between noise and output power. The EDFA also can be structured as a two-stage amplifier, where the first stage is optimized for low noise and the second stage for high pumping efficiency. A second isolator may be placed at the input of the amplifier to avoid the ASE noise, which exits at the input and the output of the amplifier, from being reflected back into the amplifier or from entering a preceding amplifier stage. For a packaged EDFA, see Fig. 3.19.

Figure 3.19 A packaged variable-gain two-stage erbium-doped fiber amplifier with single-mode fiber pigtails for the input, output, and interstage access (15 cm × 10 cm × 1.8 cm). *Source*: Reprinted by permission from Finisar Corporation.

Bandwidth. A typical EDFA has a wavelength bandwidth of around 35 nm reaching from 1.530 to 1.565-μm, corresponding to a frequency bandwidth of over 4,000 GHz. This is more bandwidth than that of the fastest p–i–n detectors and we do not need to worry about the bandwidth limitations introduced by the EDFA. In practice, the amplifier bandwidth is artificially reduced to BW_O with an optical filter to minimize the optical output noise.

Where does this huge bandwidth come from? The bandwidth depends on the available energy differences, ΔE, between the excited and ground state of the erbium ions. After all, the wavelength of the emitted photon λ must satisfy $hc/\lambda = \Delta E$. For a theoretical two-state system (excited and ground state), there would be only a single energy difference, resulting in a very narrow amplifier bandwidth. However for erbium ions embedded in silica glass, the two energy states broaden into two *energy bands*, resulting in a range of energy differences and thus a much wider bandwidth (see Fig. 3.18(b)) [1, 39]. The amplifier response can be broadened and flattened further by adding dopants besides the erbium (codopants).

Responsivity. The power gain, G, of an optical amplifier is the ratio of the output signal power (P_S) to the input signal power (P). The gain of an EDFA is typically in the range $G = 10$ to 1,000, corresponding to 10 to 30 dB. The current generated by the p–i–n photodetector, i_{OA}, expressed as a function of the optical power at the *input* of the preamplifier, P, is

$$i_{OA} = \mathcal{R}GP, \tag{3.18}$$

where $\mathcal{R}$ is the responsivity of the p–i–n photodetector. For $G = 100$ and $\mathcal{R} = 0.8$ A/W, the total responsivity of the preamplified p–i–n detector becomes an impressive 80 A/W. Whereas the APD improved the responsivity by about one order of magnitude ($M = 10$), the optically preamplified p–i–n detector can improve the responsivity by about two orders of magnitude ($G = 100$) relative to an unaided p–i–n photodetector.

The gain of an EDFA depends among other things on the pump power and the length of the erbium-doped fiber. A higher pump power results in more erbium ions being in the excited state. To obtain a gain larger than one, the rate of stimulated emissions must exceed the rate of absorptions. This point is reached when more than about 50% of the erbium ions are in the excited state and is known as *population inversion*. Figure 3.20 shows how the gain increases with pump power and eventually saturates when almost all erbium ions are in their excited state. A longer fiber permits the photon avalanche to grow larger, but when most of the pump power is used up, the percentage of excited erbium ions falls below 50% and the erbium-doped fiber turns into an attenuator. Thus for a given pump power there is an optimum fiber length.

Because the gain depends sensitively on the pump power, EDFA modules typically incorporate a microcontroller, which adjusts the power of the pump laser based on feedback. The EDFA also can perform automatic gain control (AGC) in the optical domain [40]. To that end, a small amount of light is split off from the amplified output signal and the pump power is controlled such that the average power of this light sample remains constant.

ASE Noise. As we said earlier, the EDFA not only amplifies the desired input signal, but also produces an optical noise known as ASE noise. The PSD of this noise, S_{ASE}, is nearly white across the huge bandwidth of the EDFA.[3] Thus, we

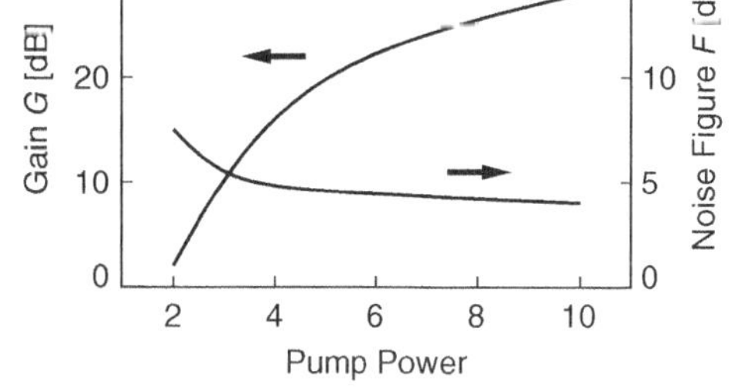

Figure 3.20 EDFA gain and noise figure as a function of the pump power (normalized to the power necessary for population inversion) [1].

3 In the following, S_{ASE} always refers to the noise PSD in *both* polarization modes, that is, $S_{ASE} = 2S'_{ASE}$, where S'_{ASE} is the noise PSD in a single (e.g., the signal's) polarization mode.

can calculate the optical noise power that reaches the photodetector as $P_{ASE} = S_{ASE} \cdot BW_O$, where BW_O is the (noise) bandwidth of the optical filter in front of the p–i–n detector. Clearly, to keep P_{ASE} low, we want to use a narrow optical filter.

How does the photodetector convert this optical noise into an electrical noise? If you thought that it was odd that optical signal *power* is converted to a proportional electrical signal *current*, wait until you hear this: because the photodetector responds to the intensity, which is proportional to the *square* of the optical field, the optical noise gets converted into *two* electrical beat-noise components. Roughly speaking, we get the terms corresponding to the expansion $(\text{signal} + \text{noise})^2 = (\text{signal})^2 + 2(\text{signal} \times \text{noise}) + (\text{noise})^2$. The first term is the desired electrical signal, the second term is known as the *signal–spontaneous beat noise*, and the third term is known as the *spontaneous–spontaneous beat noise*. A detailed analysis reveals the two electrical noise terms as [1, 41, 42]

$$\overline{i_{n,OA}^2} = 2\mathcal{R}^2 P_S S_{ASE} \cdot BW_n + \mathcal{R}^2 S_{ASE}^2 \cdot BW_O \cdot BW_n, \tag{3.19}$$

where P_S is the optical signal power incident on the p–i–n detector ($P_S = GP$). Furthermore, it is assumed that $BW_O \gg BW_n$, that the optical-filter loss is negligible, and that no polarization filter is used between the amplifier and the p–i–n detector.[4]

The first term of Eq. (3.19), the signal–spontaneous beat noise, usually is the dominant term. This noise component is proportional to the signal power P_S (and thus also to P). Note that a signal-independent optical noise density S_{ASE} generates a *signal-dependent* noise term in the electrical domain! Furthermore, this noise term is *not* affected by the optical filter bandwidth BW_O, but the electrical bandwidth BW_n does matter. The second term of Eq. (3.19), the spontaneous–spontaneous beat noise, may be closer to your expectations. Similar to the signal component, this noise current component is proportional to the optical noise power. Moreover, the optical filter bandwidth does have a limiting effect on the spontaneous–spontaneous beat noise component.

In addition to the ASE noise in Eq (3.19), optically preamplified p–i–n detectors also produce shot noise and possibly multipath interference noise. For high-gain amplifiers, the shot noise component is small and usually can be neglected (cf. Eq. (I.4)). Multipath interference (MPI) noise occurs in amplifiers with spurious reflections (e.g., from the WDM coupler, isolator, or fiber ends). The reflected waves interfere with the signal wave producing an undesirable

4 A polarization filter between the amplifier and the p–i–n detector is not normally used as this would require a polarization-controlled input signal. If it is used, however, the spontaneous–spontaneous beat noise term in Eq. (3.19) is divided by two [41]. (The signal–spontaneous beat noise term remains unaffected, because, with or without filter, only the noise component with the same polarization as the signal participates in the beating.)

intensity noise that depends on the phase noise in the signal [43]. [→ Problem 3.6.]

SNR and OSNR. Let us calculate the electrical *signal-to-noise ratio* (SNR) of the optically preamplified p–i–n detector. For an optical continuous-wave signal with power P_S and ASE noise power P_{ASE} incident on the photodetector, the electrical signal power is $I_{OA}^2 = \mathcal{R}^2 P_S^2$ and the electrical noise power, $\overline{i_{n,OA}^2}$, is given by Eq. (3.19) with $S_{ASE} = P_{ASE}/BW_O$. Dividing I_{OA}^2 by $\overline{i_{n,OA}^2}$ yields

$$SNR = \frac{P_S^2/P_{ASE}^2}{P_S/P_{ASE} + \frac{1}{2}} \cdot \frac{BW_O}{2BW_n}. \tag{3.20}$$

The ratio P_S/P_{ASE} in this equation is known as the *optical signal-to-noise ratio* (OSNR) measured in the optical bandwidth BW_O at the output of the EDFA. If the OSNR is much larger than $\frac{1}{2}$ (or −3 dB), we can neglect the $\frac{1}{2}$ in the denominator, which is due to the spontaneous–spontaneous beat noise, and we end up with the surprisingly simple result [1, 44]

$$SNR = \frac{OSNR^2}{OSNR + \frac{1}{2}} \cdot \frac{BW_O}{2BW_n} \approx OSNR \cdot \frac{BW_O}{2BW_n}. \tag{3.21}$$

For example, for a receiver with $BW_n = 7.5$ GHz, an OSNR of 14.7 dB measured in a 0.1-nm bandwidth (12.5 GHz at $\lambda = 1.55$ μm) translates into an electrical SNR of 13.9 dB. Incidentally, the result in Eq. (3.21), which we derived for a continuous-wave signal, also holds for an (ideal) NRZ-modulated signal. In Section 4.6, we use this result to analyze optically amplified transmission systems. [→ Problem 3.7.]

Noise Distribution. Although the amplitude distribution of the ASE noise in the optical domain (optical field) is Gaussian, the distribution of the ASE beat noise in the electrical domain is non-Gaussian [45]. In particular, when the optical power P_S is small, resulting in mostly spontaneous–spontaneous beat noise, the current distribution is strongly asymmetric and non-Gaussian. When the optical power is increased, making the signal–spontaneous beat noise dominant, the distribution becomes more Gaussian [46]. Figure 3.21 illustrates the distribution (probability density functions, PDF) of the p–i–n detector current for the case of a weak and strong optical signal. Note that the total detector current (signal + noise) must be positive, explaining the asymmetric shapes.

The electrical ASE beat-noise can be described well by the non-central *chi-square distribution* [45, 47]. Given n independent random variables x_i with a Gaussian distribution (and unit variance), the random variable $y = \sum_{i=1}^{n} x_i^2$ has a chi-square distribution with n degrees of freedom (the chi-square distribution is central if the mean of the Gaussian is zero and non-central if it is nonzero) [48]. The degree of freedom of the electrical beat-noise

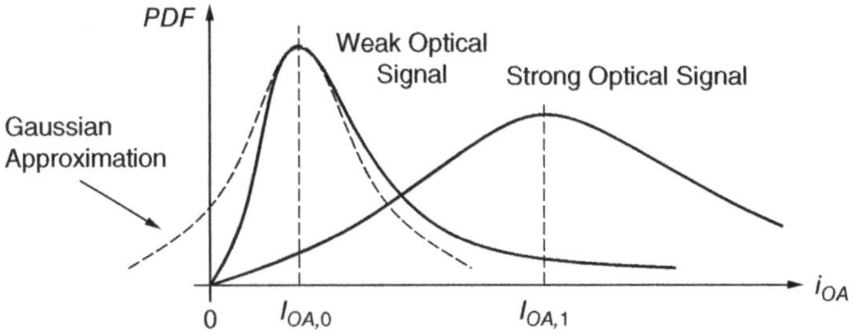

Figure 3.21 Distribution of the p–i–n detector current for a weak and strong optical signal with ASE noise (from an optical amplifier).

distribution is determined by the ratio of the optical to electrical bandwidth: $n = BW_O/BW_n$ (for unpolarized noise). The chi-square distribution is particularly non-Gaussian for small values of n. (The related Rayleigh and Rician distributions describe the random variable $\sqrt{y}$ for $n = 2$ when the mean of the Gaussian is zero and nonzero, respectively [48].)

Noise Figure of an Optical Amplifier. Just like electrical amplifiers, optical amplifiers are characterized by a noise figure F. Typical values for an EDFA noise figure are 4 to 5 dB (see Fig. 3.20). But how is the noise figure of an *optical* amplifier defined?

In an electrical system, the noise figure is defined as the ratio of the "total output noise power" to the "fraction of the output noise power due to the thermal noise of the source resistance." Usually, this source resistance is 50 Ω. Now, an optical amplifier doesn't get its signal from a 50-Ω source, and so the definition of its noise figure cannot be based on thermal 50-Ω noise. What fundamental noise is it based on? The quantum (shot) noise of the optical source!

The noise figure of an optical amplifier is defined as the ratio of the "total output noise power" to the "fraction of the output noise power due to the quantum (shot) noise of the optical source." The output noise power is measured with a perfectly efficient p–i–n photodetector ($\eta = 1$) and is quantified as the detector's mean-square noise current.[5] If we write the total output noise power as $\overline{i^2_{n,OA}}$ and the fraction that is due to the source as $\overline{i^2_{n,OA,S}}$, then the noise figure is $F = \overline{i^2_{n,OA}}/\overline{i^2_{n,OA,S}}$.

Figure 3.22 illustrates the various noise quantities introduced earlier. In Fig. 3.22(a), an ideal photodetector is illuminated directly by a noise-free (coherent) continuous-wave source and produces the DC current I_{PIN} and the mean-square shot-noise current $\overline{i^2_{n,PIN}} = 2qI_{PIN} \cdot BW_n$. In Fig. 3.22(b),

5 An equivalent definition for the noise figure of an optical amplifier is the ratio of the "input SNR" to the "output SNR," where both SNRs are measured in the electrical domain with an ideal photodetector ($\eta = 1$) and where the input SNR is based on shot noise only.

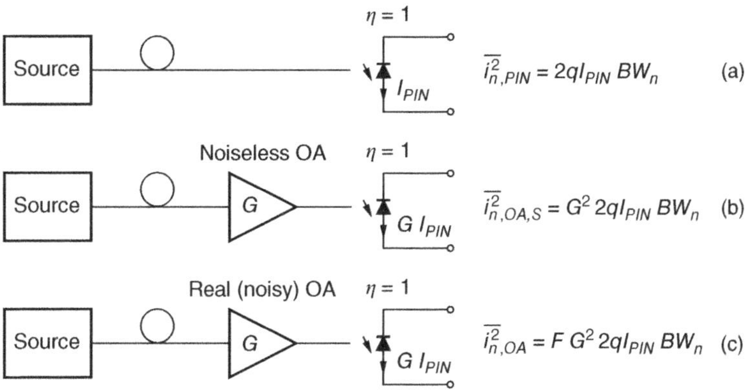

Figure 3.22 Definition of the noise figure for an optical amplifier: (a) source noise, (b) output noise due to the source, and (c) total output noise.

the signal from the same optical source is amplified with a noiseless, deterministic amplifier with gain G. This amplifier multiplies every photon from the source into exactly G photons. The ideal photodetector now produces the DC current $I_{OA} = GI_{PIN}$ and the mean-square shot-noise current $\overline{i^2_{n,OA,S}} = G^2 \cdot 2qI_{PIN} \cdot BW_n$ (cf. Problem 3.5). Note that this quantity represents the "output noise power due to the quantum (shot) noise of the optical source." In Fig. 3.22(c), we replaced the noiseless amplifier with a real amplifier with gain G and noise figure F. According to the noise figure definition, the ideal photodetector now produces a mean-square noise current that is F times larger than in Fig. 3.22(b):

$$\overline{i^2_{n,OA}} = FG^2 \cdot 2qI_{PIN} \cdot BW_n, \tag{3.22}$$

where I_{PIN} is the current produced by an ideal photodetector receiving the same amount of light as the optical preamplifier (cf. Fig. 3.22(a)).

With the definition of the noise figure in hand, we can now write the noise current of an optically preamplified p–i–n detector with preamplifier gain G, noise figure F, and p–i–n detector quantum efficiency η. This noise current is almost given by Eq. (3.22), except that the latter equation assumes an ideal detector with $\eta = 1$. To take the reduced quantum efficiency of a *real* detector into account, we have to multiply $\overline{i^2_{n,OA}}$ in Eq. (3.22) by η^2 and replace the ideal I_{PIN} in Eq. (3.22) by I_{PIN}/η. The resulting mean-square noise current is

$$\overline{i^2_{n,OA}} = \eta FG^2 \cdot 2qI_{PIN} \cdot BW_n, \tag{3.23}$$

where $I_{PIN} = \mathcal{R}P$ is the current produced by the real p–i–n photodetector receiving the same amount of light as the optical preamplifier.

Like for the unamplified p–i–n photodetector and the APD, the noise of the optically preamplified p–i–n detector is signal power dependent and therefore the noise current for zeros and ones is different. Given a DC-balanced NRZ signal with average power $\overline{P}$ and high extinction, we find with Eq. (3.23):

$$\overline{i^2_{n,OA,0}} \approx 0 \qquad \text{and} \tag{3.24}$$

$$\overline{i^2_{n,OA,1}} = \eta F G^2 \cdot 4q\mathcal{R}\overline{P} \cdot BW_n. \tag{3.25}$$

If incomplete extinction, spontaneous–spontaneous beat noise, and dark current are taken into account, $\overline{i^2_{n,OA,0}} = (\eta \tilde{F} G^2 \cdot 2q\mathcal{R}P_0 + \mathcal{R}^2 S^2_{ASE} \cdot BW_O + 2qI_{DK}) \cdot BW_n$.

If we compare the noise expression Eq. (3.14) for the APD with Eq. (3.23) for the optically preamplified p–i–n detector, we find that the excess noise factor F of the APD plays the same role as the product ηF of the optically preamplified p–i–n detector.

Noise Figure and ASE Noise. In Eq. (3.19), we expressed the electrical noise in terms of the optical ASE noise and in Eq. (3.23), we expressed the electrical noise in terms of the amplifier's noise figure. Now let us combine the two equations and find out how the noise figure is related to the ASE noise. Solving Eq. (3.23) for F and using Eqs. (3.19) and (3.3), we find

$$F = \frac{S_{ASE}}{G} \cdot \frac{\lambda}{hc} + \frac{S^2_{ASE} \cdot BW_O}{2G^2 P} \cdot \frac{\lambda}{hc}. \tag{3.26}$$

The first term is due to the signal–spontaneous beat noise, and the second term is due to the spontaneous–spontaneous beat noise. Note that this noise figure depends on the input power P and becomes infinite for $P \to 0$. The reason for this behavior is that when the signal power goes to zero, we are still left with the spontaneous–spontaneous beat noise, whereas the quantum (shot) noise due to the source does go to zero. Equation (3.26) can be extended to include shot noise terms (cf. Eq. (I.5)), but for a gain much larger than one, these terms are very small. [→ Problem 3.8.]

Sometimes it is convenient to define another type of noise figure $\tilde{F}$ that corresponds to just the first term of Eq. (3.26) [42]:

$$\tilde{F} = \frac{S_{ASE}}{G} \cdot \frac{\lambda}{hc}. \tag{3.27}$$

This noise figure is known as the *signal–spontaneous beat noise limited noise figure* [43] or the *optical noise figure*.[6] It has the advantages of being easier to measure [43] and being independent of the input power. For an OSNR much larger than −3 dB and a gain much larger than 0 dB, the two noise figures become approximately equal. Nevertheless, the fact that there are two similar but not identical noise figure definitions can be a source of confusion.

---

6 Sometimes the definition of $\tilde{F}$ includes the signal shot-noise term: $\tilde{F} = S_{ASE}/G \cdot \lambda/(hc) + 1/G$ [43] (cf. Eq. (I.5)).

A detailed analysis of the ASE noise physics reveals the following expression for its PSD [1, 36, 43]:

$$S_{ASE} = 2(G-1) \cdot \frac{N_2}{N_2 - N_1} \cdot \frac{hc}{\lambda}, \tag{3.28}$$

where N_1 is the number of erbium ions in the ground state and N_2 is the number of erbium ions in the excited state. The stronger the amplifier is "pumped," the more erbium ions are in the excited state. Thus for a strongly pumped amplifier, we have $N_2 \gg N_1$ (full population inversion). Combining Eq. (3.27) for the optical noise figure with Eq. (3.28) and assuming $G \gg 1$, we find the following simple approximation for the EDFA noise figure(s) [1, 36, 39]:

$$F \approx \tilde{F} \approx 2 \cdot \frac{N_2}{N_2 - N_1}. \tag{3.29}$$

This equation implies that increasing the pump power decreases the noise figure until it bottoms out at the theoretical limit $F = 2$ (or 3 dB), in agreement with the plot in Fig. 3.20.

Fiber Optical Parametric Amplifier and Noise Figures Below 3 dB. Why is the minimum noise figure of an EDFA two and not one? An amplifier with $F = 1$ does not require that there is zero ASE noise ($S_{ASE} = 0$), which would be quite impossible, but only that the ASE noise equals the amplified quantum (shot) noise. So, $F < 2$ does not seem totally impossible.

Is this a flaw of the EDFA? No, it is a fundamental limit for phase-insensitive amplifiers dictated by the uncertainty principle [49]. An amplifier with high gain and $F = 1$ would permit us to amplify the optical field from the quantum domain to the classical domain and measure the in-phase and quadrature components of the field accurately and simultaneously, in violation of the uncertainty principle [50].

An OOK receiver does not rely on phase information and an optical amplifier that amplifies the in-phase component but suppresses the quadrature component, a so-called *phase-sensitive amplifier*, is sufficient. Now, such an amplifier can reach $F = 1$, because it does not bring the phase information into the classical domain. Interestingly, the *fiber optical parametric amplifier* (FOPA) can be configured such that it amplifies only the field component that is phase aligned with the pump field [49]. Phase-sensitive FOPAs with sub-3-dB noise figures have been demonstrated in the lab.

Raman Amplifier and Negative Noise Figures. We conclude this section with a brief look at the *Raman amplifier* and its noise figure. This amplifier is named after the Indian physicist Chandrasekhara Venkata Raman, who discovered the Raman effect and was awarded the Nobel Prize in 1930. Raman amplification in optical fiber was first observed and measured in the early 1970s (Roger Stolen and Erich Ippen).

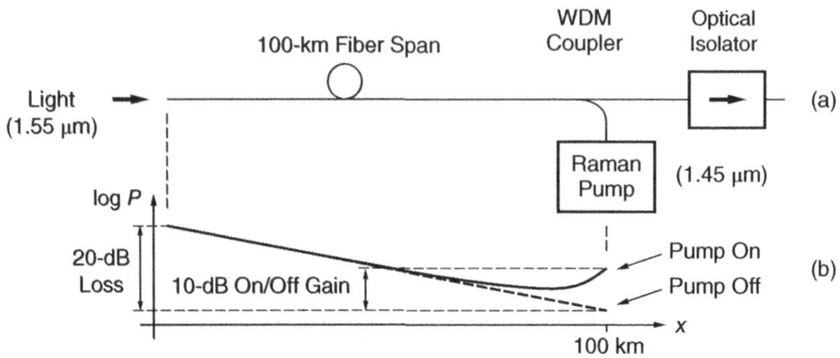

Figure 3.23 Distributed Raman amplifier: (a) structure and (b) signal-power profile.

Figure 3.23(a) shows the basic distributed Raman amplifier [51, 52]. Unlike an EDFA, which provides gain at the point where the amplifier is located, this amplifier provides distributed gain *in* the transmission fiber itself. The fiber span is pumped from the receiver end with a strong laser (1 W or so) at a frequency that is about 13 THz above (100 nm below) the signal to be amplified. The signal-power profile depicted in Fig. 3.23(b) shows that most of the amplification occurs in the last 30 km of the span where the pump power is highest. In our example, the 100-km span has a loss of 20 dB when the Raman pump is off. This loss reduces to 10 dB when the Raman pump is on, thus providing an on/off gain of 10 dB.

The Raman amplification in the transmission fiber can be explained as follows. Stimulated by the incident signal photons, some of the pump photons "decay" into a (lower-energy) signal photon and a molecular vibration (a phonon) as illustrated in Fig. 3.24. Like in the EDFA, the resulting signal photon is coherent to the stimulating one, thus amplifying the signal. This process is known as *stimulated Raman scattering* (cf. Section 2.3). The available vibrational (phonon) energy states in silica glass range from about $h \cdot 10$ THz to $h \cdot 15$ THz (see Fig. 3.24(b)), resulting in a huge amplifier bandwidth of about 5,000 GHz. This bandwidth can be made even larger by using multiple pump lasers with different wavelengths [53].

How large is the noise figure of a Raman amplifier? To answer this question, let us compare a fiber span followed by an EDFA with the same fiber span followed by a Raman pump. A fiber span with loss $1/G$ or gain G, where $G < 1$, has a noise figure of $1/G$. The same fiber span followed by a lumped optical amplifier with noise figure F has a combined noise figure of $1/G \cdot F$ (cf. Eq. (I.6)). Both facts can be proven with the noise-figure definition given earlier. For example, our 100-km fiber span with 20-dB loss followed by an EDFA with a noise figure of 5 dB has a total noise figure of $20 + 5 = 25$ dB. [→ Problem 3.9.]

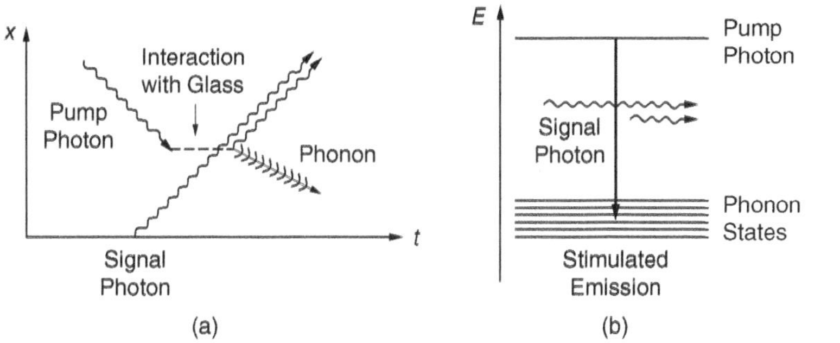

Figure 3.24 Stimulated Raman scattering: (a) space–time diagram and (b) energy diagram.

Now let us replace the EDFA with a Raman pump. With the pump off, the noise figure is 20 dB, equal to the loss. With the pump on, the loss reduces to 10 dB and ideally the noise figure also should reduce to 10 dB. But just like the EDFA, the Raman amplifier generates ASE noise and the actual noise figure is worse than that; let us say, it is 18 dB. Now, comparing the distributed Raman amplifier with the lumped EDFA amplifier, it becomes evident that the former behaves like the latter with an effective noise figure of −2 dB. Distributed Raman amplifiers can achieve negative noise figures, at least in a differential sense!

Raman amplifiers are of great practical significance, especially when used as in-line amplifiers in ultra-long-haul transmission systems. Their distributed gain permits roughly a doubling of the span length when compared to an EDFA-only solution (e.g., from 40 to 80 km) [54], cutting the number of optical amplifiers needed in half. Moreover, the bandwidth of an advanced Raman amplifier (with multiple pump wavelengths) is about twice that of an EDFA (70 nm vs 35 nm) [54], permitting a doubling of the number of WDM channels.

3.4 Integrated Photodetectors

Optoelectronic Integration. The speed and noise performance of a receiver depends critically on the total capacitance at its input node (cf. Chapter 6). The bonding pads, ESD protection circuits, and the TIA and photodetector packages all contribute to this capacitance. Bringing the photodetector as close as possible to the TIA and avoiding separate packages has the benefit of reducing this capacitance. In addition, such an integration results in a more compact system, potentially reducing the cost and improving the reliability. The photodetector and the TIA circuit can be integrated in a number of different ways:

- *Detector and circuit chip in the same package.* For high-speed optical receivers, it is common practice to *copackage* the photodetector and the

TIA into a receiver optical subassembly (ROSA), usually in the form of a small coaxial module (cf. Fig. 1.3). Similarly, a photonic integrated circuit (PIC) containing the photodiodes and a coherent receiver front-end [55, 56] can be copackaged with the TIA chips into a multi-chip module (MCM). A digital processor chip may be copackaged with an array of photodetectors and lasers (VCSELs) into an MCM to form a high-performance processor module with optical inputs and outputs [57]. This method of integration permits the technology for the photodetectors and the technology for the circuit chip to be chosen and optimized independently. For example, the photodetector can be based on an InP substrate, whereas the TIA circuit is implemented in a standard silicon technology.

- *Detector chip mounted on the circuit chip.* Pushing the integration one step further, the photodetector can be mounted on the circuit chip and can be connected by means of flip-chip bonding [58–62] or wire bonding [63], thus eliminating the substrate traces of the MCM. This method still permits an independent choice of the photodetector and circuit technologies, as in the previous case. For example, in [60] an array of AlGaAs p–i–n photodetectors were flip-chip bonded on a standard CMOS chip.
- *Detector and circuit monolithically integrated in an OEIC technology.* The next step in integration is to put the photodetector and the receiver circuits on the same chip, resulting in a so-called *optoelectronic integrated circuits* (OEIC) [5, 64]. The simplest form of an OEIC is the *p–i–n FET*, which combines a p–i–n photodetector and an FET on the same substrate. An OEIC technology is usually based on a circuit technology. Extra fabrication steps are then added to implement good-quality photodetectors (and possibly other components such as lasers, modulators, and waveguides).

  Adding photodetectors to a circuit technology involves compromises. For example, to fabricate long-wavelength (1.3 or 1.55-μm) photodetectors in a silicon circuit technology, it is necessary to incorporate a low-bandgap material such as germanium. However, germanium layers grown on silicon are prone to defects due to a lattice mismatch of about 4% between the two materials [65]. Another problem is that the maximum temperature permitted to anneal the germanium is limited in order not to degrade the performance of the silicon devices [65]. (If the germanium is deposited on top of the circuit technology's metal stack, the maximum processing temperature is limited to about 450°C [57].) Yet another problem arises if the photodetector is thicker than the circuit devices (which may be necessary to obtain a decent quantum efficiency). This lack of planarity makes it difficult to contact the devices with the metal layers [65]. Even for short-wavelength (0.85-μm) applications, where silicon can act as the absorber, incorporating fast photodetectors with high responsivity and low capacitance is challenging.

  Nevertheless, the promise of compactness, high performance, low cost, and improved reliability has resulted in the development of many OEIC technologies and the design of many integrated receivers. OEICs based on InP circuit technologies [66–68], GaAs circuit technologies [69–71], bipolar or BiCMOS circuit technologies [72–76], and CMOS technologies [27, 57, 65, 77, 78] have all been reported. Of particular interest are detectors that are compatible with CMOS technology and we discuss two examples (Ge detector and Ge/Si APD) shortly.

- *Detector and circuit monolithically integrated in a standard circuit technology.* Naturally, it would be simpler and more cost effective, if the photodetector could be implemented in a standard, that is, unmodified, circuit technology. Indeed, this is possible if the necessary performance compromises and wavelength restrictions are acceptable.
  For example, in a standard MESFET (or HFET) technology the Schottky diode formed between the gate metal and the semiconductor can be used to implement a so-called *metal–semiconductor–metal* (MSM) *photodetector* [79]. In a standard bipolar (or HBT) technology the base–collector junction may serve as a photodiode [80]. Similarly, in a standard CMOS technology the n-well to p-substrate or the p^+ (drain/source) to n-well junctions may serve as a photodiode [81]. We discuss some of these photodetectors (MSM and CMOS junctions) shortly.

CMOS Compatible Germanium Photodetector. The structure of a germanium waveguide p–i–n photodetector suitable for the 1.55-μm wavelength and compatible with SOI-CMOS technology is illustrated in Fig. 3.25.

Selective heteroepitaxy is used to grow the germanium detector on top of an SOI CMOS wafer. Although the 4% lattice mismatch between germanium and silicon can result in a high concentration of dislocations and dark current, careful processing and device design can minimize the impact of the dislocations.

The device described in [78] achieves a responsivity of 0.9 A/W at the 1.55-μm wavelength and a 1-V reverse bias. The dark current at room temperature and a 1-V reverse bias is about 300 nA. Although this dark current is much larger than that of an InGaAs/InP photodetector, it is tolerable

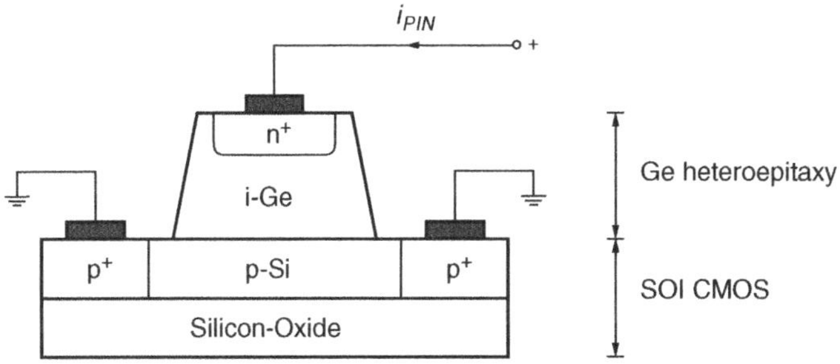

Figure 3.25 CMOS compatible germanium p–i–n photodetector [78].

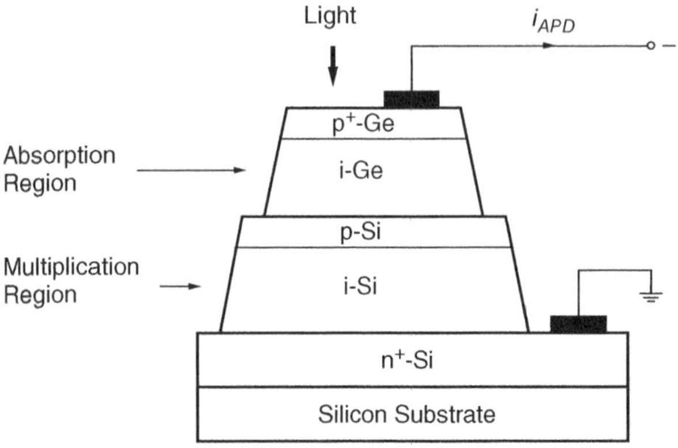

Figure 3.26 CMOS-compatible germanium/silicon APD [27].

for coherent receivers, where typical average photocurrents are 1mA or higher [56].

CMOS-Compatible Germanium/Silicon APD. The structure of a germanium/silicon APD suitable for the 1.3-μm wavelength and compatible with CMOS technology is illustrated in Fig. 3.26.

While InP-based optoelectronic devices are often superior to silicon-based devices, one of the exceptions is the area of APDs. As we have discussed in Section 3.2, a silicon multiplication region has important advantages over an InP or InAlAs multiplication region. The lower ionization-coefficient ratio k_A of silicon leads to a higher gain-bandwidth product and a lower noise.

The Ge/Si APD reported in [27] achieves a gain-bandwidth product of 340 GHz. The maximum bandwidth is 11.5 GHz for gains up to 20. The effective ionization-coefficient ratio that determines the excess noise factor (cf. Eq (3.17)) is $k_A = 0.09$. The thermal coefficient of the breakdown voltage is 0.05%/°C. This low thermal coefficient (lower than that for an InP- or InAlAs-based APD) is another advantage of the Ge/Si APD.

Metal–Semiconductor–Metal Photodetector (MESFET/HFET Technology). The MSM photodetector has the advantage that is can be implemented in an unmodified MESFET or HFET technology [79]. Nevertheless, specialized OEIC technologies also make use of this detector because of its simplicity (only one or two additional processing steps) and planarity [57, 69, 70, 73].

The operation of a basic MSM photodetector based on GaAs technology, which is suitable for the 0.85-μm wavelength, is illustrated in Fig. 3.27. Metal in direct contact with a semiconductor forms a *Schottky diode* [5]. Two interdigitated metal electrodes on top of the GaAs semiconductor form two back-to-back Schottky diodes, which constitute the MSM photodetector.

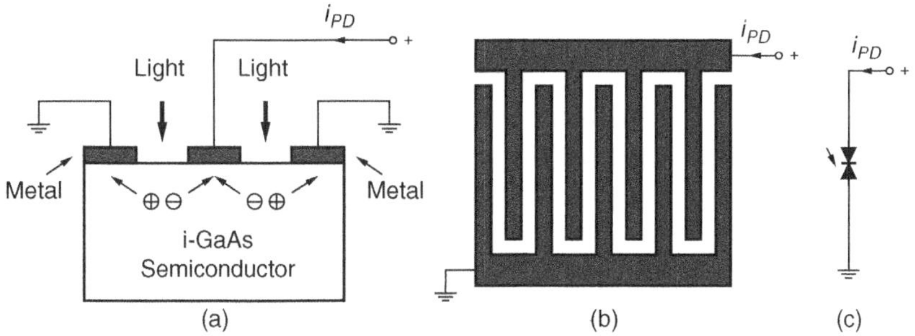

Figure 3.27 MSM photodetector: (a) cross-sectional view, (b) top view (on a different scale and with more fingers), and (c) circuit symbol.

When a bias voltage is applied across the MSM detector one diode becomes forward biased and the other diode becomes reverse biased. Unlike for a p–i–n photodetector, the polarity of the bias voltage doesn't matter, as is obvious from the MSM structure's symmetry. Light that impinges on the exposed semiconductor between the metal electrodes creates electron–hole pairs, provided that the photon energy exceeds the bandgap energy of the semiconductor. Like in a p–i–n photodetector, the electron–hole pairs get separated by the electric field in the reverse-biased Schottky diode and produce the photocurrent i_{PD}.

An obvious drawback of the MSM photodetector is that the light impinging on the metal is blocked and does not contribute to the photocurrent [4, 8, 82]. Thus, an MSM photodetector with an electrode width equal to the electrode spacing, as shown in Fig. 3.27, has a quantum efficiency (and responsivity) that is only about half of that of a p–i–n photodetector. Less shadowing and a better efficiency can be obtained by increasing the electrode spacing relative to the width, but this measure also increases the carrier transit time and thus reduces the bandwidth of the detector. (Moreover, too much spacing results in excessive carrier recombination and fewer carriers being picked up by the electrodes, reducing the quantum efficiency.) In principle, shadowing can be avoided by using transparent conductors, but this option requires special processing steps [4].

An advantage of the MSM photodetector is its low capacitance. For a given detector area, an MSM photodetector has only about 30% of the capacitance of a comparable p–i–n detector [82]. The reason for this reduction is that the MSM detector operates on the fringe capacitance between the metal fingers, whereas the p–i–n detector relies on the area capacitance between the p-layer and the n-layer. This lower capacitance can make up for some or all of the sensitivity lost because of the lower responsivity [82]. If low capacitance is not a priority, the area of the MSM photodetector can be made large (e.g., $75 \times 75\ \mu m^2$)

to simplify the coupling of the light from the fiber to the detector and to reduce the cost of the system [4].

The transit-time limited speed of the MSM photodetector is primarily given by the spacing between the metal fingers and the carrier velocities, which in turn are a function of the applied bias voltage. MSMs also tend to have low-field regions resulting in diffusion-limited tails. (The bandwidth calculation for an MSM is more complicated than that for a p–i–n structure because carriers generated far away from the surface see a weaker electric field, move well below the saturation velocity, and have a longer way to travel to the electrodes [83].) A bandwidth of 105 GHz was achieved for a small ($15 \times 10\ \mu m^2$) GaAs MSM photodetector with a finger spacing of 0.5 μm [84].

The MSM photodetector can be made sensitive to long wavelengths by using a narrow-bandgap semiconductor such as germanium or InGaAs. However, the resulting lower Schottky barrier between the metal and semiconductor leads to a larger dark current. To control this dark current, a thin layer (either a lightly doped implant or an epitaxial layer of a wide-bandgap material) can be sandwiched between the absorbing semiconductor and the metal [4, 82].

CMOS N-Well to P-Substrate Photodiode. Although the bandgap of silicon is narrow enough to absorb light at the 0.85-μm wavelength, the *indirect* nature of the bandgap results in a large absorption length, measuring about 15 μm [72]. For comparison, the absorption length of GaAs is only 1 μm [72] and that of germanium is 0.3 μm [65] at the same 0.85-μm wavelength. Thus, for a silicon photodetector to have a good quantum efficiency, a thick absorption layer is required, which unfortunately results in a long transit time for the photogenerated carriers.

When we restrict ourselves to building the photodetector in a standard CMOS technology, there are additional challenges. The junction between the lightly doped n-well and p-substrate provides the thickest depletion layer in a CMOS technology that can be used to absorb photons. A CMOS photodetector based on this junction is shown in Fig. 3.28. Even then, the thickness of the depletion layer, outlined with thin dashed lines, is only about 3 μm at 3 V [85], far less than the absorption length. As a result, the quantum efficiency of this photodiode is limited to about 37% [85]. Even worse, the light that didn't get absorbed in the depletion layer penetrates into the substrate where it creates electron–hole pairs outside of the drift field. The resulting minority carriers diffuse around aimlessly and, if not lost to recombination, contribute to a slow current tail (cf. Section 3.1). (In a silicon on insulator [SOI] technology, the insulator below the n-well can be used to block the carriers from diffusing back into the drift field [72].)

In a practical implementation of this photodetector, the single n-well region schematically shown in Fig. 3.28 is replaced by several parallel n-well stripes, which are all connected together. Such a finger layout increases the junction's

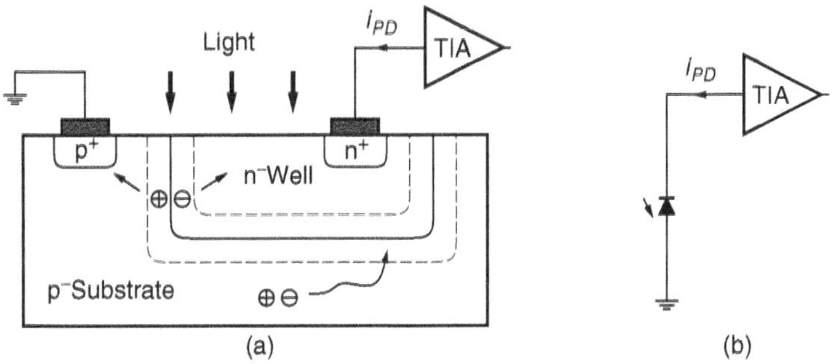

Figure 3.28 CMOS n-well to p-substrate photodiode: (a) cross-sectional view and (b) circuit view.

side-wall depletion region [86], reduces the contact resistance [81, 87], and for small finger widths can increase the bandwidth by a small amount [81, 88].

Unfortunately, the relatively thin absorption layer is not the only factor limiting the responsivity of this CMOS photodiode. Reflection and refraction of the light in the dielectric stack above the silicon attenuates the incident light even before it has a chance to reach the photodiode [85, 89]. The dielectric stack consist of the field oxide, the dielectrics used between the metal layers, and the capping layer. This stack limits the transmission of the incident light to about 40 to 70% [89]. Combining this figure with the 37% (high-frequency) quantum efficiency mentioned before yields an overall quantum efficiency of only about 20%, which corresponds to a paltry responsivity of 0.14 A/W at the 0.85-μm wavelength. Technology scaling makes the responsivity problem even worse: more metal layers result in a less transparent dielectric stack, increased doping levels and lower supply voltages result in a thinner and less effective depletion layer [89]. The deposition of silicide over CMOS active regions to facilitate low-resistance contacts presents another impediment. If not blocked over the photodiode, this silicide layer can absorb up to 95% of the incident light [90].

The bandwidth of the n-well/p-substrate photodiode is mostly determined by the minority carriers that are produced outside of the depletion layer, namely the holes in the n-well above and the electrons in the p-substrate below [81, 91]. Of those minority carriers, the electrons created deep in the substrate are the most troublesome. Typical bandwidth values are between 1 and 10 MHz [81, 92], well below what is needed for a Gb/s optical receiver. But finally there is some good news: the frequency roll-off of the responsivity at high frequencies is relatively slow. Typical values range from −3 to −5 dB/decade [81], compared with −20 dB/decade for a first-order low-pass filter. This slow roll-off can be compensated quite effectively with a simple analog equalizer. Integrated 3-Gb/s receivers with an n-well/p-substrate photodiode followed by such an equalizer have been demonstrated [81, 87]. In [81], the equalizer, which needs to have a slow frequency "roll-up," has been realized by summing the outputs of four frequency-staggered first-order high-pass filters.

CMOS P^+ to N-Well Photodiode. Another CMOS junction suitable for photodetection is the p^+ (drain or source) to n-well junction, shown in Fig. 3.29. Because the p^+ material is more heavily doped than the p-substrate, the resulting depletion layer is thinner and the responsivity lower than that of the n-well/p-substrate photodiode. On the plus side, the n-well/p-substrate photodiode can now be used to collect the carriers that are generated below the p^+/n-well photodiode thus removing the slow current tail from the upper photodiode [93]. Figure 3.29 illustrates how the photocurrent collected by the n-well/p-substrate photodiode is dumped to ground, whereas the photocurrent collected by the p^+/n-well photodiode is fed to the TIA.

In a practical implementation, the single p^+ region, schematically shown in Fig. 3.29, is replaced by several parallel p^+ stripes (fingers) located in a joint n-well [93]. The advantages of this geometry are similar to those given for the n-well/p-substrate photodetector. In a deep n-well technology, where it is possible to create an isolated p-well, the structure shown in Fig. 3.29 can be "inverted" to produce an n^+/p-well photodiode located inside the deep n-well [94].

The p^+/n-well photodiode and the n-well/p-substrate photodiode have been compared to each other extensively [81, 92]. It has been found that the p^+/n-well photodiode has a bandwidth of about 1 to 3 GHz, an improvement of almost three orders of magnitude over the n-well/p-substrate photodiode. However, the responsivity of the p^+/n-well photodiode is about 10 to 20 times smaller than that of the n-well/p-substrate photodiode. Moreover, owing to the thinner depletion layer, the capacitance per area of the p^+/n-well photodiode is about twice that of the n-well/p-substrate photodiode.

An integrated 1-Gb/s receiver using a p^+/n-well photodiode has been reported in [93]. The photodiode by itself has a responsivity ranging from 0.01 to 0.04 A/W at the 0.85-μm wavelength, where the larger value corresponds to a bias voltage near the junction breakdown (10 V). A larger reverse bias

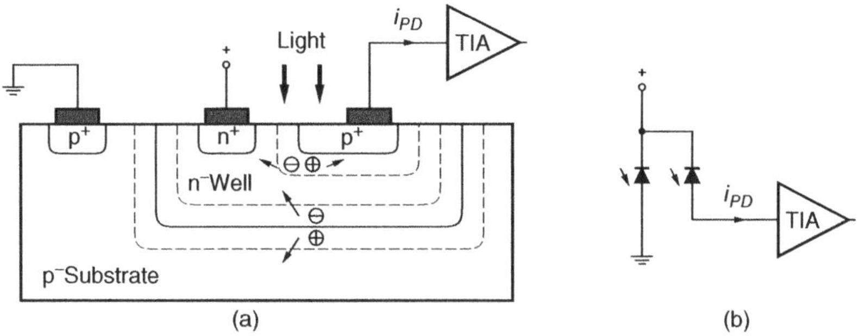

Figure 3.29 CMOS p^+ to n-well photodiode: (a) cross-sectional view and (b) circuit view.

voltage increases the thickness of the depletion layer and hence the fraction of photons absorbed there. An integrated 2.5-Gb/s receiver with a similar photodiode type biased at 6 V has been presented in [94]. Both receivers reach the stated bit rate without the use of an equalizer.

CMOS Spatially Modulated Light (SML) Detector. An innovative approach is taken by the *spatially modulated light* (SML) *detector* [85, 91]. This detector, which is shown schematically in Fig. 3.30, is based on the CMOS n-well/p-substrate photodiode. But instead of exposing all the n-well fingers to the light, every second finger is blocked by a layer of metal (usually layer-2 metal). All the illuminated junctions are connected together and all the dark junctions are connected together. Figure 3.30(a) shows the cross-sectional view of two such n-well/p-substrate junctions, one illuminated and one dark. The idea is that the troublesome carriers generated deep in the substrate have an approximately equal chance of diffusing to an illuminated junction or a dark junction. Thus, the slow current tail appears at both junctions and can be suppressed by subtracting the photocurrent of the dark junctions $i_{PD,D}$ from the photocurrent of the illuminated junctions $i_{PD,I}$. A differential TIA, which we discuss in Section 7.2, can be used to perform the subtraction $i_{PD,I} - i_{PD,D}$. The currents from the illuminated and dark junctions, $i_{PD,I}$ and $i_{PD,D}$, are sometimes called the *immediate* and the *deferred* currents, respectively. Conveniently, the indices I and D permit either interpretation.

After the current subtraction, the bandwidth of the SML detector is typically in the range from 0.6 to 1 GHz [86, 88, 92], not far from that of a p$^+$/n-well photodiode. This bandwidth is believed to increase with technology scaling because more closely spaced n-well/p-substrate photodiodes will pick up the diffusing carriers more evenly [91]. The SML detector's responsivity is typically in the range from 0.05 to 0.07 A/W at the 0.85-μm wavelength [86, 88, 92], better

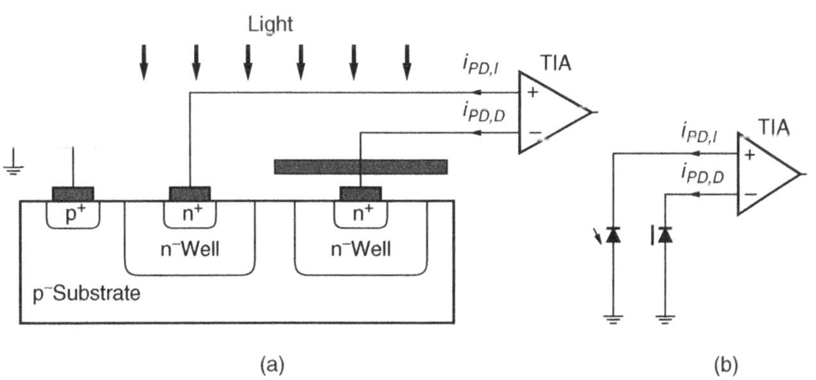

Figure 3.30 CMOS spatially modulated light detector: (a) cross-sectional view and (b) circuit view.

than that of a p^+/n-well photodiode. The capacitance per photodiode area is similar to that of the n-well/p-substrate photodiode and lower than that of the p^+/n-well photodiode. The matched photodiode capacitances combined with the fact that the signal is represented by the difference of two currents results in a superior noise immunity (cf. Section 7.2). Overall, the SML detector is a very attractive solution for integrated CMOS receivers.

The alternating pattern of dark and illuminated n-well stripes described before can be replaced by other geometrical patterns in order to increase the side-wall depletion region (better responsivity) and to collect the diffusing carriers more evenly between the two junctions (higher bandwidth). It has been found that a checker-board pattern of dark and illuminated photodiodes can improve the bandwidth at the expense of a lower responsivity [86, 95, 96]. An integrated 10-Gb/s receiver based on such an SML detector biased at 14.2 V has been reported in [95, 96]. The bandwidth can be boosted further by following the SML detector with an equalizer [86, 88, 97]. An integrated 8.5-Gb/s receiver with adaptive equalizer operating from a single 1.5-V supply has been reported in [97].

3.5 Detectors for Phase-Modulated Optical Signals

How can we detect phase-modulated optical signals? The difficulty is, of course, that all the photodetectors we discussed so far are insensitive to the optical phase and only respond to the light intensity. The solution is to let the received optical signal interfere with another optical signal in a so-called *interferometer*. If the phases of the two optical signals are aligned, constructive interference results in a high intensity; if the phases of the two optical signals are 180° apart, destructive interference results in a low intensity. The interferometer makes the phase modulation visible to the intensity detector.

Detection of DPSK. Let us focus first on the *differential* phase-shift keying (DPSK) modulation format (cf. Chapter 1), which has the advantage that the phase of the received bits can be demodulated by taking the phase of the previous bit as a reference [98, 99].

A detector for DPSK signals is shown schematically in Fig. 3.31 and the corresponding RZ-DPSK waveforms are shown in Fig. 3.32. Usually (but not necessarily) the input signal is optically preamplified. Then, the optical path branches into two arms, the delay of the first arm is one bit period longer than the delay of the second arm. At the end of the two arms the current and previous bits are available simultaneously and are brought to interfere in a 2×2 coupler. The upper output port produces the difference of the two optical signals and thus lights up when the DPSK signal encodes a one (180° phase difference). The lower output port produces the sum and thus lights up when

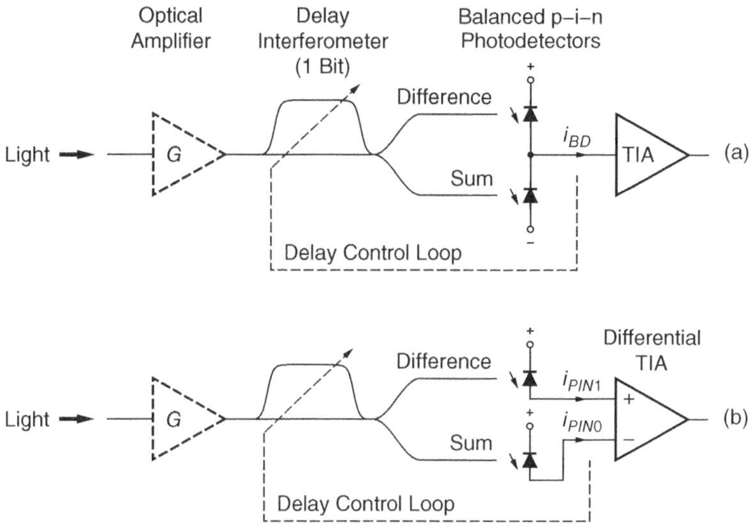

Figure 3.31 Self-coherent DPSK demodulator followed by a balanced photodetector connected to (a) a single-ended and (b) a differential TIA.

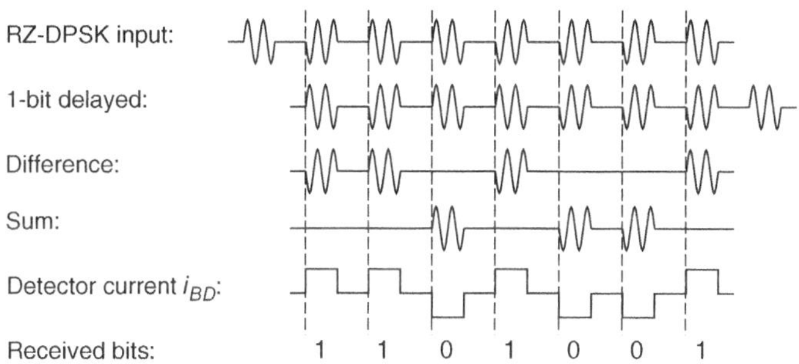

Figure 3.32 Detection of an optical RZ-DPSK signal.

the DPSK signal encodes a zero (no phase difference). This *delay interferometer* is followed by two p–i–n photodetectors, which convert the optical intensities into proportional currents. This arrangement is know as a *balanced detector*. Although it is sufficient to detect only one of the two outputs (the two outputs are complementary), detecting both outputs results in a better performance (lower required OSNR). Finally, the two currents are subtracted to produce the bipolar output current $i_{BD} = i_{PIN1} - i_{PIN0}$.

In Fig. 3.31(a) the subtraction is done by means of Kirchhoff's current law [100], whereas in Fig. 3.31(b) the subtraction is done by means of a differential TIA [101–103]. The single-ended-TIA arrangement has the advantage of requiring only one connection from the detector to the TIA.

The differential-TIA arrangement, however, may achieve a lower capacitance per input (only one photodetector per input) and thus a higher bandwidth [102]. Moreover, the differential-TIA arrangement requires only a single photodetector bias voltage, which can be lower than in the case of two stacked photodetectors.

The delay interferometer must be very precise. For two 1,550-nm light waves to be phase aligned within a couple of degrees at the coupler, the relative length of the two arms must be accurate to about 10 nm or better. To meet this high accuracy, the delay of one arm is made tunable and a feedback mechanism controls the delay (indicated by the dashed delay control loop in Fig. 3.31) [100]. The delay can be fine tuned by either a fiber heater or a piezoelectric fiber stretcher. The error signal for the feedback control can be derived, for example, from the DC current through one of the photodiodes [100].

Signal and Noise. Neglecting losses in the delay interferometer, we find with Eq. (3.18) that the current in each photodetector swings between approximately 0 and $\mathcal{R}GP$ where $\mathcal{R}$ is the responsivity of the p–i–n detectors, G is the gain of the optical preamplifier, and P is the on power of the optical input signal ($P = 2\overline{P}$ for 50% RZ-DPSK).

The balanced detector subtracts the currents of the two photodetectors and produces the *bipolar* current i_{BD} swinging between $-\mathcal{R}GP$ and $+\mathcal{R}GP$ (cf. Fig. 3.32).

The balanced DPSK detector generates the same amount of noise for the zeros and the ones. The amplitude distribution of the noise from a balanced detector with optical amplifier(s) is significantly non Gaussian, especially in its tails [104]. The noise currents from the individual p–i–n photodiodes, $i_{n,PIN0}$ and $i_{n,PIN1}$, have chi-square distributions, as we know from Section 3.3. The distribution of the "dark" photodiode is strongly non-Gaussian, whereas the illuminated photodiode is closer to Gaussian. The total noise from the balanced detector, $i_{n,BD}$, is the difference of the two individual and independent noise currents, $i_{n,PIN1} - i_{n,PIN0}$. Hence the distribution of $i_{n,BD}$ is the convolution of the distribution of $i_{n,PIN1}$ with the distribution of $-i_{n,PIN0}$.

Detection of DQPSK. A more accurate term for DPSK would be differential *binary* phase-shift keying (DBPSK) because the phase difference between two successive symbols takes on one of *two* values (0° or 180°). DPSK can be generalized to more than two phase values. In particular, for four values the format is known as differential *quadrature* phase-shift keying (DQPSK).

A detector for the DQPSK format with two delay interferometers and two balanced detectors is shown in Fig. 3.33 [105, 106]. The incoming optical signal is split into two identical copies, each one feeding a seperate delay interferometer. Because each DQPSK symbol encodes two bits, the differential delay must now be set to two bit periods (one symbol period). Moreover, instead of fine

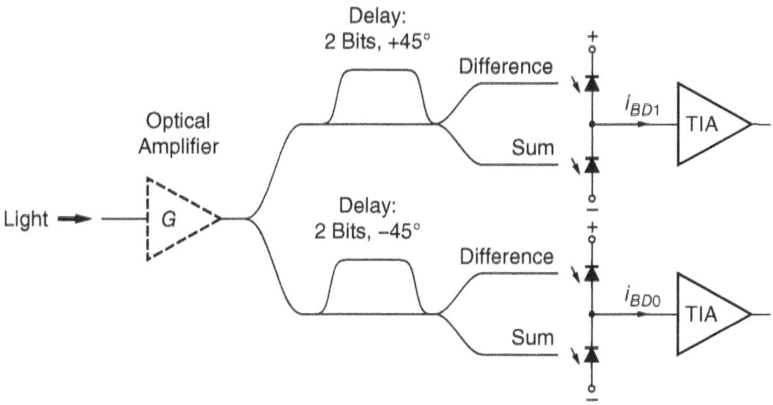

Figure 3.33 Self-coherent DQPSK demodulator followed by two balanced detectors.

tuning the interferometer arms for a 0° phase difference, one interferometer must be tuned for +45° and the other must be tuned for −45°.

How are the four phase differences of the DQPSK signal, 0°, 90°, 180°, and 270°, detected with this arrangement? Before attempting to answer this question, let us go back to the DBPSK detector in Fig. 3.31 and look at it in a more general way. Figure 3.34(a) shows all possible phase differences between the previous and the current symbol, ϕ, as points on a circle. The 0° phase, corresponding to a zero bit, is marked by a white dot and the 180° phase, corresponding to a one bit, is marked by a black dot. We know that for these two points the balanced detector produces a negative and a positive current, i_{BD}, respectively. What happens for the phases in between? For the entire right half circle, the lower photodiode receives more light than the upper one and the total output current is negative; conversely, for the entire left half circle, the

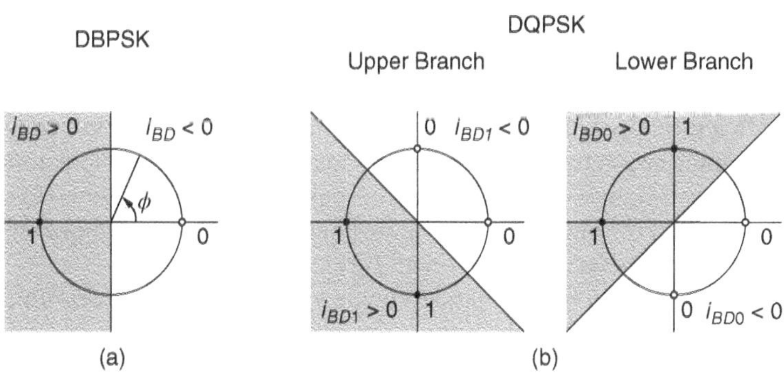

Figure 3.34 Detector currents as a function of the phase difference between successive symbols: (a) DBPSK and (b) DQPSK.

upper photodiode receives more light than the lower one and the total output current is positive.

Now we are ready to analyze the DQPSK detector. Figure 3.34(b) shows the polarity of the output currents from the upper and lower balanced detectors as a function of the symbol phase difference. Because the interferometers are offset by $+45^\circ$ and -45°, the decision boundaries are rotated by 45° counter-clockwise and 45° clockwise, respectively. As a result, each of the four phases is assigned a unique two-bit code. The 0° and 180° phases, corresponding to the DBPSK constellation, are assigned 00 and 11 and the new 90° and 270° phases are assigned 01 and 10, respectively (note that this is a Gray code).

Detection of BPSK and QPSK. How can we demodulate and detect phase-modulated optical signals that do *not* encode the information differentially? Examples of such modulation formats are *binary phase-shift keying* (BPSK) and *quadrature phase-shift keying* (QPSK).

The key idea is to use a local-oscillator (LO) laser that provides a reference against which the phase of the received signal can be measured. A receiver that makes use of interference between the received signal and an LO is known as a *coherent receiver* (cf. Fig. 3.9). In contrast, the DBPSK and DQPSK receivers discussed earlier make use of interference between the received signal and (a delayed copy of) itself and are known as *self-coherent* receivers.

Some advanced modulation formats, such as *dual-polarization quadrature phase-shift keying* (DP-QPSK), encode information in both polarization states, requiring a detector that can recover both polarization states.

Coherent Detector with Phase and Polarization Diversity. Figure 3.35 shows the basic structure of a detector that can recover the phase and polarization information. The received optical signal and the LO wave each pass through a polarization splitter separating the x-polarized component (upper path) from the y-polarized component (lower path). Then, the received signal and the LO wave pass through an array of four interferometers converting phase information to intensity information. Two of the interferometers have equal input path lengths, extracting the in-phase information, and the other two have path lengths that differ by a 90° phase shift, extracting the quadrature information. The four interferometers are followed by four balanced photodetectors providing electrical signals for the in-phase x-polarized component (IX), quadrature x-polarized component (QX), in-phase y-polarized component (IY), and quadrature y-polarized component (QY).

If the LO laser in Fig. 3.35 is phase (and frequency) locked to the transmitting laser, the coherent detector directly outputs the demodulated baseband signal. The phase locking can be achieved with an *optical* phase-locked loop (PLL). This approach, which is known as a *homodyne receiver*, however, is difficult to implement in practice [1, 107].

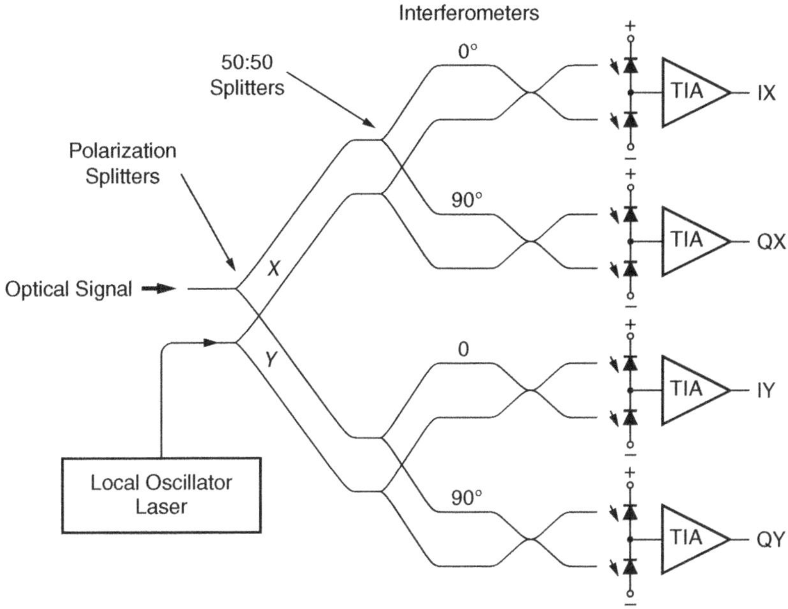

Figure 3.35 Coherent detector with phase and polarization diversity.

A better approach is to use a free-running LO laser operating at a frequency that is slightly different from the transmitter frequency (cf. Fig. 3.9). The coherent detector then outputs a phase-modulated electrical signal at a lower frequency known as the *intermediate frequency* (IF). The IF is given by the difference between the transmitter and LO frequency. This approach is known as a *heterodyne receiver* or, if the IF is so low that it falls within the signal bandwidth, as an *intradyne receiver* [1, 107]. The demodulation of the IF signal is done in the electrical domain, preferably with a digital signal processor (DSP) after A/D conversion (cf. Fig. 1.4) [108].

Figure 3.36 shows a packaged coherent receiver suitable for 100-Gb/s DP-QPSK. A PIC implementing a coherent receiver front-end including the fiber couplers, splitters, interferometers, and photodetectors is described in [56].

Signal and Noise: Single Photodetector. Before analyzing the signal and noise of a coherent receiver with balanced detectors, as shown in Fig. 3.35, let us go back to the simpler situation of a coherent receiver with a single photodetector, as illustrated in Fig. 3.9.

Because of the square-law characteristic of the photodetector, we expect to get three electrical beat components from the two optical sources: $(\text{LO} + \text{signal})^2 = (\text{LO})^2 + 2(\text{LO} \times \text{signal}) + (\text{signal})^2$. Working out the details

Figure 3.36 A packaged coherent receiver with four TIAs suitable for 100-Gb/s DP-QPSK (2.5 cm × 2.2 cm × 0.4 cm). One fiber pig tail is for the input signal, the other is for the local-oscillator laser. *Source*: Reprinted by permission from Finisar Corporation.

and assuming that the polarization of the LO and the signal are aligned, we find [1]

$$i_{PIN,\text{het}} = \mathcal{R}(P_{LO} + 2\sqrt{P_{LO}P}\sin(\Delta\omega t + \Delta\phi) + P), \tag{3.30}$$

where P is the received power, P_{LO} is the LO power, $\Delta\omega$ is the (angular) intermediate frequency (IF), and $\Delta\phi$ is the phase difference between the received signal and the LO.

The first term, $\mathcal{R}P_{LO}$, represents a DC current. For example, with an LO power of 2 mW and $\mathcal{R} = 0.8$ A/W, a constant current of 1.6 mA flows through the detector. With the usual large LO power, the third term, $\mathcal{R}P$, is small compared with the first term ($P \ll P_{LO}$) and can be neglected. The second term is the desired IF signal that contains the phase information $\Delta\phi$. This term increases with P_{LO}, which means that a larger LO power results in a stronger IF signal. In a sense, the LO provides gain, but we cannot compare this gain directly with the gain of an APD or an optical preamplifier because the LO gain acts on $\sqrt{P}$ whereas the APD gain and the optical-preamplifier gain act on P.

Assuming that the received optical signal is noise free (no ASE noise from optical amplifiers), the noise current after detection is mostly shot noise due to the LO laser. A secondary noise component is the relative intensity noise (RIN), also due to the LO laser. Neglecting the dark current of the detector, the mean-square noise current is [1]

$$\overline{i^2_{n,PIN,\text{het}}} = 2q\mathcal{R}P_{LO} \cdot BW_n + 2RIN \cdot \mathcal{R}^2P^2_{LO} \cdot BW_n, \tag{3.31}$$

where RIN is the RIN noise of the LO. Note that this noise current does *not* depend on the received signal strength P. With the usual large LO power, the noise distribution is close to Gaussian.

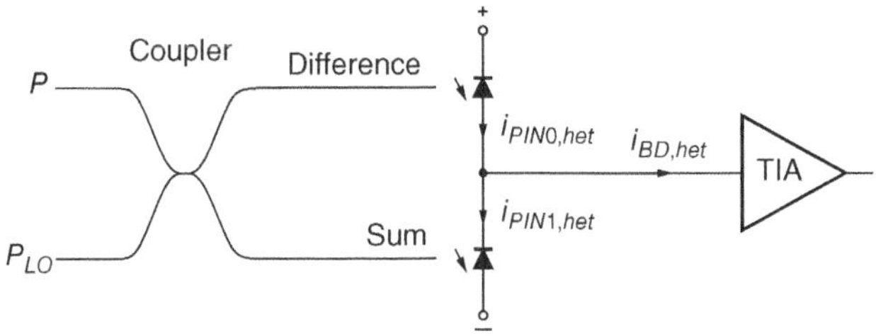

Figure 3.37 Coherent receiver with balanced photodetector.

Signal and Noise: Balanced Photodetector. Now, let us move on to the balanced detector shown in Fig. 3.37. The optical coupler produces the sum and difference of the fields from the LO and the received signal. Each output provides half of the total optical power. The currents from the two photodetectors are subtracted at the common node producing the difference $i_{BD,\text{het}} = i_{PIN0,\text{het}} - i_{PIN1,\text{het}}$.

Given perfect symmetry between the two optical paths and the two detectors, the balanced output current contains only the desired second term of Eq. (3.30):

$$i_{BD,\text{het}} = 2\mathcal{R}\sqrt{P_{LO}P}\sin(\Delta\omega t + \Delta\phi). \tag{3.32}$$

The first and third terms of Eq. (3.30) make the same current contributions to both detectors and thus cancel out. Similarly, the RIN noise current from Eq. (3.31) appears equally at both detectors and cancels out [1, 12]. The two shot-noise currents, however, are uncorrelated and do not cancel:

$$\overline{i^2_{n,BD,\text{het}}} = 2q\mathcal{R}P_{LO} \cdot BW_n. \tag{3.33}$$

The balanced detector thus has the important advantages of getting rid of the RIN noise and using the signal and LO power efficiently.

The subtraction of the two photocurrents can be done by means of Kirchhoff's current law, as shown in Fig. 3.37 [109], or by means of a differential TIA (cf. Fig. 3.31(b)) [110]. The pros and cons are similar to those discussed earlier for the DPSK detector.

Common-Mode Rejection. In practice, the symmetry of a balanced detector is not perfect, which means that the RIN noise and the DC current are not completely canceled. The asymmetry is quantified by the common-mode rejection ratio (CMRR), which is determined by applying equal optical signals to both detectors and measuring the ratio [111, 112]

$$CMRR = \frac{i_{PIN0} - i_{PIN1}}{i_{PIN0} + i_{PIN1}} = \frac{i_{BD}}{i_{PIN0} + i_{PIN1}}. \tag{3.34}$$

Typically, $CMRR < 0.1$ or less than −20 dB. (Note that with the definition Eq. (3.34) a smaller value for $CMRR$ corresponds to a better rejection.) An imbalance in power or a mismatch in delay both result in a nonzero CMRR. A power imbalance can result from a deviation in the 50/50 splitting ratio of the coupler or unequal responsivities of the detectors. A delay mismatch can result from differences in the path length from the output of the coupler to the two photodetectors [113].

Advantages of Coherent Detection. The self-coherent detector for DQPSK in Fig. 3.33 and the coherent detector for DP-QPSK in Fig. 3.35 both use interference to demodulate the phase information. But there is an important difference: The self-coherent detector compares the phase of a received symbol to a noisy reference (a previously received symbol), whereas the coherent detector compares the phase of a received symbol to an essentially noise-free reference (the local oscillator). For this reason the coherent receiver requires less OSNR to operate at the desired BER.

We introduced the coherent detector with phase and polarization diversity as a means for detecting phase modulated signal. However, this is only one of its advantages over the simple intensity detector. Other important advantages are:

- The phase and polarization information that is now available in the electrical domain permits the near perfect compensation of propagation impairments, such as chromatic dispersion and polarization mode dispersion [108, 114].
- The coherent detector can be tuned to a specific wavelength by tuning the LO laser while achieving a high level of signal rejection of neighboring channels without the need for optical filters [108].
- The coherent detector features a high optical sensitivity (minimum signal power that can be detected; cf. Section 4.4), which extends the reach of unamplified transmission systems [113].

3.6 Summary

The most common photodetectors for optical receivers are:

- The p–i–n photodetector, which has a typical responsivity in the range of 0.6 to 0.9 A/W (at $\lambda = 1.3$ to 1.55 μm). Vertically illuminated p–i–n detectors are suitable for speeds up to about 10 Gb/s and edge-coupled p–i–n detectors (waveguide and traveling-wave detectors) are suitable for speeds from 40 to more than 100 Gb/s. The primary noise mechanism is shot noise.
- The avalanche photodetector (APD), which has a typical total responsivity in the range of 5 to 20 A/W (at $\lambda = 1.3$ to 1.55 μm). Fast APDs are suitable for 10 Gb/s but do not provide much internal gain at 40 Gb/s. The primary noise mechanism is avalanche noise, quantified by the excess noise factor F.
- The optically preamplified p–i–n detector, which can have a total responsivity of several hundreds A/W. The optical preamplifier usually is an

erbium-doped fiber amplifier (EDFA). This detector is suitable for applications from 10 to more than 100 Gb/s, with the p–i–n detector being the speed-limiting device. The primary noise mechanism of the optical preamplifier is amplified spontaneous emission (ASE), quantified by the noise figure F. The p–i–n detector converts the optical ASE noise into two electrical noise components: signal–spontaneous and spontaneous–spontaneous beat noise.
- The coherent detector, which in its basic form consists of a local-oscillator laser, a beam combiner, and a p–i–n detector. When adding phase and polarization diversity it can demodulate advanced modulation formats such as DP-QPSK. This detector is suitable for applications up to more than 100 Gb/s, with the p–i–n detector being the speed-limiting device. The primary noise mechanism is shot noise due to the local oscillator.

The first three detectors are characterized by:

- A photocurrent i_{PD} that is proportional to the received optical power P. Therefore, a 3-dB change in optical power results in a 6-dB change in current.
- A signal-dependent noise current. Specifically, the mean-square noise current $\overline{i_{n,PD}^2}$ grows proportional to the signal current i_{PD} (neglecting dark-current noise and spontaneous–spontaneous beat noise). Received one bits contain more noise than zero bits; the noise is not additive.
- A non-Gaussian noise distribution with a steep roll-off toward zero (left side). The skew is most pronounced for small optical powers.

In contrast, the coherent detector generates a current that is proportional to the square root of the received optical power $\sqrt{P}$. Under typical conditions, the noise is signal independent and approximately Gaussian.

Optoelectronic integrated circuits (OEIC) combine optical devices, such as detectors, with electronic circuits, such as TIAs. Of particular interest are optical devices that are compatible with CMOS technology (silicon photonics). Examples of detectors that are suitable for integration with a circuit technology are:

- The germanium detector, which is compatible with CMOS technology.
- The metal–semiconductor–metal (MSM) detector, which can be implemented, in a standard MESFET or HFET technology.
- The spatially modulated light (SML) detector, which can be implemented in a standard CMOS technology.

To make an intensity detector sensitive to optically phase-modulated signals, the phase information must be converted to intensity information by means of optical interference. A differential phase-shift keying (DPSK) signal can be detected with a delay interferometer followed by a balanced detector.

A phase-shift keying (PSK) or quadrature amplitude modulation (QAM) signal can be detected with a coherent detector with phase diversity. A signal utilizing both polarization modes (e.g., DP-QPSK) can be detected with a coherent detector with phase and polarization diversity.

Problems

3.1 **Optical versus Electrical dBs.** A p–i–n photodetector in a 1.55-μm transmission system converts the received optical signal to an electrical signal. By how many dBs is the latter signal attenuated if we splice an additional 40 km of standard SMF into the system?

3.2 **Power Conservation in the Photodiode.** The p–i–n photodetector produces a current that is proportional to the received optical power P. When this current passes through a resistor, it produces a voltage drop that is also proportional to the received optical power. Thus, the electrical power dissipated in the resistor is proportional to P^2. We conclude that for large values of P, the electrical power will exceed the received optical power! Is energy conservation violated?

3.3 **Sensitivity of an Antenna.** An antenna is exposed to the signal power P and thermal background radiation at the temperature T. (a) Calculate the power level P at which the rms signal from the antenna becomes equal to the rms value of the noise in the bandwidth BW_n (sensitivity at $SNR = 1$). Tip: the noise from the antenna is equal to that of the resistor R_{ANT} (the radiation resistance) at temperature T [115]. (b) Evaluate the sensitivity of a receiver operating at frequency $f = 1$ GHz with noise bandwidth $BW_n = 100$ MHz and background temperature $T = 300$ K. (c) Compare the sensitivity in (a) to the fundamental sensitivity limit given in [17]:

$$P = \left[\frac{hf}{\exp(hf/kT) - 1} + \frac{hf}{2}\right] BW_n. \tag{3.35}$$

3.4 **Shot Noise Versus Thermal Noise.** The current generated in a p–i–n photodetector consists of a stream of randomly arriving electrons and thus exhibits shot noise. Does the current from a battery loaded by a resistor also exhibit shot noise?

3.5 **Amplified Shot Noise.** An APD with deterministic amplification (every primary carrier generates precisely M secondary carriers) produces the mean-square noise $\overline{i_{n,APD}^2} = M^2 \cdot 2qI_{PIN} \cdot BW_n$ (Eq. (3.14)). Now, we could argue that the DC current produced by the APD is MI_{PIN} and thus the associated shot noise should be $\overline{i_{n,APD}^2} = 2q \cdot (MI_{PIN}) \cdot BW_n$. What is wrong with this argument?

3.6 **Optically Preamplified p–i–n Detector.** Extend Eq. (3.19) for $\overline{i_{n,OA}^2}$ to include the shot noise due to the signal power, shot noise due to the ASE power, and shot noise due to the detector dark current.

3.7 **Optical Signal-to-Noise Ratio.** Equations (3.20) and (3.21) state the relationship between SNR and OSNR for a continuous-wave signal with power P_S. How does this expression change for a DC-balanced ideal NRZ-modulated signal with high extinction and an average power $\overline{P}_S$?

3.8 **Noise Figure of an Optical Amplifier.** (a) Extend Eq. (3.26) for the noise figure F to include the shot noise due to the signal power, shot noise due to the ASE power, and shot noise due to the detector dark current. (b) How large is the noise figure for $S_{ASE} = 0$, assuming $I_{DK} \ll \mathcal{R}P_S$? (c) How large must S_{ASE} be for $F = 1$, assuming $P_{ASE} \ll P_S$ and $I_{DK} \ll \mathcal{R}P_S$?

3.9 **Noise Figure of a Fiber.** (a) Calculate the noise figure F of an optical fiber with loss $1/G_1$ ($G_1 < 1$). (b) Calculate the noise figure F of an optical system consisting of an optical fiber with loss $1/G_1$ ($G_1 < 1$) followed by an EDFA with gain G_2 and noise figure F_2. (c) Calculate the noise figure F of an optical system with n segments, where each segment consists of an optical fiber with loss G ($G > 1$) followed by an EDFA with compensating gain G and noise figure F_2.

References

1 G. P. Agrawal. *Fiber-Optic Communication Systems.* John Wiley & Sons,Inc., Hoboken, NJ, 4th edition, 2010.

2 J. M. Senior. *Optical Fiber Communications: Principles and Practice.* Prentice Hall, Harlow, England, 3rd edition, 2009.

3 K. Kato. Ultrawide-band/high-frequency photodetectors. *IEEE Trans. Microwave Theory Tech.*, MTT-47(7):1265–1281, 1999.

4 P. R. Berger. MSM photodiodes. *IEEE Potentials*, 15(2):25–29, 1996.

5 S. M. Sze and K. K. Ng. *Physics of Semiconductor Devices.* John Wiley & Sons, Inc., Hoboken, NJ, 3nd edition, 2007.

6 J. E. Bowers and C. A. Burrus Jr. Ultrawide-band long-wavelength p-i-n photodetectors. *J. Lightwave Technol.*, LT-5(10):1339–1350, 1987.

7 M. Achouche, V. Magnin, J. Harari, D. Carpentier, E. Derouin, C. Jany, and D. Decoster. Design and fabrication of a p-i-n photodiode with high

responsivity and large alignment tolerances for 40-Gb/s applications. *IEEE Photonics Technol. Lett.*, 18(4):556–558, 2006.

8 S. B. Alexander. *Optical Communication Receiver Design*. SPIE Press, copublished with IEE, Bellingham, WA, 1997.

9 Y. Ota, R. G. Swartz, J. S. Schafer, M. M. Banu, A. E. Dunlop, W. C. Fischer, and T. J. Gabara. Low cost, low power digital optical receiver module for 50 Mb/s passive optical network. *Int. J. High Speed Electron. Syst.*, 7(4):471–489, 1996.

10 H. Ransijn. Receiver and transmitter IC design, May 2001. CICC'2001 Ed. Session 3-2.

11 Y. M. Greshishchev. Front-end circuits for optical communications. In *Proceedings of ISSCC'2001 Tutorial*, February 2001.

12 M. M.-K. Liu. *Principles and Applications of Optical Communications*. Irwin, McGraw-Hill, Chicago, 1996.

13 Daghighian, H. Personal communications, 2015.

14 B. L. Kasper, O. Mizuhara, and Y.-K. Chen. High bit-rate receivers, transmitters, and electronics. In I. P. Kaminow and T. Li, editors, *Optical Fiber Telecommunications IVA*, pages 784–851. Academic Press, San Diego, CA, 2002.

15 K. S. Giboney, R. L. Nagarajan, T. E. Reynolds, S. T. Allen, R. P. Mirin, M. J. W. Rodwell, and J. E. Bowers. Travelling-wave photodetectors with 172-GHz bandwidth and 76-GHz bandwidth-efficiency product. *IEEE Photonics Technol. Lett.*, 7(4):412–414, 1995.

16 J. D. Kraus. *Antennas*. McGraw Hill, New York, 2nd edition, 1988.

17 B. M. Oliver. Thermal and quantum noise. *Proc. IEEE*, 53(5):436–454, 1965.

18 H. Melchior, M. B. Fisher, and F. R. Arams. Photodetectors for optical communication systems. *Proc. IEEE*, 58(10):1466–1486, 1970.

19 A. E. Siegman. The antenna properties of optical heterodyne receivers. *Proc. IEEE*, 54(10):1350–1356, 1966.

20 W. V. Sorin. Noise sources in optical measurements. In D. Derickson, editor, *Fiber Optic Test and Measurement*, pages 597–613. Prentice Hall PTR, Upper Saddle River, NJ, 1998.

21 D. Johns and K. Martin. *Analog Integrated Circuit Design*. John Wiley & Sons, Inc., New York, 1997.

22 H. Ito, S. Kodama, Y. Muramoto, T. Furuta, T. Nagatsuma, and T. Ishibashi. High-speed and high-output InP-InGaAs unitraveling-carrier photodiodes. *IEEE J. Sel. Top. Quantum Electron.*, 10(4):709–727, 2004.

23 S. Itakura, K. Sakai, T. Nagatsuka, E. Ishimura, M. Nakaji, H. Otsuka, K. Mori, and Y. Hirano. High-current backside-illuminated photodiode array module for optical analog links. *J. Lightwave Technol.*, LT-28(6):965–971, 2010.

24 J. C. Campbell. Recent advances in telecommunications avalanche photodiodes. *J. Lightwave Technol.*, LT-25(1):109–121, 2007.
25 Maxim Integrated Products. Low-noise APD bias circuit. Maxim Application Note 1831, December 2002.
26 J. Williams. Bias voltage and current sense circuits for avalanche photodiodes. Linear Technology Corp., Application Note 92, November 2002.
27 Y. Kang, H.-D. Liu, M. Morse, M. J. Paniccia, M. Zadka, S. Litski, G. Sarid, A. Pauchard, Y.-H. Kuo, H.-W. Chen, W. S. Zaoui, J. E. Bowers, A. Beling, D. C. McIntosh, X. Zheng, and J. C. Campbell. Monolithic germanium/silicon avalanche photodiodes with 340 GHz gain-bandwidth product. *Nat. Photonics Lett.*, 3:59–63, 2009.
28 K. Shiba, T. Nakata, T. Takeuchi, T. Sasaki, and K. Makita. 10 Gbit/s asymmetric waveguide APD with high sensitivity of −30 dBm. *Electron. Lett.*, 42(20):1177–1178, 2006.
29 G. S. Kinsey, J. C. Campbell, and A. G. Dentai. Waveguide avalanche photodiode operating at 1.55 μm with a gain-bandwidth product of 320 GHz. *IEEE Photonics Technol. Lett.*, 13(8):842–844, 2001.
30 T. Nakata, T. Takeuchi, K. Makita, Y. Amamiya, T. Kato, Y. Suzuki, and T. Torikai. High-sensitivity 40-Gb/s reciever with a wideband InAlAs waveguide avalanche photodiode. In *European Conference on Optical Communication (ECOC)*, Copenhagen, Denmark, 2002.
31 S. D. Personick. Receiver design for optical fiber systems. *Proc. IEEE*, 65(12):1670–1678, 1977.
32 R. J. McIntyre. The distribution of gains in uniformly multiplying avalanche photodiodes: theory. *IEEE Trans. Electron Devices*, ED-19(6):703–713, 1972.
33 R. J. McIntyre. A new look at impact ionization—part I: a theory of gain, noise, breakdown probability, and frequency response. *IEEE Trans. Electron Devices*, ED-46(8):1623–1631, 1999.
34 T. V. Muoi. Receiver design for high-speed optical-fiber systems. *J. Lightwave Technol.*, LT-2(3):243–267, 1984.
35 H. J. R. Dutton. *Understanding Optical Communications*. Prentice Hall PTR, Upper Saddle River, NJ, 1998.
36 R. Ramaswami, K. N. Sivarajan, and G. Sasaki. *Optical Networks: A Practical Perspective*. Morgan Kaufmann Publishers, San Francisco, CA, 3rd edition, 2010.
37 B. Mason, J. M. Geary, J. M. Freund, A. Ougazzaden, C. Lentz, K. Glogovsky, G. Przybylek, L. Peticolas, F. Walters, L. Reynolds, J. Boardman, T. Kercher, M. Rader, D. Monroe, L. Ketelsen, S. Chandrasekhar, and L. L. Buhl. 40Gb/s photonic integrated receiver with −17 dBm sensitivity. In *Optical Fiber Communication Conference (OFC)*, Anaheim, CA, 2002.

38 A. K. Srivastave and Y. Sun. Advances in erbium-doped fiber amplifiers. In I. P. Kaminow and T. Li, editors, *Optical Fiber Telecommunications IVA*, pages 174–212. Academic Press, San Diego, CA, 2002.

39 J. L. Zyskind, J. A. Nagel, and H. D. Kidorf. Erbium-doped fiber amplifiers for optical communications. In I. P. Kaminow and T. L. Koch, editors, *Optical Fiber Telecommunications IIIB*, pages 13–68. Academic Press, San Diego, CA, 1997.

40 D. A. Fishman and B. S. Jackson. Transmitter and receiver design for amplified lightwave systems. In I. P. Kaminow and T. L. Koch, editors, *Optical Fiber Telecommunications IIIB*, pages 69–114. Academic Press, San Diego, CA, 1997.

41 J. L. Rebola and A. V. T. Cartaxo. *Q*-factor estimation and impact of spontaneous–spontaneous beat noise on the performance of optically preamplified systems with arbitrary optical filtering. *J. Lightwave Technol.*, LT-21(1):87–95, 2003.

42 R. C. Steele, G. R. Walker, and N. G. Walker. Sensitivity of optically preamplified receivers with optical filtering. *IEEE Photonics Technol. Lett.*, 3(6):545–547, 1991.

43 D. M. Baney. Characterization of erbium-doped fiber amplifiers. In D. Derickson, editor, *Fiber Optic Test and Measurement*, pages 519–595. Prentice Hall PTR, Upper Saddle River, NJ, 1998.

44 E. L. Goldstein and L. Eskildsen. Erbium-doped fiber amplifiers for multi-wavelength lightwave networks: Impact of the non-flat gain spectrum. In T. P. Lee, editor, *Current Trends in Optical Amplifiers and Their Applications*, pages 37–54. World Scientific, Singapore, 1996.

45 D. Marcuse. Calculation of bit-error probability for a lightwave system with optical amplifiers and post-detection Gaussian noise. *J. Lightwave Technol.*, LT-9(4):505–513, 1991.

46 V. W. S. Chan. Coding and error correction in optical fiber communications systems. In I. P. Kaminow and T. L. Koch, editors, *Optical Fiber Telecommunications IIIA*, pages 42–62. Academic Press, San Diego, CA, 1997.

47 P. A. Humblet and M. Azizoğlu. On the bit-error rate of lightwave systems with optical amplifiers. *J. Lightwave Technol.*, LT-9(11):1576–1582, 1991.

48 J. G. Proakis and M. Salehi. *Digital Communications.* McGraw Hill, New York, 5th edition, 2008.

49 W. Imajuku and A. Takada. Noise figure of phase-sensitive parametric amplifier using a Mach-Zehnder interferometer with lossy Kerr media and noisy pump. *IEEE J. Quantum Electron.*, 39(6):799–812, 2003.

50 H. A. Haus. Optimum noise performance of optical amplifiers. *IEEE J. Quantum Electron.*, 37(6):813–823, 2001.

51 S. Namiki, K. Seo, N. Tsukiji, and S. Shikii. Challenges of Raman amplification. *Proc. IEEE*, 94(5):1024–1035, 2006.
52 K. Rottwitt and A. J. Stentz. Raman amplification in lightwave communication systems. In I. P. Kaminow and T. Li, editors, *Optical Fiber Telecommunications IVA*, pages 213–257. Academic Press, San Diego, CA, 2002.
53 S. Namiki and Y. Emori. Ultrabroad-band Raman amplifiers pumped and gain-equalized by wavelength-division-multiplexed high-power laser diodes. *IEEE J. Sel. Top. Quantum Electron.*, 7(1):3–16, 2001.
54 G. Charlet and S. Bigo. Upgrading WDM submarine systems to 40-Gbit/s channel bitrate. *Proc. IEEE*, 94(5):935–951, 2006.
55 C. Doerr, L. Chen, D. Vermeulen, T. Nielsen, S. Azemati, S. Stulz, G. McBrien, X.-M. Xu, B. Mikkelsen, M. Givehchi, C. Rasmussen, and S.-Y. Park. Single-chip silicon photonics 100-Gb/s coherent transceiver. In *Optical Fiber Communication Conference (OFC)*, San Francisco, CA, 2014.
56 C. R. Doerr, P. J. Winzer, Y.-K. Chen, S. Chandrasekhar, M. S. Rasras, L. Chen, T.-Y. Liow, K.-W. Ang, and G.-Q. Lo. Monolithic polarization and phase diversity coherent receiver in silicon. *J. Lightwave Technol.*, LT-28(4):520–525, 2010.
57 I. A. Young, E. Mohammed, J. T. S. Liao, A. M. Kern, S. Palermo, B. A. Block, M. R. Reshotko, and P. L. D. Chang. Optical I/O technology for tera-scale computing. *IEEE J. Solid-State Circuits*, SC-45(1):235–248, 2010.
58 B. Analui, D. Guckenberger, D. Kucharski, and A. Narasimha. A fully integrated 20-Gb/s optoelectronic transceiver implemented in a standard 0.13-μm CMOS SOI technology. *IEEE J. Solid-State Circuits*, SC-41(12):2945–2955, 2006.
59 P. C. P. Chen, A. M. Pappu, Z. Fu, W. Wattanapanitch, and A. B. Apsel. A 10 Gb/s optical receiver in 0.25 μm silicon-on-sapphire CMOS. In *Proceedings of IEEE International Symposium on Circuits and Systems*, pages 193–196, Seattle, WA, May 2008.
60 A. Emami-Neyestanak, D. Liu, G. Keeler, N. Helman, and M. Horowitz. A 1.6Gb/s, 3mW CMOS receiver for optical communication. In *Symposium on VLSI Circuits Digest of Technical Papers*, pages 84–87, June 2002.
61 K. W. Goossen, J. E. Cunningham, G. Zhang, and J. A. Walker. Very large arrays of flip-chip bonded 1.55 μm photodetectors. *J. Lightwave Technol.*, LT-16(6):1056–1061, 1998.
62 A. Narasimha, B. Analui, Y. Liang, T. J. Sleboda, S. Abdalla, E. Balmater, S. Gloeckner, D. Guckenberger, M. Harrison, R. G. M. P. Koumans, D. Kucharski, A. Mekis, S. Mirsaidi, D. Song, and T. Pinguet. A fully integrated 4 × 10-Gb/s DWDM optoelectronic transceiver implemented in a standard 0.13 μm CMOS SOI technology. *IEEE J. Solid-State Circuits*, SC-42(12):2736–2744, 2007.

63 S. Palermo, A. Emami-Neyestanak, and M. Horowitz. A 90 nm CMOS 16 Gb/s transceiver for optical interconnects. *IEEE J. Solid-State Circuits*, SC-43(5):1235–1246, 2008.

64 M. Dagenais, R. F. Leheny, H. Temkin, and P. Bhattacharya. Applications and challenges of OEIC technology: a report on the 1989 Hilton Head workshop. *J. Lightwave Technol.*, LT-8(6):846–862, 1990.

65 S. J. Koester, C. L. Schow, L. Schares, G. Dehlinger, J. D. Schaub, F. E. Doany, and R. A. John. Ge-on-SOI-detector/Si-CMOS-amplifier receivers for high-performance optical-communication applications. *J. Lightwave Technol.*, LT-25(1):46–57, 2007.

66 K. Takahata, Y. Muramoto, H. Fukano, K. Kato, A. Kozen, S. Kimura, Y. Imai, Y. Miyamoto, O. Nakajima, and Y. Matsuoka. Ultrafast monolithic receiver OEIC composed of multimode waveguide p-i-n photodiode and HEMT distributed amplifier. *IEEE J. Sel. Top. Quantum Electron.*, 6(1):31–37, 2000.

67 S. van Waasen, A. Umbach, U. Auer, H.-G. Bach, R. M. Bertenburg, G. Janssen, G. G. Mekonnen, W. Passenberg, R. Reuter, W. Schlaak, C. Schramm, G. Unterbörsch, P. Wolfram, and F.-J. Tegude. 27-GHz bandwidth high-speed monolithic integrated optoelectronic photoreceiver consisting of a waveguide fed photodiode and an InAlAs/InGaAs-HFET traveling wave amplifier. *IEEE J. Solid-State Circuits*, SC-32(9):1394–1401, 1997.

68 R. H. Walden. A review of recent progress in InP-based optoelectronic integrated circuit receiver front-ends. In K.-C. Wang, editor, *High-Speed Circuits for Lightwave Communications*, pages 319–330. World Scientific, Singapore, 1999.

69 A. Jayakumar, M. S. Bustos, D. Cheskis, S. J. Pietrucha, M. Bonelli, S. Al-Kuran, and N. Scheinberg. 3-V MSM-TIA for gigabit Ethernet. *IEEE J. Solid-State Circuits*, SC-35(9):1271–1275, 2000.

70 C. Takano, K. Tanaka, A. Okubora, and J. Kasahara. Monolithic integration of 5-Gb/s optical receiver block for short-distance communication. *IEEE J. Solid-State Circuits*, SC-27(10):1431–1433, 1992.

71 O. Wada, T. Hamaguchi, S. Miura, M. Makiuchi, K. Nakai, H. Horimatsu, and T. Sakurai. AlGaAs/GaAs p-i-n photodiode/preamplifier monolithic photoreceiver integrated on semi-insulating GaAs substrate. *Appl. Phys. Lett.*, 46(10):981–983, 1985.

72 M. Ghioni, F. Zappa, V. P. Kesan, and J. Warnock. A VLSI-compatible high-speed silicon photodetector for optical data link applications. *IEEE Trans. Electron Devices*, ED-43(7):1054–1060, 1996.

73 G. N. Nasserbakht, J. W. Adkisson, B. A. Wooley, J. S. Harris Jr., and T. I. Kamins. A monolithic GaAs-on-Si receiver front end for optical interconnect systems. *IEEE J. Solid-State Circuits*, SC-28(6):622–630, 1993.

74 J. Sturm, M. Leifhelm, H. Schatzmayr, S. Groiß, and H. Zimmermann. Optical receiver IC for CD/DVD/blue-laser application. *IEEE J. Solid-State Circuits*, SC-40(7):1406–1413, 2005.

75 R. Swoboda, J. Knorr, and H. Zimmermann. A 5-Gb/s OEIC with voltage-up-converter. *IEEE J. Solid-State Circuits*, SC-40(7):1521–1526, 2005.

76 R. Swoboda and H. Zimmermann. 11Gb/s monolithically integrated silicon optical receiver for 850nm wavelength. In *ISSCC Digest of Technical Papers*, pages 240–241, February 2006.

77 D. Kucharski, D. Guckenberger, G. Masini, S. Abdalla, J. Witzens, and S. Sahni. 10Gb/s 15mW optical receiver with integrated germanium photodetector and hybrid inductor peaking in 0.13 μm SOI CMOS technology. In *ISSCC Digest of Technical Papers*, pages 360–361, February 2010.

78 T.-Y. Liow, K.-W. Ang, Q. Fang, J.-F. Song, Y.-Z. Xiong, M.-B. Yu, G.-Q. Lo, and D.-L. Kwong. Silicon modulators and germanium photodetectors on SOI: monolithic integration, compatibility, and performance optimization. *IEEE J. Sel. Top. Quantum Electron.*, 16(1):307–315, 2010.

79 J. Choi, B. J. Sheu, and O. T.-C. Chen. A monolithic GaAs receiver for optical interconnect systems. *IEEE J. Solid-State Circuits*, SC-29(3):328–331, 1994.

80 K. D. Pedrotti, R. L. Pierson Jr., N. H. Sheng, R. B. Nubling, C. W. Farley, and M. F. Chang. High-bandwidth OEIC receivers unsing heterojunction bipolar transistors: design and demonstration. *J. Lightwave Technol.*, LT-11(10):1601–1614, 1992.

81 S. Radovanović, A.-J. Annema, and B. Nauta. A 3-Gb/s optical detector in standard CMOS for 850-nm optical communication. *IEEE J. Solid-State Circuits*, SC-40(8):1706–1717, 2005.

82 D. L. Rogers. Integrated optical receivers using MSM detectors. *J. Lightwave Technol.*, LT-9(12):1635–1638, 1991.

83 J. B. D. Soole and H. Schumacher. Transit-time limited frequency response of InGaAs MSM photodetectors. *IEEE Trans. Electron Devices*, ED-37(11):2285–2291, 1990.

84 B. J. van Zeghbroeck, W. Patrick, J.-M. Halbout, and P. Vettiger. 105-GHz bandwidth metal-semiconductor-metal photodiode. *IEEE Electron Device Lett.*, EDL-9(10):527–529, 1988.

85 C. Rooman, D. Coppée, and M. Kuijk. Asynchronous 250-Mb/s optical receivers with integrated detector in standard CMOS technology for optocoupler applications. *IEEE J. Solid-State Circuits*, SC-35(7):953–958, 2000.

86 W.-Z. Chen, S.-H. Huang, G.-W. Wu, C.-C. Liu, Y.-T. Huang, C.-F. Chiu, W.-H. Chang, and Y.-Z. Juang. A 3.125 Gbps CMOS fully integrated optical receiver with adaptive analog equalizer. In *Proceedings of IEEE Asian Solid-State Circuits Conference (ASSCC)*, pages 396–399, November 2007.

87 Y. Dong and K. W. Martin. A high-speed fully-integrated POF receiver with large-area photo detectors in 65 nm CMOS. *IEEE J. Solid-State Circuits*, SC-47(9):2080–2092, 2012.

88 T. S.-C. Kao, F. A. Musa, and A. C. Carusone. A 5-Gbit/s CMOS optical receiver with integrated spatially modulated light detector and equalization. *IEEE Trans. Circuits Syst. I*, CASI-57(11):2844–2857, 2010.

89 F. Tavernier and M. S. J. Steyaert. High-speed optical receivers with integrated photodiode in 130 nm CMOS. *IEEE J. Solid-State Circuits*, SC-44(10):2856–2867, 2009.

90 A. C. Carusone, H. Yasotharan, and T. Kao. CMOS technology scaling considerations for multi-Gbps optical receivers with integrated photodetectors. *IEEE J. Solid-State Circuits*, SC-46(8):1832–1842, 2011.

91 J. Genoe, D. Coppée, J. H. Stiens, R. A. Vounckx, and M. Kuijk. Calculation of the current response of the spatially modulated light CMOS detector. *IEEE Trans. Electron Devices*, ED-48(9):1892–1902, 2001.

92 C. Hermans and M. S. J. Steyaert. A high-speed 850-nm optical receiver front-end in 0.18-μm CMOS. *IEEE J. Solid-State Circuits*, SC-41(7):1606–1614, 2006.

93 T. K. Woodward and A. V. Krishnamoorthy. 1 Gb/s CMOS photoreceiver with integrated detector operating at 850 nm. *Electron. Lett.*, 34(12):1252–1253, 1998.

94 W.-Z. Chen and S.-H. Huang. A 2.5 Gbps CMOS fully integrated optical receiver with lateral pin detector. In *Proceedings of IEEE Custom Integrated Circuits Conference*, pages 293–296, September 2007.

95 S.-H. Huang and W.-Z. Chen. A 10-Gbps CMOS single chip optical receiver with 2-d meshed spatially-modulated light detector. In *Proceedings of IEEE Custom Integrated Circuits Conference*, pages 129–132, September 2009.

96 S.-H. Huang, W.-Z. Chen, Y.-W. Chang, and Y.-T. Huang. A 10-Gb/s OEIC with meshed spatially-modulated photo detector in 0.18-μm CMOS technology. *IEEE J. Solid-State Circuits*, SC-46(5):1158–1169, 2011.

97 D. Lee, J. Han, G. Han, and S. M. Park. An 8.5-Gb/s fully integrated CMOS optoelectronic receiver using slope-detection adaptive equalizer. *IEEE J. Solid-State Circuits*, SC-45(12):2861–2873, 2010.

98 A. H. Gnauck and P. J. Winzer. Optical phase-shift-keyed transmission. *J. Lightwave Technol.*, LT-23(1):115–130, 2005.

99 P. J. Winzer and R.-J. Essiambre. Advanced optical modulation formats. *Proc. IEEE*, 94(5):952–984, 2006.

100 E. A. Swanson, J. C. Livas, and R. S. Bondurant. High sensitivity optically preamplified direct detection DPSK receiver with active delay-line stabilization. *IEEE Photonics Technol. Lett.*, 6(2):263–265, 1994.

101 H. Fukuyama, T. Itoh, T. Furuta, K. Kurishima, M. Tokumitsu, and K. Murata. Two-channel InP HBT differential automatic-gain-controlled transimpedance amplifier IC for 43-Gbit/s DQPSK photoreceiver. In *Compound Semiconductor Integrated Circuit Symposium (CSICS)*, pages 1–4, Monterey, CA, October 2008.

102 J. H. Sinsky, A. Adamiecki, A. Gnauck, C. A. Burrus Jr., J. Leuthold, O. Wohlgemuth, S. Chandrasekhar, and A. Umbach. RZ-DPSK transmission using a 42.7-Gb/s integrated balanced optical front end with record sensitivity. *J. Lightwave Technol.*, LT-22(1):180–185, 2004.

103 J. S. Weiner, A. Leven, V. Houtsma, Y. Baeyens, Y.-K. Chen, P. Paschke, Y. Yang, J. Frackoviak, W.-J. Sung, A. Tate, R. Reyes, R. F. Kopf, and N. G. Weimann. SiGe differential transimpedance amplifier with 50-GHz bandwidth. *IEEE J. Solid-State Circuits*, SC-38(9):1512–1517, 2003.

104 H. Kim and P. J. Winzer. Nonlinear phase noise in phase-coded transmission. In *Optical Fiber Communication Conference (OFC)*, Anaheim, CA, 2005.

105 C. R. Doerr, D. M. Gill, A. H. Gnauck, L. L. Buhl, P. J. Winzer, M. A. Cappuzzo, A. Wong-Foy, E. Y. Chen, and L. T. Gomez. Monolithic demodulator for 40-Gb/s DQPSK using a star coupler. *J. Lightwave Technol.*, LT-24(1):171–174, 2006.

106 R. A. Griffin and A. C. Carter. *Optical Differential Quadrature Phase-Shift Key (oDQPSK) for High Capacity Optical Transmission*. Optical Fiber Communication Conference (OFC), Anaheim, CA, 2002.

107 A. Leven, N. Kaneda, U.-V. Koc, and Y. K. Chen. Feed-forward phase and frequency estimation in coherent digital and analog photonic links using digital signal processing. In *Microwave Symposium*, Honolulu, HI, June 2007.

108 K. Roberts, S. H. Foo, M. Moyer, M. Hubbard, A. Sinclair, J. Gaudette, and C. Laperle. High capacity transport—100G and beyond. *J. Lightwave Technol.*, LT-33(3):563–578, 2015.

109 J. Zhang, J. Verbist, B. Moeneclaey, J. van Weerdenburg, R. van Uden, H. Chen, J. van Campenhout, C. Okonkwo, X. Yin, J. Bauwelinck, and G. Roelkens. Compact low-power-consumption 28-Gbaud QPSK/16-QAM integrated silicon photonics/electronic coherent receiver. *IEEE Photonics J.*, 8(1):1–10, 2016.

110 A. Awny, R. Nagulapalli, D. Micusik, J. Hoffmann, G. Fischer, D. Kissinger, and A. C. Ulusoy. A dual 64Gbaud 10kΩ 5% THD linear differential transimpedance amplifier with automatic gain control in 0.13 μm BiCMOS technology for optical fiber coherent receivers. In *ISSCC Digest of Technical Papers*, pages 406–407, February 2016.

111 OIF. Implementation Agreement for Integrated Dual Polarization Micro-Intradyne Coherent Receivers – IA # OIF-DPC-MRX-01.0. Optical Internetworking Forum, Fremont, CA, March 2015.

112 Y. Painchaud, M. Poulin, M. Morin, and M. Têtu. Performance of balanced detection in a coherent receiver. *OSA Opt. Express*, 17(5):3659–3672, 2009.

113 B. Zhang, C. Malouin, and T. J. Schmidt. Design of coherent receiver optical front end for unamplified applications. *OSA Opt. Express*, 20(3):3225–3234, 2012.

114 S. J. Savory, G. Gavioli, R. I. Killey, and P. Bayvel. Electronic compensation of chromatic dispersion using a digital coherent receiver. *OSA Opt. Express*, 15(5):2120–2126, 2007.

115 L. W. Couch II. *Digital and Analog Communication Systems*. Prentice Hall, Upper Saddle River, NJ, 7th edition, 2006.

4

Receiver Fundamentals

In this chapter, we examine the optical receiver at the system level and discuss how its performance is affected by the noise and bandwidth of receiver circuits such as the TIA. We start by analyzing how noise in the receiver causes bit errors. This leads to the definition of *receiver sensitivity*. We calculate and compare the sensitivity of different types of receivers (p–i–n, APD, optically preamplified, coherent, and analog receivers). We discuss how to measure the bit-error rate (BER) and the sensitivity. Then, we introduce the concept of noise bandwidth and apply it to calculate the total input-referred noise. Next, we define the *required optical signal-to-noise ratio* (required OSNR), which is important for receivers in amplified lightwave systems. After that, we introduce the concept of power penalty, which is useful to quantify receiver impairments such as intersymbol interference (ISI). Finally, we look at the trade-off between noise and ISI in NRZ and RZ receivers and draw conclusions about the best choice of the receiver's bandwidth and frequency response.

Additional receiver impairments are discussed in Appendix C: *Timing Jitter* and Appendix D: *Nonlinearity*. The mitigation of ISI is discussed in Appendix E: *Adaptive Equalizers* and Appendix F: *Decision Point Control*. The correction of bit errors is addressed in Appendix G: *Forward Error Correction*.

4.1 Receiver Model

The basic receiver model used for this chapter is shown in Fig. 4.1. It consists of (i) a photodetector model, (ii) a linear-channel model that comprises the transimpedance amplifier (TIA), the main amplifier (MA), and optionally a low-pass filter, and (iii) a decision circuit with the threshold voltage V_{DTH}.

This basic model can be extended in a number of ways: an adaptive equalizer can be added to the linear channel to compensate for signal distortions (see Appendix E). The threshold of the decision circuit can be made adaptive to optimize the BER performance (see Appendix F). A multilevel decision circuit

Analysis and Design of Transimpedance Amplifiers for Optical Receivers, First Edition. Eduard Säckinger.

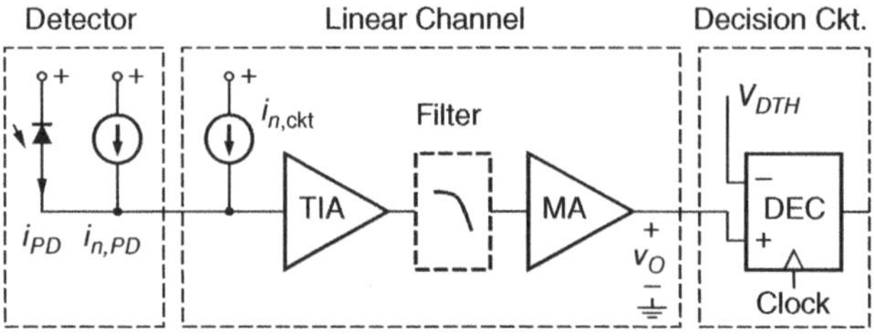

Figure 4.1 Basic receiver model.

can be used to demodulate higher-order modulation formats or to obtain soft-decision information. Forward error correction (FEC) can be used to lower the raw BER from the decision circuit (see Appendix G). In coherent heterodyne (or intradyne) receivers, a demodulator is needed to translates the intermediate-frequency (IF) signal at the output of the linear channel to a baseband signal.

In DSP-based implementations, the analog signal at the output of the linear channel is digitized and processed in the digital domain (equalization, carrier recovery, etc.). The analog decision circuit shown in Fig. 4.1 is then replaced by numerical comparators inside the DSP.

Photodetector. The detector model consist of a signal current source i_{PD} and a noise current source $i_{n,PD}$. The characteristics of these two current sources were discussed in Chapter 3 for the p–i–n photodetector, the avalanche photodetector (APD), the optically preamplified p–i–n detector, and the coherent detector. Note that in our model $i_{n,PD}$ also includes noise contributions from the transmitter and optical amplifiers, if present.

Linear Channel. The linear channel can be modeled with the complex frequency response $H(f)$ that relates the amplitude and phase of the output voltage v_O to those of the input current i_{PD}. This frequency response can be decomposed into a product of three frequency responses: one for the TIA, one for the filter, and one for the MA.

The noise characteristics of the linear channel are modeled by a single equivalent noise current generator $i_{n,ckt}$ at the input of the channel.[1] This noise is called the *circuit noise* or *amplifier noise* to distinguish it from the detector noise. (Sometimes the term *thermal noise* is used, but because bipolar circuits also produce shot noise, this is not quite accurate.) In practice, the circuit noise is determined almost exclusively by the TIA, which is the first element of the linear channel.

1 A consequence of modeling the linear-channel noise with a single noise current generator, as opposed to a noise current *and* a noise voltage generator, is that $i_{n,ckt}$ becomes dependent on the photodetector impedance, in particular its capacitance (cf. Chapter 6).

Whereas the detector noise, $i_{n,PD}$, is approximately *white* (frequency-independent spectrum), the circuit noise, $i_{n,\mathrm{ckt}}$, usually is *not white.* In Section 6.3, we calculate the PSD of the input-referred TIA noise (Eqs. (6.45) and (6.48)) and we find that its two main components are a constant part (white noise) and a part that increases with frequency like f^2. This is the case regardless whether the receiver front-end is implemented with an FET or BJT. The PSD of the circuit noise current therefore can be written in the general form

$$I^2_{n,\mathrm{ckt}}(f) = \alpha_0 + \alpha_2 f^2 + \ldots, \tag{4.1}$$

where α_0 and α_2 are parameters.

Whereas the detector noise in direct-detection receivers is *nonstationary* (rms value varies with time) and its amplitude distribution is *non-Gaussian,* especially for small optical powers, the circuit noise is *stationary* and its amplitude distribution is close to *Gaussian.*

How appropriate is a *linear* model for the TIA and MA, in particular when the MA is implemented as a limiting amplifier, which becomes strongly nonlinear for large input signals? Fortunately, for noise and sensitivity calculations, the signal levels are small enough such that nonlinear effects can be ignored.

Decision Circuit. The last block in our receiver model, the decision circuit, compares the output voltage from the linear channel, v_O, with a decision threshold voltage, V_{DTH}. If the output voltage is larger than the threshold, a one bit is detected; if it is smaller, a zero bit is detected. The result of the comparison is sampled, usually by the clock from a clock-recovery circuit, and held during the bit period. The accurate placement of the sampling instant at the center of the bit period (more precisely, at the center of the eye opening) is important. In contrast to the linear channel, the decision circuit is *nonlinear.*

The decision circuit can be implemented as a voltage comparator followed by a flip-flop or as a flip-flop with differential inputs. In DSP-based implementations, the digitized and equalized baseband signal is compared against a numerical threshold to decide the bit value.

4.2 Noise and Bit-Error Rate

The voltage v_O at the output of the linear channel is a superposition of the desired *signal* voltage v_S and the undesired *noise* voltage v_n: $v_O = v_S + v_n$. The noise voltage v_n originates from the receiver circuits, the detector, optical amplifiers, and so forth. Occasionally, the instantaneous noise voltage $v_n(t)$ becomes so large compared with the received signal $v_S(t)$ that a decision error or *bit error* occurs. In this section, we first calculate the rms value of the output noise voltage, v_n^{rms}, and then derive the bit-error rate, *BER*, resulting from this noise.

Output Noise. The output noise voltage PSD can be written as a sum of two components, one caused by the circuits and one caused by the detector. Let us start with the circuit noise, which is stationary and therefore easier to deal with. Given the input-referred circuit noise current PSD $I^2_{n,\text{ckt}}(f)$ and the frequency response of the linear channel $H(f)$, we can easily calculate the output noise voltage PSD:

$$V^2_{n,\text{ckt}}(f) = |H(f)|^2 \cdot I^2_{n,\text{ckt}}(f). \tag{4.2}$$

To keep the equations "lean," we omit indices distinguishing input and output quantities. We know from our model that a current indicates an input signal to the linear channel and a voltage indicates an output signal. Integrating the (one-sided) noise voltage PSD in Eq. (4.2) over the bandwidth of the decision circuit, BW_D, gives us the total mean-square noise voltage due to the receiver circuits experienced by the decision circuit:

$$\overline{v^2_{n,\text{ckt}}} = \int_0^{BW_D} |H(f)|^2 \cdot I^2_{n,\text{ckt}}(f)\, df. \tag{4.3}$$

Next, we have to deal with the detector noise, which may be nonstationary. We can visualize a time-dependent input noise current PSD, $I^2_{n,PD}(f,t)$, as a two-dimensional warped surface floating above the time and frequency coordinates. It can be shown [1] that this time-dependent spectrum results in the following PSD at the output of the linear channel:

$$V^2_{n,PD}(f,t) = H(f) \cdot \int_{-\infty}^{\infty} I^2_{n,PD}(f, t - t') \cdot h(t') \cdot e^{j\,2\pi f\,t'}\, dt', \tag{4.4}$$

where $h(t)$ is the impulse response of the linear channel. This means that the spectrum does not only get "shaped" along the frequency axis, but it also gets "smeared out" along the time axis! Figure 4.2 illustrates this operation for a white input noise spectrum that has a "101" time dependence. After passing through the linear channel, the noise spectrum rolls off at high frequencies and the noise, which at the input was fully contained within the two one bits, now spills over into the zero bit.

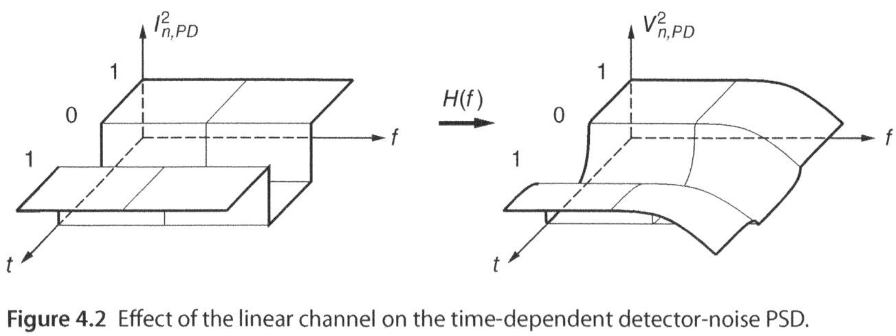

Figure 4.2 Effect of the linear channel on the time-dependent detector-noise PSD.

This spillover effect potentially complicates the noise calculations because the output noise during the nth bit period (or, more generally, the nth symbol period) depends not only on the input noise during this same period, but also on the input noise during the preceding and (possibly) the succeeding bit periods. Fortunately, the spilled noise usually decays rapidly within each bit period such that it is negligible at the sampling instant. It can be shown (cf. solution to Problem 4.1 on p. 502) that if the duration of the impulse response $h(t)$ is short compared to the bit period, Eq. (4.4) simplifies to $V^2_{n,PD}(f,t) = |H(f)|^2 \cdot I^2_{n,PD}(f,t)$. This equation has the same form as Eq. (4.2), except for the added time dependence. Integrating $V^2_{n,PD}(f,t)$ up to BW_D results in the following total mean-square output noise voltage due to the photodetector

$$\overline{v^2_{n,PD}}(t) = \int_0^{BW_D} |H(f)|^2 \cdot I^2_{n,PD}(f,t)\, df. \tag{4.5}$$

This approximate detector-noise equation provides excellent accuracy in most applications [2]. (To further analyze the noise spillover effect [2–4] it is common to consider only the worst case: If all preceding and succeeding bits are one, i.e., if they are in their noisy state, the noise at the sampling instant can be calculated uniquely; cf. Eqs. (I.9) and (I.10).) [→ Problems 4.1 and 4.2.]

The rms noise voltage at the output of the linear channel due to both noise sources is obtained by adding the (uncorrelated) mean-square noise voltages of Eqs. (4.3) and (4.5) under the square root:

$$\begin{aligned} v_n^{rms}(t) &= \sqrt{\overline{v^2_{n,PD}}(t) + \overline{v^2_{n,\text{ckt}}}} \\ &= \sqrt{\int_0^{BW_D} |H(f)|^2 [I^2_{n,PD}(f,t) + I^2_{n,\text{ckt}}(f)]\, df}. \end{aligned} \tag{4.6}$$

The latter equation is illustrated by Fig. 4.3. The input-referred noise current PSD, $I^2_n(f,t) = I^2_{n,PD}(f,t) + I^2_{n,\text{ckt}}(f)$, which increases with frequency as a result of the f^2 component of the circuit noise, is shaped by the frequency response $|H(f)|^2$, producing the output noise voltage PSD $V^2_n(f,t)$, which rolls off rapidly at high frequencies. The shaded area under the curve corresponds to the integral, the square root of which represents the total output-referred rms noise voltage $v_n^{rms}(t)$. Because the output spectrum rolls off rapidly, the precise value of the upper integration bound (BW_D) is uncritical and can be set to infinity.

For binary modulation formats (e.g., NRZ or RZ), assuming ideal waveforms and neglecting noise spillover, the time-dependent output rms noise voltage $v_n^{rms}(t)$ at the sampling instant can be described by just two values: $v_{n,0}^{rms}$ when a zero is received and $v_{n,1}^{rms}$ when a one is received.

Signal, Noise, and Bit-Error Rate. Now that we have derived the output rms noise voltage, how is it related to the bit-error rate? Figure 4.4 illustrates the situation

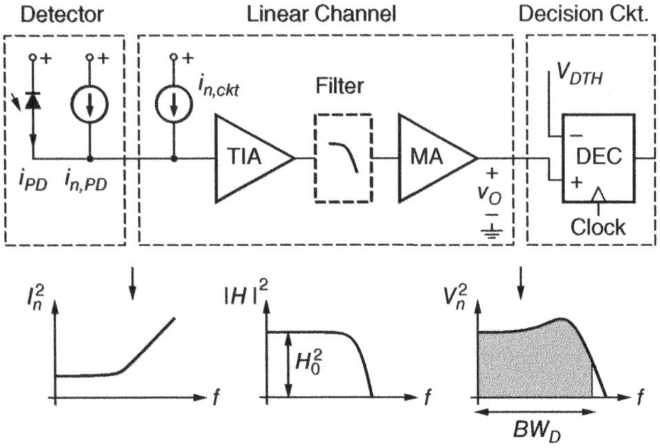

Figure 4.3 Calculation of the total output-referred noise.

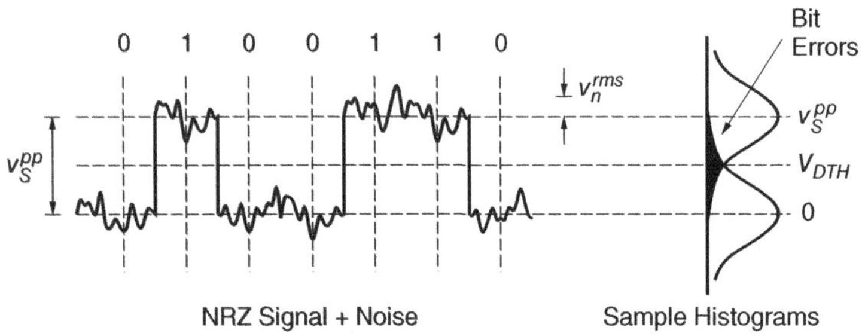

Figure 4.4 Relationship between signal, noise, and bit-error rate.

at the input of the decision circuit, where we have the non-return-to-zero (NRZ) signal $v_S(t)$ characterized by the peak-to-peak voltage v_S^{pp} and the noise $v_n(t)$ characterized by the rms voltage v_n^{rms}. For now, we assume that the NRZ signal is free of distortions (intersymbol interference) and that the noise is Gaussian[2] and signal independent; later, we generalize. The noisy signal is sampled at the center of each bit period (vertical dashed lines), producing the statistical distributions shown on the right-hand side. Both distributions are Gaussian and have a standard deviation that is equal to the rms value of the noise voltage, v_n^{rms}, which we calculated in Eq. (4.6).

The decision circuit determines whether a bit is a zero or a one by comparing the sampled output voltage v_O with the decision threshold voltage V_{DTH}, which is located at the midpoint between the zero and one levels. Setting the threshold voltage to the crossover point of the two distributions results in the fewest bit

2 Non-Gaussian detector noise is often (but not always) dominated by other approximately Gaussian noise sources. The limitations of the Gaussian model are discussed in Section 4.4.

errors (assuming zeros and ones are equally probable). Now we can define the *bit-error rate* (BER) as the probability that a zero is misinterpreted as a one or that a one is misinterpreted as a zero.[3]

Given the above model, we can derive a mathematical expression for the BER. The error probabilities are given by the shaded areas under the Gaussian tails. The area of the tails has to be weighted by $\frac{1}{2}$ before summing because zeros and ones are assumed to occur with probability $\frac{1}{2}$. The two tails are symmetric and thus have the same area. So, we can calculate just the tail above the decision threshold and take it with a weight of one:

$$BER = \int_{V_{DTH}}^{\infty} \frac{1}{v_n^{rms}} \cdot \text{Gauss}\left(\frac{v_O}{v_n^{rms}}\right) dv_O, \tag{4.7}$$

where Gauss(x) is the normalized Gaussian distribution (average = 0, standard deviation = 1, area = 1). The expression under the integral is a Gaussian with a standard deviation equal to v_n^{rms} centered at zero. Setting the decision threshold voltage to the midpoint, $V_{DTH} = v_S^{pp}/2$, and introducing the normalized variable $x = v_O/v_n^{rms}$, we can rewrite the above equation in the more elegant form:

$$BER = \int_{Q}^{\infty} \text{Gauss}(x)\, dx \qquad \text{with} \qquad Q = \frac{v_S^{pp}}{2v_n^{rms}}. \tag{4.8}$$

The lower bound of this integral, is known as the Q parameter or the *Personick Q* in honor of Stewart Personick who introduced this notation in 1973 [3].[4] The Personick Q is a measure of the ratio between signal and noise, but there are some subtle differences between Q and the common meaning of signal-to-noise ratio (SNR), which we discuss in Section 4.3. The Gaussian integral in the above equation can be expanded and approximated as follows:

$$\begin{aligned}\int_{Q}^{\infty} \text{Gauss}(x)\, dx &= \frac{1}{\sqrt{2\pi}} \int_{Q}^{\infty} e^{-\frac{x^2}{2}}\, dx \\ &= \frac{1}{2}\,\text{erfc}\left(\frac{Q}{\sqrt{2}}\right) \approx \frac{1}{\sqrt{2\pi}} \cdot \frac{\exp(-Q^2/2)}{Q}.\end{aligned} \tag{4.9}$$

The approximation on the far right is correct within 10% for $Q > 3$. The precise numerical values for the integral are listed in Table 4.1 for some round BER numbers.

For example, if the signal swing at the decision circuit is 500 mV peak-to-peak and the noise is 40 mV rms, we find $Q = 6.25$ with the right part of Eq. (4.8).

3 Unfortunately, the term *bit-error rate* is misleading because it could be interpreted as bit errors per time interval. A more accurate term would be *bit-error probability* or *bit-error ratio*, however, because the term *bit-error rate* is near universally accepted, we stick with it.
4 Note that the Personick Q is different from the Q function, $Q(x)$, used in some texts [5, 6]. In fact, the Personick Q corresponds to the argument, x, of the Q function (cf. Appendix A).

Table 4.1 Numerical relationship between Q and bit-error rate.

Q	*BER*	Q	*BER*
0.0	$\frac{1}{2}$	5.612	10^{-8}
1.282	10^{-1}	5.998	10^{-9}
2.326	10^{-2}	6.361	10^{-10}
3.090	10^{-3}	6.706	10^{-11}
3.719	10^{-4}	7.034	10^{-12}
4.265	10^{-5}	7.349	10^{-13}
4.753	10^{-6}	7.651	10^{-14}
5.199	10^{-7}	7.941	10^{-15}

Consulting Table 4.1, we can predict a BER of just over 10^{-10}. With the same swing, but 25% less noise (30 mV), we obtain $\mathcal{Q} = 8.33$, corresponding to a BER of better than 10^{-15}! This example illustrates that a small change in noise can cause the BER to change over several orders of magnitude.

An alternative and maybe more intuitive derivation of Eq. (4.8) goes as follows. With the decision threshold set to the midpoint, the distance from the center of either Gaussian to the beginning of the tail that we want to integrate is $v_S^{pp}/2$ (see Fig. 4.4). Normalizing this distance to the standard deviation of the Gaussian permits us to integrate a normalized Gaussian, that is, one with a standard deviation of one. Now referring to Eq. (4.8), the BER expression on the left is just the integral of the normalized Gaussian and the Personick Q expression on the right is just the normalized distance to the tail.

A Generalization: Unequal Noise Distributions. Now, we drop the assumption that the noise is signal independent. We know that in many receivers the noise on the ones is larger than the noise on the zeros. Given the simplified noise model introduced earlier, the rms noise alternates between the values $v_{n,0}^{rms}$ and $v_{n,1}^{rms}$, depending on whether the received bit is a zero or a one. In terms of the noise statistics, we now have two different Gaussians, a narrow one with standard deviation $v_{n,0}^{rms}$ for the zeros and a wide one with standard deviation $v_{n,1}^{rms}$ for the ones. This situation is illustrated in Fig. 4.5. The distribution for the zeros is taller than the one for the ones because both probability distributions must have a total area of one. The crossover point is now located below the midpoint and hence the optimum decision threshold voltage is shifted downward as well. The error tails are no longer symmetric.

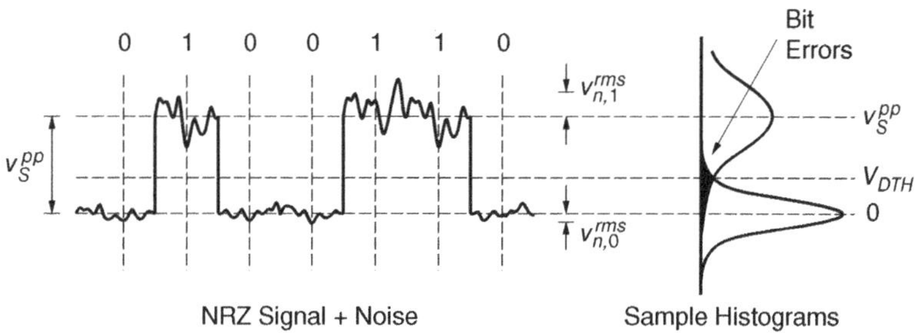

Figure 4.5 Relationship between signal, noise, and bit-error rate for unequal noise distributions.

By determining the optimum decision threshold, integrating the two error tails, and summing them with a weight of $\frac{1}{2}$ each, it can be shown (cf. solution to Problem 4.3 on p. 503) that the BER is [7]

$$BER = \int_Q^{\infty} \text{Gauss}(x)\, dx \qquad \text{with} \qquad Q = \frac{v_S^{pp}}{v_{n,0}^{rms} + v_{n,1}^{rms}}. \tag{4.10}$$

For example, with a signal swing of 500 mV peak-to-peak, a noise of 10 mV rms for the zeros and a noise of 70 mV rms for the ones, we find $Q = 6.25$, corresponding to a BER of just over 10^{-10}. As expected, Eq. (4.10) simplifies to Eq. (4.8) when the noise distributions are equal, $v_n^{rms} = v_{n,0}^{rms} = v_{n,1}^{rms}$. [→ Problem 4.3.]

Another Generalization: Noise Corrupted Threshold. So far, we assumed that the decision threshold is located at its optimum value, that is, where the two probability distributions cross over. Although this is a reasonable assumption for continuous-mode receivers, it is not accurate for burst-mode receivers [8–11]. The problem arises because the burst-mode receiver must acquire a new threshold for each individual burst based on a short and noisy preamble (cf. Chapter 1). Figure 4.6 illustrates the situation for two consecutive bursts, each with a two-bit preamble consisting of a one followed by a zero. The noise on the zeros and ones is assumed to be equal. The threshold, V_{DTH}, is established by sampling the preamble bits and taking the average (shown with the up and down arrows). Because these bits are noisy, the threshold fluctuates from burst to burst. In our example, the threshold for the first burst is too low and the threshold for the second burst is just about right due to a lucky coincidence of the noise pattern on the preamble bits. The fluctuating threshold, in turn, results in a fluctuating BER. The system BER is found by averaging the BER over many bursts.

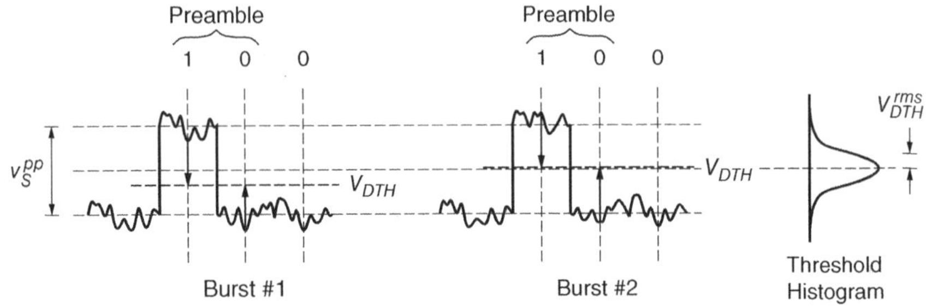

Figure 4.6 Decision-threshold acquisition from a noisy two-bit preamble.

To calculate the average BER, we first determine the probability distribution of the decision threshold (shown on the far right of Fig. 4.6), then we calculate the BER for each possible threshold and, finally, we determine the weighted average of these BERs using the threshold distribution as the weighting function [8]. It can be shown [9] that this calculation is equivalent to the following, simpler procedure: First, we convolve the noise distribution with the threshold distribution resulting in a wider noise distribution and then we use this new noise distribution together with a noise-free threshold to calculate the BER. Given the distributions for the noise and threshold, the variance of their convolution is simply the sum of the individual variances.

How large is the standard deviation of the threshold distribution, V_{DTH}^{rms}? If we take only one sample of the signal, it is equal to v_n^{rms}. But as we average over more preamble bits, the threshold becomes more sharply defined and its standard deviation shrinks to $v_n^{rms}/\sqrt{n}$, where n is the number of preamble bits [8]. Summing the variances of the noise and the threshold and inserting the result into Eq. (4.8) yields the average BER for a burst-mode receiver:

$$BER = \int_Q^{\infty} \text{Gauss}(x)\,dx \qquad \text{with} \qquad Q = \frac{1}{\sqrt{1+1/n}} \cdot \frac{v_S^{pp}}{2v_n^{rms}}. \tag{4.11}$$

For example, with a signal swing of 500 mV peak-to-peak, a noise of 40 mV rms, and a two-bit preamble, we find $Q = 5.10$, corresponding to a BER of just over 10^{-7}. Compared with a continuous-mode receiver having the same signal and noise, the BER is a thousand times larger, which suggests the use of a preamble with more than two bits.

A continuous-mode receiver extracts the decision threshold from a large number of received bits, which corresponds to a burst-mode receiver with a very long preamble, that is, $n \to \infty$. In this case, Eq. (4.11) simplifies to Eq. (4.8), as expected. Equation (4.11) can be generalized for burst-mode receivers with APDs, that is, for signals with unequal noise on the zeros and ones [9, 10].

4.3 Signal-to-Noise Ratio

Any measure of signal strength divided by any measure of noise may be called *signal-to-noise ratio* (SNR). In this sense, the Q parameter is an SNR. However to avoid confusion, we define SNR here as the *mean-free average signal power* divided by the *average noise power*. (Caution: Not all authors define SNR in the same way.)

The SNR can be calculated in the continuous-time domain, before the signal is sampled by the decision circuit, or in the sampled domain (cf. Fig. 4.7). The Q parameter, in contrast, is always calculated in the sampled domain. In general, the continuous-time and sampled SNR are not the same (although they happen to coincide for an ideal NRZ signal).

Now, let us calculate the SNR for an ideal NRZ signal with unequal noise on the zeros and ones. The mean-free average signal power is calculated as $\overline{v_S^2(t)} - \overline{v_S(t)}^2$, which is $(v_S^{pp}/2)^2$ for a DC-balanced (same number of zeros and ones), ideal NRZ signal with swing v_S^{pp} (cf. Appendix A.1). The noise power is calculated as $\overline{v_n^2(t)}$, which can be written as $\frac{1}{2}(\overline{v_{n,0}^2} + \overline{v_{n,1}^2})$, given equal probabilities for zeros and ones. Thus, the SNR follows as

$$SNR = \frac{(v_S^{pp})^2}{2\left(\overline{v_{n,0}^2} + \overline{v_{n,1}^2}\right)}. \tag{4.12}$$

SNR and Q. Comparing Eqs. (4.10) and (4.12), we realize that we cannot generally convert Q into SNR, or vice versa, unless we have additional knowledge of the noise ratio $v_{n,0}^{rms}/v_{n,1}^{rms}$ (cf. Eq. (I.11)). There are two important special cases: (i) if the noise on the zeros and ones are equal (additive noise) and (ii) if the noise on the ones is much larger than that on the zeros (multiplicative noise):

$$SNR = Q^2, \quad \text{if} \quad v_{n,1}^{rms} = v_{n,0}^{rms} \tag{4.13}$$

$$SNR = \tfrac{1}{2}Q^2, \quad \text{if} \quad v_{n,1}^{rms} \gg v_{n,0}^{rms}. \tag{4.14}$$

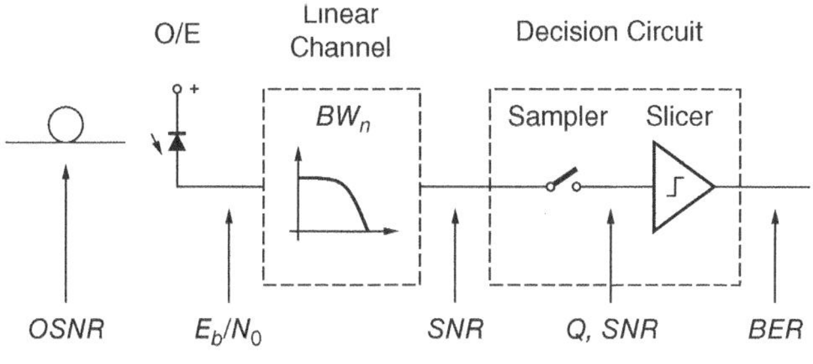

Figure 4.7 Various performance measures in an optical receiver.

For example, to achieve a BER of 10^{-12} ($Q = 7.03$), we need an SNR of 16.9 dB in the first case and 13.9 dB in the second case. [→ Problems 4.4 and 4.5.]

When expressing Q in decibels, should we use 10 log Q or 20 log Q? The discussion of the electrical SNR suggests 20 log Q (= 10 log Q^2). But in the next section, when we discuss optical sensitivity, we find that an equally strong argument can be made for 10 log Q (e.g., see Eq. (4.21)). To avoid confusion, it is best to express Q on a linear scale or else to clarify the used dB-conversion rule. Sometimes decibels resulting from the 20 log Q rule are called *electrical dBs* and those resulting from the 10 log Q rule are called *optical dBs*. [→ Problem 4.6.]

SNR for TV Signals. Although our focus is on digital transmission systems, a comparison with analog transmission systems is instructive and provides some perspective. A good example is the CATV/HFC system, where multiple TV signals are combined into a single analog broadband signal, which is then transmitted over an optical fiber (cf. Chapter 1 and Appendix A). To provide a good picture quality, this analog signal must have an SNR in the range of 27 to 51 dB, depending on the modulation format (cf. Appendix A.4 and A.5). This SNR is *much* higher than the 14 to 17 dB typically needed for an NRZ signal!

To be more precise, we should have used the term *carrier-to-noise ratio* (CNR) instead of SNR in the previous paragraph. Cable-television engineers use the term CNR for RF-modulated signals such as the TV signals in an SCM system and reserve the term SNR for baseband signals such as the NRZ signal [12].

SNR per Bit: E_b/N_0. There is yet another SNR-like quantity called E_b/N_0, often pronounced "ebno." Sometimes this quantity is referred to as *SNR per bit* [13]. E_b/N_0 is mostly used in wireless applications, but occasionally, it appears in the optical communication literature, especially when error-correcting codes are discussed. (We use it in Appendix G.) What is it about? E_b is the average energy per information bit and N_0 is the (one-sided) noise PSD. The E_b/N_0 concept only applies to situations where the noise is white, so that the noise PSD can be characterized by the single number N_0. This situation is most closely approximated at the *input* of the receiver, as shown in Fig. 4.7, before any filtering is performed, and the noise can be assumed to be approximately white.

The SNR of a (finite) signal with white noise is always zero. Meaningful SNR and Q values can be calculated only after the white noise is bandlimited with a low-pass filter. In contrast, the E_b/N_0 of a (finite) signal with white noise does have a meaningful nonzero value.

How is E_b related to the voltage swing v_S^{pp} at the output of the linear channel? The energy per bit is the average signal power multiplied by the bit period. Let us assume that the linear channel has a midband gain of one and does not attenuate the signal power. Then, we find the average energy per information bit at the input of the channel as $E_b = (\overline{v_S^2(t)} - \overline{v_S(t)}^2) \cdot T'$, where T' is the duration of the information bit. For a DC-balanced, ideal NRZ signal with swing v_S^{pp}, this is $E_b = (v_S^{pp}/2)^2 \cdot T'$ (cf. Appendix A.1).

Why this emphasis on *information* bit? Because the transmission system may use a coding scheme such as 8B/10B, where groups of 8 information bits are coded into 10 channel bits before they are transmitted over the fiber. In this case, the period of an information bit is somewhat longer than the period of a channel bit. Mathematically, we can write $T' = T/r = 1/(rB)$, where B is the channel bit rate and r is the so-called *code rate*. For example, the 8B/10B code has the code rate $r = 0.8$. Using the code rate, we can rewrite $E_b = (v_S^{pp}/2)^2/(rB)$.

Next, how is N_0 related to the rms noise v_n^{rms} at the output of the linear channel? Assuming additive white noise with the PSD N_0 at the input, we find the mean-square noise voltage at the output as $(v_n^{rms})^2 = N_0 \cdot BW_n$, where BW_n is the noise bandwidth of the linear channel. Thus, we have $N_0 = (v_n^{rms})^2/BW_n$.

Dividing E_b by N_0 reveals the following relationship with Q:

$$\frac{E_b}{N_0} = \left(\frac{v_S^{pp}}{2v_n^{rms}}\right)^2 \cdot \frac{BW_n}{rB} = Q^2 \cdot \frac{BW_n}{rB}. \tag{4.15}$$

In words, E_b/N_0 is equal to Q^2 scaled by the ratio of the noise bandwidth and the information bit rate. The latter ratio is related to the spectral efficiency of the used modulation scheme. Thus, the main difference between E_b/N_0 and Q (or SNR) is that E_b/N_0 takes the spectral efficiency of the modulation scheme into account. In texts on communication systems and forward error correction, it is often assumed that a matched-filter receiver is used. For NRZ modulation, this implies that the noise bandwidth is half of the bit rate, $BW_n = B/2$ (cf. Section 4.9), leading to the simpler relationship

$$\frac{E_b}{N_0} = \frac{Q^2}{2r}. \tag{4.16}$$

For example, to achieve a BER of 10^{-12} ($Q = 7.03$) with 8B/10B coding ($r = 0.8$), we need $E_b/N_0 = 14.9$ dB. [→ Problem 4.7.]

For a DC-balanced NRZ signal with no coding, signal-independent white noise, and a matched-filter receiver, we find $10 \log E_b/N_0 = 20 \log Q - 3$ dB. But these are a lot of assumptions and, as in the case of SNR, we have to be careful when converting E_b/N_0 to Q!

Comparison. Let us put all the SNR-like quantities that we discussed in perspective (see Fig. 4.7). The primary performance measure for a digital communication system is the BER, which is measured at the output of the decision circuit. To design and test the system, we would like to relate the BER to quantities that can be measured in the analog domain before the decision circuit. The first and most direct predictor of the BER is the Q parameter, which is determined from the sampled values at the input of the decision circuit. If we assume that the noise is Gaussian, that the two signal levels occur with equal probability, and that the decision threshold is set to its optimum value, we can easily calculate the BER from Q. In particular, we do *not* need to make any assumptions about the spectral distribution and additiveness of the noise, neither does the shape and duration of the data pulses matter (e.g., NRZ vs RZ modulation). This makes Q an excellent performance measure.

The sampled SNR measured at the input of the decision circuit is another, less direct predictor of the BER. To calculate the BER from the sampled SNR, we need additional information about the ratio of the noise on the zeros and ones. To calculate the BER from the continuous-time SNR, we need additional information about the shape of the data pulses. To calculate the BER from the E_b/N_0 at the input of the receiver, we need additional information about the receiver's noise bandwidth and code rate. We discuss the relationship between the optical signal-to-noise ratio OSNR and BER in Section 4.6.

4.4 Sensitivity

Rather than asking "What is the bit-error rate given a certain signal strength?," we could ask the other way round, "What is the minimum signal strength needed to achieve a given bit-error rate?" This minimum signal strength, when referred back to the input of the receiver, is known as the *sensitivity*. The sensitivity is one of the key characteristics of optical receivers. It tells us to what level the transmitted signal can become attenuated by the fiber and still be detected reliably by the receiver.

In the following, we define sensitivity in the electrical and the optical domain. Then, we calculate and compare the optical sensitivity of different types of receivers. Finally, we discuss the measurement of BER and sensitivity.

Electrical Sensitivity. The *electrical receiver sensitivity*, i_{sens}^{pp}, is defined as the minimum peak-to-peak signal current, i_S^{pp}, at the input of the receiver necessary to achieve a specified BER ($i_{\text{sens}}^{pp} = i_S^{pp}$ *@ BER*).

Let us relate this sensitivity to the noise of the receiver. The current swing i_S^{pp} at the input of the linear channel causes the output voltage swing $v_S^{pp} = H_0 i_S^{pp}$, where H_0 is the low-frequency value of $H(f)$ or, if the linear channel has a low-frequency cutoff, its midband value (see Fig. 4.3). We know from Eq. (4.8) that this swing and the output noise v_n^{rms} determine the BER: $Q = v_S^{pp}/(2v_n^{rms})$. Thus, solving this equation for v_S^{pp} and dividing the result by H_0, we find the electrical sensitivity

$$i_{\text{sens}}^{pp} = \frac{2Q\, v_n^{rms}}{H_0}. \tag{4.17}$$

Let us define the *input-referred* rms noise current as the rms noise voltage at the output of the linear channel divided by H_0:

$$i_n^{rms} = \frac{v_n^{rms}}{H_0}. \tag{4.18}$$

Then, we can rewrite the electrical sensitivity from Eq. (4.17) in the more compact form

$$i_{\text{sens}}^{pp} = 2\mathcal{Q}\, i_n^{rms}. \tag{4.19}$$

For example, given an input-referred rms noise current of 1 μA and a required BER of 10^{-12}, the electrical sensitivity is $14.07 \cdot 1\ \mu\text{A} = 14.07\ \mu\text{A}$.

In situations with unequal amounts of noise on the zeros and ones, we can use Eq. (4.10) to obtain the more general electrical sensitivity expression

$$i_{\text{sens}}^{pp} = \mathcal{Q}(i_{n,0}^{rms} + i_{n,1}^{rms}), \tag{4.20}$$

where $i_{n,0}^{rms} = v_{n,0}^{rms}/H_0$ and $i_{n,1}^{rms} = v_{n,1}^{rms}/H_0$ are the input-referred rms noise for zeros and ones, respectively. Note that this equation simplifies to Eq. (4.19) for the case of equal noise, $i_n^{rms} = i_{n,0}^{rms} = i_{n,1}^{rms}$.

Optical Sensitivity. The *optical receiver sensitivity*, $\overline{P}_{\text{sens}}$, is defined as the minimum optical power, averaged over time, $\overline{P}_S$, necessary to achieve a specified BER ($\overline{P}_{\text{sens}} = \overline{P}_S$ @ *BER*).

How does the optical sensitivity relate to the noise of the receiver? For an ideal NRZ signal, we have $\overline{i}_S = i_S^{pp}/2$, and with Eq. (3.3), we find $\overline{P}_S = i_S^{pp}/(2\mathcal{R})$, where $\mathcal{R}$ is the responsivity of the photodetector. With Eq. (4.19), we can express the optical sensitivity as

$$\overline{P}_{\text{sens}} = \frac{\mathcal{Q}\, i_n^{rms}}{\mathcal{R}}, \tag{4.21}$$

or, with Eq. (4.20), more generally as

$$\overline{P}_{\text{sens}} = \frac{\mathcal{Q}(i_{n,0}^{rms} + i_{n,1}^{rms})}{2\mathcal{R}}. \tag{4.22}$$

For example, given an input-referred rms noise current of 1 μA, a responsivity of 0.8 A/W, and a required BER of 10^{-12}, the optical sensitivity is $7.03 \cdot 1\ \mu\text{A}/(0.8\ \text{A/W}) = 8.79\ \mu\text{W}$, corresponding to −20.6 dBm.

Note that the optical sensitivity is based on the *average* signal value, whereas the electrical sensitivity is based on the *peak-to-peak* signal value. Thus, the optical sensitivity depends on the pulse width of the signal. For an ideal NRZ signal the average value is $\overline{i}_S = i_S^{pp}/2$, but for an ideal return-to-zero (RZ) signal with 50% duty cycle the average value is only $\overline{i}_S = i_S^{pp}/4$. This means that given an RZ and an NRZ receiver with identical electrical sensitivities, the optical sensitivity of the RZ receiver is 3 dB better than that of the NRZ receiver. (When comparing receivers with the same finite bandwidth, the RZ advantage is less than 3 dB because the RZ signal becomes more distorted than the NRZ signal. In Section 4.8, we discuss the sensitivity of receivers with a finite bandwidth.)

Optical Sensitivity with Ideal Detector. The optical sensitivity of a receiver with an *ideal photodetector* is written as $\eta\overline{P}_{\text{sens}}$. This notation, which refers to the quantum efficiency η, signifies that only the fraction η of the received power $\overline{P}$ is detected by the "internal" ideal detector. Hence $\eta\overline{P}$ is also known as the *detected optical power* [14]. This measure is useful to compare the electrical performance of different receivers while excluding the quantum efficiency of the photodetector from the comparison. With Eqs. (4.21) and (3.3), we can express this sensitivity as

$$\eta\overline{P}_{\text{sens}} = \frac{hc}{\lambda q} \cdot \mathcal{Q}\, i_n^{rms}, \tag{4.23}$$

or, with Eq. (4.22), more generally as

$$\eta\overline{P}_{\text{sens}} = \frac{hc}{\lambda q} \cdot \frac{\mathcal{Q}(i_{n,0}^{rms} + i_{n,1}^{rms})}{2}. \tag{4.24}$$

For example, given an input-referred rms noise current of 1 μA, a wavelength of 1.55 μm, and a required BER of 10^{-12}, the optical sensitivity of a receiver with an ideal photodetector is $7.03 \cdot 1$ μA/(1.25 A/W) = 5.62 μW, corresponding to −22.5 dBm.

OMA Sensitivity. Some standards, such as 10-Gigabit Ethernet IEEE 802.3ae-2002, define the optical sensitivity as the minimum *peak-to-peak* optical power, rather than the averaged optical power, necessary to achieve a specified BER ($P_{\text{sens}}^{OMA} = P_S^{pp}$ @ BER). The peak-to-peak optical power, that is, the difference between the power of the ones and the power of the zeros, is known as the *optical modulation amplitude* (OMA) [15]. Thus, this sensitivity measure is referred to as the OMA sensitivity. With Eq. (4.19), we can express the OMA sensitivity as

$$P_{\text{sens}}^{OMA} = \frac{2\mathcal{Q}\, i_n^{rms}}{\mathcal{R}}, \tag{4.25}$$

or, with Eq. (4.20), more generally as

$$P_{\text{sens}}^{OMA} = \frac{\mathcal{Q}(i_{n,0}^{rms} + i_{n,1}^{rms})}{\mathcal{R}}. \tag{4.26}$$

For example, given an input-referred rms noise current of 1 μA, a responsivity of 0.8 A/W, and a required BER of 10^{-12}, the OMA sensitivity is $2 \cdot 7.03 \cdot 1$ μA/(0.8 A/W) = 17.58 μW, corresponding to −17.6 dBm.

Compared with the average-based optical sensitivity, $\overline{P}_{\text{sens}}$, the OMA sensitivity corresponds more closely to the electrical sensitivity, i_{sens}^{pp}. The OMA measure was introduced to grant more flexibility to the transmitter design. An OMA specification permits the transmitter to be designed with a low or high extinction ratio, as long as it does not overload the receiver and meets eye safety. In contrast, an average-power specification usually comes together with a particular extinction-ratio requirement.

Sensitivity Degradations. All of the aforementioned sensitivity expressions assume an optimally set decision threshold (the same assumption as for Eqs. (4.8) and (4.10)). Moreover, the average-based sensitivity expressions ($\overline{P}_{\text{sens}}$ and $\eta\overline{P}_{\text{sens}}$, but not P_{sens}^{OMA}) assume a high extinction ratio, that is, a negligible optical power level for the zeros.

In Section 4.7, we discuss how a decision-threshold offset and a finite extinction ratio degrade the sensitivity. The (average-based) optical receiver sensitivity defined in regulatory standards usually assumes the worst permissible extinction ratio. [→ Problem 4.8.]

Reference Bit-Error Rates. When specifying an (electrical or optical) receiver sensitivity, we must do so with respect to a reference BER. The most common reference BER is 10^{-12}, which corresponds to $Q = 7.03$. Older standards, such as SONET OC-48, used higher reference BERs, such as 10^{-10}. Component manufacturers usually aim at lower BERs, such as 10^{-15} ($Q = 7.94$), to meet the required system BER.

Advanced receivers make use of forward error correction (FEC) to digitally correct decision errors (cf. Appendix G). A typical FEC decoder can turn a bit stream with $BER = 10^{-4}$ into a corrected bit stream with $BER = 10^{-12}$. In this case, the system BER can be satisfied with a BER of only 10^{-4} at the decision circuit. Thus, the sensitivity of receivers with FEC is often specified for a reference BER of 10^{-4} ($Q = 3.72$) or even 10^{-3} ($Q = 3.09$) before FEC.

Dynamic Range. As we have seen, bit errors occur when the received signal is so *weak* that it becomes corrupted by the receiver's noise. Bit errors also occur when the received signal is very *strong*. Strong signals can overload the receiver resulting in excessive pulse-width distortion and jitter (cf. Section 5.2). Thus, besides the minimum signal level, known as the sensitivity or the *sensitivity limit*, there is a maximum signal level, known as the *overload limit*, beyond which the required BER cannot be met. In analogy to the sensitivity, we define the *input overload current*, i_{ovl}^{pp}, as the maximum peak-to-peak signal current for which a specified BER can just be met. Similarly, we define the *optical overload power*, $\overline{P}_{ovl}$, as the maximum time-averaged optical power for which a specified BER can just be met.

The *dynamic range* of a receiver extends from the sensitivity limit to the overload limit and is defined as the ratio of these two limits. (In some applications the dynamic range ends before the overload limit is reached; cf. Section 5.3.)

Thus, the electrical dynamic range is $i_{out}^{pp}/i_{sens}^{pp}$, whereas the optical dynamic range is $\overline{P}_{out}/\overline{P}_{sens}$. When expressed in decibels, the electrical dynamic range is $20 \log(i_{out}^{pp}/i_{sens}^{pp})$, whereas the optical dynamic range is $10 \log(\overline{P}_{out}/\overline{P}_{sens})$.

Power Budget. Clearly, a receiver with a better sensitivity permits a longer reach (assuming fiber dispersion is not a limiting factor). How long the optical link can be made is best discussed in terms of the *power budget* (or *link budget*) [2, 7]. The difference between the transmitted power (in dBm) and the receiver sensitivity (in dBm) is the *system gain* that we can spend on various system losses, such as the fiber loss. For example, with a 1-mW transmitter (0 dBm) and a receiver with a sensitivity of −24 dBm, we have a system gain of 24 dB. We may consider spending this gain on a 120-km long single-mode fiber (SMF) with 0.2 dB/km having a total loss of 24 dB. Unfortunately, there are other losses and margins that we have to budget for and not all the gain can be spent on the fiber loss alone. Table 4.2 shows an example of a more realistic power budget. It includes losses for fiber connectors, a penalty for fiber dispersion, a margin for component aging, and finally a system (or safety) margin. After all these deductions, a gain of 16 dB is left to spend on the fiber loss, which translates into a link length of 80 km.

How significant is a 1-dB difference in receiver sensitivity? The aforementioned example shows that every dB of sensitivity gained extends the link by about 5 km. This extension is a significant fraction of the total link length and therefore system designers care about small sensitivity improvements (or degradations) such as 0.05 dB.

The power-budget example in Table 4.2 is for an unamplified optical link. Much longer links can be built when optical in-line amplifiers are used. These amplifiers add to the system gain, making it relatively easy to meet the power budget. However, they also add noise requiring us to meet another kind of budget, the OSNR budget, which we discuss in Section 4.6.

Table 4.2 Power budget for an unamplified optical link.

Parameter	Symbol	Value
Transmitter power	$\overline{P}_{out}$	0.0 dBm
Fiber connector loss (at transmitter)		−0.5 dB
Fiber attenuation (80 km of SMF at 1.55 μm)		−16.0 dB
Fiber dispersion penalty		−1.0 dB
Fiber connector loss (at receiver)		−0.5 dB
Margin for aging		−3.0 dB
System margin		−3.0 dB
Receiver sensitivity (at actual extinction ratio)	$\overline{P}_{sens}$	−24.0 dBm

Sensitivity Analysis Based on Circuit Noise Only. For the following initial calculations, we ignore the detector noise and use Eq. (4.21) with $i_n^{rms} = i_{n,ckt}^{rms}$ to estimate the sensitivity based on the circuit noise alone. Later we improve on these calculations.

The sensitivity of a receiver with a p–i–n photodetector comes out as [4, 16]

$$\overline{P}_{\text{sens},PIN} = \frac{Q\, i_{n,\text{ckt}}^{rms}}{\mathcal{R}}. \tag{4.27}$$

The APD with avalanche gain M has an M times higher responsivity (cf. Eq. (3.11)), which leads to an M times better receiver sensitivity:

$$\overline{P}_{\text{sens},APD} = \frac{1}{M} \cdot \frac{Q\, i_{n,\text{ckt}}^{rms}}{\mathcal{R}}. \tag{4.28}$$

Similarly, the optically preamplified p–i–n detector with amplifier gain G has a G times higher responsivity (cf. Eq. (3.18)), which leads to a G times better receiver sensitivity:

$$\overline{P}_{\text{sens},OA} = \frac{1}{G} \cdot \frac{Q\, i_{n,\text{ckt}}^{rms}}{\mathcal{R}}. \tag{4.29}$$

The input-referred rms noise current $i_{n,\text{ckt}}^{rms}$, which appears in all of the aforementioned sensitivity expressions, is defined in Eq. (4.18) and its calculation is discussed in Section 4.5.

Let us plug in some numbers. With the values $i_{n,\text{ckt}}^{rms} = 250$ nA at 2.5 Gb/s, 1 μA at 10 Gb/s, 4 μA at 40 Gb/s, $\mathcal{R} = 0.8$ A/W, $M = 10$, $G = 100$ (cf. Chapter 3), and $BER = 10^{-12}$, we find the sensitivity values listed in Table 4.3. We see how the sensitivities improve in proportion to the detector responsivities as we go from the p–i–n detector to the APD and finally to the optically preamplified p–i–n detector (OA + p–i–n). [→ Problem 4.9.]

Sensitivity Analysis Including Detector Noise. Next, we repeat the sensitivity calculations, but this time taking the photodetector noise, $\overline{i_{n,PD}^2}(t)$, into account. This exercise reveals the significance of the detector noise relative to the circuit

Table 4.3 Approximate receiver sensitivities at $BER = 10^{-12}$ for various photodetectors. Only the circuit noise is taken into account.

Parameter	Symbol	2.5 Gb/s	10 Gb/s	40 Gb/s
Circuit noise (rms)	$i_{n,\text{ckt}}^{rms}$	250 nA	1 μA	4 μA
Signal (pp) @ $BER = 10^{-12}$	i_{sens}^{pp}	3.5 μA	14.1 μA	56.3 μA
Sensitivity (p–i–n)	$\overline{P}_{\text{sens},PIN}$	−26.6 dBm	−20.6 dBm	−14.5 dBm
Sensitivity (APD)	$\overline{P}_{\text{sens},APD}$	−36.6 dBm	−30.6 dBm	−24.5 dBm
Sensitivity (OA + p–i–n)	$\overline{P}_{\text{sens},OA}$	−46.6 dBm	−40.6 dBm	−34.5 dBm

noise. Now, because the detector noise is signal dependent, we have to consider two different noise values, one for the zeros and one for the ones. The total input-referred mean-square noise current (detector noise plus circuit noise) can be written as $\overline{i_{n,0}^2} = \overline{i_{n,PD,0}^2} + \overline{i_{n,\text{ckt}}^2}$ for the zeros and $\overline{i_{n,1}^2} = \overline{i_{n,PD,1}^2} + \overline{i_{n,\text{ckt}}^2}$ for the ones.

The input-referred mean-square noise current of the detector, $\overline{i_{n,PD}^2}$, is defined as the mean-square noise voltage due to the detector at the output of the linear channel divided by H_0^2 (cf. Eq. (4.18)). We see in Section 4.5 that this is equivalent to $\overline{i_{n,PD}^2} = I_{n,PD}^2 \cdot BW_n$, where BW_n is the noise bandwidth of the linear channel. (The same expression does not work for the frequency-dependent circuit noise PSD.)

With the p–i–n detector noise expressions $\overline{i_{n,PIN,0}^2} \approx 0$ and $\overline{i_{n,PIN,1}^2} = 4q\mathcal{R}\overline{P} \cdot BW_n$ from Eqs. (3.8) and (3.9), we find the total noise on the zeros, $i_{n,0}^{rms} = i_{n,\text{ckt}}^{rms}$, and ones, $i_{n,1}^{rms} = \sqrt{4q\mathcal{R}\overline{P}_{\text{sens}} \cdot BW_n + (i_{n,\text{ckt}}^{rms})^2}$. Inserting these expressions into $\overline{P}_{\text{sens}} = Q(i_{n,0}^{rms} + i_{n,1}^{rms})/(2\mathcal{R})$ from Eq. (4.22) and solving for $\overline{P}_{\text{sens}}$ reveals the sensitivity of a p–i–n receiver [17]:

$$\overline{P}_{\text{sens},PIN} = \frac{Q\, i_{n,\text{ckt}}^{rms}}{\mathcal{R}} + \frac{Q^2 q \cdot BW_n}{\mathcal{R}}. \tag{4.30}$$

The first term of this equation is due to the circuit noise and is identical to Eq. (4.27); the second term is due to the shot noise of the p–i–n photodetector and is new. Using the APD noise expressions Eqs. (3.15) and (3.16) instead of the p–i–n noise expressions, we find the sensitivity of an APD receiver [4, 16, 17]:

$$\overline{P}_{\text{sens},APD} = \frac{1}{M} \cdot \frac{Q\, i_{n,\text{ckt}}^{rms}}{\mathcal{R}} + F \cdot \frac{Q^2 q \cdot BW_n}{\mathcal{R}}. \tag{4.31}$$

Finally, the sensitivity of a receiver with an optically preamplified p–i–n detector can be derived from Eqs. (3.24) and (3.25):

$$\overline{P}_{\text{sens},OA} = \frac{1}{G} \cdot \frac{Q\, i_{n,\text{ckt}}^{rms}}{\mathcal{R}} + \eta F \cdot \frac{Q^2 q \cdot BW_n}{\mathcal{R}}. \tag{4.32}$$

Note that Eq. (4.32), like the equations it was derived from, only takes the signal–spontaneous beat noise into account; for a more accurate expression that also accounts for the spontaneous–spontaneous beat noise, see Eq. (I.15) or [18].

Comparing the aforementioned three sensitivity expressions, we observe that the first term, due to the circuit noise, is suppressed with increasing detector gain (p–i–n → APD → OA + p–i–n). The second term, due to the detector noise, increases with the detector noise figure (or excess noise factor) and thus is larger for the APD and the OA + p–i–n detector than for the simple p–i–n photodetector. [→ Problem 4.10.]

Table 4.4 Receiver sensitivities at $BER = 10^{-12}$ for various photodetectors. Circuit and detector noise are taken into account.

Parameter	Symbol	2.5 Gb/s	10 Gb/s	40 Gb/s
Circuit noise (rms)	$i^{rms}_{n,\text{ckt}}$	250 nA	1 μA	4 μA
Sensitivity (p–i–n)	$\overline{P}_{\text{sens},PIN}$	−26.5 dBm	−20.5 dBm	−14.5 dBm
Sensitivity (APD)	$\overline{P}_{\text{sens},APD}$	−34.8 dBm	−28.8 dBm	−22.8 dBm
Sensitivity (OA + p–i–n)	$\overline{P}_{\text{sens},OA}$	−42.3 dBm	−36.2 dBm	−30.2 dBm

Let us evaluate these equations for some example values. For the APD, we use $F = 6$ corresponding to 7.8 dB; for the OA + p–i–n, we use $\eta = 0.64$ and $F = 3.16$ corresponding to 5 dB (cf. Chapter 3). With these numbers and assuming that the noise bandwidth is 75% of the bit rate (see Sections 4.5 and 4.8 for a justification), we obtain the sensitivity numbers shown in Table 4.4.

When comparing these numbers with the approximations in Table 4.3, we note the following:

- The sensitivity of the p–i–n receiver hardly changed at all. This means that the shot noise contributed by the p–i–n photodetector is negligible compared with the circuit noise.
- The sensitivity of the APD receiver degraded by about 1.8 dB. This means that the avalanche noise makes a significant contribution to the total noise.
- The sensitivity of the receiver with optically preamplified p–i–n detector degraded by about 4.3 dB. This means that the noise of the optical amplifier has a larger impact than the circuit noise.

Table 4.5 lists sensitivity numbers reported in the literature that were measured under back-to-back conditions. Compared with our example numbers in Table 4.4, the 10-Gb/s numbers are fairly close while the 40-Gb/s numbers

Table 4.5 Receiver sensitivities at $BER = 10^{-12}$ for various photodetectors. Measured in a back-to-back configuration.

Bit rate (Gb/s)	Photodetector	Sensitivity (dBm)	Reference
10	p–i–n	−21.5	[19]
10	APD	−27.7	[19]
10	OA + p–i–n	−35.5	[20]
40	p–i–n	−9.1	[21]
40	APD	−17.8	[22]
40	OA + p–i–n	−28.5	[23]

are several decibels worse, indicating that our 40-Gb/s example values were too optimistic.

Effects of Non-Gaussian Detector Noise. At the outset of our discussion, we made the assumption that the noise follows a Gaussian distribution. Although this is a good assumption for the circuit noise, the photodetector noise in direct-detection receivers typically has a distinctly non-Gaussian distribution (cf. Chapter 3). The deviation from the Gaussian distribution becomes most pronounced at the sensitivity limit where the optical power levels are small. Therefore, the sensitivity results in Eqs. (4.30)–(4.32) lose in accuracy when the detector noise becomes a significant fraction of the circuit noise.

Our analysis of the unamplified p–i–n receiver is good because the Gaussian circuit noise dominates the non-Gaussian detector noise, but the analysis of the APD receiver and the optically preamplified p–i–n receiver needs to be revisited. In those cases, the noise at the decision circuit must be taken as a superposition of the non-Gaussian detector noise and the Gaussian circuit noise. The total bit-error-determining noise distribution is obtained by *convolving* the two noise distributions.

A non-Gaussian analysis of the APD receiver reveals that the Gaussian approximation estimates the sensitivity within about 1 dB, but underestimates the optimum decision threshold (finds it too far below the midpoint) [24].

A non-Gaussian analysis of the optically preamplified p–i–n receiver finds that the Gaussian approximation estimates the sensitivity pretty accurately, within about 0.3 dB, but significantly underestimates the optimum decision threshold [25]. The closeness of the sensitivity approximation is due to a lucky mathematical coincidence [26]; the actual chi-square distributions are not close to the Gaussian distributions as is evident from the discrepancy in the optimum decision threshold. Moreover, the mathematical coincidence does not extend to optically amplified receivers with balanced detectors (cf. Section 3.5). For such receivers a non-Gaussian noise treatment is mandatory to avoid errors of several decibels [26].

Optimum APD Gain. We observed in Section 3.2 that there must be an optimum APD gain, because the avalanche gain M can be increased only at the expense of a higher excess noise factor F (cf. Eq. (3.17)). Now with Eq. (4.31) for the APD receiver sensitivity in our possession, we can derive a mathematical expression for the *optimum APD gain*, M_{opt}, that yields the best sensitivity. Intuitively, if M is chosen too low, the first term in Eq. (4.31) containing $1/M$ limits the sensitivity; conversely, if M is chosen too high, the second term containing F limits the sensitivity.

Combining Eqs. (4.31) and (3.17) and solving for the M that minimizes $\overline{P}_{\text{sens},APD}$, we find [2, 4, 16]

$$M_{\text{opt}} = \sqrt{\frac{i^{rms}_{n,\text{ckt}}}{Qk_A q \cdot BW_n} - \frac{1-k_A}{k_A}}. \tag{4.33}$$

We see that M_{opt} increases with increasing circuit noise, $i^{rms}_{n,\text{ckt}}$, which makes sense because the APD gain helps to suppress this noise. Furthermore, M_{opt} decreases with increasing k_A, which means that the optimum gain for a relatively noisy detector, such as an InGaAs APD ($k_A \approx 0.6$), is smaller than that for a comparatively quiet detector, such as a silicon APD ($k_A \approx 0.03$).

When taking the non-Gaussian distribution of the APD noise into account, the optimum gain is somewhat lower than given in Eq. (4.33) [24]. The noise corrupted decision threshold in burst-mode receivers (cf. Section 4.2) also affects the optimum gain. Specifically, the optimum gain is lower for short preambles but approaches the optimum gain of continuous-mode receivers for long preambles [10].

Interestingly, there is no corresponding optimum gain for an optically preamplified p–i–n detector. As we know, the noise figure F of the optical amplifier *decreases* with increasing gain G (cf. Fig. 3.20). Therefore, a higher gain always improves the sensitivity and there is no (finite) optimum gain.

Sensitivity of Coherent Receivers. So far we have been focusing on direct-detection receivers. Now, let us analyze a coherent heterodyne (or intradyne) receiver for binary phase-shift keying (BPSK) (cf. Section 3.5). BPSK, like NRZ, is a two-valued format, but unlike NRZ, the optical power of a BPSK signal is constant while the optical phase flips back and forth between 0° and 180° (cf. Chapter 1).

To demodulate the IF signal from the coherent receiver front-end, we recover the IF carrier and multiply (mix) it with the incoming IF signal. This method is known as synchronous or coherent demodulation. It can be shown [7] that for a coherent BPSK receiver with synchronous demodulation the BER relates to the SNR as $Q = \sqrt{SNR}$. Note that this is the same relationship as for a direct-detection NRZ receiver with additive noise (cf. Eq. (4.13)).

The signal and noise of a coherent receiver with balanced detector was discussed in Section 3.5. From Eq. (3.32), the mean-square value of the sinusoidal IF signal is $\overline{i_S^2} = \frac{1}{2}(2\mathcal{R}\sqrt{P_{LO}\overline{P}_S})^2 = 2\mathcal{R}^2 P_{LO}\overline{P}_S$, where P_{LO} is the power of the local oscillator (LO). From Eq. (3.33), the mean-square noise of the detector is $2q\mathcal{R}P_{LO} \cdot BW_n$, where it was assumed that the incoming optical signal is free of ASE noise (no optical amplifiers) and the relative intensity noise (RIN) from the LO laser is fully suppressed by the balanced detector. Including the circuit noise, the total noise becomes $\overline{i_n^2} = \overline{i_{n,\text{ckt}}^2} + 2q\mathcal{R}P_{LO} \cdot BW_n$. Dividing the signal power by the noise power reveals [7]

$$SNR = \frac{2\mathcal{R}^2 P_{LO}\overline{P}_S}{\overline{i_{n,\text{ckt}}^2} + 2q\mathcal{R}P_{LO} \cdot BW_n}. \tag{4.34}$$

Using $Q = \sqrt{SNR}$ and solving for $\overline{P}_S$, we find the sensitivity of the heterodyne receiver for a BPSK signal:

$$\overline{P}_{\text{sens,het}} = \frac{1}{2P_{LO}} \cdot \left(\frac{Q\, i_{n,\text{ckt}}^{rms}}{\mathcal{R}}\right)^2 + \frac{Q^2 q \cdot BW_n}{\mathcal{R}}. \tag{4.35}$$

Comparing this sensitivity to that of the simple p–i–n receiver in Eq. (4.30), we see that the first term due to the circuit noise is suppressed by P_{LO}, while the second term due to the shot noise is the same. Thus, by choosing a strong enough LO, we can make the first term small compared with the second term. When this condition is met, the coherent receiver is said to be shot-noise limited.

Evaluating Eq. (4.35) for the example values $\mathcal{R} = 0.8$ A/W, $BW_n = 0.75B$, and $P_{LO} = 2$ mW, we obtain the sensitivities shown in Table 4.6. The coherent receiver turns out to be even more sensitive than the optically preamplified p–i–n receiver (cf. Table 4.4).

In practice, the common-mode rejection (CMRR) of the balanced detector is not perfect (cf. Eq. (3.34)) and some RIN noise from the LO laser (cf. Eq. (3.31)) leaks through. This effect adds another term to the sensitivity expression in Eq. (4.35) that is proportional to P_{LO}. As a result, the sensitivity does not improve indefinitely with increasing LO power, but reaches an optimum, which occurs when the residual RIN noise is balanced with the circuit noise [27]. Other sensitivity degradations not accounted for in Eq. (4.35) are caused by the intrinsic loss when phase and polarization diversity is used and the excess loss of the interferometers [27].

Coherent receivers for higher-order modulation formats such as QPSK or 16-QAM require a higher SNR than the BPSK receiver in our example and thus have a lower sensitivity [28].

Table 4.6 Receiver sensitivity at $BER = 10^{-12}$ for coherent heterodyne detection of a BPSK signal. Circuit and detector shot noise are taken into account.

Parameter	Symbol	2.5 Gb/s	10 Gb/s	40 Gb/s
Circuit noise (rms)	$i_{n,\text{ckt}}^{rms}$	250 nA	1 μA	4 μA
Sensitivity (coherent)	$\overline{P}_{\text{sens,het}}$	−47.0 dBm	−40.3 dBm	−32.2 dBm

Sensitivity of Analog Receivers. To add some perspective, let us have a brief look at analog receivers and what mechanisms determine their sensitivity.

Assuming a transmitter with sinusoidal modulation and a receiver with a p–i–n photodetector, we can calculate the CNR as follows (cf. Section 4.2). The average photocurrent is $\mathcal{R}\overline{P}_S$. The amplitude of the AC photocurrent is $m\mathcal{R}\overline{P}_S$, where m is the *modulation index*. Because the AC signal is a sine wave, its mean-square value is $\overline{i_S^2} = \frac{1}{2}(m\mathcal{R}\overline{P}_S)^2$. As for the noise, we have to account for three major components: the input-referred circuit noise $\overline{i_{n,\mathrm{ckt}}^2}$, the shot noise of the photodetector, which follows from Eq. (3.7) as $\overline{i_{n,PIN}^2} = 2q\mathcal{R}\overline{P}_S \cdot BW_n$, and the RIN noise of the laser in the transmitter, $\overline{i_{n,RIN}^2} = 2RIN \cdot \mathcal{R}^2\overline{P}_S^2 \cdot BW_n$ [7]. Dividing the signal power by the sum of the three noise powers reveals [7]:

$$CNR = \frac{\frac{1}{2}m^2\mathcal{R}^2\overline{P}_S^2}{\overline{i_{n,\mathrm{ckt}}^2} + 2q\mathcal{R}\overline{P}_S \cdot BW_n + 2RIN \cdot \mathcal{R}^2\overline{P}_S^2 \cdot BW_n}. \tag{4.36}$$

The CNR given in Eq. (4.36) can be approximated by three asymptotes, each one corresponding to one dominating noise component. If circuit noise dominates, we have $CNR \approx m^2\mathcal{R}^2\overline{P}_S^2/(2\overline{i_{n,\mathrm{ckt}}^2})$; if shot noise dominates, we have $CNR \approx m^2\mathcal{R}\overline{P}_S/(4q \cdot BW_n)$; and if RIN noise dominates, we have $CNR \approx m^2/(4RIN \cdot BW_n)$. Figure 4.8 shows the total CNR (solid line) together with the three asymptotes (dashed lines) for a typical analog CATV application ($m = 5\%$, $\mathcal{R} = 0.9$ A/W, $i_{n,\mathrm{ckt}}^{rms} = 12$ nA, $BW_n = 4$ MHz, and $RIN = -153$ dB/Hz). Each asymptote has a different dependence on the received optical power $\overline{P}_S$: the CNR due to the circuit noise increases with $\overline{P}_S^2$, the CNR due to the shot noise increases with $\overline{P}_S$, and the CNR due to the RIN noise is independent of $\overline{P}_S$. We can see that as we increase $\overline{P}_S$, the RIN noise, which increases proportional to the signal, becomes the ultimate limit for the CNR.

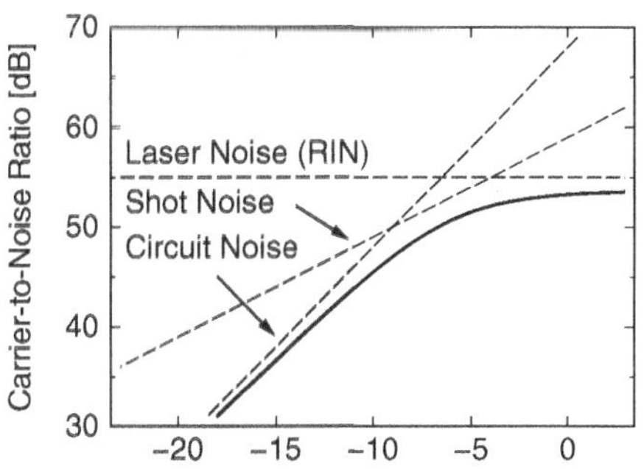

Figure 4.8 CNR as a function of received optical power for an analog CATV receiver.

The sensitivity of an analog receiver is the minimum optical power necessary to achieve the required CNR. In the case of an analog CATV application, a CNR of about 50 dB is required (cf. Appendix A) and based on Fig. 4.8 an input power of at least −6 dBm is necessary. Thus, the receiver operates in the shot-noise or RIN noise dominated regime. For more information on analog receivers, see Refs [12, 29].

BER Plots. The sensitivity and dynamic range of a digital optical receiver is determined by measuring the BER as a function of the received power and comparing it against the reference BER. A graph of the BER against the received optical power is known as a *BER plot*.

What do we expect these plots to look like? For a p–i–n receiver and an NRZ signal, we expect the curves to follow the theoretical result in Eq. (4.30), as long as the receiver is operated far away from its overload limit. Thus, the Q value calculated from the measured BER should be a linear function of the received optical power $\overline{P}_S$ (shot noise is assumed to be negligible). It is convenient to plot the BER values on a graph paper (coordinate system) that represents this relationship by a straight line. One possibility, illustrated in Fig. 4.9, is to choose $\overline{P}_S$ for the x-axis and $-Q$ for the y-axis (the minus sign makes the *BER* go up with y). To make the graph more legible, we label the y-axis with BER values rather than Q values. This results in what may look like a logarithmic *BER* scale, but it is not logarithmic. Alternatively, if we prefer to represent the power in dBm rather than in mW, we can choose 10 log $\overline{P}_S$ (the power in dB) for the x-axis and −10 log Q for the y-axis [30]. The linear relationship between Q and $\overline{P}_S$ is still represented by a straight line in these logarithmic coordinates. Figure 4.10 shows an example of that type of BER plot. Note that both BER-plot examples show the same BER data. The second BER-plot type is more common because its logarithmic x-axis can represent a larger power range.

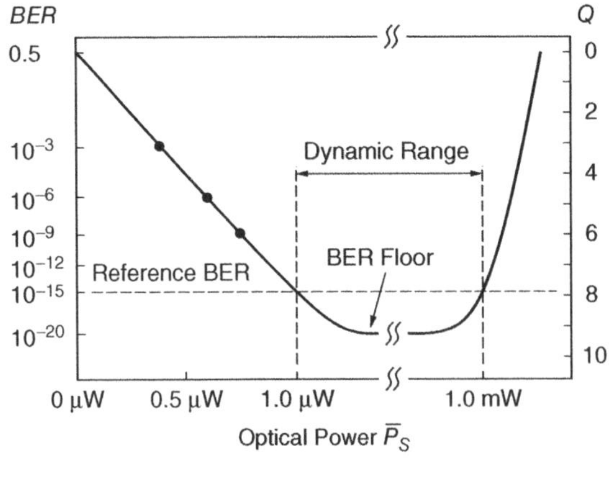

Figure 4.9 BER plot in the linear ($\overline{P}_S$, $-Q$) coordinate system. For small power values, the plot follows a straight line down to $BER = 10^{-15}$, where the sensitivity is 1 µW.

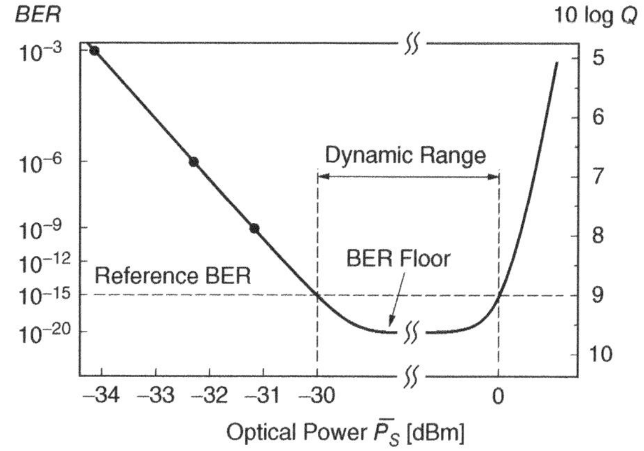

Figure 4.10 BER plot in the (10 log $\overline{P}_S$, −10 log Q) coordinate system. For small power values, the plot follows a straight line down to $BER = 10^{-15}$, where the sensitivity is −30 dBm.

Data points plotted in this way can easily be *extrapolated* down to very low BERs. For example, Figs 4.9 and 4.10 show data points at $BER = 10^{-3}$, 10^{-6}, and 10^{-9} that can be extrapolated down to 10^{-15}, revealing a sensitivity of 1 μW (or −30 dBm) at this reference BER. Such extrapolations are very convenient because they avoid time-consuming low-BER measurements. However, the extrapolation is only correct if the following conditions are met: (i) the receiver noise must closely follow a Gaussian distribution over a range of many sigmas, (ii) the noise must not depend on the received power level, and (iii) the signal must not become distorted at high received power levels.

If the noise does increase with the received power level, the BER curves on our graph paper deviate from a straight line, becoming flatter with increasing power. Such a deviation happens for APD receivers and optically preamplified receivers and is a consequence of the Q^2 term in Eqs. (4.31) and (4.32). In the extreme case, the BER curve flattens out completely, that is, even for high received power levels, the BER never goes below a certain value. This minimum BER value is known as the *BER floor*. Note that a BER floor limits the validity of BER extrapolations.

At very high power levels, the receiver front-end tends to overload, which means that the signal distortions become so severe that the BER *increases* rapidly with $\overline{P}_S$. Figures 4.9 and 4.10 show BER plots with an optical overload power of 1 mW (or 0 dBm). The dynamic range, defined by the sensitivity and overload limits, is also indicated in the plots.

BER plots can be helpful to diagnose problems with the receiver. For example, from a few measured data points, we can see if they are on a straight line or not. A divergence from a straight line may indicate a BER floor. Moreover, from the slope of the plot in the linear coordinate system, we can infer the circuit noise $i^{rms}_{n,ckt}$, assuming $\mathcal{R}$ is known and the detector noise is negligible.

The BER plot depends on the modulation format used. In our earlier example NRZ modulation was assumed. For higher-order modulation formats, such as

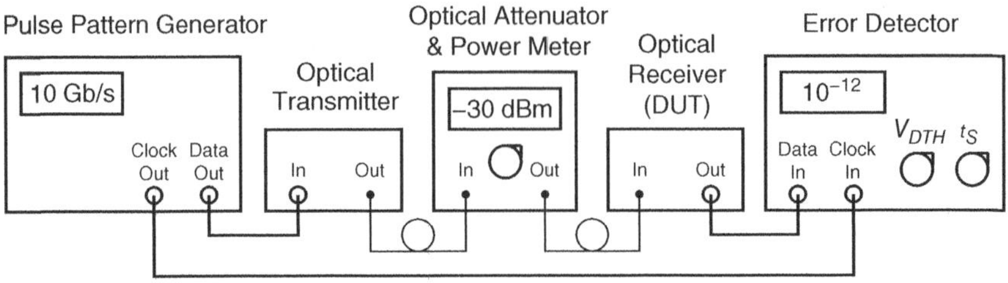

Figure 4.11 Bit-error rate test set (BERT) to measure the BER of an optical receiver.

4-PAM, QPSK, and 16-QAM, the BER for a given optical power is higher than that of a two-valued format, such as NRZ or BPSK.

BER Measurement. Figure 4.11 shows a typical measurement setup to characterize an optical receiver. The setup consists of a pulse pattern generator, an optical transmitter, an optical attenuator with power meter, the device under test (DUT), and an error detector. The pulse pattern generator and the error detector together are known as a *bit-error rate test set* (BERT).

The pulse pattern generator produces a pattern, such as a pseudorandom bit sequence (PRBS), in the form of an electrical NRZ signal. This signal is converted to an optical signal by means of a high-quality optical transmitter (a.k.a. electrical-to-optical converter). The optical signal then passes through an adjustable attenuator with a built-in power meter that measures the optical output power. The attenuated optical signal is fed to the receiver to be characterized, that is, to the DUT. The electrical output signal from the receiver is checked with an error detector, a test instrument that detects bit errors by comparing the received bit sequence with the expected sequence. To synchronize the error detector with the incoming bits, the error detector can recover the clock from the received signal or it can get the bit clock directly from the pulse pattern generator (as shown in Fig. 4.11). The sampling instant t_S (phase offset) and the decision threshold voltage V_{DTH} at the error detector must be adjusted until the lowest BER is found. Modern error detectors can find the optimum decision point automatically.

To make a BER plot, we proceed as follows. With the attenuator, we set the optical power for the DUT to a few decibels below the expected receiver sensitivity and measure the BER. This measurement can be done quickly because the BER is high (much higher than the reference BER). Then, we increase the optical power by a small amount (e.g., 0.2 dB) and measure the BER again. We repeat this step until we reach BER values below the reference BER (or we run out of patience). Next, we take measurements near the overload limit. Finally, we plot the measured BER values against the set power values on the appropriate graph paper.

Low BER Issues. At very low BERs, such as 10^{-15}, it becomes extremely time consuming to obtain an accurate BER reading. For example, to collect 10 errors at a BER of 10^{-15} and a bit rate of 10 Gb/s, we have to wait an average of 10^6 seconds, which amounts to about 12 days! Even then, we have to wonder if 10 errors are enough for an accurate reading. Assuming the error count during a given time interval follows the Poisson distribution, we can estimate the uncertainty after collecting n errors as $\sqrt{n}$. (A Poisson distribution with mean n has a standard deviation of $\sqrt{n}$.) Thus, for 10 collected errors, the uncertainty is 3.2 or 32%; for 100 errors, this uncertainty shrinks to 10%, and so forth.

In some situations, it may be too tedious to collect even a single error. Fortunately, it is often not necessary to determine the exact BER value, but it is sufficient to be reasonably certain that the BER is *below* a certain value. For example, we may not care if the exact BER is $0.5 \cdot 10^{-12}$ or 10^{-15}, but we would like to have 95% confidence that the BER is below 10^{-12}. Again making use of the Poisson distribution, it can be shown [31] that receiving $3n$ bits *without* an error is sufficient to conclude with a 95% confidence level that the BER is below $1/n$. For example, if we measured $3 \cdot 10^{12}$ bits without an error, we are 95% sure that $BER < 10^{-12}$.

4.5 Noise Bandwidths and Personick Integrals

Total Input-Referred Noise. We saw in Section 4.4 that the input-referred rms noise current, i_n^{rms}, plays a key role in determining the receiver sensitivity. Can this noise be calculated from the input-referred noise current PSD $I_n^2(f)$? It is tempting to integrate the input-referred noise PSD over all frequencies and write

$$\overline{i_n^2} \stackrel{?}{=} \int_0^{\infty} I_n^2(f)\, df. \tag{4.37}$$

But this cannot be right, because the integral does not converge if $I_n^2(f)$ contains the usual white or f^2-noise components (or both), leading to an unbounded noise current. Maybe we should integrate only up to the receiver's bandwidth:

$$\overline{i_n^2} \stackrel{?}{=} \int_0^{BW} I_n^2(f)\, df. \tag{4.38}$$

At least, now we get a finite noise current. But note that the result is very sensitive to the upper bound of the integration because it is located in the rising part of the spectrum. Moreover, should that upper bound be the 3-dB bandwidth or the noise bandwidth or something else? Unfortunately, all of our guesses are wrong!

Let us take a step back. Equation (4.18) defines the total input-referred noise as the total output noise divided by the midband value of $H(f)$. In squared form, this is $\overline{i_n^2} = \overline{v_n^2}/H_0^2$. In turn, the mean-square output noise is obtained by integrating the output noise PSD. Working our way backward, the output noise PSD is obtained by shaping the input-referred noise PSD with $|H(f)|^2$ (cf. Eq. (4.6)). Putting these steps together, we find the correct equation for the total input-referred mean-square noise:

$$\overline{i_n^2} = \frac{1}{H_0^2} \int_0^{BW_D} |H(f)|^2 I_n^2(f)\, df, \tag{4.39}$$

where $H(f)$ is the frequency response of the linear channel, H_0 is its midband value, BW_D is the bandwidth of the decision circuit, and $I_n^2(f) = I_{n,PD}^2(f) + I_{n,ckt}^2(f)$ is the input-referred noise current PSD of the detector and receiver circuits combined.

Noise Bandwidths. Equation (4.39) can be simplified, if we make suitable assumptions about the noise spectrum. We know from Section 4.1 that the input-referred noise current PSD can be approximated as (cf. Eq. (4.1))

$$I_n^2(f) = \alpha_0 + \alpha_2 f^2. \tag{4.40}$$

Parameter α_0 describes the white part of the spectrum (due to the circuit and detector noise), and parameter α_2 describes the f^2-noise part (we are neglecting possible $1/f$ and f-noise for now). Now, we plug this PSD into Eq. (4.39):

$$\overline{i_n^2} = \frac{1}{H_0^2} \int_0^{BW_D} |H(f)|^2 (\alpha_0 + \alpha_2 f^2)\, df, \tag{4.41}$$

expand it as

$$\overline{i_n^2} = \alpha_0 \cdot \frac{1}{H_0^2} \int_0^{BW_D} |H(f)|^2\, df + \alpha_2 \cdot \frac{1}{H_0^2} \int_0^{BW_D} |H(f)|^2 f^2\, df, \tag{4.42}$$

and rewrite it in the form (we will see why in a moment)

$$\overline{i_n^2} = \alpha_0 \cdot BW_n + \frac{\alpha_2}{3} \cdot BW_{n2}^3, \tag{4.43}$$

where

$$BW_n = \frac{1}{H_0^2} \int_0^{BW_D} |H(f)|^2\, df, \tag{4.44}$$

$$BW_{n2}^3 = \frac{3}{H_0^2} \int_0^{BW_D} |H(f)|^2 f^2\, df. \tag{4.45}$$

The bandwidths BW_n and BW_{n2} depend *only* on the receiver's frequency response $|H(f)|$ and the decision circuit's bandwidth BW_D. The latter is uncritical as long as it is larger than the receiver bandwidth and the receiver

Table 4.7 Numerical values for BW_n and BW_{n2}.

$H(f)$	BW_n	BW_{n2}
1st-order low pass	$1.57 \cdot BW_{3dB}$	∞
2nd-order low pass, crit. damped ($Q = 0.500$)	$1.22 \cdot BW_{3dB}$	$2.07 \cdot BW_{3dB}$
2nd-order low pass, Bessel ($Q = 0.577$)	$1.15 \cdot BW_{3dB}$	$1.78 \cdot BW_{3dB}$
2nd-order low pass, Butterworth ($Q = 0.707$)	$1.11 \cdot BW_{3dB}$	$1.49 \cdot BW_{3dB}$
Brick-wall low pass	$1.00 \cdot BW_{3dB}$	$1.00 \cdot BW_{3dB}$
Rectangular (impulse response) filter	$0.500 \cdot B$	∞
NRZ to full raised-cosine filter	$0.564 \cdot B$	$0.639 \cdot B$

response rolls off rapidly. Because these conditions usually are met and to keep things simple, we subsequently use $BW_D \to \infty$. Numerical values for the bandwidths BW_n and BW_{n2} for some simple receiver responses are listed in Table 4.7 (cf. Eqs. (H.8) and (H.9)). As soon as these bandwidths and the noise parameters α_0 and α_2 are known, we can easily calculate the total input-referred noise current with Eq. (4.43).

Why did we choose this peculiar form for Eq. (4.43)? Because it leads to a neat interpretation of the bandwidths BW_n and BW_{n2}. If we were to integrate the input-referred noise current PSD Eq. (4.40) up to the bandwidth BW, as suggested by the incorrect Eq. (4.38), we would get

$$\overline{i_n^2} = \alpha_0 \cdot BW + \frac{\alpha_2}{3} \cdot BW^3. \tag{4.46}$$

By comparing this result with Eq. (4.43), we realize that the latter equation can be interpreted as the result of integrating the white-noise component of the input-referred noise current PSD up to BW_n and the f^2-noise component up to BW_{n2}. This interpretation is illustrated by Fig. 4.12. You probably guessed it: BW_n is the *noise bandwidth* (a.k.a., *noise equivalent bandwidth*) of the receiver's frequency response [32, 33]; BW_{n2} could be called the second-order noise bandwidth, because it plays the same role as the (zeroth-order) noise bandwidth BW_n, when we replace the white (f^0) noise with f^2 noise [34, 35].

Now let us go back and see how far off a simple integration of the input-referred noise current PSD up to the 3-dB bandwidth would be. Integrating up to the 3-dB point means that we set $BW_n = BW_{n2} = BW_{3dB}$. Consulting Table 4.7, we find that this is the right thing to do in the case of the brick-wall low-pass response, but in all other cases, we incur an error. For example, in the case of the second-order Butterworth response, we under-estimate the white-noise power by a factor 1.11 and the f^2-noise power by a factor $1.49^3 = 3.33$. If we integrate the input-referred noise current PSD up to the noise bandwidth BW_n, we get the correct white-noise power contribution,

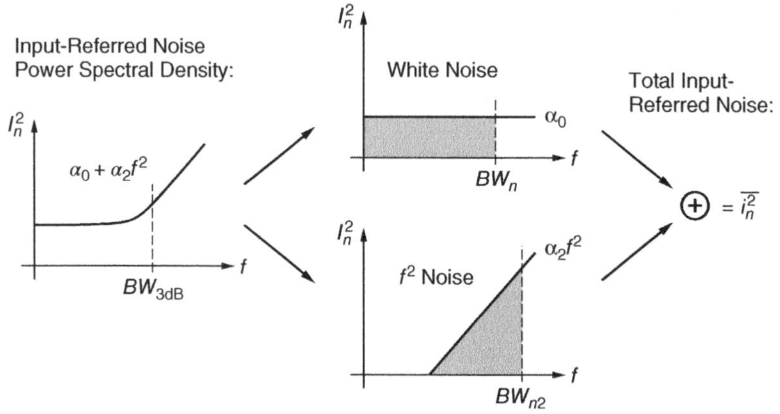

Figure 4.12 Interpretation of BW_n and BW_{n2} as integration bounds.

but we still underestimate the f^2-noise power by a factor $(1.49/1.11)^3 = 2.43$. These are significant differences and, if ignored, leave us scratching our heads over why the measured receiver sensitivity is not as good as the predicted one!

Earlier (in Chapter 3 and Section 4.4), we calculated the mean-square noise currents of various photodetectors by multiplying the noise current PSD with the noise bandwidth of the receiver: $\overline{i_{n,PD}^2} = I_{n,PD}^2 \cdot BW_n$. We now understand that this is correct because the detector noise is *white* ($\alpha_2 = 0$). In Chapter 6, we calculate the input-referred mean-square noise current of transimpedance amplifiers, $\overline{i_{n,TIA}^2}$. Since the latter amplifiers have a colored input-referred noise spectrum ($\alpha_2 \neq 0$), both terms in Eq. (4.43) must be taken into account.

Personick Integrals. For an electrical engineer, the noise bandwidths BW_n and BW_{n2} have an intuitive meaning, and this is why we introduced them first. In the optical receiver literature, however, we often find the so-called *Personick integrals* instead of the noise bandwidths (named after the same Personick who gave us the Personick Q) [2–4, 14, 16, 36]. These integrals are designated with I_1, I_2, and I_3 and are defined such that the input-referred mean-square noise current can be written as

$$\overline{i_n^2} = \alpha_0 \cdot I_2 B + \alpha_2 \cdot I_3 B^3, \tag{4.47}$$

where B is the bit rate. Thus, by comparing this equation with Eqs. (4.43)–(4.45), the second and third Personick integrals can be identified as

$$I_2 = \frac{1}{H_0^2 B} \int_0^\infty |H(f)|^2 \, df = \frac{BW_n}{B}, \tag{4.48}$$

$$I_3 = \frac{1}{H_0^2 B} \int_0^\infty |H(f)|^2 \left(\frac{f}{B}\right)^2 df = \frac{1}{3}\left(\frac{BW_{n2}}{B}\right)^3. \tag{4.49}$$

In other words, the Personick integrals I_2 and I_3 are normalized noise bandwidths. For example, if the receiver has a second-order Butterworth frequency response with a 3-dB bandwidth equal to $\frac{2}{3}$ of the bit rate (we justify this choice in Section 4.8), we find with the help of Table 4.7 the values $I_2 = 0.740$ and $I_3 = 0.329$ for the Personick integrals (cf. Eqs. (H.10) and (H.11)).

What happened to the *first* Personick integral? This integral, I_1, as well as a sum $\sum_1$ (also introduced by Personick) relate to the noise spillover that occurs when the nonstationary detector noise passes through the linear channel. If we neglect this effect, as we did in Section 4.2, I_1 and $\sum_1$ are not needed.

A receiver model often used for theoretical considerations transforms ideal NRZ pulses at the input into pulses with a full raised-cosine spectrum at the output (we discuss this frequency response in Section 4.9). The Personick integrals for this case have the often-seen values $I_2 = 0.564$ and $I_3 = 0.087$ (corresponding to the last entry of Table 4.7). Most practical receivers, however, have I_2 and I_3 values that are larger than that [37].

Generalized Noise Bandwidths and Personick Integrals. It is straightforward to generalize the concept of noise bandwidths for an input-referred noise current PSD of the general form $I_n^2(f) = \sum_{i=0}^{k} \alpha_i f^i$. Now there are $k+1$ noise bandwidths, $BW_{n0}, BW_{n1}, BW_{n2}, \ldots, BW_{nk}$, one for each power of f [35]:

$$BW_{ni}^{\,i+1} = \frac{i+1}{H_0^2} \int_0^{\infty} |H(f)|^2 f^i \, df, \tag{4.50}$$

where, of course, $BW_{n0} = BW_n$. Similarly, the concept of Personick integrals can be generalized for an arbitrary spectrum in the form of a power series [38–40].

In particular, the noise bandwidth BW_{n1}, which relates to an input-referred noise PSD that increases linearly with frequency, is of significance in systems with strong flicker noise (cf. Section 6.3). The corresponding Personick integral is designated with I_f [39] (although $I_{2.5}$ would have been a more logical choice) and is defined as:

$$I_f = \frac{1}{H_0^2 B} \int_0^{\infty} |H(f)|^2 \frac{f}{B} \, df = \frac{1}{2} \left(\frac{BW_{n1}}{B} \right)^2. \tag{4.51}$$

4.6 Optical Signal-to-Noise Ratio

Cascade of Optical Amplifiers. Ultra-long-haul fiber links without electrical regenerators became possible with the development of reliable optical in-line amplifiers in the 1990s. These amplifiers are inserted into the fiber link at regular intervals to boost the optical signal. Let us consider a 10,000-km long fiber link connecting two continents as an example. With a fiber loss of 0.2 dB/km, we need to compensate a total link loss of 2,000 dB. One way to

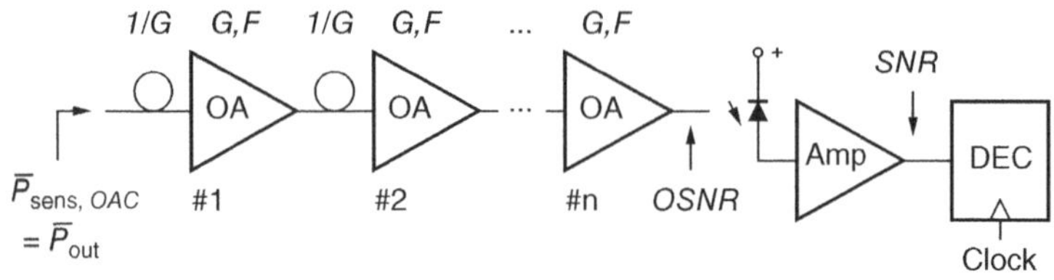

Figure 4.13 A cascade of optical amplifiers followed by a p–i–n receiver.

accomplish this is by inserting 200 optical amplifiers with gain $G = 10$ dB at intervals of 50 km (see Fig. 4.13). Although the optical signal at the output of this amplified link is as strong as at the input, the accumulated noise from the 200 amplifiers is a potential problem.

In the following, we analyze the BER of the aforementioned link. We take two approaches: in the first (and unconventional) approach, we regard the whole fiber link including all the amplifiers as part of the *detector*, which leads to an analysis very similar to our previous sensitivity calculations. In the second approach, we give up the notion of sensitivity and do the calculations in terms of OSNR. The two approaches are equivalent and lead to the same results.

First, let us consider the whole 10,000-km link, including all the optical amplifiers to be part of the detector. In other words, the input of our extended receiver is at the transmitter's end. To derive the "sensitivity" of this system, we regard the whole amplified link as a single equivalent optical amplifier with gain G' and noise figure F'. If we assume that the fiber loss is exactly balanced by the optical amplifier gain G, we have $G' = 1$. It can be shown (cf. solution to Problem 3.9 on p. 501) that the noise figure of a fiber with loss G is G. With the amplifier noise figure F, the total noise figure of the link becomes $F' = nGF$, where n is the number of amplifiers and G is the gain of each amplifier (cf. Eq. (I.7)). Substituting G' and F' into Eq. (4.32) for the sensitivity of an optically preamplified p–i–n receiver yields

$$\overline{P}_{\text{sens},OAC} = \frac{\mathcal{Q}\, i_{n,\text{ckt}}^{rms}}{\mathcal{R}} + \eta n G F \cdot \frac{\mathcal{Q}^2 q \cdot BW_n}{\mathcal{R}}. \tag{4.52}$$

For example, with 200 amplifiers, $G = 10$ dB, and $F = 5$ dB, we get the compound noise figure $F' = 38$ dB. The "sensitivity" $\overline{P}_{\text{sens},OAC}$ for a 10-Gb/s system ($BW_n = 7.5$ GHz, $i_{n,\text{ckt}}^{rms} = 1$ μA, $\mathcal{R} = 0.8$ A/W, $\eta = 0.64$, $BER = 10^{-12}$) comes out as −5.3 dBm, essentially regardless of the circuit noise $i_{n,\text{ckt}}^{rms}$. This "sensitivity" is very low, but recall that it refers to the far end of the fiber link. So all this means is that the transmitter must launch a power of at least $\overline{P}_{\text{out}} = -5.3$ dBm into the fiber to get the desired BER.

We can draw two conclusions from this example. First, it is indeed possible to send bits over a 10,000-km fiber link with 200 erbium-doped fiber amplifiers (EDFAs) and receive them with a low BER such as 10^{-12}! This fact is also demonstrated by commercial transpacific undersea cables such as the 8,600-km long TPC-5, Segment J connecting the United Stated with Japan, which contains 260 EDFA-type repeaters spaced 33 km apart [41].

The second conclusion is that the concept of receiver sensitivity loses its meaning in situations where many in-line amplifiers contribute most of the system noise. Remember, receiver sensitivity is the minimum optical power required to achieve a certain BER based on detector and circuit noise but no incoming optical noise. In optically amplified long-haul and ultra-log-haul systems, the concept of sensitivity must be replaced by the concept of required OSNR, which we discuss next.

Optical Signal-to-Noise Ratio. OSNR, like SNR, can be defined in more than one way. To avoid confusion, let us define it here as the *average optical signal power* divided by the *average optical noise power*, $\overline{P}/P_{ASE}$. The average optical noise power is measured in the optical (noise) bandwidth BW_O and can be written as $P_{ASE} = S_{ASE} \cdot BW_O$, where S_{ASE} is the optical noise PSD in both polarization modes (cf. Section 3.3).

The designer of optically amplified transmission systems is interested in the minimum OSNR required at the receiver rather than the sensitivity of the receiver. The transmit power, the spacing of the optical amplifiers, their gain and noise figure are then selected such that the required OSNR at the receiver is met. To illustrate this method, let us repeat the previous calculation but now thinking in terms of OSNR.

First, we want to know what OSNR is required at the receiver to meet the desired BER. This OSNR is known as the *required OSNR*, $OSNR_{\text{req}}$. To keep things simple at this point, we consider a direct-detection receiver with signal–spontaneous beat noise only (no spontaneous–spontaneous beat noise and no circuit noise) and an NRZ signal with zero optical power for the zeros (high extinction). With these assumptions, the input-referred noise for the zeros is $i_{n,0}^{rms} = 0$ and the noise for the ones is $i_{n,1}^{rms} = \mathcal{R}\sqrt{4\overline{P} \cdot P_{ASE}} \cdot \sqrt{BW_n/BW_O}$, where we have used Eq. (3.19) and $P_{ASE} = S_{ASE} \cdot BW_O$. The signal swing for an ideal NRZ signal at the photodetector is $i_S^{pp} = 2\mathcal{R}\overline{P}$. Inserting these expression into $Q = i_S^{pp}/(i_{n,0}^{rms} + i_{n,1}^{rms})$ and solving for $OSNR_{\text{req}} = \overline{P}/P_{ASE}$, we find [42, 43]

$$OSNR_{\text{req}} = Q^2 \cdot \frac{BW_n}{BW_O}. \tag{4.53}$$

For example, a 10-Gb/s receiver with $BW_n = 7.5$ GHz operating at the 1.55-μm wavelength needs an OSNR of 14.7 dB measured in a 0.1-nm optical bandwidth ($BW_O = 12.5$ GHz at $\lambda = 1.55$ μm) to achieve a BER of 10^{-12}.

Next, we want to know what OSNR we get at the end of a chain of n amplifiers. The optical signal power at the end is $\overline{P}_{\text{out}}$, the same as the launch power of the transmitter because the amplifier gain and the fiber loss are balanced. The optical noise power at the end of the chain is the sum of the noise from all n amplifiers, $P_{ASE} = nS_{ASE} \cdot BW_O$, because the gain from each EDFA output to the link output is one. Thus, we have [44]

$$OSNR = \frac{\overline{P}_{\text{out}}}{nS_{ASE} \cdot BW_O} \approx \frac{\overline{P}_{\text{out}}}{nGF \cdot hc/\lambda \cdot BW_O}, \tag{4.54}$$

where, on the right-hand side, we expressed S_{ASE} in terms of the amplifier noise figure, $F \approx \tilde{F}$, using Eq. (3.27). If we transform Eq. (4.54) into the log domain and specialize it for $\lambda = 1.55$ μm and a BW_O corresponding to 0.1 nm, we obtain the engineering rule [45]

$$OSNR\ (\text{dB}) \approx 58\ \text{dB} + \overline{P}_{\text{out}}\ (\text{dBm}) - G\ (\text{dB}) - F\ (\text{dB}) - 10\ \log n. \tag{4.55}$$

For example, with the familiar values $n = 200$, $F = 5$ dB, and $G = 10$ dB, we find that we need $\overline{P}_{\text{out}} = -5.3$ dBm to achieve the required OSNR of 14.7 dB in a 0.1-nm optical bandwidth. This is same transmit power that we found earlier with the extended receiver argument! [→ Problem 4.11.]

OSNR Budget. The aforementioned example glosses over several issues that are important in practical systems. Similar to our power budget in Table 4.2, we need to account for system penalties and margins [44, 46, 47]. Table 4.8 shows a more realistic OSNR budget. After calculating the OSNR in the first six lines with Eq. (4.55), starting with a transmit power of 4 dBm, we deduct various OSNR penalties and margins. The OSNR penalties account for propagation

Table 4.8 OSNR budget for an amplified optical link.

Parameter	Symbol	Value
Transmitter power	$\overline{P}_{\text{out}}$	4.0 dBm
Constant in Eq. (4.55)		+58.0 dB
Amplifier gain (or span loss)	G	−10.0 dB
Amplifier noise figure	F	−5.0 dB
Number of amplifiers (10 log n in dB)	n	−23.0 dB
OSNR according to Eq. (4.55)	$OSNR$	24.0 dB
Penalty for dispersion, nonlinearity, crosstalk, etc.		−3.0 dB
Penalty for manufacturing and time variations		−3.0 dB
Margin for aging and repair		−1.0 dB
System margin		−1.0 dB
Worst-case OSNR at the receiver	$OSNR$	16.0 dB

impairments (fiber dispersion, fiber nonlinearity, and crosstalk in WDM systems), manufacturing variations, and time variations. The margins account for aging, repair, and added system reliability. After these deductions we obtain the worst-case OSNR at the receiver. To recover the data from the received optical signal at the desired BER, this OSNR must match (or exceed) $OSNR_{req}$.

In practice, the required OSNR is also quite a bit higher than suggested by the simple Eq. (4.53). A finite extinction ratio and spontaneous–spontaneous beat noise both degrade the required OSNR. Forward error correction (FEC) can be used to counteract these degradation (cf. Appendix G). The 16-dB OSNR at the bottom of Table 4.8 is just about enough for a practical 10-Gb/s NRZ receiver with FEC to operate at $BER = 10^{-12}$. Measured values for the required OSNR are listed in Table 4.11.

As a practical matter it is important to ascertain that the transmission system operates with the margins it has been designed for. For our example in Table 4.8 we expect a received OSNR of 18 dB at the beginning of the system life, slowly degrading to 17 dB over its lifetime. Although it is possible to measure the received OSNR to confirm the margin, it is more accurate and more direct to measure the Q parameter with the receiver itself [44, 46]. This measurement can be performed by sweeping the decision threshold of the receiver while observing the BER [48].

To bring the link analysis in line with this practice, the OSNR budget in Table 4.8 can be recast in terms of Q parameters, in this context often called Q factors [44, 46] (cf. the table on p. 505). When these Q factors are expressed in decibels using the conversion rule 20 log Q, OSNR penalties become equal to Q penalties and OSNR margins become equal to Q margins because Q^2 is proportional to $OSNR$ (cf. Eq. (4.53)). The Q-factor budget starts out the same way as the OSNR budget, but then the OSNR value in line 6 is converted to the so-called *ideal* Q factor. After all penalties and margins are deducted from the ideal Q factor, the resulting worst-case Q factor must match (or exceed) the value required for the desired BER. [→ Problem 4.12.]

OSNR Limit and BER Plots. To make a connection between the receiver sensitivity and the required OSNR, let us consider a graph that shows the received OSNR on the x-axis and the received optical power, $\overline{P}_S$, on the y-axis. Such a graph contains a region for which the BER is less that the reference BER. Figure 4.14 shows an example region for which $BER < 10^{-12}$.

On the far right, where the OSNR is high, the graph represents the situation of an unamplified transmission system. Imagine that we follow a vertical line, located in this region, from low to high power levels. The BER starts out high, but decreases with increasing power. When we pass the sensitivity limit, it falls below 10^{-12}. Eventually the BER reaches a minimum (BER floor) and then starts to increase again. When we cross the overload limit, the BER climbs back

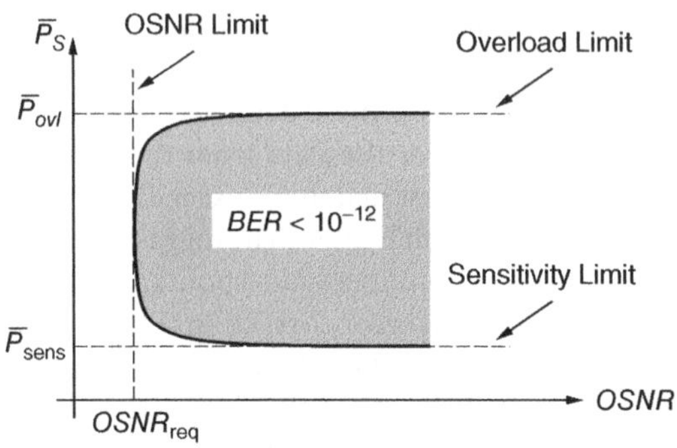

Figure 4.14 OSNR, sensitivity, and overload limits.

above 10^{-12}. Of course, this BER profile is just what we discussed in Section 4.4 and illustrated with the BER plots in Figs 4.9 and 4.10.

Now, if we move our vertical line further and further to the left, which means that we decrease the OSNR (increase the optical noise), the corresponding BER plot changes its shape. The BER floor rises and the dynamic range shrinks, up to the point where it becomes zero. At this point, we have reached the *OSNR limit*. Beyond this limit, we cannot meet the BER, regardless of the optical power available (cf. Eq. (I.16)). It is near this OSNR limit that transmission systems with many in-line amplifiers operate. [→ Problem 4.13.]

In Section 4.4 we became accustomed to plot the BER against the received optical power (cf. Figs 4.9 and 4.10). For amplified systems, however, we are interested in a plot that shows the BER against the received OSNR [49]. An example of such a plot is shown in Fig. 4.15. Whereas the conventional BER plot corresponds to a vertical slice through Fig. 4.14, this plot corresponds to a horizontal slice through the same figure. The required OSNR for a given reference BER can be read directly from this BER plot. From Fig. 4.15, we find $OSNR_{req} = 20.7$ dB at $BER = 10^{-12}$.

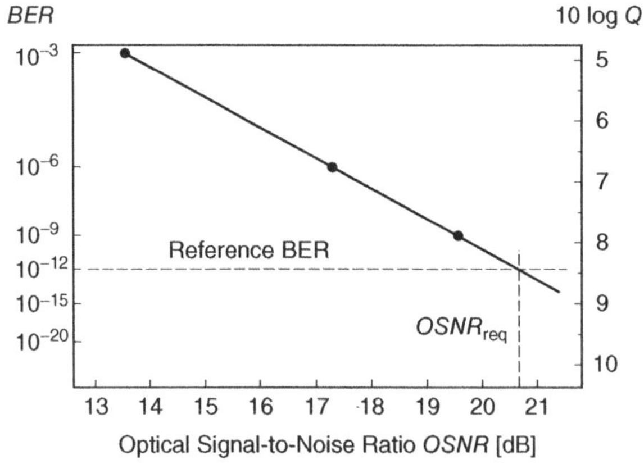

Figure 4.15 BER versus OSNR plot in the (10 log *OSNR*, −10 log *Q*) coordinate system.

Equation (4.53) suggests that for an NRZ direct-detection receiver Q is proportional to $\sqrt{OSNR}$. As before, we can take advantage of this knowledge and choose a coordinate system (or graph paper) in which this relationship is represented by a straight line. Choosing 10 log $OSNR$ (OSNR in dB) for the x-axis and -10 log Q for the y-axis does the trick (cf. Fig. 4.15). The main differences between this BER plot and the conventional BER plot are that the x-axis represents OSNR instead of power, that the expected slope of the plot is halved (because of the square root), and that the plot has no rising part (a high OSNR does not cause overload).

Noise-Loaded BER Measurement. To measure the BER as a function of the OSNR, we need to extend the setup in Fig. 4.11 by adding an optical noise source with variable output power. This extra noise is added to the signal from the transmitter with an optical coupler. The resulting noisy optical signal passes through an attenuator with power meter feeding the receiver under test [47]. By adjusting the power of the optical noise relative to the transmitted signal, we can set the OSNR and by adjusting the attenuator in front of the receiver we can set the received optical power. Thus, we can measure the BER for any combination of $OSNR$ and $\overline{P}_S$, that is, for any point in the plane of Fig. 4.14. The BER measured in this way is known as the *noise-loaded BER*.

Optical Reference Bandwidth. Earlier in this section, we defined OSNR as the optical signal power divided by the optical ASE noise power in the bandwidth BW_O. Clearly, the choice of this bandwidth affects the OSNR value. With the ASE noise being approximately white, a doubling of this bandwidth reduces the OSNR by 3 dB. In practice, a 0.1-nm optical bandwidth, which corresponds to $BW_O = 12.5$ GHz at $\lambda = 1.55$ μm, has become a common choice, regardless of the bit rate [50].

Figure 4.16 illustrates the spectrum of the NRZ signal and the noise relative to the 0.1-nm optical bandwidth. Whereas the signal spectrum of a 2.5 Gb/s

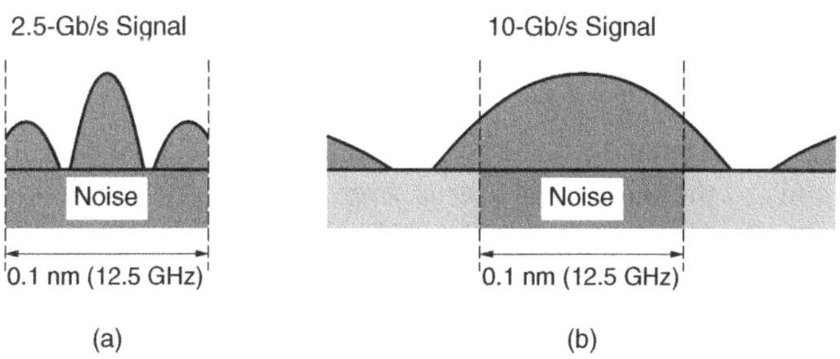

Figure 4.16 Definition of OSNR in a 0.1-nm optical bandwidth for (a) a 2.5 Gb/s and (b) a 10 Gb/s NRZ signal.

Table 4.9 Required optical signal-to-noise ratio (in 0.1 nm). Only signal–spontaneous beat noise is taken into account and high extinction is assumed (Eq. (4.53)).

Parameter	Symbol	10 Gb/s	40 Gb/s	160 Gb/s
OSNR w/o FEC	$OSNR_{req}$ @ $BER = 10^{-12}$	14.7 dB	20.7 dB	26.8 dB
OSNR with FEC	$OSNR_{req}$ @ $BER = 10^{-4}$	9.2 dB	15.2 dB	21.2 dB

signal fits mostly into the 0.1-nm bandwidth, a large part of the 10-Gb/s spectrum (and an even larger part of the 40-Gb/s spectrum) is truncated. To avoid this truncation, the signal power for the OSNR calculation is defined as the *total* power, not restricted to the 0.1-nm bandwidth; only the noise power is restricted to the 0.1-nm optical (noise) bandwidth. Thus, OSNR is not so much a signal-to-noise ratio but rather a signal-to-noise *density* ratio [50].

Bit-Rate Dependence. Let us evaluate the simple expression for $OSNR_{req}$ in Eq. (4.53) for three different bit rates and two different reference BERs to get a better feeling for the numerical values. The results are shown in Table 4.9. For this example we assume that a system without error correction requires $BER = 10^{-12}$, whereas a system with error correction is happy with $BER = 10^{-4}$ at the decision circuit, which is then boosted to $BER = 10^{-12}$ after error correction. As before, the noise bandwidth of the receiver is taken to be $0.75B$.

Table 4.9 makes clear that the required OSNR depends on the bit rate, increasing by 6 dB for every quadrupling in speed. This dependence is a direct consequence of the convention to measure OSNR in a *fixed* optical bandwidth, regardless of the bit rate. In contrast, the required electrical SNR at the decision circuit is bit-rate independent and is measured over a bandwidth that is large compared with the receiver bandwidth.

Required-OSNR Degradations. The required OSNR, as calculated with Eq. (4.53), is rather optimistic. In practice, the received signal has a finite extinction ratio, that is, the zeros have a non-zero power. This results in a reduced signal swing and in signal–spontaneous beat noise for the zeros. Both effects push the required OSNR to a higher value (cf. Eq. (I.17)). [→ Problem 4.14.]

Another effect neglected in Eq. (4.53) is the spontaneous–spontaneous beat noise, which adds to the noise of the zeros, even (and especially) when the extinction is high. When including this effect, the required OSNR goes up further [6, 44, 46] (cf. Eqs. (I.18) and (I.19)). Table 4.10 lists example values calculated for an extinction ratio of 10 and taking spontaneous–spontaneous beat noise into account. To further improve the accuracy of the required-OSNR calculations, the non-Gaussian noise statistics must be taken into account [51, 52]. [→ Problems 4.15 and 4.16.]

Table 4.10 Required optical signal-to-noise ratio (in 0.1 nm). Signal–spontaneous and spontaneous–spontaneous beat noise are taken into account and $ER = 10$ (Eq. (I.19)).

Parameter	Symbol	10 Gb/s	40 Gb/s	160 Gb/s
OSNR w/o FEC	$OSNR_{req}$ @ $BER = 10^{-12}$	18.5 dB	24.5 dB	30.5 dB
OSNR with FEC	$OSNR_{req}$ @ $BER = 10^{-4}$	13.1 dB	19.0 dB	25.0 dB

Table 4.11 Required optical signal-to-noise ratio (in 0.1 nm). Measured in a back-to-back configuration.

Bit rate (Gb/s)	BER	Required OSNR (dB)	Reference
10	10^{-12}	22.4	[20]
10	10^{-12}	20.6	[20]
10	10^{-4}	15.6	[20]
10	10^{-4}	15.9	[20]
40	10^{-9}	24.0	[43]
160	10^{-9}	28.0	[43]

Table 4.11 lists required-OSNR values reported in the literature that were measured under back-to-back conditions. These values are a couple of decibels higher than those in Table 4.10.

Required OSNR and Modulation Formats. The required OSNR also depends on the modulation format. The NRZ format and a direct-detection receiver were assumed thus far. In comparison, the RZ format requires less OSNR because the ones carry more optical power for a given *average* signal power. In practice, the RZ format requires about 1 to 3 dB less OSNR than NRZ [53]. (Interestingly, if an optical matched filter is used, the RZ format has no OSNR advantage over the NRZ format [54].)

The differential phase-shift keying (DPSK) format requires about 3 dB less OSNR when compared with the NRZ-OOK format, provided a receiver with balanced detector is used [55] (cf. Section 3.5). This improvement is due to the steep inner tails of the non-Gaussian noise distribution associated with this modulation format [26].

Higher-order modulation formats, such as 4-PAM, QPSK, and 16-QAM, require more OSNR than two-valued formats, such as NRZ or BPSK [56].

Required OSNR and Electrical Bandwidth. Equation (4.53) suggests that the required OSNR can be minimized by making the noise bandwidth, BW_n, of the receiver as small as possible. This makes sense, because a receiver with a smaller noise bandwidth picks up less noise from the optical amplifiers. However, there are limits to how small the receiver bandwidth can be made without causing undue signal distortions. We discuss this trade-off in Section 4.8 when we study the impact of the receiver bandwidth.

Another requirement for a good receiver is that its noise bandwidth must remain independent of the received optical power. In amplified systems the received power can become quite large and a bandwidth increase at high power levels would be detrimental to the required OSNR.

In summary, for an unamplified transmission system, a high receiver *sensitivity* is the primary goal. To this end, the receiver must be designed to have a low *input-referred noise current*, $i_{n,\mathrm{ckt}}^{rms}$. For an optically amplified transmission system, a low *required OSNR* at the receiver is the primary goal. To this end, the receiver must be designed for a small and signal-independent *noise bandwidth*, BW_n.

4.7 Power Penalty

The power-penalty concept is useful to quantify impairments in the receiver, the transmitter, and the fiber. Moreover, it permits us to derive specifications for the building blocks that make up a communication system.

Definition. Imagine a comprehensive model of an optical communication system that accurately predicts the BER based on many system parameters: the transmit power, extinction ratio, fiber loss, fiber dispersion, fiber nonlinearity, detector responsivity, detector noise, circuit noise, receiver intersymbol interference (ISI), decision threshold, and so forth. Such a model would be very complex and unwieldy to deal with in practice. In contrast, the simple system model that we developed in Section 4.4 predicts the BER based on just a few major system parameters: the transmit power, fiber loss, detector responsivity, detector noise, and circuit noise. Clearly, this simple model cannot be very accurate because it makes many idealizing assumptions such as an infinite extinction ratio, no fiber dispersion, no fiber nonlinearity, no receiver ISI, an optimum decision threshold, and so forth.

Fortunately, there is a straightforward way to improve on the simple model with first-order "corrections" that quantify the neglected impairments without making the model overly complex. To do that, we choose *one* impairment that is not modeled by the simple model (say a deviation from the optimum decision threshold) and ask: "By how much do we have to increase the transmit power to overcome this impairment?" This power increment is the *power penalty* due to the impairment under consideration. We repeat this procedure for other impairments and in this way obtain power penalties for all of them.

Table 4.12 Examples of impairments leading to power penalties.

Transmitter:	Extinction ratio
	Relative intensity noise
Fiber:	Dispersion
	Nonlinearity
Detector:	Dark current
TIA/MA:	Intersymbol interference
	Crosstalk
	Nonlinearity
CDR:	Decision-threshold offset
	Decision-threshold noise
	Sampling-time offset
	Sampling-time jitter

In the presence of multiple impairments, we can multiply the power penalties (or add their dB values) to obtain an estimate of the total power penalty. This is a convenient and practical way of analyzing a transmission system, but we have to keep in mind that by considering only one impairment at a time we neglect possible interactions between them.

In summary, the power penalty PP for a particular impairment is defined as the increase in average transmitted power necessary to achieve the same BER as in the absence of the impairment. Power penalties usually are expressed in decibels using the conversion rule 10 log PP. Table 4.12 gives examples of impairments in various parts of the optical communication system. Subsequently, we calculate the power penalty for some of these impairments.

Example 1: Decision-Threshold Offset. To illustrate the power-penalty concept, let us make an example and calculate the power penalty for the case when V_{DTH} is not at its optimum value, that is, when we have a decision-threshold offset. For this example we assume an equal amount of noise on the zeros and ones. Figure 4.17(a) shows the situation where the threshold is at its optimum value, V_{DTH}, located at the midpoint in between the zero and one levels. Figure 4.17(b) shows the situation where the decision threshold voltage is too high. We can write the errant threshold voltage as

$$V'_{DTH} = V_{DTH} + \delta v_S^{pp}, \tag{4.56}$$

where V_{DTH} is the optimum threshold and δ is the threshold offset relative to the signal swing. The high threshold causes many ones to be misinterpreted as zeros, thus significantly increasing the BER.

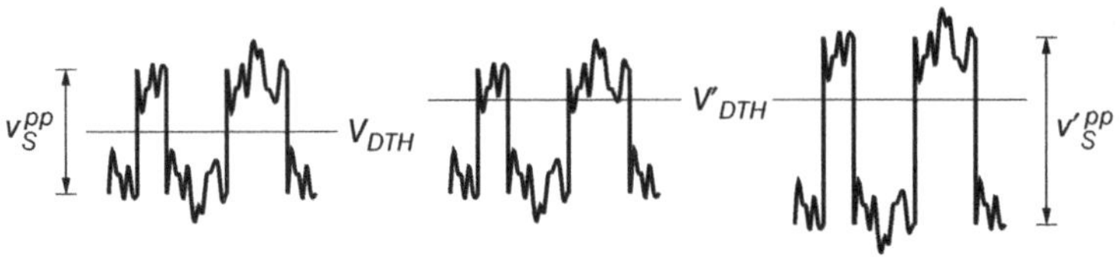

Figure 4.17 Power penalty due to decision-threshold offset: (a) without offset, (b) with offset, and (c) with offset and increased signal swing to restore the original BER.

To restore the original BER, we need to increase the signal swing as shown in Fig. 4.17(c). Specifically, we must increase the signal swing from v_S^{pp} to $v_S^{\prime pp}$ such that the difference between the one level and the decision threshold becomes the same as in Fig. 4.17(a). Assuming an AC coupled system where the signal swings from $-v_S^{pp}/2$ to $+v_S^{pp}/2$ (as opposed to from 0 to v_S^{pp}) and thus grows symmetrically around the zero level, the new swing becomes

$$v_S^{\prime pp} = v_S^{pp} + 2\delta v_S^{pp} = (1 + 2\delta)v_S^{pp}. \tag{4.57}$$

With this increased swing, the probability of misinterpreting a one as a zero becomes identical to that in Fig. 4.17(a), approximately restoring the BER. (The probability of misinterpreting a zero as a one becomes very small in Fig. 4.17(c). Thus, the overall BER is now only about half of the original BER in Fig. 4.17(a), but because BER is an exponentially "steep" function of swing, this difference does not alter the power penalty by much.)

In summary, to restore the original BER, we need to increase the signal voltage swing at the decision circuit by $v_S^{\prime pp}/v_S^{pp} = 1 + 2\delta$. Assuming linearity in the fiber and the receiver, increasing the (transmitted) optical power by the same amount will do just that. Thus, the power penalty for the decision-threshold offset δ is

$$PP = 1 + 2\delta. \tag{4.58}$$

For example, a 10% decision-threshold offset causes a power penalty of 0.79 dB ($PP = 1.2$).

Now let us see how we can use the power-penalty idea to find receiver specifications. If we solve Eq. (4.58) for δ, we get

$$\delta = \frac{PP - 1}{2}. \tag{4.59}$$

This means that if the largest acceptable power penalty is PP, we must control the decision threshold to a precision better than the δ given in Eq. (4.59). For example, given a maximum acceptable power penalty of 0.05 dB ($PP = 1.0116$), the decision-threshold offset must be less than 0.58% of the signal swing.

Example 2: Dark Current. In Chapter 3, we briefly discussed the detector's dark current and how it interferes with the received signal. Now we have the necessary tool to quantify this impairment. The dark current by itself just adds an offset to the signal but leaves its swing unchanged. As long as the receiver can compensate for this offset, there is no power penalty for this effect. However, the *noise* associated with the dark current will add to the receiver noise and cause a power penalty. Let us calculate it!

According to Eq. (3.7), the dark current I_{DK} of a p–i–n detector produces the shot noise

$$\overline{i_{n,DK}^2} = 2qI_{DK} \cdot BW_n. \tag{4.60}$$

This noise power adds to the receiver noise, which we assume is dominated by the circuit noise $\overline{i_{n,\text{ckt}}^2}$. (Neglecting the detector noise overestimates the power penalty somewhat.) So, the dark-current noise increases the noise power by

$$\frac{\overline{i_{n,\text{ckt}}^2} + \overline{i_{n,DK}^2}}{\overline{i_{n,\text{ckt}}^2}} = 1 + \frac{2qI_{DK} \cdot BW_n}{\overline{i_{n,\text{ckt}}^2}}. \tag{4.61}$$

To overcome this additional noise we must increase the signal swing at the decision circuit in proportion to the increase in rms noise. This preserves the Q parameter and thus the BER. The power penalty follows as

$$PP = \sqrt{1 + \frac{2qI_{DK} \cdot BW_n}{\overline{i_{n,\text{ckt}}^2}}}. \tag{4.62}$$

With the example values for our 10 Gb/s receiver ($i_{n,\text{ckt}}^{rms} = 1\ \mu\text{A}$, $BW_n = 7.5$ GHz) and a worst-case dark current of 2 nA, we find the power penalty to be 0.000 0104 dB ($PP = 1.000\,0024$). As expected from our simple argument in Chapter 3, it is exceedingly small.

To find a specification for the dark current we ask: "What is the maximum dark current consistent with a given power penalty?" Solving Eq. (4.62) for I_{DK} reveals [14]

$$I_{DK} < (PP^2 - 1) \cdot \frac{\overline{i_{n,\text{ckt}}^2}}{2q \cdot BW_n}. \tag{4.63}$$

With the same example values as before, we find that the dark current must be less than 9.7 μA to keep the power penalty below 0.05 dB ($PP = 1.0116$). For a p–i–n detector this is easily satisfied.

The aforementioned example illustrates that dark current is no problem for high-speed p–i–n receivers. For APD receivers, the noise due to the dark current is given by $\overline{i_{n,DK}^2} = FM^2 \cdot 2qI_{DK} \cdot BW_n$, where I_{DK} is now the *primary* dark current of the APD (cf. Eq. (3.14)). This noise is significantly larger than for the

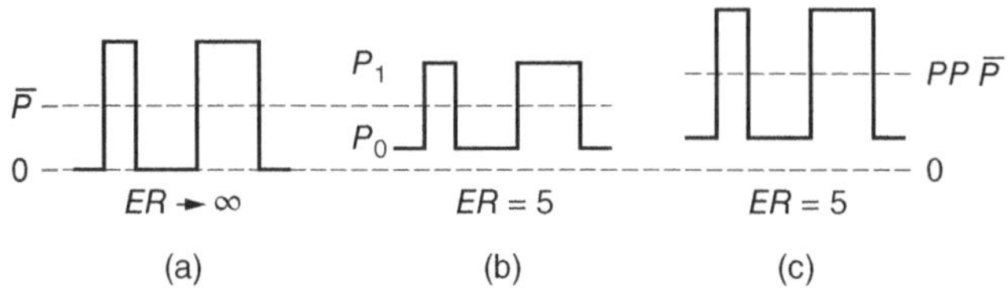

Figure 4.18 Power penalty due to finite extinction ratio: (a) with infinite ER, (b) with $ER = 5$, and (c) with $ER = 5$ and increased average power (1.5×) to restore the original signal swing.

p–i–n receiver and may present a practical impairment (cf. Eq. (I.22)). Dark current also is important in low bit-rate receivers (< 10 Mb/s) where the circuit noise is much lower than in our example and Eq. (4.63) requires a correspondingly smaller dark current. For more information on the dark current and its impact on the receiver performance, see Ref. [14]. [→ Problem 4.17.]

Example 3: Extinction Ratio. As we know, optical transmitters do not go *completely* dark when a zero is transmitted. This effect is quantified by the *extinction ratio* (ER), which is defined as

$$ER = \frac{P_1}{P_0}, \tag{4.64}$$

where P_0 is the optical power emitted for a zero and P_1 is the power for a one.[5] Thus, an ideal transmitter would have an infinite ER. The ER usually is expressed in decibels using the conversion rule 10 log ER.

How large is the power penalty caused by a finite ER? Figure 4.18(a) and (b) illustrates how decreasing the ER reduces the optical signal swing (optical modulation amplitude), $P_1 - P_0$, while the average power $\overline{P} = (P_1 + P_0)/2$ is kept constant. To restore the original signal swing, we have to increase the average transmitted power, as shown in Fig. 4.18(c). The power penalty PP due to a finite extinction ratio can be derived easily as (see solution to Problem 4.8 on p. 504 or [7])

$$PP = \frac{ER + 1}{ER - 1}. \tag{4.65}$$

For example, an extinction ratio of 10 dB ($ER = 10$) causes a power penalty of 0.87 dB ($PP = 1.22$), while an extinction ratio of 6 dB ($ER = 4$) causes a power penalty of 2.2 dB ($PP = 1.67$).

System Dependence of Power Penalties. Examples 1 to 3 tacitly assumed that we are dealing with an *unamplified* optical transmission system. In our first example (decision-threshold offset), we assumed that the act of increasing the

5 Sometimes ER is defined as P_0/P_1. Here, we follow [57] and use P_1/P_0, which results in $ER > 1$.

transmit power does not introduce new impairments, such as an increase in system noise. Whereas the noise in unamplified transmission systems with p–i–n detectors is approximately constant, it is *not* constant in systems with optical amplifiers, APDs, or both. In those systems, the noise level increases with increasing transmit power, leading to a larger power penalty than given in Eq. (4.58).

In the second example (dark current), we assumed that the circuit noise dominates the system noise. Although this is true for unamplified transmission systems with p–i–n detectors, in optically amplified transmission systems most of the noise originates from the optical amplifiers, leading to a smaller power penalty than given in Eq. (4.62).

In the third example (extinction ratio), like in the first example, the act of increasing the transmit power to compensate for the finite extinction ratio adds noise in amplified systems. This necessitates a larger power increase to compensate for this noise, leading to a larger power penalty than given in Eq. (4.65). Clearly, we need to know the type of transmission system when calculating power penalties.

4.8 Intersymbol Interference and Bandwidth

Now, we turn to the question of how large the receiver bandwidth should be made. To get a feeling for the answer, consider the following dilemma. If we make the receiver bandwidth wide enough such that the signal waveform remains undistorted, the signal picks up a lot of noise, which translates into a low receiver sensitivity (cf. Section 4.4). Alternatively, if we make the receiver bandwidth narrow such that much of the noise gets filtered out, the signal suffers from a distortion known as *intersymbol interference* (ISI). Like noise, ISI negatively impacts the sensitivity. ISI reduces the signal swing in a bit-pattern dependent way, with the "01010101 …" pattern typically resulting in the lowest swing. We can conclude from this line of reasoning that there must be an *optimum receiver bandwidth* for which the sensitivity (or the required OSNR) is best.

As a rule of thumb, the optimum 3-dB bandwidth for NRZ receivers is about 60 to 70% of the bit rate, B [2, 17]:

$$BW_{3dB} \approx 0.6B \dots 0.7B \tag{4.66}$$

or $BW_{3dB} \approx \frac{2}{3}B$ to pick a specific number in this range. Later in this section, we justify this rule and also discuss the optimum bandwidth for RZ receivers.

Intersymbol Interference. Before taking on the general case of a signal with distortion and noise, let us first study the simpler case of a signal with distortion due to the finite receiver bandwidth alone. If we apply an ideal NRZ signal to the

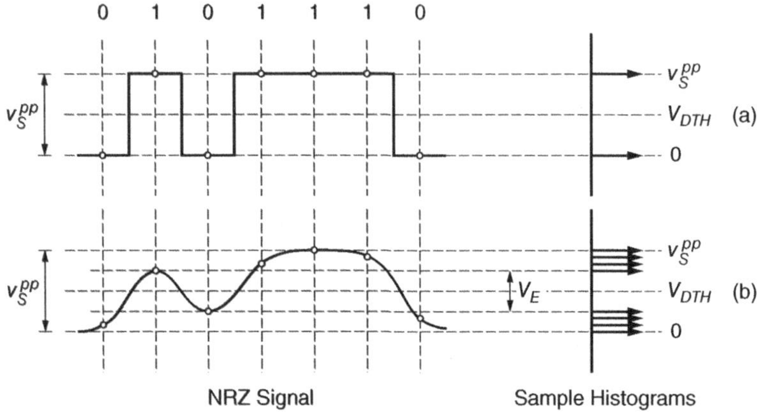

Figure 4.19 NRZ signal at the decision circuit of (a) an infinite-bandwidth receiver and (b) a finite-bandwidth receiver.

input of an *infinite-bandwidth* receiver,[6] we obtain, not surprisingly, an undistorted NRZ signal at the output, as shown in Fig. 4.19(a). The histogram of the sampled signal values, shown on the right-hand side, consists of two lines (Dirac pulses), one for the zero level and one for the one level. Now, applying the same ideal NRZ signal to the input of a *finite-bandwidth* receiver, we obtain a (linearly) distorted NRZ signal at the output, as shown in Fig. 4.19(b). The sampled values of this distorted signal do not only depend on the current bit, but also on the bits, or more generally the symbols, before and after the current bit (or symbol), in other words we have intersymbol interference.

Upon closer inspection of the waveform in Fig. 4.19(b), we find that in this example each sample value depends on three bits: the bit being sampled, the bit before that, and the bit after that. For example, if the sampled bit is a one, the four bit sequences "... 010 ...," "... 011 ...," "... 110 ...," and "... 111 ..." result in slightly different sample values for the middle bit. In the "111" case, we get the largest value. In the "010" case, which represents an isolated one, we get the lowest value. In the "011" and "110" cases, we get intermediate values, with the latter being somewhat larger. Clearly, if the amount of ISI depends on only the *two* adjacent bits, the two histograms have at most four lines, as shown in Fig. 4.19(b). In the general case where the amount of ISI depends on n bits, each histogram has 2^n lines, where some lines may fall on top of each other.

How can we quantify the distortion resulting from the finite bandwidth? The *vertical eye opening*, V_E, is defined as the difference between the smallest sampled value for a one and the largest sampled value for a zero. The smaller the vertical eye opening, the more distorted the signal is. For the signal in Fig. 4.19(b),

6 With *infinite bandwidth* we mean that the amplitude response is flat and the phase response is linear up to frequencies well above the bit rate.

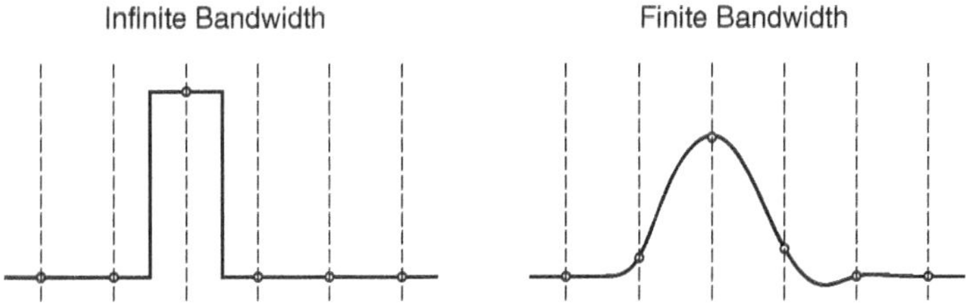

Figure 4.20 Single-bit response.

V_E is half of v_S^{pp}, that is, the vertical eye opening is 50%. As the name suggests, V_E appears in the eye diagram as the vertical opening at the sampling instant. For a discussion of the eye diagram and its openings, see Appendix B.

Single-Bit Response. Analyzing ISI for a random bit sequence, as we did in Fig. 4.19, can be tricky, especially when the ISI depends on many bits. So, let us take a step back and see what happens to a single one bit embedded in a long string of zeros, that is, to the sequence ... 00000100000 ... An example, consistent with the waveforms in Fig. 4.19, is shown in Fig. 4.20. The response on the right is known as the *single-bit response* (SBR).

In this example, the finite bandwidth has three effects on the signal samples: The one-bit pulse is attenuated by about 25%, about 8% of the one-bit pulse spills over into the preceding bit slot, and about 17% spills over into the succeeding bit slot. All the other signal samples are zero. The spilling into a bit slot before the main pulse (the cursor) is known as *precursor ISI* and the spilling into a bit slot after the main pulse is known as *postcursor ISI*.

Knowing the SBR of a linear channel is enough to figure out the ISI for any bit sequence. All we have to do is to superimpose time-shifted copies of the SBR. Going back to Fig. 4.19, we can do the following analysis: the first sample consists of the precursor ISI of the succeeding one pulse (8%), the second sample consists of the attenuated one pulse (75%), the third sample consists of the postcursor ISI of the preceding pulse *plus* the precursor ISI of the succeeding pulse (17% + 8%), the fourth sample consists of the attenuated one pulse *plus* the precursor ISI of the succeeding pulse (75% + 8%), the fifth sample consists of the attenuated one pulse *plus* the postcursor ISI of the preceding pulse *plus* the precursor ISI of the succeeding pulse (75% + 17% + 8%), and so forth.

You may have noticed that the three contributions to the fifth sample add up to 100%. This is so because this sample is in the middle of three consecutive ones. For an NRZ signal with all ones, there is no difference between the infinite- and finite-bandwidth cases. Thus, the pulse attenuation has to be counterbalanced by all the pre- and postcursor ISI taken together. In the case of an RZ signal, however, the pulse train corresponding to an all-one sequence does

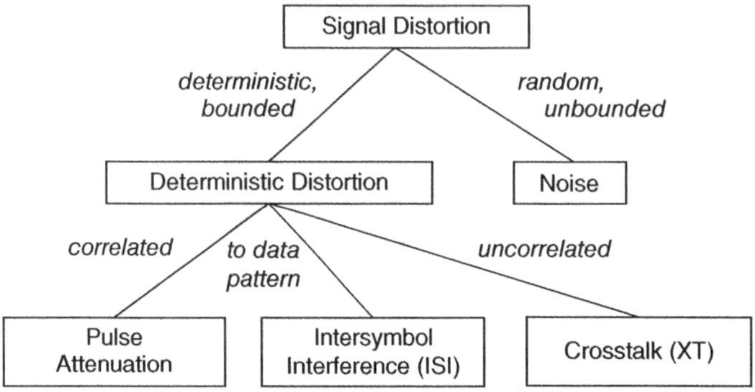

Figure 4.21 Relationship between noise, intersymbol interference, and crosstalk.

depend on the bandwidth. There, the pulse attenuation is larger than the sum of all ISI components.

ISI, Noise, and Crosstalk. Both ISI and noise distort the signal and result in eye closure. But there are important differences. Whereas random noise is unpredictable, ISI is deterministic, that is, it can be calculated, if the bit sequence and the transfer function (or SBR) are known. Whereas random noise has a Gaussian histogram, ISI has a histogram consisting of discrete lines. Whereas random noise is unbounded (as a result of its Gaussian distribution), ISI is bounded (only a finite number of bits have a finite impact on the current bit). Figure 4.21 shows how we can divide signal distortions into deterministic distortions, such as ISI, and random distortions (noise).

Apart from the finite receiver bandwidth that our discussion focused on, there are several other mechanisms that cause ISI. For example, a low-frequency cutoff that is insufficiently low (e.g., because the AC-coupling capacitors are too small) causes baseline wander. Other culprits are insufficient phase linearity (or, equivalently, excessive group-delay variation) and reflections on circuit board traces and cables due to impedance mismatch. Besides these linear distortions, a receiver produces nonlinear distortions when driven into overload.

All forms of ISI are correlated to the data pattern they interfere with. In contrast, *crosstalk* (XT), another type of signal distortion, is *not* correlated to the data pattern. Crosstalk originates, for example, from another communication channel, from a switching power supply, or from switching activity in a digital block on the same chip. Like ISI, crosstalk is deterministic and bounded and as such its effect on the signal can be quantified by the reduction of the vertical eye opening, V_E.

Example: Butterworth Receiver. Figure 4.22 illustrates signal distortions in a 10-Gb/s receiver with a second-order Butterworth response (characterized by

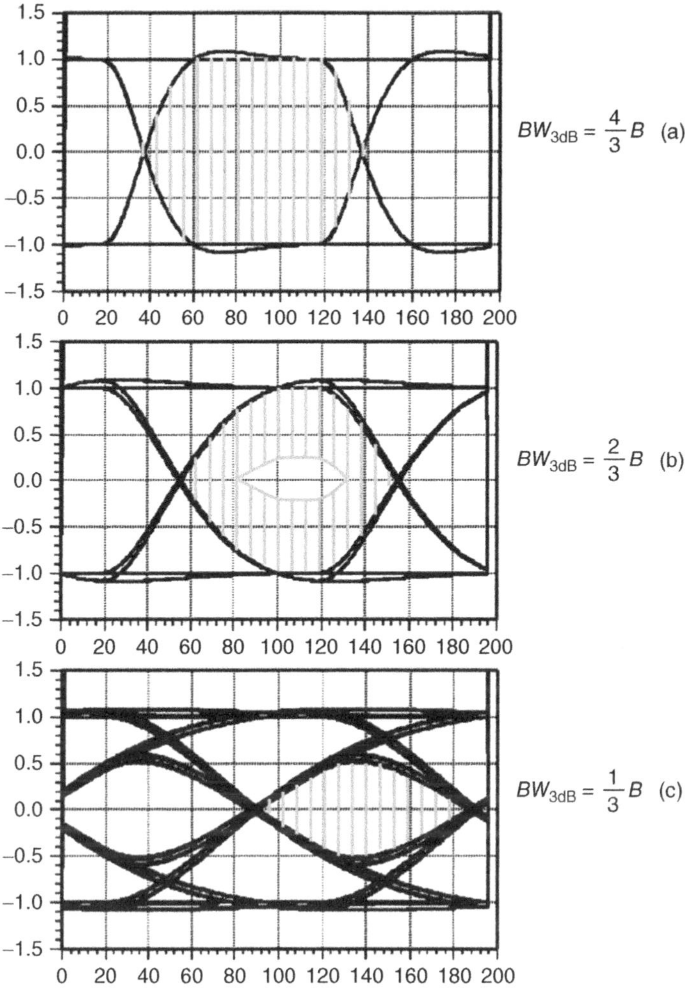

Figure 4.22 ISI and noise in the Butterworth receiver.

a maximally flat amplitude response). For this example, we assume again that the received input signal is an ideal NRZ waveform. The output waveforms for three different receiver bandwidths are shown from top to bottom in the form of eye diagrams (cf. Appendix B).

For now, we ignore the gray stripes in the eye diagrams (which symbolize noise). Figure 4.22(a) shows the eye diagram for a wideband receiver with a 3-dB bandwidth of $\frac{4}{3}$ the bit rate, that is, about twice the optimum bandwidth given in Eq. (4.66). As expected, we get a clean eye with little distortion. If we sample this eye around the 90-ps mark, the eye opening is 100%. (The slight overshoot does not impact the eye opening.) Figure 4.22(b) shows the eye diagram for a receiver with the optimum bandwidth, $\frac{2}{3}$ of the bit rate. This eye too is clean and has an opening of 100%, if sampled near 110 ps. Finally, Fig. 4.22(c) shows the eye diagram of a narrowband receiver with only half the optimum bandwidth. In this case, we observe severe distortions resulting in a partially closed eye. In particular, for the optimum sampling instant at 140 ps, the eye opening is only about 50%. Note that for this sampling instant, we get roughly four distinct values for the one level and four distinct values for the zero level. Thus, we have a situation similar to that in Fig. 4.19(b) and we can conclude that the ISI depends mostly on *two* adjacent bits. If we look more closely, however, we see that the four traces in the eye are really four bands with a fine structure, indicating that there are additional bits that contribute a small amount of ISI.

Power Penalty Due to ISI. ISI alone does not cause any bit errors in the receiver as long as the eye has some opening left ($V_E > 0$) and the decision threshold is located in that opening. Once we include noise, however, errors do occur and then ISI has the unfortunate effect of increasing the BER. To bring the BER back down, we need to increase the power. In other words, ISI causes a power penalty. How large is this penalty?

Figure 4.23(a) shows the histograms of the sampled signal values for the case of noise only (copied from Fig. 4.4). Figure 4.23(b) shows the histograms for the case of ISI, but no noise (copied from Fig. 4.19). In the case of ISI *and* noise, the original noise histograms are broadened (smeared out) and shifted closer together, as shown in Fig. 4.23(c). This results in a larger overlap of the two histograms and thus a larger BER. Mathematically, the composite distributions in Fig. 4.23(c) are obtained by convolving the corresponding distributions in Fig. 4.23(a) and (b). In our example, the composite distributions consist of a superposition of four shifted Gaussians each. To find the power penalty, we must calculate the shaded areas under the tails of Fig. 4.23(c) and determine by how much we have to increase v_S^{pp} to bring this area back to what it was in Fig. 4.23(a).

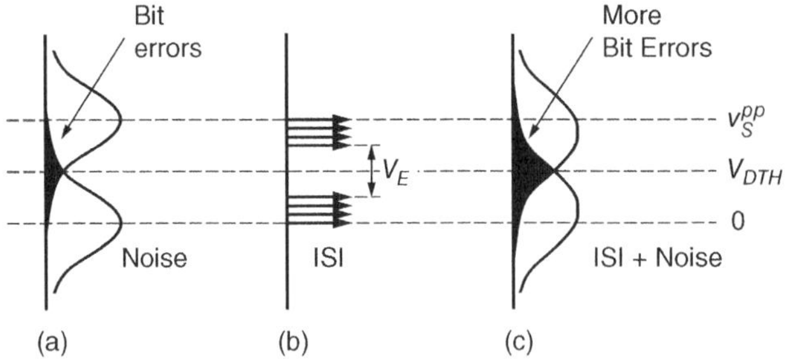

Figure 4.23 Histograms of a signal with (a) noise only, (b) ISI only, and (c) ISI *and* noise.

The power-penalty calculation outlined earlier is rather complicated and a simple but useful approximation would be welcome. Rather than calculating the power penalty for a random bit sequence, let us take the worst-case bit sequence, usually "01010101 ..." Now, the two distributions in Fig. 4.23(b) simplify to just two lines: one for the zeros and one for the ones. These two worst-case lines are separated by V_E, the vertical eye opening for random data. Thus, we end up in the same situation as for the infinite-bandwidth case, except that the signal swing is reduced from v_S^{pp} to V_E. The convolutions to calculate the composite distributions are now trivial. All the ISI does, in this worst-case scenario, is shift the Gaussians closer together such that the spacing of their mean values reduces from v_S^{pp} to V_E (their standard deviations remain the same). Everything we learned about calculating the BER for an undistorted waveform with noise (cf. Section 4.2) is still valid in the presence of ISI. All we have to do is replace v_S^{pp} with V_E:

$$BER \approx \int_Q^{\infty} \text{Gauss}(x)\, dx \qquad \text{with} \qquad Q = \frac{V_E}{v_{n,0}^{rms} + v_{n,1}^{rms}}. \tag{4.67}$$

Figure 4.24 illustrates the calculation of the power penalty using the worst-case approximation. In Fig. 4.24(a), the signal from the infinite-bandwidth receiver without distortion has a vertical eye opening of v_S^{pp}. In Fig. 4.24(b), the eye opening of the finite-bandwidth receiver is reduced to V_E because of ISI. To restore the original BER (for the worst-case waveform), we must increase the full signal swing of the distorted signal to $v_S'^{pp}$ such that its vertical eye opening V_E' becomes equal to v_S^{pp}, as shown in Fig. 4.24(c). The power penalty due to ISI is given by the ratio $v_S'^{pp}/v_S^{pp}$, which equals $v_S'^{pp}/V_E'$, and thus

$$PP = \frac{v_S^{pp}}{V_E}. \tag{4.68}$$

Because this power penalty is based on the worst-case bit sequence it somewhat overestimates the actual penalty.

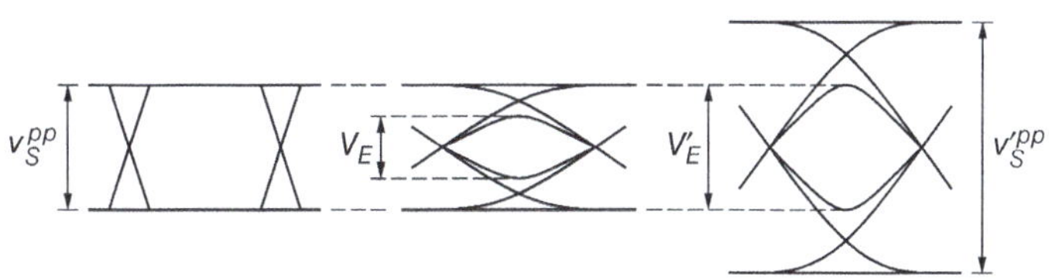

Figure 4.24 Power penalty due to ISI: (a) infinite-bandwidth receiver, (b) finite-bandwidth receiver, and (c) finite-bandwidth receiver with increased signal swing to restore the original BER.

To illustrate Eq. (4.68), let us estimate the power penalties for our Butterworth receiver by inspecting the vertical eye openings in Fig. 4.22. For the bandwidths $BW_{3\text{dB}} = \frac{4}{3}B$ and $BW_{3\text{dB}} = \frac{2}{3}B$, we find no power penalty ($PP = 1$) and for the bandwidth $BW_{3\text{dB}} = \frac{1}{3}B$, we find a 3-dB power penalty ($PP = 2$). The same values can be found analytically from the step response of the Butterworth transfer function (cf. Eq. (I.25)). [→ Problem 4.18.]

Separating ISI from Noise. Histograms captured with an oscilloscope contain a mixture of noise, ISI, and crosstalk. How can we experimentally determine the amount of noise, as measured by $v_{n,0}^{rms}$ and $v_{n,1}^{rms}$, and the amount of deterministic signal distortions, as measured by V_E? Here is a choice of three methods:

- *Two-Tests Method*: First, test the system with an all-zero (or all-one) data pattern to determine the noise. Then, test the same system with a repetitive data pattern to determine the vertical eye opening. Triggering on the pattern clock and using the averaging feature of the oscilloscope suppresses the noise in the latter measurement. The weakness of this method is that crosstalk, if present, is erroneously included in the noise measurements, exaggerating the noise values.
- *Tail-Fit Method*: Measure the histograms of the zeros and ones (containing all distortion components) and fit two (independent) Gaussians to their inner tails (see Fig. 4.25(a)). The standard deviations of the two Gaussians provide an estimate for the noise on the zeros and ones. The separation of the centers of the two Gaussians provides an estimate for the vertical eye opening. The weakness of this method is that low-probability deterministic distortions are counted as noise.

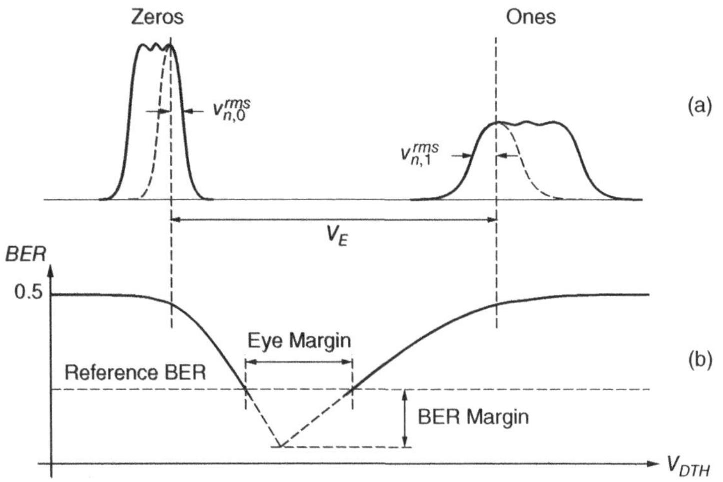

Figure 4.25 (a) Histograms (solid) with Gaussians fitted to the inner tails (dashed); (b) vertical bathtub curve (solid) with a model-based extrapolation (dashed).

- *BERT-Scan Method*: Measure the BER while sweeping the decision threshold, producing a so-called *vertical bathtub curve*, as shown in Fig. 4.25(b) (cf. Appendix B). Then, determine V_E, $v_{n,0}^{rms}$, and $v_{n,1}^{rms}$ by fitting a parametrized mathematical model to the bathtub curve [48]. Roughly speaking, the opening near the top of the bathtub curve corresponds to the vertical eye opening, whereas the slopes to the left and right relate to the noise on the zeros and ones, respectively. The weakness of this method is similar to that listed for the tail-fit method.

Having found V_E, $v_{n,0}^{rms}$, and $v_{n,1}^{rms}$ by one of the aforementioned methods, we can use Eq. (4.67) to estimate the BER. Comparing the estimated BER with the required (reference) BER can provide valuable information about the system margin (see Fig. 4.25(b)) [48].

Optimum Bandwidth for NRZ Receivers. Now we are ready to combine the basic noise-based sensitivity from Eq. (4.21) with the power penalty due to ISI to obtain the actual receiver sensitivity:

$$\overline{P}_{\text{sens}} = PP(BW_{3\text{dB}}) \cdot \frac{Q \cdot i_n^{rms}(BW_{3\text{dB}})}{\mathcal{R}}. \tag{4.69}$$

Both, the power penalty, PP, and the rms noise current, i_n^{rms}, depend on the receiver bandwidth, $BW_{3\text{dB}}$. Specifically, for a fixed white input-referred noise, the rms noise current increases proportional to $\sqrt{BW_{3\text{dB}}}$ and we can rewrite Eq. (4.69) as $\overline{P}_{\text{sens}} \propto PP(BW_{3\text{dB}}) \cdot \sqrt{BW_{3\text{dB}}}$. A graphical representation of this expression, where we have used the power-penalty values of our Butterworth receiver, is shown in Fig. 4.26. The basic noise-based sensitivity is represented by the dashed line sloping up from $P_S - 3.0$ dB to $P_S - 1.5$ dB and to P_S for each doubling of the bandwidth (following the square-root law), where P_S is an arbitrary reference value. The power penalty due to ISI is indicated by the arrow. The actual receiver sensitivity is shown with the solid line. As expected, the best sensitivity is reached near (actually a little bit below) the bandwidth $\frac{2}{3}B$.

That $\frac{2}{3}B$ is about the optimum bandwidth can also be seen directly from the eye diagram. Going back to Fig. 4.22, we now add noise to the eye diagrams.

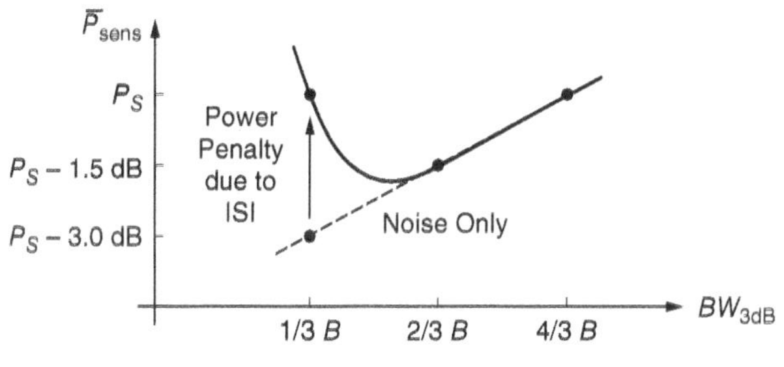

Figure 4.26 Sensitivity as a function of receiver bandwidth.

The gray stripes represent the peak value of the noise voltage. But, what exactly is meant with the *peak* value, isn't Gaussian noise unbounded? Yes, but the trick here is to define the peak value of the noise voltage based on the required BER. For example, if we are ready to accept a BER of 10^{-12}, then we do not care about noise voltages exceeding $7.03v_n^{rms}$ (because this happens only with a probability of 10^{-12}), and so we can take the peak value of the noise voltage as $7.03v_n^{rms}$ or more generally as Qv_n^{rms}.

Now, as we decrease the bandwidth in steps of 2× from the wideband to the optimum-bandwidth to the narrowband receiver in Fig. 4.22, the peak value of the noise reduces in steps of $\sqrt{2}\times$ from 1.00 to 0.71 to 0.50 V. For clarity, only the noise inside the eye is shown; in reality, of course, noise is present on both sides of the signal traces. We observe that whereas the eyes for the wide and narrowband receivers are completely closed by the noise, the eye for the optimum-bandwidth receiver is open at the center. To recover the received data at the desired BER, we must make a decision in the open part of the eye, that is, the decision threshold and sampling instant must define a point in the open part of the eye. In our example, this is only possible for the receiver with the optimum bandwidth: $BW_{3dB} = \frac{2}{3}B$.

Note that we have been using the term "eye opening" in two different senses. Eye opening in the first sense, for which we used the symbol V_E, refers to the eye with ISI but *without* noise. Eye opening in the second sense (as used in the previous paragraph) refers to the eye with ISI *and* noise and is given by $V_E - 2Qv_n^{rms}$. To avoid confusion, we use the term (vertical) eye opening only when referring to V_E and use the term (vertical) eye *margin* instead when referring to $V_E - 2Qv_n^{rms}$ (cf. Appendix B). To recover a bit stream at the desired BER, the vertical eye margin for that BER must be larger than zero.

Although $BW_{3dB} \approx \frac{2}{3}B$ results in the best sensitivity according to Fig. 4.26, it is possible to build a practical receiver with a much smaller bandwidth. From the graph we can estimate that a receiver with $BW_{3dB} = \frac{1}{3}B$ is about 1.5 dB less sensitive than the optimum receiver. This narrowband approach is attractive for optically preamplified receivers at very high speeds where a small loss in sensitivity is acceptable, if in return the receiver can be built from lower speed electronic components. In [58], a 40-Gb/s receiver front-end with a 3-dB bandwidth of only 20 GHz (but good phase linearity up to 35 GHz) has been reported.

For the analysis in Fig. 4.26 we assumed that the rms noise current increases proportional to $\sqrt{BW_{3dB}}$. This is approximately true for optically amplified systems where the input-referred receiver noise is dominated by white detector noise. However, for unamplified systems where the receiver noise is dominated by the circuit noise, the rms noise grows more like $BW_{3dB}^{1.5}$ (cf. Section 6.4). In that case the optimum bandwidth is pushed to a somewhat lower value.

The detailed numerical analysis in [59] finds that the optimum electrical bandwidth for optically preamplified NRZ receivers is in the range from $0.6B$ to

0.8B. In ultradense WDM systems, the optical filters that are used to separate the wavelengths become so narrow that the optimum electrical bandwidth shifts toward a higher value close to the full bit rate B [60].

Optimum Bandwidth for RZ Receivers. What about the optimum receiver bandwidth for a 50%-RZ signal? Observing that such an RZ signal at the bit rate B is like an NRZ signal at the bit rate $2B$ where every second bit is forced to zero, we might guess that the optimum bandwidth is about twice that for an NRZ signal, that is, $BW_{3dB} \approx \frac{4}{3}B$. This guess is close, but not fully accurate because an NRZ signal with every second bit forced to zero (let us call this an RZ-like pattern) and a regular random NRZ signal are affected differently by a finite-bandwidth receiver. Given the same channel response, the power penalty for an RZ-like pattern is smaller than that for a random pattern. This is so because the eye closure from the bottom is *less severe* for an RZ-like pattern: the intervening zeros protect the zero data bits from ISI. (Note that the eye diagram for an RZ-like pattern can be constructed from the eye diagram for a random pattern by deleting all traces that are not consistent with the condition that every second bit must be a zero. Deleting traces always increases the eye opening.) Because of this difference in power penalty, the best trade-off between ISI and noise occurs at a bandwidth somewhat below $\frac{4}{3}B$.

To get a more quantitative understanding, we repeat our sensitivity analysis for the 50%-RZ case. First, we simulate the eye diagram at the output of the linear channel for different bandwidths and measure the vertical eye openings, V_E. Then, we put these eye openings in relation to those for a infinite-bandwidth channel, v_S^{pp}, to calculate the power penalties. In contrast to the NRZ case, pulse attenuation now plays a more prominent role: even for an all-one bit sequence, which corresponds to a square wave at frequency B, the band limited RZ signal does not reach the full swing v_S^{pp}. Finally, we use these penalties to correct the noise-based sensitivity, which is 3 dB better for 50%-RZ than for NRZ (cf. Section 4.4). Figure 4.27 shows the result of this procedure for our familiar Butterworth receiver.[7]

The figure shows that the power penalty for RZ is *larger* than that for NRZ given the same bandwidth and bit rate. This is consistent with the fact that the RZ signal has a wider spectrum than the NRZ signal (cf. Appendix A) and thus is truncated more severely by the receiver. In the time domain, the larger power penalty can be attributed to the fact that the RZ signal suffers from pulse attenuation in addition to ISI. (ISI by itself is often smaller for RZ than for NRZ.) Although the power penalty is larger for RZ, it is not so large that it eliminates the 3-dB advantage inherent to the 50%-RZ format. At the bandwidth $\frac{2}{3}B$, for

7 The small *negative* power penalty for large bandwidths is due to a consistent overshoot for the ones. In contrast to the NRZ signal where overshoot occurred for only *some* ones and thus did not contribute to the vertical eye opening, it does occur for *every* one in the RZ signal and thus does increase the vertical eye opening.

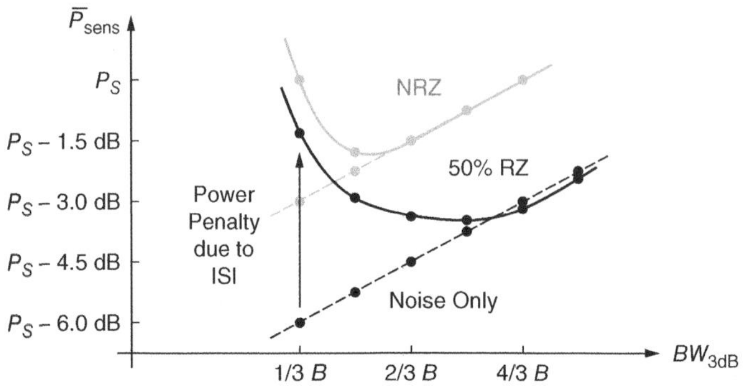

Figure 4.27 Sensitivity as a function of receiver bandwidth: RZ versus NRZ format.

example, the power penalty for RZ is 1.2 dB and that for NRZ is 0 dB, still leaving a sensitivity advantage of 3 dB − 1.2 dB = 1.8 dB for the RZ format.

What is the optimum bandwidth for a 50%-RZ receiver? Compared with the plot for the NRZ receiver, the sensitivity minimum is significantly broader. Any bandwidth between $\frac{2}{3}B$ and $\frac{4}{3}B$ results in a very similar sensitivity.

The detailed numerical analysis in [59] finds the optimum electrical bandwidth for optically preamplified 33%-RZ receivers to be in the range from $0.8B$ to $1.2B$.

Bandwidth Allocation. As we know, the receiver consists of a cascade of building blocks: photodetector, TIA, an optional filter, MA, and so forth. The overall receiver bandwidth can be estimated by adding the reciprocal-square bandwidths of the individual blocks: $1/BW^2 \approx 1/BW_1^2 + 1/BW_2^2 + \ldots$ Thus, each individual block must have a bandwidth that is *larger* than the desired receiver bandwidth.

There are several strategies for allocating bandwidths to the individual blocks such that the desired overall bandwidth is met:

- A precise filter, located after the TIA, controls the bandwidth and frequency response of the receiver. A fourth-order Bessel–Thomson filter, which exhibits good phase linearity, is a common choice. All other blocks are designed for a significantly larger bandwidth. This strategy is typical for low-speed receivers.
- The TIA controls the receiver bandwidth and no filter is used. All other blocks are designed for a significantly larger bandwidth. This strategy relaxes the bandwidth requirement for the TIA, permitting a higher transimpedance and lower noise (we study this trade-off in Section 6.2), but the receiver's frequency response is less well controlled.
- All blocks determine the receiver bandwidth jointly. This strategy is typical for high-speed receivers when overdesigning the blocks is not an option.

4.9 Frequency Response

In the previous section, we assumed that the receiver has a second-order Butterworth frequency response. This response is maximally flat (in amplitude) and is fairly straightforward to implement. But is it the optimum frequency response?

A theoretical analysis shows that the optimum frequency response depends on many factors, such as the shape of the received pulses (i.e., the amount of ISI in the received signal), the spectrum of the input-referred noise, the sampling jitter in the decision circuit, the bit estimation technique used, and so forth. Figure 4.28 shows a decision tree distinguishing the most important cases.

If the NRZ pulses at the input of the receiver are well shaped, specifically if the pulses are broadened by less than 14% of the bit interval ($1/B$), the *matched-filter response* is the best choice [1]. For now let us assume that the input-referred noise is white and the sampling jitter is negligible. Then, the matched filter is defined by an impulse response $h(t)$ that is proportional to (or matched to) a time-reversed copy of the received pulses $x(t)$, that is, $h(t) \propto x(T - t)$, where T is the duration of the received pulses. This definition implies that the matched filter's frequency response matches the spectral shape of the received pulses. It can be shown that in the absence of ISI, the matched-filter response maximizes the sampled signal-to-noise ratio and thus results in the lowest BER [1, 61].

For an undistorted NRZ signal, the matched filter is given by $h(t) \propto x(T - t) = x(t)$, where $x(t)$ is a rectangular pulse starting at $t = 0$ and ending at $T = 1/B$, hence this filter is known as the *rectangular filter*. We discuss this case as well as a possible implementation (integrate and dump) shortly.

If the input-referred noise spectrum is not white or if the decision circuit exhibits sampling jitter, the concept of matched filtering can be generalized to take these effects into account [5, 36]. In Fig. 4.28 this case is referred to as *modified matched filter*.

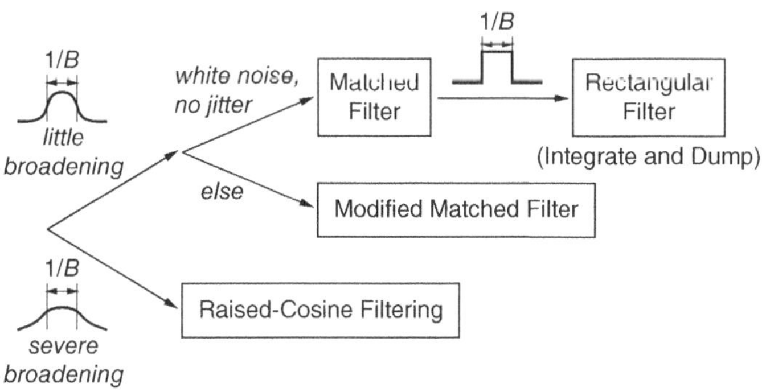

Figure 4.28 Decision tree to determine the optimum receiver response.

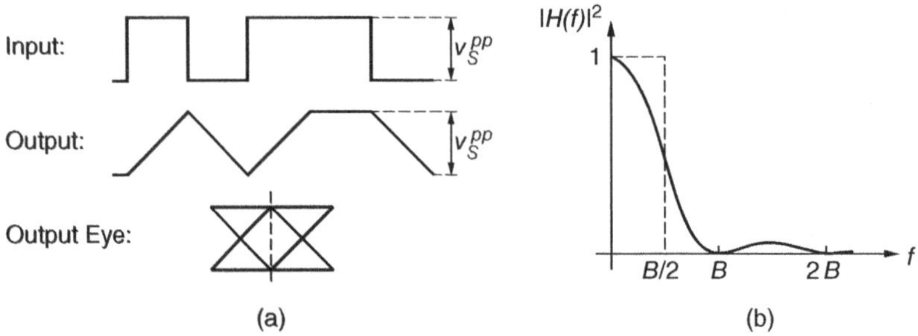

Figure 4.29 Rectangular-filter receiver: (a) waveforms and (b) squared frequency response.

If the received pulses are significantly broadened, they overlap with each other, that is, they suffer from ISI. For such pulses, the matched-filter response may not be suitable because it exacerbates the ISI problem by further broadening the pulses. For severely broadened pulses (more than 20% of the bit interval), *raised-cosine filtering* often is the best choice [1]. Raised-cosine filtering is defined as the transformation of the (broadened) input pulses into pulses with a raised-cosine spectrum (Fourier transform). Note that the receiver itself does *not* have a raised-cosine response, except in the theoretical case when the received pulses are Dirac pulses. Pulses with a raised-cosine spectrum have the desirable property that they are zero at $t = nT$ for all n except 0 with $T = 1/B$ and thus superimpose to form a signal that is free of ISI [1, 61]. We clarify this approach with an example shortly.

Why do we not use raised-cosine filtering in all cases? Because the sampled signal-to-noise ratio of a raised-cosine receiver is not as good as that of a matched-filter receiver. Hence, if ISI in the received signal is weak, matched filtering is preferable. In fact, when making a *joint* decision on the received bit sequence by using a maximum likelihood sequence estimator (MLSE) (cf. Appendix E) the optimum receiver response is always the matched-filter (or modified matched-filter) response, regardless of the pulse broadening.

Rectangular Filter. Let us make a simple example to illustrate the concept of matched filtering. Consider that we receive an undistorted NRZ signal with white noise. As we said, the matched filter for this case is the rectangular filter.

In the time domain, this filter convolves the received signal with a pulse of duration $T = 1/B$. This pulse is the matched filter's impulse response, $h(t)$. Figure 4.29(a) shows how this convolution results in a triangular output signal. Note that despite of the slow edges, the output signal reaches its full value and is free of ISI when sampled at the instant of maximum eye opening (dashed line in the eye diagram).

In the frequency domain, the filter has a low-pass characteristic that can be obtained by taking the Fourier transform of its impulse response: a rectangular pulse starting at $t = 0$ and ending at $T = 1/B$. The frequency response, when normalized for a DC gain of one, turns out to be (cf. Eq. (I.28))

$$H(f) = \frac{\sin(\pi f/B)}{\pi f/B} \cdot e^{-j\,\pi f/B}. \tag{4.70}$$

The squared frequency response $|H(f)|^2$, which matters for noise calculations, is plotted in Fig. 4.29(b) on a linear scale. The noise bandwidth of this filter is $BW_n = B/2$ (indicated by the dashed line) and the 3-dB bandwidth is $BW_{3dB} = 0.443B$. [→ Problem 4.19.]

The combination of a small noise bandwidth and the absence of ISI is what makes this an ideal receiver response. However, the triangular eye shape implies that we have to sample *exactly* at the center of the eye to avoid ISI. Any sampling-time offset or sampling-time jitter results in a power penalty.

What response does the matched-filter approach give for an RZ receiver? The matched filter for 50%-RZ pulses is again a rectangular filter, but this time with half the pulse width, $T = 0.5/B$. The matched-filter bandwidth and thus the receiver bandwidth becomes twice that of the NRZ case. Like the NRZ matched filter, the RZ matched filter introduces no ISI and causes no signal attenuation.

Integrate and Dump. As we have seen, the rectangular filter convolves the received signal, $x(t)$, with the rectangular pulse, $h(t)$. In the case of NRZ, $h(t)$ has the value $1/T$ on the interval from $t = 0$ to T and is zero everywhere else. (The area under the pulse is one corresponding to a DC gain of one.) We can thus write the output signal $y(t)$ from the filter as

$$y(t) = \int_{-\infty}^{\infty} h(t - t') \cdot x(t')\, dt' = \frac{1}{T} \int_{t-T}^{t} x(t')\, dt'. \tag{4.71}$$

The second form can be interpreted as the moving average of $x(t)$ computed over the time interval T. The decision circuit in the receiver samples the output signal, $y(t)$, at the instant of maximum eye opening, which occurs at the end of each bit period, $t = nT$. Thus the sampled signal for the nth bit is

$$y(nT) = \frac{1}{T} \int_{(n-1)T}^{nT} x(t')\, dt'. \tag{4.72}$$

This expression suggests that the rectangular filter can be replaced by a circuit that integrates the received signal $x(t)$ over the bit period T. Figure 4.30 compares the output waveforms from the rectangular filter and from the bit integrator. Although the waveforms are different, the samples $y(nT)$ at the end of each bit period (indicated by the circles) are the same. Note that the bit integrator needs to start the integration from $y = 0$ at the beginning of each bit period, and thus the integrator must be reset immediately after each integration. Hence, this method is called *integrate and dump* [1, 61]. In practice, the instantaneous reset can be avoided by using two integrators operating in a

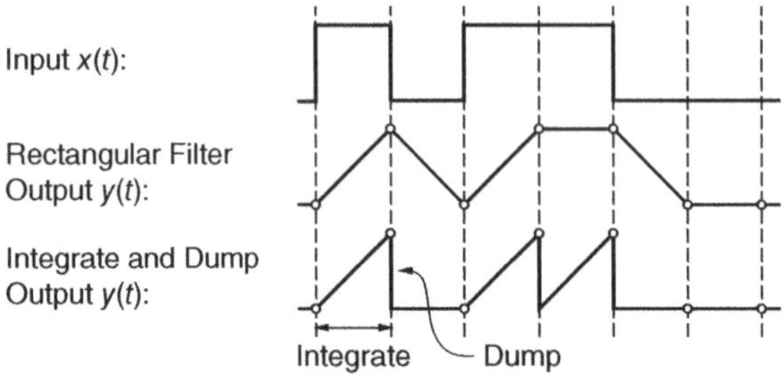

Figure 4.30 Rectangular filter versus integrate and dump.

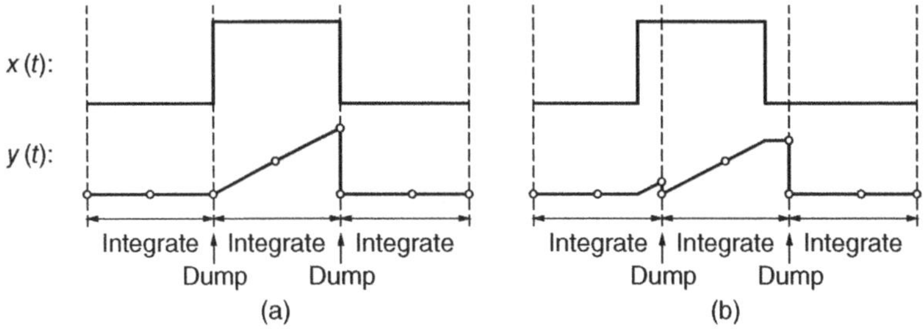

Figure 4.31 Integrate-and-dump waveforms for (a) a phase-aligned clock and (b) a misaligned clock.

ping-pong fashion: while one integrator integrates the input signal, the other integrator can be sampled and reset in the course of a full bit period.

The integrate-and-dump approach lends itself well to monolithic integration. Implementations of optical integrate-and-dump receivers have been reported in [62, 63]. For implementations of other serial integrate-and-dump receivers, see Refs [64–68].

Clock Recovery for Integrate-and-Dump Receivers. The integrate-and-dump receiver requires a clock signal that defines the time interval during which the received signal is integrated. This clock signal must be phase aligned with the incoming bits. How can we obtain such a clock signal?

One solution is to sample the analog output signal from the integrator at the beginning, the middle, and the end of the bit period and to infer the clock phase error by comparing the samples [63]. Figure 4.31 shows the waveforms of an integrate-and-dump receiver with the correct and an incorrect clock phase. If the three samples do not lie on a straight line, the clock is not properly aligned and its phase must be shifted.

Raised-Cosine Filtering. Let us make a simple example to illustrate the concept of raised-cosine filtering. We want to calculate the frequency response that transforms undistorted NRZ pulses into pulses with a full raised-cosine spectrum (this frequency response is called "NRZ to full raised-cosine filter" in Table 4.7).

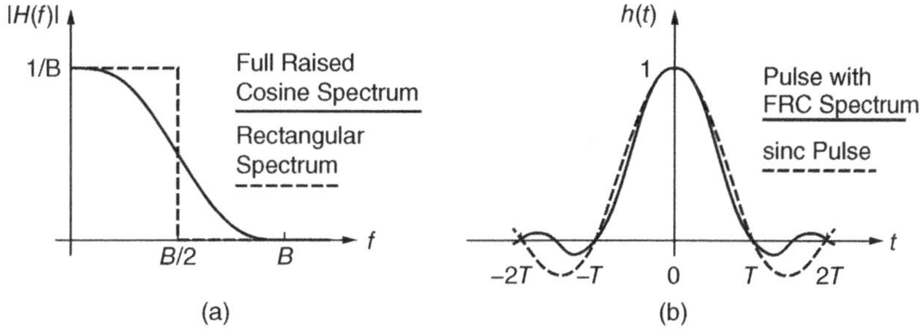

Figure 4.32 (a) Raised-cosine spectra and (b) corresponding time-domain pulses.

The full raised-cosine spectrum is defined as

$$H_{FRC}(f) = \frac{1 + \cos(\pi f/B)}{2B} \quad \text{for} \quad f < B \tag{4.73}$$

and $H_{FRC}(f) = 0$ for $f \geq B$. This spectrum and the corresponding time-domain pulse (the inverse Fourier transform) are shown in Fig. 4.32 (solid lines). The time-domain pulse has the desirable properties that it is one at $t = 0$ and zero at $t = nT$ for all $n \neq 0$ with $T = 1/B$. Thus, the raised-cosine spectrum guarantees that the data signal, which is a superposition of time-shifted copies of the aforementioned pulse, is free of ISI (if sampled precisely at nT). The full raised-cosine spectrum belongs to a family of raised-cosine spectra that differ in their roll-off characteristics but are all ISI free. The roll off is quantified by the *excess bandwidth* measured relative to the *Nyquist bandwidth* $B/2$. The full raised-cosine spectrum has 100% excess bandwidth. Also illustrated in Fig. 4.32 is the case with 0% excess bandwidth (dashed lines). For a full discussion of the raised-cosine spectra see [1, 13, 61].

The desired frequency response of the receiver is obtained by dividing the full raised-cosine spectrum by the spectrum of the incoming NRZ pulses. The spectrum (Fourier transform) of an ideal NRZ pulse centered at $t = 0$ and swinging from zero to one is (cf. Eq. (I.28))

$$H_{NRZ}(f) = \frac{\sin(\pi f/B)}{\pi f} \tag{4.74}$$

leading to the receiver's frequency response[8]

$$H(f) = \frac{H_{FRC}(f)}{H_{NRZ}(f)} = \frac{1 + \cos(\pi f/B)}{2} \cdot \frac{\pi f/B}{\sin(\pi f/B)} \quad f < B. \tag{4.75}$$

8 Equation (4.75) requires the filter to respond instantaneously (the input and output pulses are both centered at $t = 0$). To make the filter realizable, the output pulse can be delayed by an arbitrary amount of time t_{delay}. This modification adds the linear phase factor $\exp(-j2\pi f t_{\text{delay}})$ to the frequency response in Eq. (4.75) but leaves the magnitude unchanged.

The noise bandwidth of this response is $BW_n = 0.564B$ and its 3-dB bandwidth is $BW_{3dB} = 0.580B$.

Compared with the matched-filter receiver for the same ideal NRZ pulse, the noise bandwidth of the raised-cosine receiver is about 13% larger. Because both receivers produce an ISI-free output signal, the raised-cosine receiver is suboptimal in this case. As we pointed out earlier, raised-cosine filtering is most attractive when the received pulses are significantly broadened.

What bandwidth does the raised-cosine approach give for an RZ receiver? We might expect a somewhat wider bandwidth than for the NRZ receiver. However, going through the math we find that the 3-dB bandwidth must be *reduced* from $0.580B$ for NRZ to $0.394B$ for 50%-RZ pulses [16]. How can this be explained? Recall that the raised-cosine approach forces the same output pulse shape no matter whether the input pulse shape is NRZ or RZ. Therefore, the RZ receiver has to broaden the pulses more than the NRZ receiver, which explains the narrower bandwidth for the RZ receiver.

In practice, receivers for severely broadened pulses use equalizers. The adaptive equalizer automatically adjusts the receiver response according to the incoming pulse shapes in an effort to minimize ISI. For an introduction to adaptive equalizers, see Appendix E.

4.10 Summary

The basic receiver model consists of a photodetector, a linear channel (comprising the transimpedance amplifier, main amplifier, equalizer, etc.), and a decision circuit. The noise at the output of the linear channel originates from the receiver circuits (mostly the transimpedance amplifier) and the detector (and optical amplifiers, if present).

If the noise's instantaneous value becomes too large, the decision circuit makes a bit error. For Gaussian noise, a simple mathematical expression can be found that relates the bit-error rate (BER) to the signal swing and the rms value of the noise on the zeros and ones. To obtain a low BER, a large signal-to-noise ratio (SNR), or more accurately, a large Q parameter (Personick Q) is necessary.

The electrical receiver sensitivity is the minimum input signal (peak-to-peak) required to achieve a specified BER. For a high sensitivity, a low input-referred rms noise is required. The optical receiver sensitivity is the minimum optical input power (averaged over time or peak-to-peak) required to achieve a specified BER.

- The sensitivity of a p–i–n receiver is mostly determined by the circuit noise (detector shot noise is small).
- The sensitivity of an APD receiver is determined by the circuit noise and the detector's avalanche noise.
- The sensitivity of an optically preamplified p–i–n receiver is determined by the noise from the optical amplifier (ASE noise) and the circuit noise.
- The sensitivity of a coherent receiver with balanced detector is mostly determined by the shot noise from the local oscillator. Circuit noise and relative-intensity noise (RIN) from the local oscillator are secondary.
- The sensitivity of an analog receiver is mostly determined by the shot noise and the RIN noise from the transmitter. Circuit noise is secondary.

The optical sensitivity and the optical overload power of a digital receiver can be determined from the BER plot, a graph of BER vs received power.

The input-referred rms noise current of a receiver can be calculated by integrating the noise PSD at the output of the linear channel and then referring it back to the input. Alternatively, noise bandwidths or Personick integrals (normalized noise bandwidths) can be used to compute the input-referred rms noise current from the input-referred noise current PSD.

In transmission systems that contain many optical in-line amplifiers, the optical amplifier noise dominates the circuit noise. The concept of optical sensitivity then must be replaced by the concept of required optical signal-to-noise ratio (OSNR). The required OSNR is the minimum OSNR at the input of the receiver that is required to achieve a specified BER.

Impairments in transmission systems, such as a finite extinction ratio, intersymbol interference (ISI), or a decision-threshold offset, can be quantified by power penalties. The power penalty specifies the increase in optical power necessary to overcome a particular impairment. Power penalties depend on the type of transmission system (e.g., unamplified vs amplified).

The optimum receiver bandwidth is determined by a trade-off between noise and ISI. Common bandwidths are

- 60 to 70% of the bit rate for conventional non-return-to-zero (NRZ) receivers.
- 80 to 120% of the bit rate for return-to-zero (RZ) receivers.

The optimum frequency response depends on the shape of the received pulses, the bit estimation technique, and other factors. Well-known responses are

- The matched-filter response, which maximizes the SNR.
- The raised-cosine filtering response, which eliminates the ISI.

For rectangular pulses (e.g., NRZ), the matched filter can be implemented with an integrate-and-dump circuit. In practice, adaptive equalizers are used to remove the ISI.

Problems

4.1 Filtered Detector Noise. Assume that the impulse response of the linear channel, $h(t)$, is zero everywhere except on the time interval $[0 \ldots \xi]$. Under what conditions can Eq. (4.4) be simplified to $V^2_{n,PD}(f,t) = |H(f)|^2 \cdot I^2_{n,PD}(f,t)$?

4.2 Noise Spillover. (a) Assume that the detector noise spectrum is white and the decision circuit bandwidth is infinite. Derive the exact expression for the total mean-square output noise, $\overline{v^2_{n,PD}}(t)$, from Eq. (4.4). (b) Assume the following: an ideal NRZ signal with $I^2_{n,PD}(t) = 0$ for zeros and $I^2_{n,PD}(t) = I^2_{n,PD}$ for ones, a linear channel with a first-order low-pass response that has the 3-dB bandwidth BW_{3dB}, and sampling of the output signal at the end of the bit periods of the input signal. Calculate the worst-case mean-square output noise for zeros, $\overline{v^2_{n,PD,0}}$, and for ones, $\overline{v^2_{n,PD,1}}$, at the sampling instant.

4.3 BER for the Case of Unequal Noise. Derive Eq. (4.10) for the BER when noise on the zeros and ones is unequal. Approximate the optimum decision-threshold voltage assuming that it is strongly determined by the exponential factors of the Gaussians.

4.4 SNR Requirement for the Case of Unequal Noise. Let ξ be the ratio between the rms noise on the zeros and ones of an ideal NRZ signal. Assume Gaussian noise distributions and equal probabilities for zeros and ones. Generalize the relationship between SNR and Q given in Eqs. (4.13) and (4.14) for an arbitrary ξ.

4.5 SNR Requirement for Finite-Slope NRZ Signal. (a) Derive the (continuous-time) SNR requirement for an NRZ signal with linear slopes, where each slope occupies the fraction $\xi < 1$ of a bit period, such that the signal can be detected at a specified BER. Assume additive Gaussian noise and equal probabilities for zeros and ones. (b) What is the SNR value for $\xi = 0.3$ and $BER = 10^{-12}$? (c) What is the requirement for the *sampled* SNR given a finite-slope NRZ signal?

4.6 SNR Definitions. An ideal 50%-RZ signal swings between 0 and 1 V (zeros and ones have equal probabilities). The rms noise on the 0-V level is 20 mV and the rms noise on the 1-V level is 140 mV. Seven students are asked to calculate the SNR of this signal. Each student comes up with a different answer: (a) 7.96 dB, (b) 13.98 dB, (c) 15.57 dB, (d) 15.92 dB, (e) 16.82 dB, (f) 16.99 dB, and (g) 22.84 dB. Who is right?

4.7 E_b/N_0 **and SNR.** (a) Derive the relationship between E_b/N_0 at the input of the linear channel and SNR at the output of the linear channel. Assume that the linear channel bandlimits the noise but does not attenuate the signal. (b) Under which circumstances become E_b/N_0 and SNR identical?

4.8 **Sensitivity for Finite Extinction Ratio.** Equation (4.22) gives the receiver sensitivity for an optical signal with high extinction ratio. Generalize this equation assuming that the power for the zeros is $P_0 = P_1/ER$ instead of 0, where P_1 is the power for the ones and ER is the extinction ratio.

4.9 **Sensitivity of p–i–n Receiver.** Engineers use the following rule to estimate the sensitivity of a p–i–n receiver (cf. [69])

$$\overline{P}_{\text{sens},PIN}\ [\text{dBm}] \approx -21.53\ \text{dBm} + 10\log(i^{rms}_{n,\text{ckt}}\ [\mu\text{A}]) - 10\log(\mathcal{R}\ [\text{A/W}]). \quad (4.76)$$

Explain the origin of this equation. What is the meaning of −21.53 dBm?

4.10 **Sensitivity of Optically Preamplified Receiver.** The sensitivity in Eq. (4.32) takes only circuit and signal–spontaneous beat noise into account. Derive a more precise expression that includes the effect of spontaneous–spontaneous beat noise. (a) Write $\overline{P}_{\text{sens},OA}$ as a function of S_{ASE}. (b) Write $\overline{P}_{\text{sens},OA}$ as a function of the optical noise figure $\tilde{F}$, which is defined in Eq. (3.27).

4.11 **Sensitivity vs OSNR.** Show that the sensitivity analysis leading up to Eq. (4.52) is equivalent to the OSNR analysis based on Eqs. (4.53) and (4.54).

4.12 **Q-factor Budget.** (a) Redo the budget in Table 4.8 using Q factors instead of OSNR values. Assume that the receiver requires an OSNR of 16 dB at $BER = 10^{-4}$ and that the receiver's Q is proportional to $\sqrt{OSNR}$. (b) What value of Q should be observed at the beginning of life?

4.13 **Sensitivity of p–i–n Receiver with Finite OSNR.** Calculate the sensitivity of a p–i–n receiver assuming that the received optical signal contains noise from in-line amplifiers. The latter noise is specified in terms of the OSNR. Neglect shot and spontaneous–spontaneous beat noise. (a) What is the sensitivity in the limit $OSNR \to \infty$? (b) What minimum $OSNR$ is required given a high received power?

4.14 **Required OSNR for Receiver with Finite Extinction.** Equation (4.53) gives the required OSNR for an optical signal with high extinction ratio. Generalize this equation for the case when the power for the zeros is $P_0 = P_1/ER$ instead of 0, where P_1 is the power for the ones and ER is the extinction ratio.

4.15 **Required OSNR for Receiver with Spontaneous–Spontaneous Beat Noise.** Equation (4.53) takes only signal–spontaneous beat noise into account. (a) Derive an expression that includes the effect of spontaneous–spontaneous beat noise, assuming high extinction. (b) Same as (a), but for the finite extinction ratio ER. Tip: Write Q as a function of $OSNR$.

4.16 **Required OSNR for Receiver with Optical Preamplifier.** An optically preamplified receiver has the sensitivity $\overline{P}_{\text{sens},OA}$ and the preamplifier has the noise figure F. Estimate the required OSNR for this receiver.

4.17 **Power Penalty due to APD Dark Current.** (a) Calculate the power penalty due to the primary dark current in an APD receiver. Assume that the APD gain is set to its optimum value. To keep things simple, make the following approximations: assume that the avalanche noise equals the circuit noise at the optimum APD gain, neglect the second term in Eq. (3.17), and neglect the second term under the root in Eq. (4.33). (b) Derive a specification for the dark current given the maximum acceptable power penalty PP.

4.18 **Power Penalty due to ISI.** Calculate the power penalty incurred when passing an ideal NRZ waveform through (a) a first-order low-pass filter and (b) a second-order Butterworth low-pass filter. Assume that the filtered signal is sampled at the bit boundaries of the input signal. (c) How large are the power penalties for $BW_{3\text{dB}} = \frac{1}{3}B, \frac{2}{3}B$, and $\frac{4}{3}B$ for each filter?

4.19 **Rectangular-Filter Response.** The rectangular filter has an impulse response $h(t)$, which is one in the interval from 0 to $1/B$ and zero everywhere else. Calculate the frequency response of this filter (magnitude and phase).

References

1 M. M.-K. Liu. *Principles and Applications of Optical Communications.* Irwin, McGraw-Hill, Chicago, 1996.
2 S. B. Alexander. *Optical Communication Receiver Design.* SPIE Press, copublished with IEE, Bellingham, WA, 1997.

3 S. D. Personick. Receiver design for digital fiber optic communication systems. *Bell Syst. & Tech. J.*, 52(6):843–886, 1973.
4 R. G. Smith and S. D. Personick. Receiver design for optical fiber communication systems. In H. Kressel, editor, *Topics in Applied Physics: Semiconductor Devices for Optical Communication*, volume 39. Springer-Verlag, Berlin, Germany, 1982.
5 L. W. Couch II. *Digital and Analog Communication Systems.* Prentice Hall, Upper Saddle River, NJ, 7th edition, 2006.
6 R. Ramaswami, K. N. Sivarajan, and G. Sasaki. *Optical Networks: A Practical Perspective.* Morgan Kaufmann Publishers, San Francisco, CA, 3rd edition, 2010.
7 G. P. Agrawal. *Fiber-Optic Communication Systems.* John Wiley & Sons, Inc., Hoboken, NJ, 4th edition, 2010.
8 C. A. Eldering. Theoretical determination of sensitivity penalty for burst mode fiber optic receivers. *J. Lightwave Technol.*, LT-11(12):2145–2149, 1993.
9 P. Menéndez-Valdés. Performance of optical direct receivers using noise corrupted decision threshold. *J. Lightwave Technol.*, LT-13(11):2202–2214, 1995.
10 P. Ossieur, X.-Z. Qiu, J. Bauwelinck, and J. Vandewege. Sensitivity penalty calculation for burst-mode receivers using avalanche photodiodes. *J. Lightwave Technol.*, LT-21(11):2565–2575, 2003.
11 C. Su, L.-K. Chen, and K.-W. Cheung. Theory of burst-mode receiver and its applications in optical multiaccess networks. *J. Lightwave Technol.*, LT-15(4):590–606, 1997.
12 W. Ciciora, J. Farmer, D. Large, and M. Adams. *Modern Cable Television Technology: Video, Voice, and Data Communications.* Morgan Kaufmann, San Francisco, CA, 2nd edition, 2004.
13 J. G. Proakis and M. Salehi. *Digital Communications.* McGraw Hill, New York, 5th edition, 2008.
14 T. V. Muoi. Receiver design for high-speed optical-fiber systems. *J. Lightwave Technol.*, LT-2(3):243–267, 1984.
15 M. Dudek. OMA proposal. IEEE 802.3ae 10Gb/s Task Force, November 2000.
16 B. L. Kasper. Receiver design. In S. E. Miller and I. P. Kaminow, editors, *Optical Fiber Telecommunications II*, pages 689–722. Academic Press, San Diego, CA, 1988.
17 R. G. Smith and B. L. Kasper. Optical receivers. In J. D. Gibson, editor, *The Communications Handbook*, pages 789–802. CRC Press, Boca Raton, FL, 1997.
18 B. L. Kasper, O. Mizuhara, and Y.-K. Chen. High bit-rate receivers, transmitters, and electronics. In I. P. Kaminow and T. Li, editors, *Optical Fiber Telecommunications IVA*, pages 784–851. Academic Press, San Diego, CA, 2002.

19 L. D. Tzeng, O. Mizuhara, T. V. Nguyen, K. Ogawa, I. Watanabe, K. Makita, M. Tsuji, and K. Taguchi. A high-sensitivity APD receiver for 10-Gb/s system applications. *IEEE Photonics Technol. Lett.*, 8(9):1229–1231, 1996.

20 K. Ogawa, L. D. Tzeng, Y. K. Park, and E. Sano. Advances in high bit-rate transmission systems. In I. P. Kaminow and T. L. Koch, editors, *Optical Fiber Telecommunications IIIA*, pages 336–372. Academic Press, San Diego, CA, 1997.

21 K. Krishnamurthy, J. Xu, X. Shou, R. Vetury, S. Jaganathan, Y.-Z. Liu, R. Pullela, R. Yu, and M. J. W. Rodwell. 40 Gb/s TDM system transceiver prototype in InP technology. In Y. Liu and H. Yang, editors, *High-Speed Optical Transceivers, Integrated Circuits Designs and Optical Devices Techniques*, pages 167–189. World Scientific, Singapore, 2006.

22 T. Nakata, T. Takeuchi, K. Makita, Y. Amamiya, T. Kato, Y. Suzuki, and T. Torikai. High-sensitivity 40-Gb/s reciever with a wideband InAlAs waveguide avalanche photodiode. In *European Conference on Optical Communication (ECOC)*, Copenhagen, Denmark, 2002.

23 Y. Miyamoto, M. Yoneyama, T. Otsuji, K. Yonenaga, and N. Shimizu. 40-Gbit/s TDM transmission technologies based on ultra-high-speed IC's. *IEEE J. Solid-State Circuits*, SC-34(9):1246–1253, 1999.

24 J. Conradi. A simplified non-Gaussian approach to digital optical receiver design with avalanche photodiodes: theory. *J. Lightwave Technol.*, LT-9(8):1019–1026, 1991.

25 B. Chan and J. Conradi. On the non-Gaussian noise in erbium-doped fiber amplifiers. *J. Lightwave Technol.*, LT-15(4):680–687, 1997.

26 H. Kim and P. J. Winzer. Nonlinear phase noise in phase-coded transmission. In *Optical Fiber Communication Conference (OFC)*, Anaheim, CA, 2005.

27 B. Zhang, C. Malouin, and T. J. Schmidt. Design of coherent receiver optical front end for unamplified applications. *OSA Opt. Express*, 20(3):3225–3234, 2012.

28 K. Kikuchi. Coherent optical communications: historical perspectives and future directions. In M. Nakazawa, K. Kikuchi, and T. Miyazaki, editors, *High Spectral Density Optical Communication Technologies*, pages 11–49. Springer, Berlin Heidelberg, 2010.

29 M. R. Phillips and T. E. Darcie. Lightwave analog video transmission. In I. P. Kaminow and T. L. Koch, editors, *Optical Fiber Telecommunications IIIA*, pages 523–559. Academic Press, San Diego, CA, 1997.

30 Circadiant Systems. Straight-line BER plots. Circadiant Technical Brief TB010, May 2003.

31 M. Müller, R. Stephens, and R. McHugh. Total jitter measurement at low probability levels, using optimized BERT scan method. In *DesignCon East*, Worcester, MA, September 2005.

32 P. R. Gray, P. J. Hurst, S. H. Lewis, and R. G. Meyer. *Analysis and Design of Analog Integrated Circuits.* John Wiley & Sons, Inc., New York, 5th edition, 2009.

33 D. Johns and K. Martin. *Analog Integrated Circuit Design.* John Wiley & Sons, Inc., New York, 1997.

34 E. Säckinger. *Broadband Circuits for Optical Fiber Communication.* John Wiley & Sons, Inc., Hoboken, NJ, 2005.

35 E. Säckinger. On the noise optimum of FET broadband transimpedance amplifiers. *IEEE Trans. Circuits Syst. I*, CASI-59(12):2881–2889, 2012.

36 A. Buchwald and K. Martin. *Integrated Fiber-Optic Receivers.* Kluwer Academic Publishers, Boston, MA, 1995.

37 J. J. Morikuni, A. Dharchoudhury, Y. Leblebici, and S. M. Kang. Improvements to the standard theory for photoreceiver noise. *J. Lightwave Technol.*, LT-12(7):1174–1184, 1994.

38 J. L. Gimlett and N. K. Cheung. Dispersion penalty analysis for LED/single-mode fiber transmission systems. *J. Lightwave Technol.*, LT-4(9):1381–1392, 1986.

39 B. L. Kasper and J. C. Campbell. Multigigabit-per-second avalanche photodiode lightwave receivers. *J. Lightwave Technol.*, LT-5(10):1351–1364, 1987.

40 K. Ogawa. Considerations for optical receiver design. *IEEE J. Sel. Areas Commun.*, SAC-1(3):524–532, 1983.

41 W. C. Barnett, H. Takahira, J. C. Baroni, and Y. Ogi. The TPC-5 cable network. *IEEE Commun. Mag.*, 34(2):36–40, 1996.

42 V. L. da Silva, Y. L. Barberio, O. T. Blash, L. J. Button, K. Ennser, L. L. Hluck, A. J. Lucero, M. Rukosueva, S. Tsuda, and R. J. Whitman. Capacity upgrade for non-zero dispersion-shifted fiber based systems. In *National Fiber Optics Engineers Conference (NFOEC)*, 1999.

43 R.-J. Essiambre, G. Raybon, and B. Mikkelsen. Pseudo-linear transmission of high-speed TDM signals: 40 and 160 Gb/s}. In I. P. Kaminow and T. Li, editors, *Optical Fiber Telecommunications IVB*, pages 232–304. Academic Press, San Diego, CA, 2002.

44 N. S. Bergano. Undersea communication systems. In I. P. Kaminow and T. Li, editors, *Optical Fiber Telecommunications IVB*, pages 154–197. Academic Press, San Diego, CA, 2002.

45 J. L. Zyskind, J. A. Nagel, and H. D. Kidorf. Erbium-doped fiber amplifiers for optical communications. In I. P. Kaminow and T. L. Koch, editors, *Optical Fiber Telecommunications IIIB*, pages 13–68. Academic Press, San Diego, CA, 1997.

46 N. S. Bergano. Undersea amplified lightwave systems design. In I. P. Kaminow and T. L. Koch, editors, *Optical Fiber Telecommunications IIIA*, pages 302–335. Academic Press, San Diego, CA, 1997.

47 D. A. Fishman and B. S. Jackson. Transmitter and receiver design for amplified lightwave systems. In I. P. Kaminow and T. L. Koch, editors, *Optical Fiber Telecommunications IIIB*, pages 69–114. Academic Press, San Diego, CA, 1997.

48 N. S. Bergano, F. W. Kerfoot, and C. R. Davidson. Margin measurements in optical amplifier systems. *IEEE Photonics Technol. Lett.*, 5(3):304–306, 1993.

49 Circadiant Systems. BER vs. OSNR. Circadiant Technical Brief TB007, February 2003.

50 Circadiant Systems. OSNR. Circadiant Technical Brief TB001, December 2002.

51 P. A. Humblet and M. Azizo glu. On the bit-error rate of lightwave systems with optical amplifiers. *J. Lightwave Technol.*, LT-9(11):1576–1582, 1991.

52 D. Marcuse. Derivation of analytical expressions for the bit-error probability in lightwave systems with optical amplifiers. *J. Lightwave Technol.*, LT-8(12):1816–1823, 1990.

53 P. J. Winzer and R.-J. Essiambre. Advanced optical modulation formats. *Proc. IEEE*, 94(5):952–984, 2006.

54 G. Bosco, R. Gaudino, and P. Poggiolini. An exact analysis of RZ versus NRZ sensitivity in ASE noise limited optical systems. In *European Conference on Optical Communication (ECOC), Amsterdam*, The Netherlands, 2001.

55 A. H. Gnauck and P. J. Winzer. Optical phase-shift-keyed transmission. *J. Lightwave Technol.*, LT-23(1):115–130, 2005.

56 P. Bayvel, C. Behrens, and D. S. Millar. Digital signal processing (DSP) and its application in optical communication systems. In I. P. Kaminow, T. Li, and A. E. Willner, editors, *Optical Fiber Telecommunications VIB*, pages 163–212. Academic Press, 2013.

57 Telcordia Technologies. Synchronous optical network (SONET) transport systems: common generic criteria, GR-253-CORE, Issue 3. Telcordia Technologies (formerly Bellcore), Piscataway, NJ, September 2000.

58 J. Müllrich, H. Thurner, E. Müllner, J. F. Jensen, W. E. Stanchina, M. Kardos, and H.-M. Rein. High-gain transimpedance amplifier in InP-based HBT technology for receiver in 40-Gb/s optical-fiber TDM links. *IEEE J. Solid-State Circuits*, SC-35(9):1260–1265, 2000.

59 P. J. Winzer, M. Pfennigbauer, M. M. Strasser, and W. R. Leeb. Optimum filter bandwidths for optically preamplified NRZ receivers. *J. Lightwave Technol.*, LT-19(9):1263–1273, 2001.

60 I. Lyubomirsky, S. Shetty, J. Roman, and M. Y. Frankel. Optimum 10-Gb/s NRZ receiver bandwidths for ultradense WDM transmission systems. *IEEE Photonics Technol. Lett.*, 14(6):870–872, 2002.

61 S. Haykin and M. Moher. *Communication Systems*. John Wiley & Sons, Inc., Hoboken, NJ, 5th edition, 2009.

62 R. P. Jindal. Silicon MOS amplifier operation in the integrate and dump mode for gigahertz band lightwave communication systems. *J. Lightwave Technol.*, LT-8(7):1023–1026, 1990.

63 A. E. Stevens. *An Integrate-and-Dump Receiver for Fiber Optic Networks.* PhD thesis, Columbia University, New York, 1995.

64 J. Savoj and B. Razavi. A CMOS interface circuit for detection of 1.2Gb/s RZ data. In *ISSCC Digest of Technical Papers*, pages 278–279, February 1999.

65 S. Sidiropoulos and M. Horowitz. A 700-Mb/s/pin CMOS signaling interface using current integrating receivers. *IEEE J. Solid-State Circuits*, SC-32(5):681–690, 1997.

66 M. S. J. Steyaert, W. Dehaene, J. Craninckx, M. Walsh, and P. Real. A CMOS rectifier-integrator for amplitude detection in hard disk servo loops. *IEEE J. Solid-State Circuits*, SC-30(7):743–751, 1995.

67 F. Yang, J. H. O'Neill, D. Inglis, and J. Othmer. A CMOS low-power multiple 2.5–3.125-Gb/s serial link macrocell for high IO bandwidth network ICs. *IEEE J. Solid-State Circuits*, SC-37(12):1813–1821, 2002.

68 J. L. Zerbe, P. S. Chau, C. W. Werner, T. P. Thrush, H. J. Liaw, B. W. Garlepp, and K. S. Donnelly. 1.6 Gb/s/pin 4-PAM signaling and circuits for a multidrop bus. *IEEE J. Solid-State Circuits*, SC-36(5):752–760, 2001.

69 H. Ransijn. Receiver and transmitter IC design, May 2001. CICC'2001 Ed. Session 3-2.

5

Transimpedance Amplifier Specifications

In this chapter we examine the main specifications of the TIA: the transimpedance, the input overload current, the maximum input current for linear operation, the bandwidth, the phase linearity, the group-delay variation, the jitter, the input-referred noise current, and the crosstalk (for multichannel TIAs). We discuss the measurement of the transimpedance and the noise and calculate the power penalty due to crosstalk. We conclude with a table showing the data of commercial parts for illustration.

5.1 Transimpedance

Definition. Figure 5.1 defines the input current i_I and the output voltage v_O of single-ended and differential TIAs. Note that in the differential cases, Figs 5.1(b) and (c), v_O is defined as the voltage difference $v_O = v_{OP} - v_{ON}$ and i_I is defined as the current difference $i_I = i_{IP} - i_{IN}$. All output voltages are defined under terminated conditions, that is, a TIA designed to drive a 50-Ω load must be terminated with 50-Ω resistors. The input current i_I is the current from the photodetector. More precisely, it is the photocurrent generated by the *intrinsic* photodetector [1]. At low and intermediate frequencies, however, there is little difference between these two currents. We return to this distinction later in this section.

At low frequencies, the *transimpedance*, Z_T, is defined as the output voltage change, Δv_O, per input current change, Δi_I (see Fig. 5.2)

$$Z_T = \frac{\Delta v_O}{\Delta i_I}. \tag{5.1}$$

The term *transimpedance* derives from the older term *transfer impedance*, which indicates that the voltage and current defining the impedance are measured at two different ports. In contrast, the *driving point impedance* is defined as the ratio of the voltage and current measured at the same port.

Analysis and Design of Transimpedance Amplifiers for Optical Receivers, First Edition. Eduard Säckinger.

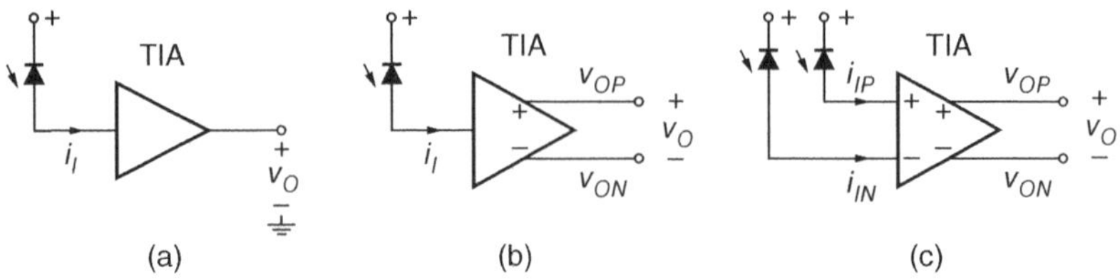

Figure 5.1 Input and output signals of (a) a single-ended, (b) a differential-output, and (c) a fully differential TIA.

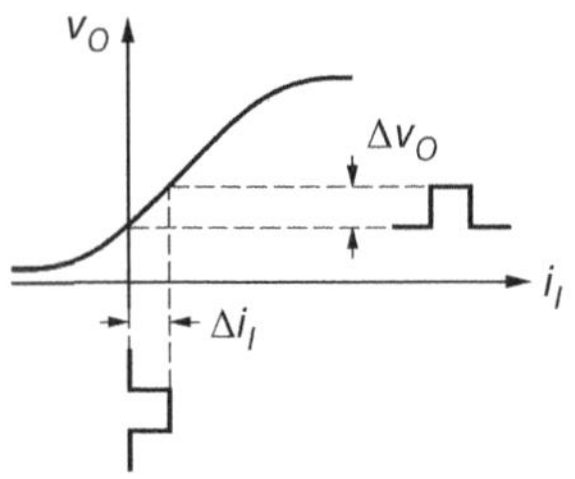

Figure 5.2 Input current change Δi_I and output voltage change Δv_O for a single-ended TIA.

The transimpedance is specified either in units of Ω or dBΩ. In the latter case, the value is calculated as $20\log(Z_T/\Omega)$. For example, 1 kΩ corresponds to 60 dBΩ.

For TIAs with differential outputs it is important to distinguish between the differential and the single-ended transimpedance. With our definition $v_O = v_{OP} - v_{ON}$, we obtain the differential transimpedance. In contrast, if we define $v_O = v_{OP}$ or $v_O = v_{ON}$, we obtain the single-ended transimpedance, which is only half of the differential transimpedance (assuming a constant output common-mode voltage).

The small-signal transimpedance must be determined for an input signal swing Δi_I that is small enough such that the transfer function can be regarded as linear (see Fig. 5.2). For larger input signals, the transfer function becomes nonlinear (compressive), causing the effective transimpedance to drop (cf. Section 5.3). For very large input signals, the output signal gets corrupted by excessive pulse-width distortion and jitter (cf. Section 5.2).

Frequency Response. At high frequencies, the output voltage swing Δv_O diminishes relative to Δi_I and the output voltage no longer follows the input current instantaneously. Thus, the low-frequency transimpedance defined in Eq. (5.1) must be generalized to the frequency dependent and complex quantity $Z_T(f) = |Z_T(f)| \cdot \exp(j\phi(f))$, where $|Z_T(f)|$ is the magnitude of the transimpedance and $\phi(f)$ is the phase shift from the input to the output signal. Equivalently, this complex quantity can be expressed as

$$Z_T(f) = \frac{V_o}{I_i}, \tag{5.2}$$

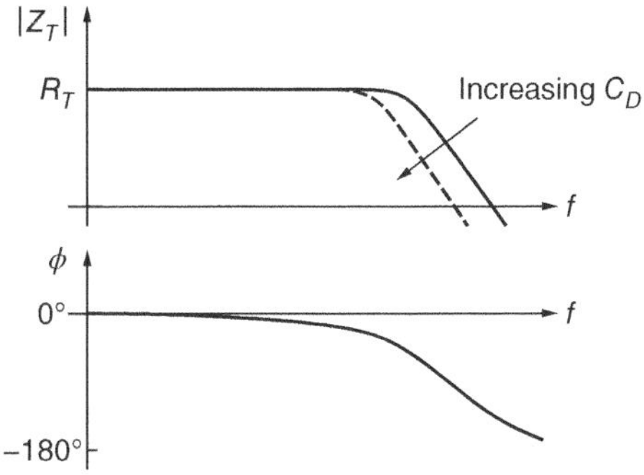

Figure 5.3 Frequency response (magnitude and phase) of a TIA.

where V_o is the output voltage phasor and I_i is the input current phasor. At low frequencies, Z_T becomes real, that is, the phase shift goes to zero, as illustrated in Fig. 5.3. (Here we assumed that $Z_T(f)$ does not have a low-frequency cutoff.) When Z_T is real, the transimpedance is also known as the *transresistance*, R_T.

At low frequencies, all of the photocurrent enters the TIA and there is no distinction between the intrinsic photocurrent and the current flowing through the TIA's input terminal. At high frequencies, however, photodetector parasitics and packaging parasitics make these two currents different. In particular, the photodetector capacitance, C_D, shunts some of the photocurrent to ground. (How much depends on C_D and the TIA's input impedance.) In this case, the transimpedance response, $Z_T(f)$, based on the intrinsic photocurrent is more relevant than that based on the TIA's input-terminal current [1]. When reporting $Z_T(f)$, the associated C_D (or better, the photodetector/package model) must be stated along with it. As illustrated with the dashed line in Fig. 5.3, a larger C_D typically reduces the transimpedance at high frequencies (i.e., it reduces the bandwidth).

Typical transimpedance values for some commercial parts are listed in Table 5.1. In general, a larger transimpedance is better because it relaxes the gain and noise requirements of the subsequent main amplifier. In Section 6.2, we see that the maximum achievable transimpedance is limited to a value that depends on the bandwidth (bit rate), the speed of the technology, the TIA topology, and so forth. In particular, the maximum transimpedance diminishes for increasing bandwidths, assuming everything else remains the same.

Most high-speed test equipment, such as network analyzers, are designed to measure circuits in a 50-Ω environment. However, the transimpedance of a TIA must be measured for a capacitive source impedance, as presented by the photodetector, rather than a 50-Ω source impedance. There are two practical methods to experimentally determine the transimpedance: using a photodetector simulator or using a network analyzer (NWA) together with some math.

Measurement with Photodetector Simulator. The measurement setup with a photodetector simulator is shown in Fig. 5.4 [2–5]. The simulator generates the

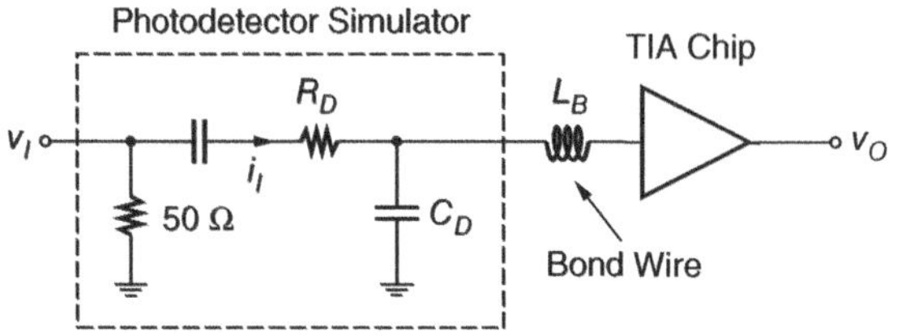

Figure 5.4 Measurement of the transimpedance with a photodetector simulator.

"photocurrent" i_I with the resistor R_D, which has a large value around 1 to 2 kΩ to approximate a current source. The simulator further models the photodetector capacitance with the capacitor C_D. The photodetector simulator is connected to the TIA chip with a bond wire L_B similar to what would be used to connect the actual photodetector. The "photocurrent" of the simulator can be controlled with the input voltage v_I. Assuming that the TIA's input impedance is much smaller than R_D, the simulated photocurrent is given by $i_I = v_I/R_D$. Thus, by measuring the voltage gain v_O/v_I of the photodetector simulator plus the TIA and multiplying the gain by R_D, we obtain the transimpedance Z_T. The voltage gain can be measured, for example, with a network analyzer (hence the 50-Ω termination resistor at the input).

Measurement with Network Analyzer. Another method to experimentally determine the transimpedance is to first characterize the TIA by itself as a (linear) black box using a network analyzer, as shown in Fig. 5.5, and then to mathematically combine the transfer functions of the photodetector/package model (intrinsic photodetector plus parasitics) with the TIA black-box model, as shown in Fig. 5.6. The black-box model usually is in the form of 2-port S parameters and hence this method is also known as the S-parameter method.

After measuring the S parameters S_{11} and S_{21}, the transimpedance can be calculated as [1]

$$Z_T(f) = \frac{H_{PD}(f)S_{21}(f)R_0}{[1 - S_{11}(f)] + Y_D(f)[1 + S_{11}(f)]R_0}, \tag{5.3}$$

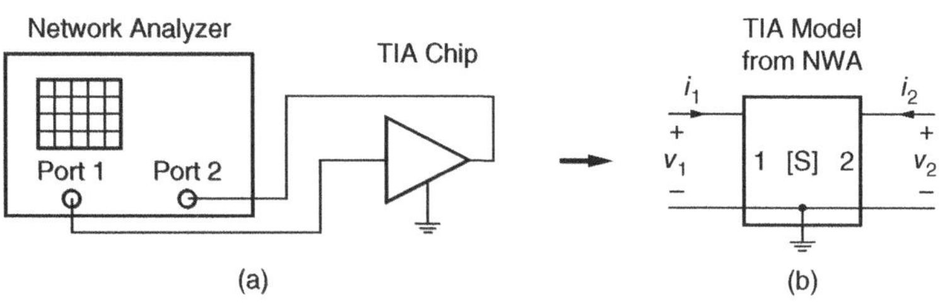

Figure 5.5 (a) Measurement of a TIA with a network analyzer and (b) the resulting S-parameter black-box model.

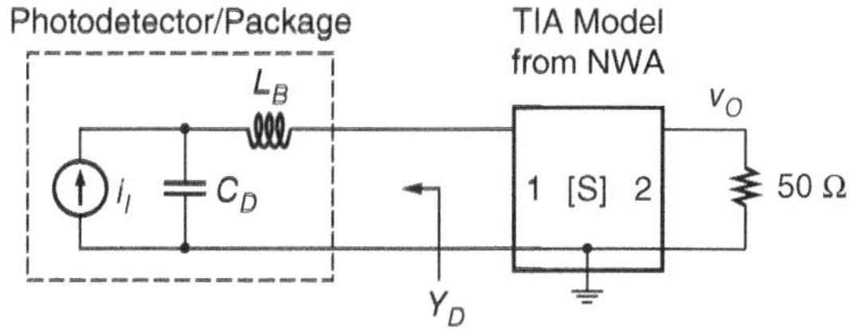

Figure 5.6 Calculation of the transimpedance from the photodetector/package model and the measured S parameters.

where R_0 is the characteristic impedance relative to which the S parameters were measured (usually 50 Ω), Y_D is the output admittance of the photodetector (including parasitics) as seen from the TIA input (see Fig. 5.6), and H_{PD} is the current referral function referring the intrinsic photodetector current, i_I, to the input of the TIA. Equation (5.3) follows easily from the facts that the input impedance of the black box is $(1 + S_{11})/(1 - S_{11})R_0$ and that the voltage gain from the input to the output (loaded with R_0) is $S_{21}/(1 + S_{11})$. For an arbitrary linear photodetector model, it can be shown that its output admittance is given by $Y_D = \mathcal{C}/\mathcal{D}$ and its current referral function by $H_{PD} = 1/\mathcal{D}$, where $\mathcal{C}$ and $\mathcal{D}$ are the ABCD parameters (chain parameters) of the network located between the intrinsic photodetector and the TIA (cf. Eq. (I.32)). [→ Problem 5.1.]

An approximation of Eq. (5.3), often encountered in the literature [5–7], is

$$Z_T(f) \approx \frac{S_{21}(f)R_0}{1 - S_{11}(f)}. \tag{5.4}$$

This expression is valid for frequencies that are low enough such that the output admittance of the photodetector can be neglected, $Y_D \approx 0$, and the current into the TIA input closely matches the intrinsic photodetector current, $H_{PD} \approx 1$.

Let us illustrate Eq. (5.3) for the simple photodetector/package model shown in Fig. 5.6. The capacitance C_D models the detector and the inductor L_B models a bond wire. The output admittance of the photodetector plus package is found to be $Y_D = sC_D/(1 + s^2L_BC_D)$ and the current referral function is $H_{PD} = 1/(1 + s^2L_BC_D)$. Inserting these two expressions into Eq. (5.3) results in

$$Z_T = \frac{S_{21}R_0}{(1 - \omega^2L_BC_D)(1 - S_{11}) + j\omega C_D(1 + S_{11})R_0}. \tag{5.5}$$

While the photodetector-simulator method requires a more elaborate measurement setup, it presents the TIA with a realistic input impedance and therefore guarantees stability for a correctly designed TIA. In contrast, the S-parameter method, can only be applied to TIAs that are stable for a 50-Ω source impedance in addition to the photodetector impedance. If stability is not an issue and measurements are done carefully, both the photodetector-simulator and the S-parameter method give comparable results [5].

5.2 Input Overload Current

Definition. Figure 5.7 depicts an idealized current waveform, $i_I(t)$, generated by a single-ended photodetector. This current is unipolar, that is, it is always positive. Its magnitude can be measured either as the signal's peak-to-peak value i_I^{pp} or its average value $\bar{i}_I$. In the case of a non-return-to-zero (NRZ) signal with DC balance (same number of zeros and ones) and high extinction ($i_I \approx 0$ for zeros), the peak-to-peak value is twice the average value: $i_I^{pp} = 2\bar{i}_I$. The current from a balanced detector, such as the one in Fig. 3.37, is bipolar, that is, it swings around zero. Its magnitude is measured zero to peak or peak to peak; its average value is zero.

For large input signals, TIAs tend to produce pulse-width distortion and jitter, as sketched in Fig. 5.8. This effect causes the bit-error rate (BER) of the receiver to increase rapidly with received power (cf. Figs 4.9 and 4.10). Eventually, the BER will exceed the maximum acceptable value, such as 10^{-12}, and the TIA is said to be overloaded. The *input overload current*, i_{ovl}^{pp}, is the maximum input current, measured peak to peak, for which the BER, jitter, or another performance criterion is just met. For a discussion of jitter and pulse-width distortion, see Appendix C.

The dynamic range of the TIA extends from its sensitivity i_{sens}^{pp} to its overload current i_{ovl}^{pp} and is defined as the ratio $i_{ovl}^{pp}/i_{sens}^{pp}$. (Here we assumed that the dynamic range is not limited by nonlinear distortions, cf. Section 5.3.) Expressed in decibels, the electrical dynamic range is $20\log(i_{ovl}^{pp}/i_{sens}^{pp})$.

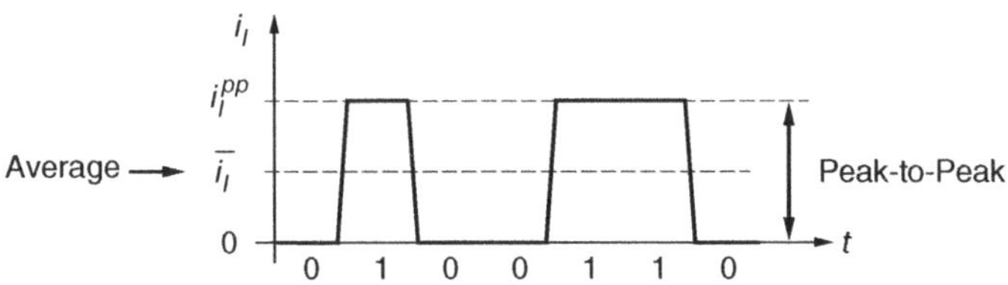

Figure 5.7 TIA input signal current: peak-to-peak value and average value.

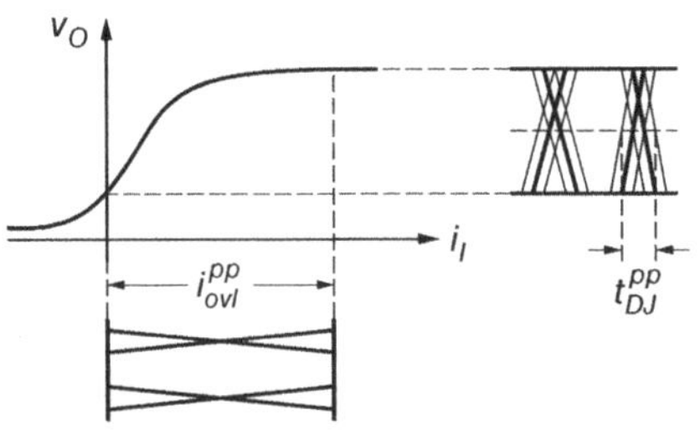

Figure 5.8 Pulse-width distortion and jitter as a result of input overload.

Typical Values. The input overload currents for some commercial parts are listed in Table 5.1. The overload requirement of a TIA can be estimated from the largest expected optical power, $\overline{P}_{ovl}$, at the receiver as $i_{ovl}^{pp} \approx 2\mathcal{R}\overline{P}_{ovl}$, where $\mathcal{R}$ is the photodetector's responsivity and a DC-balanced NRZ signal with high extinction was assumed. If the largest expected optical power is given as an optical modulation amplitude (OMA) instead of an average value, then $i_{ovl}^{pp} \approx \mathcal{R}P_{ovl}^{pp}$. Assuming a responsivity of 0.8 A/W, we find the following input overload currents:

Standard	$\overline{P}_{ovl}$ (dBm)	P_{ovl}^{OMA} (dBm)	i_{ovl}^{pp} (mA)
SONET OC-192 short reach	0.0		1.6
SONET OC-192 long reach	−3.0		0.8
100GBASE-SR10	2.4	3.0	1.6
100GBASE-LR4/ER4	4.5	4.5	2.3

5.3 Maximum Input Current for Linear Operation

Depending on the application, the useful dynamic range of the TIA ends before reaching the overload limit. For example, a receiver with equalizer or a receiver for an advanced modulation format (e.g., 4-PAM, DP-QPSK) requires a certain amount of linearity to function correctly. The *maximum input current for linear operation*, i_{lin}^{pp}, which is illustrated in Fig. 5.9 is always smaller than the overload current i_{ovl}^{pp}. For a discussion of nonlinearity (gain compression, total harmonic distortion, intermodulation distortions, and composite distortions), see Appendix D.

Definition. Several definitions for i_{lin}^{pp} are in use. For example, it can be defined as the peak-to-peak input current for which the transimpedance drops by 1 dB (about 11%) below its small-signal value, that is, the 1-dB compression point. Or it can be defined as the peak-to-peak sinusoidal input current for which the total harmonic distortion (THD) reaches 1% or 5%. The relevant definition depends on the application of the TIA. For PAM applications the level separation mismatch matters [8] and therefore a compression-point criterion

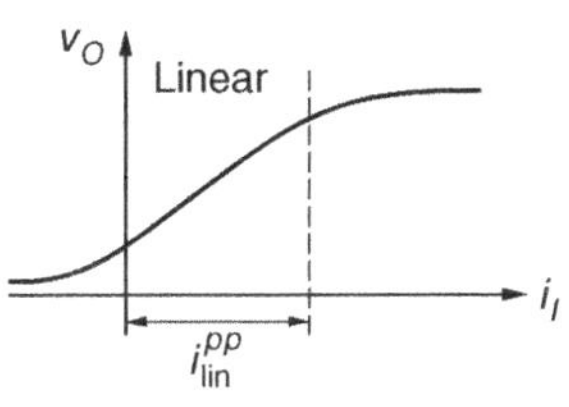

Figure 5.9 Maximum input current for linear operation.

is a good choice. For DP-QPSK applications, a THD criterion has been used [9]. For multicarrier applications composite distortions matter and therefore a CSO, CTB, or MTPR criterion is used (see Appendix D).

A TIA with a fixed transimpedance, R_T, must be able to deliver an output voltage swing larger than $R_T i^{pp}_{\text{lin}}$. This output-swing requirement can be relaxed, if the transimpedance is made adaptive such that its value reduces for large input signals. We explore this idea further in Section 7.4.

Typical Values. The maximum input currents for linear operation for some commercial parts are listed in Table 5.1.

For digital receivers that perform linear signal processing, such as equalization, the requirements for i^{pp}_{lin} are similar to those for i^{pp}_{ovl} discussed before. In applications, where the TIA's output signal is fed directly into a limiting amplifier or slicer, i^{pp}_{lin} is uncritical and can be as small as a few microamps.

For 4-PAM applications, a *level separation mismatch ratio*, defined as 3 × (minimum level separation)/(top to bottom level separation), of more than 92% is commonly required [8]. To meet this goal, the gain compression at i^{pp}_{lin} must remain less than 1.3 dB.

For 100-Gb/s DP-QPSK receivers, the agreement [9] requires $THD < 5\%$ for an AC current of 0.36 mA_{pp} and a DC current of 1.3 mA per photodetector.

For analog CATV/HFC receivers, the CSO and CTB distortions for the maximum input signal must stay below −51 dBc (see Appendix D).

5.4 Bandwidth

Definition. The TIA's *bandwidth*, $BW_{3\text{dB}}$, is defined as the (upper) frequency at which the transimpedance $|Z_T(f)|$ drops by 3 dB below its midband value (see Fig. 5.10). More precisely, this bandwidth is called the 3-dB bandwidth to distinguish it from other types of bandwidths such as the phase bandwidth or the noise bandwidth. The bandwidth depends among other things on the photodetector impedance, in particular its capacitance C_D (cf. Section 5.1). Hence the photodetector capacitance must be stated along with the bandwidth.

As we know from Section 4.8, the bandwidth determines the amount of intersymbol interference (ISI) introduced into the signal and the amount of noise picked up by the signal. The optimum bandwidth is given by a trade-off between these two mechanisms. For this reason it is important to specify a minimum bandwidth (to limit ISI) as well as a maximum bandwidth (to limit noise).

In optically amplified transmission systems, the receiver bandwidth directly impacts the required OSNR (cf. Section 4.6) and thus must be well controlled.

In addition to the 3-dB bandwidth, the maximum permitted peaking in the frequency response must be specified. When designing a TIA, it is often possible to improve the 3-dB bandwidth at the expense of more peaking

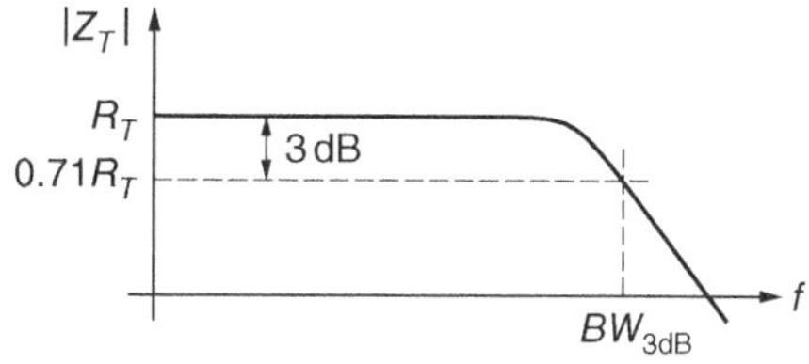

Figure 5.10 Definition of the 3-dB bandwidth.

(cf. Section 6.2). Usually, a flat frequency response (no peaking), as shown in Fig. 5.10, is required to keep signal distortions low.

Typical Values. Typical bandwidth values for some commercial parts are listed in Table 5.1. The optimum TIA bandwidth depends on many factors including the modulation format, the bandwidth allocation strategy chosen for the receiver, and the amount of optical filtering performed. These factors were discussed in Section 4.8. For a conventional NRZ receiver, operating at the bit rate B, the TIA bandwidth is chosen around $0.6B$ to $0.7B$, assuming that the TIA sets the receiver bandwidth. If the receiver bandwidth is set in another way (e.g., with an electrical filter), a wider TIA bandwidth around $0.9B$ to $1.2B$ is chosen:

Speed (Gb/s)	BW_{3dB} (GHz)
2.5	1.5 … 3
10	6 … 12
40	24 … 48

5.5 Phase Linearity and Group-Delay Variation

Definition. The 3-dB bandwidth specification alone does not say anything about the phase of $Z_T(f)$. Even if the frequency response $|Z_T(f)|$ is flat up to sufficiently high frequencies, distortions in the form of data-dependent jitter may occur if the phase linearity of $Z_T(f)$ is insufficient. The *phase linearity* (or, more accurately, the *phase distortion*), $\Delta\phi$, is defined as the largest deviation of the phase from the linear phase within the bandwidth of interest [2, 10], usually the 3-dB bandwidth. To determine the phase linearity, the phase is plotted as a function of linear frequency (not the usual logarithmic frequency), as illustrated in Fig. 5.11(a) (cf. Fig. H.6 on p. 491).

The *group delay* (a.k.a. *envelope delay*), τ, is related to the phase, ϕ, as $\tau(\omega) = -d\phi/d\omega$, where ϕ is in radians and $\omega = 2\pi f$. Thus, a linear phase corresponds to a constant group delay. The *group-delay variation*, $\Delta\tau$, is defined as the largest deviation of the group delay from the constant group delay within the bandwidth of interest (see Fig. 5.11(b)).

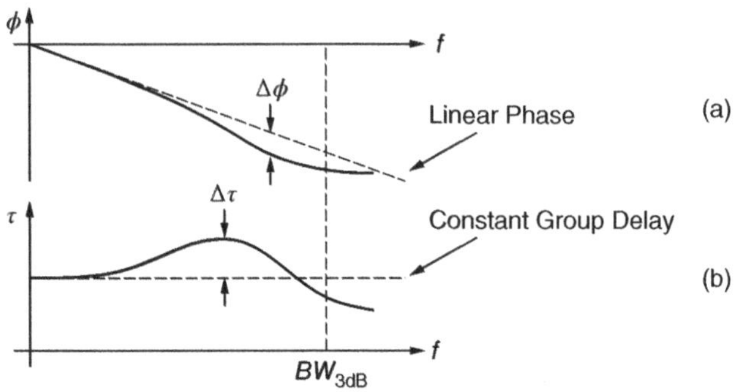

Figure 5.11 Definition of (a) the phase linearity and (b) the group-delay variation.

Can phase linearity and group-delay variation be related to each other? Only as an inequality. For a phase that wiggles around the linear-phase trajectory the group-delay variation can be made arbitrarily large by choosing rapid wiggles (large derivative), while the phase linearity itself can be kept small by limiting the magnitude of the wiggles. However, for a given phase linearity $\Delta\phi$, the group-delay variation must be at least $\Delta\tau \geq \Delta\phi/(2\pi BW_{3\text{dB}})$. The equality holds when the phase error builds up evenly throughout the whole bandwidth (cf. Appendix H).

Typical Values. Typically, a phase linearity, $\Delta\phi$, of less than 15° [2] to 20° [11, 12] is required over the bandwidth of interest to limit the generation of data-dependent jitter. Similarly, the group-delay variation, $\Delta\tau$, should be less than 10% of the bit period (0.1 UI) over the bandwidth of interest. The corresponding values are listed in the following table. Note that for $BW_{3\text{dB}} = \frac{2}{3}B$, a phase linearity of 15 to 20° implies a group-delay variation of at least 0.063 to 0.083 UI, making the two recommendations consistent.

Speed (Gb/s)	$\Delta\phi$	$\Delta\tau$ (ps)
2.5	<20°	<40
10	<20°	<10
40	<20°	<2.5

5.6 Timing Jitter

Good phase linearity and low group-delay variation guarantee low data-dependent jitter for *small* input signals. Remember that phase linearity and group-delay variation are derived from the *small-signal* quantity $Z_T(f)$.

However, to limit jitter generation over the full range of input signal swings, an explicit jitter specification is needed in addition to the phase linearity and group-delay variation. A discussion of jitter and its types (deterministic jitter, random jitter, etc.) is given in Appendix C.

Definition. To determine the jitter introduced by a TIA, an ideal (for simulations) or low-jitter (for measurements) data signal is applied to the input of the TIA. The deterministic jitter at the output of the TIA is then separated from the random jitter using one of the methods discussed in Appendix C. This procedure is repeated for a number of input current levels that cover the dynamic range from i^{pp}_{sens} to i^{pp}_{ovl}. Often the deterministic jitter is seen to increase with input current, reaching the maximum for i^{pp}_{ovl} [13].

Typical Values. Typical jitter values for some commercial parts are listed in Table 5.1. The deterministic jitter, measured peak to peak, t^{pp}_{DJ}, should be less than 10% of the bit period (0.1 UI). This specification is consistent with that given in Section 5.5 because a group-delay variation of 0.1 UI typically results in a linear (small-signal) jitter that is much less than 0.1 UI (cf. Appendix H), leaving a comfortable margin for nonlinear (large-signal) jitter.

Speed (Gb/s)	t^{pp}_{DJ} (ps)
2.5	<40
10	<10
40	<2.5

5.7 Input-Referred Noise Current

The input-referred noise current is one of the most critical TIA parameters. This is especially true in unamplified direct-detection systems, where the noise of the TIA often dominates all other noise sources (the noise from the photodetector, main amplifier, etc.) and therefore determines the sensitivity of the receiver. We discussed the impact of circuit noise on the performance of different types of receivers in Section 4.4.

Definition. Figure 5.12(left) shows a noiseless TIA with an imagined noise current generator, $i_{n,TIA}$, at the input. This noise current is chosen such that, together with the noiseless TIA, it reproduces the output noise of the actual noisy TIA shown in Fig. 5.12(right). The noise current $i_{n,TIA}$ is known as the *input-referred noise current* or the *equivalent input noise current*.

Several points must be made about this definition: First, the value of the input-referred noise current depends on the source impedance, that is, the photodetector impedance. The latter is determined mostly by the photodetector

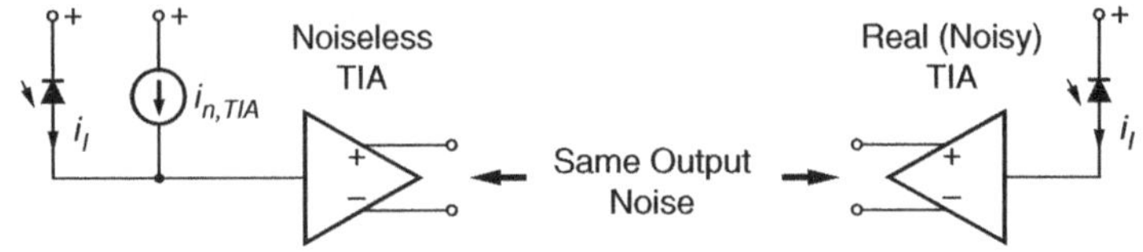

Figure 5.12 Definition of the input-referred noise current.

capacitance C_D and package parasitics. As we see in Section 6.3, a larger photodetector capacitance generally means more input-referred high-frequency noise current. Hence it is important to state the photodetector capacitance (or better the photodetector/package model) along with the input-referred noise current.

Second, it is possible to model the noise properties of a TIA in a source-impedance independent way by placing a noise current generator *and* a partially correlated noise voltage generator at the input of the noiseless TIA. We discuss this noise model, which we call the four-parameter model because it requires four (real) parameters for its description, in Section 6.5. The important point here is that the input-referred noise current in the four-parameter model is *different* from the input-referred noise current $i_{n,TIA}$ defined here and the two noise currents must not be confused!

Third, the input-referred noise current must be referred to the intrinsic photodetector such that it can be meaningfully compared to the intrinsic photocurrent signal [1]. Because of photodetector parasitics and package parasitics, this noise current is somewhat different from the noise current referred to the input of the TIA, especially at high frequencies (cf. Section 5.1). Hence it is important to state the photodetector/package model (e.g., C_D, L_B) along with the input-referred noise current.

Fourth, for TIAs with differential outputs the input-referred noise current must be chosen such that it reproduces the *differential* output noise. In general, the input-referred noise current that reproduces the differential noise and the input-referred noise current that reproduces the single-ended noise are not the same. They are equal only if all the output noise is contained in the differential mode (and the common mode is noise free), that is, if the two single-ended noise voltages are fully correlated and of opposite polarity. For example, simulations of a particular TIA design yielded an input-referred rms noise current that was 30% higher when reproducing the single-ended instead of the differential output noise. [→ Problem 5.2.]

The input-referred noise current, $i_{n,TIA}$, in Fig. 5.12 can be quantified in a number of different ways:

- **Input-Referred Noise Current PSD.** The power spectral density (PSD) of the input-referred noise current, $I^2_{n,TIA}(f)$, or the *input-referred noise current PSD* for short, is illustrated with an example in Fig. 5.13. The square root of

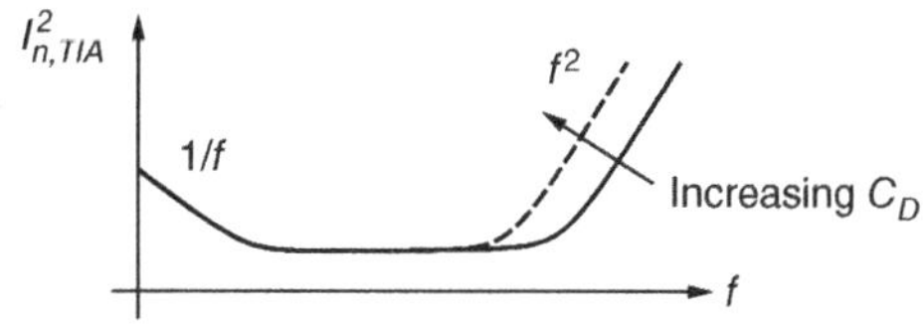

Figure 5.13 Typical input-referred noise current PSD.

this quantity, is known as the root spectral density of the input-referred noise current, $I_{n,TIA}(f)$, or the *input-referred noise current density* for short. Noise current PSDs are measured in pA^2/Hz and noise current densities are measured in $\text{pA}/\sqrt{\text{Hz}}$. The input-referred noise current PSD typically consists of a frequency-independent part, an f^2 part at high frequencies, and possibly a $1/f$ part at low frequencies. Frequency-independent noise is also known as *white noise*. Continuing the color analogy, f^2 noise, f noise, $1/f$, noise, and $1/f^2$ noise are referred to as violet, blue, pink, and red noise, respectively.

In Section 6.3, we analyze the input-referred noise current PSD quantitatively. Because this PSD is *not white*, it cannot be characterized by a single number, but a graph must be provided instead. To compare the noise performance of different TIAs, it is necessary to look at the whole PSD up to at least the TIA's noise bandwidths, which are larger than the 3-dB bandwidth (cf. Section 4.5). But, even knowing the whole PSD is not enough to infer the TIA's sensitivity.

- **Input-Referred RMS Noise Current.** Another measure of the input-referred noise current that relates directly to the sensitivity and can be expressed by a single number (in nA or μA) is its rms noise value, $i^{rms}_{n,TIA}$. As we discussed at length in Section 4.5, the *input-referred rms noise current*, or the *total input-referred noise current*, is determined by dividing the rms output noise voltage by the TIA's midband transimpedance value. The rms output noise voltage, in turn, is obtained by integrating the output-referred noise voltage PSD and taking the square root of it. Thus, we have

$$i^{rms}_{n,TIA} = \frac{v^{rms}_{n,TIA}}{R_T} = \frac{1}{R_T}\sqrt{\int_0^{f_u} |Z_T(f)|^2\, I^2_{n,TIA}(f)\, df}, \tag{5.6}$$

where $v^{rms}_{n,TIA}$ is the rms output noise, $|Z_T(f)|$ is the frequency response of the transimpedance, and R_T is its midband value (cf. Eq. (4.39)). For analytical calculations, the integration can be carried out to infinity ($f_u \to \infty$); for simulations (and measurements), it is usually enough to integrate up to about 2× the TIA's noise bandwidths ($f_u > 2\max\{BW_n, BW_{n2}\}$), after which the contributions to the rms output noise become very small. See Section 4.5 for a discussion of the noise bandwidths. The square of the input-referred rms noise current is known as the *input-referred mean-square noise current*, $\overline{i^2_{n,TIA}}$, and is measured in units of nA^2 or μA^2.

As we discussed in Section 4.4, the input-referred rms noise current, $i^{rms}_{n,TIA}$, directly determines the (electrical) sensitivity of the TIA as (cf. Eq. (4.19))

$$i^{pp}_{\text{sens}} = 2Q\, i^{rms}_{n,TIA} \tag{5.7}$$

and therefore this noise measure is a good metric to compare different TIAs (designed for the same bit rate). When the TIA noise dominates the receiver circuit noise, $i^{rms}_{n,TIA}$ can also serve as an estimate for the receiver's noise, $i^{rms}_{n,\text{ckt}} \approx i^{rms}_{n,TIA}$, which, in turn, determines the electrical sensitivity of the receiver.

- **Averaged Input-Referred Noise Current Density.** The input-referred noise current can also be described by the *averaged input-referred noise current density.* This quantity is defined as the input-referred rms noise current, $i^{rms}_{n,TIA}$, divided by the square root of the TIA's noise bandwidth[1] [15]:

$$I^{\text{avg}}_{n,TIA} = \frac{i^{rms}_{n,TIA}}{\sqrt{BW_n}} = \sqrt{\frac{\int_0^{f_u} |Z_T(f)|^2\, I^2_{n,TIA}(f)\, df}{\int_0^{f_u} |Z_T(f)|^2\, df}}, \tag{5.8}$$

where BW_n is the noise bandwidth. The expression on the right follows from Eq. (5.6) and the definition of the noise bandwidth Eq. (4.44). It is important to realize that $I^{\text{avg}}_{n,TIA}$ is a *weighted* average (weighted by $|Z_T(f)|^2$) and therefore cannot be obtained by simply averaging $I^2_{n,TIA}(f)$ or $I_{n,TIA}(f)$ over the TIA's bandwidth. Such simple averages frequently underestimates the true averaged noise value.

What is the meaning of the averaged input-referred noise current density? It can be interpreted as the noise current density of the white noise source that must be applied to the input of a noise-free TIA to reproduce the rms output noise (but not its spectral distribution) of the real (noisy) TIA. This fact follows easily from the definition in Eq. (5.8) and is illustrated in Fig. 5.14. This interpretation also suggests a simple procedure to determine the averaged input-referred noise current density. If we apply a white noise current source to the input of the real (noisy) TIA and adjust its amplitude until the

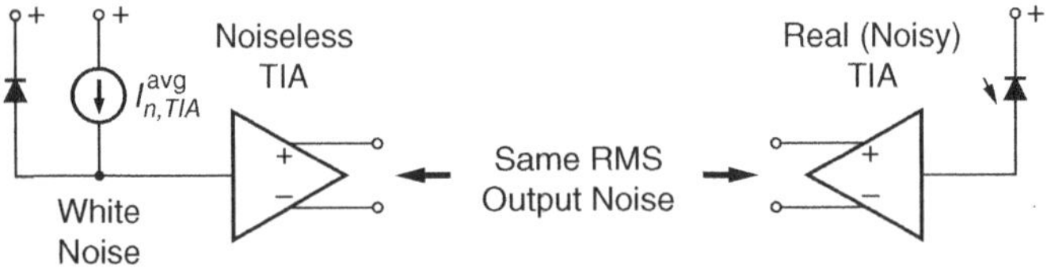

Figure 5.14 Interpretation of the averaged input-referred noise current density.

1 Sometimes the 3-dB bandwidth is used instead of the noise bandwidth [14]. However, the interpretation of $I^{\text{avg}}_{n,TIA}$ illustrated in Fig. 5.14 is valid only if the noise bandwidth is used.

TIA's output noise power is twice that without the noise source, then the noise current density of the applied white noise current equals the averaged input-referred noise current density of the TIA [16].

Typical input-referred rms noise current values for some commercial parts are listed in Table 5.1. Note that the rms noise tends to increase with the bit rate. We examine the dependence of $i_{n,TIA}^{rms}$ on the bit rate in Section 6.4. [→ Problem 5.3.]

Measurement. The input-referred noise current generator cannot be measured directly because it exists only in the mind of the imaginative engineer. To determine the input-referred noise, we have to measure the output noise, which is a real noise quantity, and then mathematically refer it back to the input using the separately measured transimpedance function $|Z_T(f)|$. In practice, we can measure the output noise voltage PSD with a *spectrum analyzer*. If the noise floor of the spectrum analyzer is too high, a broadband low-noise amplifier can be inserted between the TIA and the spectrum analyzer [4, 15], as shown in Fig. 5.15.

Because the measured noise depends on the source impedance, the TIA's input must be terminated with a (dark) photodetector or an equivalent impedance. To find the input-referred noise current PSD, $I_{n,TIA}^2(f)$, from the measured output noise voltage PSD, we first subtract the noise voltage PSDs due to the low-noise amplifier and the spectrum analyzer and then refer the remaining noise back through the low-noise amplifier and the TIA to the input.

The input-referred rms noise current and the averaged input-referred noise current density can be calculated from $I_{n,TIA}^2(f)$ using Eqs. (5.6) and (5.8), respectively.

If we are not interested in the spectral information, but only in the rms noise, we can replace the spectrum analyzer in Fig. 5.15 with an oscilloscope to measure the rms output noise voltage [15, 17, 18]. Dividing the de-embedded rms output noise voltage by the midband amplifier gain and the midband transimpedance, R_T, yields the input-referred rms noise current.

Relationship to the Noise Figure. Can the input-referred noise current be determined from a noise-figure measurement? In principle yes, but we need to

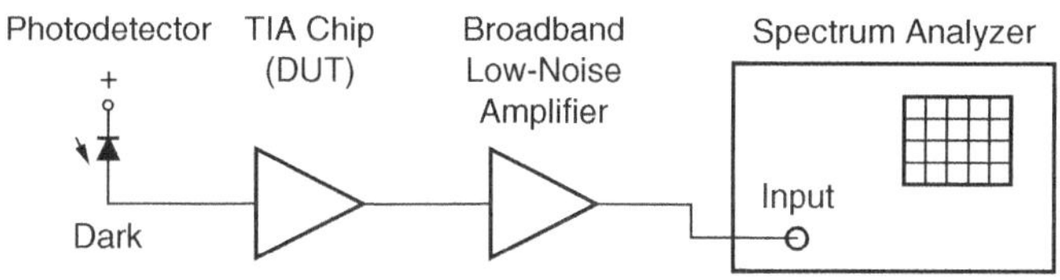

Figure 5.15 Measurement of the TIA's output noise voltage PSD.

determine the noise figure F for the source impedance Z_D presented by the photodetector. Most noise-figure meters, however, measure F only for a 50-Ω source impedance and, unfortunately, we cannot mathematically transform a 50-Ω noise figure to another source impedance. To find the noise figure for an arbitrary source impedance, we need to measure the noise figure for (at least) four separate source impedances, which requires an elaborate test set with impedance tuners. From these four measurements we can then infer four noise parameters (usually, the minimum noise figure, $F_{\min}$, the real and imaginary part of the optimum source admittance, Y_{opt}, and the noise resistance, R_n; cf. Section 6.5), which, in turn, permit us to calculate the noise figure for an arbitrary source impedance [19, 20]. (In practice, more than four impedance points are usually measured in order to be able to calculate the noise parameters with sufficient accuracy [21]. Conversely, if we have *a priori* knowledge about the noise mechanisms in the device under test, we can get away with fewer than four measurements [21, 22].)

After we have found the (spot) noise figure $F(f)$ for the photodetector source impedance Z_D, we can proceed to calculate the input-referred noise current PSD as follows:

$$I_{n,TIA}^2(f) = F(f) \cdot 4kT\text{Re}\left\{\frac{1}{Z_D(f)}\right\} = F(f) \cdot 4kTG_D(f). \tag{5.9}$$

By definition, the noise figure (first factor) multiplies the noise current PSD generated by the source impedance (second factor) to produce the input-referred noise current PSD of the TIA plus the source impedance. Clearly, a purely reactive photodetector, $G_D = 0$, produces no noise and therefore results in an undefined noise figure ($F \rightarrow \infty$). Thus, the noise-figure method of determining the input-referred noise current is only applicable if the photodetector is lossy. An example of a lossy photodetector model is shown in Fig. 5.16 (cf. Fig. 3.4(a)). For this model, we have $Z_D(\omega) = (1 + sR_{PD}C_{PD})/(sC_{PD})$ and the noise current generating conductance follows as $G_D(\omega) = \text{Re}\{1/Z_D(\omega)\} = (\omega C_{PD})^2 R_{PD}/(1 + \omega^2 R_{PD}^2 C_{PD}^2)$.

Equation (5.9) expresses the equivalent noise current PSD at the input of the TIA. But we know that this is not quite what we want: The noise current

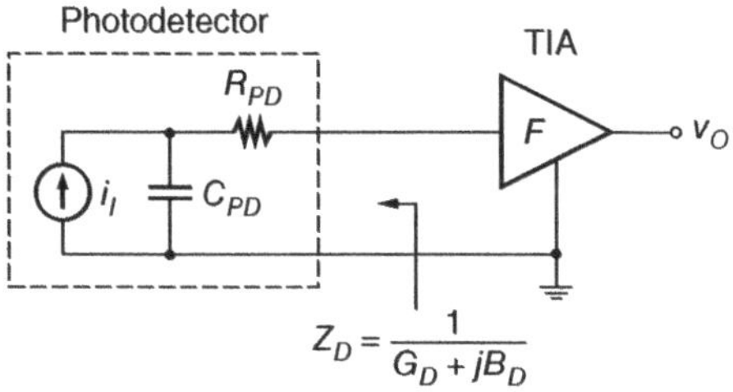

Figure 5.16 Lossy photodetector model followed by a TIA with noise figure F measured for the source impedance Z_D.

should be referred to the intrinsic photodetector. This flaw can easily be corrected by dividing Eq. (5.9) by $|H_{PD}|^2$, where H_{PD} is the current referral function introduced in Section 5.1. Remember, H_{PD} refers the intrinsic photodetector current, i_I, to the input of the TIA. Thus we have

$$I^2_{n,TIA} = \frac{F \cdot 4kTG_D}{|H_{PD}|^2} = F \cdot 4kTG_{Di}, \tag{5.10}$$

where $G_{Di} = G_D/|H_{PD}|^2$ is known as the *photodetector intrinsic conductance* [1]. Continuing our example from Fig. 5.16, we find the referral function $|H_{PD}|^2 = 1/(1 + \omega^2 R^2_{PD} C^2_{PD})$ and therefore $G_{Di} = (\omega C_{PD})^2 R_{PD}$ [1, 23]. For an arbitrary linear photodetector model, it can be shown that its intrinsic conductance is given by $G_{Di} = \text{Re}\{C^* D\}$, where C and D are the ABCD parameters (chain parameters) of the network located between the intrinsic photodetector and the TIA [1] (cf. Eq. (I.34)). [→ Problem 5.4.]

5.8 Crosstalk

Multichannel receivers (e.g., 4 × 25 Gb/s) or receivers for parallel optical interconnects often make use of multichannel TIA chips. Because the TIAs on those chips are located in close proximity to each other, crosstalk through the power supply, substrate, mutual-wire capacitance, or mutual-wire inductance is an important concern [24–28].

Definition. Figure 5.17 shows an array of receiver channels with crosstalk. To quantify crosstalk, we pick a channel, the so-called victim channel, and analyze how it is disturbed by the other channels, known as aggressor channels. Without crosstalk, the victim channel's output signal is v_S. With crosstalk from the other channels (v_{AS1} and v_{AS2} in Fig. 5.17), the output signal becomes $v_S + v_X$,

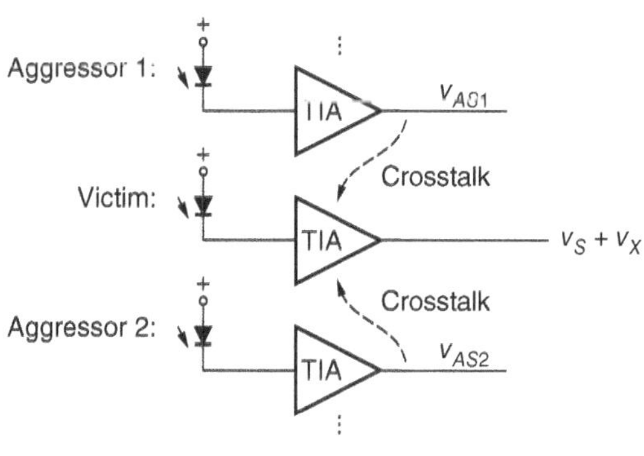

Figure 5.17 Crosstalk interference from adjacent channels.

where v_X is the total crosstalk voltage. A similar analysis can be made for every receiver channel on the chip.

Electrical crosstalk from the nth aggressor channel ($n = 1, 2$ in Fig. 5.17) to a particular victim channel is defined as

$$XT_n = \frac{v_{Xn}^{pp}}{v_{ASn}^{pp}}, \tag{5.11}$$

where v_{Xn} is the crosstalk voltage due to the nth aggressor channel and v_{ASn} is the output signal of the nth aggressor channel. Electrical crosstalk usually is expressed in decibels using the conversion rule $20 \log XT$.

The total worst-case crosstalk voltage appearing in the victim channel, v_X, can be derived from Eq. (5.11) by summing the individual contributions in a peak-to-peak manner:

$$v_X^{pp} = \sum_n v_{Xn}^{pp} = \sum_n XT_n \cdot v_{ASn}^{pp}, \tag{5.12}$$

where the summation is over all the aggressor channels.

In the special case where N dominant aggressor channels have the same swing v_{AS}^{pp} and the same crosstalk XT to the victim channel while all the other channels produce negligible crosstalk, the sum can be simplified to the product

$$v_X^{pp} = N \cdot XT \cdot v_{AS}^{pp}. \tag{5.13}$$

Power Penalty. To overcome the BER degradation due to crosstalk, more optical power must be transmitted. In other words, crosstalk can be quantified by a power penalty (or sensitivity penalty), as discussed in Section 4.7. Figure 5.18 illustrates the derivation of the power penalty due to crosstalk [27]. The clean eye in Fig. 5.18(a) is degraded by crosstalk to that in Fig. 5.18(b). To restore the vertical eye opening to that of the clean eye (and thus to approximately restore the BER), we need to increase the signal swing from v_S^{pp} to $v_S'^{pp} = v_S^{pp} + v_X^{pp}$, as shown in Fig. 5.18(c). (It is assumed here that the act of increasing the signal swing does not significantly increase the crosstalk.) Thus, the power penalty is

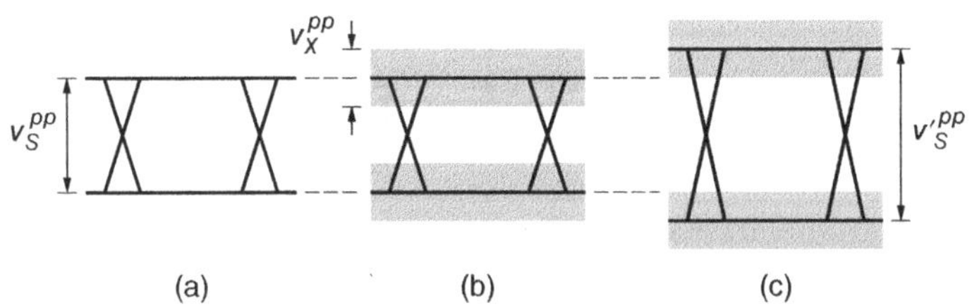

Figure 5.18 Power penalty due to crosstalk: (a) without crosstalk, (b) with crosstalk, and (c) with crosstalk and increased signal swing to restore the original BER.

$PP = (v_S^{pp} + v_X^{pp})/v_S^{pp} = 1 + v_X^{pp}/v_S^{pp}$. Inserting Eq. (5.13) for v_X^{pp} yields

$$PP = 1 + N \cdot XT \cdot \frac{v_{AS}^{pp}}{v_S^{pp}}. \tag{5.14}$$

Power penalties usually are expressed in decibels using the conversion rule $10 \log PP$ (cf. Section 4.7). For example, with two dominant aggressor channels (the adjacent channels), a per-channel crosstalk of −20 dB, and equal swings in the aggressor and victim channels, we have $PP = 1 + 2 \cdot 0.1 \cdot 1 = 1.2$ corresponding to a power penalty of 0.79 dB.

Finally, we can solve Eq. (5.14) for XT to obtain an expression for the largest permissible crosstalk that complies with a given power penalty:

$$XT \leq \frac{PP - 1}{N} \cdot \frac{v_S^{pp}}{v_{AS}^{pp}}. \tag{5.15}$$

For example, given a power penalty of 0.5 dB ($PP = 1.122$), five aggressor channels, and a 10× higher swing for the aggressor channels than the victim channel, the per-channel crosstalk must be kept below $0.122/5 \cdot 0.1 = 0.00244$, corresponding to less than −52.3 dB [27].

Typical Values. The quad 25-Gb/s TIA chip in [28] achieves a crosstalk of less than −17 dB over the whole bandwidth and a power penalty of less than 0.8 dB. The power penalty of the quad 25-Gb/s TIA chip in [25] is less than 0.15 dB. The crosstalk of the dual 64-Gb/s TIA chip in [29] is below −34 dB over the whole bandwidth.

5.9 Product Examples

Table 5.1 summarizes typical specifications for some commercial TIA chips. The numbers have been taken from data sheets that were available at the time of writing. Besides the parts shown, many more are available from several manufacturers. For up-to-date product information, please contact the manufacturer directly.

All listed TIA parts have differential outputs and thus the *differential* transimpedance is shown under R_T. The maximum input current for linear operation, i_{lin}^{pp}, is specified for the linearity criterion shown at the bottom of the table, which varies from part to part.

The PHY1090 and ONET8511T are linear TIAs for applications that require electronic dispersion compensation (EDC) such as 10-Gigabit Ethernet LRM receivers. The other parts are limiting TIAs for applications without equalization. The ASNT6123-BD contains two independent TIAs for dual-channel applications.

Table 5.1 High-speed TIA products.

Company & product	Speed (Gb/s)	R_T (kΩ)	BW_{3dB} (GHz)	t_{DJ}^{pp} (ps)	$i_{n,TIA}^{rms}$ (μA)	i_{lin}^{pp} (μA)	i_{ovl}^{pp} (mA)	Power (mW)	Technology
Maxim MAX3864	2.5	2.8	2.0	24	0.49	40[a)]	2.0	112	SiGe Bipolar
Maxim PHY1097	2.5	25.0	1.7	20	0.15		4.4	158	
Maxim MAX3797	10.3	5.5	7.7	6	1.25		1.0	105	SiGe Bipolar
Maxim PHY1090	10.3	2.0	6.0	5	<1.10	600[b)]	2.0	149	
Hittite HMC960	11.3	1.3	7.5	<10	1.00	300	3.0	304	
TI ONET8551T	11.3	10.0	9.0	10	0.90		2.5	92	
TI ONET8511T	11.3	5.5	8.0	8	1.00	1,400[c)]	2.0	152	
Adsantec ASNT6123-BD	25.0		18.0	<2			1.5	495	SiGe BiCMOS
Adsantec ASNT6121	43.0	4.0	35.0	<2			2.0	475	SiGe BiCMOS
Hittite HMC6590	43.0	4.0	39.0		3.30		4.0	310	SiGe BiCMOS

a) Measured at 5% gain compression.
b) Measured at 2% THD; increases to 2 mA at 6.5% THD.
c) Measured at 5% THD.

5.10 Summary

The main specifications of the transimpedance amplifier (TIA) are as follows:

- The *transimpedance*, which we want to be as large as possible to relax the gain and noise requirements for the subsequent main amplifier.
- The *input overload current*, which must be large enough to avoid harmful pulse-width distortion and jitter when the maximum optical signal power is received.
- The *maximum input current for linear operation*, which must be large enough to avoid harmful signal distortions (signal compression, harmonic, and intermodulation distortions) when the maximum optical signal power is received. This specification is important for receivers using linear signal processing (e.g., equalization), an advanced modulation format (e.g., 4-PAM, DP-QPSK), or a linear modulation scheme (e.g., CATV/HFC).
- The *3-dB bandwidth*, which for NRZ modulation is between $0.6B$ and $1.2B$, where B is the bit rate, depending on the bandwidths of the other components in the receive path.
- The *phase linearity* (or, more accurately, the phase distortion) and the related *group-delay variation*, which both must be kept small over the bandwidth of interest to minimize (small-signal) jitter and other signal distortions.
- The *jitter*, which must be kept low for signals ranging from the sensitivity limit to the overload limit.
- The *input-referred noise current*, which must be as small as possible to obtain high sensitivity, especially in p–i–n receivers. The TIA noise performance can be specified in terms of the input-referred noise current PSD, the averaged input-referred noise current density, or the input-referred rms noise current. The latter quantity determines the sensitivity.
- The *crosstalk* in multichannel TIA chips, which must be kept small to avoid sensitivity degradations (power penalties).

Besides the specifications discussed in this chapter there are several other important TIA specifications. Examples are the power-supply current, the power-supply rejection, the input bias voltage at the photodetector, the output voltage swing, and the S parameter of the output (S_{22}). For TIAs with differential outputs, the differential output voltage swing, the common-mode output voltage, and the mixed-mode S parameters of the outputs (S_{dd22}, S_{cc22}, S_{dc22}, and S_{cd22}) must be specified. If the TIA contains a voltage regulator or filter for the photodetector bias, a received signal strength indicator (RSSI), or an automatic gain control (AGC) loop, the parameters associated with these features must be specified as well.

Problems

5.1 Transimpedance and S Parameters. A photodetector (including extrinsic and packaging parasitics) can be modeled as a cascade of an intrinsic photodetector (an ideal current source) and a two-port network. (a) Derive Z_T in terms of the ABCD parameters of this two-port network and the S parameters of the subsequent TIA. (b) Demonstrate that Eq. (5.3) is correct and find expressions for Y_D and H_{PD} in terms of ABCD parameters. (c) Find the ABCD parameters, Y_D, and H_{PD} for the photodetector model shown in Fig. 5.6.

5.2 Input-Referred RMS Noise Current. A TIA with differential outputs has a differential transimpedance of 1 kΩ and an rms output noise voltage of 1 mV at each output terminal. (a) Assuming that the noise at the two outputs is uncorrelated, how much noise is in the differential mode and how much noise is in the common mode? (b) What is the input-referred rms noise current required to reproduce the single-ended output noise and what is the input-referred rms noise current required to reproduce the differential output noise?

5.3 TIA Dynamic Range. A 10-Gb/s NRZ receiver must be able to handle optical input signals in the range from −19 to +3 dBm at $BER < 10^{-12}$. Assume that the photodetector responsivity is 0.8 A/W and that the optical signal has high extinction. (a) What is the optical and electrical dynamic range of this receiver? (b) What input overload current and input-referred rms noise current should the TIA have? (c) What averaged input-referred noise current density should the TIA have, if its noise bandwidth is 7.5 GHz?

5.4 Photodetector Intrinsic Conductance. A photodetector (including extrinsic and packaging parasitics) can be modeled as a cascade of an intrinsic photodetector (an ideal current source) and a two-port network. (a) Express the photodetector intrinsic conductance G_{Di} in terms of the ABCD parameters of the two-port network. (b) Find the ABCD parameters and G_{Di} for the two-port network shown in Fig. 5.16.

References

1 A. Leven, R. Reuter, and Y. Baeyens. Unified analytical expressions for transimpedance and equivalent input noise current of optical receivers. *IEEE Trans. Microwave Theory Tech.*, MTT-48(10):1701–1706, 2000.

2 J. Müllrich, H. Thurner, E. Müllner, J. F. Jensen, W. E. Stanchina, M. Kardos, and H.-M. Rein. High-gain transimpedance amplifier in InP-based HBT technology for receiver in 40-Gb/s optical-fiber TDM links. *IEEE J. Solid-State Circuits*, SC-35(9):1260–1265, 2000.

3 M. Neuhäuser, H.-M. Rein, H. Wernz, and A. Felder. 13 Gbit/s Si bipolar preamplifier for optical front ends. *Electron. Lett.*, 29(5):492–493, 1993.

4 S. M. Park and H.-J. Yoo. 1.25-Gb/s regulated cascode CMOS transimpedance amplifier for Gigabit Ethernet applications. *IEEE J. Solid-State Circuits*, SC-39(1):112–121, 2004.

5 C. Q. Wu, E. A. Sovero, and B. Massey. 40-GHz transimpedance amplifier with differential outputs using InP–InGaAs heterojunction bipolar transistors. *IEEE J. Solid-State Circuits*, SC-38(9):1518–1523, 2003.

6 D. Li, G. Minoia, M. Repossi, D. Baldi, E. Temporiti, A. Mazzanti, and F. Svelto. A low-noise design technique for high-speed CMOS optical receivers. *IEEE J. Solid-State Circuits*, SC-49(6):1437–1447, 2014.

7 J. S. Weiner, A. Leven, V. Houtsma, Y. Baeyens, Y.-K. Chen, P. Paschke, Y. Yang, J. Frackoviak, W.-J. Sung, A. Tate, R. Reyes, R. F. Kopf, and N. G. Weimann. SiGe differential transimpedance amplifier with 50-GHz bandwidth. *IEEE J. Solid-State Circuits*, SC-38(9):1512–1517, 2003.

8 M. Bassi, F. Radice, M. Bruccoleri, S. Erba, and A. Mazzanti. A 45Gb/s PAM-4 transmitter delivering 1.3V_{ppd} output swing with 1V supply in 28nm CMOS FDSOI. In *ISSCC Digest of Technical Papers*, pages 66–67, February 2016.

9 OIF. Implementation Agreement for Integrated Dual Polarization Micro-Intradyne Coherent Receivers – IA # OIF-DPC-MRX-01.0. Optical Internetworking Forum, Fremont, CA, March 2015.

10 M. Neuhäuser, H.-M. Rein, and H. Wernz. Low-noise, high-gain Si-bipolar preamplifiers for 10 Gb/s optical-fiber links—design and realization. *IEEE J. Solid-State Circuits*, SC-31(1):24–29, 1996.

11 K. Ogawa, L. D. Tzeng, Y. K. Park, and E. Sano. Advances in high bit-rate transmission systems. In I. P. Kaminow and T. L. Koch, editors, *Optical Fiber Telecommunications IIIA*, pages 336–372. Academic Press, San Diego, CA, 1997.

12 L. D. Tzeng, O. Mizuhara, T. V. Nguyen, K. Ogawa, I. Watanabe, K. Makita, M. Tsuji, and K. Taguchi. A high-sensitivity APD receiver for 10-Gb/s system applications. *IEEE Photonics Technol. Lett.*, 8(9):1229–1231, 1996.

13 A. Maxim. A 54 dBΩ + 42 dB 10 Gb/s SiGe transimpedance-limiting amplifier using bootstrap photodiode capacitance neutralization and vertical threshold adjustment. *IEEE J. Solid-State Circuits*, SC-42(9):1851–1864, 2007. This paper may contain falsified information, see the June 2008 issue.

14 B. Razavi. *Design of Integrated Circuits for Optical Communications.* McGraw-Hill, New York, 2003.

15 J. S. Weiner, J. S. Lee, A. Leven, Y. Baeyens, V. Houtsma, G. Georgiou, Y. Yang, J. Frackoviak, A. Tate, R. Reyes, R. F. Kopf, W.-J. Sung, N. G. Weimann, and Y.-K. Chen. An InGaAs–InP HBT differential transimpedance amplifier with 47-GHz bandwidth. *IEEE J. Solid-State Circuits*, SC-39(10):1720–1723, 2004.

16 R. G. Smith and S. D. Personick. Receiver design for optical fiber communication systems. In H. Kressel, editor, *Topics in Applied Physics: Semiconductor Devices for Optical Communication*, volume 39. Springer-Verlag, Berlin, Germany, 1982.

17 J.-Y. Dupuy, F. Jorge, M. Riet, V. Nodjiadjim, H. Aubry, and A. Konczykowska. 59-dBΩ 68-GHz variable gain-bandwidth differential linear TIA in 0.7–μm InP DHBT for 400-Gb/s optical communication systems. In *Compound Semiconductor Integrated Circuit Symposium (CSICS)*, pages 1–4, New Orleans, LA, October 2015.

18 H. Mohammadnezhad, A. K. Bidhendi, M. M. Green, and P. Heydari. A low-power BiCMOS 50 Gbps G_m-boosted dual-feedback transimpedance amplifier. In *Proceedings of IEEE Bipolar/BiCMOS Circuits and Technology Meeting*, pages 161–164, October 2015.

19 T. H. Lee. *The Design of CMOS Radio-Frequency Integrated Circuits*. Cambridge University Press, Cambridge, UK, 2nd edition, 2003.

20 D. M. Pozar. *Microwave Engineering*. John Wiley &Sons, Inc., Hoboken, NJ, 4th edition, 2012.

21 S. Asgaran, M. J. Deen, C.-H. Chen, G. A. Rezvani, Y. Kamali, and Y. Kiyota. Analytical determination of MOSFET's high-frequency noise parameters from NF_{50} measurements and its application in RFIC design. *IEEE J. Solid-State Circuits*, SC-42(5):1034–1043, 2007.

22 R. P. Jindal. Transimpedance preamplifier with 70-dB AGC range in fine-line NMOS. *IEEE J. Solid-State Circuits*, SC-23(3):867–869, 1988.

23 M. S. Park and R. A. Minasian. Ultra-low-noise and wideband-tuned optical receiver synthesis and design. *J. Lightwave Technol.*, LT-12(2):254–259, 1994.

24 S. Goswami, T. Copani, A. Jain, H. Karaki, B. Vermeire, H. J. Barnaby, G. Fetzer, R. Vercillo, and S. Kiaei. A 96Gb/s-throughput transceiver for short-distance parallel optical links. In *ISSCC Digest of Technical Papers*, pages 230–231, February 2008.

25 G. Kalogerakis, T. Moran, T. Nguyen, and G. Denoyer. A quad 25Gb/s 270mW TIA in 0.13μm BiCMOS with <0.15dB crosstalk penalty. In *ISSCC Digest of Technical Papers*, pages 116–117, February 2013.

26 C. Kromer, G. Sialm, C. Berger, T. Morf, M. L. Schmatz, F. Ellinger, D. Erni, G.-L. Bona, and H. Jäckel. A 100-mW 4 × 10 Gb/s transceiver in 80-nm CMOS for high-density optical interconnects. *IEEE J. Solid-State Circuits*, SC-40(12):2667–2679, 2005.

27 T. Nagahori, K. Miyoshi, Y. Aizawa, Y. K. Y. Nukada, N. Kami, and N. Suzuki. An analog front-end chip set employing an electro-optical mixed design on SPICE for 5-Gb/s/ch parallel optical interconnection. *IEEE J. Solid-State Circuits*, SC-36(12):1984–1991, 2001.

28 T. Takemoto, F. Yuki, H. Yamashita, Y. Lee, T. Saito, S. Tsuji, and S. Nishimura. A compact 4 × 25-Gb/s 3.0 mW/Gb/s CMOS-based optical receiver for board-to-board interconnects. *J. Lightwave Technol.*, LT-28(23):3343–3350, 2010.

29 A. Awny, R. Nagulapalli, D. Micusik, J. Hoffmann, G. Fischer, D. Kissinger, and A. C. Ulusoy. A dual 64Gbaud 10kΩ 5% THD linear differential transimpedance amplifier with automatic gain control in 0.13μm BiCMOS technology for optical fiber coherent receivers. In *ISSCC Digest of Technical Papers*, pages 406–407, February 2016.

6

Basic Transimpedance Amplifier Design

We start our exploration of TIA topologies with the low- and high-impedance front-ends. These simple front-ends illustrate important design trade-offs and motivate the need for more sophisticated TIA topologies. Then, we move on to the popular shunt-feedback architecture, which we study in great detail. We calculate the transimpedance, input impedance, and output impedance. We analyze the stability and the transimpedance limit of single- and multistage implementations. Next, we derive the noise performance of TIAs with FET and BJT front-ends. We explain Ogawa's noise factor and its relationship to induced gate noise. After that, we discuss the noise optimization of TIAs with FET and BJT front-ends by means of device sizing and biasing. We examine the impact of constraints, such as a constant gain-bandwidth product, on the noise optimum. Finally, we investigate noise-matching networks and their properties.

6.1 Low- and High-Impedance Front-Ends

The term TIA in its wide sense refers to any circuit that converts a current into a voltage. This includes the simple resistor! In this sense, the *low-impedance front-end* and the *high-impedance front-end* are simple forms of TIAs.

Low-Impedance Front End. A simple low-impedance front-end is shown in Fig. 6.1(a). A 50-Ω load resistor converts the photodetector current i_I into a proportional voltage following Ohm's law. This voltage is then brought to the output with a unity-gain buffer that has a high-impedance input (capacitive loading only). Clearly, the output voltage (at low frequencies) is $v_O = 50\ \Omega \times i_I$ and thus the transimpedance is $R_T = 50\ \Omega$. We cannot beat this transimpedance amplifier in terms of simplicity!

The low-impedance front-end has a respectable bandwidth. For example, given a photodetector capacitance of $C_D = 0.15$ pF and a buffer input capacitance of $C_I = 0.15$ pF, the bandwidth is 11 GHz. The relatively low

Analysis and Design of Transimpedance Amplifiers for Optical Receivers, First Edition. Eduard Säckinger.

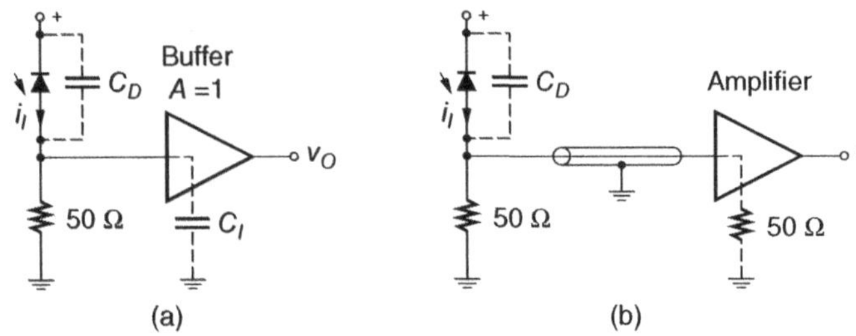

Figure 6.1 Low-impedance front-end with (a) high-impedance buffer and (b) 50-Ω amplifier.

transimpedance value and the resulting small output voltage is a disadvantage. Another drawback is the significant thermal noise current associated with the small 50-Ω resistor. The input-referred noise current density of this front-end is white and given by $I_{n,res} = \sqrt{4kT/R}$, which is about 18 pA/$\sqrt{\text{Hz}}$ for $R = 50$ Ω. (To keep things simple at this point, we are neglecting noise contributions from the buffer and subsequent circuits, which would be important in practice [cf. Eq. (I.36)].)

Figure 6.1(b) shows an easy way to construct a low-impedance front-end in the lab. A photodetector module with a built-in 50-Ω load resistor is connected to an off-the-shelf broadband low-noise amplifier with 50-Ω inputs and outputs. Because of the 50-Ω terminations on both sides, we can use a standard 50-Ω cable to connect the two modules without having to worry about reflections. The (low-frequency) transimpedance of this arrangement is $R_T = 25\ \Omega \times A$, where A is the voltage gain of the amplifier. The 25-Ω factor is due to the 50-Ω load resistor and the 50-Ω amplifier input termination, which are connected in parallel. For a noise-free amplifier ($F = 0$ dB), the input-referred noise current density is again 18 pA/$\sqrt{\text{Hz}}$. The bandwidth, however, increases because the photodetector capacitance is now in parallel to a 25-Ω resistance. In our example with $C_D = 0.15$ pF, the bandwidth increases to 42 GHz (assuming that the amplifier does not limit the bandwidth). [→ Problem 6.1.]

The low-impedance front-end is useful in situations where circuit noise is not a primary concern. A 40-Gb/s receiver with an optically preamplified p–i–n detector that is directly connected to a limiting amplifier (LA) has been described in [1]. With a powerful optical amplifier and a photodetector with a high saturation current it is even possible to generate enough swing to directly drive the receiver's clock and data recovery circuit (CDR) without the need of an LA [2] (cf. Section 3.1). For example, given an optical power of +10 dBm, a photodetector responsivity of $\mathcal{R} = 0.8$ A/W, and $R_T = 25$ Ω, the output signal from the low-impedance front-end is about 400 mV_{pp}. This approach is of interest for high-speed receivers when sufficiently fast TIAs and MAs are not yet available.

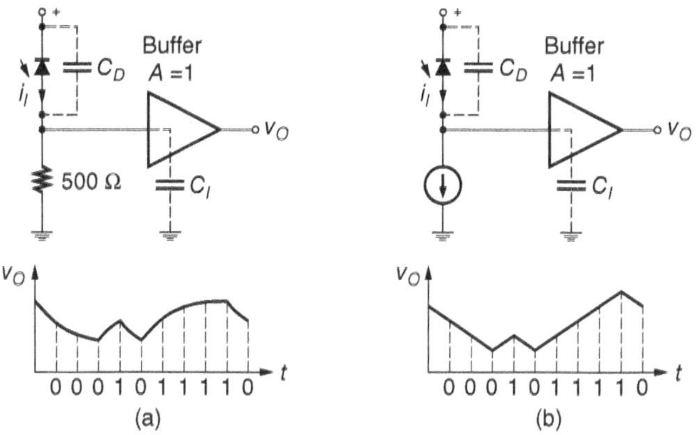

Figure 6.2 (a) High-impedance front-end and (b) integrating front-end.

High-Impedance Front-End. To get around the noise problem of the low-impedance front-end, we may consider increasing the value of the load resistor, which brings us to the high-impedance front-end shown in Fig. 6.2(a). Assuming a load resistor of 500 Ω, the transimpedance increases to $R_T = 500\ \Omega$, a reasonable value for a 10-Gb/s TIA. The noise improves as expected, and is now down to 5.8 $\text{pA}/\sqrt{\text{Hz}}$. (Again, we are neglecting important noise contributions from the buffer and subsequent circuits. But even these noise contributions are reduced as a result of the higher transimpedance.)

Unfortunately, the bandwidth of this high-impedance front-end is reduced to a mere 1.1 GHz given our example values, way too little for a 10-Gb/s TIA. As a result, the output signal suffers from a lot of ISI, as shown in Fig. 6.2(a). An equalizer with a zero at 1.1 GHz and a pole at around 7 GHz can be used to boost high frequencies and reduce the ISI [3, 4].

Another problem of the high-impedance front-end is its low input overload current and small dynamic range [3, 5, 6]. Assuming a peak-to-peak input current of 2 mA from the photodetector, the output voltage swing becomes 1 V for a long run of zeros followed by a long run of ones, whereas the swing for the low-impedance front-end under the same conditions is only 100 mV. This large swing may overload the input stage of the subsequent equalizer. In addition, the large swing modulates the reverse bias voltage of the photodetector, which in the case of an APD may lead to an undesirable gain modulation.

If the maximum permissible output voltage swing is $v_{O,ovl}^{pp}$, then the input overload current is given by $i_{ovl}^{pp} = v_{O,ovl}^{pp}/R$, where R is the load resistor. Thus, as we increase R to reduce the noise, the overload current decreases proportional to $1/R$. The sensitivity i_{sens}^{pp}, however, is proportional to i_n^{rms} (cf. Section 4.4) and therefore, the sensitivity improves only proportional to $1/\sqrt{R}$. Combining these two observations, we conclude that the dynamic range shrinks with increasing R.

High-impedance front-ends were used in early high-sensitivity receivers [3, 5, 6]. These receivers typically operated in the 10 to 100 Mb/s range and used load resistors in the mega Ohms. More recently, high-impedance front-ends followed by an FFE or DFE (cf. Appendix E) found application in low-power receivers operating at 10 Gb/s and beyond [7, 8].

Integrating Front-End. Replacing the load resistor with a current source, as illustrated in Fig. 6.2(b), turns the high-impedance front-end into an *integrating front-end*. This front-end can be viewed as the limiting case of the high-impedance front-end when $R \rightarrow \infty$. Whereas the transimpedance and the noise are further improved, the output signal becomes more distorted (it is the integral of the received signal) and can assume very large values after long runs of zeros or ones. As a result, the input overload current for a random NRZ signal becomes very small.

Nevertheless, with some modifications the integrating front-end can be made practical. If we restrict the input signal to be DC balanced and to have short runs (e.g., by 8B/10B encoding it, cf. Chapter 1), the output signal swing becomes bounded and the input overload current assumes a practical value. The output signal can then be equalized by taking the difference between a sample at the end of the bit period and a sample at the beginning of the bit period (a differentiator compensating for the integration) [9]. Such integrating and double-sampling front-ends are a good match for parallel optical interconnect applications because they achieve high density and consume little power [10] (low noise is of secondary importance in these applications). Another way of making the integrating front-end practical is to add a switch to discharge $C_D + C_I$ at the end of every bit period [11]. This leads to the integrate-and-dump front-end, which we discussed in Section 4.8 [11]. The dump switch eliminates the ISI and solves the overload problem for long runs by periodically forcing the output voltage to zero.

Summary and Outlook. In summary, the low-impedance front-end offers a large bandwidth and a high overload current, whereas the high-impedance front-end offers a lower noise. Is there a way to get a large bandwidth, a high input overload current, and low noise all at the same time? Yes, by using the shunt-feedback topology, which we discuss next.

6.2 Shunt-Feedback TIA

The block diagram of the basic *shunt-feedback TIA* is shown in Fig. 6.3(a). The inverting voltage amplifier A can be implemented in many different ways (see Chapter 9 for examples). The simplest realization consist of a single common-source MOSFET stage, as shown in Fig. 6.3(b).

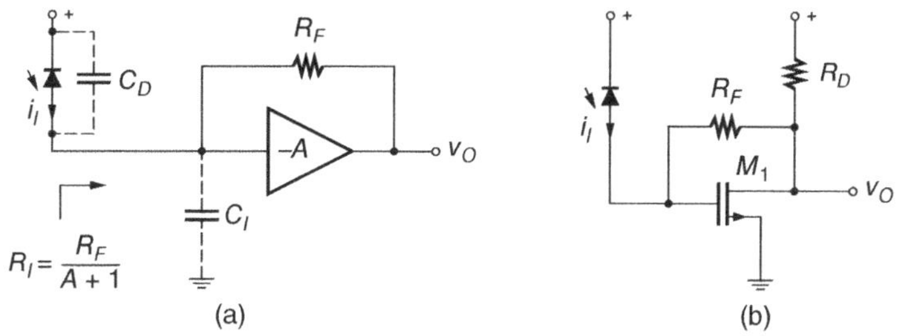

Figure 6.3 Basic shunt-feedback transimpedance amplifier: (a) block diagram and (b) simple transistor-level implementation with a single MOSFET.

Simple Analysis. For now, let us assume that the voltage amplifier has a high gain, such that it maintains a virtual ground at its input. Then, all of the current from the photodetector, i_I, flows into R_F, and the amplifier's output must respond such that its input remains at virtual ground. The output voltage necessary for this is $v_O = -R_F \cdot i_I$. Hence, the transimpedance is R_F. Note that the virtual ground also keeps the reverse bias voltage of the photodetector constant, a prerequisite for good overload behavior.

Let us make the model of the voltage amplifier one step more realistic. We drop the assumption of a high gain and use the finite value $-A$ instead, which implies that the input is no longer a precise virtual ground. Furthermore, we take the input impedance as $1/(sC_I)$, that is, as purely capacitive (see Fig. 6.3(a)), a good assumption for an amplifier with an FET input stage such as the one shown in Fig. 6.3(b). Finally, we take the output impedance as zero and the bandwidth as infinite. Obviously, the last two assumptions are not realistic and need to be revisited later. Because the photodetector capacitance, C_D, and the input capacitance, C_I, appear in parallel (from an AC point of view), we can combine them into a single (total) capacitance $C_T = C_D + C_I$. Given this amplifier model, we find the frequency-dependent transimpedance as

$$Z_T(s) = -R_T \cdot \frac{1}{1 + s/\omega_p}, \tag{6.1}$$

where

$$R_T = \frac{A}{A+1} R_F, \tag{6.2}$$

$$\omega_p = \frac{A+1}{R_F C_T}. \tag{6.3}$$

Equation (6.2) confirms our earlier observation that the transimpedance R_T is approximately equal to the feedback resistor R_F, given a gain A that is much larger than unity. The 3-dB bandwidth of the TIA follows from Eq. (6.3) as

$$BW_{3dB} = \frac{\omega_p}{2\pi} = \frac{A+1}{2\pi R_F C_T}. \tag{6.4}$$

This means that the bandwidth is $A+1$ times *larger* than that of a high-impedance front-end with load resistor R_F and total capacitance C_T. Thus, the shunt-feedback TIA benefits from the low noise of a large R_F without the drawback of a slow response!

The bandwidth improvement can also be understood by analyzing the circuit's closed-loop input resistance, R_I. For $A = 0$, that is, in the absence of the voltage amplifier, $R_I = R_F$. With the amplifier present ($A > 0$), the feedback action reduces the input resistance to (cf. Fig. 6.3(a))

$$R_I = \frac{R_F}{A+1}. \tag{6.5}$$

Because the pole ω_p is determined by R_I and C_T, it speeds up by the same factor $A+1$ that the input resistance decreases.

The low input resistance of the shunt-feedback TIA also leads to a high input overload current. For a given input current swing, the input voltage swing is $A+1$ times smaller than that of the high-impedance front-end. Conversely, given the maximum permissible input voltage swing $v_{I,ovl}^{pp}$, the input overload current is $A+1$ times larger:

$$i_{ovl}^{pp} = (A+1)\frac{v_{I,ovl}^{pp}}{R_F}. \tag{6.6}$$

Alternatively, we can express the input overload current in terms of the maximum permissible output voltage swing $v_{O,ovl}^{pp}$:

$$i_{ovl}^{pp} = \frac{A+1}{A} \cdot \frac{v_{O,ovl}^{pp}}{R_F} = \frac{v_{O,ovl}^{pp}}{R_T}. \tag{6.7}$$

The actual overload current is given by either Eq. (6.6) or (6.7), whichever expression is smaller.

The voltage swings v_I^{pp} and v_O^{pp} are limited by a number of mechanisms. For example, in a voltage amplifier with bipolar junction transistors (BJTs or HBTs), a large voltage swing can cause the base–collector diode of a critical transistor to become forward biased (BJT saturation), resulting in severe signal distortions. Moreover, a large voltage swing may push a bias current source in the voltage amplifier out of its operating range, resulting in a reduced bias current and thus a slow response. Finally, a large input voltage swing may reduce the reverse bias of the photodetector to the point where its response becomes too slow.

In summary, the shunt-feedback TIA features a high transimpedance ($R_T \approx R_F$) and a low input-referred noise current PSD ($I_{n,TIA}^2 = 4kT/R_F$ plus the noise contributions from the voltage amplifier), similar to those of the high-impedance front-end. Its bandwidth and input overload current, however, are better than those of the high-impedance front-end by a factor $A+1$. But nothing is free! To realize all these advantages, we need a voltage amplifier with the necessary gain, bandwidth, and low noise. Next, we consider the effects of a voltage amplifier with finite bandwidth.

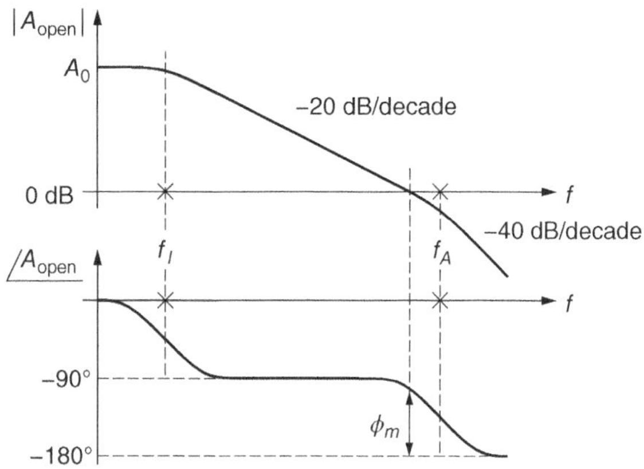

Figure 6.4 Open-loop frequency response (magnitude and phase) of a TIA with a single-pole voltage amplifier. The input pole is at $f_I = 1/(2\pi R_F C_T)$ and the amplifier pole is at $f_A = 1/(2\pi T_A)$.

Effects of Finite Amplifier Bandwidth. Next, let us replace our infinite-bandwidth amplifier with a more realistic single-pole amplifier, which is a good approximation for a single-stage amplifier. The transfer function of this amplifier is $A(s) = A_0/(1 + sT_A)$, where A_0 is the DC gain and T_A is the time constant of the pole. The 3-dB bandwidth of this amplifier, not to be confused with the 3-dB bandwidth of the TIA, is given by $f_A = 1/(2\pi T_A)$. The amplifier's input impedance is $1/(sC_I)$ and its output impedance is zero, as before. Now, the open-loop frequency response, $|A_{open}(\omega)|$, has two poles, as shown in Fig. 6.4. The high-frequency pole at $f_A = 1/(2\pi T_A)$ is due to the voltage amplifier and the low-frequency pole at $f_I = 1/(2\pi R_F C_T)$ is due to the low-pass filter formed by R_F and C_T. In a second-order system like this, we have to watch out for undesired peaking in the closed-loop frequency response.

Given the aforementioned voltage amplifier model with a finite bandwidth, we find the closed-loop transimpedance as [12]

$$Z_T(s) = -R_T \cdot \frac{1}{1 + s/(\omega_0 Q) + s^2/\omega_0^2}, \tag{6.8}$$

where

$$R_T = \frac{A_0}{A_0 + 1} R_F, \tag{6.9}$$

$$\omega_0 = \sqrt{\frac{A_0 + 1}{R_F C_T T_A}}, \tag{6.10}$$

$$Q = \frac{\sqrt{(A_0 + 1) R_F C_T T_A}}{R_F C_T + T_A}. \tag{6.11}$$

In these equations, R_T is the transimpedance at DC and did not change from the case with infinite bandwidth (Eq. (6.2)), ω_0 is the (angular) frequency of the pole pair, and Q is the quality factor of the pole pair,[1] which controls the peaking (not

1 The quality factor Q is related to the damping factor ζ as $Q = 1/(2\zeta)$.

to be confused with the Q-factor from Chapter 4). For $Q > 0.5$, the two poles become conjugate complex and thus no longer can be identified as originating from the voltage amplifier or the input node, they are simply *system* poles.

It is worth spending a few minutes to study the impact of the pole quality factor Q on the frequency response and the time-domain behavior. Two values are of particular interest:

- For $Q = 1/\sqrt{3} = 0.577$, we obtain the so-called *Bessel–Thomson response* or *Bessel response* for short. W. A. Thomson first studied filters with this response and published his findings in 1952. More than 100 years earlier the mathematician Friedrich W. Bessel (1784–1846) developed functions (Bessel functions) that ultimately proved important in describing this response [13]. The Bessel response is characterized by a maximally flat group delay, $\tau(f)$, resulting in a small group-delay variation $\Delta\tau$ (0.058 UI for $BW_{3dB} = \frac{2}{3}B$). Moreover, it has no peaking in the amplitude response $|Z_T(f)|$ and produces only a negligible amount of overshoot (0.54%) and jitter (0.002 UI for $BW_{3dB} = \frac{2}{3}B$) in the time domain.
- For $Q = 1/\sqrt{2} = 0.707$, we obtain the so-called *Butterworth response*, named after S. Butterworth who first studied filters with this response and published his findings in 1930 [13]. It is characterized by a maximally flat amplitude response, $|Z_T(f)|$. However, the Butterworth response does have a small amount of peaking in the group delay, $\tau(f)$, resulting in a slightly larger group-delay variation $\Delta\tau$ (0.070 UI for $BW_{3dB} = \frac{2}{3}B$). Moreover, it produces more overshoot (4.8%) and jitter (0.016 UI for $BW_{3dB} = \frac{2}{3}B$) in the time domain than the Bessel response.

For larger values of Q, amplitude and group-delay peaking in the frequency domain as well as overshoot and jitter in the time domain become progressively worse. Appendix H discusses four second-order low-pass transfer functions with $Q = 1/2, 1/\sqrt{3}, 1/\sqrt{2}$, and 1 in the frequency and time domain.

Amplifier Bandwidth Requirements. Let us assume that we chose the Butterworth response for our TIA design. This choice keeps phase linearity, group-delay variation, and jitter well within the limits discussed in Chapter 5 and has been well studied in the literature [12, 14]. By setting Eq. (6.11) to $1/\sqrt{2}$ and assuming a reasonably large gain such that $A_0^2 \gg 1$, we find that the bandwidth of the voltage amplifier has to be[2]

$$f_A = \frac{2A_0}{2\pi R_F C_T}. \tag{6.12}$$

The interpretation of this equation is that the two open-loop poles shown in Fig. 6.4 must be spaced apart by a factor $2A_0$. Equivalently, the voltage amplifier

2 For small gains: $f_A = (A_0 + \sqrt{A_0^2 - 1})/(2\pi R_F C_T)$.

bandwidth (f_A) must be a factor two larger than the unity-gain frequency of the open-loop response ($A_0/[2\pi R_F C_T]$). [→ Problem 6.2.]

If we had chosen the Bessel response with its somewhat better phase linearity, group-delay variation, and jitter, the open-loop poles had to be spaced *further* apart, namely by a factor $3A_0 + 1$ (cf. Eq. (I.37)). Given the same A_0, R_F, and C_T, a voltage amplifier with about 50% more bandwidth and a correspondingly higher power dissipation would be required [15]. The Butterworth response appears to be a pretty good compromise in terms of performance and amplifier requirements.

That a better dynamic stability (lower Q) demands a larger pole spacing (and thus a faster voltage amplifier) can also be understood in terms of the phase margin ϕ_m. The latter is measured at the unity-gain frequency of the open-loop response, as shown in Fig. 6.4. If the pole spacing is equal to A_0, the high-frequency pole is located just about at the unity-gain frequency and causes a 45° phase lag. If we further assume $A_0 \gg 1$ such that the phase lag due to the low-frequency pole is still a full 90° at the unity-gain frequency, the phase margin is approximately 45°. Now increasing the pole spacing beyond A_0, the phase lag due to the high-frequency pole reduces and the phase margin asymptotically increases to 90°. The larger the phase margin, the better the stability.

TIA Bandwidth. The 3-dB bandwidth of the TIA response in Eq. (6.8) is $BW_{3dB} = \omega_0/(2\pi)$, where ω_0 is given by Eq. (6.10) and a Butterworth response was assumed (see Appendix H). Inserting the amplifier time constant necessary for the Butterworth response, $T_A = R_F C_T/(2A_0)$ (from Eq. (6.12)), we arrive at the following TIA bandwidth:

$$BW_{3dB} = \frac{\sqrt{2A_0(A_0+1)}}{2\pi R_F C_T}. \tag{6.13}$$

Comparing this equation with Eq. (6.4) for the TIA with an infinite-bandwidth voltage amplifier, we find that the TIA's bandwidth *increased* by about a factor $\sqrt{2}$. In other words, if we start out with a TIA that has a very wideband voltage amplifier and then reduce its bandwidth (e.g., by loading it with a capacitor [16]) to its optimum value, the TIA's bandwidth improves by about 40%.

How can a system get faster by making one of its components slower? A qualitative explanation for this effect is that a voltage amplifier with the optimum bandwidth introduces a phase shift near the TIA's roll off turning the mostly negative feedback into more of a positive feedback, which in turn locally boosts the gain and pushes the bandwidth out. Another explanation of the same effect is that the feedback loop around the finite-bandwidth voltage amplifier synthesizes an input impedance with an inductive or negative-capacitive component, which in turn tunes out some of the bandwidth-limiting capacitance C_T. We analyze the TIA's input impedance in more detail later in this section.

Transimpedance Limit. Let us say we are only interested in TIA designs free of amplitude peaking, which includes Butterworth, Bessel, and critically damped designs. Under this condition, given by $Q \leq 1/\sqrt{2}$, it can be shown that the 3-dB bandwidth of the second-order system in Eq. (6.8) is bounded by $BW_{3dB} \leq \omega_0/(2\pi)$ (see Appendix H). When applying this bound to Eq. (6.10) and expressing R_F in terms of R_T using Eq. (6.9), we arrive at the following inequality, which is known as the *transimpedance limit* [17, 18]:

$$R_T \leq \frac{A_0 f_A}{2\pi C_T \cdot BW_{3dB}^2}. \tag{6.14}$$

The expression $A_0 f_A$ in this equation represents the gain-bandwidth product of the single-pole voltage amplifier.

For a single-transistor voltage amplifier, the gain-bandwidth product can be written as $A_0 f_A \approx g_m/(2\pi C_L)$, where g_m is the transconductance and C_L is the load capacitance of the transistor. The related technology parameter f_T can be written as $f_T = g_m/(2\pi \tilde{C}_I)$, where $\tilde{C}_I$ is the input capacitance of the amplifying transistor under shorted-output conditions. We use the tilde to distinguish the input capacitance under shorted-output conditions, which does not contain a Miller term (e.g., $\tilde{C}_I = C_{gs} + C_{gd}$ for a FET) from the full input capacitance under active-output conditions (e.g., $C_I = C_{gs} + C_{gd}(A_0 + 1)$). Thus, the gain-bandwidth product is roughly proportional to the technology parameter $f_T : A_0 f_A \approx \tilde{C}_I/C_L \cdot f_T$.

For noise reasons, which we discuss in Section 6.4, the input capacitance $\tilde{C}_I$ is made to approximately match the photodetector capacitance C_D. Thus, the total capacitance, C_T, is roughly proportional to the photodetector capacitance: $C_T = C_D + C_I \approx 2C_D$.

Going back to Eq. (6.14), we conclude that for a given technology (fixed $A_0 f_A$) and a given photodetector (fixed C_T), the attainable transimpedance drops with the *square* of the desired TIA bandwidth. For example, if we want to quadruple the bit rate (quadruple BW_{3dB}) without using a faster technology and photodetector, the transimpedance will degrade by a factor *sixteen*. Even with a technology that is four times faster, and a photodetector with half the capacitance, the transimpedance still degrades by a factor two. This is the reason why higher bit-rate TIAs generally have lower transimpedance values.

Figure 6.5 graphically illustrates Eq. (6.14). The transimpedance limit divides the design space (BW_{3dB}, R_T) into two regions, one with design points that are unrealizable and one with design points that may be realizable. The location of the limit is determined by the expression $A_0 f_A/C_T$, which is approximately proportional to the technology parameter f_T/C_D. With increasing value, the limit moves toward the upper right-hand corner (dashed line in Fig. 6.5). It can be shown that for all practical purposes the points on the transimpedance limit *can* be realized [18]. That is, for a given $A_0 f_A$, C_T, and BW_{3dB}, we can find values for A_0, f_A, and R_F that result in the maximum transimpedance

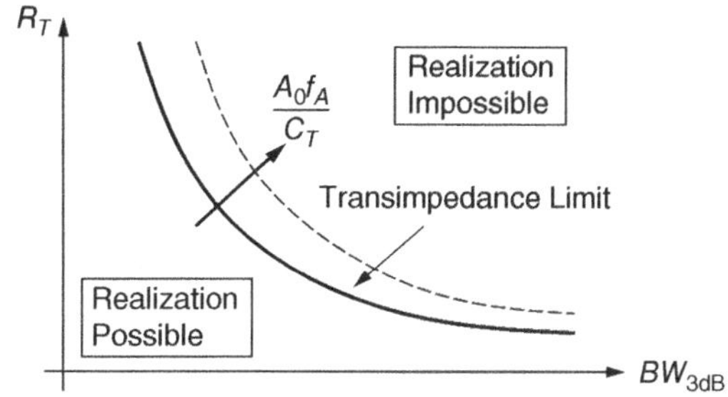

Figure 6.5 The transimpedance limit.

permitted by Eq. (6.14). Thus, Eq. (6.14) provides a useful tool for estimating the transimpedance before the design is started. [→ Problem 6.3.]

In deriving the transimpedance limit Eq. (6.14) we required $Q \leq 1/\sqrt{2}$. What if we had chosen the more general stability criterion $Q \leq Q_0$? In that case, the right-hand side of Eq. (6.14) had to be multiplied by $\rho^2(Q_0)$, where $\rho(Q)$ is given by Eq. (H.3) in Appendix H. For example, with the more conservative choice $Q \leq 1/\sqrt{3}$, the transimpedance limit is reduced by 38.2%, whereas with the more aggressive choice $Q \leq 1$, the transimpedance limit is increased by 61.8%. But in either case the form of the limit remains the same.

In deriving the transimpedance limit Eq. (6.14) we assumed the basic shunt-feedback topology of Fig. 6.3(a) with a single-pole voltage amplifier. What if we consider a multipole voltage amplifier or another topology such as a TIA with post amplifier (see Section 7.1), a common-base/gate TIA (see Section 7.5), or a current-mode TIA (see Section 8.2)? These alternatives *do* result in different transimpedance limits, as we discuss later [18]. The transimpedance limit in Eq. (6.14) is *not* a fundamental limit!

It is instructive to compare Eq. (6.14) with the corresponding equation for the low-impedance front-end. The transimpedance of the latter front-end is simply given by $R_T = 1/(2\pi C_T \cdot BW_{3dB})$. Thus, as long as the gain-bandwidth product of the voltage amplifier is larger than the required front-end bandwidth, $A_0 f_A > BW_{3dB}$, the shunt-feedback TIA achieves a higher transimpedance than the low-impedance front-end.

Transimpedance-Bandwidth Product? The gain-bandwidth product, $A_0 \times BW_{3dB}$, is often used as a figure of merit for single-stage voltage amplifiers. This product is a measure of the technology's speed and the designer's skill and is roughly independent of the particular gain and bandwidth values. Designing a single-stage amplifier with 20-dB gain and 10-GHz bandwidth is about as hard as designing a 26-dB, 5-GHz amplifier; they both have a gain-bandwidth product of 100 GHz.

In analogy to single-stage amplifiers, it is tempting to use the transimpedance-bandwidth product, $R_T \times BW_{3dB}$, as a figure of merit for TIAs. For example, a TIA with a 1-kΩ transimpedance and a 10-GHz bandwidth yields the impressive 10 THzΩ product. However, Eq. (6.14) shows that this product is *not* independent of the particular transimpedance and bandwidth values (even though the TIA is build around a single-stage voltage amplifier) and therefore is not a very good figure of merit. For example, according to Eq. (6.14), a TIA with a 4-kΩ transimpedance and a 5-GHz bandwidth could be designed in the same technology (same $A_0 f_A$ and C_T) as the 1-kΩ, 10-GHz TIA, yet it achieves twice the transimpedance-bandwidth product. It is easier to obtain a high transimpedance than a high bandwidth! The transimpedance-bandwidth product of a TIA obeying Eq. (6.14) is inversely proportional to the bandwidth and thus grows indefinitely for small bandwidths.

An analogous situation occurs for multistage voltage amplifiers. It is well known that the gain-bandwidth product is *not* a valid basis for comparing multistage amplifiers [19]. For example, increasing the load resistance in both stages of a two-stage amplifier by n increases the overall gain by n^2 yet reduces the bandwidth only by n. Hence, the gain-bandwidth product increases by a factor n.

Amplifier Gain. When discussing the transimpedance limit in Fig. 6.5, we assumed that the voltage gain can be traded arbitrarily against the bandwidth as long as their product, $A_0 f_A$, does not exceed the capability of the technology. However, for resistively loaded gain stages, like the one shown in Fig. 6.3(b), the gain is related to the DC voltage drop across the load resistor, V_R, which must remain less than about half of the supply voltage. Thus, there is a limit to how large the gain can be made.

Using the square-law FET model, the voltage gain is found to be $A_0 = 2V_R/(V_{GS} - V_{TH})$, where $V_{GS} - V_{TH}$ is the FET's overdrive voltage. For example, given a power-supply voltage of 1 V, $V_R < 0.5$ V, and $V_{GS} - V_{TH} = 0.3$ V, the gain is limited to 3.3×. In nanoscale technologies the square-law to linear-law transition occurs at very small overdrive voltages (≈ 0.2 V for a 90-nm n-MOSFET [20]), further reducing the maximum gain. An active load instead of or in addition to the load resistor can be used to mitigate the gain limitation.

For BJT or HBT stages, the voltage gain is given by $A_0 = V_R/V_T$, where V_T is the thermal voltage, which is about 25 mV at room temperature. Given the same limit on the voltage drop as before, $V_R < 0.5$ V, the maximum gain is about 20×, which is much better than in the FET case. In advanced SiGe and InP HBTs the parasitic emitter resistance becomes significant, reducing the gain to $A_0 \approx V_R/2V_T$ [20].

Numerical Examples. To get a better feeling for numerical values, we want to illustrate the foregoing theory with a 10-Gb/s TIA design example. Figure 6.6 shows the familiar shunt-feedback TIA annotated with some example values. The photodetector and amplifier input capacitance are 0.15 pF each, the voltage amplifier has a low-frequency gain of 14 dB (5×), and the feedback resistor is 600 Ω (cf. Table 6.1).

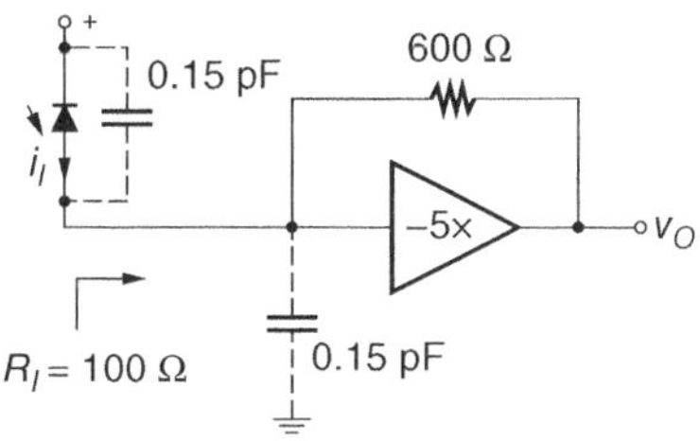

Figure 6.6 Example values for a 10-Gb/s shunt-feedback TIA.

Table 6.1 Parameters and performance of our shunt-feedback TIA examples.

Parameter	Symbol	10 Gb/s	40 Gb/s
Gain-bandwidth product	$A_0 f_A$	44 GHz	177 GHz
Detector capacitance incl. parasitics	C_D	0.15 pF	75 fF
Voltage amplifier DC gain	A_0	5	5
Voltage amplifier 3-dB bandwidth	f_A	8.8 GHz	35.4 GHz
Voltage amplifier input capacitance	C_I	0.15 pF	75 fF
Feedback resistor	R_F	600 Ω	300 Ω
Transimpedance	R_T	500 Ω	250 Ω
Input impedance	R_I	100 Ω	50 Ω
TIA 3-dB bandwidth	$BW_{3\text{dB}}$	6.85 GHz	27.4 GHz

From these data, we can easily calculate the transimpedance of the TIA at low frequencies:

$$R_T = \frac{5}{5+1} \cdot 600\ \Omega = 500\ \Omega, \tag{6.15}$$

which is equal to 54 dBΩ. This is somewhat lower than R_F, as expected. The TIA bandwidth, assuming a Butterworth response, turns out to be

$$BW_{3\text{dB}} = \frac{\sqrt{2 \cdot 5 \cdot (5+1)}}{6.28 \cdot 600\ \Omega \cdot 0.3\ \text{pF}} = 6.85\ \text{GHz}, \tag{6.16}$$

which is suitable for a 10-Gb/s receiver in which the TIA sets the bandwidth. For comparison, the high-impedance front-end with the same transimpedance (500 Ω) and the same total capacitance (0.3 pF), which we discussed in Section 6.1, has a bandwidth of only 1.1 GHz.

To achieve the assumed Butterworth response, the voltage amplifier bandwidth must be

$$f_A = \frac{2 \cdot 5}{6.28 \cdot 600\ \Omega \cdot 0.3\ \text{pF}} = 8.84\ \text{GHz}, \tag{6.17}$$

corresponding to $T_A = 18$ ps. Thus, we need a technology in which we can realize a single-stage amplifier with the gain-bandwidth product

$$A_0 f_A = 5 \cdot 8.84\ \text{GHz} = 44.2\ \text{GHz}. \tag{6.18}$$

Table 6.2 Performance of our 10-Gb/s shunt-feedback TIA, low-impedance front-end (Low-*Z*), and high-impedance front-end (High-*Z*) examples.

Parameter	Symbol	TIA	Low-*Z*	High-*Z*
Transimpedance	R_T	500 Ω	50 Ω	500 Ω
3-dB bandwidth	BW_{3dB}	6.85 GHz	11 GHz	1.1 GHz
Overload current	i_{ovl}^{pp}	2 mA	4 mA	0.4 mA
Noise current	$I_{n,TIA}$	5.3 pA$\sqrt{\text{Hz}}$	18 pA$\sqrt{\text{Hz}}$	5.8 pA$\sqrt{\text{Hz}}$

The TIA's input impedance at low frequencies is

$$R_I = \frac{600\ \Omega}{5+1} = 100\ \Omega, \tag{6.19}$$

which means that for a maximum permissible input voltage swing, $v_{I,ovl}^{pp}$, of 0.2 V, the input overload current, i_{ovl}^{pp}, is 2 mA. Under the same condition, the overload current of the 500-Ω high-impedance front-end is five times lower: $i_{ovl}^{pp} = 0.4$ mA.

Finally, the input-referred noise current density of our TIA due to R_F is 5.3 pA/$\sqrt{\text{Hz}}$. (To keep things simple at this point, we are neglecting noise contributions from the voltage amplifier. We discuss them in Section 6.3.) This value is comparable with that of the 500-Ω high-impedance front-end.

Table 6.1 summarizes the parameters of our TIA example. Table 6.2 summarizes the performance results of our TIA example and compares them with the low-impedance and high-impedance front-end examples from Section 6.1. Note that all overload currents are based on $v_{I,ovl}^{pp} = 0.2$ V and that all noise spectral densities are based on resistor noise only.

What transimpedance would we expect for a 40-Gb/s TIA realized in a technology that permits a 177-GHz gain-bandwidth product connected to a detector that has a 75-fF capacitance? Assuming the TIA bandwidth needs to be 27.4 GHz and the input capacitance matches the detector capacitance, the transimpedance limit makes the following prediction

$$R_T \leq \frac{177\ \text{GHz}}{6.28 \cdot 0.15\ \text{pF} \cdot (27.4\ \text{GHz})^2} = 250\ \Omega. \tag{6.20}$$

Working the example out yields the values shown in Table 6.1. Compared with our 10-Gb/s example, all bandwidths are four times larger and all capacitances and resistances are half as large. Because we chose a Butterworth response for our example, the worked-out transimpedance is at the upper limit of the range predicted by Eq. (6.20).

Input Impedance. We know that the input impedance of the TIA at low frequencies is given by $R_I = R_F/(A_0 + 1)$, but how does this impedance change for higher frequencies and how is it impacted by the bandwidth of the voltage amplifier? The input impedance plays an important role in situations where matching is required. For example, when a transmission line is used to connect the photodetector to the TIA and matching between the characteristic impedance of the transmission line and the TIA's input impedance is required to avoid reflections [21]. Examining the input impedance also yields insights into why reducing the voltage-amplifier's bandwidth can extend the TIA's bandwidth.

The input impedance is easily derived from the transimpedance by noticing that the former is *the input voltage divided by the input current* and the latter is *the output voltage divided by the input current*. Thus, dividing the transimpedance in Eq. (6.8) by the voltage gain $-A(s) = -A_0/(1 + sT_A)$ reveals the input impedance

$$Z_I(s) = \frac{Z_T(s)}{-A(s)} = \frac{R_F}{A_0 + 1} \cdot \frac{1 + sT_A}{1 + s/(\omega_0 Q) + s^2/\omega_0^2}, \tag{6.21}$$

where the expressions for ω_0 and Q are the same as in Eqs. (6.10) and (6.11), respectively. The low-frequency value of Z_I equals $R_I = R_F/(A_0 + 1)$, as expected. Furthermore, Z_I has the same poles as Z_T and a new zero at $1/T_A$.

To gain more insight into the impedance function $Z_I(s)$, it is useful to find an equivalent *RLC* network. The capacitance C_T, which appears directly in parallel with the TIA input port, should be a parallel part of this network. Separating this capacitance out yields

$$Z_I(s) = \frac{1}{sC_T} \Bigg\| \left(\frac{R_F}{A_0 + 1} \cdot \frac{1 + sT_A}{1 + sT_A/(A_0 + 1)} \right). \tag{6.22}$$

The impedance in parallel with C_T has the low-frequency value $R_I = R_F/(A_0 + 1)$. Above the frequency $f_A = 1/(2\pi T_A)$, which corresponds to the open-loop high-frequency pole, the impedance starts to rise like that of an inductor. Eventually, the impedance tops out at the value R_F. An *RLC* network representation of the entire input impedance is shown in Fig. 6.7(a). (The expressions for the network components can be obtained by calculating the impedance of the network and comparing it with Eq. (6.22).)

For an infinite-bandwidth voltage amplifier ($T_A = 0$), the input impedance consists simply of C_T in parallel with $R_I = R_F/(A_0 + 1)$. This input impedance

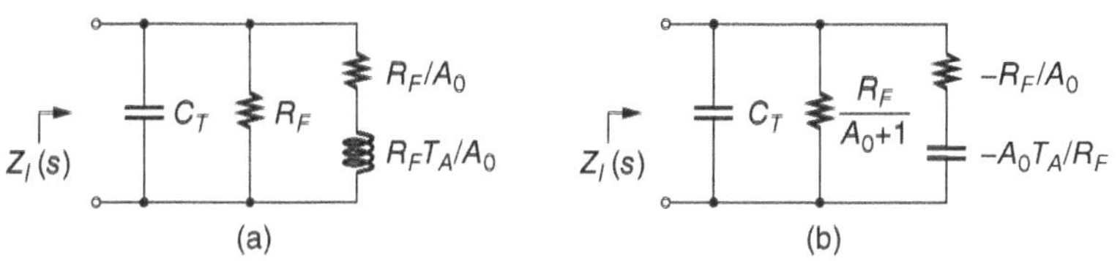

Figure 6.7 Equivalent networks for the TIA's input impedance: (a) with inductive component and (b) with negative-capacitive component.

leads to the simple expression for the TIA's bandwidth given in Eq. (6.4). For a finite-bandwidth voltage amplifier ($T_A > 0$), however, an inductive component with the value $L = R_F T_A / A_0$ appears. If chosen appropriately, this inductance can tune out some of the capacitance, extending the bandwidth of the TIA from Eq. (6.4) to Eq. (6.13), corresponding to about 40%.

The network in Fig. 6.7(a) is not the only one that mimics the input impedance of the TIA given by Eq. (6.22). An interesting alternative is the network shown in Fig. 6.7(b), which features a negative capacitance. For a Butterworth design, the value of the negative capacitance becomes $-C_T/2$ (using $T_A = R_F C_T / (2A_0)$ from Eq. (6.12)). The partial cancellation of C_T by this negative capacitance offers yet another explanation for the aforementioned bandwidth extension. So, which one is it: does the input impedance have an inductive or a negative-capacitive component? There is no answer to this question, both networks in Fig. 6.7 model the input impedance equally well.

In general, the input impedance should be made just low enough to meet the bandwidth requirement. For example, in our 10-Gb/s TIA with $C_T = 0.3$ pF, $R_I = 100\ \Omega$ was sufficient to meet the required bandwidth. However, there are circumstances, where a 50-Ω input impedance is either natural or required. Many 40-Gb/s TIAs do have an input impedance near 50 Ω, simply as a result of the typical value for C_T and the bandwidth requirement (cf. Table 6.1). In a receiver with balanced detection (cf. Section 3.5), the two photodetectors may be physically spaced apart to accommodate two fibers and therefore cannot both be brought very close to the TIA [21]. In this case, transmission lines are needed to connect the photodetectors to the TIA. To avoid reflections (more importantly, double reflections), the input impedance of the TIA must be matched to the characteristic impedance of the transmission lines, usually 50 Ω. The inductive or negative-capacitive component of the input impedance helps to improve the broadband input matching (i.e., the input return loss or S_{11}) [22].

Shunt-feedback TIAs with 50-Ω inputs have applications beyond the amplification of photocurrents. They have been used as general-purpose low-noise broadband amplifiers [23–25], as input stages to low-noise broadband amplifiers [26–28], and as low-noise amplifiers (LNA) in broadband wireless receivers [22, 29].

Effects of Nonzero Amplifier Output Impedance. So far, we assumed that the voltage amplifier's output impedance is zero, however, in practical implementations, such as the circuit in Fig. 6.3(b), this is not the case. To study the effect of a nonzero output impedance, we model the voltage amplifier as shown in the box of Fig. 6.8(a). An ideal voltage amplifier with gain $-A_0'$ is followed by an RC low-pass filter that realizes a bandwidth-limiting pole ($T_A' = R_L C_L$) and the nonzero output impedance $R_L/(1 + sR_L C_L)$ [30]. The amplifier model in Fig. 6.8(b) is the Norton equivalent of the model in Fig. 6.8(a) and behaves

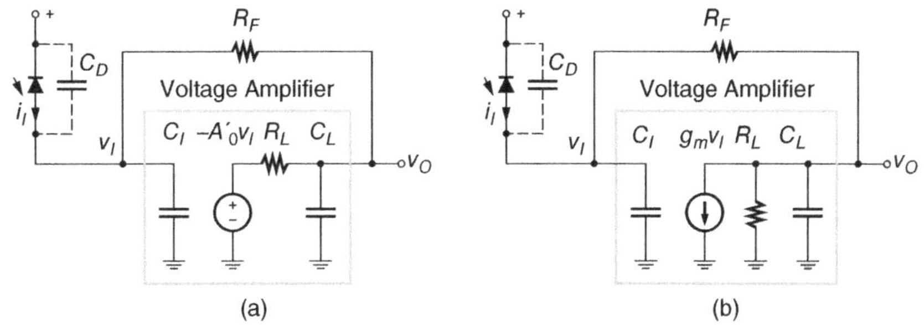

Figure 6.8 TIA with a voltage amplifier model that has a nonzero output impedance. The amplifier models in (a) and (b) are equivalent when $A'_0 = g_m R_L$.

exactly the same way if $A'_0 = g_m R_L$ is satisfied. The reason for introducing this alternative view is that it corresponds directly to the single-transistor circuit in Fig. 6.3(b): g_m corresponds to the transconductance of M_1, R_L to the drain resistor R_D in parallel with $1/g_o$ of M_1, and C_L to the total load capacitance at the output node including C_{db} of M_1. Thus, the subsequent discussion also applies to this transistor circuit. (For a generalized discussion that also includes C_{gd} of M_1 see the solution to Problem 6.5 on p. 512.)

Calculating the transimpedance of either of the two circuits in Fig. 6.8 results in [30]

$$Z_T(s) = -R_T \cdot \frac{1}{1 + s/(\omega_0 Q) + s^2/\omega_0^2}, \tag{6.23}$$

where

$$R_T = \frac{A'_0 R_F - R_L}{A'_0 + 1}, \tag{6.24}$$

$$\omega_0 = \sqrt{\frac{A'_0 + 1}{R_F C_T T'_A}}, \tag{6.25}$$

$$Q = \frac{\sqrt{(A'_0 + 1) R_F C_T T'_A}}{(R_F + R_L) C_T + T'_A}. \tag{6.26}$$

In these expressions, $A'_0 = g_m R_L$ and $T'_A = R_L C_L$ are the *unloaded* amplifier gain and *unloaded* amplifier time constant, respectively. It is important to make this distinction, because connecting the feedback resistor R_F to the output alters both the gain and the time constant. Comparing this result with Eqs. (6.8)–(6.11) for the zero-output-impedance case, we find the following:

- Both transimpedance expressions have two poles and no zero.
- The DC transresistance, R_T, is reduced by the factor $1 - R_L/(A'_0 R_F)$, if we identify A_0 with A'_0.

- The pole frequency, ω_0, remains unchanged, if we identify A_0 with A'_0 and T_A with T'_A.
- The Q factor is reduced by the factor $R_F/(R_F + R_L)$, if we identify A_0 with A'_0 and T_A with T'_A and assume $R_F C_T \gg T_A$.

In fact, the expressions are so similar, that for a given A_0, T_A, and R_L, we can find values for A'_0 and T'_A that yield the exact same $Z_T(s)$ and $Z_I(s)$ (cf. Eq. (I.38)). [→ Problem 6.4.]

Combining Eqs. (6.24) and (6.25), we easily find the transimpedance limit for the TIA with nonzero output impedance:

$$R_T \leq \frac{A'_0 f'_A}{2\pi C_T \cdot BW_{3dB}^2} \left(1 - \frac{R_L}{A'_0 R_F}\right). \tag{6.27}$$

This expression consists of two parts: the transimpedance limit for the case with zero output impedance and a factor that represents the loss in transimpedance due to the nonzero output resistance, which is always smaller than one. The loss in transimpedance increases with R_L, but remains negligible as long as $R_L \ll A'_0 R_F$ or, equivalently, $g_m R_F \gg 1$. Note that the loss factor on the right-hand side can also be written as $1 - 1/(g_m R_F)$. [→ Problem 6.5.]

Instead of the amplifier models in Fig. 6.8, we could have chosen an ideal single-pole voltage amplifier with a series resistor at the output. Although the latter model has a load-independent pole, the results are similar to what we have found here (cf. Eqs. (I.47)–(I.52)). For either model, the effects of the output impedance become negligible when $R_L \ll R_F$. [→ Problem 6.6.]

Numerical Example. To illustrate the effects of a nonzero amplifier output impedance, let us take the values $C_T = 0.3$ pF, $g_m = 50$ mS, $R_L = 120\ \Omega$, $C_L = 0.18$ pF, and $R_F = 580\ \Omega$ (see Table 6.3). The (unloaded) gain-bandwidth product of the voltage amplifier is $A'_0 f'_A = g_m/(2\pi C_L) = 44$ GHz, which has been chosen to match that of the 10-Gb/s example in Table 6.1. The unloaded gain $A'_0 = 6$ and the unloaded time constant $T'_A = 21.6$ ps have been chosen larger than the A_0 and T_A of our previous example to compensate for the loading effect of R_F and the nonzero R_L.

With Eqs. (6.25), (6.26), and (H.2), we find $Q = 0.7$ and $BW_{3dB} = 6.8$ GHz, that is, after closing the loop, the frequency response and bandwidth are essentially the same as in our previous example. With Eq. (6.24), we find the DC transimpedance as $R_T = 480\ \Omega$, which is about 4% lower than in our example with zero output impedance. This drop in transimpedance is the result of the nonzero output impedance and can also be found from the transimpedance limit. The first factor in the transimpedance limit Eq. (6.27) has the same value as in our previous example because we kept $A'_0 f'_A$, C_T, and BW_{3dB} the same. The second factor evaluates to about 0.96, consistent with the observed drop in transimpedance.

Table 6.3 Parameters and performance of our 10-Gb/s shunt-feedback TIA example with nonzero amplifier output impedance.

Parameter	Symbol	Value
Gain-bandwidth product	$A_0' f_A'$	44 GHz
Detector capacitance incl. parasitics	C_D	0.15 pF
Amplifier DC gain (unloaded)	A_0'	6
Amplifier 3-dB bandwidth (unloaded)	f_A'	7.4 GHz
Input capacitance	C_I	0.15 pF
Transconductance	g_m	50 mS
Load resistance	R_L	120 Ω
Load capacitance	C_L	0.18 pF
Feedback resistor	R_F	580 Ω
Transimpedance	R_T	480 Ω
Input impedance	R_I	100 Ω
Output impedance	R_O	17.1 Ω
TIA 3-dB bandwidth	BW_{3dB}	6.8 GHz

Output Impedance. For a voltage amplifier with zero output impedance, the TIA's output impedance is, of course, also zero. But what happens for the more realistic voltage amplifier of Fig. 6.8? Interestingly, the TIA's output impedance is *lower* than that of its constituent voltage amplifier alone. Calculating the output impedance yields

$$Z_O(s) = \frac{R_L}{A_0' + 1} \cdot \frac{1 + sR_F C_T}{1 + s/(\omega_0 Q) + s^2/\omega_0^2}, \tag{6.28}$$

where the expressions for ω_0 and Q are the same as in Eqs. (6.25) and (6.26), respectively. At low frequencies, the output resistance is $R_O = R_L/(A_0' + 1)$, that is, $A_0' + 1$ times smaller than that of the voltage amplifier alone.

As in the case of the input impedance, it is instructive to find an equivalent network for Z_O. We start by separating out the parallel components R_L and C_L

$$Z_O(s) = R_L \left\| \frac{1}{sC_L} \right\| \left(\frac{1}{g_m} \cdot \frac{1 + sR_F C_T}{1 + sC_T/g_m} \right). \tag{6.29}$$

Thus, the impedance in parallel with R_L and C_L has the low-frequency value $1/g_m$. Above the frequency $f_I = 1/(2\pi R_F C_T)$, which corresponds to the open-loop low-frequency pole, the impedance starts to rise like that of an inductor. Eventually, the impedance tops out at R_F. This behavior can also be inferred directly from Fig. 6.8(b): At low frequencies, C_I and C_D can be neglected and R_F shorts the input to the output, resulting in the output resistance $1/g_m$; at high frequencies, C_I and C_D short the input to ground, shutting down the

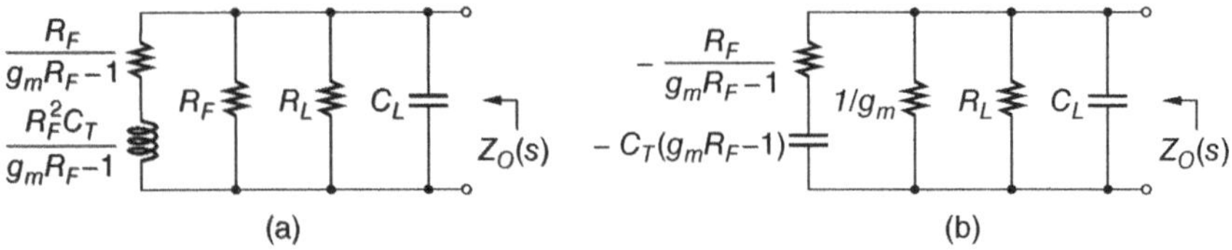

Figure 6.9 Equivalent networks for the TIA's output impedance: (a) with inductive component and (b) with negative-capacitive component.

voltage-controlled current source (transistor) and making the resistor R_F appear between the output and ground. Figure 6.9 shows *RLC* network representations of the entire output impedance. Again, there is a choice of using inductive or negative-capacitive components.

Numerical Example. For our 10-Gb/s TIA example in Table 6.3, the low-frequency output impedance is $R_O = 120\ \Omega/(6+1) = 17.1\ \Omega$. The inductive effect starts to kick in at $1/(6.28 \cdot 580\ \Omega \cdot 0.3\ \text{pF}) = 915$ MHz increasing the output impedance like that of an inductor for almost a decade until the bandwidth of the TIA (6.8 GHz) is reached. Thereafter, the output impedance peaks and rolls off to zero as a result of the load capacitance C_L. This output impedance is plotted in Fig. 6.10.

For comparison, the transimpedance and input impedance of the same TIA are also plotted in Fig. 6.10. The input-impedance expressions for this case are given in the solution to Problem 6.4(b) on p. 512. The low-frequency input impedance is $R_I = (R_F + R_L)/(A_0' + 1) = (580\ \Omega + 120\ \Omega)/(6+1) = 100\ \Omega$ and the inductive effect starts to be felt only at $f_z = 1/(2\pi \cdot R_L \parallel R_F \cdot C_L) = 1/(6.28 \cdot 120\ \Omega \parallel 580\ \Omega \cdot 0.18\ \text{pF}) = 8.9$ GHz (cf. Eq. (I.40)). Because this frequency is already beyond the TIA's bandwidth, it results mostly in a broadening of the bandwidth but little increase in the input impedance. At high frequencies, the input impedance rolls off to zero as a result of the capacitance C_T.

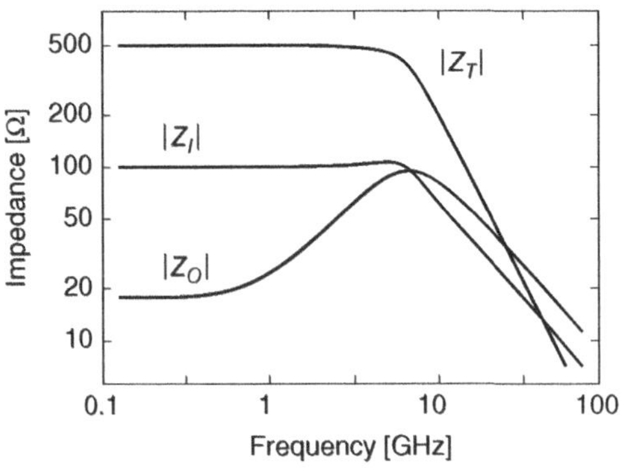

Figure 6.10 Output impedance $|Z_O|$, input impedance $|Z_I|$, and transimpedance $|Z_T|$ as a function of frequency for our 10-Gb/s TIA example.

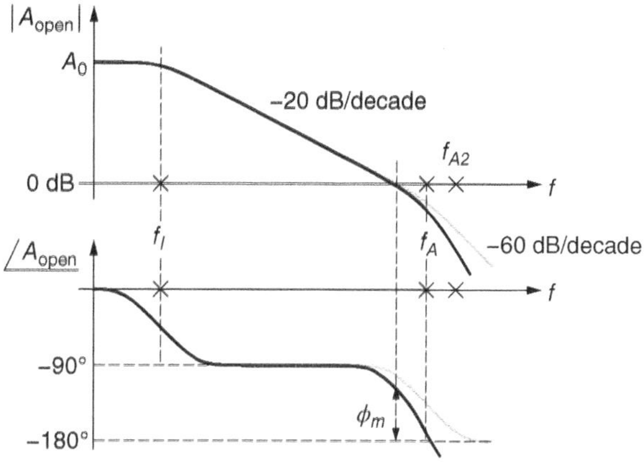

Figure 6.11 Open-loop frequency response (magnitude and phase) of a TIA with a voltage amplifier that has one dominant pole at $f_A = 1/(2\pi T_A)$ and one nondominant pole at $f_{A2} = 1/(2\pi T_{A2})$.

Voltage Amplifiers with Nondominant Poles. The single-pole model that we assumed for our finite-bandwidth single-stage voltage amplifier often is still too simplistic. In practice, additional poles are caused by cascode transistors, buffers, and level shifters. Moreover, inductive broadbanding techniques add poles and zeros to extend the bandwidth. In the following, let us discuss the impact of a single nondominant pole in the voltage amplifier. Figure 6.11 shows the open-loop frequency response, which now has three real poles: the input pole at $f_I = 1/(2\pi R_F C_T)$, the dominant amplifier pole at $f_A = 1/(2\pi T_A)$, and the new nondominant amplifier pole at $f_{A2} = 1/(2\pi T_{A2})$.

The effect of the new pole is an additional phase shift at the unity-gain frequency, which reduces the phase margin and thus the dynamic stability. The additional phase shift at frequency f is given by $\tan^{-1}(2\pi T_{A2} f)$ and with the unity-gain frequency at about $A_0/(2\pi R_F C_T)$, the phase margin is reduced by

$$\Delta\phi_m \approx \tan^{-1}\left(\frac{A_0\,T_{A2}}{R_F C_T}\right). \tag{6.30}$$

For example, a nondominant amplifier pole that is located a factor $4A_0$ above the input pole results in a phase-margin degradation of $\tan^{-1}(1/4) = 14°$.

What can we do to improve the stability in the presence of a nondominant amplifier pole? One approach is to place the dominant amplifier pole, f_A, at a higher frequency than required by Eq. (6.12) such that it causes less phase lag at the unity-gain frequency and the phase margin becomes sufficient again.

Another technique is to add a small capacitor C_F in parallel to the feedback resistor R_F as shown in Fig. 6.12. This capacitor introduces a zero at $1/(2\pi R_F C_F)$ in the open-loop response, which can be used to compensate the nondominant amplifier pole. This zero is known as a *phantom zero* because it disappears in the closed-loop response [31, 32].

A feedback capacitor helps to improve the stability regardless of the number of nondominant poles. While this method could also be used to stabilize a

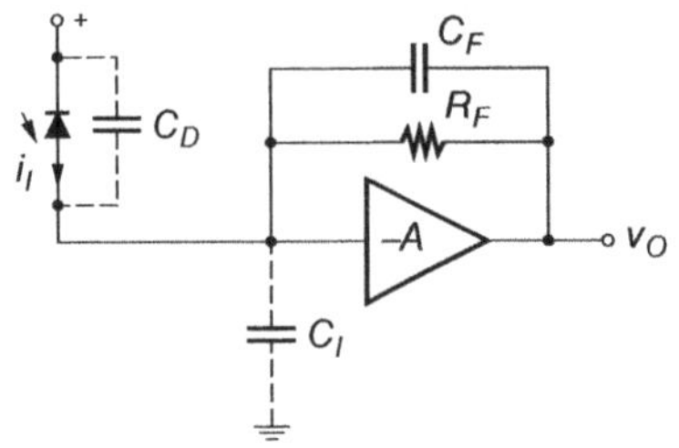

Figure 6.12 Shunt-feedback TIA with feedback capacitor.

TIA without nondominant amplifier poles (cf. Eq. (I.59)), it has the drawback of lowering the transimpedance limit (cf. Eq. (I.61)). In the case of a single amplifier pole, it is better to stabilize the TIA by reducing the gain and increasing the bandwidth of the voltage amplifier.

A feedback capacitor is also useful for a TIA with variable feedback resistor. By varying the feedback capacitor and the feedback resistor in concert, stability and constant bandwidth can be maintained for all gain settings (cf. Section 7.4). [→ Problem 6.7.]

Multistage Voltage Amplifier. Up to now we have considered only single-stage voltage amplifiers. Are there situations where it is beneficial to use a multistage amplifier as part of a shunt-feedback TIA? A multistage amplifier can achieve a much higher gain than a single-stage amplifier while loosing comparatively little in terms of bandwidth. The higher gain permits a larger feedback resistor, leading to a higher transimpedance and less noise from the feedback resistor. However, a multistage amplifier has multiple poles that create phase lags. These phase shifts add up and reduce the phase margin and the stability. To analyze this situation quantitatively, we derive the (approximate) transimpedance limit for a TIA with multiple identical single-pole stages, $A(s) = [A_0/(1+sT_A)]^n$, under the constraints of a given gain-bandwidth product per stage and good dynamic stability.

We can use the bandwidth expression Eq. (6.13) to estimate the bandwidth of the multistage TIA, if we replace A_0 by A_0^n for the total DC gain of the n-stage amplifier and assume $A_0^n \gg 1$:

$$BW_{3\text{dB}} \approx \frac{\sqrt{2A_0^n}}{2\pi R_F C_T}. \tag{6.31}$$

With more than two poles, stability can no longer be described by a single Q factor. Hence, we use the phase margin ϕ_m as the stability measure [33, 34]. The phase margin without any amplifier poles (only the pole due to $R_F C_T$) is approximately 90°. With n identical amplifier poles, this margin is degraded by n times the amount given in Eq. (6.30):

$$\phi_m \approx 90^\circ - n \cdot \tan^{-1}\left(\frac{A_0^n T_A}{R_F C_T}\right), \tag{6.32}$$

where all poles have the same time constant T_A. The bandwidth per stage, $f_A = 1/(2\pi T_A)$, required to achieve the phase margin $\phi_m \geq \Phi_m$ becomes

$$f_A \geq \frac{BW_{3dB}}{\sqrt{2}\,\tan\left(\frac{90^\circ - \Phi_m}{n}\right)}, \tag{6.33}$$

where we used Eq. (6.31). Given the gain-bandwidth product $A_0 f_A$, the gain of each stage is limited to

$$A_0 \leq \sqrt{2}\,\tan\left(\frac{90^\circ - \Phi_m}{n}\right) \cdot \frac{A_0 f_A}{BW_{3dB}}. \tag{6.34}$$

Combining this result with Eq. (6.31) and using $R_T \approx R_F$ for $A_0^n \gg 1$, finally yields the (approximate) transimpedance limit for the multistage TIA [18]:

$$R_T \leq \sqrt{2^{n+1}}\,\tan^n\left(\frac{90^\circ - \Phi_m}{n}\right) \cdot \frac{(A_0 f_A)^n}{2\pi C_T \cdot BW_{3dB}^{n+1}}. \tag{6.35}$$

A similar result was found in [34]. Several of the approximations made are pessimistic and the exact limit tends to be higher. Nevertheless, for $n = 1$, this limit assumes the same form as the single-stage transimpedance limit in Eq. (6.14) and for $\Phi_m = \tan^{-1}(2) \approx 63.4^\circ$, the two equations become identical.

Equation (6.35) shows that the trade-off between transimpedance and bandwidth becomes more and more lopsided as we add more stages. For example, doubling the bandwidth of a 3-stage TIA results in a 16-fold drop in transimpedance (making the transimpedance-bandwidth product even less meaningful).

How does this transimpedance limit compare to that of a single-stage TIA? On one hand, the first factor in Eq. (6.35) drops rapidly with increasing n (see Table 6.4), reducing the transimpedance. On the other hand, the second factor is larger by a factor $(A_0 f_A/BW_{3dB})^{n-1}$, which boosts the transimpedance. As a consequence, the transimpedance improves only if $(A_0 f_A/BW_{3dB})^{n-1}$ is large enough to overcome the first factor. Let us call the ratio $A_0 f_A/BW_{3dB}$ the *bandwidth headroom* because it is a measure of how close the TIA's target

Table 6.4 First factor of Eq. (6.35) for $\Phi_m = \tan^{-1}(2)$ and minimum bandwidth headroom needed for an n-stage TIA to outperform an $(n-1)$-stage TIA.

Number of stages n	First factor of Eq. (6.35)	Minimum $A_0 f_A/BW_{3dB}$
1	1.0	–
2	0.158	6.34
3	0.0151	10.42
4	0.00104	14.55
5	0.0000556	18.68

bandwidth is to the capability of the technology. In our 10-Gb/s example with $A_0 f_A = 44$ GHz and $BW_{3dB} = 6.85$ GHz (see Table 6.1), the bandwidth headroom is 6.4. In this case, the transimpedance for a one or two-stage design is about the same and for more than two stages it is significantly lower, making the single-stage design the best choice (cf. Fig. 7.2 in Chapter 7). Table 6.4 shows that if the bandwidth headroom is less than 6.34, that is, the TIA operates relatively close to the speed limit of the technology, there is no advantage of using multiple stages. However, if the technology is fast compared with the TIA speed, a multistage TIA can boost the transimpedance [35]. For example, if we were to realize our 10-Gb/s example TIA in a technology with $A_0 f_A = 88$ GHz, Table 6.4 predicts that a three-stage TIA leads to the highest transimpedance.

When discussing the transimpedance limit Eq. (6.35), we assumed that the voltage gain of each stage can be traded arbitrarily against its bandwidth as long as the product, $A_0 f_A$, does not exceed the capability of the technology. However, as we pointed out earlier, the gain of each stage may be limited by the supply voltage and headroom considerations. In such gain-limited situations, a multistage amplifier may be the preferred solution even if Eq. (6.35) does not exceed Eq. (6.14).

Verification by Simulation. Hopefully, the analysis in this section provided some understanding of the trade-offs involved when designing a shunt-feedback TIA. However, this simplified analysis does not replace detailed transistor-level simulations that include all parasitics. Especially at 10 Gb/s and beyond, bond-wire inductance and packaging parasitics play an important role. Only careful simulation over all relevant PVT corners can confirm whether the transimpedance, bandwidth, amplitude peaking, group-delay variations, etc. are all within specifications.

6.3 Noise Analysis

In Section 4.4, we emphasized the importance of the input-referred noise current and its impact on receiver sensitivity. Now, we want to analyze this noise quantity for the shunt-feedback TIA. First, we study the input-referred noise current PSD, $I^2_{n,TIA}(f)$, for the case of an FET and BJT front-end. This analysis provides an explanation for the white and f^2-noise components, which we have encountered earlier. Then, we turn to the *total* input-referred noise current, $i^{rms}_{n,TIA}$, which determines the sensitivity. Finally, we illustrate the noise expressions with numerical examples.

Input-Referred Noise Current PSD of an FET Front-End. Figure 6.13 shows the familiar shunt-feedback TIA with the input stage (or front-end) implemented with an FET in common-source configuration. We discuss complete transistor-level circuits in Chapter 9, but for the following simplified noise analysis, it is

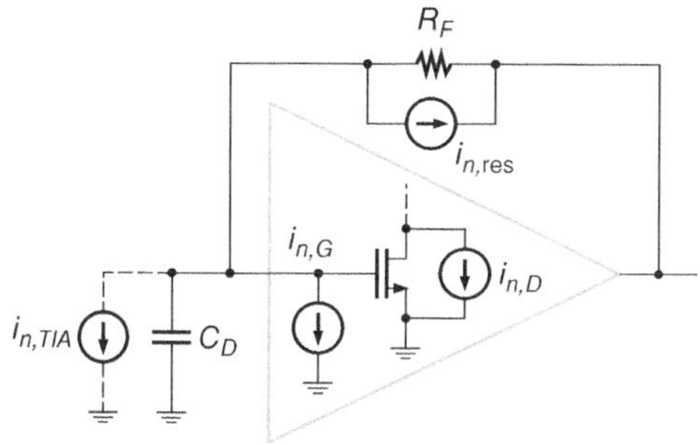

Figure 6.13 Calculation of the equivalent input noise current generator $i_{n,TIA}$ for a TIA with FET front-end.

sufficient to consider just the input transistor. The most significant device noise sources are shown in the figure: the thermal noise of the feedback resistor, $i_{n,\text{res}}$, the thermal noise of the FET, $i_{n,D}$, and the shot noise due to the gate current, $i_{n,G}$. As we know, the effect of all these noise sources can be described by a single imagined equivalent noise current generator, $i_{n,TIA}$, at the input of the TIA (shown with dashed lines).

The input-referred noise current PSD of the TIA can be broken into two major components: the noise from the feedback resistor (or resistors, in a differential implementation) and the noise from the amplifier front-end. Because they are uncorrelated, we can write

$$I_{n,TIA}^2(f) = I_{n,\text{res}}^2(f) + I_{n,\text{front}}^2(f). \tag{6.36}$$

In high-speed receivers, the noise contribution from the amplifier front-end typically is larger than the contribution from the feedback resistor. In low-speed receivers, however, the resistor noise may become dominant. We discuss the ratio of these two noise components further in Section 6.4.

Feedback-Resistor Noise. The noise current PSD of the feedback resistor is white (frequency independent) and given by the well-known thermal-noise equation:

$$I_{n,\text{res}}^2(f) = \frac{4kT}{R_F}. \tag{6.37}$$

This noise current contributes directly to the input-referred TIA noise in Eq. (6.36) because $i_{n,\text{res}}$ has the same effect on the TIA output as $i_{n,TIA}$ does. Note that in Section 6.2 this was the *only* noise source that we considered.

Gate Shot Noise. The shot noise generated by the gate current, I_G, is white and has the well-known noise current PSD

$$I_{n,G}^2 = 2qI_G. \tag{6.38}$$

It also contributes directly to the input-referred TIA noise and is uncorrelated to any of the other noise sources. MOSFETs in 0.13-μm and larger technologies have extremely small gate currents, but in nanoscale technologies, gate leakage

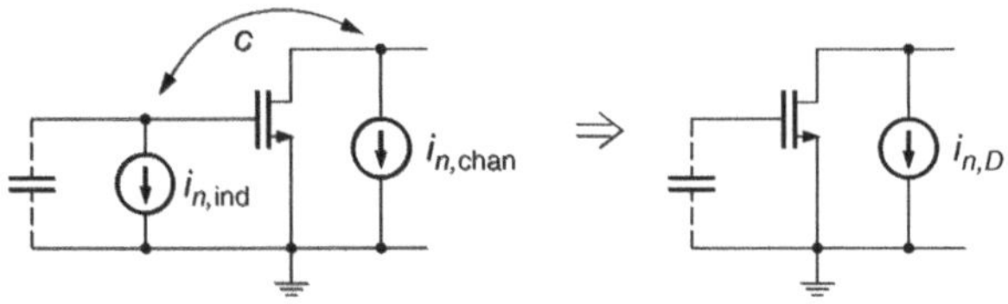

Figure 6.14 When driven by a capacitive source, the FET's channel noise generator and partially correlated induced gate noise generator in van der Ziel's noise model (left) can be combined into a single drain noise generator (right).

due to direct tunneling through the thin gate oxide increases rapidly [36]. Metal–semiconductor FETs (MESFETs) and heterostructure FETs (HFETs) can have gate currents up to 1 μA [4, 37]. Nevertheless, in high-speed TIAs, shot noise due to the gate current usually is negligible.

Channel Noise and Induced Gate Noise. The thermal fluctuations in the FET channel result in a *channel noise current* $i_{n,\text{chan}}$ and due to the capacitive coupling between the channel and the gate also in an *induced gate noise current* $i_{n,\text{ind}}$ (see Fig. 6.14) [38, 39]. For long-channel devices in saturation and strong inversion, the channel noise current PSD is found theoretically to be $I^2_{n,\text{chan}} = 4kT\frac{2}{3}g_{d0}$ [38]. The symbol g_{d0} stands for the drain–source conductance at zero drain–source voltage while maintaining the gate–source voltage of normal, saturated operation. Thus, to evaluate the noise of an FET we need to determine a small-signal parameter for an operating point that *differs* from that of normal operation. To get around this difficulty we may use the approximation $g_{d0} \approx g_m + g_{mb} + g_o$, where all the quantities on the right-hand side refer to the normal, saturated operating point, g_{mb} is the substrate transconductance, and g_o is the drain–source (output) conductance [40–42]. If the body effect and channel-length modulation can be neglected, we can further simplify to $g_{d0} \approx g_m$. Although the latter approximation is often taken for granted, it may be far off (by a factor two or more) for FETs with short channel lengths, high overdrive voltages, or both [43].

There are a number of mechanisms including hot electrons [44, 45], velocity saturation [46, 47], and weak avalanche [48] that increase the noise of short-channel devices significantly over the noise of their long-channel counterparts. To account for these and other mechanisms (e.g., substrate-resistance noise [49]), the channel noise current PSD is commonly written as [41, 48, 50, 51],

$$I^2_{n,\text{chan}} = 4kT\gamma g_{d0}, \tag{6.39}$$

where γ is a fudge factor known as the *channel noise factor*. For MOSFETs, the value of γ is typically in the range $\frac{2}{3}$ to 3, where smaller channel lengths and larger drain–source voltages tend to result in larger values [41].

The induced gate noise current PSD increases with frequency (violet noise) and can be written as [48, 50, 51]

$$I_{n,\mathrm{ind}}^2 = 4kT\delta\frac{(2\pi f C_{gs})^2}{5g_{d0}}, \tag{6.40}$$

where δ is known as the *gate noise factor*. For long-channel devices in saturation and strong inversion, the theoretical value of δ is $\frac{4}{3}$. For short-channel devices, δ increases similar to γ [48]. Moreover, because the induced gate noise and the channel noise have a common physical origin they are partially correlated, an effect quantified by the correlation coefficient c [39, 52]. (Caution: Not all authors use the same polarity conventions to define c [53].)

***Ogawa's Noise Factor* Γ.** Interestingly, if the source impedance is capacitive, often a reasonable approximation for optical receivers, the induced gate noise current produces a *white* gate noise voltage, which is then translated into a *white* drain noise current by the FET's transconductance (ignoring the feedforward path through C_{gd}). It is a common practice in the optical receiver literature to refer the induced gate noise current to the drain side and combine it with the (partially correlated) channel noise, thus arriving at a FET model with a single white drain noise generator, $i_{n,D}$, as shown in Fig. 6.14 [53]. For this reason, it was sufficient to have only the single noise generator $i_{n,D}$ instead of the two noise generators $i_{n,\mathrm{chan}}$ and $i_{n,\mathrm{ind}}$ in Fig. 6.13. The PSD of the drain noise current generator, $i_{n,D}$, is commonly written as

$$I_{n,D}^2 = 4kT\Gamma g_m, \tag{6.41}$$

where

$$\Gamma = \gamma\frac{g_{d0}}{g_m} + \langle\text{induced gate noise terms}\rangle. \tag{6.42}$$

Γ is known as *Ogawa's noise factor*, in honor of Kinichiro Ogawa, who analyzed it first for GaAs MESFETs [54] and later for MOSFETs [55]. The first term is due to the channel noise and follows easily from Eq. (6.39). The remaining terms are due to the induced gate noise and are somewhat harder to calculate [53] (see Eq. (I.68) for the full expression).

The noise factor Γ is usually larger than γ. For MOSFETs the multiplier $g_{d0}/g_m \approx 1 + (g_{mb} + g_o)/g_m$ typically results in an increase of 10 to 100%, whereas the induced gate noise terms typically result in a *decrease* of 10 to 15%, depending on the ratio $C_{gs}/\tilde{C}_T$ (cf. Eq. (I.69)). The decreasing effect of the induced gate noise is explained by the particular nature of the partial correlation: Whereas the uncorrelated portion of the induced gate noise adds to the channel noise, the correlated portion of the induced gate noise *cancels* a somewhat larger amount of the channel noise, leading to a small net noise reduction [53]. Typical values for n-MOSFETs are $\Gamma \approx 2.0$ for a channel length of 0.18 μm and $\Gamma \approx 1.0$ for $L = 0.5$ μm [53]. For 1-μm GaAs MESFETs, $\Gamma = 1.1$

to 1.75 [4, 54] and for silicon junction FETs (JFETs), $\Gamma \approx 0.7$ [4] have been reported. [→ Problem 6.8.]

While the aforementioned FET noise model is very convenient for hand calculations, it is also important to be aware of its limitations. Strictly speaking, it can be applied only when the source impedance is *purely capacitive*. In practice, there is always some resistive component due to R_F or a biasing resistor. In this case the noise model is only valid at frequencies high enough such that the capacitive component dominates the resistive one. Fortunately, as we shall see shortly, at low frequencies, where the model accuracy is lacking, the FET noise is usually dominated by other noise sources, such as the feedback-resistor noise. When an inductor, such as a bond wire, is part of the input network, the noise model is only valid if the resonance associated with the inductor is well above the frequency band of interest. When the inductor is used for noise matching (cf. Section 6.5), however, the resonance is in the band of interest and the induced gate noise can no longer be accurately accounted for with Ogawa's noise factor Γ. Finally, the noise model in Eq. (6.41) becomes invalid in the linear regime (for $V_{DS} \to 0$, g_m vanishes, but the noise does not) [40, 42, 47]. In contrast, the model based on Eqs. (6.39) and (6.40) is valid in the saturated *and* linear regimes if γ and δ are taken to be bias dependent.

Input-Referral Function. Let us go back to calculating the equivalent input-referred noise current in Fig. 6.13. Unlike $i_{n,\mathrm{res}}$ and $i_{n,G}$, the drain noise generator, $i_{n,D}$, is *not* located directly at the input of the TIA and we have to apply a transformation to find its contribution to the input-referred TIA noise. A straightforward way to do this transformation is to calculate the transfer function from the current $i_{n,D}$ to the output voltage of the TIA and divide that by the transfer function from the current $i_{n,TIA}$ to the output voltage. Equivalently, but easier, we can calculate the current $i_{n,TIA}$ that results in the *same* output voltage as the current $i_{n,D}$ and then divide the phasor of the first current by the second. Still easier, we can calculate the currents $i_{n,TIA}$ and $i_{n,D}$ that when applied simultaneously result in *zero* output voltage and then divide the phasor of the first current by the second and multiply the result by −1. (The zero output voltage condition simplifies the calculation a lot.) In any case, we call the resulting transformation an *input-referral function* and use the notation $H^{-1}(s)$ to indicate that the referral occurs in the direction *opposite* to the signal flow.

The input-referral function from the drain noise current generator to the equivalent input noise current generator in Fig. 6.13 is [37, 56]

$$H_D^{-1}(s) = -\frac{sR_F\tilde{C}_T + 1}{g_m R_F(1 - sC_{gd}/g_m)} \approx -\frac{sR_F\tilde{C}_T + 1}{g_m R_F}, \tag{6.43}$$

where we neglected feedforward through C_{gd} in the second form. The total capacitance has a tilde on it because it does *not* contain a Miller term:

$\tilde{C}_T = C_D + \tilde{C}_I$. As before, C_D is the photodetector capacitance (including parasitic capacitances in parallel to it) and $\tilde{C}_I = C_{gs} + C_{gd}$. Note that the referral function from the input noise current generator to the drain noise current generator, $H_D(s)$, has a low-pass characteristic and hence the input-referral function, $H_D^{-1}(s)$, has a *high-pass* characteristic.

It is important to distinguish between transfer functions and referral functions. The referral function answers the question: For a given current injected at the input, what current must be alternatively injected at the drain to have the same effect on the output voltage? In contrast, the transfer function answers the question: For a given current injected at the input, what current does appear at the drain? Whereas the transfer function depends on the load at the drain and does contain a Miller term of the form $A_0 C_{gd}$, the referral function does *not* depend on the load and does *not* contain a Miller term, as can be seen from (6.43).

Input-Referred Drain Noise. Referring the white drain noise current PSD, $I_{n,D}^2$, from Eq. (6.41) back to the input using $H_D^{-1}(s)$ from Eq. (6.43) yields

$$\begin{aligned} I_{n,\text{in},D}^2(f) &= |H_D^{-1}(j2\pi f)|^2 \cdot I_{n,D}^2 = \frac{(2\pi f R_F \tilde{C}_T)^2 + 1}{(g_m R_F)^2} \cdot 4kT\Gamma g_m \\ &= 4kT\Gamma \frac{(2\pi \tilde{C}_T)^2}{g_m} f^2 + \frac{4kT\Gamma}{g_m R_F^2}. \end{aligned} \tag{6.44}$$

And here we meet the long-awaited f^2-noise component. We now understand that it arises from a white-noise generator and a high-pass function referring it to the input. [→ Problem 6.9.]

Putting It All Together. We now have everything we need to write down the complete input-referred noise current PSD of the TIA with FET front-end [37, 56]:

$$I_{n,TIA}^2(f) = \frac{4kT}{R_F} + 2qI_G + 4kT\Gamma \frac{(2\pi \tilde{C}_T)^2}{g_m} f^2 + \dots, \tag{6.45}$$

where the first term is due to the feedback-resistor noise, Eq. (6.37), the second term is due to the gate shot noise, Eq. (6.38), and the third term is due to the thermal FET noise (channel noise and induced gate noise), Eq. (6.44). We have neglected the second term of Eq. (6.44), which is small compared with the feedback-resistor noise if $g_m R_F \gg \Gamma$. (However, for small values of R_F, this noise may become significant.)

Figure 6.15(a) illustrates the thermal FET noise component and the feedback-resistor noise component graphically. The FET noise from Eq. (6.44) starts out below the feedback-resistor noise (by a factor $g_m R_F / \Gamma$) and then begins to rise at the frequency $f_I = 1/(2\pi R_F \tilde{C}_T)$, corresponding to the zero in Eq. (6.43). The f^2-noise corner of the overall input-referred noise

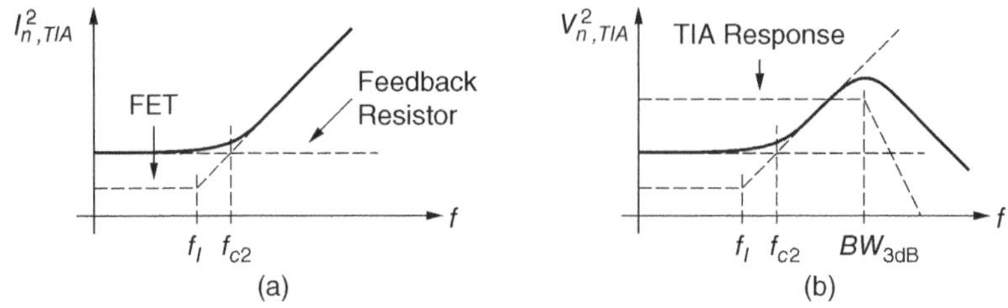

Figure 6.15 (a) Input-referred noise current PSD of a TIA with FET front-end and (b) the corresponding output-referred noise voltage PSD.

(solid line) is at the somewhat higher frequency $f_{c2} = \sqrt{g_m R_F/\Gamma + 1}/(2\pi R_F \tilde{C}_T)$ (cf. Eq. (I.74)). Interestingly, the 3-dB bandwidth of the TIA, given by $BW_{3dB} = \sqrt{2A_0(A_0+1)}/(2\pi R_F C_T)$ for a Butterworth response (cf. Eq. (6.13)), is often *higher* than the f^2-noise corner. As a result, the output-referred noise voltage PSD may have a "hump," as shown in Fig. 6.15(b). [→ Problem 6.10.]

Second-Order Effects. The noise analysis presented so far can be extended in a number of ways:

Besides the white thermal noise, the FET also produces $1/f$ *noise* (a.k.a. *flicker noise*), which contributes to the TIA's input-referred noise [37]. In analogy to the white noise, which when referred back to the input results in an f^2 noise component, the $1/f$ noise, when referred back to the input, results in an f noise component (cf. Eq. (I.75)). Because $1/f$ noise is confined to relatively low frequencies, it has little impact on high-speed TIAs. (Typically, the $1/f$-noise corner frequency of GaAs MESFETs and HFETs is between 10 and 100 MHz, that of MOSFETs is between 1 and 10 MHz, and that of JFETs is below 100 kHz [4, 37]). [→ Problem 6.11.]

Gate, drain, and source access resistances to the intrinsic device also produce thermal noise. In particular, the gate resistance R_g of a polysilicon gate with minimum length and a large width can become significant in nanoscale technologies. To account for that, we have to add the input-referred noise current term $4kTR_g(2\pi C_D)^2 f^2$ (cf. the subsequent discussion of the base resistance in BJTs). However, the gate resistance usually can be made small enough with a finger-structure layout such that this noise becomes negligible.

The FET's load resistor (or active load) and subsequent gain stages also contribute to the TIA's input-referred noise current. Taking the noise due to the load resistor R_L into account has the same effect as increasing the Γ factor from Γ to $\Gamma + 1/(g_m R_L)$ [4, 57]. For example, for a 10× voltage gain, $g_m R_L = 10$, Γ effectively increases by about 3 to 15%. Thus, if the gain of the first stage is sufficiently large, the noise of the load resistor (and the subsequent stages) is negligible.

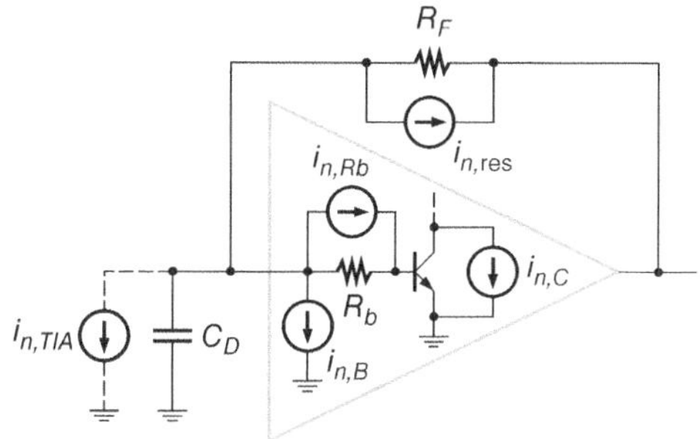

Figure 6.16 Calculation of the equivalent input noise current generator $i_{n,TIA}$ for a TIA with bipolar front-end.

When feedforward through the gate–drain capacitance C_{gd} is considered (first form of Eq. (6.43)), the input-referral function is augmented by a (right half plane) pole at g_m/C_{gd} [58]. As a result, the f^2 noise does not increase indefinitely, but tops out at this pole frequency.

A lossy photodetector, as shown in Fig. 3.4(a) or 5.16, also contributes noise. Usually R_{PD} is small enough for this noise to be negligible, but if it is not, we can account for it by adding the input-referred noise current term $4kTR_{PD}(2\pi C_{PD})^2 f^2$ [59] (or more generally, $4kTG_{Di}$, cf. Section 5.7).

Input-Referred Noise Current PSD of a BJT Front-End. A TIA with a BJT common-emitter front-end and its major noise sources is shown in Fig. 6.16. The BJT generates shot noise due to the collector current, $i_{n,C}$, and shot noise due to the base current, $i_{n,B}$. At high frequencies, these two noise generators are partially correlated [60], but in the following calculations we neglect this relatively small effect. The BJT further generates thermal noise due to its base resistance (base-spreading and contact resistance), $i_{n,Rb}$, which is typically much larger than the corresponding gate-resistance noise of a FET. The noise from the feedback resistor, $i_{n,\text{res}}$, is the same as for the FET front-end.

The shot noise generated by the base current, I_B, is given by $I^2_{n,B} = 2qI_C/\beta$, where I_C is the collector current and β is the DC current gain of the BJT ($I_B = I_C/\beta$). This white noise current contributes directly to the input-referred TIA noise. Then, we have the shot noise generated by the collector current, which is $I^2_{n,C} = 2qI_C$. This noise current must be referred to the input. If we neglect R_b, the input-referral function is the same as in Eq. (6.43) and we can write

$$\begin{aligned} I^2_{n,\text{in},C}(f) &= |H_C^{-1}(j2\pi f)|^2 \cdot I^2_{n,C} \approx \frac{(2\pi f R_F \tilde{C}_T)^2 + 1}{(g_m R_F)^2} \cdot 2qI_C \\ &= 2qI_C \frac{(2\pi \tilde{C}_T)^2}{g_m^2} f^2 + \frac{2qI_C}{(g_m R_F)^2}. \end{aligned} \quad (6.46)$$

Note how the white shot noise is transformed into a f^2-noise component, just like in the FET case. Finally, we have the thermal noise generated by the base resistance, which is given by $I^2_{n,Rb} = 4kT/R_b$. This noise current, too, must be referred to the input. In this case, the input-referral function is $H_{Rb}^{-1}(s) = -(sR_bC_D + R_b/R_F)$, where C_D is the photodetector capacitance (including parasitic capacitances in parallel to it). Thus, the noise contribution due to R_b is

$$\begin{aligned} I^2_{n,\text{in},Rb}(f) &= |H_{Rb}^{-1}(j2\pi f)|^2 \cdot I^2_{n,Rb} = \left[(2\pi f R_b C_D)^2 + \frac{R_b^2}{R_F^2}\right] \cdot \frac{4kT}{R_b} \\ &= 4kTR_b(2\pi C_D)^2 f^2 + \frac{4kTR_b}{R_F^2}. \end{aligned} \tag{6.47}$$

Note that the base resistance is giving rise to another f^2-noise component.

In summary, we can write the input-referred noise current PSD of the TIA with BJT front-end as [37, 61]

$$I^2_{n,TIA}(f) = \frac{4kT}{R_F} + \frac{2qI_C}{\beta} + 2qI_C\frac{(2\pi \tilde{C}_T)^2}{g_m^2}f^2 + 4kTR_b(2\pi C_D)^2 f^2 + \dots, \tag{6.48}$$

where the first term is due to the feedback-resistor noise, the second term is due to the base shot noise, the third term is due to the collector shot noise, and the fourth term is due to the base resistance. We have neglected the second term of $I^2_{n,\text{in},C}$ in Eq. (6.46), which is small compared with the base shot noise if $(g_m R_F)^2 \gg \beta$, and we also have neglected the second term of $I^2_{n,\text{in},Rb}$ in Eq. (6.47), which is small compared with the noise from the feedback resistor if $R_F \gg R_b$.

Equation (6.48) can be extended by including $1/f$ noise [37], but as for the FET front-end, only very low bit-rate receivers are affected by it. Another extension is to include the parasitic emitter resistance, which in advanced SiGe and InP HBTs is significant ($\approx 1/g_m$). The noise due to the collector load resistor R_L can be accounted for by increasing the collector shot noise term in Eq. (6.48) by a factor $1 + 2/(g_m R_L)$. Thus, as long as the input stage has sufficient gain, this noise and the noise from subsequent stages are negligible.

Throughout this section, we assumed that the TIA is implemented as a single-ended circuit, that is, that there is only one feedback resistor and one input transistor. A differential TIA (e.g., see Fig. 9.5) has more noise sources that must be taken into account.

Equations (6.45) and (6.48) show that, regardless of whether the TIA has an FET or BJT front-end, the input-referred noise current PSD is composed of white-noise terms and f^2-noise terms, justifying the form $I^2_{n,TIA}(f) = \alpha_0 + \alpha_2 f^2$, which we introduced in Section 4.1.

Input-Referred RMS Noise Current and Noise Bandwidths. Having discussed the input-referred noise current PSD, we now turn to the *total* input-referred noise current also known as the input-referred rms noise current. This is the quantity that determines the sensitivity of the TIA (cf. Eq. (5.7)). The total noise can be obtained from the noise PSD by weighting it with the TIA's transfer function (squared) and integrating it across all frequencies, as given by Eq. (5.6). Alternatively, noise bandwidths (or Personick integrals) can be used to do the same calculation, as discussed in Section 4.5.

Let us review the use of noise bandwidths. If the input-referred noise current PSD can be decomposed into a white-noise and an f^2-noise part, $I_{n,TIA}^2 = \alpha_0 + \alpha_2 f^2$, then the input-referred mean-square noise current is

$$\overline{i_{n,TIA}^2} = \alpha_0 \cdot BW_n + \frac{\alpha_2}{3} \cdot BW_{n2}^3, \tag{6.49}$$

where BW_n is the noise bandwidth of the TIA for white noise and BW_{n2} is the noise bandwidth for f^2 noise. The noise bandwidths are found from the transfer function of the TIA by evaluating the integrals in Eqs. (4.44) and (4.45).

For the second-order TIA model given by Eqs. (6.8)–(6.11) the noise-bandwidth integrals can be solved analytically [62] (cf. Appendix H) resulting in

$$BW_n = \frac{A_0 + 1}{4(R_F C_T + T_A)}, \tag{6.50}$$

$$BW_{n2}^3 = \frac{3(A_0 + 1)^2}{16\pi^2 (R_F C_T + T_A) R_F C_T T_A}. \tag{6.51}$$

Now, to calculate the input-referred mean-square noise current of the TIA with FET front-end, we separate the noise current PSD from Eq. (6.45) into a white and f^2-noise part and apply Eq. (6.49):

$$\overline{i_{n,TIA}^2} \approx \left(\frac{4kT}{R_F} + 2qI_G\right) BW_n + \frac{1}{3}\left(4kT\Gamma\frac{(2\pi \tilde{C}_T)^2}{g_m}\right) BW_{n2}^3. \tag{6.52}$$

The square root of this expression is the input-referred rms noise current.

Repeating the same procedure for the TIA with BJT front-end using the noise current PSD from Eq. (6.48), we find:

$$\begin{aligned}\overline{i_{n,TIA}^2} \approx &\left(\frac{4kT}{R_F} + \frac{2qI_C}{\beta}\right) BW_n \\ &+ \frac{1}{3}\left(2qI_C\frac{(2\pi \tilde{C}_T)^2}{g_m^2} + 4kTR_b(2\pi C_D)^2\right) BW_{n2}^3.\end{aligned} \tag{6.53}$$

Numerical Example: FET Front-End. To illustrate the foregoing theory, let us calculate the input-referred noise current of our 10-Gb/s TIA example in Table 6.1. The input-referred noise current PSD of a single-ended shunt-feedback TIA with FET front-end is given by Eq. (6.45). To evaluate this expression numerically, we use the following typical FET parameters in addition to those listed in Table 6.1: $\Gamma = 2$, $I_G = 0$, $\tilde{C}_I = 0.1$ pF, $g_m = 50$ mS, and $T = 300$ K, which implies $f_T = g_m/(2\pi \tilde{C}_I) = 80$ GHz (cf. Table 6.5).

Table 6.5 Parameters and noise performance of our FET shunt-feedback TIA examples.

Parameter	Symbol	10 Gb/s	40 Gb/s
Transition frequency	f_T	80 GHz	318 GHz
Ogawa's noise factor	Γ	2	2
Gate current	I_G	0	0
Input capacitance (without Miller)	$\tilde{C}_I$	0.1 pF	50 fF
Transconductance	g_m	50 mS	100 mS
Noise bandwidth (white noise)	BW_n	7.58 GHz	30.3 GHz
Noise bandwidth (f^2 noise)	BW_{n2}	10.22 GHz	40.9 GHz
Input-referred rms noise current	$i^{rms}_{n,TIA}$	889 nA	2.51 μA
Averaged inp.-ref. noise curr. density	$I^{avg}_{n,TIA}$	10.2 pA/$\sqrt{\text{Hz}}$	14.4 pA/$\sqrt{\text{Hz}}$

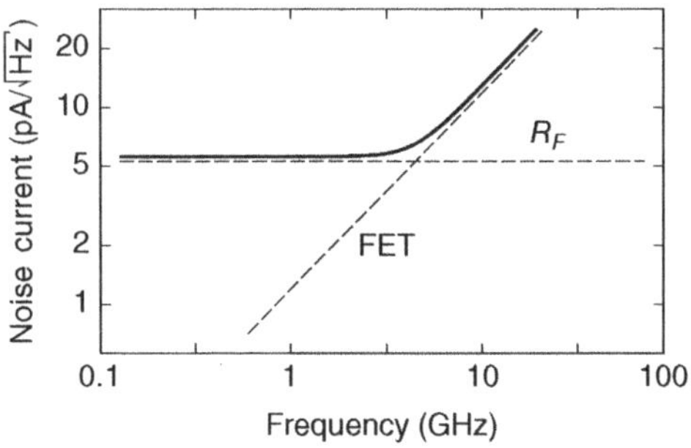

Figure 6.17 Input-referred noise current density of our 10-Gb/s FET TIA example.

The resulting input-referred noise current density is plotted in Fig. 6.17 with a solid line and the contributions from the feedback resistor R_F (600 Ω) and the FET are shown with dashed lines. We see that at low frequencies, the noise from the feedback resistor dominates, bringing the total spectral density to about 5.3 pA/$\sqrt{\text{Hz}}$. But at frequencies above about 4.2 GHz, the f^2-noise due to the FET dominates making a significant contribution to the total noise, as we will see shortly.

Next, to calculate the input-referred rms noise current we need the noise bandwidths for our 10-Gb/s TIA example. Using Eqs. (6.50) and (6.51) we find $BW_n = 7.58$ GHz and $BW_{n2} = 10.22$ GHz. Note that these noise bandwidths are significantly larger than the 3-dB bandwidth, which was 6.85 GHz. In fact, $BW_n/BW_{3\text{dB}} = 7.58/6.85 = 1.11$ and $BW_{n2}/BW_{3\text{dB}} = 10.22/6.85 = 1.49$, exactly what we would expect based on Table 4.7.

Now, we can calculate the input-referred rms noise current using Eq. (6.52):

$$i^{rms}_{n,TIA} \approx \sqrt{(458\text{ nA})^2 + (762\text{ nA})^2} = 889\text{ nA}. \tag{6.54}$$

The feedback resistor (first term) contributes about 27% and the FET (second term) contributes about 73% to the total noise (in terms of mean-square current).

The averaged input-referred noise current density is obtained by dividing the total input-referred noise current by the square-root of the noise bandwidth, resulting in $I_{n,TIA}^{avg} = 10.2\ \text{pA}/\sqrt{\text{Hz}}$. This number is nearly twice that of the feedback-resistor noise alone ($5.3\ \text{pA}/\sqrt{\text{Hz}}$).

How much more noisy is our 40-Gb/s example in Table 6.1 compared to the 10-Gb/s example? Assuming that we have a four times faster technology at our disposal, we find the noise numbers shown in Table 6.5. The averaged input-referred noise current density went up by a factor $\sqrt{2}$, while the input-referred rms noise current went up by a factor $2\sqrt{2}$ (both terms in Eq. (6.52) increased by a factor of 8). The increase in rms noise can be decomposed into two factors: a factor 2 due to the quadrupling of the bandwidth and a factor $\sqrt{2}$ due to the higher noise density. The second factor depends on the speed of the available technology. For a less than four times faster technology, the input-referred noise current density would be higher.

Numerical Example: BJT Front-End. The input-referred noise current PSD of a single-ended shunt-feedback TIA with BJT front-end is given by Eq. (6.48). To evaluate this expression numerically, we use the following typical BJT parameters in addition to those listed in Table 6.1: $\beta = 100$, $R_b = 40\ \Omega$, $\tilde{C}_I = 0.1$ pF, $g_m = 50$ mS, and $T = 300$ K, which implies $I_C = g_m kT/q = 1.3$ mA and $f_T = g_m/(2\pi\tilde{C}_I) = 80$ GHz (cf. Table 6.6).

The resulting input-referred noise current density is plotted in Fig. 6.18. At low frequencies, the noise from the feedback resistor (R_F) dominates again.

Table 6.6 Parameters and noise performance of our BJT shunt-feedback TIA examples.

Parameter	Symbol	10 Gb/s	40 Gb/s
Transition frequency	f_T	80 GHz	318 GHz
DC current gain	β	100	100
Base resistance	R_b	40 Ω	20 Ω
Input capacitance (without Miller)	$\tilde{C}_I$	0.1 pF	50 fF
Transconductance	g_m	50 mS	100 mS
Noise bandwidth (white noise)	BW_n	7.58 GHz	30.3 GHz
Noise bandwidth (f^2 noise)	BW_{n2}	10.22 GHz	40.9 GHz
Input-referred rms noise current	$i_{n,TIA}^{rms}$	771 nA	2.18 μA
Averaged inp.-ref. noise curr. density	$I_{n,TIA}^{avg}$	$8.9\ \text{pA}/\sqrt{\text{Hz}}$	$12.5\ \text{pA}/\sqrt{\text{Hz}}$

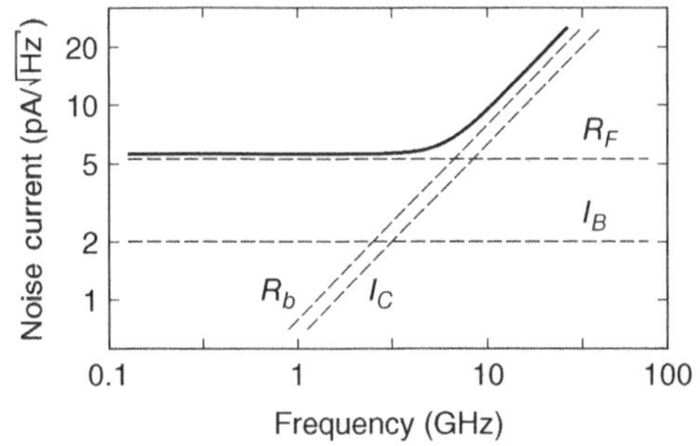

Figure 6.18 Input-referred noise current density of our 10-Gb/s BJT TIA example.

At high frequencies, the noise from the base resistance (R_b) now plays the most important role.

Using the same noise bandwidths as in our FET example, we find the input-referred rms noise current with Eq. (6.53) as

$$i_{n,TIA}^{rms} \approx \sqrt{(458\ \text{nA})^2 + (177\ \text{nA})^2 + (381\ \text{nA})^2 + (458\ \text{nA})^2} = 771\ \text{nA}, \tag{6.55}$$

where the terms from left to right are due to R_F, I_B, I_C, and R_b. The feedback-resistor contribution (first term) is again dominated by the transistor contributions (second to fourth term). The largest transistor contribution is from the base resistance, which is as large as the feedback-resistor contribution.

Low-Impedance Front-End Revisited. In Section 6.1, we discussed the noise of low-impedance front-ends. We wrote down the contribution from the 50-Ω load resistor, but we did not know the contribution from the subsequent buffer amplifier. Interestingly, the noise equations that we just derived for the shunt-feedback TIA also hold for the low-impedance front-end shown in Fig. 6.1(a), if we re-interpret the feedback resistor R_F as the 50-Ω load resistor and the voltage amplifier in the feedback loop as the buffer amplifier. In general, adding feedback to an amplifier does not affect the input-referred noise, except for the noise contributed by the resistive part of the feedback network [63].

For a low-impedance front-end with our example amplifier as a buffer, the input-referred noise current density is as shown in Fig. 6.17 (or Fig. 6.18), except that the noise contribution from R_F goes up from 5.3 to 18 pA/$\sqrt{\text{Hz}}$, making the low-impedance front-end significantly noisier than the shunt-feedback TIA. (However, in severely power-constrained systems, the amplifier front-end noise may become so high that the advantage of the shunt-feedback TIA over the low-impedance front-end all but vanishes [64].)

6.4 Noise Optimization

Having derived analytical expressions for the input-referred rms noise current, we now have all the pieces in place to tackle the noise optimization problem.

The optimization procedures for TIAs with FET and BJT front-ends are somewhat different and we discuss them separately. After the formal treatment, we illustrate the optimization procedures with numerical examples. We conclude with a discussion of the questions: "How does noise (and the resulting sensitivity) scale with the bit rate?" and "Does reducing the photodetector capacitance or increasing the photodetector responsivity have more impact on the sensitivity?"

Noise Optimization of a TIA with FET Front-End. The input-referred mean-square noise current of a (single-ended) TIA with an FET front-end can be written as (cf. Eq. (6.52))

$$\overline{i^2_{n,TIA}} = \frac{4kT}{R_F} \cdot BW_n + 2qI_G \cdot BW_n + \frac{4kT\Gamma[2\pi(C_D + \tilde{C}_I)]^2}{3g_m} \cdot BW^3_{n2} + \dots, \tag{6.56}$$

where we have expanded $\tilde{C}_T = C_D + \tilde{C}_I$, C_D is the photodetector capacitance (with all parasitic capacitances in parallel to it), and $\tilde{C}_I$ is the input capacitance of the FET under shorted-output conditions ($\tilde{C}_I = C_{gs} + C_{gd}$, no Miller term).

As we know, the first term of Eq. (6.56) can be minimized by making R_F as large as possible. To that end, a fast technology (high f_T) resulting in a high transimpedance limit is desirable.

The second term suggests the use of an FET with a low gate-leakage current, I_G. Specifically, I_G should be small enough such that this noise term can be neglected when compared with the feedback-resistor noise term. This condition is fulfilled when $I_G \ll 2V_T/R_F \approx 50$ mV$/R_F$, where $V_T = kT/q$ is the thermal voltage. For example, for $R_F = 600\ \Omega$, we need $I_G \ll 83\ \mu$A, which is easily satisfied.

The third term increases with C_D. Thus we should minimize the photodetector capacitance as well as the pad, ESD protection, and all other capacitances in parallel to it. The third term also increases with the input capacitance, $\tilde{C}_I = C_{gs} + C_{gd}$. However, simply minimizing $\tilde{C}_I$ is not desirable because $\tilde{C}_I$ and the transconductance, g_m, which appears in the denominator, are related by means of the technology parameter f_T as $g_m = 2\pi f_T \tilde{C}_I$. Instead, we should minimize the entire expression $(C_D + \tilde{C}_I)^2/g_m$. First, a technology with a high f_T should be selected because this implies a low $\tilde{C}_I/g_m$ ratio. Second, for a given technology and f_T, we can rewrite the third term as

$$\overline{i^2_{n,TIA,3}} = \frac{(C_D + \tilde{C}_I)^2}{\tilde{C}_I}\xi \qquad \text{where} \qquad \xi = \frac{8\pi kT\Gamma}{3f_T} \cdot BW^3_{n2}. \tag{6.57}$$

Taking the derivative with respect to the variable $\tilde{C}_I$ and setting the result to zero, we find the noise minimum at

$$\tilde{C}_I = C_D. \tag{6.58}$$

This means that the minimum is reached when the front-end FET is sized such that its input capacitance, $\tilde{C}_I = C_{gs} + C_{gd}$, matches C_D, the photodetector

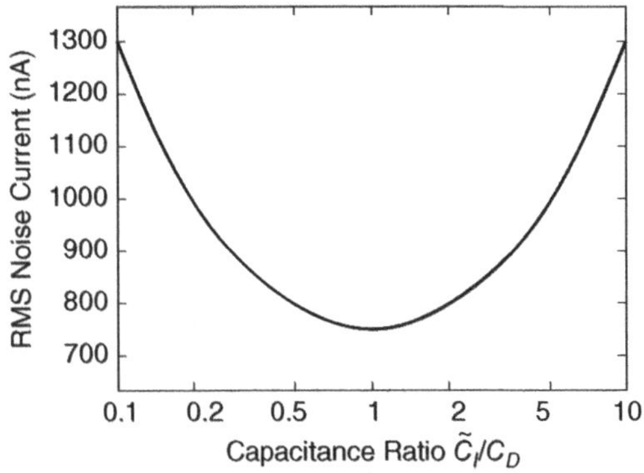

Figure 6.19 Input-referred rms noise current due to the third term of Eq. (6.56) versus input capacitance normalized to the detector capacitance for our 10-Gb/s TIA example.

capacitance with all parasitic capacitances in parallel to it [4, 6, 56, 65]. For this optimum input capacitance the mean-square noise current becomes $\overline{i^2_{n,TIA,3}} = 4C_D\xi$.

Intuitively, we can explain the existence of a minimum as follows: Starting with a very small $\tilde{C}_I$, it helps to increase $\tilde{C}_I$ because this also increases g_m, which suppresses the noise ($C_D + \tilde{C}_I$ remains approximately constant in this regime). But at some point, the total capacitance at the input node, $\tilde{C}_T = C_D + \tilde{C}_I$, becomes so large that it hurts the noise performance more than g_m helps (a large $\tilde{C}_T$ hurts quadratically, whereas a large g_m helps only linearly). Thus, an optimum exists somewhere in the middle.

Numerical Example. Figure 6.19 illustrates the noise contribution from the third term of Eq. (6.56) for our 10-Gb/s TIA example with the values from Tables 6.1 and 6.5. Clearly, the minimum is reached for $\tilde{C}_I/C_D = 1$. The numerical value of ξ only determines the magnitude of the noise, but does not affect where the minimum occurs.

Variables, Constants, and Constraints. When optimizing an expression with multiple parameters, it is important to be clear about which parameters are kept constant, which parameters are varied, and what constraints apply to the parameters that are varied. The optimization result depends on these assumptions [66].

Let us choose a fixed channel length, L, and a fixed drain current density, I_D/W, then we vary the channel width, W, to optimize the noise. As a consequence of the fixed current density, the drain current I_D varies with the channel width and thus $I_D \propto \tilde{C}_I$. Moreover, the transconductance g_m also varies with the channel width and thus $g_m \propto \tilde{C}_I$ or more explicitly $g_m = 2\pi f_T \tilde{C}_I$. (Doubling the channel width is like putting two equally biased FETs in parallel to each other.) In contrast, f_T, which is determined by the ratio of the transconductance and the input capacitance, is fixed. The gate–source bias voltage, V_{GS}, is also fixed, which is good because it avoids potential headroom issues when varying the channel width. In summary, we effectively have a single variable: $\tilde{C}_I \propto W \propto I_D \propto g_m$ and the following constants: L, I_D/W, f_T, and V_{GS}.

What values should we choose for the fixed channel length and the fixed drain current density? A short channel length is necessary for a high f_T, but it limits the maximum voltage gain. Often, the minimum channel length is chosen. A drain current density for which the FET operates near its maximum f_T or near its minimum noise figure is often a good choice. For nanoscale MOSFETs, the maximum f_T is reached for a current density of 0.2 to 0.5 mA/μm and the minimum noise figure is reached for a current density of 0.15 to 0.2 mA/μm [20]. Lower current densities may be chosen for low-power applications.

There are a couple of more constants and constraints that must be considered. For a given technology and f_T, the gain-bandwidth product of a single-stage amplifier, $A_0 f_A$, is fixed. This condition constrains how A_0 and f_A can vary during the optimization. Furthermore, the bandwidth, BW_{3dB}, and the stability, Q, of the TIA must stay fixed (which implies that the noise bandwidths stay fixed too). After all, the TIA needs a certain bandwidth to support the desired bit rate. We will see shortly that these conditions force certain parameters (e.g., A_0, f_A, R_F) to vary with $\tilde{C}_I$ in a particular way.

Finally, we consider the parameters C_D, Γ, and I_G as fixed. Strictly speaking, the noise factor Γ does depend somewhat on C_{gs} [53], but usually the dependence is weak enough, especially for short-channel MOSFETs with small correlation coefficients, such that this effect can be neglected. Similarly, the leakage current I_G increases with the gate area and thus $\tilde{C}_I$. However, in high-speed receivers the gate shot noise is very small compared with the feedback-resistor noise, such that this effect can be neglected.

Power Considerations. The constant current-density constraint implies that smaller FETs consume less power. From Fig. 6.19, we see that the noise minimum is relatively shallow and by choosing a somewhat smaller than-optimum $\tilde{C}_I/C_D$ ratio we can save a fair amount of power while sacrificing little in terms of noise performance [67, 68].

For example, for $\tilde{C}_I/C_D = 0.5$ the power dissipation of the input FET is cut in half, while the rms noise current increases by only 6.1%, corresponding to a 0.26 dB loss in optical sensitivity. (An additional benefit is that the reduced input capacitance leads to a higher transimpedance, as we will see shortly.)

Effect of Feedback-Resistor Noise on the Optimum. What happens if we add R_F to the list of fixed parameters? Then, the feedback-resistor noise (first term in Eq. (6.56)) is not part of the optimization problem and $\tilde{C}_I = C_D$ is the optimum for the entire noise expression in Eq. (6.56). But is it possible to vary $\tilde{C}_I$ and simultaneously keep R_F and the frequency response ($BW_{3\,dB}$ and Q) fixed, or have we imposed too many constraints? Using our TIA model with a single-pole voltage amplifier as a reference, we find that it is indeed possible to adjust the amplifier parameters A_0 and f_A as a function of $\tilde{C}_I$ such that the frequency response remains fixed (cf. Eqs. (I.76) and (I.77)). The problem with adding the constant R_F constraint is that it forces the gain-bandwidth product, $A_0 f_A$, to vary with $\tilde{C}_I$ thus violating our constant gain-bandwidth constraint. [→ Problem 6.12.]

It is easy to see that if we keep the open-loop input pole, $f_I \approx 1/[2\pi R_F(C_D + \tilde{C}_I)]$, and the open-loop output pole, $f_A = 1/(2\pi T_A)$, and the open-loop DC gain, A_0, fixed, the entire frequency response (open- or closed-loop) is fixed. Thus $A_0 f_A$, BW_{3dB}, and Q all remain constant, as desired. But now, R_F must become a variable. It must vary in concert with $\tilde{C}_I$ such that f_I remains fixed:

$$R_F = \frac{1}{2\pi f_I (C_D + \tilde{C}_I)}. \tag{6.59}$$

If we increase the FET's width (and thus $\tilde{C}_I$), we must reduce R_F according to Eq. (6.59). Moreover, increasing the FET's width (and thus g_m) requires that we reduce the load resistor R_L to keep A_0 $(= g_m R_L)$ constant and increase C_L to keep f_A $(= 1/(2\pi R_L C_L))$ constant.

With the first noise term in Eq. (6.56) now being dependent on $\tilde{C}_I$ the optimization problem changes to

$$\overline{i^2_{n,TIA}} = (C_D + \tilde{C}_I)\xi_1 + \frac{(C_D + \tilde{C}_I)^2}{\tilde{C}_I}\xi_2, \tag{6.60}$$

where

$$\xi_1 = 8\pi kT f_I \cdot BW_n \qquad \text{and} \qquad \xi_2 = \frac{8\pi kT\Gamma}{3 f_T} \cdot BW^3_{n2}. \tag{6.61}$$

The first term is due to the feedback-resistor's noise and the familiar second term is due to the FET's noise. All optimization constraints are now incorporated into the noise expression and only a single optimization variable, $\tilde{C}_I$, remains.

Taking the derivative with respect to the variable $\tilde{C}_I$ and setting the result to zero, we find the noise minimum at

$$\tilde{C}_I = \psi C_D \qquad \text{where} \qquad \psi = \frac{1}{\sqrt{1 + \xi_1/\xi_2}}. \tag{6.62}$$

Thus, the optimum input capacitance is somewhat *smaller* than C_D [33, 35, 66–68]. This is not surprising: A smaller input capacitance permits a larger and less noisy feedback resistor for the same open-loop input pole frequency. Thus lower capacitance solutions are preferred, shifting the curve in Fig. 6.19 to the left by the factor ψ. Of course, for $\xi_1 = 0$ (no resistor noise), the minimum reverts back to its old location, $\psi = 1$.

Let us bring the expression for ψ into a more practical form. Given our TIA model with a single-pole voltage amplifier Eqs. (6.10) and (6.11) we can calculate f_I as a function of $A_0 f_A$, BW_{3dB}, and Q. For $Q = 1/\sqrt{2}$, we find $f_I \approx BW_{3dB}^2/(A_0 f_A)$ (cf. Eq. (I.84) on p. 518). Next, we express BW_n and BW_{n2} in terms of BW_{3dB} and Q (cf. Eqs. (H.8) and (H.9)). Then, using these results with Eqs. (6.61) and (6.62) and assuming that BW_{3dB} is not too close to $A_0 f_A$, we can approximate the optimum as (cf. Eq. (I.89)) [66]

$$\psi \approx \sqrt{\frac{\Gamma A_0 f_A}{\Gamma A_0 f_A + f_T}}. \tag{6.63}$$

Note that this optimum is (approximately) independent of BW_{3dB} and Q. [→ Problem 6.13.]

Noise after Optimization; Figure of Merit. How large is the noise current after the TIA has been optimized? Inserting Eq. (6.62) into Eq. (6.60) and simplifying the result yields $\overline{i_{n,TIA}^2} = (1 + 1/\psi)^2 C_D \xi_2$. With Eq. (6.61) and assuming a second-order Butterworth response ($Q = 1/\sqrt{2}$) we can write more explicitly

$$\overline{i_{n,TIA}^2} = \sqrt{8}\pi^2 kT \left(1 + \frac{1}{\psi}\right)^2 \frac{\Gamma C_D}{f_T} \cdot BW_{3dB}^3. \tag{6.64}$$

This result suggests that the expression $f_T/(\Gamma C_D)$ can serve as a figure of merit [4]. Technologies for which this figure of merit is high yield low-noise TIAs.

We further conclude from Eq. (6.64) that the optimized mean-square noise current scales with the third power of the bandwidth, BW_{3dB}^3. So, what happened to the noise of the feedback resistor that in Eq. (6.56) used to scale with the first power of the bandwidth? Remember, we optimized under the constraint of a constant gain-bandwidth product. The transimpedance limit of a TIA with a single-pole voltage amplifier Eq. (6.14) tells us that under this condition the transimpedance scales like BW_{3dB}^{-2}. In turn, R_F also scales approximately like BW_{3dB}^{-2} (assuming BW_{3dB} is not too close to $A_0 f_A$ such that $A_0/[A_0 + 1] \approx 1$). Therefore, the noise of the feedback resistor now scales (approximately) with the third power of the bandwidth, just like the noise of the FET.

It can be shown that, after noise optimization, the fraction of the mean-square noise current originating from the FET is ψ (cf. Eq. (I.82)) and that originating from the feedback resistor is $1 - \psi$.

Numerical Example. Let us evaluate Eq. (6.63) for our 10-Gb/s FET TIA example in Tables 6.1 and 6.5:

$$\psi \approx \sqrt{\frac{2 \cdot 44\text{ GHz}}{2 \cdot 44\text{ GHz} + 80\text{ GHz}}} = 0.72. \tag{6.65}$$

With the detector capacitance $C_D = 0.15$ pF, we find that the FET should be sized for $\tilde{C}_I = 0.72 \cdot 0.15\text{ pF} = 0.11$ pF. This is very close to the value used in our example (0.1 pF) and we can consider it noise optimized. For our 40-Gb/s FET TIA example, we find the same optimum value $\psi = 0.72$.

After noise optimization, we expect about 72% of the noise to originate from the FET and 28% from the feedback resistor. This is in good agreement with what we observed for our example in Section 6.3.

Second-Order Effects. In our discussion we have dealt with the noise of the front-end FET and the feedback resistor. Although these are the dominant noise sources, the FET's load resistor and subsequent stages also make some contribution to the input-referred noise, especially in high-speed TIAs where gains are low. How does the inclusion of these noise sources alter the sizing recommendations for the FET?

As we know from Section 6.3, the thermal noise of the load resistor, R_L, can be accounted for by increasing Γ from Γ to $\Gamma + 1/(g_m R_L)$. From Eqs. (6.63) we conclude that the inclusion of this noise source pushes ψ to a slightly higher value and thus the optimum FET size becomes slightly larger.

If we include the noise of the stages that follow the shunt-feedback TIA, the optimum shifts slightly in the direction that results in a higher transimpedance, that is, toward a smaller FET size [65]. This is so because a higher transimpedance can better suppress the noise from the subsequent stages.

A more significant shift of the optimum toward a smaller FET size occurs when we include the post amplifier's bandwidth-limiting effect. Adding a post amplifier to the TIA has only a minuscule effect on the input-referred noise of the combination, but it typically does have a significant effect on the bandwidth of the combination. For example, in the case that the post amplifier and the TIA have the same second-order Butterworth response, the 3-dB bandwidth of the combination is only about 80% of the 3-dB bandwidth of the TIA alone [69]. Similarly, it can be shown that the noise bandwidths BW_n and BW_{n2} shrink to 75% and 63%, respectively, when adding such a post amplifier. With most of the input-referred FET noise located at high frequencies and the noise bandwidths shrinking, the importance of the white R_F noise increases relative to the FET noise. In response, the optimum shifts in the direction that results in a larger R_F, that is, toward a smaller FET size.

If we add a series inductor to the feedback resistor, the optimum also shifts toward a smaller FET size [20].

Noise Optimization of a TIA with BJT Front-End. The input-referred mean-square noise current of a (single-ended) TIA with a BJT front-end can be written as (cf. Eq. (6.53))

$$\begin{aligned}\overline{i^2_{n,TIA}} = \frac{4kT}{R_F} \cdot BW_n + \frac{2qI_C}{\beta} \cdot BW_n + \frac{2qI_C[2\pi(C_D + \tilde{C}_I)]^2}{3g_m^2} \cdot BW_{n2}^3 \\ + \frac{4kTR_b(2\pi C_D)^2}{3} \cdot BW_{n2}^3 + \dots,\end{aligned} \tag{6.66}$$

where we have expanded $\tilde{C}_T = C_D + \tilde{C}_I$, C_D is the photodetector capacitance (with all parasitic capacitances in parallel to it) and $\tilde{C}_I$ is the input capacitance of the BJT under shorted-output conditions ($\tilde{C}_I = C_{be} + C_{bc}$, no Miller term).

Before we can optimize this expression, we must be clear about which parameters are kept constant, which parameters are varied, and what constraints apply to the parameters that are varied. Analogous to the FET case, we choose a fixed emitter width, W_E, and a fixed collector current density, I_C/A_E, then we vary the length of the emitter stripe, L_E, to optimize the noise. As a consequence of the fixed emitter width, the base resistance, R_b, varies inversely with the emitter length and thus $R_b \propto 1/\tilde{C}_I$. More explicitly, we can write $R_b = 1/(2\pi f_{Rb}\tilde{C}_I)$, where f_{Rb} is the pole frequency due to the base resistance, which is independent of L_E. As a consequence of the fixed current density, the collector current, I_C, varies with the emitter area and thus $I_C \propto \tilde{C}_I$. More explicitly, we can write $I_C = 2\pi f_T V_T \tilde{C}_I$, where V_T is the thermal voltage. Moreover, the transconductance, g_m, also varies with $\tilde{C}_I$: $g_m = 2\pi f_T \tilde{C}_I$. In contrast, f_T and V_{BE} remain fixed. In summary, we effectively have a single variable: $\tilde{C}_I \propto A_E \propto L_E \propto 1/R_b \propto I_C \propto g_m$ and the following constants: W_E, I_C/A_E, f_T, V_{BE}, C_D, and β.

What values should we choose for the fixed emitter width and the fixed collector current density? A narrow emitter stripe helps to keep the base resistance small. Often, the minimum emitter width is chosen. A collector current density for which the BJT operates near its maximum f_T or near its minimum noise figure is often a good choice. For a typical SiGe HBT, the maximum f_T is reached for a current density of 15 to 20 mA/μm^2 and the minimum noise figure is reached for a current density of 1.5 to 8 mA/μm^2 [20]. Lower current densities may be chosen for low-power applications.

Now, we can recast Eq. (6.66) into a form that takes the aforementioned constraints into account. Substituting $I_C = 2\pi f_T V_T \tilde{C}_I$, $g_m = 2\pi f_T \tilde{C}_I$, and $R_b = 1/(2\pi f_{Rb}\tilde{C}_I)$ into Eq. (6.66) results in

$$\overline{i^2_{n,TIA}} = \frac{4kT}{R_F} \cdot BW_n + \frac{4\pi kT f_T \tilde{C}_I}{\beta} \cdot BW_n + \frac{4\pi kT(C_D + \tilde{C}_I)^2}{3 f_T \tilde{C}_I} \cdot BW_{n2}^3 + \frac{8\pi kT C_D^2}{3 f_{Rb} \tilde{C}_I} \cdot BW_{n2}^3 + \cdots \quad (6.67)$$

Note that the third term (collector shot noise) has the same form as the third term in the noise equation for the FET front-end (drain noise). Thus, the optimum input capacitance that minimizes the third term is straightforward to determine. From our study of the FET front-end we know that the expression $(C_D + \tilde{C}_I)^2/\tilde{C}_I$ reaches its minimum for $\tilde{C}_I = C_D$.

The optimum input capacitance that minimizes the *entire* noise expression Eq. (6.67) is different from that for the third term alone. The second term (base shot noise), pulls the optimum toward a smaller size, especially in low-speed receivers. The fourth term (base resistance noise), pulls the optimum toward a larger size. The first term (feedback-resistor noise) pulls the optimum toward a smaller size, if a constant gain-bandwidth product is imposed. Next, we quantify these three effects.

Effect of Feedback-Resistor Noise, Base Shot Noise, and Base Resistance Noise on the Optimum. To keep the gain-bandwidth product, $A_0 f_A$, the TIA's bandwidth, BW_{3dB}, and its stability, Q, constant during the optimization of $\tilde{C}_I$, the feedback resistor must vary with $\tilde{C}_I$ as (cf. Eq. (6.59))

$$R_F = \frac{1}{2\pi f_I (C_D + \tilde{C}_I)}. \tag{6.68}$$

Inserting R_F into the first noise term of Eq. (6.67), the noise optimization problem becomes

$$\overline{i^2_{n,TIA}} = \xi_0 + \tilde{C}_I \xi_1 + \frac{(C_D + \tilde{C}_I)^2}{\tilde{C}_I}\xi_2 + \frac{C_D^2}{\tilde{C}_I}\xi_3, \tag{6.69}$$

where

$$\xi_0 = 8\pi kT f_I C_D \cdot BW_n \quad \text{and} \quad \xi_1 = 4\pi kT\left(2f_I + \frac{f_T}{\beta}\right) \cdot BW_n, \tag{6.70}$$

$$\xi_2 = \frac{4\pi kT}{3f_T} \cdot BW_{n2}^3 \quad \text{and} \quad \xi_3 = \frac{8\pi kT}{3f_{Rb}} \cdot BW_{n2}^3. \tag{6.71}$$

The first term is due to the C_D part of the feedback-resistor noise. The second term is due to the $\tilde{C}_I$ part of the feedback-resistor noise and the base shot noise. The third term is due to the collector shot noise. Finally, the fourth term is due to the base resistance noise. All optimization constraints are now incorporated into the noise expression and only a single optimization variable, $\tilde{C}_I$, remains.

Taking the derivative with respect to the variable $\tilde{C}_I$ and setting the result to zero, we find the noise minimum at

$$\tilde{C}_I = \psi C_D \quad \text{where} \quad \psi = \sqrt{\frac{\xi_2 + \xi_3}{\xi_1 + \xi_2}}. \tag{6.72}$$

Like in the FET case, ψ describes how much smaller the optimum $\tilde{C}_I$ is compared with C_D. Together with our optimization constraints, ψ determines not only the optimum device size but also the optimum bias current. Note that the result in Eq. (6.72) has the same form as in the FET case, but the particular expression for ψ and its dependence on BW_{3dB} are different.

Following the same method as for the FET front-end, that is, using our TIA model with a single-pole amplifier to calculate f_I, assuming BW_{3dB} is not too close to $A_0 f_A$, and setting $Q = 1/\sqrt{2}$, we can approximate ψ as

$$\psi \approx \left(\frac{2f_T + f_{Rb}}{f_{Rb}}\right)^{1/2} \cdot \left(\frac{f_T^2}{\beta\, BW_{3dB}^2} + \frac{2f_T + A_0 f_A}{A_0 f_A}\right)^{-1/2}. \tag{6.73}$$

In contrast to the FET case, the optimum now *does* depend on the TIA's bandwidth, $BW_{3\text{dB}}$.

To gain more insight into Eq. (6.73) we approximate it with two asymptotes, one valid in the low-speed regime (where the first term in the parenthesis dominates) and one valid in the high-speed regime (where the second term in parenthesis dominates). For the low-speed regime, we have

$$\psi \approx \sqrt{\frac{(2f_T + f_{Rb})\beta}{f_{Rb}} \cdot \frac{BW_{3\text{dB}}}{f_T}} \quad \text{(low speed).} \tag{6.74}$$

Thus, ψ increases linearly with the TIA's bandwidth until it saturates in the high-speed regime at a value near one

$$\psi \approx \sqrt{\frac{(2f_T + f_{Rb})A_0 f_A}{(2f_T + A_0 f_A) f_{Rb}}} \quad \text{(high speed).} \tag{6.75}$$

Why are the optimum BJT size and bias current smaller for low-speed TIAs? Because the base shot noise becomes dominant. For a constant size and bias current, the noise components due to the feedback resistor, collector current, and base resistance all scale like $BW_{3\text{dB}}^3$, while the noise due to the base current scales like $BW_{3\text{dB}}$ (cf. Eq. (6.67) and $R_F \propto BW_{3\text{dB}}^{-2}$), making it the dominant noise at low speeds. By making the BJT size and bias current proportional to $BW_{3\text{dB}}$, as recommended by Eq. (6.74), the base shot noise is reduced to scale like $BW_{3\text{dB}}^2$ while other terms are increased to scale like $BW_{3\text{dB}}^2$. Overall, the noise performance is improved by reducing the dominant noise source.

Numerical Example. Let us evaluate Eq. (6.73) for our 10-Gb/s BJT TIA example in Tables 6.1 and 6.6. With $f_{Rb} = 1/(6.28 \cdot 40\ \Omega \cdot 0.1\ \text{pF}) = 40$ GHz, we can approximate the optimum as

$$\psi \approx \sqrt{\frac{(160 + 40)/40}{80^2/(100 \cdot 6.85^2) + (160 + 44)/44}} = 0.91. \tag{6.76}$$

With the detector capacitance $C_D = 0.15$ pF, we find that the BJT should be sized and biased for $\tilde{C}_I = 0.91 \cdot 0.15\ \text{pF} = 0.14$ pF, which is about 40% larger than the value used in our example (0.1 pF). If we were to increase the BJT by 40%, the rms noise would reduce from 771 to 759 nA, which is probably not worth the additional power consumption. For our 40-Gb/s BJT TIA example, we find the same optimum value $\psi = 0.91$.

Alternative Noise Optimization Procedures. In the aforementioned noise optimization procedure, we varied the device geometry and the bias current in sync ($L_E \propto I_C$), thus keeping the current density at a fixed value. While this is a reasonable thing to do in an integrated circuit design, it is not possible in a design with discrete transistors where the device geometry is a given. In the latter case, which has been analyzed in [4], only the bias current can be varied to optimize the noise. With a fixed device geometry, the fourth term in Eq. (6.66), which is due to the base resistance, is not part of the optimization problem. In contrast, with a variable device geometry, all terms in Eq. (6.66) can be balanced against each other.

Noise optimization can also be approached as a two-dimensional problem where the geometry and the bias current are treated as two independent variables. This approach is suitable when using numerical methods, but for an analytical treatment, a single variable is preferred.

In [20], the TIA noise is optimized under a fixed current-density constraint, as we did, but the optimum geometry is expressed in terms of HBT noise parameters.

Low-Noise Circuit Techniques. The noise optimization techniques discussed in this section dealt with device sizes, bias currents, and component values, but took the basic shunt-feedback topology as a given. Altering the circuit topology opens up another degree of freedom that can be exploited to optimize the noise performance. Let us look at some possibilities.

Inserting a suitable passive network in between the photodetector and the TIA input can reduce the noise. This technique is known as *noise matching* and we discuss it in Section 6.5.

Noise can also be reduced by subtracting two voltages (or currents) with the same noise waveform but complementary signal waveforms. This technique is known as *noise canceling* [70–72] and we encounter an example in Fig. 7.24.

Another circuit technique is to replace the noisy feedback resistor with another feedback element that is less noisy. In Section 8.1, we discuss capacitive, optical, and active feedback as possible candidates.

Finally, the TIA can be designed for a bandwidth that is lower than the target bandwidth and then be followed by an equalizer that restores the target bandwidth [73]. This technique is reminiscent of the high-impedance front-end, which also achieves low noise by sacrificing bandwidth and then restores the bandwidth with an equalizer.

Scaling of Noise and Sensitivity with Bit Rate. How does the input-referred rms noise current of a TIA scale with the bit rate? This is an interesting question because it is closely related to the question of how the sensitivity of a p–i–n receiver scales with the bit rate. What sensitivity can we expect for a receiver operating at 10, 40, or 160 Gb/s?

From Eq. (6.64), we know that the input-referred mean-square noise current of an optimized TIA scales like BW_{3dB}^3. At least, this is the case as long as the base or gate current can be neglected (i.e., the BJT operates in the high-speed regime). The input-referred rms noise current thus scales like $BW_{3dB}^{1.5}$ or, with the receiver bandwidth being proportional to the bit rate B, like $B^{1.5}$.

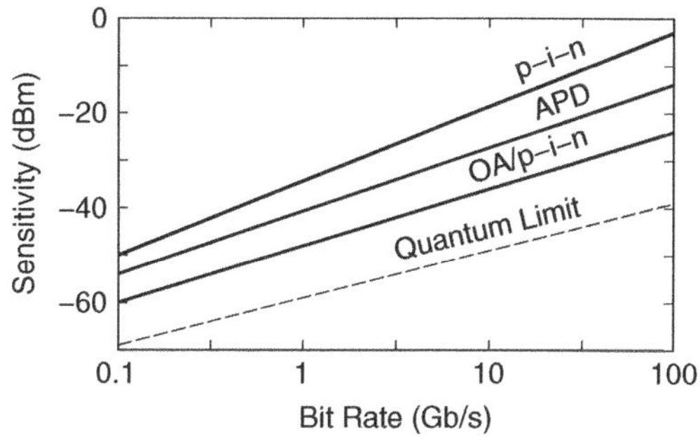

Figure 6.20 Scaling of optical receiver sensitivity (at $BER = 10^{-9}$) with bit rate [74].

The aforementioned scaling law was derived under the assumption of a given technology (fixed f_T, $A_0 f_A$, and C_D). However, TIAs for higher bit rates typically are fabricated in faster technologies and use lower capacitance detectors (higher f_T, $A_0 f_A$, and lower C_D). The figure of merit $f_T/(\Gamma C_D)$ indicates that this results in a lower noise (cf. Eq. (6.64)). For our TIA examples in Tables 6.1, 6.5, and 6.6, we assumed $f_T \propto A_0 f_A \propto B$, and $C_D \propto B^{-0.5}$ and found that the rms noise current scales like $B^{0.75}$. In practice, the technology may not scale as aggressively, and the rms noise ends up scaling somewhere between $B^{0.75}$ and $B^{1.5}$.

Assuming that the rms noise scales like B and that the detector's responsivity is bit-rate independent, the optical sensitivity of a p–i–n receiver degrades by 10 dB for every decade of speed increase. In practice, faster photodetectors tend to have a lower quantum efficiency and thus a lower responsivity (cf. Chapter 3), resulting in a slope of more than 10 dB per decade. The curves in Fig. 6.20, which are based on the experimental receiver-sensitivity data presented in [74], show a slope of about 15.8 dB per decade for the p–i–n receiver.

For a receiver with an APD or an optically preamplified p–i–n detector, the sensitivity is determined jointly by the TIA noise and the detector noise (cf. Eqs. (4.31) and (4.32)). In the extreme case when the detector noise dominates the TIA noise, we can conclude from Eqs. (4.30)–(4.32) that the optical sensitivity scales like B, corresponding to a slope of 10 dB per decade. In practice, the TIA contributes some noise and the scaling law is somewhere between B and $B^{1.5}$, corresponding to a slope of 10 to 15 dB per decade. The experimental data in Fig. 6.20 confirms this expectation: we find a slope of about 13.5 dB per decade for APD receivers and 12 dB per decade for optically preamplified p–i–n receivers.

Capacitance versus Responsivity of Photodetector. Integrated photodetectors often feature a lower capacitance than stand-alone photodetectors because their parasitic capacitances are smaller (cf. Section 3.4). At the same time, integrated photodetectors often suffer from a lower responsivity than optimized stand-alone photodetectors. Thus, the following question arises: How do these two quantities trade against each other? If C_D is reduced by 20% and $\mathcal{R}$ is reduced by 20%, does the overall receiver sensitivity improve or degrade?

To answer this question we recall from Eq. (4.27) that the optical sensitivity of a receiver (for which the TIA noise dominates) is given by

$$\overline{P}_{\text{sens}} = \frac{\mathcal{Q}\sqrt{\overline{i^2_{n,TIA}}}}{\mathcal{R}}. \tag{6.77}$$

Clearly, the sensitivity is inversely proportional to the responsivity, $\overline{P}_{\text{sens}} \propto 1/\mathcal{R}$, but its dependence on the detector capacitance C_D is less obvious.

To find the latter dependence, we first have to determine how the noise varies with C_D. We recall from Eq. (6.52) that the input-referred mean-square noise current of a TIA with FET front-end is given by

$$\overline{i^2_{n,TIA}} = \frac{4kT}{R_F} \cdot BW_n + 2qI_G \cdot BW_n + \frac{4kT\Gamma(2\pi\tilde{C}_T)^2}{3g_m} \cdot BW^3_{n2} + \ldots. \tag{6.78}$$

When varying C_D, we must keep $BW_{3\text{dB}}$, Q, f_T, etc. fixed to be fair. Applying the constant-bandwidth constraint to the first term in Eq. (6.78), we find that it is proportional to $\tilde{C}_T$. (A larger capacitance $\tilde{C}_T$ implies a smaller and noisier resistor R_F.) For a noise-optimized TIA, which implies $C_D \propto \tilde{C}_T$ (cf. Eq. (6.62)), the first term is also proportional to C_D. A larger leakage current I_G in the second term implies a larger FET with a larger input capacitance $\tilde{C}_I$. Thus, for a noise-optimized TIA ($C_D \propto \tilde{C}_I$) the second term is also proportional to C_D. For a fixed f_T, the third term is proportional to $\tilde{C}_T^2/\tilde{C}_I$, which, for a noise-optimized TIA, is once more proportional to C_D. With all three terms proportional to C_D, we can conclude that the entire mean-squared noise current of the TIA is proportional to the detector capacitance [75].

Plugging this result into Eq. (6.77), we find that the optical sensitivity of a receiver scales like [75]

$$\overline{P}_{\text{sens}} \propto \frac{\sqrt{C_D}}{\mathcal{R}}. \tag{6.79}$$

We can now answer the question about what happens when C_D and $\mathcal{R}$ are both reduced by 20%: $\overline{P}_{\text{sens}}$ goes up by $\sqrt{0.8}/0.8 = 1.118$, that is, the sensitivity degrades by about 12%. In conclusion, reducing C_D is good, but not as effective as increasing $\mathcal{R}$.

6.5 Noise Matching

Apart from the transistor size and bias optimizations discussed in the previous section, the noise performance of a receiver can also be improved by *noise matching* the photodetector to the receiver front-end. In the following,

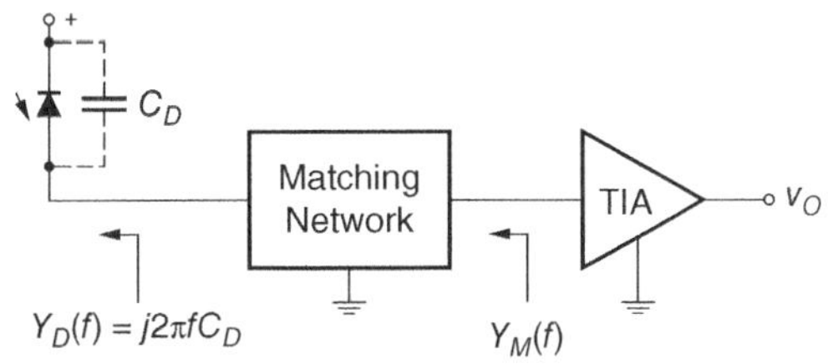

Figure 6.21 A noise-matching network transforms the admittance seen by the TIA input from $Y_D(f)$ to $Y_M(f)$.

we revisit the FET front-end and apply some simple noise-matching networks to it. Then, we discuss noise matching in a more general setting.

Noise-Matching Network. Figure 6.21 shows the block diagram of a receiver with a noise-matching network. The admittance of the photodetector, Y_D, is transformed with a passive network into another admittance, Y_M, for which the TIA produces less noise. Given the simple capacitive photodetector model, the former admittance is $Y_D(f) = j2\pi f C_D$. Finding a more optimal $Y_M(f)$ and the associated network is the objective of this section.

There is however a catch: the noise-matching network also changes the frequency response of the overall transimpedance $Z_T(f)$. As a result, the bandwidth, low-frequency cutoff, group-delay variation, and so on of the TIA may no longer meet the requirements. Thus the noise-matching network must meet two criteria:

- The noise-matching network must minimize the noise current referred to the intrinsic photodetector (i.e., "through" the noise-matching network) in the bandwidth of interest.
- The combination of noise-matching network and TIA must exhibit a flat frequency response $Z_T(f)$ from the intrinsic photodetector to the TIA's output in the bandwidth of interest.

FET Front-End Connected to an Arbitrary Source Admittance. As a first step in exploring the idea of noise matching, let us again calculate the input-referred noise current of a FET front-end, but this time assuming an *arbitrary* source admittance Y_S (see Fig. 6.22). We already know that minimizing the

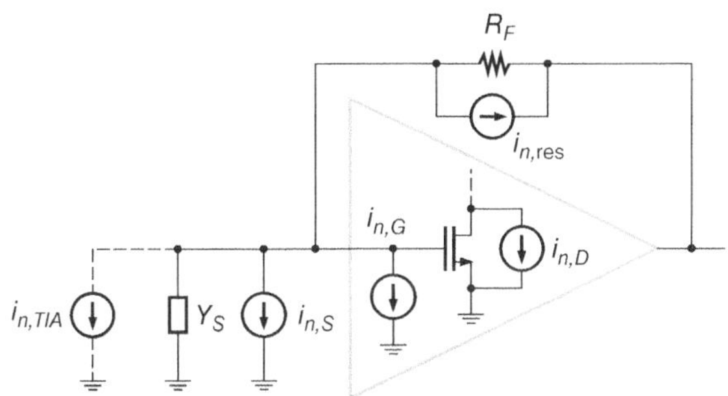

Figure 6.22 Calculation of the equivalent input noise current generator $i_{n,TIA}$ for a TIA with FET front-end driven by an arbitrary source admittance $Y_S(f)$.

photodetector capacitance, C_D, improves the TIA's noise performance. Now, let us find out which general admittance function $Y_S(f)$ results in the lowest TIA noise.

There are a number of differences compared with our calculation in Section 6.3. First, the source, which may have a resistive component, now itself generates the noise current $i_{n,S}$. If we decompose the source admittance into a real and imaginary part, $Y_S = G_S + jB_S$, the corresponding noise current PSD amounts to $I^2_{n,S} = 4kT|G_S|$. Second, the input-referral function for the drain noise, Eq. (6.43), must be generalized for the arbitrary source admittance Y_S, resulting in $H_D^{-1}(s) = -(Y_S + 1/R_F + s\tilde{C}_I)/g_m$ (where we again neglected feedforward through C_{gd}). Finally, we can no longer rely on Ogawa's noise factor Γ to account for the induced gate noise because this method works only for capacitive sources. To keep things simple, we ignore the induced gate current here and write $I^2_{n,D} = I^2_{n,\text{chan}} = 4kT\gamma g_{d0}$. Referring all noise sources to the input yields

$$I^2_{n,TIA}(f) = 4kT|G_S(f)| + \frac{4kT}{R_F} + 2qI_G + 4kT\gamma g_{d0}\,\frac{[G_S(f) + 1/R_F]^2 + [B_S(f) + 2\pi f\tilde{C}_I]^2}{g_m^2} + \cdots \quad (6.80)$$

The first term is due to the real (resistive) part of the source admittance, the second term is due to the feedback resistor, the third term is due to the gate shot noise, and the last term is due to the FET's channel noise. Note that if we let $G_S(f) = 0$ and $B_S(f) = 2\pi f C_D$, we get back our familiar result for the capacitive photodetector (less the effect of the induced gate noise).

To find the optimum source admittance $\tilde{Y}_{\text{opt}}(f) = \tilde{G}_{\text{opt}}(f) + j\tilde{B}_{\text{opt}}(f)$, we take the derivative of Eq. (6.80) with respect to G_S and B_S and set them both to zero. This results in the optimum (cf. Eq. (I.109))

$$\tilde{G}_{\text{opt}}(f) = 0 \quad \text{and} \quad \tilde{B}_{\text{opt}}(f) = -2\pi f\tilde{C}_I \quad (6.81)$$

under the condition that $g_m^2 R_F \geq 2\gamma g_{d0}$, which is normally satisfied. Figure 6.23(a) shows the trajectory of this frequency-dependent optimum in

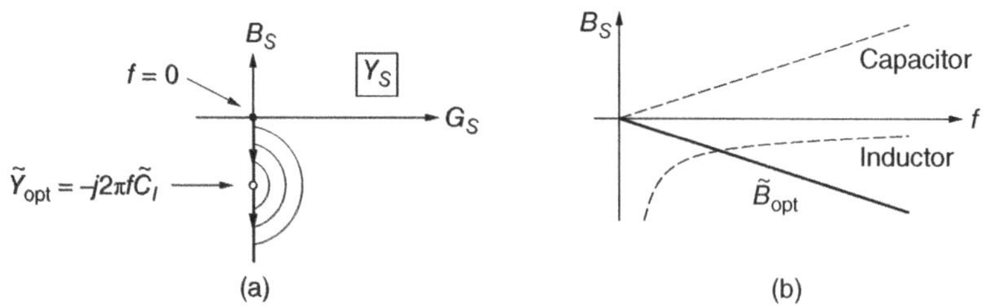

Figure 6.23 (a) Contours of constant input-referred noise current PSD in the source-admittance plane and (b) optimum susceptance as a function of frequency.

the source-admittance plane. The contours of constant input-referred noise current PSD, $I^2_{n,TIA}(f) = \text{const.}$, plotted for one particular frequency, show how the noise increases as we move away from the optimum. When the frequency is varied, these contours move along with the optimum.

Figure 6.23(b) shows the optimum susceptance $\tilde{B}_{\text{opt}}$ as a function of frequency and compares it with the susceptance of the capacitive photodetector as well as an inductor. Clearly, a smaller photodetector capacitance brings the function closer to the optimum, but cannot reach it. A negative capacitance in parallel to the photodetector would be ideal, but it cannot be implemented with a passive network (an active implementation is possible [76], but it adds noise of its own). An inductive source has the same sign as $\tilde{B}_{\text{opt}}$, but its frequency dependence does not match $\tilde{B}_{\text{opt}}(f)$, yielding the minimum noise for only one frequency.

Noise-matching networks typically achieve a perfect match for only a couple of discrete frequencies. For this reason, broadband noise matching is hard to attain. Noise matching is well-suited for narrow-band receivers, such as a subcarrier multiplexing (SCM) receiver operating over a frequency span of about one octave [77, 78]. In this narrow-band context, a front-end with a noise-matching network is known as a *resonant front-end* or a *tuned front-end*.

It is well known from the theory of linear noisy two-ports that there is an optimum source admittance, usually designated Y_{opt}, that results in the minimum *noise figure* [50, 79, 80]. Is this optimum admittance the same as $\tilde{Y}_{\text{opt}}$ discussed earlier? Not quite, and this is why we used the tilde! Whereas the optimum susceptances happen to be the same, $B_{\text{opt}} = \tilde{B}_{\text{opt}}$, the optimum conductances are different, $G_{\text{opt}} \neq \tilde{G}_{\text{opt}}$. Moreover, whereas the contours of constant noise figure are (non-concentric) circles in the admittance plane, the contours of constant input-referred noise current PSD are concentric circle segments, as shown in Fig. 6.23(a). We continue to discuss the relationship between the optimum input-referred noise current PSD and the optimum noise figure toward the end of this section.

Noise Matching with a Shunt Inductor. Knowing that the optimum source admittance is on the inductive side, we may try the simple noise-matching network shown in Fig. 6.24, consisting of the shunt inductor L_P. The inductor transforms the photodetector admittance $Y_D = j\omega C_D$ to the new admittance

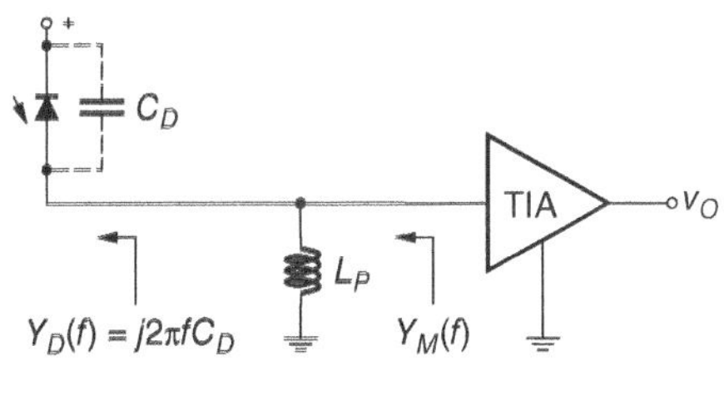

Figure 6.24 Shunt inductor as a noise-matching network.

$Y_M = j[\omega C_D - 1/(\omega L_P)]$, where $\omega = 2\pi f$. The shunt inductor subtracts from the photodetector susceptance, thus moving the source susceptance in the direction of the noise optimum.

Perfect noise matching is reached for $Y_M(\omega) = \tilde{Y}_{\text{opt}}(\omega) = -j\omega\tilde{C}_I$, which requires the inductor value [77, 78]

$$L_P = \frac{1}{\omega^2(C_D + \tilde{C}_I)}. \tag{6.82}$$

The fact that this optimum inductance is frequency dependent means that a perfect match can be obtained only for one particular frequency. At this frequency, Eq. (6.80) evaluates to $I^2_{n,TIA} = 4kT/R_F$, assuming $I_G R_F \ll 2V_T$ and $g^2_m R_F \gg \gamma g_{d0}$. Thus at resonance, the parallel inductor removes the effect of the FET's channel noise entirely, leaving only the feedback-resistor noise.

If we had done the calculations with the induced gate noise included, the optimum shunt inductor would have come out slightly larger to compensate for the correlated part of the induced gate noise (see Eq. (I.99) for the inductor value). Similarly, the minimum noise would have come out slightly higher because it includes the uncorrelated part of the induced gate noise (see Eq. (I.100)). [→ Problem 6.14.]

The main drawback of the shunt-inductor network is that it gives the TIA a passband response, which makes it unsuitable for receivers of broadband signals, such as NRZ.

Noise Matching with a Series Inductor. Let us try the series inductor shown in Fig. 6.25. This simple noise-matching network results in a low-pass response that is suitable for NRZ receivers. The inductor may be realized with a bond wire of the appropriate length. The series inductor L_S transforms the photodetector admittance to $Y_M(\omega) = j\omega C_D/(1 - \omega^2 L_S C_D)$. If we require $Y_M(\omega) = \tilde{Y}_{\text{opt}}(\omega)$, we find the inductor value [77, 78]

$$L_S = \frac{1}{\omega^2 C_D} + \frac{1}{\omega^2 \tilde{C}_I}. \tag{6.83}$$

Done? No, this inductor minimizes the noise referred to node x in Fig. 6.25, *not* the noise referred to the photodetector at node y. (In the case of the shunt-inductor network, nodes x and y were identical.) To do it right, we must

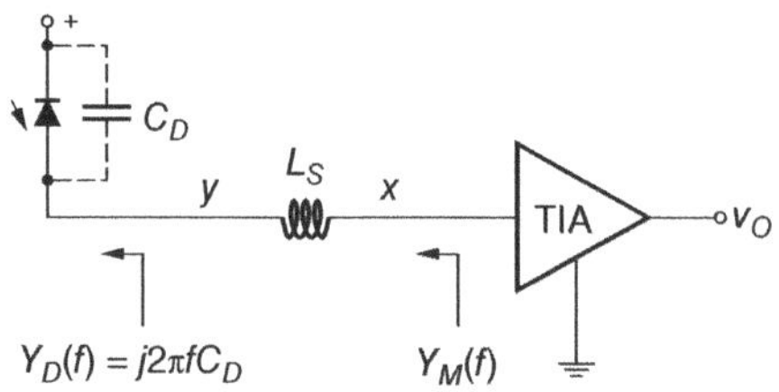

Figure 6.25 Series inductor as a noise-matching network.

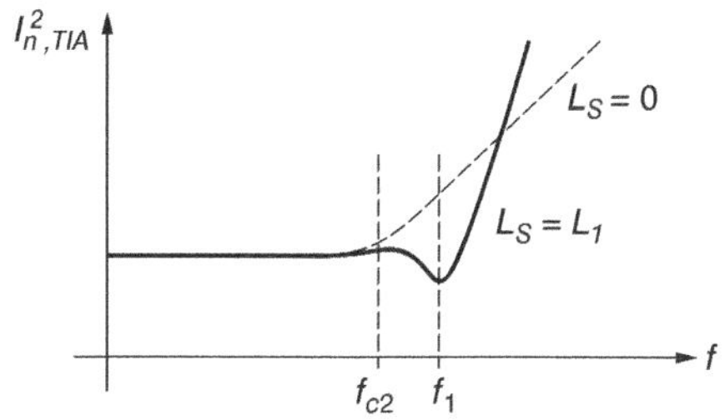

Figure 6.26 Input-referred noise current PSD with (solid line) and without (dashed line) the series inductor L_S for noise matching at the frequency f_1.

go back to Eq. (6.80) and multiply it with $|H_M^{-1}|^2$, where $H_M^{-1}(s) = 1 + s^2 L_S C_D$ is the input-referral function referring the noise current generator at node x backwards through the matching network to node y. Then, we set Y_S, the source admittance seen by the TIA in Eq. (6.80), to Y_M and find the optimum inductor value by taking the derivative with respect to L_S and setting it to zero. After some work, we find the optimum series inductor value

$$L_S = \frac{1}{\omega^2 C_D} \frac{\omega^2 (C_D + \tilde{C}_I)\tilde{C}_I + G^2}{\omega^2 \tilde{C}_I^2 + G^2}, \tag{6.84}$$

where $G^2 = g_m^2/(\gamma g_{d0} R_F) + 1/R_F^2$. Note that if the ω^2 terms dominate the G^2 terms, the expression simplifies to Eq. (6.83).

Like in the shunt-inductor case, the noise can be optimized only at a single frequency. What is a good frequency to pick? Because the input-referred noise current PSD is largest above the f^2-noise corner, f_{c2}, optimizing for a frequency in that region has the most impact. Figure 6.26 shows the input-referred noise current PSD with and without a series inductor. The series inductor $L_S = L_1$ is chosen to minimize the noise at frequency f_1, which is located somewhat above f_{c2}. The inductor clearly helps to reduce the noise current PSD at f_1, but does it optimize the TIA's sensitivity? To determine the sensitivity, we need to weight and integrate the noise current PSD as prescribed by Eq. (5.6). Unfortunately, this calculation is complicated by the fact that the TIA's frequency response (the weighting function in the integral) also depends on L_S. In general, computer simulations are needed to determine the TIA's sensitivity and to find the optimum value(s) for the matching component(s). In Section 7.7, we continue the discussion on how L_S affects the TIA's frequency response and bandwidth.

Numerical Example. As an illustration of the foregoing theory, let us add a series inductor to our 10-Gb/s FET TIA example in Table 6.5. With the f^2-noise corner at 4.2 GHz, we pick $f_1 = 10$ GHz for the frequency where we want to minimize the noise. Inserting f_1 and the numerical values of our example (with $\gamma g_{d0}/g_m = 2$) into Eq. (6.84), we find $L_S = 2.88$ nH. (The approximation Eq. (6.83) would have overestimated the value as $L_S = 4.23$ nH.) Running computer simulations of the TIA example confirms that the input-referred

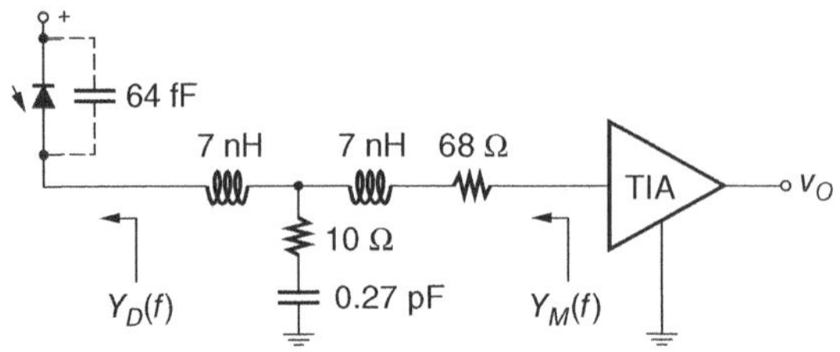

Figure 6.27 Example noise-matching network for a 10-Gb/s NRZ receiver [81].

noise current density dips at 10 GHz down to 5.4 pA/$\sqrt{\text{Hz}}$, significantly lower than the 14 pA/$\sqrt{\text{Hz}}$ without the inductor (cf. Fig. 6.26). However, inserting the inductor also changes the frequency response $Z_T(f)$, which now has almost twice the bandwidth (11.1 GHz) and about 1.7 dB of peaking.

Noise Matching with RLC Network. In practice, noise-matching networks are often more complex than the simple shunt and series inductors discussed so far. Figure 6.27 shows the low-pass noise-matching network proposed in [81] for a 10-Gb/s NRZ receiver. It employs three reactive elements to obtain noise matching across a wide range of frequencies and two small resistors to flatten the frequency response. While these resistors add some unwanted noise, a practical design needs to balance the noise performance against other performance measures such as frequency response ripple and group-delay distortion.

In the following, we broaden our view of noise matching. Our goal is to come up with an expression for the input-referred noise current PSD that is valid for an arbitrary front-end with an arbitrary noise-matching network.

Four-Parameter Noise Model. It is well known that the noise of a general linear two-port can be modeled by adding two partially correlated noise generators to the noiseless linear model [50, 80, 82]. There are several equivalent ways of doing that, but most often a *noise voltage generator*, v_{n4}, and a *noise current generator*, i_{n4}, are added at the input of the noiseless two-port, as shown in Fig. 6.28(a).

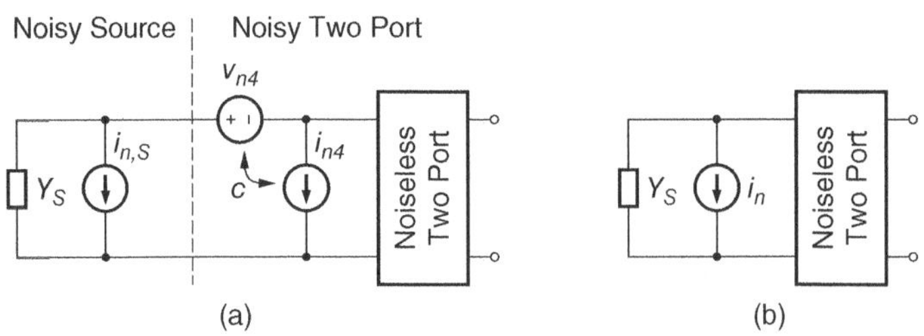

Figure 6.28 (a) General four-parameter noise model of a linear two-port connected to a noisy source and (b) one-parameter noise model of the two-port and the source.

The partial correlation between the two noise generators is specified by the complex *correlation coefficient* c, which is defined as [80]

$$c = \frac{\overline{\underline{V}^*_{n4}\underline{I}_{n4}}}{\sqrt{\overline{|\underline{V}_{n4}|^2}\ \overline{|\underline{I}_{n4}|^2}}}, \tag{6.85}$$

where $\underline{V}_{n4}$ and $\underline{I}_{n4}$ are the phasors of v_{n4} and i_{n4} in a narrow bandwidth around the frequency of interest, the star denotes the complex conjugate, and an overbar indicates the average over different noise waveform realizations. (We underline phasors to distinguish them from root spectral densities.) Note that without the averaging overbars, Eq. (6.85) simply equals $\exp(j\phi)$, where ϕ is the phase difference between the two phasors. Thus, for fully correlated noise sources we have $|c| = 1$ and $\arg(c) = \phi$. For uncorrelated noise sources, however, the ensemble average of $\underline{V}^*_{n4}\underline{I}_{n4}$ vanishes and we have $c = 0$. The signs of $\mathrm{Re}\{c\}$ and $\mathrm{Im}\{c\}$ depend on the chosen reference polarities for the noise generators and whether $\underline{V}_{n4}$ or $\underline{I}_{n4}$ is conjugated in Eq. (6.85): here we follow the convention in [80].

The noise model in Fig. 6.28(a), like all general linear two-port noise models, is based on four real parameters. To fully specify the model, we need to provide the two PSDs V^2_{n4} and I^2_{n4} and the two components $\mathrm{Re}\{c\}$ and $\mathrm{Im}\{c\}$, all of which may be a function of frequency. All four parameters depend on the noisy two-port only. In particular, they do *not* depend on the source or load admittances. In fact, they permit the calculation of noise voltages and currents for *arbitrary* source and load admittances.

Relationship to Input-Referred Noise Current. At this point it is important to emphasize that the noise current i_{n4} is *different* from what we have called the input-referred noise current throughout this book. The noise model just introduced has four parameters, which are independent of the source admittance. In contrast, the noise model used so far to describe the TIA has only one parameter (the input-referred noise current), which *does* depend on the source admittance. To avoid confusion, we call the new model *four-parameter noise model* and use the subscript $n4$ instead of n for its noise current and noise voltage generators. Figure 6.28(a) and (b) shows the four-parameter noise model and the one-parameter noise model side by side. If we make the (not uncommon) mistake of calculating or simulating the noise current i_{n4} and then use it with Eqs. (5.6) and (5.7) to determine the TIA's or the receiver's sensitivity, we get an entirely wrong (and unrealistically optimistic) result. To be clear, we use the term *input-referred noise current* only for i_n in the one-parameter noise model.

Of course, if we have the four-parameter noise model and a particular source admittance, we can calculate the corresponding input-referred noise current. Moving the noise current generator of the source admittance, $i_{n,S}$, to the right side of v_{n4} and converting the series connection of Y_S and v_{n4} into its Norton equivalent parallel connection of Y_S and a noise current generator, we find the phasor relationship

$$\underline{I}_n = \underline{I}_{n,S} + \underline{I}_{n4} + Y_S \underline{V}_{n4}, \tag{6.86}$$

where $\underline{I}_n$ is the input-referred noise current phasor in the one-parameter model. If i_{n4} and v_{n4} are uncorrelated, the corresponding input-referred noise current PSD follows easily as $I_n^2 = I_{n,S}^2 + I_{n4}^2 + |Y_S|^2 V_{n4}^2$. If i_{n4} and v_{n4} are correlated, however, there is an additional noise term. We can find the full expression by considering the mean-square-magnitude of the noise current phasor

$$\overline{|\underline{I}_n|^2} = \overline{|\underline{I}_{n,S}|^2} + \overline{|\underline{I}_{n4} + Y_S \underline{V}_{n4}|^2}. \tag{6.87}$$

Using the complex identity $|z_1 + z_2|^2 = |z_1|^2 + 2\,\mathrm{Re}\{z_1^* z_2\} + |z_2|^2$, we can expand

$$\overline{|\underline{I}_n|^2} = \overline{|\underline{I}_{n,S}|^2} + \overline{|\underline{I}_{n4}|^2} + 2\,\mathrm{Re}\{Y_S \overline{\underline{I}_{n4}^* \underline{V}_{n4}}\} + |Y_S|^2 \overline{|\underline{V}_{n4}|^2}. \tag{6.88}$$

With the definition Eq. (6.85), we can introduce the correlation coefficient c

$$\overline{|\underline{I}_n|^2} = \overline{|\underline{I}_{n,S}|^2} + \overline{|\underline{I}_{n4}|^2} + 2\,\mathrm{Re}\{c^* Y_S\} \sqrt{\overline{|\underline{I}_{n4}|^2}\ \overline{|\underline{V}_{n4}|^2}} + |Y_S|^2 \overline{|\underline{V}_{n4}|^2}. \tag{6.89}$$

Making use of the fact that mean-square-magnitude noise phasors are proportional to their PSDs and that $I_{n,S}^2 = 4kT|\mathrm{Re}\{Y_S\}|$, we can rewrite in terms of PSDs

$$I_n^2 = 4kT|\mathrm{Re}\{Y_S\}| + I_{n4}^2 + 2\,\mathrm{Re}\{c^* Y_S\} I_{n4} V_{n4} + |Y_S|^2 V_{n4}^2. \tag{6.90}$$

where I_{n4}^2 and V_{n4}^2 are PSDs and I_{n4} and V_{n4} are root spectral densities. The third noise term is the correlation term that we were looking for. Equation (6.90) clearly shows that I_{n4}^2 is but one of four components contributing to the total input-referred noise current PSD, I_n^2.

Four-Parameter Noise Model without Correlation Coefficient. Calculating with partially correlated noise phasors can be tricky. To avoid partial correlations, the noise current generator i_{n4} can be split into two parallel generators: a *correlated noise current generator*, i_{n4c}, and an *uncorrelated noise current generator*, i_{n4u}, as shown in Fig. 6.29. Whereas the phasor of the first noise current source is fully correlated to the phasor of v_{n4} by means of the complex *correlation admittance* $Y_c = G_c + jB_c$, the latter noise current source is fully uncorrelated.

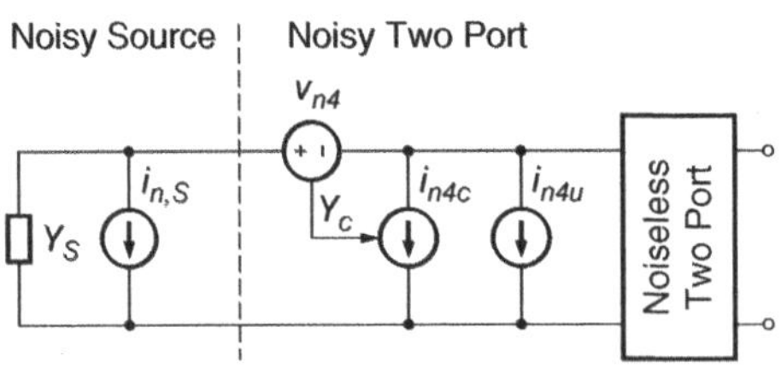

Figure 6.29 General four-parameter noise model of a linear two-port containing only fully correlated and fully uncorrelated noise generators.

This two-port noise model has again four noise parameters: the PSDs V_{n4}^2 and I_{n4u}^2 and the components G_c and B_c.

Relating these four parameters to the input-referred noise current in our one-parameter model is now much easier than before. The phasor equation becomes

$$\begin{aligned} \underline{I}_n &= \underline{I}_{n,S} + \underline{I}_{n4u} + \underline{I}_{n4c} + Y_S\underline{V}_{n4} \\ &= \underline{I}_{n,S} + \underline{I}_{n4u} + Y_c\underline{V}_{n4} + Y_S\underline{V}_{n4} \\ &= \underline{I}_{n,S} + \underline{I}_{n4u} + (Y_c + Y_S)\underline{V}_{n4}. \end{aligned} \tag{6.91}$$

Realizing that all three terms in the last form are uncorrelated and inserting $I_{n,S}^2 = 4kT|\text{Re}\{Y_S\}|$, the PSD follows as [80]

$$I_n^2 = 4kT|\text{Re}\{Y_S\}| + I_{n4u}^2 + |Y_S + Y_c|^2 V_{n4}^2, \tag{6.92}$$

which is the corresponding result to Eq. (6.90).

Naturally, the parameters of the two-port noise model in Fig. 6.29 can be related to those of our previous model in Fig. 6.28(a). Starting with the definition of Y_c and rewriting it in terms of c, we find [80]

$$Y_c = \frac{\underline{I}_{n4c}}{\underline{V}_{n4}} = \frac{\overline{\underline{I}_{n4c}\underline{V}_{n4}^*}}{\overline{\underline{V}_{n4}\underline{V}_{n4}^*}} = \frac{\overline{\underline{I}_{n4}\underline{V}_{n4}^*}}{\overline{|\underline{V}_{n4}|^2}} = c\sqrt{\frac{\overline{|\underline{I}_{n4}|^2}}{\overline{|\underline{V}_{n4}|^2}}} = c\frac{I_{n4}}{V_{n4}}, \tag{6.93}$$

thus Y_c and c have the same phase, $\arg(Y_c) = \arg(c)$. The uncorrelated portion of the noise current PSD in Fig. 6.29 then becomes

$$I_{n4u}^2 = I_{n4}^2 - I_{n4c}^2 = I_{n4}^2 - |Y_c|^2 V_{n4}^2 = I_{n4}^2 - \left|c\frac{I_{n4}}{V_{n4}}\right|^2 V_{n4}^2 = (1 - |c|^2)I_{n4}^2. \tag{6.94}$$

The fully correlated portion follows as $I_{n4c}^2 = |c|^2 I_{n4}^2$. Thus, the magnitude of c describes the fraction of the partially correlated noise current that is fully correlated to the noise voltage. As a check, we can insert Eqs. (6.93) and (6.94) into Eq. (6.92) and, after some work, we recover Eq. (6.90).

Four-Parameter Noise Model Based on the Noise Figure. Equations (6.90) and (6.92) are two different parametrization of the same relationship between I_n^2 and Y_S. Yet another way to write the same relationship is [50, 79]

$$I_n^2 = \left(F_{\min} + \frac{R_n}{\text{Re}\{Y_S\}}|Y_S - Y_{\text{opt}}|^2\right) \cdot 4kT\,\text{Re}\{Y_S\}. \tag{6.95}$$

This form is of practical interest because it expresses the input-referred noise current PSD as a multiple of the noise current PSD of the source admittance. The multiplier, the expression in the parenthesis, is the noise figure. Again, there are four noise parameters: the *noise resistance*, R_n, the *minimum noise figure*, $F_{\min}$, and the complex source admittance for which the minimum noise figure is reached, $Y_{opt} = G_{opt} + jB_{opt}$. Comparing Eq. (6.95) with Eq. (6.92), we can relate these noise parameters to those of the model in Fig. 6.29 as follows: $R_n = V_{n4}^2/(4kT)$, $G_{opt}^2 = I_{n4u}^2/V_{n4}^2 + G_c^2$, $B_{opt} = -B_c$, and $F_{min} = 1 + 2R_n(g_{opt} + G_c)$ [50].

Example. Let us make an example to illustrate the foregoing theory. Equation (6.80) expresses the input-referred noise current PSD of a TIA with FET front-end (neglecting induced gate noise) for an arbitrary source admittance Y_S. By comparing this equation to the general form in Eq. (6.92), we find the four noise parameters of that TIA:

$$I_{n4u}^2 = \frac{4kT}{R_F} + 2qI_G, \qquad V_{n4}^2 = \frac{4kT\gamma g_{d0}}{g_m^2},$$
$$G_c = 1/R_F, \qquad B_c = 2\pi f\tilde{C}_I. \tag{6.96}$$

In this example, i_{n4u} and v_{n4} both have white noise spectra. Interestingly, $Y_c \neq 0$, which means that $i_{n4c} \neq 0$ and thus that i_{n4} and v_{n4} are partially correlated.

Let us try to understand where this correlation comes from. Wouldn't it be enough to have i_{n4u}, which models the feedback-resistor noise and the gate shot noise, and v_{n4}, which models the channel noise divided by g_m? If the input port is shorted and thus v_{n4} drives the gate directly, this is indeed enough. But if the input port is open ($Y_S = 0$), v_{n4} is disconnected and the channel noise is no longer modeled. Thus, we need to introduce a noise current generator, i_{n4c}, that drops the equivalent noise voltage v_{n4} across the input admittance $Y_c = 1/R_F + s\tilde{C}_I$. Now, the channel noise is modeled for the shorted and open cases, but what about an arbitrary source admittance? It can be shown that if the model includes the v_{n4} noise voltage generator needed for the shorted case *and* the i_{n4c} noise current generator needed for the open case and both generators are fully correlated, the channel noise is modeled correctly for any source admittance. Thus in this example, the correlation arises because the channel noise is modeled with two noise generators that are both located on the gate side.

If we had included the induced gate noise in our example, I_{n4u}^2 would increase by the uncorrelated part of the induced gate noise (see Eq. (I.102)) and B_c would reduce somewhat due to the correlated part of the induced gate noise (see Eq. (I.103)). [→ Problem 6.15.]

General Noise Matching. Now with Eq. (6.92) in hand, we can revisit the question about which source admittance minimizes the input-referred noise current PSD in a more general setting. Taking the derivative with respect to G_S and B_S and setting them both to zero yields the optimum (cf. Eqs. (I.105) and (I.106))

$$\tilde{G}_{opt}(f) = 0 \qquad \text{and} \qquad \tilde{B}_{opt}(f) = -B_c \tag{6.97}$$

under the condition that $|G_c| \leq 2kT/V_{n4}^2$, which is normally satisfied. Thus for any front-end, the input-referred noise current PSD becomes smallest when the source admittance is purely reactive and its susceptance equals the negative correlation susceptance (assuming the correlation conductance is sufficiently small). In the case of a TIA with FET front-end this means $\tilde{Y}_{\text{opt}}(f) = -j2\pi f \tilde{C}_I$, which is consistent with Eq. (6.81). [→ Problem 6.16.]

As we know, the optimum source admittance in Eq. (6.97) is only relevant when the input and output nodes of the noise-matching network are identical. For a general (two-port) noise-matching network, we must refer the noise to the input of the noise-matching network before minimizing it. This is done by multiplying Eq. (6.92) with the input-referral function $|H_M^{-1}|^2$. Still, we are not completely done. For a general photodetector model, in which the photocurrent source does not connect directly to the output node, we need one more input-referral function to move the equivalent noise generator from the input of the noise-matching network to the intrinsic photodetector: $|H_{PD}^{-1}|^2$ (cf. Section 5.7). Using Eq. (6.92) with Y_M as the source admittance and applying the two input-referral functions finally yields the general input-referred noise current PSD for a TIA with a noise-matching network:

$$I_{n,TIA}^2 = \frac{4kT|\text{Re}\{Y_M\}| + I_{n4u}^2 + |Y_M + Y_c|^2 V_{n4}^2}{|H_{PD}|^2 \cdot |H_M|^2}. \tag{6.98}$$

The admittance Y_M and the referral function $|H_M|^2$ can further be expressed in terms of the ABCD parameters (chain parameters) of the noise-matching network (cf. Eqs. (I.111) and (I.112)):

$$Y_M = \frac{AY_D + C}{BY_D + D} \quad \text{and} \quad |H_M|^2 = \frac{1}{|BY_D + D|^2}, \tag{6.99}$$

where Y_D is the admittance of the photodetector.

For example, for the simple noise-matching network consisting of only a parallel inductor L_P (cf. Fig. 6.24), the ABCD parameters are:

$$\begin{bmatrix} A & B \\ C & D \end{bmatrix} = \begin{bmatrix} 1 & 0 \\ 1/(sL_P) & 1 \end{bmatrix}. \tag{6.100}$$

With $Y_D = sC_D$ we find the familiar expression $Y_M = j[\omega C_D - 1/(\omega L_P)]$ and $|H_M|^2 = 1$. Similarly, for the series inductor L_S (cf. Fig. 6.25), the ABCD parameters are:

$$\begin{bmatrix} A & B \\ C & D \end{bmatrix} = \begin{bmatrix} 1 & sL_S \\ 0 & 1 \end{bmatrix}. \tag{6.101}$$

With $Y_D = sC_D$ we find the familiar expressions $Y_M = j\omega C_D/(1 - \omega^2 L_S C_D)$ and $|H_M|^2 = 1/(1 - \omega^2 L_S C_D)^2$. [→ Problem 6.17.]

Finally, the photodetector admittance Y_D and the referral function $|H_{PD}|^2$ can be expressed in terms of the ABCD parameters of the photodetector model (cf. Section 5.1, Eq (I.32)):

$$Y_D = \frac{C}{D} \quad \text{and} \quad |H_{PD}|^2 = \frac{1}{|D|^2}. \tag{6.102}$$

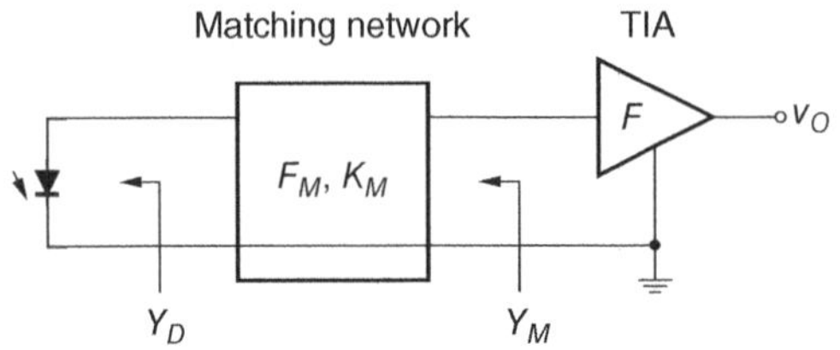

Figure 6.30 Noise-figure view of a TIA with noise-matching network.

For the simple photodetector model consisting of a photocurrent source with parallel capacitor C_D, the ABCD parameters are

$$\begin{bmatrix} A & B \\ C & D \end{bmatrix} = \begin{bmatrix} 1 & 0 \\ sC_D & 1 \end{bmatrix} \tag{6.103}$$

resulting in $Y_D = sC_D$ and $H_{PD} = 1$, as expected.

This concludes our noise-matching tour. We started from special cases, went on to the general case, and then came back to our special cases.

The Noise-Figure View. Given a photodetector with a nonzero resistive component G_D, which a real photodetector always has, it is possible to discuss the TIA's noise in terms of its noise figure (cf. Section 5.7). While this is not the usual approach, it provides another view point that may be helpful to RF engineers who feel more comfortable with noise figures than with input-referred noise currents.

Figure 6.30 shows a general block diagram of a TIA that is coupled to the photodetector through a noise-matching network. The TIA is characterized by the noise figure F and the noise-matching network is characterized by the noise figure F_M and the available power gain (or loss, if less than one) K_M.

The total noise figure of the TIA plus the matching network can be found with the well-known Friis noise formula [83, 84] as $F_{tot} = F_M + (F - 1)/K_M$. Inserting this total noise figure into Eq. (5.9), we find the input-referred noise current PSD in terms of the noise figures F and F_M:

$$I^2_{n,TIA} = \left(F_M + \frac{F - 1}{K_M} \right) \cdot 4kTG_D. \tag{6.104}$$

(This is the noise referred to Y_D; if we want the noise referred to the intrinsic photodetector, we must replace G_D with G_{Di}, as in Eq. (5.10).)

Next, let us assume that the matching network is passive and lossless, as in the case of the simple shunt- or series-inductor network. Then, $K_M = 1$ (no power gain) and $F_M = 1$ (no noise) and the aforementioned equation simplifies to $I^2_{n,TIA} = F \cdot 4kTG_D$. Thus, the noise figure of the TIA alone, F, and the noise figure of the TIA combined with the passive, lossless noise-matching network, F_{tot}, are the same. Can that be right? When we calculated with noise current PSDs, we had to use the input-referral function H_M^{-1} to refer the noise from the

TIA back to the photodetector, but now that we calculate with noise figures, input referral seems unnecessary. The resolution of this apparent contradiction is that the noise figure describes the noise relative to the noise of the source admittance and the source admittances for F and F_{tot} are not the same. The source admittance for F is Y_M and the source admittance for F_{tot} is Y_D. Thus, although the noise figures are the same, the equivalent noise current PSDs are not. In fact, the ratio of the noise-generating source conductances G_M and G_D equals the referral function $|H_M|^2$:

$$|H_M|^2 = \frac{\text{Re}\{Y_M\}}{\text{Re}\{Y_D\}} = \frac{G_M}{G_D}. \tag{6.105}$$

It can be shown that this property holds for any network that is reciprocal and lossless (cf. Eqs. (I.113)–(I.115)). Thus, the noise current PSD approach and the noise figure approach are equivalent. [→ Problem 6.18.]

From the aforementioned reasoning, we conclude that minimizing the noise figure of the TIA, F, also minimizes the input-referred noise current, $I^2_{n,TIA}$, as long as the matching network is passive and lossless ($K_M = F_M = 1$). It is well known that the noise figure is minimized for the source admittance $Y_{\text{opt}} = G_{\text{opt}} + jB_{\text{opt}}$, where [50, 79, 80]

$$G_{\text{opt}} = \sqrt{\frac{I^2_{n4u}}{V^2_{n4}} + G_c^2} \quad \text{and} \quad B_{\text{opt}} = -B_c. \tag{6.106}$$

This optimum source admittance has a positive, nonzero real part, G_{opt}. What does that mean? Should we add a resistor to the matching network to make $Y_M = Y_{\text{opt}}$? No, this would violate our premise that the matching network is lossless, making $F_M > 1$, $F_{\text{tot}} \neq F$, and thus Y_{opt} no longer the optimum for the whole system. What $G_{\text{opt}} > 0$ means is that we should use a network that losslessly *transforms* any existing real part of Y_D into G_{opt}. But if we have $G_D \rightarrow 0$, this is not possible and the optimum source admittance becomes $\tilde{Y}_{\text{opt}} = 0 + jB_{\text{opt}}$, bringing us back to Eq. (6.97).

6.6 Summary

Circuits for converting an input current to an output voltage include the low-impedance front-end, the high-impedance front-end, the integrating front-end, and the shunt-feedback TIA:

- The low-impedance front-end is simple, has a large bandwidth, and a high overload current, but suffers from a low transimpedance and high noise.
- The high-impedance front-end has a higher transimpedance and better noise performance, but suffers from a low overload current and needs an equalizer to achieve a large bandwidth.

- The integrating front-end is similar to the high-impedance front-end, but replaces the load resistor with a current source.
- The shunt-feedback TIA simultaneously achieves a large bandwidth, a high transimpedance, a high overload current, and low noise, but consumes more power than the other approaches.

The dynamic stability of a shunt-feedback TIA, as measured by the pole quality factor or the phase margin, is critical. TIAs are often designed for a Butterworth or Bessel–Thomson response.

The maximum possible transimpedance of a TIA is limited by the speed of the circuit technology ($A_0 f_A$), the photodetector capacitance (C_D), and the required TIA bandwidth ($BW_{3\text{dB}}$). Expressions for the maximum transimpedance (the transimpedance limit) are listed in Table 6.7 for some TIA topologies. Shunt-feedback TIAs with multistage voltage amplifiers are of interest when the technology is much faster than the TIA's target bandwidth ($A_0 f_A \gg BW_{3\text{dB}}$) or when the gain of a single stage is limited.

The input and output impedance of shunt-feedback TIAs have an inductive or negative-capacitive component. The inductive input component helps to tune out some of the photodetector capacitance, thus extending the bandwidth.

Noise of a TIA with FET front-end: The input-referred noise current PSD has a white part, which is mostly due to the feedback-resistor noise and the gate shot noise, and an f^2 part, which is mostly due to the channel and induced gate noise. The channel noise and the induced gate noise can be jointly characterized by Ogawa's noise factor Γ.

Noise of a TIA with BJT front-end: The input-referred noise current PSD has a white part, which is mostly due to the feedback-resistor noise and the base

Table 6.7 Summary of transimpedance limit expressions (LZ = low-impedance front-end, SFB = shunt feedback).

TIA topology	Transimpedance expression
LZ (shunt resistor)	$= \dfrac{1}{2\pi C_T \cdot BW_{3\,\text{dB}}}$
SFB TIA, single-stage amp	$\leq \dfrac{A_0 f_A}{2\pi C_T \cdot BW_{3\,\text{dB}}^2}$
SFB TIA, single-stage amp, $R_L \neq 0$	$\leq \dfrac{A_0' f_A'}{2\pi C_T \cdot BW_{3\,\text{dB}}^2}\left(1 - \dfrac{R_L}{A_0' R_F}\right)$
SFB TIA, single-stage amp, $C_F \neq 0$	$\leq \dfrac{A_0 f_A}{2\pi (C_T + C_F) \cdot BW_{3\,\text{dB}}^2}$
SFB TIA, multistage amp	$\leq \sqrt{2^{n+1}} \tan^n\left(\dfrac{90° - \phi_m}{n}\right) \cdot \dfrac{(A_0 f_A)^n}{2\pi C_T \cdot BW_{3\,\text{dB}}^{n+1}}$

shot noise, and an f^2 part, which is mostly due to the collector shot noise and the base resistance noise.

The total input-referred noise current, which determines the sensitivity, is found from the input-referred noise current PSD with the help of two noise bandwidths, one for white noise (BW_n) and one for f^2 noise (BW_{n2}).

Noise optimization of a TIA with FET front-end: Choose a circuit technology and a photodetector for which $f_T/(\Gamma C_D)$ is large. For low-speed TIAs, a low gate-leakage current and a low $1/f$ noise corner frequency are also important. Then, choose a channel length (often the minimum length) and a drain current density (often one that brings f_T close to its maximum) and size the width of the front-end FET such that $(C_{gs} + C_{gd})/C_D$ equals Eq. (6.63), usually around 0.7. To save power at the expense of a small noise penalty, the FET's width can be made somewhat smaller than that.

Noise optimization of a TIA with BJT front-end: Choose a circuit technology and a photodetector for which f_T/C_D is large and R_b is small. For low-speed TIAs, a high current gain β is also important. Then, choose an emitter width (often the minimum width) and a collector current density (often one that brings f_T close to its maximum) and size the emitter length of the front-end BJT such that $(C_{be} + C_{bc})/C_D$ equals Eq. (6.73), usually around 0.9 for high-speed receivers. To save power at the expense of a small noise penalty, the BJT's emitter length can be made somewhat smaller than that.

Besides the transistor size and bias current optimizations, the noise can be reduced by inserting a noise-matching network between the photodetector and the TIA. Popular noise-matching networks are the shunt and series inductor. While noise-matching networks can reduce the noise at specific frequencies, broadband noise matching is difficult to achieve.

Problems

6.1 **Low-Impedance Front-End.** A low-impedance front-end as in Fig. 6.1(b) has an amplifier with gain $A = 40$ dB, noise figure $F = 2$ dB, and noise bandwidth $BW_n = 10$ GHz. (a) How large is the transimpedance of this arrangement? (b) How large is the input-referred rms noise current? (c) How does the optical sensitivity of this front-end compare with a TIA front-end with $i_{n,TIA}^{rms} = 1.0$ μA?

6.2 **Open-Loop Pole Spacing.** For a two-pole TIA, the open-loop pole spacing, $R_F C_T / T_A$, must be $2A_0$ to obtain a Butterworth response after closing the loop (Eq. (6.12)). (a) Given an arbitrary closed-loop Q value, what is the required open-loop pole spacing? (b) What is the required pole spacing for a Bessel response? (c) What is the required pole spacing for a critically damped response?

6.3 Transimpedance Limit. In the available technology, we can realize amplifier stages with a gain-bandwidth product of 44 GHz. Given $C_T = 0.3$ pF and the requirement that the TIA bandwidth must be 70% of the bit rate, what transimpedance values do we expect for the basic shunt-feedback TIA with a single-stage voltage amplifier operating at 2.5, 10, and 40 Gb/s?

6.4 Single-Transistor Shunt-Feedback TIA. (a) Find the transformation that maps A_0 and T_A of the zero-output-impedance TIA model described by Eqs. (6.8)–(6.11) to A_0' and T_A' of the single-transistor TIA model described by Eqs. (6.23)–(6.26) for a given load resistor R_L. (b) Calculate the input impedance $Z_I(s)$ of the single-transistor TIA model in Fig. 6.8. (c) Show that this input impedance matches that of the zero-output-impedance TIA in Eq. (6.21), if A_0' and T_A' are chosen according to the transformation found in (a).

6.5 Single-Transistor Shunt-Feedback TIA with C_F. (a) Calculate the transimpedance $Z_T(s)$ of the single-transistor shunt-feedback TIA shown in Fig. 6.8 extended with a capacitor C_F in parallel to R_F. (b) How is the transimpedance limit Eq. (6.14) affected by R_L and C_F?

6.6 Voltage Amplifier with Nonzero Output Resistance. (a) Calculate the transimpedance $Z_T(s)$ of the shunt-feedback TIA shown in Fig. 6.3(a) assuming a single-pole voltage amplifier, $A_0/(1 + sT_A)$, with zero output impedance followed by the series resistance R_L. (b) How is the transimpedance limit Eq. (6.14) affected by R_L? (c) Calculate the output impedance $Z_O(s)$ of the same TIA circuit.

6.7 Feedback Capacitance. (a) Calculate the transimpedance $Z_T(s)$ of the TIA with a feedback capacitor shown in Fig. 6.12. Assume a voltage amplifier model with a single pole (time constant T_A) and zero output impedance. (b) What value for C_F do we need to obtain a Butterworth response (assume $C_F \ll C_T$)? (c) What is the 3-dB bandwidth of the TIA given a Butterworth response (assume $T_A \ll R_F(C_T + A_0C_F)$)? (d) How is the transimpedance limit Eq. (6.14) affected by C_F?

6.8 Ogawa's Noise Factor Γ. The correlation between the induced gate noise current generator $i_{n,\text{ind}}$ and the channel noise current generator $i_{n,\text{chan}}$ is described by the correlation coefficient c [39]:

$$c = \frac{\overline{-\underline{I}^*_{n,\text{ind}}\underline{I}_{n,\text{chan}}}}{\sqrt{\overline{|\underline{I}_{n,\text{ind}}|^2}\;\overline{|\underline{I}_{n,\text{chan}}|^2}}}, \tag{6.107}$$

where $\underline{I}_{n,\text{ind}}$ and $\underline{I}_{n,\text{chan}}$ are the phasors of $i_{n,\text{ind}}$ and $i_{n,\text{chan}}$ in a narrow bandwidth around the frequency of interest, the star denotes the complex conjugate, and an overbar indicates the average over different noise waveform realizations. Phasors are underlined to distinguish them from root spectral densities and the noise current source polarities are as shown in Fig. 6.14. (The minus sign in Eq. (6.107) arises from the fact that the reference polarity of the channel noise generator in Fig. 6.14 is the reverse of that in [39]. Caution: Not all authors define c in the same way [53].) (a) Given the source capacitance C_D and ignoring feedforward through C_{gd}, refer the induced gate noise current to the drain side and combine it with the channel noise. (b) Find the full expression for Ogawa's Noise Γ factor. (c) Assuming $\delta = 2\gamma$, $c = j0.395$, and $g_{d0} = g_m$, how large is the contribution of the induced gate noise relative to the channel noise for $C_{gs}/\tilde{C}_T = \frac{1}{2}, \frac{1}{3}$, and $\frac{1}{4}$?

6.9 **Input-Referral Function.** (a) Show that the three definitions of the input-referral function given in the main text are all equivalent. (b) Calculate the transfer functions from $i_{n,TIA}$ and from $i_{n,D}$ to the output of the TIA for the circuit in Fig. 6.13. Assume a MOS model with the parameters C_{gs}, C_{gd}, and g_m (but $g_o = C_{db} = 0$), a resistive drain load R_D (no load capacitance), and an ideal voltage follower between the drain node and the output of the TIA. Divide the transfer functions and show how the Miller factors cancel out in the input-referral function.

6.10 **TIA f^2-Noise Corner.** The "f^2-noise corner" occurs at the frequency where the white noise and f^2 noise are equally strong. (a) Derive an expression for the f^2-noise corner frequency of a TIA with a MOSFET front-end. (b) How is this corner frequency related to the bit rate?

6.11 **TIA with $1/f$ Noise.** Assume that the white channel noise and the $1/f$ noise can be written in the combined form $I_{n,D}^2 = 4kT\Gamma g_m(1 + f_c/f)$, where f_c is the $1/f$-noise corner frequency. Calculate the input-referred noise current spectrum of a TIA with a MOSFET front-end.

6.12 **Noise Optimum of FET Front-End with Constant R_F.** (a) Given a shunt-feedback TIA with a single-pole voltage amplifier and values for R_F, C_D, and C_I, find the DC gain A_0 and amplifier pole f_A necessary to achieve a desired BW_{3dB} and Q. (b) For what range of BW_{3dB} are the solutions for A_0 and f_A valid? Assume that the maximum realizable gain-bandwidth product is $A_0 f_A$(max).

6.13 **Noise Optimum of FET Front-End with Constant $A_0 f_A$.** (a) Show that after minimizing Eq. (6.60), the noise fraction due to the FET is ψ. (b) Given a shunt-feedback TIA with a single-pole voltage amplifier and values for $A_0 f_A$, C_D, and C_I, find the input pole f_I and amplifier pole f_A

necessary to achieve a desired BW_{3dB} and Q. (c) For what range of BW_{3dB} are the solutions for f_I and f_A valid? Assume that the feedback resistor must remain below $R_F(\max)$. (d) Derive Eq. (6.63).

6.14 **Induced Gate Noise and Noise Matching with Shunt Inductor.** (a) Calculate the input-referred noise current PSD of a TIA with FET front-end driven by the arbitrary source admittance Y_S while taking the induced gate noise current into account. See Problem 6.8 for a description of the induced gate noise current. (b) Assuming c is purely imaginary, find the optimum source admittance $\tilde{Y}_{\text{opt}}$. (c) Find the optimum shunt inductor L_P for the noise-matching circuit shown in Fig. 6.24. (d) Find the residual input-referred noise current after noise matching.

6.15 **Noise Parameters of a FET Front-End with Induced Gate Noise.** Calculate the noise parameters V_{n4}^2, I_{n4u}^2, and Y_c for a FET front-end as we did in Eq. (6.96), but now taking the induced gate noise current into account. See Problem 6.8 for a description of the induced gate noise current.

6.16 **Optimum Source Admittance.** (a) Find the source admittance that minimizes the input-referred noise current PSD, $\tilde{Y}_{\text{opt}}$, for the general noise model in Eq. (6.92). (b) Find the source admittance that minimizes the noise figure, Y_{opt}, for the general noise model in Eq. (6.92). (c) Specialize the results for the case of a TIA with a FET front-end.

6.17 **Noise Matching and ABCD Parameters.** Prove the expressions for Y_M and H_M in Eq. (6.99).

6.18 **Reciprocal and Lossless Matching Network.** Prove Eq. (6.105) assuming a reciprocal and lossless matching network. Tip: For a reciprocal network $\mathcal{AD} - \mathcal{BC} = 1$ and for a lossless network $\mathcal{A}, \mathcal{D}$ are real and $\mathcal{B}, \mathcal{C}$ are imaginary [85].

References

1 M. Reinhold, C. Dorschky, E. Rose, R. Pullela, P. Mayer, F. Kunz, Y. Baeyens, T. Link, and J.-P. Mattia. A fully integrated 40-Gb/s clock and data recovery IC with 1:4 DEMUX in SiGe technology. *IEEE J. Solid-State Circuits*, SC-36(12):1937–1945, 2001.

2 H. Ito, S. Kodama, Y. Muramoto, T. Furuta, T. Nagatsuma, and T. Ishibashi. High-speed and high-output InP-InGaAs unitraveling-carrier photodiodes. *IEEE J. Sel. Top. Quantum Electron.*, 10(4):709–727, 2004.

3 J. E. Goell. An optical repeater with high-impedance input amplifier. *Bell Syst. Tech. J.*, 53(4):629–643, 1974.

4 B. L. Kasper. Receiver design. In S. E. Miller and I. P. Kaminow, editors, *Optical Fiber Telecommunications II*, pages 689–722. Academic Press, San Diego, CA, 1988.

5 T. V. Muoi. Receiver design for high-speed optical-fiber systems. *J. Lightwave Technol.*, LT-2(3):243–267, 1984.

6 R. G. Smith and S. D. Personick. Receiver design for optical fiber communication systems. In H. Kressel, editor, *Topics in Applied Physics: Semiconductor Devices for Optical Communication*, volume 39. Springer-Verlag, Berlin, Germany, 1982.

7 M. H. Nazari and A. Emami-Neyestanak. A 24-Gb/s double-sampling receiver for ultra-low-power optical communication. *IEEE J. Solid-State Circuits*, SC-48(2):344–357, 2013.

8 J. Proesel, A. Rylyakov, and C. Schow. Optical receivers using DFE-IIR equalization. In *ISSCC Digest of Technical Papers*, pages 130–131, February 2013.

9 A. Emami-Neyestanak, D. Liu, G. Keeler, N. Helman, and M. Horowitz. A 1.6Gb/s, 3mW CMOS receiver for optical communication. In *Symposium on VLSI Circuits Digest of Technical Papers*, pages 84–87, June 2002.

10 S. Palermo, A. Emami-Neyestanak, and M. Horowitz. A 90 nm CMOS 16 Gb/s transceiver for optical interconnects. *IEEE J. Solid-State Circuits*, SC-43(5):1235–1246, 2008.

11 M. Georgas, J. Orcutt, R. J. Ram, and V. Stojanović. A monolithically-integrated optical receiver in standard 45-nm SOI. *IEEE J. Solid-State Circuits*, SC-47(7):1693–1702, 2012.

12 M. Abraham. Design of Butterworth-type transimpedance and bootstrap-transimpedance preamplifiers for fiber-optic receivers. *IEEE Trans. Circuits Syst.*, CAS-29(6):375–382, 1982.

13 P. Starič and E. Margan. *Wideband Amplifiers*. Springer, Dordrecht, The Netherlands, 2006.

14 M. B. Das, J.-W. Chen, and E. John. Designing optoelectronic integrated circuit (OEIC) receivers for high sensitivity and maximally flat frequency response. *J. Lightwave Technol.*, LT-13(9):1876–1884, 1995.

15 W.-Z. Chen and R.-M. Gan. 1.8V variable gain transimpedance amplifiers with constant damping factor for burst-mode optical receiver. In *IEEE Radio Frequency Integrated Circuits (RFIC) Symposium*, pages 691–694, Long Beach, CA, June 2005.

16 F.-T. Chien and Y.-J. Chan. Bandwidth enhancement of transimpedance amplifier by a capacitive-peaking design. *IEEE J. Solid-State Circuits*, SC-34(8):1167–1170, 1999.

17 S. S. Mohan, M. D. M. Hershenson, S. P. Boyd, and T. H. Lee. Bandwidth extension in CMOS with optimized on-chip inductors. *IEEE J. Solid-State Circuits*, SC-35(3):346–355, 2000.

18 E. Säckinger. The transimpedance limit. *IEEE Trans. Circuits Syst. I*, CASI-57(8):1848–1856, 2010.

19 E. M. Cherry. Impedance mismatching in wide-band transistor amplifier design. *IEEE Trans. Circuit Theory*, 17(1):131–132, 1970.

20 S. Voinigescu. *High-Frequency Integrated Circuits*. Cambridge University Press, 2013.

21 J. S. Weiner, J. S. Lee, A. Leven, Y. Baeyens, V. Houtsma, G. Georgiou, Y. Yang, J. Frackoviak, A. Tate, R. Reyes, R. F. Kopf, W.-J. Sung, N. G. Weimann, and Y.-K. Chen. An InGaAs–InP HBT differential transimpedance amplifier with 47-GHz bandwidth. *IEEE J. Solid-State Circuits*, SC-39(10):1720–1723, 2004.

22 B. Razavi. Cognitive radio design challenges and techniques. *IEEE J. Solid-State Circuits*, SC-45(8):1542–1553, 2010.

23 M.-C. Chiang, S.-S. Lu, C.-C. Meng, S.-A. Yu, S.-C. Yang, and Y.-J. Chan. Analysis, design, and optimization of InGaP–GaAs HBT matched-impedance wide-band amplifiers with multiple feedback loops. *IEEE J. Solid-State Circuits*, SC-37(6):694–701, 2002.

24 R. G. Meyer and R. A. Blauschild. A 4-terminal wide-band monolithic amplifier. *IEEE J. Solid-State Circuits*, SC-16(6):634–638, 1981.

25 K.-Y. Toh, R. G. Meyer, D. C. Soo, G. M. Chin, and A. M. Voshchenkov. Wide-band, low-noise, matched-impedance amplifiers in submicrometer MOS technology. *IEEE J. Solid-State Circuits*, SC-22(6):1031–1040, 1987.

26 T. O. Dickson, R. Beerkens, and S. P. Voinigescu. A 2.5-V 45-Gb/s decision circuit using SiGe BiCMOS logic. *IEEE J. Solid-State Circuits*, SC-40(4):994–1003, 2005.

27 T. O. Dickson, K. H. K. Yau, T. Chalvatzis, A. M. Mangan, E. Laskin, R. Beerkens, P. Westergaard, M. Tazlauanu, M.-T. Yang, and S. P. Voinigescu. The invariance of characteristic current densities in nanoscale MOSFETs and its impact on algorithmic design methodologies and design porting of Si(Ge) (Bi)CMOS high-speed building blocks. *IEEE J. Solid-State Circuits*, SC-41(8):1830–1845, 2006.

28 W. Pöhlmann. A silicon-bipolar amplifier for 10 Gbit/s with 45 dB gain. *IEEE J. Solid-State Circuits*, SC-29(5):551–556, 1994.

29 J.-H. C. Zhan and S. S. Taylor. A 5GHz resistive-feedback CMOS LNA for low-cost multi-standard applications. In *ISSCC Digest of Technical Papers*, pages 200–201, February 2006.

30 F. Y. Liu, D. Patil, J. Lexau, P. Amberg, M. Dayringer, J. Gainsley, H. F. Moghadam, X. Zheng, J. E. Cunningham, A. V. Krishnamoorthy, E. Alon, and R. Ho. 10-Gbps, 5.3-mW optical transmitter and receiver circuits in 40-nm CMOS. *IEEE J. Solid-State Circuits*, SC-47(9):2049–2067, 2012.

31 E. H. Nordholt. *The Design of High-Performance Negative-Feedback Amplifiers*. Elsevier, Amsterdam, The Netherlands, 1983.

32 H. Ransijn. Receiver and transmitter IC design, May 2001. CICC'2001 Ed. Session 3-2.

33 I. Hehemann, W. Brockherde, H. Hofmann, A. Kemna, and B. J. Hosticka. A single-chip optical CMOS detector with *in-situ* demodulating and integrating readout for next-generation optical storage systems. *IEEE J. Solid-State Circuits*, SC-39(4):629–635, 2004.

34 D. M. Pietruszynski, J. M. Steininger, and E. J. Swanson. A 50-Mbit/s CMOS monolithic optical receiver. *IEEE J. Solid-State Circuits*, SC-23(6):1426–1433, 1988.

35 M. Ingels and M. S. J. Steyaert. A 1-Gb/s, 0.7-μm CMOS optical receiver with full rail-to-rail output swing. *IEEE J. Solid-State Circuits*, SC-34(7):971–977, 1999.

36 A.-J. Annema, B. Nauta, R. van Langevelde, and H. Tuinhout. Analog circuits in ultra-deep-submicron CMOS. *IEEE J. Solid-State Circuits*, SC-40(1):132–143, 2005.

37 S. B. Alexander. *Optical Communication Receiver Design*. SPIE Press, copublished with IEE, Bellingham, WA, 1997.

38 A. van der Ziel. Thermal noise in field-effect transistors. *Proc. IRE*, 50(8):1808–1812, 1962.

39 A. van der Ziel. Gate noise in field-effect transistors at moderately high frequencies. *Proc. IEEE*, 51(3):461–467, 1963.

40 R. M. Fox. Comments on circuit models for MOSFET thermal noise. *IEEE J. Solid-State Circuits*, SC-28(2):184–185, 1993.

41 R. P. Jindal. Compact noise models for MOSFETs. *IEEE Trans. Electron Devices*, ED-53(9):2051–2061, 2006.

42 G. Nicollini, D. Pancini, and S. Pernici. Simulation-oriented noise model for MOS devices. *IEEE J. Solid-State Circuits*, SC-22(6):1209–1212, 1987.

43 Z. Deng and A. M. Niknejad. On the noise optimization of CMOS common-source low-noise amplifiers. *IEEE Trans. Circuits Syst. I*, CASI-58(4):654–667, 2011.

44 W. Baechtold. Noise behavior of GaAs field-effect transistors with short gate lengths. *IEEE Trans. Electron Devices*, ED-19(5):674–680, 1972.

45 G. Knoblinger, P. Klein, and M. Tiebout. A new model for thermal channel noise of deep-submicron MOSFETS and its application in RF-CMOS design. *IEEE J. Solid-State Circuits*, SC-36(5):831–837, 2001.

46 A. J. Scholten, H. J. Tromp, L. F. Tiemeijer, R. van Langevelde, R. J. Havens, P. W. H. de Vreede, R. F. M. Roes, P. H. Woerlee, A. H. Montree, and D. B. M. Klaassen. Accurate thermal noise model for deep-submicron CMOS. In *Proceedings of International Electron Devices Meeting*, pages 155–158, December 1999.

47 B. Wang, J. R. Hellums, and C. G. Sodini. MOSFET thermal noise modeling for analog integrated circuits. *IEEE J. Solid-State Circuits*, SC-29(7):833–835, 1994.

48 A. J. Scholten, L. F. Tiemeijer, R. van Langevelde, R. J. Havens, A. T. A. Zegers-van Duijnhoven, and V. C. Venezia. Noise modeling for RF CMOS circuit simulation. *IEEE Trans. Electron Devices*, ED-50(3):618–632, 2003.

49 D. K. Shaeffer, A. R. Shahani, S. S. Mohan, H. Samavati, H. R. Rategh, M. del Mar Hershenson, M. Xu, C. P. Yue, D. J. Eddleman, and T. H. Lee. A 115-mW, 0.5-μm CMOS GPS receiver with wide dynamic-range active filters. *IEEE J. Solid-State Circuits*, SC-33(12):2219–2231, 1998.

50 T. H. Lee. *The Design of CMOS Radio-Frequency Integrated Circuits*. Cambridge University Press, Cambridge, UK, 2nd edition, 2003.

51 D. K. Shaeffer and T. H. Lee. A 1.5-V, 1.5-GHz CMOS low noise amplifier. *IEEE J. Solid-State Circuits*, SC-32(5):745–759, May 1997. Also see the Corrections in the June 2005 and October 2006 issues.

52 M. Shoji. Analysis of high-frequency thermal noise of enhancement mode MOS field-effect transistors. *IEEE Trans. Electron Devices*, ED-13(6):520–524, 1966.

53 E. Säckinger. On the excess noise factor Γ of a FET driven by a capacitive source. *IEEE Trans. Circuits Syst. I*, CASI-58(9):2118–2126, 2011.

54 K. Ogawa. Noise caused by GaAs MESFETs in optical receivers. *Bell Syst. Tech. J.*, 60(6):923–928, 1981.

55 K. Ogawa, B. Owen, and H. J. Boll. A long-wavelength optical receiver using a short-channel Si-MOSFET. *Bell Syst. Tech. J.*, 62(5):1181–1188, 1983.

56 D. Johns and K. Martin. *Analog Integrated Circuit Design*. John Wiley &Sons, Inc., New York, 1997.

57 B. L. Kasper and J. C. Campbell. Multigigabit-per-second avalanche photodiode lightwave receivers. *J. Lightwave Technol.*, LT-5(10):1351–1364, 1987.

58 R. P. Jindal. General noise considerations for gigabit-rate NMOSFET front-end design for optical-fiber communication systems. *IEEE Trans. Electron Devices*, ED-34(2):305–309, 1987.

59 J. E. Goell. Input amplifiers for optical PCM receivers. *Bell Syst. Tech. J.*, 53(9):1771–1793, 1974.

60 J. C. J. Paasschens, R. J. Havens, and L. F. Tiemeijer. Modelling the correlation in the high-frequency noise of (hetero-junction) bipolar transistors using charge-partitioning. In *Proceedings of Bipolar/BiCMOS Circuits and Technology Meeting*, pages 221–224, September 2003.

61 R. G. Meyer and R. A. Blauschild. A wide-band low-noise monolithic transimpedance amplifier. *IEEE J. Solid-State Circuits*, SC-21(4):530–533, 1986.

62 J. J. Morikuni, A. Dharchoudhury, Y. Leblebici, and S. M. Kang. Improvements to the standard theory for photoreceiver noise. *J. Lightwave Technol.*, LT-12(7):1174–1184, 1994.

63 C. D. Motchenbacher and F. C. Fitchen. *Low-Noise Electronic Design.* John Wiley &Sons, Inc., New York, 1973.

64 B. S. Leibowitz, B. E. Boser, and K. S. J. Pfister. A 256-element CMOS imaging receiver for free-space optical communication. *IEEE J. Solid-State Circuits*, SC-40(9):1948–1956, 2005.

65 A. A. Abidi. Gigahertz transresistance amplifiers in fine line NMOS. *IEEE J. Solid-State Circuits*, SC-19(6):986–994, 1984.

66 E. Säckinger. On the noise optimum of FET broadband transimpedance amplifiers. *IEEE Trans.Circuits Syst. I*, CASI-59(12):2881–2889, 2012.

67 A. A. Abidi. On the noise optimum of gigahertz FET transimpedance amplifiers. *IEEE J. Solid-State Circuits*, SC-22(6):1207–1209, 1987.

68 A. A. Abidi. On the choice of optimum FET size in wide-band transimpedance amplifiers. *J. Lightwave Technol.*, LT-6(1):64–66, 1988.

69 E. Säckinger. *Broadband Circuits for Optical Fiber Communication.* John Wiley &Sons, Inc., Hoboken, NJ, 2005.

70 M. Atef and H. Zimmermann. Optical receiver using noise cancelling with an integrated photodiode in 40 nm CMOS technology. *IEEE Trans.Circuits Syst. I*, CASI-60(7):1929–1936, 2013.

71 L. Belostotski. No noise is good noise. *IEEE Microwave Mag.*, 17(8):28–40, 2016.

72 F. Bruccoleri, E. A. M. Klumperink, and B. Nauta. Wide-band CMOS low-noise amplifier exploiting thermal noise canceling. *IEEE J. Solid-State Circuits*, SC-39(2):275–282, 2004.

73 D. Li, G. Minoia, M. Repossi, D. Baldi, E. Temporiti, A. Mazzanti, and F. Svelto. A low-noise design technique for high-speed CMOS optical receivers. *IEEE J. Solid-State Circuits*, SC-49(6):1437–1447, 2014.

74 J. L. Zyskind, J. A. Nagel, and H. D. Kidorf. Erbium-doped fiber amplifiers for optical communications. In I. P. Kaminow and T. L. Koch, editors, *Optical Fiber Telecommunications IIIB*, pages 13–68. Academic Press, San Diego, CA, 1997.

75 D. L. Rogers. Integrated optical receivers using MSM detectors. *J. Lightwave Technol.*, LT-9(12):1635–1638, 1991.

76 L. Belostotski, A. Madanayake, and L. T. Bruton. Wideband LNA with an active $-C$ element. *IEEE Microwave Wireless Compon. Lett.*, 22(10):524–526, 2012.

77 T. E. Darcie, B. L. Kasper, J. R. Talman, and C. A. Burrus Jr. Resonant p-i-n-FET receivers for lightwave subcarrier systems. *J. Lightwave Technol.*, LT-6(4):582–589, 1988.

78 S. D. Greaves and R. T. Unwin. The design of tuned front-end GaAs MMIC optical receivers. *IEEE Trans. Microwave Theory Tech.*, MTT-44(4):591–597, 1996.

79 D. M. Pozar. *Microwave Engineering*. John Wiley &Sons, Inc., Hoboken, NJ, 4th edition, 2012.

80 H. Rothe and W. Dahlke. Theory of noisy fourpoles. *Proc. IRE*, 44(6):811–818, 1956.

81 M. S. Park and R. A. Minasian. Ultra-low-noise and wideband-tuned optical receiver synthesis and design. *J. Lightwave Technol.*, LT-12(2):254–259, 1994.

82 R. Müller. *Rauschen (Halbleiter-Elektronik 15)*. Springer-Verlag, Berlin, 1979.

83 C. Bowick. *RF Circuit Design*. Newnes, Boston, MA, 1982.

84 B. Razavi. *RF Microelectronics*. Prentice-Hall, Inc., Upper Saddle River, NJ, 1998.

85 A. Leven, R. Reuter, and Y. Baeyens. Unified analytical expressions for transimpedance and equivalent input noise current of optical receivers. *IEEE Trans. Microwave Theory Tech.*, MTT-48(10):1701–1706, October 2000.

7

Advanced Transimpedance Amplifier Design I

The basic shunt-feedback TIA from the previous chapter can be enhanced in a number of ways to meet the needs of practical applications. In the following, we discuss shunt-feedback TIAs with post amplifier, differential inputs and outputs, DC input current control, and variable or adaptive transimpedance. Then, we turn to another class of TIAs, the so-called feedforward TIAs, which includes the common-base and common-gate TIA. After that, we analyze the regulated-cascode TIA. Finally, we discuss the application of inductive broadband techniques, ultimately leading to the distributed amplifier front-end.

7.1 TIA with Post Amplifier

In Section 6.2, we discussed the use of a voltage amplifier as part of the shunt-feedback TIA. Now, we discuss the use of a voltage amplifier as a *post amplifier*, that is, an amplifier that follows the basic TIA as shown in Fig. 7.1. The post amplifier comes in various forms. It can be just an output buffer to drive an off-chip load [1–3], it can be a single-stage amplifier [4–6], or it can be a multistage high-gain amplifier [7–9]. The last case amounts to a TIA with a main amplifier (MA) integrated on the same chip. The combination of a TIA and a limiting amplifier is known as a *limiting transimpedance amplifier* [10]. In the following, we discuss the trade-offs associated with the design of a TIA with post amplifier.

Transimpedance and Bandwidth. The addition of a post amplifier with voltage gain A_1 boosts the overall transimpedance by a factor A_1. Thus for the circuit in Fig. 7.1, we have

$$R_T = A_1 \cdot \frac{A_0 R_F}{A_0 + 1}, \tag{7.1}$$

where A_0 is, as usual, the low-frequency value of $A(s)$.

How large can we make the post-amplifier gain and how does the post amplifier impact the overall bandwidth? Let us take our 10-Gb/s TIA example

Analysis and Design of Transimpedance Amplifiers for Optical Receivers, First Edition. Eduard Säckinger.

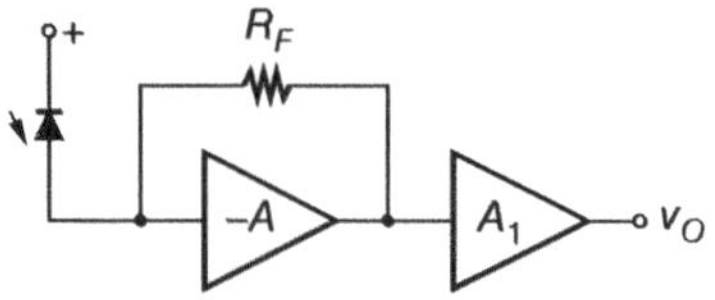

Figure 7.1 TIA with post amplifier.

from the previous chapter (Table 6.1) to illustrate the situation. There, the basic shunt-feedback TIA had a bandwidth of 6.85 GHz and a transimpedance of 500 Ω. In the technology that we assumed for the example ($A_0 f_A = 44$ GHz), we can build a single-stage post amplifier with a gain of two and a bandwidth of 22 GHz. Thus, with this post amplifier the transimpedance increases to 1 kΩ while the bandwidth shrinks only very little from the original 6.85 GHz to about 6.5 GHz. We could obtain an even higher overall transimpedance by increasing the gain of the post-amplifier stage to about three and redesigning the shunt-feedback TIA for a slightly larger bandwidth that compensates for the bandwidth shrinkage caused by the post amplifier.

The bandwidth of a cascade of n identical second-order Butterworth stages is given by [11, 12]

$$BW_{3dB} = \sqrt[4]{\sqrt[n]{2} - 1} \cdot BW_S, \tag{7.2}$$

where BW_S is the 3-dB bandwidth of the individual stages. For example, if we follow a (second-order Butterworth) shunt-feedback TIA with a bandwidth of 8.54 GHz by a (second-order Butterworth) post amplifier with the same bandwidth, the overall bandwidth is 6.85 GHz, corresponding to a bandwidth shrinkage of about 20%.

Transimpedance Limit. The maximum transimpedance that can be obtained with a (single-stage) TIA followed by an n-stage post amplifier and an overall bandwidth of BW_{3dB} is given by the transimpedance limit [11, 12]

$$R_T \leq \sqrt[4]{\left(\sqrt[n+1]{2} - 1\right)^{n+2}} \cdot \frac{(A_0 f_A)^{n+1}}{2\pi C_T \cdot BW_{3dB}^{n+2}}, \tag{7.3}$$

where it was assumed that the TIA has a flat response ($Q \leq 1/\sqrt{2}$), that all post-amplifier stages have a second-order Butterworth response with the same bandwidth as the TIA, and that each post-amplifier stage has a gain-bandwidth product of $A_0 f_A$ (see solution to Problem 7.1 on p. 523 for a derivation). For $n = 0$ (no post amplifier), this limit is identical to the transimpedance limit in Eq. (6.14), as we would expect. For $n > 0$, the first factor decreases with n (see Table 7.1) while the second factor increases by $(A_0 f_A / BW_{3dB})^n$ relative to Eq. (6.14). Fortunately, in most situations the second factor grows faster than the first one shrinks, leading to a net gain in transimpedance over a TIA without post amplifier. Table 7.1 shows that if the bandwidth headroom, $A_0 f_A / BW_{3dB}$,

Table 7.1 First factor of Eq. (7.3) and minimum bandwidth headroom needed for an n-stage post amplifier to outperform an $(n-1)$-stage post amplifier.

Number of stages n	First factor of Eq. (7.3)	Minimum $A_0 f_A / BW_{3dB}$
0	1.0	–
1	0.516	1.94
2	0.260	1.99
3	0.125	2.08
4	0.0573	2.18

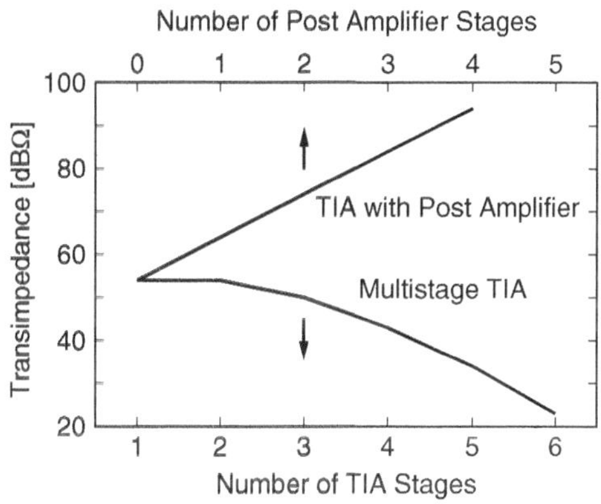

Figure 7.2 Transimpedance limit of TIA with post amplifier versus multistage TIA for our 10-Gb/s example ($A_0 f_A = 44$ GHz, $C_T = 0.3$ pF, and $BW_{3dB} = 6.85$ GHz).

is more than about two, adding a post-amplifier stage helps to boost the transimpedance, nearly regardless of how many stages were there already. [→ Problem 7.1.]

For our 10-Gb/s example in Table 6.1 and a single-stage post amplifier, the transimpedance limit comes out as $0.516 \times (44/6.85) \times 500\ \Omega = 1.65\ \text{k}\Omega$, corresponding to a boost of 3.3× over a TIA without post amplifier. For a two-stage post amplifier the transimpedance limit goes up to $0.260 \times (44/6.85)^2 \times 500\ \Omega = 5.34\ \text{k}\Omega$, corresponding to a boost of 10.7×, and so on. This nearly exponential growth in transimpedance is plotted in Fig. 7.2 (upper curve).

Comparison with Multistage TIA. It is interesting to compare the transimpedance limit in Eq. (7.3) with that for a multistage TIA in Eq. (6.35). The two forms are similar, but the first factor in Eq. (6.35) is much smaller than that in Eq. (7.3), especially for large n. This implies that it is much harder to boost the transimpedance by adding gain stages *within* the feedback loop than adding them *after* the feedback loop. The reason for this difference is that adding gain stages within the loop is constrained by stability considerations, whereas adding gain stages after the loop is not.

Figure 7.2 compares the two transimpedance limits for the case of our 10-Gb/s TIA example. Because $A_0 f_A / BW_{3dB}$ is only 6.4, a single gain stage within the loop is the best choice (cf. Table 6.4). Another comparison for the larger ratio $A_0 f_A / BW_{3dB} = 20$ can be found in [12].

Noise Considerations. Given that it is so easy to boost the transimpedance with a post amplifier, should we still bother to optimize the transimpedance of the shunt-feedback section? Yes, because we also want to minimize the input-referred noise current. A larger transimpedance in the shunt-feedback section implies a larger feedback resistor, R_F, and thus a lower noise contribution from it. In contrast, boosting the transimpedance with a post amplifier keeps the input-referred noise approximately constant. [→ Problem 7.2.]

In the extreme case when all gain is located in the post amplifier, the circuit degenerates into a noisy low-impedance front-end. Thus, regardless of whether there is a post amplifier or not, we should make R_F as large as possible, or at least large enough such that the feedback-resistor noise becomes small when compared with other noise sources.

Power Consumption, Coupling, and Stability. Figure 7.3(a) shows the classical partitioning of the receive chain into a TIA chip and an MA chip. A standard 50-Ω transmission line provides a broadband connection between the two chips. To

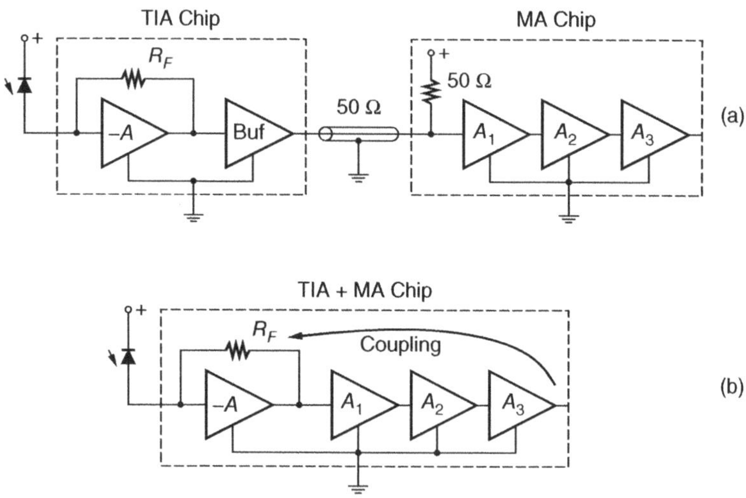

Figure 7.3 TIA and MA: (a) two-chip solution and (b) single-chip solution.

avoid reflections on the transmission line, an undesirable source of ISI, the input of the MA chip needs to have a 50-Ω termination. Consequently, the buffer on the TIA side must be able to drive a 50-Ω load. Preferably, the driver also has a 50-Ω output impedance to prevent residual reflections from the MA from being reflected back into the MA (double reflections).

The main drawbacks of this partitioning are the power dissipated in the buffer and the termination as well as the cost and board space usage resulting from the two chip solution [8]. The obvious solution is to integrate the TIA and the MA on the same chip, as shown in Fig. 7.3(b). Unfortunately, this integration brings the highly sensitive TIA input and the noisy MA output(s) in close proximity and coupling between them may result in instability and oscillations [13].

The following coupling mechanisms in an integrated TIA + MA solution can be identified:

- Capacitive coupling between metal traces, bond pads, and package pins.
- Coupling through the chip substrate.
- Coupling though the inductance and resistance of shared power and ground connections.
- Magnetic coupling between bond wires.

To minimize the coupling, the following techniques can be utilized [8, 10, 14, 15]:

- Place the input and output pads as far apart as possible, ideally on opposite sides of the chip.
- Use differential signaling and symmetric placement of the associated traces and pads. Equally coupled signals are suppressed as a common-mode disturbance.
- Use a buried-layer shield under sensitive input pads to reduce substrate coupling.
- Surround the TIA and MA blocks by shield rings (e.g., substrate contacts or isolation trenches) and separate the blocks by about 100 to 400 μm to reduce lateral substrate coupling. The separated blocks can be interconnected with on-chip microstrip lines, which provide shielding from the substrate and a well defined return path. (For the implementation of microstrip lines, shielded coplanar waveguides, and other transmission lines in CMOS see [16–18].) Impedance matching to these lines is not required because the lines tend to be short compared with the minimum signal wavelength.
- Mount the (thinned) chip with a conductive glue on a ground plane to provide a good (vertical) substrate connection to an external ground.
- Use separate power and ground pads for the TIA, MA, and output buffer blocks and use on-chip decoupling capacitors to avoid coupling through the power network.
- Use an on-chip voltage regulator for the sensitive TIA block to reduce the supply coupling.

- Keep ground bond wires short and use double bonding, if possible, to minimize their inductance. Arrange critical bond wires at right angle to minimize magnetic coupling.

The optimum partitioning of an optical receiver into individual chips is subject to debate. Besides integrating the MA function into the TIA front-end chip, as suggested earlier, the MA function could also be integrated into the subsequent mixed-signal (CMOS) chip [19].

7.2 TIA with Differential Inputs and Outputs

Differential circuits have a number of important advantages over their single-ended counterparts. Among them are the improved immunity to power-supply and substrate noise and the superior rejection of parasitically coupled signals. For these reasons, *differential TIAs* find application in noisy environments, such as systems on a chip (SOC) or multichannel receivers. Moreover, differential TIAs have low even-order distortions, which is important for linear applications (cf. Section 8.5).

Differential-input TIAs can subtract two photocurrents and thus find application in receivers with balanced detectors (cf. Section 3.5). Differential-output TIAs can produce twice the voltage swing compared with single-ended TIAs, making them an attractive solution for low-voltage systems. Moreover, differential-output TIAs facilitate the connection to differential MAs, avoiding the need for a reference voltage.

The main drawback of differential TIAs is their higher power consumption.

Balanced Differential TIA with Two Photodetectors. Figure 7.4 shows a balanced differential shunt-feedback TIA. The differential voltage amplifier at the center of the circuit senses the input differential voltage and drives the output differential voltage: $(v_{OP} - v_{ON}) = A_0(v_{IP} - v_{IN})$. Simultaneously, the differential amplifier keeps the output common-mode voltage constant, that is, independent of the input voltages: $\frac{1}{2}(v_{OP} + v_{ON}) = V_{OCM}$. Implementation examples for differential voltage amplifiers are given in Chapter 9 (e.g., Fig. 9.5).

The differential TIA in Fig. 7.4 responds to the *difference* of the two photocurrents, $(v_{OP} - v_{ON}) = R_T(i_{IP} - i_{IN})$, where the differential transimpedance is

$$R_T = \frac{\Delta(v_{OP} - v_{ON})}{\Delta(i_{IP} - i_{IN})} = \frac{A_0}{A_0 + 1} R_F. \tag{7.4}$$

Note that this is the same expression as in Eq. (6.9). The transimpedance, bandwidth, and stability analysis, which we carried out in Section 6.2, remains valid for the differential TIA, if we replace the single-ended input voltage, v_I, by the *differential* input voltage, $v_{IP} - v_{IN}$, the single-ended output voltage, v_O, by the

Figure 7.4 Balanced differential TIA with two photodetectors.

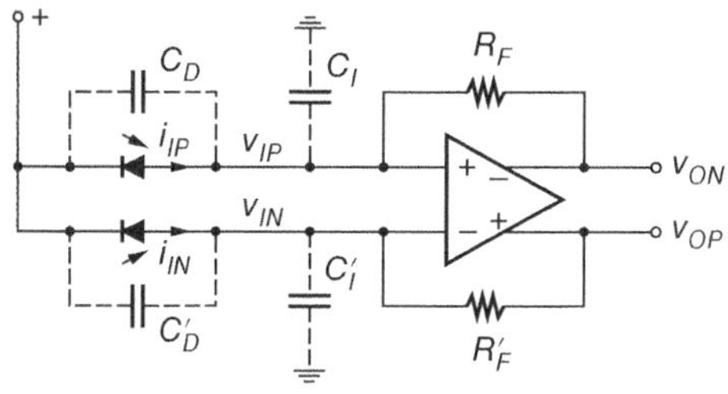

differential output voltage, $v_{OP} - v_{ON}$, the single-ended voltage-amplifier gain by the *differential* gain, $A_0 = \Delta(v_{OP} - v_{ON})/\Delta(v_{IP} - v_{IN})$, and so forth.

The two matched photodetectors in Fig. 7.4 present a balanced input impedance. Any noise on the power supply or the substrate couples equally to the noninverting and the inverting input of the voltage amplifier and thus is suppressed as a common-mode disturbance. If the basic shunt-feedback topology is enhanced with a variable feedback resistor, an inductive input network, a common-base/gate input stage, or something else, it must be done in a balanced manner to preserve the noise immunity.

When preceded by the appropriate optical front-end, the receiver in Fig. 7.4 can detect advanced modulation formats, such as quadrature phase-shift keying (QPSK) and differential quadrature phase-shift keying (DQPSK) [20–22] (cf. Section 3.5). When connected to a spatially modulated light (SML) detector, the differential TIA can subtract the deferred photocurrent from the immediate photocurrent (cf. Section 3.4).

Balanced Differential TIA with One Photodetector. Many receivers rely on only a single photodetector. Although it is possible to use the arrangement in Fig. 7.4 and keep one photodetector dark [1, 23], this may not be a cost-effective solution. We are thus looking for a way to connect a *single* photodetector to the differential TIA without compromising its superior noise immunity. A straightforward solution is shown in Fig. 7.5, where the dark photodetector is replaced with a dummy or replica capacitor C_R that closely matches the photodetector capacitance C_D [15]. For best supply-noise cancellation, the photodetector and

Figure 7.5 Balanced differential TIA with one photodetector.

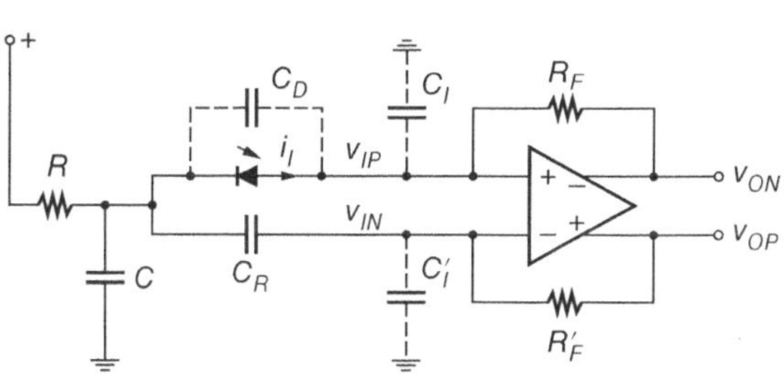

C_R must be connected to the same supply node. Moreover, an RC low-pass filter can be inserted into the supply line, as shown in Fig. 7.5 [24, 25].

In practice, however, the matching between C_R and C_D is never perfect and supply-noise coupling to the inverting and the noninverting input of the TIA is slightly different, resulting in imperfect noise cancellation. Moreover, if C_R is located on chip and C_D is located off chip, the bond wire connecting the photodetector to the chip may pick up some noise which is not canceled by the replica path.

The differential transimpedance is the same as before

$$R_T = \frac{\Delta(v_{OP} - v_{ON})}{\Delta i_I} = \frac{A_0}{A_0 + 1} R_F. \tag{7.5}$$

The single-ended input resistance seen by the photodetector, however, is larger than that of a single-ended TIA (cf. Eq. (I.117) on p. 524)

$$R_{I,se} = \frac{\Delta v_{IP}}{\Delta i_I} \approx \frac{R_F}{2}. \tag{7.6}$$

As a result, the voltage swing at the photodetector of a differential TIA typically is larger than that of a single-ended TIA. [→ Problem 7.3.]

Unbalanced Differential TIA. If noise immunity is not a primary concern, we can replace the replica capacitor C_R in Fig. 7.5 by a large capacitor, $C \to \infty$, shorting the unused input to AC ground, as shown in Fig. 7.6 [1, 26]. This large capacitor disables the AC feedback through R_F' and we end up with essentially a single-ended topology. In this configuration the noise contribution of R_F' is eliminated and the input-referred rms noise current is reduced. However, because the input capacitances are no longer balanced, power-supply and substrate noise couple differently to the two inputs, causing noise to leak into the differential mode.

The transimpedance, bandwidth, and stability analysis, which we have carried out for the single-ended TIA, remain valid for the unbalanced differential TIA in Fig. 7.6, if we replace the single-ended input voltage, v_I, by the single-ended input voltage, v_{IP}, the single-ended output voltage, v_O, by the single-ended output voltage, v_{ON}, the single-ended voltage-amplifier

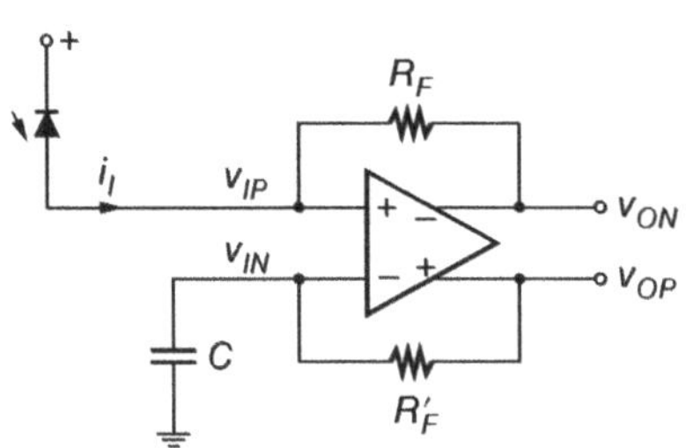

Figure 7.6 Unbalanced differential TIA with one photodetector.

gain, A_0, by *half* of the differential gain, $A_0/2 = -\Delta v_{ON}/\Delta v_{IP}$, and so forth. It follows that the single-ended transimpedance is now given by $R_{T,se} = -\Delta v_{ON}/\Delta i_I = (A_0/2)/[(A_0/2)+1] \cdot R_F$, where A_0 is the differential gain. Although only one output provides shunt feedback, both outputs are swinging in a complementary fashion due to the common-mode constraint $\frac{1}{2}(v_{OP}+v_{ON}) = V_{OCM}$. Thus, the differential transimpedance is twice the single-ended one:

$$R_T = \frac{\Delta(v_{OP}-v_{ON})}{\Delta i_I} = \frac{A_0}{A_0/2+1} R_F \approx 2R_F. \tag{7.7}$$

In comparison with the balanced TIA in Fig. 7.5, the unbalanced TIA has about twice the transimpedance and a somewhat better sensitivity (lower input-referred noise current) but it has less immunity to power-supply and substrate noise. Furthermore, its single-ended input resistance is

$$R_{I,se} = \frac{\Delta v_{IP}}{\Delta i_I} = \frac{R_F}{A_0/2+1}, \tag{7.8}$$

which is lower than that in Eq. (7.6). If a TIA that has been designed for the balanced configuration is operated in the unbalanced configuration, its pole placement becomes nonoptimal because the loop gain is cut in half by AC grounding the unused input. The resulting unbalanced configuration has a lower bandwidth and a lower quality factor [1].

Pseudo-Differential TIAs. Instead of a differential voltage amplifier, we can use two matched single-ended voltage amplifiers, an arrangement known as a *pseudo-differential amplifier*. Whereas the differential and the pseudo-differential amplifier amplify the differential mode in the same way, $(v_{OP}-v_{ON}) = A_0(v_{IP}-v_{IN})$, they treat the common mode differently. The output common-mode voltage of the differential amplifier is constant, as we know, but the output common-mode voltage of the pseudo-differential amplifier does depend on the input voltages. In other words, the differential amplifier rejects the common mode, whereas the pseudo-differential amplifier does not.

A balanced *pseudo-differential TIA* with two photodetectors is shown in Fig. 7.7. Its transimpedance is the same as of the balanced differential TIA. Similarly, its symmetry and impedance balance result in a good power supply and substrate noise rejection. However, depending on the optical inputs, this pseudo-differential TIA outputs a common-mode signal, which must be suppressed in a subsequent stage. For a more complete circuit example, see Fig. 9.6 in Chapter 9.

An unbalanced pseudo-differential TIA with one photodetector is shown in Fig. 7.8. This circuit consists of a single-ended main TIA and a matched replica TIA (a.k.a. dummy TIA) [8, 27–29]. Only the v_{ON} output swings while

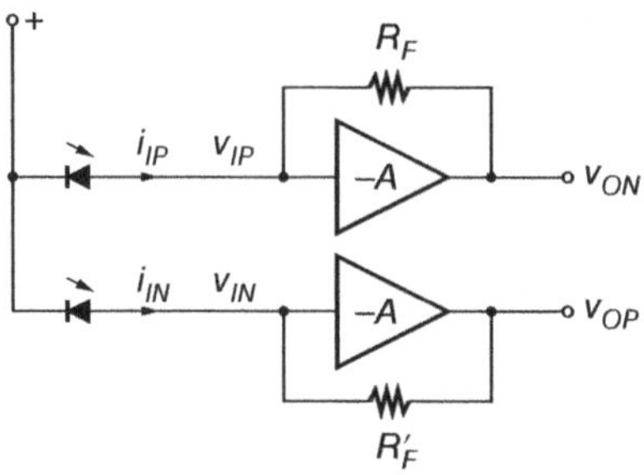

Figure 7.7 Balanced pseudo-differential TIA with two photodetectors.

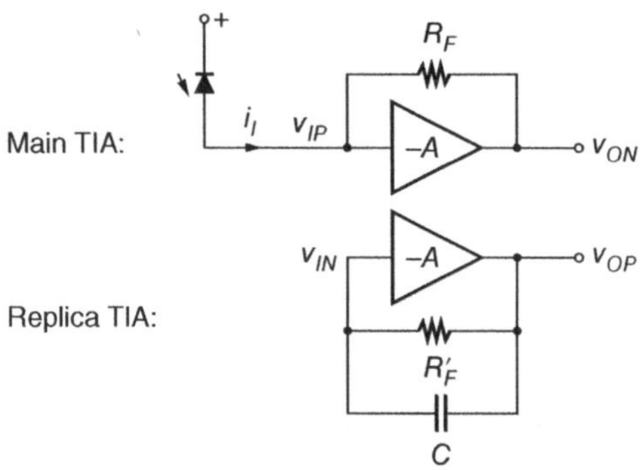

Figure 7.8 Unbalanced pseudo-differential TIA with one photodetector.

the v_{OP} output remains constant. The replica TIA simply produces a DC voltage that tracks the dark level of the signal voltage over process, voltage, and temperature.

Because the replica TIA only provides a DC output voltage, its bandwidth can be reduced with a large feedback capacitor C. This has the advantage that most of the replica TIA's noise is filtered out [8, 27, 28]. Thus, the input-referred noise current of this unbalanced pseudo-differential TIA approaches that of a single-ended TIA. The power consumption of the unbalanced pseudo-differential TIA can be reduced by scaling the bias currents in the replica TIA down (e.g., by one half) relative to the main TIA [8, 28].

From the unbalanced pseudo-differential TIA in Fig. 7.8 it is only a small step to the circuit in Fig. 7.9, which consists of a single-ended TIA followed by an RC low-pass filter to generate a reference voltage [5, 7, 30]. Again, only the v_{ON} output swings while the v_{OP} output remains essentially constant and tracks the average value of v_{ON}. By eliminating the replica TIA, the circuit in Fig. 7.9 consumes less power, however, a large, possibly external, capacitor is needed to prevent the reference voltage from drifting during long runs of zeros or ones.

Output Signal Waveforms. It is instructive to compare the waveforms of the output signals v_{OP} and v_{ON} of the various TIA topologies (see Fig. 7.10). For the

Figure 7.9 Single-ended TIA with low-pass reference voltage generator.

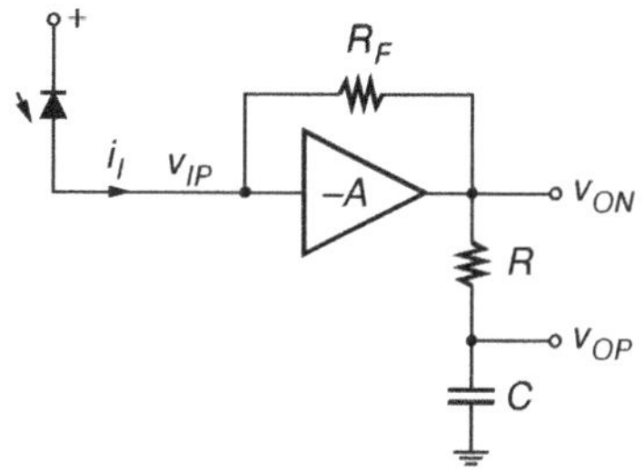

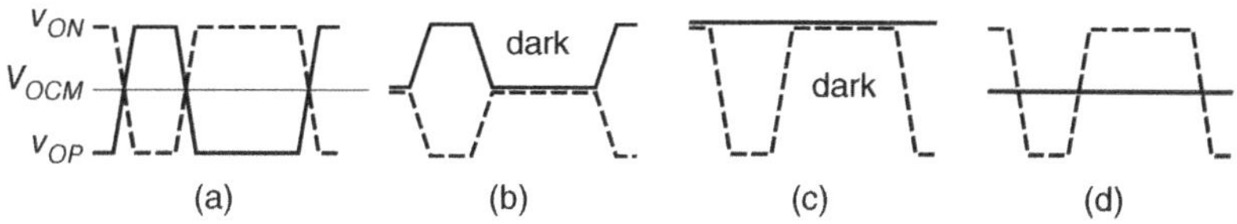

Figure 7.10 Output signals of (a) differential TIA with two photodetectors (complementary input signals), (b) differential TIA with one photodetector (no offset control), (c) pseudo-differential TIA with one photodetector (no offset control), and (d) single-ended TIA with low-pass reference voltage generator.

differential TIA with two photodetectors and complementary optical signals, the differential input current, $i_{IP} - i_{IN}$, and the differential output voltage, $v_{OP} - v_{ON}$, both swing symmetrically about zero. With the common-mode constraint $\frac{1}{2}(v_{OP} + v_{ON}) = V_{OCM}$, we find that the individual outputs swing symmetrically about the common-mode voltage V_{OCM}, as shown in Fig. 7.10(a). For the (balanced or unbalanced) differential TIA with one photodetector the situation is quite different. When the photodetector is dark, the input current is zero and hence v_{OP} and v_{ON} are both equal to the output common-mode voltage V_{OCM}. When the detector is illuminated, a current starts to flow into R_F, forcing v_{ON} (dashed line) to decrease. Meanwhile, v_{OP} (solid line) has to increase by the same amount to keep the output common-mode voltage at V_{OCM}, as illustrated in Fig. 7.10(b). For the pseudo-differential TIA with one photodetector, only the v_{ON} signal swings while v_{OP} remains at the dark level, as shown in Fig. 7.10(c). Finally, for the single-ended TIA with a low-pass reference voltage generator, shown in Fig. 7.10(d), only the v_{ON} signal swings while v_{OP} remains at the average level.

7.3 TIA with DC Input Current Control

Offset Control. The signals in Figs 7.10(b) and (c) suffer from a large systematic output offset voltage. In Fig. 7.10(b) this offset voltage limits the maximum possible output signal swing: v_{OP} can only swing in the upper half and v_{ON} can only swing in the lower half of the supply range.

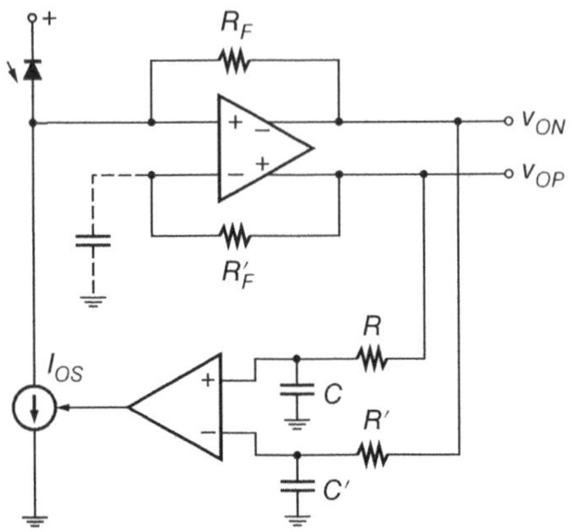

Figure 7.11 Differential TIA with automatic offset control.

Figure 7.11 shows the differential TIA from Fig. 7.5 (or Fig. 7.6) enhanced with an *automatic offset control* (AOC) circuit. The feedback control circuit determines the output offset voltage by subtracting the time-averaged (RC low-pass filtered) values of the two output signals and, in response to this difference, controls the DC current source I_{OS} (usually implemented with a MOSFET) until the output offset voltage becomes zero. In steady state, I_{OS} equals the average photocurrent from the detector and the current into R_F swings symmetrically about zero. With the offset control circuit added, the outputs swing as shown in Fig. 7.10(a) rather than in Fig. 7.10(b).

The control loop in Fig. 7.11 must be sufficiently slow to prevent the output signals from drifting when long runs of zeros or ones are received. This means that large values for R, R', C, and C' are needed, possibly requiring external capacitors. To reduce the size of the capacitors, the two grounded capacitors, C and C', can be combined into one floating capacitor half their size, resulting in the same cutoff frequency (for the differential mode) but requiring only a quarter of the chip area. Alternatively, Miller multiplication may be exploited to implement large effective capacitances with small on-chip capacitors [8, 28]. To avoid large resistors and improve the accuracy of the offset compensation loop, the RC low-pass filters can be replaced by a g_mC integrator [15]. To obtain a faster response to signal increases, the RC low-pass filters can be replaced by leaky peak detectors [31]. Finally, the AOC loop can be implemented in the digital domain with a current DAC providing the current I_{OS}.

Care must be taken to minimize the capacitive load presented by the current source I_{OS} to the TIA input and to minimize noise injection from the AOC circuit [8, 15]. To that end, we may consider moving the current source in Fig. 7.11 from the active input to the unused (and AC grounded) input and reversing its polarity to $-I_{OS}$ [26, 32]. Although this arrangement removes the output offset voltage as expected, it has the drawback that the amplifier's average input common-mode voltage becomes dependent on the received power

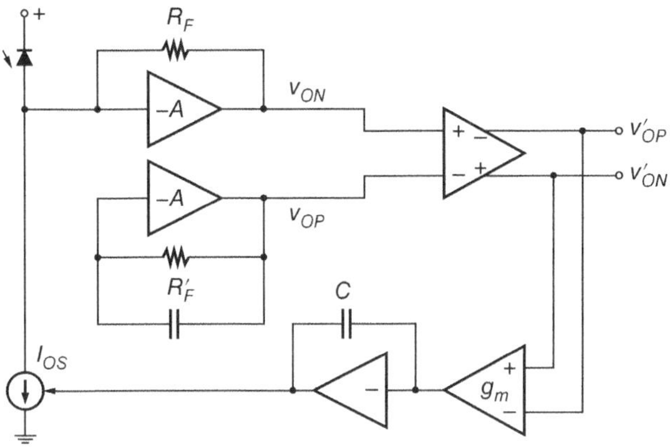

Figure 7.12 Pseudo-differential TIA with post amplifier and automatic offset control.

level. A large optical power level may push the common-mode voltage out of its permitted operating range.

Figure 7.12 shown another example of an offset control loop [8, 28]. Here, the pseudo-differential TIA from Fig. 7.8 is followed by a post amplifier, which subtracts the signals v_{OP} and v_{ON} and produces the complementary output signals v'_{OP} and v'_{ON} (i.e., the common-mode voltage $(v'_{OP} + v'_{ON})/2$ is constant). The control loop, which is implemented with a $g_m C$ integrator, forces the average output voltage of the main TIA to be identical to that of the replica TIA. Because the input current of the replica TIA is zero, the average input current of the main TIA (after I_{OS} is subtracted) is forced to zero as well. Thus, like in the case of the differential TIA, I_{OS} ends up matching the average photocurrent. With the AOC circuit added, the outputs of the pseudo-differential TIA section swing as shown in Fig. 7.10(d) rather than in Fig. 7.10(c).

The DC current I_{OS} produced by the AOC circuits in Figs 7.11 and 7.12 is proportional to the average received optical power. Therefore, a copy of this current may be used for a received signal strength indicator (RSSI) [8], a loss of signal (LOS) alarm [8], or to control the transimpedance of the TIA [33] (cf. Section 7.4).

DC Overload Control. In coherent receivers with balanced detection, the strong local-oscillator laser causes large DC currents in both photodetectors (cf. Section 3.5). While these two DC currents are essentially equal (a common-mode current) and thus ideally do not cause an output offset error, they can easily overload the TIA. For the differential TIA with two photodetectors in Fig. 7.4, a common-mode current from the detectors moves the common-mode input voltage up and for large currents it may go out of its permitted range. For the pseudo-differential TIA with two photodetectors in Fig. 7.7, a common-mode current from the detectors causes the common-mode output voltage to move down and for large currents it may cause the outputs to saturate.

To avoid these problems, control loops that remove the DC input currents from the photodetectors can be used. The basic idea is the same as the one discussed earlier for offset control, except that the primary goal now is to remove the input common-mode current (rather than the output offset voltage). A replica circuit similar to that in Fig. 7.12 can be used to determine each DC input current and to force it to zero by subtracting I_{OS} (see Fig. 9.7 in Chapter 9).

DC input current control can also be applied to single-ended TIAs with one photodetector. The goal now is to make the DC component of the output signal independent of the received power level. The average voltage drop across R_F can be taken as the input to the error amplifier thus forcing the average current into the TIA to zero [33].

7.4 TIA with Adaptive Transimpedance

How does the basic shunt-feedback TIA respond to an increasing input signal? Starting with a small input signal, the output voltage swing v_O^{pp} increases proportional to the input current swing, i_I^{pp}. Eventually, the voltage amplifier starts to compress and the TIA no longer responds linearly. This point defines the maximum input current for linear operation, i_{lin}^{pp} (cf. Section 5.3). Continuing to increase the input signal swing, the output swing ultimately reaches the overload limit of the voltage amplifier, $v_{O,ovl}^{pp}$, resulting in large signal distortions and jitter. This point defines the input overload current of the TIA, $i_{ovl}^{pp} = v_{O,ovl}^{pp}/R_T$ (cf. Section 6.2).

The maximum input current for linear operation as well as the input overload current can be pushed to higher values by reducing the transimpedance for large input currents. The block diagram of a TIA with *adaptive transimpedance* or *automatic gain control* (AGC) is shown in Fig. 7.13(a). Its response to an

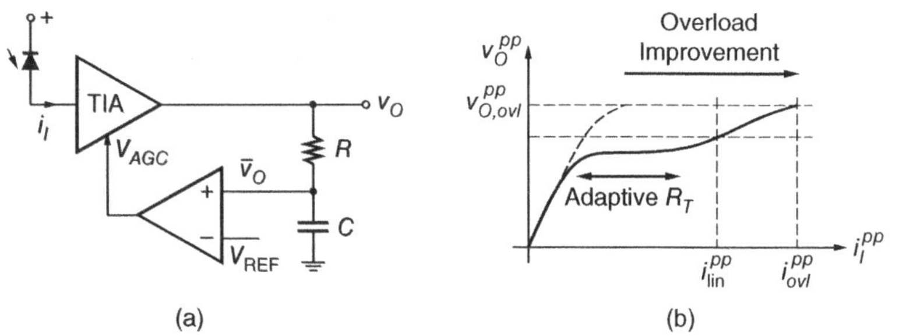

Figure 7.13 TIA with adaptive transimpedance: (a) block diagram and (b) output swing versus input swing for fixed (dashed line) and adaptive (solid line) transimpedance.

increasing input signal is plotted in Fig. 7.13(b) and compared to that of a TIA with fixed transimpedance. Starting with a small input signal, the transimpedance is set to its highest value and the output swing increases rapidly with input swing. When the output swing approaches the nonlinear limit, a control loop reduces the transimpedance such that the output swing remains below that limit. Continuing to increase the input signal swing, the output swing initially remains relatively constant up to the point where the loop cannot reduce the transimpedance any further. Then, the output swing starts to increase again, eventually resulting in nonlinear distortions and overload. However, the adaptive TIA reaches these limits for significantly larger input signals than the fixed TIA.

A large linear range is important in applications that perform linear signal processing, such as equalization and data conversion, and in applications that use higher-order modulation, such as 4-PAM and QAM. Wide dynamic range TIAs also are important in optical storage systems [34–36] and in laser radar systems [37, 38].

Control Loop. The control loop shown in Fig. 7.13(a) uses an *RC* low-pass filter to extract the average value $\overline{v}_O$ from the output signal [4, 39]. Given a DC-balanced signal with the lower level at a fixed voltage, the average value $\overline{v}_O$ is proportional to the swing v_O^{pp}. An error amplifier subtracts $\overline{v}_O$ from the reference voltage V_{REF} and outputs the AGC voltage V_{AGC} that adjusts the transimpedance. The loop reaches steady state when $\overline{v}_O \approx V_{\text{REF}}$. The *RC* filter's bandwidth must be sufficiently small to prevent the gain from drifting when long runs of zeros or ones are received. Loops with bandwidths as low as 3 Hz have been reported [40].

If the TIA also contains an AOC circuit, as discussed in Section 7.3, the offset control voltage from this circuit, which is a good measure of the received signal strength, can be reused to control the transimpedance, eliminating the need to generate a separate gain control voltage [33].

The required narrow loop bandwidth prevents the gain control mechanism from responding quickly to changes in the input signal swing. Thus, for an abrupt signal increase, the TIA may temporarily overload. To avoid this, the signal swing can be determined with two leaky peak detectors instead of the *RC* low-pass filter [7, 40–42]. This approach simultaneously gives a quick response to signal increases and a slow response to signal decreases. The latter feature also keeps the drift due to long runs of zeros or ones small. When discussing burst-mode TIAs in Section 8.4, we come back to the use of peak detectors for gain control.

The transimpedance of a TIA can be controlled by (i) varying the feedback resistor, (ii) shunting a variable amount of photodetector current away from the input, or (iii) attenuating the photodetector current by a variable amount. We discuss these three methods next.

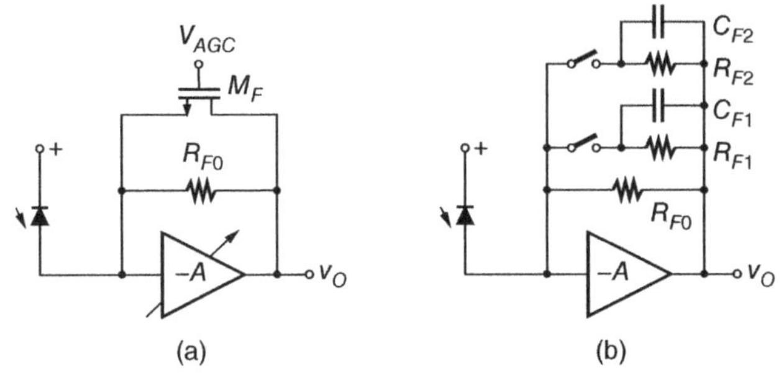

Figure 7.14 TIA with variable feedback resistor: (a) continuous control and (b) discrete control.

Variable Feedback Resistor. The transimpedance R_T can be controlled by means of the feedback resistor R_F, as is evident from the relationship

$$R_T = \frac{A_0 R_F}{A_0 + 1}. \tag{7.9}$$

The feedback resistor may be varied *continuously* in response to a control voltage (V_{AGC}) or *discretely* in response to a digital code. Figure 7.14(a) shows a TIA with continuous control, where the feedback resistor is implemented with a MOSFET, M_F, operating in the linear regime [4, 29, 43]. A fixed resistor, R_{F0}, in parallel to M_F improves the linearity and limits the maximum resistance. Figure 7.14(b) shows a TIA with discrete control, where the maximum feedback resistor, R_{F0}, is reduced by switching additional resistors, R_{F1}, R_{F2}, etc., in parallel to it [34, 44]. The latter scheme typically allows for a wider transimpedance tuning range but requires a digital controller.

Dynamic Range Improvement. As we discussed in Section 6.2, the input overload current is inversely proportional to the feedback resistor, $i_{ovl}^{pp} \propto 1/R_F$ (assuming the voltage amplifier is able to sink i_{ovl}^{pp}). The same is true for the maximum input current for linear operation, $i_{\text{lin}}^{pp} \propto 1/R_F$. The sensitivity, the lower end of the dynamic range, is proportional to the input-referred rms noise current, $i_{\text{sens}}^{pp} \propto i_{n,TIA}^{rms}$. For small values of R_F, when the feedback-resistor noise dominates, the sensitivity varies with R_F as $i_{\text{sens}}^{pp} \propto 1/\sqrt{R_F}$; for large values of R_F, when the amplifier front-end noise dominates, the sensitivity becomes independent of R_F. The optical overload and sensitivity limits following from this analysis are plotted in Fig. 7.15 as a function of R_F on a log–log scale [45].

What happens when the feedback resistor is made adaptive? For strong optical signals, R_F is reduced to prevent the high input currents from overloading the TIA. For weak optical signals, R_F is increased to reduce the noise contributed by this resistor. Figure 7.15 shows how an adaptive feedback resistor extends the dynamic range over what can be achieved with any particular fixed feedback resistor.

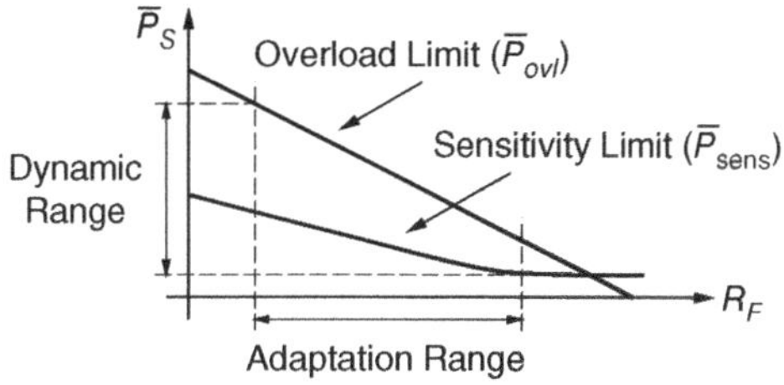

Figure 7.15 Extension of the dynamic range with an adaptive feedback resistor.

Stability and Bandwidth Control. An important consideration for TIAs with variable transimpedance is their stability. From Eq. (6.11) we see that if we reduce R_F, while keeping the voltage amplifier (T_A and A_0) fixed, the quality factor Q increases, pushing the TIA toward instability. This can be explained with the help of Fig. 6.4: If we reduce R_F, the open-loop low-frequency pole, $f_I = 1/(2\pi R_F C_T)$, speeds up, while the open-loop high-frequency pole, $f_A = 1/(2\pi T_A)$, and the loop gain, A_0, remain constant. Consequently, the open-loop pole spacing and thus the stability is reduced (cf. Eq. (6.12)).

Another concern is that the bandwidth of the TIA increases when we reduce R_F while keeping A_0 fixed. This effect can be seen from Eqs. (6.4) or (6.13). Increasing the open-loop low-frequency pole while keeping the DC loop gain constant increases the open-loop unity-gain frequency. In optically amplified transmission systems, an increase in bandwidth causes the receiver to pick up more noise from the optical amplifiers, resulting in a degraded SNR and higher bit error rate, as explained in Section 4.6 (cf. Eq. (4.53)). In unamplified transmission systems, an increase in bandwidth at high power levels usually is not an issue. The bandwidth increase causes the receiver to generate more noise, but the received signal is much stronger and the resulting SNR is sufficiently high.

One way to keep the TIA's bandwidth approximately constant is to reduce the gain of the voltage amplifier proportional to the feedback resistor, $A_0 \propto R_F$ [4, 44, 46, 47]. This approach is indicated in Fig. 7.14(a) by the arrow through the voltage amplifier. Varying A_0 proportional to R_F, results in the bandwidth of the voltage amplifier varying inversely proportional to R_F, $f_A \propto 1/A_0 \propto 1/R_F$, assuming $A_0 f_A$ is constant. As a result, f_I and f_A both become inversely proportional to R_F and the open-loop pole spacing remains fixed. Moreover, the open-loop unity-gain frequency becomes approximately independent of R_F. For the closed-loop response this means that the bandwidth remains approximately constant (cf. Eq. (I.118)) while the quality factor Q decreases (stability improves) with decreasing R_F (cf. Eq. (I.119)). Circuit techniques to obtain good tracking between A_0 and R_F have been proposed [44]. [→ Problem 7.4.]

For our 10-Gb/s TIA example in Table 6.1, reducing R_F by a factor four from 600 to 150 Ω and cutting A_0 by a factor four from 5.0 to 1.25 results in $R_T = 83$ Ω, $BW_{3dB} = 8.8$ GHz and $Q = 0.43$. We observe a small increase in bandwidth (up from 6.85 GHz) and a decrease in the quality factor (down from 0.71).

Another technique to keep the bandwidth constant is to add feedback capacitors, C_{F1}, C_{F2}, etc., when R_F is reduced [34, 43]. This approach is shown in Fig. 7.14(b). We can estimate the necessary C_F from the requirement that the open-loop low-frequency time constant including the Miller contribution from C_F, given by $R_F[C_T + (A_0 + 1)C_F]$, remains constant (cf. Eq. (I.58)). The feedback capacitors added to keep the bandwidth constant also provide good stability for all transimpedance settings. In this scheme, the voltage gain remains at its full value, keeping the input voltage swing small.

For our 10-Gb/s TIA example, reducing R_F by a factor four from 600 to 150 Ω and adding the feedback capacitor $C_F = 0.15$ pF (note that $0.3\text{ pF} + [5 + 1] \times 0.15\text{ pF} = 4 \times 0.3\text{ pF}$) results in $R_T = 125$ Ω, $BW_{3\text{dB}} = 5.9$ GHz and $Q = 0.43$. The slightly smaller value $C_F = 0.125$ pF would have given a more constant bandwidth.

Although Fig. 7.14 illustrates the gain compensation method for a TIA with continuous control and the feedback-capacitance compensation method for a TIA with discrete control, the methods do not need to be paired in this way. For example, in [43] R_F is controlled in a continuous manner and discrete feedback capacitors are switched in when the control voltage crosses predefined thresholds while in [4] R_F and the voltage-amplifier gain are controlled in discrete steps.

Equation (7.9) suggests that the transimpedance also could be controlled by varying the gain A_0 while keeping R_F fixed [48]. However, this method has the problem that very low gains, $A_0 \ll 1$, are needed to reduce the transimpedance to the values needed in practice. These very low gains lead to very small bandwidths (high input impedances) and large input voltage swings that far exceed the output voltage swing. For our 10-Gb/s TIA example, to lower the transimpedance from 500 to 125 Ω, we would have to cut the gain from 5.0 to 0.26, resulting in a bandwidth of only 1.1 GHz.

Variable Input Shunt Resistor. An alternative to the TIA with variable feedback resistor is the TIA with variable input shunt resistor, R_S. This approach is shown in Fig. 7.16. To prevent the input shunt resistor from disturbing the operating point of the TIA, it must connect to a node with a voltage that matches the quiescent voltage of the TIA input (V_1). The shunt resistor can connect to an AC

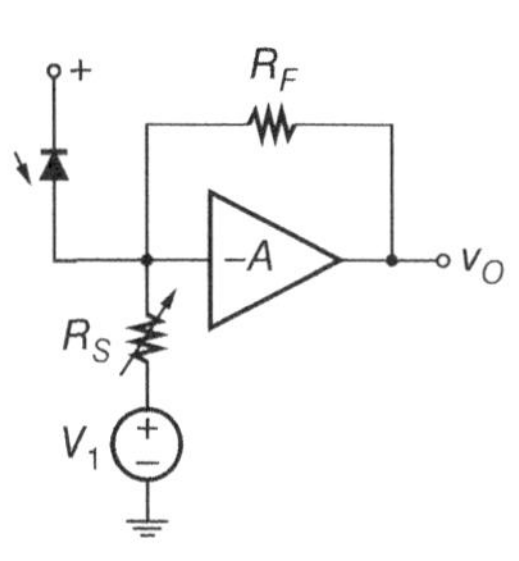

Figure 7.16 TIA with variable input shunt resistor.

ground if a parallel shunt path for the DC current is provided [33]. This scheme extends the dynamic range of the TIA as follows: For large optical signals, R_S is reduced to divert some of the photodetector current to ground, thus preventing the input current from overloading the TIA. For weak optical signals, R_S is increased such that almost all photodetector current flows into the TIA. In the latter case, the noise contribution from the shunt resistor and the associated sensitivity degradation are negligible.

The addition of R_S modifies the low-frequency transimpedance of the TIA as follows (cf. Eq. (I.122)) [48]

$$R_T = \frac{A_0 R_F}{A_0 + 1 + R_F/R_S}, \tag{7.10}$$

confirming that R_S can be used to control the transimpedance. Varying the shunt resistor has the advantage over varying the feedback resistor that it is easier to maintain stability and constant bandwidth. More specifically, if we reduce R_S, the open-loop low-frequency pole, which is given by $(R_S + R_F)/(R_S R_F C_T)$, speeds up *and* the loop gain, which is given by $A_0 R_S/(R_S + R_F)$, decreases by the same amount. Thus, for a loop gain that is significantly larger than one, the closed-loop response remains approximately constant (cf. Eq. (I.121)). When R_S becomes so small that the loop gain drops below unity, stability is guaranteed by virtue of that low loop gain. [→ Problem 7.5.]

For our 10-Gb/s TIA example in Table 6.1, we can lower the transimpedance from 500 to 125 Ω by adding the input shunt resistor $R_S = 33.3\ \Omega$, resulting in $BW_{3dB} = 9.6$ GHz and $Q = 0.53$. In this example, the loop gain dropped by a factor 19 from 5.0 to 0.26 and the open-loop low-frequency pole sped up by the same factor 19, moving past the open-loop high-frequency pole. A small feedback capacitor, $C_F \approx 45$ fF, could be added to bring the bandwidth back to 6.85 GHz.

Another advantage of the shunt-resistor control is that the voltage amplifier does not need to sink the whole overload current from the photodetector: much of this current is picked up by the shunt resistor. A downside is that a relatively large FET is needed to realize the low value of R_S. Such an FET has a correspondingly large parasitic capacitance that adds substantially to C_T.

The methods of variable feedback resistor and variable shunt resistor can also be combined to extend the control range. For example, in [29] the feedback resistor is reduced first and after this resistor has reached its lowest value, a shunt mechanism similar to Fig. 7.16 is turned on to reduce the transimpedance further.

Implementation of Voltage-Controlled Resistors. The variable resistor in the aforementioned TIAs is usually implemented with an FET operating in the linear regime. Over what range can the resistance of such an FET resistor be tuned? From the FET model for the linear regime it follows that the drain–source resistance varies approximately inversely proportional to the overdrive voltage, $R \propto 1/(V_{GS} - V_{TH})$. Thus, the resistance tuning range is given by the maximum and minimum possible overdrive voltages. The maximum overdrive voltage is limited by the supply voltage and headroom considerations whereas the minimum overdrive voltage is limited by the signal voltage from drain to source and the tolerable nonlinear distortions [29]. As a rule, the overdrive voltage needs to be at least a factor of two more than the drain–source voltage to limit the modulation of the resistance by the signal voltage. For example, for a maximum drain–source voltage of 0.1 V (a swing of 0.2 V_{pp}) and a supply voltage of 4 V, the overdrive voltage range is about 0.2 to 2 V resulting in a resistance ratio of 10:1.

To increase the resistance tuning range beyond 10:1, two (or more) FETs with different sizes can be used [29]. Drain and source of the two FETs are connected in parallel and the gates are controlled individually in a sequential manner. For example, to cover the 100:1 range from 20 kΩ to 200 Ω, the gate voltage of the small FET is first swept from low to high while the large FET is turned off, tuning the resistance from 20 to 2 kΩ; then, the gate voltage of the 10× larger FET is swept from low to high while the small FET is kept on, tuning the resistance from 2 kΩ to 200 Ω.

A fixed resistor may be connected in parallel to the voltage-controlled FET to limit the maximum resistance for low control voltages and to linearize the channel resistance for small overdrive voltages.

Variable Input Current Attenuator. Another way to increase the TIA's input overload current is to insert a *current attenuator* in between the photodetector and the TIA [37, 38, 49]. This approach is sketched in Fig. 7.17. The AC current from the photodetector, i_i, is split into two parts, one of which, i_a, flows into the TIA and the other, $i_i - i_a$, is dumped into the power rail. The split is accomplished with the differential pair Q_1 and Q_1', which is controlled by the voltage difference $V_{AGC} - V_{\text{REF}}$. For $V_{AGC} \gg V_{\text{REF}}$, Q_1 is fully turned on and virtually all signal current flows into the TIA, $i_a = i_i$ (ignoring R_E, R_C, R_C' for now); for $V_{AGC} = V_{\text{REF}}$, both transistors are equally turned on and the signal current is split in half, $i_a = \frac{1}{2} i_i$; for $V_{AGC} \ll V_{\text{REF}}$, Q_1' is fully turned on and virtually no signal current flows into the TIA, $i_a = 0$.

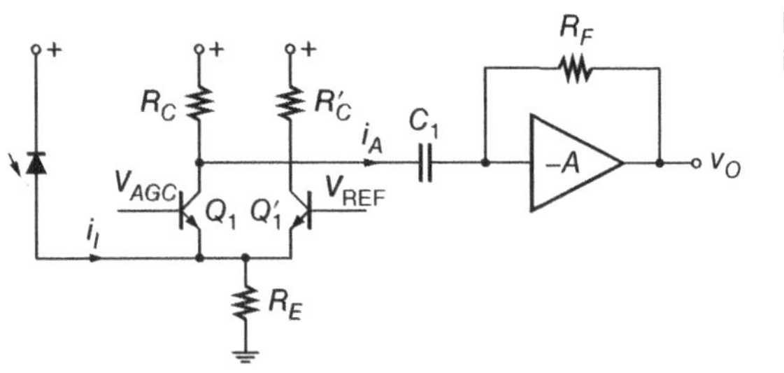

Figure 7.17 TIA with variable input current attenuator.

The bias current through R_E must be chosen significantly larger than the maximum photodetector current to prevent the transistors from current starving. Moreover, the impedance seen into the common-emitter node of Q_1 and Q_1' must be low compared to the emitter resistor, $1/(g_{m1} + g_{m1}') \ll R_E$, such that almost all signal current from the photodetector flows into the differential pair. Similarly, the input impedance of the shunt-feedback TIA must be low compared to R_C, such that almost all signal current from the attenuator flows into the TIA. With these assumptions and a high current gain for Q_1 and Q_1', the transimpedance can be written as

$$R_T = \frac{1}{1 + \exp[(V_{\mathrm{REF}} - V_{AGC})/V_T]} \cdot \frac{A_0 R_F}{A_0 + 1}, \tag{7.11}$$

where V_T is the thermal voltage. The first factor is the voltage-controlled attenuation and the second factor is the familiar transimpedance of the TIA without attenuator. To derive the first factor, we write the attenuation as $i_a/i_i = g_{m1}/(g_{m1} + g_{m1}')$ and use $g_{m1} = I_{C1}/V_T$, $g_{m1}' = I_{C1}'/V_T$, and $I_{C1} = (I_{C1} + I_{C1}')/(1 + \exp[-(V_{AGC} - V_{\mathrm{REF}})/V_T])$ [50].

One drawback of the simple attenuator in Fig. 7.17 is that the DC voltage drop across R_C varies with the control voltage. To prevent the attenuator from disturbing the TIA's operating point, AC coupling capacitor C_1 is inserted. A more elegant solution is to connect a second differential pair, controlled by $-(V_{AGC} - V_{\mathrm{REF}})$, in parallel to the outputs of the first one such that the bias currents through R_C and R_C' remain constant regardless of the control voltage [37, 38]. DC coupling between the attenuator and the TIA is then possible.

When used with a differential-input TIA, the collectors of Q_1 and Q_1' can be connected to the inverting and noninverting inputs of the TIA, respectively [37, 38]. In this configuration, the transimpedance is zero for $V_{AGC} = V_{\mathrm{REF}}$, that is, when the attenuator splits the photodetector current 1:1.

The presence of an input current attenuator increases the input-referred noise current by the amount of the attenuation. At the same time, the input current attenuator may reduce the total capacitance at the voltage amplifier input, C_T, by decoupling the photodetector capacitance. The latter effect decreases the input-referred noise current of the TIA. For large photodetector capacitances, the overall noise performance with and without the attenuator stays about the same [37, 38, 49]. Stability and bandwidth variations are usually small in this control scheme because the output impedance of the attenuator is largely independent of the control voltage.

The circuit in Fig. 7.17 shares some properties with the TIA with common-base input stage, which we discuss in Section 7.5.

Nonlinear Networks. If linearity for large input signals is not a requirement, nonlinear elements, such as diodes and transistors, can be used to increase the

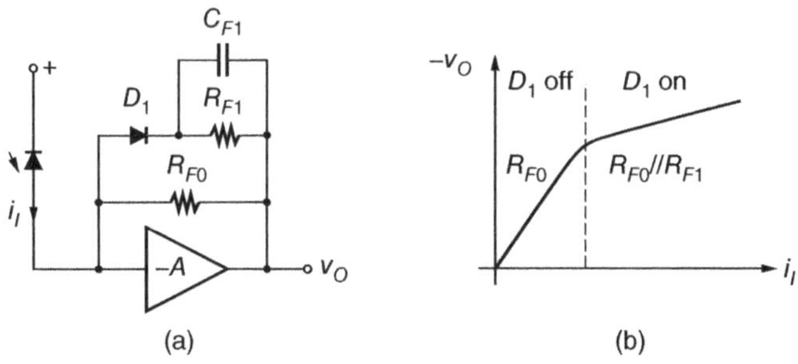

Figure 7.18 TIA with nonlinear feedback network: (a) circuit and (b) DC transfer function.

TIA's input overload current. This method avoids the need for a control loop and results in an instantaneous response to changes in signal strength.

Figure 7.18(a) shows an example where the linear feedback resistor has been replaced by a *nonlinear* feedback network [27]. For small input signals, diode D_1 is turned off and the transimpedance is determined by R_{F0}. For input signals large enough to forward biases D_1, the feedback resistance reduces to $R_{F0} \parallel R_{F1}$, thus reducing the transimpedance and preventing the TIA from overloading. Capacitor C_{F1} has the same stabilizing function as in Fig. 7.14(b). The DC transfer function for this nonlinear TIA is depicted in Fig. 7.18(b). Note that this graph, unlike the one in Fig. 7.13(b), relates instantaneous signal values.

Figure 7.19(a) shows an example of a shunt-feedback TIA with a nonlinear input shunt network. For small input signals, diode D_1 is off and the transimpedance is set by R_F. For signals large enough to turn the diode on, some input current is shunted away from R_F, resulting in a logarithmic compression. The corresponding DC transfer function is depicted in Fig. 7.19(b). The breaking point between the linear and logarithmic regime can be controlled with the voltage source V_1. More sophisticated variations of this idea using bipolar transistors have been reported in [51, 52].

In Section 8.4, we discuss the use of nonlinear TIAs in burst-mode applications.

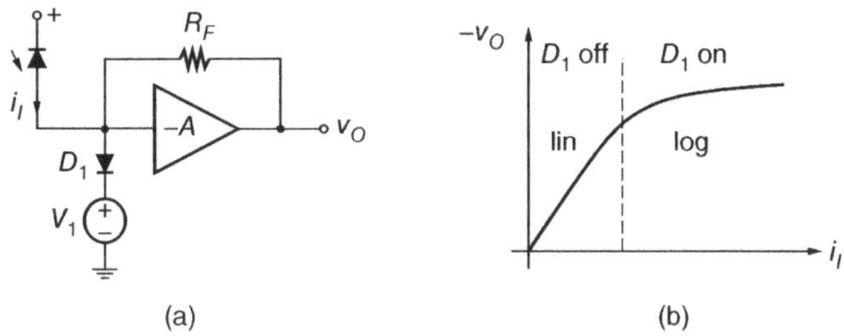

Figure 7.19 TIA with nonlinear input shunt network: (a) circuit and (b) DC transfer function.

7.5 Common-Base and Common-Gate TIAs

The bandwidth and stability of the shunt-feedback TIA depend on the photodetector capacitance, C_D. This dependence is made explicit by Eqs. (6.10) and (6.11) and can be explained as follows. If we increase C_T ($= C_D + C_I$), the open-loop low-frequency pole at $1/(R_F C_T)$ slows down, which reduces the TIA's bandwidth. If we decrease C_T, the open-loop low-frequency pole speeds up, which reduces the open-loop pole spacing and thus degrades the stability (cf. Fig. 6.4 and Eq. (6.12)). In applications where the TIA has to work with a variety of photodetectors exhibiting a range of capacitances, this dependence can be problematic.

One solution is to add a feedback capacitor C_F, such that the resulting Miller capacitance at the input dominates the photodetector capacitance: $(A_0 + 1)C_F \gg C_D$ [29]. Then, the bandwidth is mostly determined by R_F and C_F and the sensitivity to C_D is greatly reduced (cf. Eq. (I.58)). However, this approach also significantly reduces the maximum achievable transimpedance (cf. Eqs. (I.59) and (I.61)).

Another solution is to interpose a current buffer in the form of a common-base or common-gate stage in between the photodetector and the shunt-feedback TIA [53]. The input capacitance seen by the shunt-feedback TIA is now independent of C_D and, if the bandwidth of the current buffer is sufficiently high, the overall bandwidth is insensitive to C_D.

Before discussing the combination of a current buffer with a shunt-feedback TIA in more details, let us analyze the common-base (and common-gate) stage by itself. Not only is this a good warm-up exercise, but the common-base (and common-gate) stage can also serve as a simple TIA by itself. This type of TIA is known as a *feedforward TIA* [54] or an *open-loop TIA* [30]. Despite its simplicity, it is suitable for applications that do not require the lowest possible noise but demand low power consumption (e.g., optical interconnects) [54–57].

Common-Base (Common-Gate) Feedforward TIA. Figure 7.20 shows the schematic of a basic common-base feedforward TIA. The common-base stage presents a low input impedance, $R_I \approx 1/g_m$, to the photodetector, pushing the input pole to a high frequency. The AC current from the photodetector, i_i passes mostly unattenuated through Q_1 into the collector resistor R_C, where it is converted to a voltage, $v_o \approx R_C\, i_i$. Thus, the transimpedance of this TIA is approximately R_C. Note that Q_1 acts as a unity-gain current buffer (or current follower) with an input impedance that is much lower than its output impedance. The emitter resistor R_E supplies the bias current for Q_1. In some implementations, R_E is replaced by a current source transistor.

To calculate the frequency response, we make the following simplifying assumptions (for now): $R_E \gg 1/g_m$, $\beta \gg 1$, $g_o \ll 1/R_C$, and $R_b \ll 1/g_m$. We define the total capacitance at the input node $C_T = C_D + C_I$, where C_D is

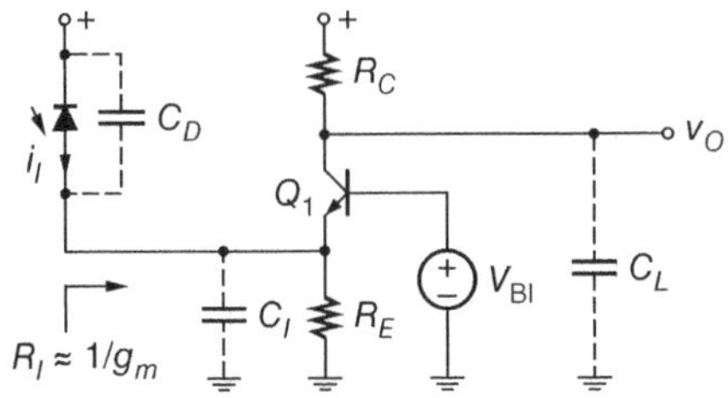

Figure 7.20 Basic common-base feedforward TIA.

the photodetector capacitance plus all parasitic capacitances in parallel to it and $C_I \approx C_{be}$. (Note that in contrast to the shunt-feedback TIA, C_I does *not* contain a Miller component.) We also define the total capacitance at the output node C_L as the sum of C_{bc}, the collector–substrate capacitance, and the external load capacitance. The resulting transfer function has two real poles, one corresponding to the input node and one corresponding to the output node:

$$Z_T(s) = \frac{R_T}{(1 + s/\omega_{p1})(1 + s/\omega_{p2})}, \tag{7.12}$$

where

$$R_T = R_C, \qquad \omega_{p1} = \frac{g_m}{C_T}, \qquad \omega_{p2} = \frac{1}{R_C C_L}. \tag{7.13}$$

The frequency response of Eq. (7.12) is shown in Fig. 7.21. Because both poles are real (and negative), stability is guaranteed for any value of C_T.

Equations (7.12) and (7.13) also hold for a common-gate feedforward TIA if g_m is replaced by $g_m + g_{mb}$, where g_{mb} is the substrate transconductance and a FET with grounded substrate is assumed.

Second-Order Effects. A finite emitter resistor R_E or a finite current gain β diverts some of the photodetector AC current away from R_C resulting in the lower transimpedance $R_T = R_C/(1 + g_m/R_E + 1/\beta)$, the lower input resistance $R_I = 1/(g_m + 1/R_E + g_m/\beta)$, and the faster input pole $\omega_{p1} = (g_m + 1/R_E + g_m/\beta)/C_T$. Conversely, a nonzero output conductance g_o increases the input resistance from $1/g_m$ to $1/g_m \cdot (1 + g_o R_C)$ [30].

A nonzero base resistance R_b increases the input impedance and for large base resistances the input impedance becomes inductive [58]. While an

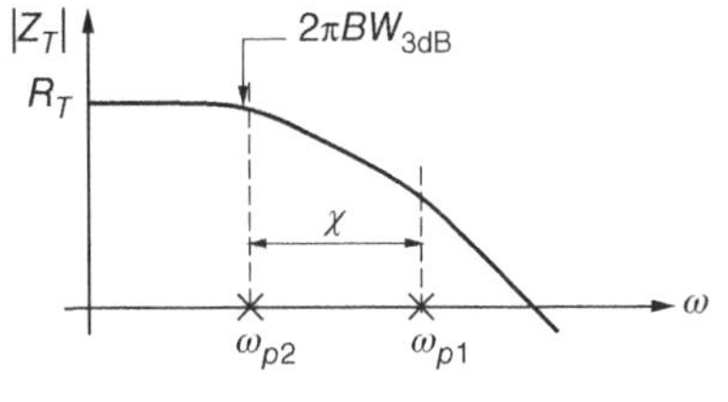

Figure 7.21 Frequency response of the common-base feedforward TIA.

inductive input speeds up the input pole, it may also degrade the TIA's stability (cf. Eq. (I.128)). [→ Problem 7.6.]

Bias Considerations. The DC output voltage of the circuit in Fig. 7.20 is $V_{CC} - I_C R_C$, where V_{CC} is the supply voltage and I_C is the collector current. To maximize the signal headroom, this output voltage should be set to about $V_{CC}/2$. Thus, the bias current essentially determines the transimpedance of the TIA, $R_T = R_C \approx V_{CC}/(2I_C)$, as well as the input resistance, $R_I = 1/g_m = V_T/I_C$. Note that the shunt-feedback TIA does not suffer from a similar trade off because R_F, which determines the transimpedance, does not need to support a bias current.

The photodetector DC current and the collector bias current in Fig. 7.20 must sum up to the bias current supplied by R_E. Thus, the collector bias current and g_m are reduced when a large optical signal is received. This undesirable effect can be eliminated by adding a control loop that removes the photodetector DC current or by making Q_1 part of a floating current mirror that imposes a constant collector current [59] (see Fig. 9.16 in Chapter 9 for an example).

3-dB Bandwidth. Let us determine the 3-dB bandwidth of the common-base TIA. In order to make the TIA's bandwidth insensitive to the photodetector capacitance, we choose the input pole ω_{p1} at a higher frequency than the output pole ω_{p2} [59]. The bandwidth is then, as indicated in Fig. 7.21, mostly determined by the dominant output pole, $BW_{3\text{dB}} \approx \omega_{p2}/(2\pi)$.

To determine the 3-dB bandwidth more accurately, we need to know the *pole spacing*, which is defined as $\chi = \omega_{p1}/\omega_{p2}$. Then, we can use Eq. (H.16) on p. 486 in Appendix H to calculate $BW_{3\text{dB}}$ as a function of ω_{p2} and χ. Table 7.2 gives numerical values for the 3-dB bandwidth (normalized to the dominant pole frequency) as a function of the pole spacing. Clearly, $BW_{3\text{dB}} \leq \omega_{p2}/(2\pi)$ for any value of χ and $BW_{3\text{dB}} = \omega_{p2}/(2\pi)$ for $\chi \to \infty$, in which case the response reduces to first order.

Table 7.2 Normalized 3-dB and noise bandwidths of Eq. (7.12) as a function of the pole spacing $\chi = \omega_{p1}/\omega_{p2}$ (for $\omega_{p1} \geq \omega_{p2}$).

Pole spacing χ	$BW_{3dB}/[\omega_{p2}/(2\pi)]$	BW_n/BW_{3dB}	BW_{n2}/BW_{3dB}
1	0.644	1.22	2.07
2	0.838	1.25	2.20
3	0.912	1.29	2.41
4	0.946	1.33	2.61
5	0.964	1.36	2.80
10	0.990	1.44	3.53
100	0.99990	1.56	7.76

The table also shows that a larger pole spacing makes the bandwidth less sensitive to variations in C_D. For example, if the pole spacing for the largest value of C_D is two, the TIA's bandwidth varies less than 16.2% (= 100% − 83.8%) over the full range of C_D. If the minimum pole spacing is three, the variation is less than 8.8%(= 100% − 91.2%).

Transimpedance Limit. To derive a transimpedance limit that can be compared to our earlier results for the shunt-feedback TIA, we write the gain-bandwidth product of the common-base stage as $A_0 f_A$. With the DC voltage gain $A_0 = g_m R_C$ and the stage bandwidth $f_A = \omega_{p2}/(2\pi)$, we can write the output pole as $\omega_{p2} = 2\pi A_0 f_A/(g_m R_C)$. Using $g_m = \omega_{p1} C_T$ from Eq. (7.13) and $\omega_{p1} = \chi\omega_{p2}$, we find $\omega_{p2} = \sqrt{2\pi A_0 f_A/(\chi R_C C_T)}$. Observing that $BW_{3\text{dB}} \le \omega_{p2}/(2\pi)$ and using $R_T = R_C$, the transimpedance limit becomes [12]

$$R_T \le \frac{A_0 f_A}{2\pi C_T \chi \cdot BW_{3\text{dB}}^2}, \tag{7.14}$$

where the equality is reached for large values of χ. Interestingly, for $\chi = 1$, that is, for two identical real poles, the transimpedance limit coincides with that of the shunt-feedback TIA, Eq. (6.14). As we increase the pole spacing to decouple the bandwidth from C_D, the maximum achievable transimpedance is linearly reduced.

Note that the inequality in Eq. (7.14) did not result from a stability condition (e.g., $Q \le Q_0$ or $\phi_m \ge \Phi_m$) as in the transimpedance limit for the shunt-feedback TIA, but from (lazily) bounding the bandwidth as $BW_{3\text{dB}} \le \omega_{p2}/(2\pi)$. When going through the trouble of calculating the exact bandwidth of Eq. (7.12) as a function of ω_{p2} and χ, the transimpedance can be written as an equality (see Eq. (I.130) on p. 525 for the expression). The exact solution is illustrated graphically in Fig. 7.22 (solid line) along with the bound of Eq. (7.14) (dashed line). The exact transimpedance reaches its maximum for $\chi = 1$ with a value equal to about 41% of the shunt-feedback limit in Eq. (6.14). In conclusion, given the same bandwidth and technology, our feedforward TIA model always has a lower transimpedance than the shunt-feedback TIA model, even in the best case when the two poles are at the same frequency. [→ Problem 7.7.]

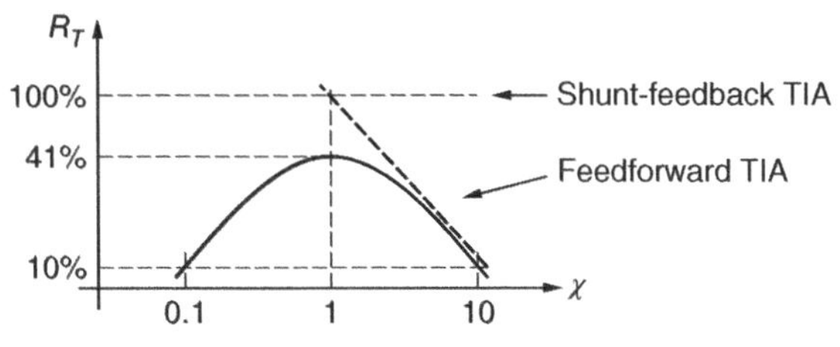

Figure 7.22 Transimpedance versus pole spacing of the feedforward TIA (on a log–log scale): exact solution Eq. (I.130) (solid line) and bound Eq. (7.4) (dashed line).

Input-Referred Noise Current PSD. To calculate the input-referred noise current PSD, we first need to find the input-referral functions for the individual device noise sources. The thermal noise current of R_E and the base shot-noise current of Q_1 contribute directly to the input-referred noise. The collector shot-noise current is input referred by $H_C^{-1}(s) = (sR_EC_T + 1)/(g_mR_E)$, where we neglected R_b. The thermal noise current of R_C is input referred by $H_{RC}^{-1}(s) = 1 + H_C^{-1}(s) \approx 1 + sC_T/g_m$, where we again neglected R_b and further assumed $g_mR_E \gg 1$. Finally, the thermal noise current of R_b is input referred by $H_{Rb}^{-1}(s) = sR_bC_D + R_b/R_E$. After referring all device noise sources to the input, we find the noise current PSD as

$$I_{n,TIA}^2(f) = \frac{4kT}{R_C} + \frac{4kT}{R_E} + \frac{2qI_C}{\beta} + 2qI_C\frac{(2\pi C_T)^2}{g_m^2}f^2 + 4kTR_b(2\pi C_D)^2f^2 + \cdots, \tag{7.15}$$

where the terms are due to the collector resistor, emitter resistor, and base current, collector current, and the base resistance, respectively, and the following simplifying assumptions were made: $(g_mR_E)^2 \gg \beta$ (the white noise due to the base current dominates that due to the collector current), $R_E \gg R_b$ (the white noise due to R_E dominates that due to R_b), and $g_mR_C \gg 2$ (the f^2 noise due to the collector current dominates that due to R_C).

Interestingly, this noise expression is very similar to that of the shunt-feedback TIA with a common-emitter input stage given in Eq. (6.48). The main differences are that R_C now plays the role of R_F and that there is a new noise term due to the bias resistor R_E (or, if a bias transistor is used, a new noise term due to that transistor). It is a well known fact that common-emitter, common-collector, and common-base configurations are all described by very similar input-referred noise expressions [49, 60].

The corresponding noise expression for the common-gate TIA essentially consists of the noise terms known from the shunt-feedback TIA with common-source input stage plus the noise current of the new biasing device (cf. Eq. (I.131)). [→ Problem 7.8.]

Noise Bandwidths and Input-Referred RMS Noise Current. To calculate the input-referred rms noise current we need to know the noise bandwidths, BW_n and BW_{n2}, in addition to the noise current PSD. For $\chi = 1$, the common-base TIA exhibits a critically damped second-order response and the corresponding noise bandwidths can be found in Table 4.7. For $\chi \to \infty$, the second-order response degenerates into a first-order response and the noise bandwidth for white noise becomes $BW_n = \pi/2 \cdot BW_{3dB}$ while the noise bandwidth for f^2 noise diverges like $BW_{n2} \to \sqrt[3]{3\pi/2 \cdot \chi} \cdot BW_{3dB}$. The noise bandwidths for an arbitrary pole spacing can be calculated with Eqs. (H.20) and (H.21) on p. 487 in Appendix H. Some noise bandwidth values (normalized to the 3-dB bandwidth) are listed in Table 7.2 and plotted in Fig. 7.23 (cf. Fig. H.3). Note

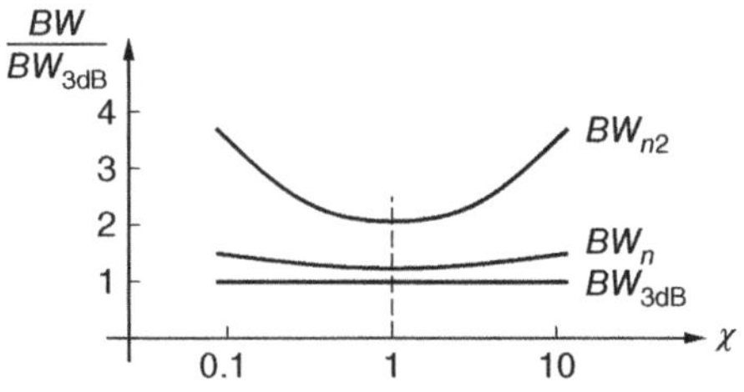

Figure 7.23 Noise bandwidths versus pole spacing of the feedforward TIA.

that both noise bandwidths increase with increasing pole spacing even though the 3-dB bandwidth is held fixed. For symmetry reasons, the noise bandwidths assume the same values for χ and $1/\chi$ and reach their minima for $\chi = 1$.

Relatively simple rms noise expressions for the common-base and the common-gate TIA can be found when using approximate noise bandwidths (see Eqs. (I.132) and (I.133) on p. 526 for the expressions). [→ Problem 7.9.]

Noise Optimization. What is the optimum transistor size that minimizes the input-referred rms noise current? Like in Section 6.4, we first choose a fixed collector-current density, I_C/A_E, (often one for which the BJT operates near its maximum f_T) and a fixed emitter width, W_E, (often the minimum width), then we vary the emitter length, L_E, to find the noise optimum. Under these constraints, the transistor size (emitter length) determines I_C, C_I, g_m, and R_b.

An important difference between the shunt-feedback and common-base TIA is that the latter has fewer degrees of freedom for noise optimization. If we want a certain TIA bandwidth and pole spacing, the location of the two poles are fixed and with a given C_D and f_T the transconductance g_m and thus the transistor size is already determined. There is no room left for noise optimization.

However, if we do not require a particular pole spacing, we can choose it such that it minimizes the input-referred noise current under the appropriate constraints. Because so many variables in the expression of the input-referred rms noise current depend on the pole spacing χ (when subjected to the constraints of a constant BW_{3dB}, $A_0 f_A$, and f_T), an analytical solution is difficult to find. Nevertheless, we can discuss the optimum in a qualitative way: If we neglect the base shot noise and the thermal noise due to R_b and R_C and assume that BW_n and BW_{n2} are independent of χ (a crude approximation), the noise optimum occurs for $C_I = C_D$, that is, the same condition that we found for the shunt-feedback TIA. Taking the base shot noise into account pulls this minimum toward a smaller value of C_I and taking the base resistance noise into account pulls the minimum toward a larger value of C_I, just like for the shunt-feedback TIA. From the transimpedance limit, illustrated in Fig. 7.22, we know that R_C reaches its maximum for $\chi = 1$. Thus, taking the thermal noise of R_C into account pulls the minimum toward $\chi = 1$. Moreover, BW_n and BW_{n2} both reach their minima for $\chi = 1$, as illustrated in Fig. 7.23, pulling the minimum further toward $\chi = 1$. From the interplay of these "forces," we may get a rough sense of where the optimum occurs.

Even if we choose the pole spacing that minimizes the noise, the common-base TIA is still noisier than the corresponding shunt-feedback TIA: First, the noise of R_C is larger than that of R_F due to the lower transimpedance limit. Second, R_E results in an additional noise term. Third, the noise bandwidths for a given 3-dB bandwidth are larger as a result of the common-base TIA's slower frequency-response roll-off. (For the combination of TIA and MA, the difference in noise bandwidths between the common-base and shunt-feedback TIA is smaller because the MA provides additional high-frequency filtering, resulting in a steeper roll-off.)

Numerical Example. Let us conclude the discussion of the common-base TIA with a numerical example (see Table 7.3). For simplicity, we choose the same BJT, operating at the same collector current, as in our 10-Gb/s shunt-feedback TIA example in Table 6.6. Given $g_m = 50$ mS, $C_D = 0.15$ pF, and $C_I = 0.1$ pF, the input resistance is $1/g_m = 20\ \Omega$ and the input pole frequency is $g_m/(2\pi C_T) = 31.8$ GHz. To obtain a TIA bandwidth of 6.85 GHz, as in our shunt-feedback example, the output pole frequency must be set to 7.17 GHz. (This can be found with Eq. (H.16).) The resulting pole spacing is 31.8 GHz/7.17 GHz = 4.44. With this large pole spacing, the TIA's bandwidth is rather insensitive to the photodetector capacitance: doubling C_D from 0.15 to 0.3 pF reduces the bandwidth by only 6%.

Given a technology that permits a gain-bandwidth product of 44 GHz, as in our 10-Gb/s shunt-feedback TIA example in Table 6.1, the voltage gain of

Table 7.3 Parameters and performance of our common-base TIA examples.

Parameter	Symbol	10 Gb/s	40 Gb/s
Transition frequency	f_T	80 GHz	318 GHz
Gain-bandwidth product	$A_0 f_A$	44 GHz	177 GHz
Detector capacitance incl. parasitics	C_D	0.15 pF	75 fF
Input capacitance	C_I	0.1 pF	50 fF
Transconductance	g_m	50 mS	100 mS
Load resistor	R_C	123 Ω	61.7 Ω
Input pole frequency	f_{p1}	31.8 GHz	127.4 GHz
Output pole frequency	f_{p2}	7.17 GHz	28.7 GHz
Pole spacing	χ	4.44	4.44
Transimpedance	R_T	123 Ω	61.7 Ω
Input impedance	R_I	20 Ω	10 Ω
3-dB bandwidth	BW_{3dB}	6.85 GHz	27.4 GHz

the common-base stage comes out as $A_0 = 44$ GHz/7.17 GHz = 6.14, and thus $R_C = A_0/g_m = 123\ \Omega$. Clearly, the 123-Ω transimpedance achieved with this topology is much inferior to the 500 Ω of our shunt-feedback example. According to the transimpedance limit Eq. (7.14), we would expect it to be lower by $\chi = 4.44$, which approximately agrees with our numbers. Therefore, to design a common-base TIA with a higher transimpedance, we must reduce the pole spacing, which can be done by reducing the size of Q_1.

To determine the rms noise current of our 10-Gb/s common-base TIA example, we first need to calculate the noise bandwidths with Eqs. (H.20) and (H.21), which yields $BW_n = 9.2$ GHz and $BW_{n2} = 18.5$ GHz (cf. Table 7.4). Note that these bandwidths are significantly larger than those of our shunt-feedback example in Table 6.6. Next, we calculate the input-referred rms noise current as

$$i_{n,TIA}^{rms} \approx \sqrt{(1{,}110\text{ nA})^2 + (195\text{ nA})^2 + (926\text{ nA})^2 + (1{,}110\text{ nA})^2} = 1{,}840\text{ nA},$$

where we neglected the noise due to the emitter resistor and the terms are due to the collector resistor, and base current, collector current, and the base resistance, respectively (same order as in Eq. (7.15)). Note that this noise current is significantly larger than that of our shunt-feedback example in Table 6.6.

To lower the input-referred rms noise current of our 10-Gb/s common-base TIA example, we can reduce the pole spacing, which can be done by reducing the size of Q_1. Shortening the emitter length by a factor two ($C_I = 50$ fF, $g_m = 25$ mS, $R_b = 80\ \Omega$) while keeping BW_{3dB}, $A_0 f_A$, and f_T fixed, reduces the pole spacing from 4.44 to 2.58. This modification increases the transimpedance (and R_C) from 123 to 228 Ω, reduces the noise bandwidths from $BW_n = 9.2$ to 8.7 GHz and from $BW_{n2} = 18.5$ to 15.9 GHz, and reduces the input-referred rms noise current from 1.84 to 1.71 μA. Moreover, the collector current is reduced from 1.3 to 0.65 mA. A potential downside of this optimization is that the TIA's bandwidth is now more sensitive to the photodetector capacitance.

Feedforward TIA with Differential Outputs. The common-base feedforward TIA shown in Fig. 7.20 is single-ended. In many applications, such as multichannel

Table 7.4 Parameters and noise performance of our common-base TIA examples.

Parameter	Symbol	10 Gb/s	40 Gb/s
DC current gain	β	100	100
Base resistance	R_b	40 Ω	20 Ω
Noise bandwidth (white noise)	BW_n	9.2 GHz	36.8 GHz
Noise bandwidth (f^2 noise)	BW_{n2}	18.5 GHz	73.9 GHz
Input-referred rms noise current	$i_{n,TIA}^{rms}$	1.84 μA	5.19 μA
Averaged inp.-ref. noise curr. density	$I_{n,TIA}^{avg}$	19.2 pA/$\sqrt{\text{Hz}}$	27.1 pA/$\sqrt{\text{Hz}}$

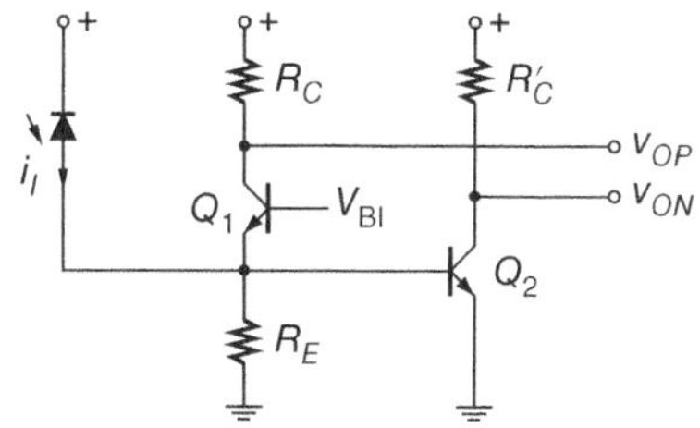

Figure 7.24 Feedforward TIA with differential outputs.

receivers, a differential topology with its superior noise immunity would be preferable (cf. Section 7.2). To this end, a pseudo-differential common-base TIA, consisting of two matched, but separate, single-ended common-base TIAs, can be used. For a circuit example of this approach, see Fig. 9.16 in Chapter 9.

Another way to obtain differential outputs with a feedforward TIA is shown in Fig. 7.24 [61]. Here, the common-base stage with Q_1 is accompanied by a common-emitter stage with Q_2. The transimpedance from i_I to the output of the common-base stage, v_{OP}, is R_C, as before. The transimpedance from i_I to the output of the common-emitter stage, v_{ON}, is $-g_{m2}/g_{m1} \cdot R'_C$. In order for the two output signals to be amplitude matched, the following condition must be satisfied: $R_C = g_{m2}/g_{m1} \cdot R'_C$.

The topology in Fig. 7.24 has another interesting property: some of the noise generated by Q_1 gets canceled in the differential output voltage, $v_{OP} - v_{ON}$. The noise current generator from the collector to the emitter of Q_1 produces noise voltages at the collector and the emitter nodes that are 180° out of phase. After the voltage inversion by Q_2, the noise voltages appear in phase at the two outputs, v_{OP} and v_{ON}. In order for the two noise voltages to be amplitude matched, the following condition must be satisfied: $R_C = g_{m2}R_E \cdot R'_C$ [62]. Looking at the differential output signal, the two signal voltages add constructively, whereas the two noise voltages cancel each other. For an insightful comparison of noise matching and noise canceling, see [63].

An implementation of a feedforward TIA with a common-gate and a common-source pair to provide differential outputs has been reported in [64].

Shunt-Feedback TIA with Common-Base (Common-Gate) Input Stage. As pointed out at the beginning of this section, the common-base/gate current buffer can also be combined with the shunt-feedback TIA. A shunt-feedback TIA with common-base input stage is shown in Fig. 7.25. (For a shunt-feedback TIA with common-gate input stage, see Fig. 9.15 in Chapter 9.) The common-base stage is now primarily loaded by the input impedance of the shunt-feedback TIA, which is $R_F/(A_0 + 1)$ at low frequencies. The collector resistor R_C is usually much larger than that or is left away entirely [53].

Compared with the common-base stage alone, the small and noisy collector resistor, R_C, is replaced by a much larger and quieter feedback resistor, R_F, while

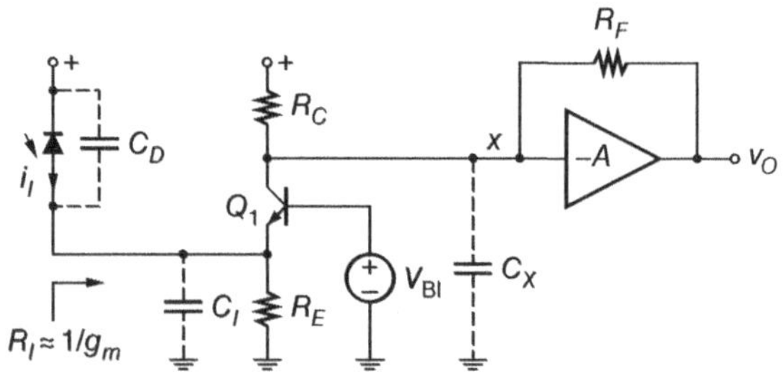

Figure 7.25 Shunt-feedback TIA with common-base input stage.

maintaining a similar resistance at node x. Compared with the shunt-feedback TIA alone, the photodetector capacitance C_D is now isolated from the critical node x, making the TIA's bandwidth and stability less dependent on C_D. A drawback of this arrangement is the power consumption, which is larger than that of the common-base TIA or the shunt-feedback TIA alone.

For a common-base stage with unity current gain and a single real input pole as in Eq. (7.12), $R_C \gg R_F/(A_0+1)$, and a single-stage shunt-feedback section as described by Eqs. (6.8)–(6.11), the transimpedance of the common-base shunt-feedback TIA is

$$Z_T(s) = -\frac{R_T}{[1+s/\omega_{p1}][1+s/(\omega_0 Q)+s^2/\omega_0^2]}, \tag{7.16}$$

where

$$R_T = \frac{A_0}{A_0+1} \cdot R_F, \quad \omega_{p1} = \frac{g_m}{C_T}, \tag{7.17}$$

$$\omega_0 = \sqrt{\frac{A_0+1}{R_F C_X T_A}}, \quad Q = \frac{\sqrt{(A_0+1)R_F C_X T_A}}{R_F C_X + T_A}, \tag{7.18}$$

$C_T = C_D + C_I$ is the total capacitance an the input node, C_X is the total capacitance at node x (including the Miller component), A_0 is the low-frequency gain of the voltage amplifier, and T_A is its time constant (cf. Section 6.2).

If the input pole of the common-base stage, ω_{p1}, is placed sufficiently high, such that it does not interfere with the frequency response of the shunt-feedback TIA section, the overall TIA bandwidth is mostly determined by the shunt-feedback section. Analogous to the pole spacing of the common-base TIA, we can define the *cutoff-frequency spacing* $\chi = \omega_{p1}/\omega_{b2}$, where $\omega_{b2}/(2\pi)$ is the 3-dB bandwidth of the shunt-feedback section. For any value of χ, the 3-dB bandwidth of the overall TIA can be bounded as $BW_{3dB} \leq \omega_{b2}/(2\pi)$, where equality is reached for large values of χ.

Transimpedance Limit. The transimpedance limit Eq. (6.14) that we derived for the shunt-feedback TIA still applies to the shunt-feedback section in Fig. 7.25. Thus, given a common-base/gate stage with unity current gain, and requiring $Q \leq 1/\sqrt{2}$ for the shunt-feedback section, the transimpedance limit is $R_T \leq 2\pi A_0 f_A/(C_X \omega_{b2}^2)$. To get an estimate for C_X, we assume, as before, that the gain-bandwidth product of the common-base/gate stage is $A_0 f_A$. Thus, $C_X = g_m/(2\pi A_0 f_A)$ and with $g_m = \omega_{p1} C_T$ from Eq. (7.17) and $\omega_{p1} = \chi \omega_{b2}$, we have $C_X = \chi \omega_{b2}/(2\pi A_0 f_A) \cdot C_T$. Observing that $BW_{3dB} \leq \omega_{b2}/(2\pi)$, the transimpedance limit for the common-base/gate shunt-feedback TIA becomes [12]

$$R_T \leq \frac{(A_0 f_A)^2}{2\pi C_T \chi \cdot BW_{3dB}^3}, \tag{7.19}$$

where the equality is reached for large values of χ and $Q = 1/\sqrt{2}$.

Comparing this limit with Eq. (6.14), we see that if the bandwidth headroom exceeds the cutoff-frequency spacing, $A_0 f_A/BW_{3dB} > \chi$, the common-base/gate shunt-feedback TIA can achieve a higher transimpedance than the basic shunt-feedback TIA. The reason for this boost is that the addition of an input stage can reduce the capacitance at node x, which permits a larger value for R_F while maintaining the same bandwidth. As a result, the transimpedance increases and the noise contributed by the feedback resistor decreases.

Input-Referred Noise Current PSD. To estimate the input-referred noise current PSD let us assume that the common-base stage has unity current gain and that $\omega_{p1} \gg \omega_{b2}$. Then, the input-referral function for the noise from the shunt-feedback section is one and the input-referred noise current PSD of the whole circuit is simply the sum of Eqs. (7.15) and (6.48) [53]:

$$\begin{aligned} I_{n,TIA}^2(f) = {} & \frac{4kT}{R_F} + \frac{4kT}{R_C} + \frac{4kT}{R_E} + \frac{2qI_{C1}}{\beta_1} + \frac{2qI_{C2}}{\beta_2} \\ & + 2qI_{C1} \frac{(2\pi C_T)^2}{g_{m1}^2} f^2 + 2qI_{C2} \frac{(2\pi \tilde{C}_X)^2}{g_{m2}^2} f^2 \\ & + 4kTR_{b1}(2\pi C_D)^2 f^2 + 4kTR_{b2}(2\pi C_X')^2 f^2 + \cdots, \end{aligned} \tag{7.20}$$

where I_{C1}, β_1, g_{m1}, and R_{b1} refer to the common-base transistor and I_{C2}, β_2, g_{m2}, and R_{b2} refer to the input transistor of the shunt-feedback section. $\tilde{C}_X$ is the total capacitance at node x under shorted-output conditions and C_X' is the capacitance contributed by the common-base stage to node x.

The noise contribution from R_F (first term) may be smaller than that of a TIA without common-base input stage because R_F may be larger, as explained earlier. However, this advantaged may easily be nullified by the five additional noise terms from the common-base stage. In many cases, adding a common-base (common-gate) input stage increases the total noise of the TIA [1], but noise equality [37, 38], and even noise improvements [53] also have been reported.

Numerical Example. As an illustration, let us combine the 10-Gb/s common-base TIA example in Table 7.3 with the 10-Gb/s shunt-feedback TIA example in Table 6.1. Assuming that $R_C \gg 100\ \Omega$ and $C_X = 0.3$ pF, the resulting shunt-feedback TIA with common-base input stage combines the high transimpedance of the shunt-feedback TIA example ($R_T = 500\ \Omega$) with the low input impedance of the common-base TIA example ($R_I = 20\ \Omega$).

To estimate the input-referred rms noise current of our common-base shunt-feedback TIA example, let us assume that $R_E \parallel R_C = R_F = 600\ \Omega$, $\tilde{C}_X = C_T = 0.25$ pF and $C'_X = C_D = 0.15$ pF. Then, the noise PSD contributions from the common-base stage and the shunt-feedback section in Eq. (7.20) happen to be equal. Given the high input pole frequency of the common-base stage, the noise bandwidths of the shunt-feedback TIA with and without the common-base input stage are approximately the same. Thus, the total input-referred rms noise current is about $\sqrt{2}\times$ that of the basic shunt-feedback TIA in Table 6.6, which evaluates to 1.09 μA. This is significantly less than the noise of the common-base TIA in Table 7.4, which was 1.84 μA. The elimination of the small load resistor at the collector and the steeper roll-off of the frequency response provided by the shunt-feedback section do make a difference.

7.6 Regulated-Cascode TIA

So far, the focus has been on common-base rather than common-gate input stages. The reason for this preference is that BJTs achieve a significantly higher g_m than FETs for a given bias current. More precisely, for a BJT $g_m = I_C/V_T$, neglecting the parasitic emitter resistance, and for an FET $g_m = 2I_D/(V_{GS} - V_{TH})$, assuming the simple quadratic model. With the thermal voltage $V_T \approx$ 25 mV and the typical overdrive voltage $V_{GS} - V_{TH} \approx 0.3$ V, the BJT's g_m is about six times larger than the FET's. (A smaller overdrive voltage helps to increase g_m at the expense of a lower f_T.) For the FET to achieve the same g_m as a BJT, it would need an about six times larger bias current. Thus, a BJT input stage is more power efficient and more effective in achieving a large pole spacing (or cutoff-frequency spacing).

Interestingly, there is an alternative way to increase the effective g_m, without increasing the FET's bias current, namely by adding a booster amplifier, as shown in Fig. 7.26(a). This arrangement is known as a *regulated cascode* (RGC) stage. Like the common-gate stage, the regulated-cascode stage can be used as a TIA by itself [54, 56] or it can be used as an input stage to a shunt-feedback TIA [65–68]. We focus here on the former variant.

The booster amplifier B senses any voltage change at the input node and controls the gate voltage to counteract it. The resulting input resistance is $B_0 + 1$ times lower than that of the simple common-gate stage: $R_I \approx 1/[(B_0 + 1)g_{m1}]$.

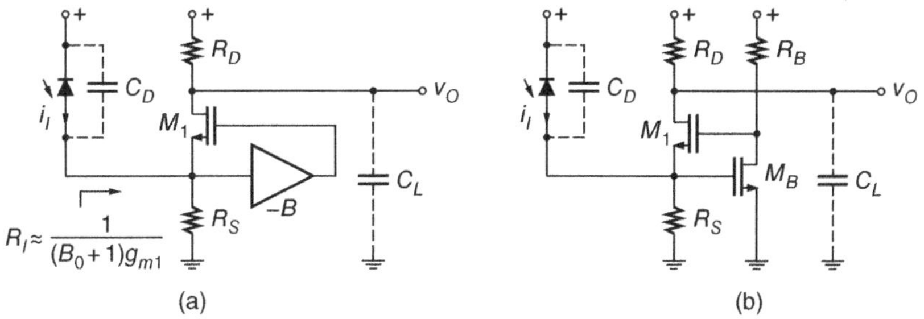

Figure 7.26 The regulated-cascode TIA: (a) block diagram and (b) a common transistor-level implementation with MOSFETs.

A common transistor-level implementation of the RGC TIA is shown in Fig. 7.26(b) [65–67]. Here, the booster amplifier is realized with a common-source MOSFET stage, M_B and R_B, which has the DC gain $B_0 = g_{mB}R_B$. Other topologies and realizations have been devised as well. For example, an RGC TIA with multistage booster amplifier [54] (see Fig. 9.17 in Chapter 9), a differential RGC TIA [69], an RGC TIA with MESFETs [70], and an RGC TIA with BJTs [71] have been reported.

For the subsequent analysis of the RGC TIA in Fig. 7.26(a) we model the booster amplifier as having a single pole, $B(s) = B_0/(1 + sT_B)$, infinite input resistance, and zero output resistance, just like we did for the voltage amplifier in the shunt-feedback TIA. The 3-dB bandwidth of this booster amplifier is $f_B = 1/(2\pi T_B)$. With the simplifying assumptions $R_S \to \infty$, $g_{mb1} = 0$, $g_{o1} = 0$, and $C_{gd1} = 0$ the transimpedance becomes

$$Z_T(s) = R_T \frac{1 + s/\omega_z}{[1 + s/(\omega_0 Q) + s^2/\omega_0^2][1 + s/\omega_{p2}]}, \tag{7.21}$$

where

$$R_T = R_D, \quad \omega_z = \frac{B_0 + 1}{T_B}, \quad \omega_{p2} = \frac{1}{R_D C_L}, \tag{7.22}$$

$$\omega_0 = \sqrt{\frac{(B_0 + 1)g_{m1}}{C_T T_B}}, \quad Q = \frac{\sqrt{(B_0 + 1)C_T g_{m1} T_B}}{C_T + B_0 C_M + g_{m1} T_B}, \tag{7.23}$$

C_L is the total load capacitance at the output node, and C_T is the total capacitance at the input node at low frequencies. C_T can be broken down into three components, $C_T = C_D + C_I + C_M$, where C_D is the photodetector capacitance with all parasitic capacitances in parallel to it, C_I is the input capacitance of the booster amplifier plus C_{sb1} ($C_I = C_{gsB} + C_{sb1}$ in Fig. 7.26(b)), and C_M is C_{gs1} plus all capacitances in parallel to it ($C_M = C_{gs1} + C_{gdB}$ in Fig. 7.26(b)).

The expressions for the low-frequency transimpedance, R_T, and the output pole, ω_{p2}, are the same as for the common-gate TIA. The real input pole of the common-gate TIA, ω_{p1}, however, is replaced by a (complex) pole pair given by ω_0, Q, and a zero at ω_z. To get a handle on this fairly complex expression, let us make the simplification $C_M = T_B = 0$. Now, ω_0 and ω_z both go to infinity, leaving only a single real input pole at $\omega_0 Q = (B_0 + 1) \cdot g_{m1}/C_T$. The first factor, $B_0 + 1$, is the DC gain of the booster amplifier plus one and the second factor, g_{m1}/C_T, is the input pole of the unboosted common-gate TIA. As expected, the presence of the booster amplifier increases the transconductance of the input FET and thus the frequency of the input pole by $B_0 + 1$. For the transistor-level RGC TIA in Fig. 7.26(b), the input pole is boosted by the corresponding factor $g_{mB}R_B + 1$ (cf. Eqs. (I.134)–(I.137)). [→ Problem 7.10.]

The zero and (complex) pole pair in Eq. (7.21) could potentially result in a peaked frequency response or even instability. From Eq. (7.22) we can see that $\omega_z > B_0/T_B = 2\pi B_0 f_B$, that is, the frequency of the zero is beyond the gain-bandwidth product of the booster amplifier. Located at such a high frequency, the zero has little impact on the overall frequency response. A large value for Q in Eq. (7.23) is another concern. If necessary, Q can be lowered by reducing B_0 and simultaneously increasing f_B, thus keeping the gain-bandwidth product $B_0 f_B$ fixed. From an open-loop perspective, this modification reduces the loop gain and speeds up the nondominant open-loop pole. Both changes help to improve the phase margin. In practice, a Q value slightly larger than $1/\sqrt{2}$ is often beneficial as it increases the RGC TIA's bandwidth and, owing to the roll-off resulting from the output pole at ω_{p2}, does not necessarily result in a peaked overall response.

3-dB Bandwidth. Assuming that the zero can be neglected and that $Q = 1/\sqrt{2}$, the 3-dB bandwidth of the RGC input node (the input cutoff frequency) is $f_{b1} = \omega_0/(2\pi)$. With Eq. (7.23) we find [12]

$$f_{b1} = \sqrt{\frac{g_{m1}}{2\pi C_T} \cdot B_0 f_B \cdot \frac{B_0 + 1}{B_0}} \approx \sqrt{\frac{g_{m1}}{2\pi C_T} \cdot B_0 f_B}. \tag{7.24}$$

The first factor under the root, $g_{m1}/(2\pi C_T)$, is the input cutoff frequency of the unboosted common-gate TIA and the second factor, $B_0 f_B$, is the gain-bandwidth product of the booster amplifier. Ignoring the third factor, which is close to one for $B_0 \gg 1$, leads to the approximation shown in the second form of Eq. (7.24).

This equation has the following interesting interpretation: Adding a booster amplifier speeds up the bandwidth of the input node to approximately the geometrical mean of its bandwidth *before* boost is applied and the gain-bandwidth product of the booster amplifier. This is a consequence of the booster gain being constrained by the gain-bandwidth product and the necessity of the booster bandwidth to be commensurate with the input cutoff frequency.

In practice, the input cutoff frequency is often higher than given by the approximation in Eq. (7.24) because Q may be larger than $1/\sqrt{2}$ and the neglected zero and the $(B_0 + 1)/B_0$ factor do have some impact.

Transimpedance Limit. An approximate transimpedance limit can be derived following the by now familiar procedure. Let us assume that the input FET M_1 and the booster amplifier B are both constrained by the same gain-bandwidth product: $g_{m1}R_D \cdot \omega_{p2}/(2\pi) = A_0 f_A = B_0 f_B$. Requiring $Q \leq 1/\sqrt{2}$ and assuming $\omega_z \to \infty$ (no peaking from the input node), the input cutoff frequency can be bounded as $f_{b1} \leq \sqrt{g_{m1}/(2\pi C_T) \cdot A_0 f_A}$ (cf. Eq. (7.24)) and the overall TIA bandwidth can be bounded as $BW_{3\text{dB}} \leq \omega_{p2}/(2\pi)$. Now, we can write the output pole as $\omega_{p2} = 2\pi A_0 f_A/(g_{m1}R_D)$. Inserting $g_{m1} \geq \omega_{b1}^2 C_T/(2\pi A_0 f_A)$ and $\omega_{b1} = \chi\omega_{p2}$, we find $\omega_{p2}^3 \leq (2\pi A_0 f_A)^2/(\chi^2 R_D C_T)$. Using $BW_{3\text{dB}} \leq \omega_{p2}/(2\pi)$ and $R_T = R_C$, the transimpedance limit of the RGC TIA is found as [12]

$$R_T \leq \frac{(A_0 f_A)^2}{2\pi C_T \chi^2 \cdot BW_{3\text{dB}}^3}, \tag{7.25}$$

where the equality is reached for $Q = 1/\sqrt{2}$ and large values of χ. Some of the assumptions and approximations (e.g., $Q \leq 1/\sqrt{2}$ and neglecting the zero) are pessimistic and the actual limit tends to be higher.

Compared with the limit of the common-gate TIA, Eq. (7.14), we gained the factor $A_0 f_A/(\chi BW_{3\text{dB}})$, which equals the ratio of the bandwidth headroom to the cutoff-frequency spacing. The reason for this improvement is that the addition of a booster amplifier permits a reduction of M_1's width without loss in bandwidth. The smaller M_1, in turn, results in a smaller load capacitance C_L, permitting a larger drain resistor, R_D, and thus a larger transimpedance.

The transimpedance limit of the shunt-feedback TIA with common-gate input stage, Eq. (7.19), improves by the same factor, $A_0 f_A/(\chi BW_{3\text{dB}})$, when the common-gate input stage is replaced by an RGC input stage [12].

Power Consumption. The RGC TIA typically consumes less power than a common-gate TIA with the same bandwidth. This power reduction can be explained and quantified as follows [54]: The addition of a booster amplifier with gain B_0 permits the width, and thus the bias current, of the input FET M_1 to be reduced by the factor $B_0 + 1$, while maintaining roughly the same input cutoff frequency f_{b1}. Assuming that the booster FET, M_B, and M_1 are biased at the same current level, the total current of the RGC TIA becomes $2/(B_0 + 1)$ times that of the common-gate TIA. For example, for $B_0 = 3$ the RGC TIA consumes only half the power of the common-gate TIA at the same bandwidth.

Input-Referred Noise Current PSD. Let us analyze the noise of the transistor-level RGC TIA in Fig. 7.26(b). To calculate the input-referral functions, we make the following simplifying assumptions: $R_S \to \infty$, $g_{mb1} = 0$, $g_{o1} = g_{oB} = 0$, and $C_{gd1} = 0$. The noise current of R_S contributes directly to the input. The drain noise current of M_1 is input referred by $H_{D1}^{-1}(s) = s[(C_D + C_I)/(B_0 + 1) + C_M]/g_{m1}$, which is valid for $f \ll (C_T + B_0 C_M)/\{2\pi R_B[(C_D + C_I)C_M + C_T C_B]\}$ where $C_I = C_{gsB} + C_{sb1}$, $C_M = C_{gs1} + C_{gdB}$, $C_B = C_{dbB}$, $C_T = C_D + C_I + C_M$, and $B_0 = g_{mB}R_B$. The noise current of R_D is input referred by $H_{RD}^{-1}(s) = 1 + H_{D1}^{-1}(s)$. Finally, the drain noise current of M_B and the noise current of R_B are both input referred by $H_B^{-1}(s) = sR_B(C_D + C_I)/(B_0 + 1)$, which is valid for $f \ll (B_0 + 1)/(2\pi R_B C_B) = 1/(2\pi\omega_z)$ (cf. Eq. (I.135)).

After referring all device noise sources to the input, we find the noise current PSD as [67, 72]

$$
\begin{aligned}
\overline{I_{n,TIA}^2}(f) = {} & \frac{4kT}{R_D} + \frac{4kT}{R_S} + 4kT\tilde{\Gamma}_1 \frac{\left[2\pi\left(\frac{C_D + C_I}{B_0 + 1} + C_M\right)\right]^2}{g_{m1}} f^2 \\
& + 4kT\tilde{\Gamma}_B \frac{B_0^2}{(B_0 + 1)^2} \frac{[2\pi(C_D + C_I)]^2}{g_{mB}} f^2 + \cdots,
\end{aligned}
\tag{7.26}
$$

where we neglected the gate shot noise and the induced gate noise of M_1 and M_B ($\tilde{\Gamma}_1 = \gamma_1 g_{d01}/g_{m1}$ and $\tilde{\Gamma}_B = \gamma_B g_{d0B}/g_{mB}$), we assumed that the f^2 noise due to M_1 dominates that due to R_D ($g_{m1}R_D \gg 1/\tilde{\Gamma}_1$) and that the noise due to M_B dominates that due to R_B ($B_0 = g_{mB}R_B \gg 1/\tilde{\Gamma}_B$). Note that this noise expression reverts to that of the common-gate TIA, if we let $B_0 = 0$.

The first two terms of Eq. (7.26) are due to the resistors R_D and R_S, respectively. Because the RGC TIA permits a larger R_D than the common-gate TIA, the noise contribution of the first term is reduced.

The third term is due to M_1. Because the RGC TIA permits an M_1 that is smaller by a factor $B_0 + 1$ than the corresponding common-gate TIA, the noise of this term is also reduced by about a factor $B_0 + 1$. (When converting a common-gate TIA with a FET characterized by g_{m1} and C_{gs1} into an RGC TIA with roughly the same bandwidth and pole spacing, $g_{m1} \to g_{m1}/(B_0 + 1)$ and $C_M \approx C_{gs1} \to C_{gs1}/(B_0 + 1)$. Thus, the numerator of the third noise term is reduced by $(B_0 + 1)^2$ and the denominator is reduced by $B_0 + 1$.)

The fourth term is due to the booster amplifier and is new. The magnitude of this noise term depends on the size of M_B and can be optimized similar to our discussion in Section 6.4. Because this term is proportional to $(C_D + C_{sb1} + C_{gsB})^2/g_{mB}$, it reaches its minimum for $C_{gsB} = C_D + C_{sb1}$. For an RGC TIA with a large value for B_0, this fourth term easily becomes the largest. For a high-speed RGC TIA with a relatively small value for B_0, the third and fourth terms usually have comparable magnitudes.

Noise Optimization. It is convenient to start the design of an RGC TIA with a common-gate TIA prototype that has the desired bandwidth and pole spacing. Then, we convert this prototype into an RGC TIA by adding a booster amplifier with gain B_0 and reducing the size of M_1 by $B_0 + 1$. The gain B_0 must be chosen such that the peaking in the overall frequency response is acceptable.

At this point, we still have the freedom to choose the size of M_B, at least within certain bounds. Considering the fourth noise term of Eq. (7.26) in isolation, the noise minimum is reached for the capacitive matching condition $C_{gsB} = C_D + C_{sb1}$, as mentioned earlier. However, when taking all noise terms into consideration, the optimum is pulled toward a smaller M_B. A smaller M_B, for example, results in a smaller C_I, which reduces the third noise term.

Numerical Example: Common-Gate TIA Prototype. Let us conclude the discussion of the RGC TIA with a numerical example (see Table 7.5). We start by designing a common-gate TIA prototype using the same FET as for our 10-Gb/s shunt-feedback TIA example in Table 6.5. For simplicity we assume $R_S \to \infty$ and $g_{mb} = g_o = 0$. Given $g_m = 50$ mS, $C_D = 0.15$ pF, and $C_I = C_{gs} + C_{sb} = 0.1$ pF $+ 0.08$ pF $= 0.18$ pF, the input resistance is $1/g_m = 20\ \Omega$ and the input pole frequency is $g_m/(2\pi C_T) = 24.1$ GHz.

Table 7.5 Parameters and performance of our 10-Gb/s common-gate TIA prototype and RGC TIA example.

Parameter	Symbol	Common-gate TIA	RGC TIA
Gain-bandwidth product	$A_0 f_A$	44 GHz	44 GHz
Detector capacitance incl. parasitics	C_D	0.15 pF	0.15 pF
Noise factor	$\tilde{\Gamma}, \tilde{\Gamma}_B$	2	2
Gate–source capacitance	C_{gs}	100 fF	25 fF
Source–bulk capacitance	C_{sb}	80 fF	20 fF
Transconductance	g_m	50 mS	12.5 mS
Load resistor	R_D	118 Ω	547 Ω
Gate–source capacitance (booster)	C_{gsB}		25 fF
Drain–bulk capacitance (booster)	C_{dbB}		20 fF
Transconductance (booster)	g_{mB}		12.5 mS
Load resistor (booster)	R_B		240 Ω
Input cutoff frequency	f_{p1}	24.1 GHz	29.5 GHz
Output pole frequency	f_{p2}	7.43 GHz	6.43 GHz
Pole spacing	χ	3.25	4.58
Transimpedance	R_T	118 Ω	547 Ω
Input impedance	R_I	20 Ω	20 Ω
3-dB bandwidth	BW_{3dB}	6.85 GHz	6.85 GHz
Inp.-ref. noise from R_D	$I_{n,in,RD}$	11.8 pA/$\sqrt{\text{Hz}}$	5.5 pA/$\sqrt{\text{Hz}}$
Inp.-ref. noise from M_1 @ 10 GHz	$I_{n,in,M1}$	16.9 pA/$\sqrt{\text{Hz}}$	7.5 pA/$\sqrt{\text{Hz}}$
Inp.-ref. noise from M_B @ 10 GHz	$I_{n,in,MB}$		15.0 pA/$\sqrt{\text{Hz}}$

To obtain a TIA bandwidth of 6.85 GHz, as for our shunt-feedback example in Table 6.1, the output pole frequency must be set to 7.43 GHz (from Eq. (H.16)). The resulting pole spacing is 24.1 GHz/7.43 GHz = 3.25. Given a technology that permits a gain-bandwidth product of 44 GHz, as in our shunt-feedback example, the voltage gain of the common-gate stage comes out as $A_0 = 44$ GHz/7.43 GHz = 5.92, and thus $R_D = A_0/g_m = 118\ \Omega$, which is also the low-frequency transimpedance R_T.

To calculate the noise of our common-gate TIA prototype, we use the solution to Problem 7.8 on p. 526. At low frequencies, the input-referred noise current density due to R_D is 11.8 pA/$\sqrt{\text{Hz}}$. At 10 GHz, the FET adds about 16.9 pA/$\sqrt{\text{Hz}}$ to this noise.

Numerical Example: Regulated-Cascode TIA. Next, we convert our 10-Gb/s common-gate TIA prototype into an RGC TIA by reducing the size of the FET by a factor four and adding a booster amplifier with $B_0 = 3$ to boost $g_{m1} = 12.5$ mS back to an effective value of 50 mS. We build the booster amplifier with a FET of the same size biased at the same current, that is, $g_{mB} = 12.5$ mS. To obtain a DC gain of three, the load resistor must be $R_B = 240\ \Omega$. To calculate the input cutoff frequency, we use the solution to Problem 7.10 on p. 526. With the scaled-down device capacitances $C_{gs1} = C_{gsB} = 0.1$ pF/4, $C_{sb1} = C_{dbB} = 0.08$ pF/4, and $C_{gd1} = C_{gdB} = 0$, we find $C_I = C_{gsB} + C_{sb1} = 45$ fF, $C_M = C_{gs1} = 25$ fF, $C_B = C_{dbB} = 20$ fF, and $C_T = C_D + C_I + C_M = 220$ fF and with Eqs. (I.136) and (I.137), we can calculate the input pole parameters as $\omega_0 = 2\pi \cdot 23.9$ GHz and $Q = 0.94$. Finally, with Eq. (H.2), we find the input cutoff frequency as $f_{b1} = 29.5$ GHz. Note that the peaking associated with the relatively high Q value results in an input cutoff frequency that is somewhat higher than that of the common-gate prototype.

A numerical analysis shows that we need to place the output pole at 6.43 GHz to obtain an overall bandwidth of 6.85 GHz. This low-frequency output pole flattens the overall response such that no peaking occurs despite the large Q value of the input poles. (The zero in Eq. (I.135) is at 133 GHz and thus far out of band.) Given a technology that permits a gain-bandwidth product of 44 GHz and $f_{p2} = 6.43$ GHz, we find $A_0 = 6.84$ and $R_D = R_T = 547\ \Omega$. (The pessimistic transimpedance limit Eq. (7.25) would have predicted less than half of that.)

The noise performance of our RGC TIA example can be calculated with Eq. (7.26). At low frequencies, the input-referred noise current density due to R_D is 5.5 pA/$\sqrt{\text{Hz}}$. This is only about half of the noise of the common-gate prototype, a direct consequence of the larger R_D. At 10 GHz, M_1 adds about 7.5 pA/$\sqrt{\text{Hz}}$ and M_B adds about 15.0 pA/$\sqrt{\text{Hz}}$ to this noise. Compared with the common-gate prototype, the noise current density due to M_1 is reduced to less than half. Even when including the additional noise due to M_B, the RGC TIA is still quieter than the common-gate prototype.

In conclusion, our RGC TIA example has a transimpedance that is about 4.6 times larger than that of the common-gate prototype, consumes only half the power, and has comparable, if not better, noise performance (cf. Table 7.5).

Other Common-Gate Input Stages. Besides the regulated-cascode stage there are several other variations of the common-gate stage that feature an input resistance of less than $1/g_m$. We briefly discuss three examples.

Instead of a feedback loop around the voltage amplifier, as in Fig. 7.25, a global feedback loop encompassing the common-gate stage and the voltage amplifier, as shown in Fig. 7.27, can be used. This feedback loop reduces the input resistance from $1/g_{m1}$ to $[1/g_{m1}] \parallel [R_F/(g_{m1}R_DA_0 + 1)]$, where we neglected R_S [73]. With the shunt-feedback resistor R_F now connecting to the input node rather than node x, the pole frequency associated with node x is lower (higher resistance) and the open-loop gain is higher (if $g_{m1}R_D > 1$), making this feedback configuration potentially slower and less stable. The configuration in Fig. 7.27 can also be realized with a common-base instead of a common-gate input stage [1, 53].

Instead of negative feedback, positive feedback and matched devices can be used to theoretically reduce the input resistance to zero. A first example is shown in Fig. 7.28(a) [35]. The current mirror consisting of M_3 and M_4 copies the drain current of M_1 into M_2. Under ideally matched conditions ($M_1 = M_2$ and $M_3 = M_4$) and zero drain–source conductance, the gate–source voltage of M_2 always equals that of M_1. Consequently, the input voltage at node x replicates the bias voltage V_{BI} at node y for any input current i_I, which is to say that the input resistance is zero. Of course, when there is device mismatch or nonzero output conductance the input impedance becomes nonzero and may even be negative (cf. Eq. (I.138)). The current mirror, consisting of M_3 and M_5, copies the input current into the load resistor R_D, producing the output voltage v_O. Unfortunately, this current mirror contributes a fair amount of input-referred noise current, making this a rather noisy circuit.

Another example, operating on the same principles, is the *translinear cross-quad* shown in Fig. 7.28(b) [74]. Instead of using a current mirror to force the collector currents of two devices to be equal, the two devices are stacked to achieve the same goal (neglecting base currents). Under ideally matched

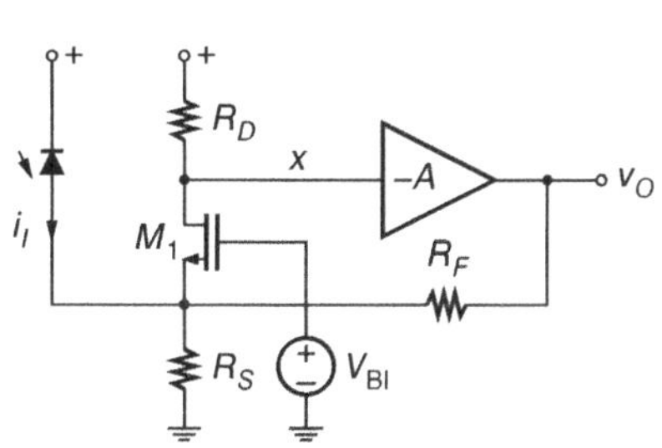

Figure 7.27 TIA with common-gate input stage and a global feedback loop.

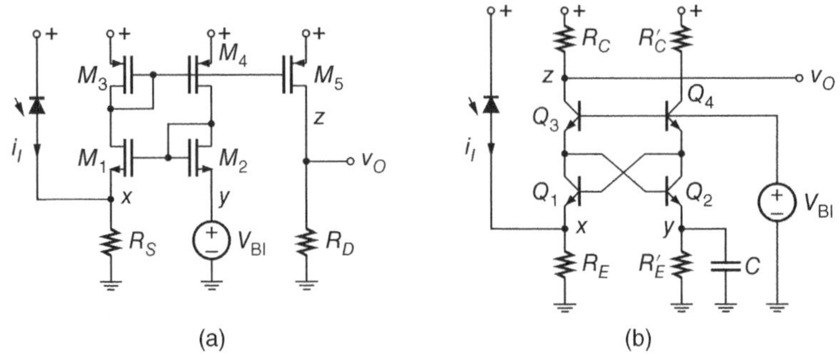

Figure 7.28 Common-gate/common-base input stages with theoretically zero input resistance: (a) V_{GS} voltage mirror and (b) translinear cross-quad.

conditions ($Q_1 = Q_3$ and $Q_2 = Q_4$), zero collector–emitter conductance, and zero base currents, we have $V_{BE1} = V_{BE3}$ and $V_{BE2} = V_{BE4}$. Due to cross coupling, the voltage at node x is $V_{\mathrm{BI}} - V_{BE4} - V_{BE1}$ and the voltage at node y is $V_{\mathrm{BI}} - V_{BE3} - V_{BE2}$. Thus, under matched conditions, these two voltages become equal. Again, the input voltage at node x replicates that at node y for any input current i_I, which means that the differential input resistance is zero. To make the single-ended input resistance at node x zero, node y can be AC grounded with a large capacitor C. This circuit has better noise properties than the previous one because it avoids the current mirrors, however, it requires a higher supply voltage because of the larger number of stacked transistors and resistors. [→ Problem 7.11.]

The common-gate/common-base stage, the regulated-cascode stage, as well as the circuits in Fig. 7.28 are all current buffers in the sense that they have a low input impedance (ideally zero), a high output impedance (ideally infinite), and an output current that equals the input current. Now, if such a current buffer has a third terminal that sets the input voltage to that of the third terminal, the buffer is known as a *current conveyor* [75]. Interestingly, the two circuits in Fig. 7.28 do have such a terminal, namely node y (node x is the current buffer input and node z is the current buffer output). (The regulated cascode can be turned into a current conveyor by making the booster-amplifier inputs differential.) Thus, these circuits are not only current buffers but also current conveyors and the associated TIAs are sometimes known as current conveyor TIAs.

7.7 TIA with Inductive Broadbanding

In Section 6.5, we used inductors to improve the noise performance of TIAs by means of noise matching. Now, we investigate the use of inductors to broaden the TIA's bandwidth.

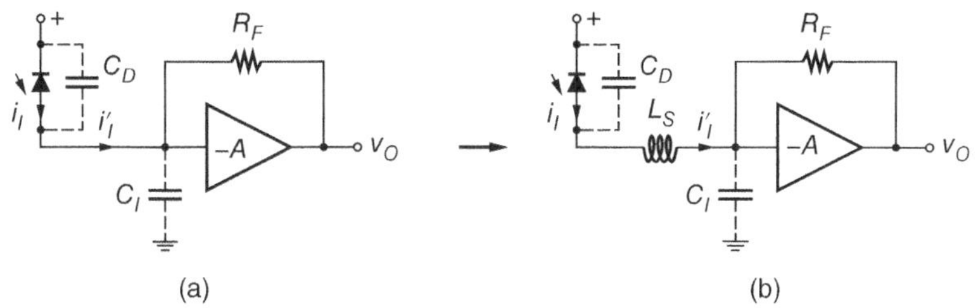

Figure 7.29 TIA (a) without and (b) with an input series inductor.

Input Series Inductor. Figure 7.29 shows a shunt-feedback TIA with and without an input series inductor.

The inductor L_S splits the formerly parallel shunt capacitors C_D and C_I apart and thus adds two new poles to the TIA's transimpedance function. To keep the following discussion simple, let us assume that the original TIA in Fig. 7.29(a) has only one pole at $\omega_p = (A_0 + 1)/(R_F C_T) = 1/(R_I C_T)$ (cf. Eq. (6.3)), that is, we neglect the pole(s) in the voltage amplifier. When adding the inductor, the following third-order transimpedance function is obtained:

$$Z_T(s) = -R_T \cdot \frac{1}{1 + sR_I C_T + s^2 L_S C_D + s^3 R_I C_I L_S C_D}, \tag{7.27}$$

where

$$R_T = \frac{A_0}{A_0 + 1} R_F, \qquad R_I = \frac{R_F}{A_0 + 1}, \qquad \text{and} \qquad C_T = C_D + C_I. \tag{7.28}$$

Here, C_D is the photodetector capacitance and C_I is the input capacitance of the TIA including all parasitic capacitances in parallel to it (e.g., the bond pad and ESD circuit capacitance). Note that if we let $L_S = 0$, two of the three poles disappear and the first-order transimpedance function of Eq. (6.1) is recovered.

Bandwidth Improvement. If we want the transimpedance response to be maximally flat (Butterworth response), the series inductor needs to be (cf. Eq. (I.141))

$$L_S = \frac{2}{3} R_I^2 C_T = \frac{2/3}{(2\pi BW_{3dB})^2 C_T}, \tag{7.29}$$

where BW_{3dB} is the original bandwidth without inductor.

With this inductor, the new bandwidth becomes twice the original bandwidth, $2BW_{3dB}$, quite a substantial improvement! Figure 7.30(a) illustrates how the inductor turns the first-order transimpedance function (dashed line) into a third-order function with twice the bandwidth (solid line).

But there is a catch: The Butterworth solution can be found only if C_D is three times larger than C_I. For the example values $C_D = C_I = 0.15$ pF and $R_I = 100\ \Omega$, there is no solution, but for $C_D = 0.225$ pF and $C_I = 0.075$ pF, we find

$L_S = 2$ nH. The reason for this difficulty is that we need to satisfy *two* conditions to force the three poles into a neat Butterworth (or Bessel, etc.) configuration, but L_S provides only *one* degree of freedom. [→ Problem 7.12.]

In practice, it is often not possible to obtain the desired (e.g., maximally flat) response, necessitating a compromise. Nevertheless, Eq. (7.29) can serve as a starting point for the required inductance. A similar equation, $L_S \approx 1/[(2\pi BW_{3dB})^2 C_D]$, was provided in [76] based on the intuition that the inductor and the photodetector capacitance should resonate near the cutoff frequency.

To gain more insight into the origin of the bandwidth extension, it is useful to analyze the ratio of the current flowing into the TIA, i_I', to the current generated in the photodetector, i_I (cf. Fig. 7.29) [76]. Figure 7.30(b) shows the frequency response of I_i'/I_i without and with the inductor. At low frequencies, the full photodetector current appears at the TIA input, that is, $I_i'/I_i = 1$ (or 0 dB). At higher frequencies and without the inductor, some of the photodetector current is shunted to ground by C_D, resulting in a roll off near the original bandwidth (dashed line). Now adding the inductor, C_D, L_S, and C_I create a resonance that improves the current transfer from the photodetector to the TIA in a frequency band near the original bandwidth (solid line). This resonance can even result in a current gain, $I_i'/I_i > 1$, an effect that can be used to compensate for bandwidth limitations elsewhere in the TIA. Note that the same signal-current enhancement that improves the bandwidth is also responsible for the reduction in input-referred noise that we discussed earlier in the context of noise matching (cf. Section 6.5).

Rise-Time Improvement. The bandwidth improvement in the frequency domain can be explained in terms of L_S resonating out (or tuning out) the undesired capacitances C_D and C_I. The corresponding rise-time improvement in the time domain can be explained in terms of L_S stagger charging the capacitances C_D and C_I, as illustrated in Fig. 7.31.

Without L_S, capacitors C_D and C_I are connected in parallel and must charge simultaneously (dashed line). With L_S present, initially all photodetector current charges C_D (C_I is temporarily isolated by L_S), then after a small delay,

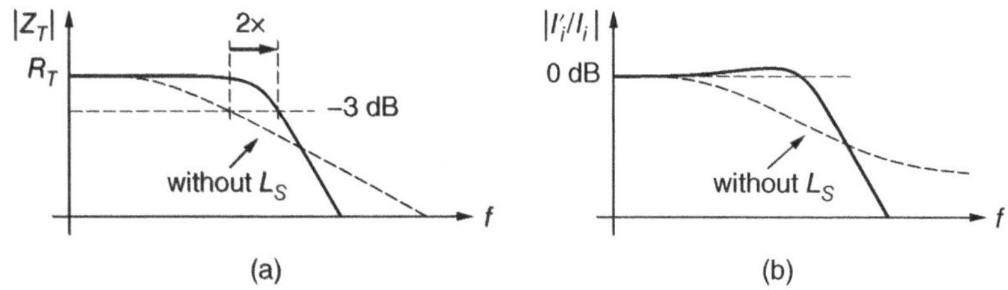

Figure 7.30 Effect of input series inductor on (a) the transimpedance and (b) the photodetector current transfer efficiency.

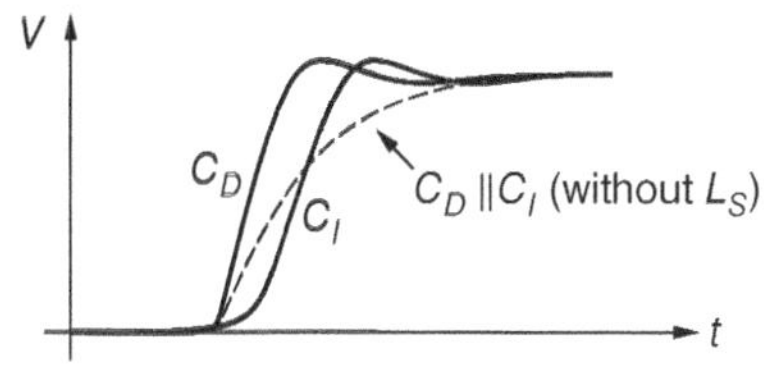

Figure 7.31 Charging of the capacitors C_D and C_I in response to a step current with (solid line) and without (dashed line) the inductor L_S.

the photodetector current charges C_I (solid lines). Because C_D and C_I are each smaller than their parallel combination, the rise times resulting from stagger charging are shorter. Note that the inductor does not improve the delay time (or latency), but it is the rise time that determines the maximum bit rate that can be processed.

In high-speed receivers, the photodetector and the TIA chip are often copackaged to minimize the interconnect parasitics (cf. Section 3.4). The bond wire that is typically used for this interconnection can play the role of L_S. Choosing the optimum length for this bond wire can result in a substantial bandwidth and noise improvement [76, 77].

Separating Capacitances with Inductors. The idea of inserting a series inductor between the photodetector and the TIA input to improve the speed can be generalized as follows: Look for parallel capacitances in the signal path and break them up by inserting small inductors. Like before, stagger charging the broken up capacitances results in faster edges. Two examples of this technique are shown in Fig. 7.32.

In Fig. 7.29(b), we separated C_D from the input capacitance of the TIA, but the latter capacitance still has (at least) two parallel components that can be broken up further: the capacitance of the front-end transistor and the capacitance of the bond pad (also the capacitance of the ESD protection circuit). Figure 7.32(a) shows how the detector capacitance C_D can be separated from the bond-pad capacitance C_B with a bond wire L_B and additionally C_B can be separated from the transistor capacitance C_I with an on-chip spiral inductor L_1 [1].

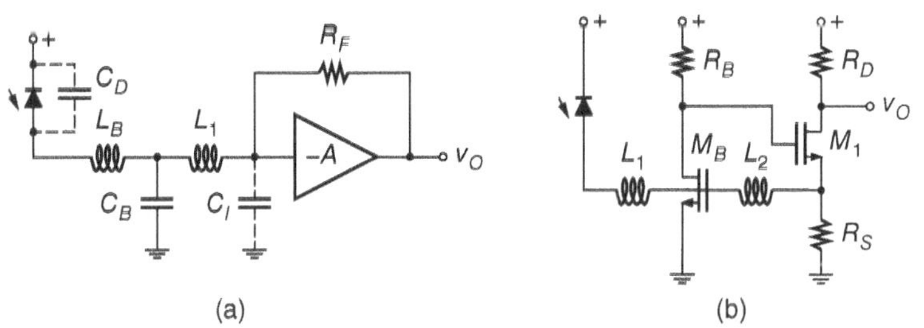

Figure 7.32 Separating capacitances with inductors in (a) a shunt-feedback TIA and (b) a regulated-cascode TIA.

In a regulated-cascode TIA, the photodetector is connected to two FETs: the source of the common-gate FET, M_1, and the gate of the booster-amplifier FET, M_B. Figure 7.32(b) shows how two on-chip spiral inductors L_1 and L_2 can be used to separate the capacitances of these two transistors from each other and from C_D [72].

In both examples, the two inductors add four new poles, making the overall response at least fifth order. To force all five poles into a Butterworth or Bessel configuration, we need to satisfy four conditions, which may be difficult, considering that the two inductors provide only two degrees of freedom. Nevertheless, solutions with acceptable amplitude ripple and group-delay variation can usually be found [1, 72].

Deferring Resistor Currents with Inductors. In the same way inductors can defer currents in capacitors, leading to stagger charging, inductors can also defer currents in resistors, speeding up the charging of capacitances at the same node. Two examples of this technique are shown in Fig. 7.33.

Figure 7.33(a) shows a shunt-feedback TIA with an inductor L_F in series with the feedback resistor [9, 78]. In response to a current step from the photodetector, the inductor L_F temporarily blocks the current from flowing into the feedback branch thus leaving more current for charging the capacitances C_D and C_I more quickly. In the frequency domain the same effect can be understood as follows: At high frequencies, the inductor increases the impedance of the feedback branch thus emphasizing high frequencies in the closed-loop response and pushing the cutoff frequency to a higher value. Besides increasing the speed, the feedback inductor also helps to reduce the input-referred noise [9].

The second example in Fig. 7.33(b) shows a common-gate TIA with a shunt peaking inductor [55]. In response to a drain current step, the inductor L_D defers the current flow into the load resistor R_D thus leaving more drain current for charging the load capacitance C_L more quickly. Equivalently, we can

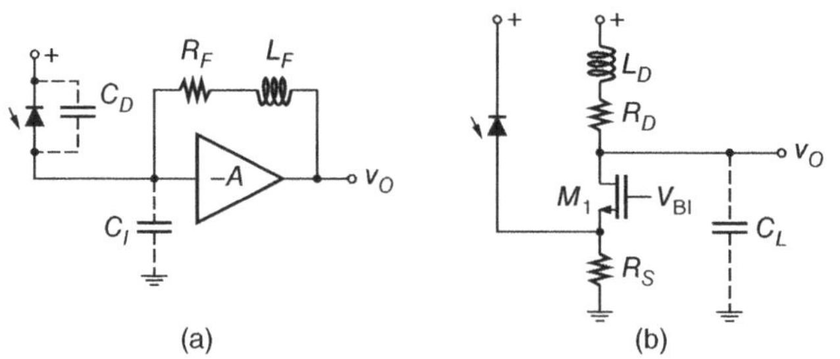

Figure 7.33 Deferring resistor currents with inductors in (a) a shunt-feedback TIA and (b) a common-gate TIA.

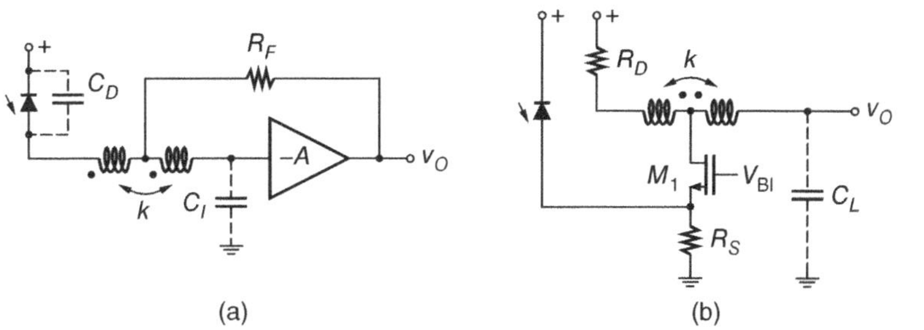

Figure 7.34 Application of T-coils in (a) a shunt-feedback TIA and (b) a common-gate TIA.

say that at high frequencies, L_D increases the load impedance counteracting the impedance decrease (and gain roll-off) due to C_L.

T-Coil Networks. The methods of separating capacitances with inductors and deferring resistor currents with inductors can be combined, often producing superior results than each method by itself. Whereas an implementation with multiple individual inductors is possible, it is more efficient to use a so-called *T-coil network*. A T-coil consists of a pair of magnetically coupled inductors (a transformer) and is equivalent to a three-inductor T network but consumes less chip area and permits one of the three inductors to be negative [79–81].

Figure 7.34(a) shows a shunt-feedback TIA with a T-coil that separates the capacitances C_D and C_I, resulting in stagger charging of those capacitances, while also deferring the current into the feedback resistor R_F [82]. Figure 7.34(b) shows a common-gate TIA with a T-coil that separates the drain capacitance from the load capacitance C_L while also deferring the current into the drain resistor R_D [83].

7.8 Distributed-Amplifier TIA

In the previous section, we saw how the speed of a circuit can be increased by breaking up parallel capacitances and inserting small inductors. But what can we do if the speed is limited by the capacitance of a *single* large transistor? We can replace the large transistor by n smaller transistors, each one n times smaller than the original one, and connect them all in parallel. This replacement does not affect the low-frequency behavior of the circuit, but it provides the opportunity to insert inductors between the individual small transistors. This is the basic idea behind a distributed amplifier.

Figure 7.35(a) shows a conventional low-impedance front-end with a FET amplifier stage. Let us assume that the input capacitance of the FET, C_I,

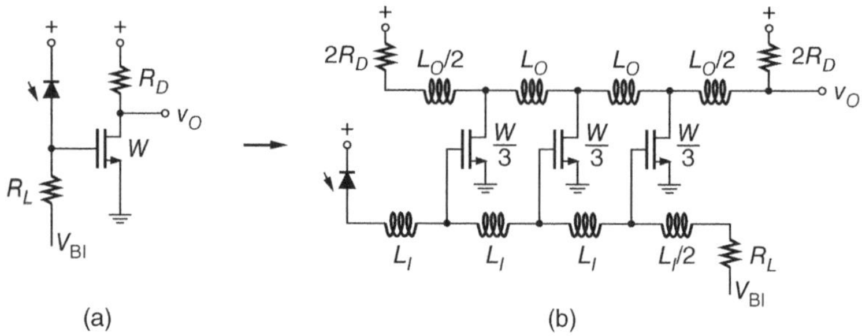

Figure 7.35 (a) Low-impedance front-end and (b) distributed amplifier front-end.

dominates the photodetector capacitance, C_D, and simply inserting an inductor between the detector and the FET does not provide sufficient bandwidth. Applying the idea outlined earlier, we break the FET of width W into three FETs ($n = 3$) of width $W/3$ and connect them in parallel with a string of input inductors L_I and a string of output inductors L_O. The resulting arrangement, shown in Fig. 7.35(b), is known as a *distributed amplifier front-end* or a *distributed preamplifier* (see [84–86] for design examples).

The low-frequency transimpedance of the lumped front-end in Fig. 7.35(a) is given by $R_T = g_m R_D R_L$. By construction, the distributed amplifier front-end must have the same R_T, where g_m now designates the transconductance of all transistors in parallel and R_D designates the value of the two output load resistors in parallel. If we define the transconductance of the individual transistor as $g'_m = g_m/n$ and the resistance of each individual output load resistor as $R'_D = 2R_D$, we can rewrite the low-frequency transimpedance as [84]

$$R_T = g_m R_D R_L = \frac{n}{2}\, g'_m R'_D R_L. \tag{7.30}$$

How large should we make L_I and L_O to obtain a flat broadband response? The answer is provided by the theory of artificial transmission lines [79, 87]: The characteristic impedance of each artificial transmission line must match its termination resistors. In the passband, the characteristic impedances of the input and output lines can be approximated as

$$Z_{TL,I} = \sqrt{\frac{L_I}{C'_I}} \quad \text{and} \quad Z_{TL,O} = \sqrt{\frac{L_O}{C'_O}}, \tag{7.31}$$

where C'_I is the input capacitance of each individual transistor, $C'_I = C_I/n$, and C'_O is the output capacitance of each individual transistor, $C'_O = C_O/n$. Requiring that $Z_{TL,I}$ and $Z_{TL,O}$ match the respective termination resistors R_L and R'_D, results in

$$L_I = R_L^2 C'_I \quad \text{and} \quad L_O = R_D'^2 C'_O. \tag{7.32}$$

Bandwidth. The theory of artificial transmission lines also provides the bandwidth of the input and output lines [79, 87]:

$$f_{\text{cutoff},I} = \frac{1}{\pi\sqrt{L_I C_I'}} \quad \text{and} \quad f_{\text{cutoff},O} = \frac{1}{\pi\sqrt{L_O C_O'}}. \tag{7.33}$$

With Eqs. (7.32) and (7.33), we find $f_{\text{cutoff},I} = 2n/(2\pi R_L C_I)$ and $f_{\text{cutoff},O} = n/(2\pi R_D C_O)$. Thus, assuming that the bandwidth of the original lumped low-impedance front-end was limited by its input pole to $1/[2\pi R_L(C_D + C_I)]$ and that $C_I \gg C_D$, the bandwidth of the distributed amplifier front-end is $2n$ times larger than that of the lumped front-end.

The bandwidth, however, is not only determined by the cutoff frequencies of the input and output transmission lines, but also by the *phase synchronization* between these two lines. Figure 7.36(a) illustrates the three different paths in our example circuit through which the signal propagates from input to output (in general, there are n such paths). Ideally, the delay through all paths is identical such that the signals at the output add up in phase. In the time domain, path differences result in smeared out edges (superimposed time-shifted copies of the edges). In the frequency domain, they manifest themselves as nulls (due to destructive interference) and a reduction in bandwidth. With the delays of the input-line and output-line sections given as [79, 87]

$$\tau_I = \sqrt{L_I C_I'} \quad \text{and} \quad \tau_O = \sqrt{L_O C_O'}, \tag{7.34}$$

phase synchronization is established when $L_I C_I' = L_O C_O'$.

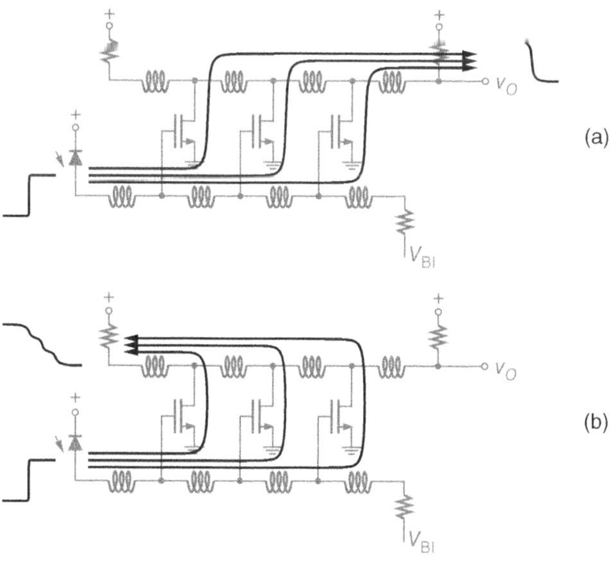

Figure 7.36 (a) Forward paths and (b) reverse paths in the distributed amplifier front-end.

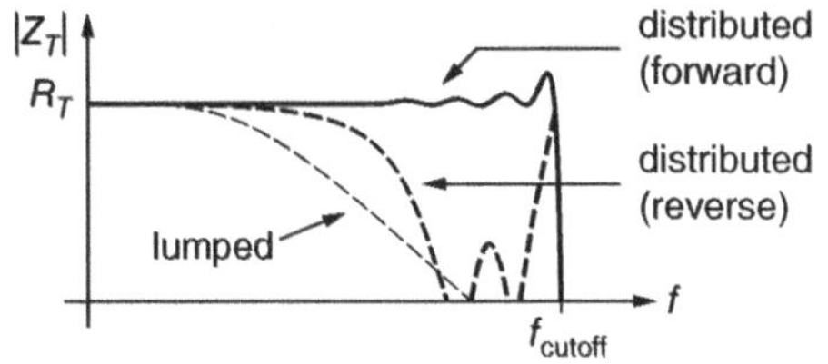

Figure 7.37 Frequency response of the forward and reverse transimpedance of a distributed amplifier front-end.

If we take the output signal from the back end of the output transmission line, we experience large delay mismatches, as shown in Fig. 7.36(b). As a result, the reverse transimpedance, that is, the transimpedance to the back end output, has deep nulls and a bandwidth that is significantly smaller than that of the forward transimpedance [84, 87]. This situation is illustrated in Fig. 7.37. The nulls occur when the phases of the reverse paths are evenly distributed around the circle and thus cancel out. Assuming phase synchronization in the forward direction, $\tau = \tau_I = \tau_O$, the first null occurs for $f = 1/(2\tau n)$. Expressed in terms of the cutoff frequency, this frequency is $f = \pi/(2n) \cdot f_{\text{cutoff}}$. Thus, the factor $2n$ that we gained in the forward bandwidth is lost in the reverse bandwidth. Interestingly, at the very high frequency $f = 1/(2\tau) = \pi/2 \cdot f_{\text{cutoff}}$, all reverse paths are again in phase and the full gain is restored (cf. Fig. 7.37).

Numerical Example. Let us make a numerical example to illustrate the foregoing theory (see Table 7.6). We start with the lumped low-impedance front-end of Fig. 7.35(a) and the values $R_L = 50\ \Omega$, $R_D = 25\ \Omega$, $g_m = 400$ mS,

Table 7.6 Parameters and performance of our lumped and distributed front-end examples.

Parameter	Symbol	Lumped TIA	Distributed TIA
Detector capacitance incl. parasitics	C_D	0.15 pF	0.15 pF
Input capacitance	C_I	1.2 pF	8×0.15 pF
Output capacitance	C_O	1.2 pF	8×0.15 pF
Transconductance	g_m	400 mS	8×50 mS
Input load resistor	R_L	50 Ω	50 Ω
Output load resistor	R_D	25 Ω	50 Ω ‖ 50 Ω
Input inductors	L_I		0.375 nH
Output inductors	L_O		0.375 nH
Input pole/cutoff frequency	$f_{\text{cutoff},I}$	2.36 GHz	42.5 GHz
Output pole/cutoff frequency	$f_{\text{cutoff},O}$	5.31 GHz	42.5 GHz
Transimpedance	R_T	500 Ω	500 Ω
Input impedance	R_I	50 Ω	50 Ω
3-dB bandwidth	BW_{3dB}	2.04 GHz	39.4 GHz

$C_D = 0.15$ pF, $C_I = 1.2$ pF, and $C_O = 1.2$ pF. We easily find the amplifier gain 400 mS $\times$ $25\ \Omega = 10$, the low-frequency transimpedance $R_T = 10 \times 50\ \Omega = 500\ \Omega$, the input pole frequency 2.36 GHz, and the output pole frequency 5.31 GHz. The lumped front-end's 3-dB bandwidth is found with Eq. (H.16) as $BW_{3dB} = 2.04$ GHz. Clearly, the bandwidth is limited by the input capacitance of the large transistor and not the photodetector.

Next, we convert the lumped front-end into a distributed front-end with $n = 8$ sections (cf. Fig. 7.35(b)). The individual transistors are characterized by $g'_m = 400$ mS/8 = 50 mS and $C'_I = C'_O = 1.2$ pF/8 = 0.15 pF (same as C_D). With Eq. (7.32) and the termination resistors $R_L = R'_D = 50\ \Omega$, we find the inductor values $L_I = L_O = 0.375$ nH. The cutoff frequency of both lines is found with Eq. (7.33) as 42.5 GHz, which is also an estimate for the forward bandwidth. The delay of each section is found with Eq. (7.34) as $\tau = 7.5$ ps. Thus the first null of the reverse response occurs at 8.33 GHz.

A circuit simulation of our distributed front-end example reveals that the 3-dB forward bandwidth is 39.4 GHz and the 3-dB reverse bandwidth is 3.7 GHz. Thus, the 3-dB forward bandwidth is about 19 times larger than that of the lumped front-end.

Noise Analysis. If the noise of the low-impedance front-end is dominated by the thermal noise of the load resistor R_L, the input-referred noise current density is white and given by $I_n = \sqrt{4kT/R_L}$, which is about 18 pA/$\sqrt{\text{Hz}}$ for $R_L = 50\ \Omega$. Is the noise of the distributed amplifier front-end the same? By construction, it must be the same at low frequencies. But note that the noise from the load resistor R_L propagates in the opposite direction than the signal. For this reason, the noise from R_L is not amplified by the forward transimpedance, but by something close to the reverse transimpedance [84]. (It is not *exactly* the reverse transimpedance because the input line is not terminated at the photodetector end and some noise gets reflected back into the forward direction.) Figure 7.38 compares a typical input-referred noise current PSD of a distributed amplifier front-end (solid line) [84] with that of a low-impedance front-end (dashed line). At low frequencies both noise densities are the same, but in the band above the reverse bandwidth and below the cutoff frequency, the noise density of the distributed amplifier front-end is lower. The width of this low-noise band increases with the number of sections, n.

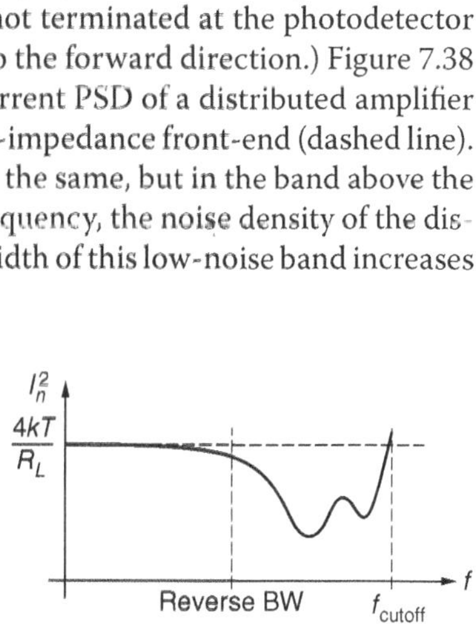

Figure 7.38 Typical input-referred noise current PSD of a distributed amplifier front-end.

A more complete noise analysis must include the noise contributions of the transistors and the termination resistors of the output line [84].

Noise Optimization. To reduce the noise of a distributed amplifier front-end, we can increase the load resistor R_L to, say, 100 Ω [84]. But to preserve impedance matching, a larger R_L also requires a larger L_I, which in turn results in a lower cutoff frequency of the input line. In our example, to match $R_L = 100\ \Omega$ we would need to increase L_I to 1.5 nH, reducing the cutoff frequency to 21.2 GHz. Thus, R_L controls a trade-off between noise performance and bandwidth.

Another noise-reduction technique is to replace R_L with an active circuit that is less noisy than a passive resistor. The input impedance of a shunt-feedback TIA can serve as such an active termination [85].

Second-Order Effects. In our discussion we neglected loss in the transmission lines, mismatch between the transmission lines and their terminations, and other effects that are important in practice.

Loss in the input as well as the output lines limits the maximum number of sections that can be gainfully used [87]. Loss is caused, by the series and substrate resistances of the inductors as well as the resistive input and output components of the transistors (e.g., R_g and g_o of a FET). Cascoded sections can be used to lower the output conductance of the transistors [87, 88] (as well as to lower the input capacitance and to improve the output-to-input isolation). Loss in the inductors results in a gain droop toward higher frequencies [89, 90]. Parasitic capacitances associated with the inductors can detune the phase synchronization and lead to additional gain droop [89, 90].

Mismatch between the artificial transmission lines and the termination resistors results in gain ripple, especially near the cutoff frequency (cf. Fig. 7.37) [87]. The half sections ($L_I/2$, $L_O/2$) at the ends of the transmission lines shown in Fig. 7.35(b) only provide approximate matching at high frequencies. Replacing them with m-derived half sections improves the high-frequency matching [79, 87, 91]. To compensate for gain degradations at low frequencies, a frequency dependent termination can be used [85, 88].

In a variation of the distributed amplifier front-end shown in Fig. 7.35(b), the small inductors that connect the individual transistors are replaced by short transmission lines, such as coplanar waveguides [85, 86]. A distributed amplifier making use of transmission lines is sometimes called a traveling-wave amplifier [87]. Other elements, such as mutually coupled inductors and T-coil networks, also have been used in place of the simple inductors [87].

Designing a distributed amplifier with a flat frequency response and a small group-delay variation is a challenge [87], especially in CMOS technology [89, 90]. Careful optimization guided by detailed circuit simulations is necessary to achieve this goal.

7.9 Summary

The basic shunt-feedback TIA can be enhanced in a number of ways to meet the needs of practical applications:

- *TIA with post amplifier.* Adding a post amplifier substantially increases the transimpedance while only modestly reducing the bandwidth. Parasitic coupling and stability are important issues when integrating a TIA and a high-gain post amplifier on the same chip.
- *TIA with differential inputs and outputs.* Differential-input TIAs can subtract the currents of two photodetectors. When used with a single photodetector, the unused TIA input can be terminated in a balanced or unbalanced manner. Balanced differential TIAs are more immune to power-supply noise and parasitically coupled signals. Unbalanced differential TIAs can provide a higher transimpedance and lower noise. Differential-output TIAs can produce twice the output voltage swing compared with single-ended TIAs. Pseudo-differential TIAs consist of two matched single-ended TIAs. Compared to differential TIAs, they lack common-mode suppression.
- *TIA with DC input current control.* DC input current control removes the average current from the photodetector current feeding only the AC current into the TIA. This provision removes the output offset voltage, maximizes the available output swing, and prevents large DC currents from overloading the TIA. A common implementation consists of a control loop that measures the DC input current (or a proxy thereof, such as the output offset voltage) and controls the amount of DC current that needs to be shunted away from the TIA input(s).
- *TIA with adaptive transimpedance.* Automatic transimpedance (or gain) control increases the input dynamic range of the TIA. A common implementation consists of a control loop that measures the output voltage swing and controls the TIA's transimpedance by means of a variable feedback resistor, a variable input shunt resistor, or a variable input attenuator. The transimpedance can be controlled in discrete steps or continuously. Bandwidth and stability must be maintained over the entire transimpedance tuning range. This requires the simultaneous adjustment of the feedback resistor and the amplifier gain or the feedback capacitance.

Feedforward TIAs, such as the common-base and common-gate TIAs, do not have an explicit feedback loop. Generally, they have a lower transimpedance and are noisier but consume less power and avoid the stability issues associated with shunt-feedback TIAs. A fast input pole relative to the output pole (a large pole spacing χ) makes the feedforward TIA's bandwidth less dependent on the photodetector capacitance, but it also reduces the maximum achievable transimpedance (cf. Table 7.7).

Table 7.7 Summary of transimpedance limit expressions (SFB = shunt feedback, CB = common base, CG = common gate, RGC = regulated cascode).

TIA topology	Transimpedance expression
SFB TIA, single-stage amplifier	$\leq \dfrac{A_0 f_A}{2\pi C_T \cdot BW_{3dB}^2}$
SFB TIA with n-stage post amplifier	$\leq \sqrt{\left(\sqrt[n+1]{\sqrt[4]{2}} - 1\right)^{n+2}} \cdot \dfrac{(A_0 f_A)^{n+1}}{2\pi C_T \cdot BW_{3dB}^{n+2}}$
CB/CG TIA (bound)	$\leq \dfrac{A_0 f_A}{2\pi C_T \chi \cdot BW_{3dB}^2}$
CB/CG TIA (equality)	$= \dfrac{\sqrt{\chi^4 + 6\chi^2 + 1} - \chi^2 - 1}{2\chi} \cdot \dfrac{A_0 f_A}{2\pi C_T \cdot BW_{3dB}^2}$
RGC TIA	$\leq \dfrac{(A_0 f_A)^2}{2\pi C_T \chi^2 \cdot BW_{3dB}^3}$
SFB TIA with CB/CG input stage	$\leq \dfrac{(A_0 f_A)^2}{2\pi C_T \chi \cdot BW_{3dB}^3}$
SFB TIA with RGC input stage	$\leq \dfrac{(A_0 f_A)^3}{2\pi C_T \chi^2 \cdot BW_{3dB}^4}$

The regulated-cascode TIA includes a booster amplifier that reduces the input resistance below that of the common-gate TIA ($<1/g_m$). The regulated-cascode TIA can be designed with smaller transistors, consumes less power, and achieves a higher transimpedance than the common-gate TIA (cf. Table 7.7).

A common-base/gate input stage or a regulated-cascode input stage can be combined with a shunt-feedback TIA to realize a large pole spacing and a high transimpedance at the expense of a higher power dissipation.

The bandwidth of a TIA can be increased by inserting a small inductor or an *LC* filter in between the photodetector and the TIA. More generally, separating parallel capacitances by inserting small inductors results in stagger charging, which is faster than simultaneous charging. Moreover, inserting small inductors in series with load or feedback resistors defers the resistor currents, leading to faster charging times. Besides improving the bandwidth, these inductive broadbanding techniques also improve the noise performance.

The distributed amplifier front-end replaces large transistors with multiple smaller transistors that are interconnected with small inductors (or short transmission lines). Converting a lumped front-end into a distributed front-end with n sections increases the bandwidth by a factor up to about $2n$.

Problems

7.1 **Transimpedance Limit of TIA with Post Amplifier.** Derive the maximum transimpedance as a function of the overall bandwidth BW_{3dB} for a shunt-feedback TIA with a single-pole voltage amplifier followed by an n-stage post amplifier. Assume that the TIA has $Q \leq 1/\sqrt{2}$ to avoid peaking, that all post-amplifier stages have a second-order Butterworth response with the same bandwidth, BW_S, as the TIA, and that each post-amplifier stage has a gain-bandwidth product of $A_0 f_A$.

7.2 **TIA Parameters.** You are reviewing a paper that describes a single-ended TIA consisting of a shunt-feedback stage and a post-amplifier stage. The shunt-feedback amplifier and the post amplifier both have a gain of 6 dB. The section on experimental results reports the transimpedance as 62.5 dBΩ and the averaged input-referred noise current density as 4 $\text{pA}/\sqrt{\text{Hz}}$. What is your comment?

7.3 **Differential TIA.** (a) Calculate the low-frequency relationship between v_{IP} and i_I of the balanced TIA shown in Fig. 7.5. From this result, derive the single-ended input resistance $R_{I,se} = \Delta v_{IP}/\Delta i_I$ "seen" by the photodetector. (b) Derive the single-ended input resistance again, this time starting from the differential input resistance given by $R_{I,d} = 2\Delta(v_{IP} - v_{IN})/\Delta(i_{IP} - i_{IN}) = 2R_F/(A_0 + 1)$ and the common-mode input resistance given by $R_{I,c} = \frac{1}{2}\Delta(v_{IP} + v_{IN})/\Delta(i_{IP} + i_{IN}) = R_F/2$.

7.4 **Variable Feedback Resistor.** A shunt-feedback TIA has a variable feedback resistor R_F (with negligible parallel capacitance). How does the TIA's quality factor Q and bandwidth BW_{3dB} vary for the following cases? Make the simplifying assumptions that $T_A \ll R_F C_T$ and $A_0 \gg 1$ to estimate Q and that $T_A \ll R_F C_T/A_0$ and $A_0 \gg 1$ to estimate BW_{3dB}. (a) The adaptation mechanism controls only R_F; A_0 and T_A remain fixed. (b) The adaptation mechanisms controls A_0 to be proportional to R_F and the gain-bandwidth product remains constant. (c) The adaptation mechanisms controls A_0 to be proportional to $\sqrt{R_F}$ and the gain-bandwidth product remains constant.

7.5 **Input Shunt Resistor.** Calculate the transimpedance $Z_T(s)$ of a TIA with input shunt resistor, as shown in Fig. 7.16. Assume a single-pole model (with time constant T_A) for the voltage amplifier. (a) Find the expressions for R_T, ω_0, and Q. (b) Does the transimpedance limit Eq. (6.14) still hold for this case?

7.6 **Common-Base TIA with Base Resistance.** Derive the transfer function Eq. (7.12) of the common-base TIA including the base resistance R_b. What is the maximum value for R_b such that the input poles have $Q \leq 1/\sqrt{2}$? What is the bandwidth due to the input poles for $Q = 1/\sqrt{2}$?

7.7 **Transimpedance Limit for Common-Base TIA.** Derive the transimpedance of the common-base TIA as a function of its bandwidth BW_{3dB} and technology, as we did in Eq. (7.14), but instead of a bound use the precise bandwidth expression. What pole spacing results in the maximum transimpedance and what is its value?

7.8 **Noise PSD of Common-Gate TIA.** Calculate the input-referred noise current PSD of a common-gate TIA with load resistor R_D and source resistor R_S.

7.9 **Total Noise of Common-Base and Common-Gate TIAs.** Calculate the total input-referred noise currents of the common-base and the common-gate TIAs assuming that the output pole ω_{p2} dominates and that the pole spacing is large.

7.10 **Regulated-Cascode TIA.** Calculate the transimpedance $Z_T(s)$ of the transistor-level regulated-cascode TIA shown in Fig. 7.26(b). Make the following simplifying assumptions: $R_S \rightarrow \infty$, $g_{mb1} = 0$, $g_{o1} = 0$, $C_{gd1} = 0$, and $g_{oB} = 0$ (or include g_{oB} in R_B).

7.11 **Current Conveyor TIA.** Calculate the input resistance R_I of the circuits shown in Fig. 7.28(a) and (b). Make the simplifying assumptions that the transistors (MOSFETs and BJTs) have zero input and zero output conductance and no body effect, that is, g_m is their only model parameter.

7.12 **Bandwidth Extension with Input Series Inductor.** (a) What conditions must be met for Eq. (7.27) to take the form of a Butterworth response? Calculate the increase in 3-dB bandwidth resulting from L_S under these conditions. (b) Repeat (a) for a Bessel response.

References

1 H. H. Kim, S. Chandrasekhar, C. A. Burrus, and J. Bauman. A Si BiCMOS transimpedance amplifier for 10-Gb/s SONET receiver. *IEEE J. Solid-State Circuits*, SC-36(5):769–776, 2001.

2 N. Scheinberg, R. J. Bayruns, and T. M. Laverick. Monolithic GaAs transimpedance amplifiers for fiber-optic receivers. *IEEE J. Solid-State Circuits*, SC-26(12):1834–1839, 1991.

3 O. Wada, T. Hamaguchi, S. Miura, M. Makiuchi, K. Nakai, H. Horimatsu, and T. Sakurai. AlGaAs/GaAs p-i-n photodiode/preamplifier monolithic photoreceiver integrated on semi-insulating GaAs substrate. *Appl. Phys. Lett.*, 46(10):981–983, 1985.

4 H. Khorramabadi, L. D. Tzeng, and M. J. Tarsia. A 1.06Gb/s, −31dBm to 0dBm BiCMOS optical preamplifier featuring adaptive transimpedance. In *ISSCC Digest of Technical Papers*, pages 54–55, February 1995.

5 K. Ohhata, T. Masuda, K. Imai, R. Takeyari, and K. Washio. A wide-dynamic-range, high-transimpedance Si bipolar preamplifier IC for 10-Gb/s optical fiber links. *IEEE J. Solid-State Circuits*, SC-34(1):18–24, 1999.

6 J. S. Weiner, J. S. Lee, A. Leven, Y. Baeyens, V. Houtsma, G. Georgiou, Y. Yang, J. Frackoviak, A. Tate, R. Reyes, R. F. Kopf, W.-J. Sung, N. G. Weimann, and Y.-K. Chen. An InGaAs–InP HBT differential transimpedance amplifier with 47-GHz bandwidth. *IEEE J. Solid-State Circuits*, SC-39(10):1720–1723, 2004.

7 W.-Z. Chen, Y.-L. Cheng, and D.-S. Lin. A 1.8-V 10-Gb/s fully integrated CMOS optical receiver analog front-end. *IEEE J. Solid-State Circuits*, SC-40(6):1388–1396, 2005.

8 A. Maxim. A 54 dBΩ + 42 dB 10 Gb/s SiGe transimpedance-limiting amplifier using bootstrap photodiode capacitance neutralization and vertical threshold adjustment. *IEEE J. Solid-State Circuits*, SC-42(9):1851–1864, 2007. This paper may contain falsified information, see the June 2008 issue.

9 H. Tran, F. Pera, D. S. McPherson, D. Viorel, and S. P. Voinigescu. 6-kΩ 43-Gb/s differential transimpedance-limiting amplifier with auto-zero feedback and high dynamic range. *IEEE J. Solid-State Circuits*, SC-39(10):1680–1696, 2004. Also see the Correction in the December 2004 issue.

10 W. Steiner, H.-M. Rein, and J. Berntgen. Substrate coupling in a high-gain 30-Gb/s SiGe amplifier—modeling, suppression, and measurement. *IEEE J. Solid-State Circuits*, SC-40(10):2035–2045, 2005.

11 J. Kim and J. F. Buckwalter. Bandwidth enhancement with low group-delay variation for a 40-Gb/s transimpedance amplifier. *IEEE Trans. Circuits Syst. I*, CASI-57(8):1964–1972, 2010.

12 E. Säckinger. The transimpedance limit. *IEEE Trans. Circuits Syst. I*, CASI-57(8):1848–1856, 2010.

13 M. Ohara, Y. Akazawa, N. Ishihara, and S. Konaka. High gain equalizing amplifier integrated circuits for a gigabit optical repeater. *IEEE J. Solid-State Circuits*, SC-20(3):703–707, 1985.

14 J. Müllrich, H. Thurner, E. Müllner, J. F. Jensen, W. E. Stanchina, M. Kardos, and H.-M. Rein. High-gain transimpedance amplifier in InP-based HBT technology for receiver in 40-Gb/s optical-fiber TDM links. *IEEE J. Solid-State Circuits*, SC-35(9):1260–1265, 2000.

15 A. Schild, H.-M. Rein, J. Müllrich, L. Altenhain, J. Blank, and K. Schrödinger. High-gain SiGe transimpedance amplifier array for a 12 × 10 Gb/s parallel optical-fiber link. *IEEE J. Solid-State Circuits*, SC-38(1):4–12, 2003.

16 C. H. Doan, S. Emami, A. M. Niknejad, and R. W. Brodersen. Millimeter-wave CMOS design. *IEEE J. Solid-State Circuits*, SC-40(1):144–155, 2005.

17 B. Kleveland, C. H. Diaz, D. Vook, L. Madden, T. H. Lee, and S. S. Wong. Exploiting CMOS reverse interconnect scaling in multigigahertz amplifier and oscillator design. *IEEE J. Solid-State Circuits*, SC-36(10):1480–1488, 2001.

18 F. Vecchi, M. Repossi, W. Eyssa, P. Arcioni, and F. Svelto. Design of low-loss transmission lines in scaled CMOS by accurate electromagnetic simulations. *IEEE J. Solid-State Circuits*, SC-44(9):2605–2615, 2009.

19 E. Säckinger and W. C. Fischer. A 3-GHz, 32-dB CMOS limiting amplifier for SONET OC-48 receivers. *IEEE J. Solid-State Circuits*, SC-35(12):1884–1888, 2000.

20 H. Fukuyama, T. Itoh, T. Furuta, K. Kurishima, M. Tokumitsu, and K. Murata. Two-channel InP HBT differential automatic-gain-controlled transimpedance amplifier IC for 43-Gbit/s DQPSK photoreceiver. In *Compound Semiconductor Integrated Circuit Symposium (CSICS)*, pages 1–4, Monterey, CA, October 2008.

21 J. H. Sinsky, A. Adamiecki, A. Gnauck, C. A. Burrus Jr., J. Leuthold, O. Wohlgemuth, S. Chandrasekhar, and A. Umbach. RZ-DPSK transmission using a 42.7-Gb/s integrated balanced optical front end with record sensitivity. *J. Lightwave Technol.*, LT-22(1):180–185, 2004.

22 J. S. Weiner, A. Leven, V. Houtsma, Y. Baeyens, Y.-K. Chen, P. Paschke, Y. Yang, J. Frackoviak, W.-J. Sung, A. Tate, R. Reyes, R. F. Kopf, and N. G. Weimann. SiGe differential transimpedance amplifier with 50-GHz bandwidth. *IEEE J. Solid-State Circuits*, SC-38(9):1512–1517, 2003.

23 Y. Dong and K. W. Martin. A high-speed fully-integrated POF receiver with large-area photo detectors in 65 nm CMOS. *IEEE J. Solid-State Circuits*, SC-47(9):2080–2092, 2012.

24 B. S. Leibowitz, B. E. Boser, and K. S. J. Pfister. A 256-element CMOS imaging receiver for free-space optical communication. *IEEE J. Solid-State Circuits*, SC-40(9):1948–1956, 2005.

25 T. Nagahori, K. Miyoshi, Y. Aizawa, Y. K. Y. Nukada, N. Kami, and N. Suzuki. An analog front-end chip set employing an electro-optical mixed

design on SPICE for 5-Gb/s/ch parallel optical interconnection. *IEEE J. Solid-State Circuits*, SC-36(12):1984–1991, 2001.
26 D. Caruth, S. C. Shen, D. Chan, M. Feng, and J. Schutt-Ainé. A 40 Gb/s integrated differential PIN+TIA with DC offset control using InP SHBT technology. In *IEEE Gallium Arsenide Integrated Circuit (GaAs IC) Symposium*, pages 59–62, October 2002.
27 S. Brigati, P. Colombara, L. D'Ascoli, U. Gatti, T. Kerekes, and P. Malcovati. A SiGe BiCMOS burst-mode 155-Mb/s receiver for PON. *IEEE J. Solid-State Circuits*, SC-37(7):887–894, 2002.
28 A. Maxim. A 10Gb/s SiGe transimpedance amplifier using a pseudo-differential input stage and a modified Cherry-Hooper amplifier. In *Symposium on VLSI Circuits Digest of Technical Papers*, pages 404–407, June 2004. This paper may contain falsified information, see the June 2008 issue.
29 R. G. Meyer and W. D. Mack. A wideband low-noise variable-gain BiCMOS transimpedance amplifier. *IEEE J. Solid-State Circuits*, SC-29(6):701–706, 1994.
30 B. Razavi. *Design of Integrated Circuits for Optical Communications*. McGraw-Hill, New York, 2003.
31 N. Ishihara, S. Fujita, M. Togashi, S. Hino, Y. Arai, N. Tanaka, Y. Kobayashi, and Y. Akazawa. 3.5-Gb/s × 4-ch Si bipolar LSI's for optical interconnections. *IEEE J. Solid-State Circuits*, SC-30(12):1493–1501, 1995.
32 J. Müllrich, T. F. Meister, M. Rest, W. Bogner, A. Schöpflin, and H.-M. Rein. 40 Gb/s transimpedance amplifier in SiGe bipolar technology for receiver in optical-fibre TDM links. *Electron. Lett.*, 34(5):452–453, 1998.
33 J. D. Yoder. Optical receiver preamplifier dynamic range enhancing circuit and method. U.S. Patent No. 5,734,300, March 1998.
34 K. Kieschnick and H. Zimmermann. High-sensitivity BiCMOS OEIC for optical storage systems. *IEEE J. Solid-State Circuits*, SC-38(4):579–584, 2003.
35 N. Tadić and H. Zimmermann. Low-power BiCMOS optical receiver with voltage-controlled transimpedance. *IEEE J. Solid-State Circuits*, SC-42(3):613–626, 2007.
36 N. Tadić, M. Zogović, W. Gaberl, and H. Zimmermann. A 78.4 dB photo-sensitivity dynamic range, 285 TΩHz transimpedance bandwidth product BiCMOS optical sensor for optical storage systems. *IEEE J. Solid-State Circuits*, SC-46(5):1170–1182, 2011.
37 T. Ruotsalainen, P. Palojärvi, and J. Kostamovaara. A current-mode gain-control scheme with constant bandwidth and propagation delay for a transimpedance preamplifier. *IEEE J. Solid-State Circuits*, SC-34(2):253–258, 1999.

38 T. Ruotsalainen, P. Palojärvi, and J. Kostamovaara. A wide dynamic range receiver channel for a pulsed time-of-flight laser radar. *IEEE J. Solid-State Circuits*, SC-36(8):1228–1238, 2001.

39 H. Ikeda, T. Ohshima, M. Tsunotani, T. Ichioka, and T. Kimura. An auto-gain control transimpedance amplifier with low noise and wide input dynamic range for 10-Gb/s optical communication systems. *IEEE J. Solid-State Circuits*, SC-36(9):1303–1308, 2001.

40 R. G. Meyer and W. D. Mack. Monolithic AGC loop for a 160 Mb/s transimpedance amplifier. *IEEE J. Solid-State Circuits*, SC-31(9):1331–1335, 1996.

41 D. M. Pietruszynski, J. M. Steininger, and E. J. Swanson. A 50-Mbit/s CMOS monolithic optical receiver. *IEEE J. Solid-State Circuits*, SC-23(6):1426–1433, 1988.

42 M. A. T. Sanduleanu and P. Manteman. A low noise, wide dynamic range, transimpedance amplifier with automatic gain control for SDH/SONET (STM16/OC48) in a 30GHz f_T BiCMOS process. In *Proceedings of European Solid-State Circuits Conference*, September 2001.

43 K. Schrödinger, J. Stimma, and M. Mauthe. A fully integrated CMOS receiver front-end for optic gigabit Ethernet. *IEEE J. Solid-State Circuits*, SC-37(7):874–880, 2002.

44 K. Phang and D. A. Johns. A CMOS optical preamplifier for wireless infrared communications. *IEEE Trans. Circuits Syst. II*, CASII-46(7):852–859, 1999.

45 B. L. Kasper, A. R. McCormick, C. A. Burrus Jr., and J. R. Talman. An optical-feedback transimpedance receiver for high sensitivity and wide dynamic range at low bit rates. *J. Lightwave Technol.*, LT-6(2):329–338, 1988.

46 W.-Z. Chen and R.-M. Gan. 1.8V variable gain transimpedance amplifiers with constant damping factor for burst-mode optical receiver. In *IEEE Radio Frequency Integrated Circuits (RFIC) Symposium*, pages 691–694, Long Beach, CA, June 2005.

47 W.-Z. Chen and R.-M. Gan. A single chip 2.5 Gbps CMOS burst mode optical receiver. In *Symposium on VLSI Circuits Digest of Technical Papers*, pages 120–121, June 2006.

48 K. Yamashita, T. Kinoshita, Y. Takasaki, M. Maeda, T. Kaji, and N. Maeda. A variable transimpedance preamplifier for use in wide dynamic range optical receivers. *IEEE J. Solid-State Circuits*, SC-21(2):324–330, 1986.

49 L. A. D. van den Broeke and A. J. Nieuwkerk. Wide-band integrated optical receiver with improved dynamic range using a current switch at the input. *IEEE J. Solid-State Circuits*, SC-28(7):862–864, 1993.

50 P. R. Gray, P. J. Hurst, S. H. Lewis, and R. G. Meyer. *Analysis and Design of Analog Integrated Circuits*. John Wiley & Sons, Inc., New York, 5th edition, 2009.

51 D. Mičušík and H. Zimmermann. A 240MHz-BW 112dB-DR TIA. In *ISSCC Digest of Technical Papers*, pages 554–555, February 2007.

52 D. Mičušík and H. Zimmermann. 130dB-DR transimpedance amplifier with monotonic logarithmic compression and high-current monitor. In *ISSCC Digest of Technical Papers*, pages 78–79, February 2008.

53 T. Vanisri and C. Toumazou. Integrated high frequency low-noise current-mode optical transimpedance preamplifiers: theory and practice. *IEEE J. Solid-State Circuits*, SC-30(6):677–685, 1995.

54 C. Kromer, G. Sialm, T. Morf, M. L. Schmatz, F. Ellinger, D. Erni, and H. Jäckel. A low-power 20-GHz 52-dBΩ transimpedance amplifier in 80-nm CMOS. *IEEE J. Solid-State Circuits*, SC-39(6):885–894, 2004.

55 P. C. P. Chen, A. M. Pappu, Z. Fu, W. Wattanapanitch, and A. B. Apsel. A 10 Gb/s optical receiver in 0.25μm silicon-on-sapphire CMOS. In *Proceedings of IEEE International Symposium on Circuits and Systems*, pages 193–196, Seattle, WA, May 2008.

56 Y. Perelman and R. Ginosar. A low-light-level sensor for medical diagnostic applications. *IEEE J. Solid-State Circuits*, SC-36(10):1553–1558, 2001.

57 C. L. Schow, F. E. Doany, C. Chen, A. V. Rylyakov, C. W. Baks, D. M. Kuchta, R. A. John, and J. A. Kash. Low-power 16 × 10 Gb/s bi-directional single chip CMOS optical transceivers operating at <5 mW/Gb/s/link. *IEEE J. Solid-State Circuits*, SC-44(1):301–313, 2009.

58 I. Darwazeh and B. Wilson. Hybrid π common base analysis. *IEEE Trans. Circuits Syst.*, CAS-37(5):655–656, 1990.

59 D. Guckenberger, J. D. Schaub, and K. T. Kornegay. A DC-coupled low-power transimpedance amplifier architecture for Gb/s communication system applications. In *IEEE Radio Frequency Integrated Circuits (RFIC) Symposium*, pages 515–518, June 2004.

60 C. D. Motchenbacher and F. C. Fitchen. *Low-Noise Electronic Design*. John Wiley & Sons, Inc., New York, 1973.

61 G. Kalogerakis, T. Moran, T. Nguyen, and G. Denoyer. A quad 25Gb/s 270mW TIA in 0.13μm BiCMOS with <0.15dB crosstalk penalty. In *ISSCC Digest of Technical Papers*, pages 116–117, February 2013.

62 F. Bruccoleri, E. A. M. Klumperink, and B. Nauta. Wide-band CMOS low-noise amplifier exploiting thermal noise canceling. *IEEE J. Solid-State Circuits*, SC-39(2):275–282, 2004.

63 L. Belostotski. No noise is good noise. *IEEE Microwave Mag.*, 17(8):28–40, 2016.

64 S. G. Kim, S. H. Jung, Y. S. Eo, S. H. Kim, X. Ying, H. Choi, C. Hong, K. Lee, and S. M. Park. A 50-Gb/s differential transimpedance amplifier in 65nm CMOS technology. In *IEEE Asian Solid-State Circuits Conference (ASSCC)*, pages 357–360, November 2014.

65 S. M. Park, J. Lee, and H.-J. Yoo. 1-Gb/s 80-dBΩ fully differential CMOS transimpedance amplifier in multichip on oxide technology for optical interconnects. *IEEE J. Solid-State Circuits*, SC-39(6):971–974, 2004.

66 S. M. Park and C. Toumazou. A packaged low-noise high-speed regulated cascode transimpedance amplifier using a 0.6 μm N-well CMOS technology. In *Proceedings of European Solid-State Circuits Conference*, September 2000.

67 S. M. Park and H.-J. Yoo. 1.25-Gb/s regulated cascode CMOS transimpedance amplifier for Gigabit Ethernet applications. *IEEE J. Solid-State Circuits*, SC-39(1):112–121, 2004.

68 B. A. Rodriguez, G. C. Temes, K. W. Martin, S. M. L. Law, R. Handy, and N. Kadekodi. An NMOS buffer amplifier. *IEEE J. Solid-State Circuits*, SC-19(1):69–71, 1984.

69 S. Goswami, J. Silver, T. Copani, W. Chen, H. J. Barnaby, B. Vermeire, and S. Kiaei. A 14mW 5Gb/s CMOS TIA with gain-reuse regulated cascode compensation for parallel optical interconnects. In *ISSCC Digest of Technical Papers*, pages 100–101, February 2009.

70 S. I. Long and J. Q. Zhang. Low power GaAs current-mode 1.2 Gb/s interchip interconnections. *IEEE J. Solid-State Circuits*, SC-32(6):890–897, 1997.

71 S. Kurtti and J. Kostamovaara. Laser radar receiver channel with timing detector based on front end unipolar-to-bipolar pulse shaping. *IEEE J. Solid-State Circuits*, SC-44(3):835–847, 2009.

72 Z. Lu, K. S. Yeo, J. Ma, M. A. Do, W. M. Lim, and X. Chen. Broad-band design techniques for transimpedance amplifiers. *IEEE Trans. Circuits Syst. I*, CASI-54(3):590–600, 2007.

73 C.-F. Liao and S.-I. Liu. 40 Gb/s transimpedance-AGC amplifier and CDR circuit for broadband data receivers in 90 nm CMOS. *IEEE J. Solid-State Circuits*, SC-43(3):642–655, 2008.

74 B. Gilbert. Current-mode circuits from a translinear viewpoint: a tutorial. In C. Toumazou, F. J. Lidgey, and D. G. Haigh, editors, *Analogue IC Design: The Current-Mode Approach*, pages 11–91. Institution of Engineering and Technology (IET), London, U.K., 1990.

75 A. S. Sedra and G. W. Roberts. Current conveyor theory and practice. In C. Toumazou, F. J. Lidgey, and D. G. Haigh, editors, *Analogue IC Design: The Current-Mode Approach*, pages 93–126. Institution of Engineering and Technology (IET), London, U.K., 1990.

76 M. Neuhäuser, H.-M. Rein, and H. Wernz. Low-noise, high-gain Si-bipolar preamplifiers for 10 Gb/s optical-fiber links-design and realization. *IEEE J. Solid-State Circuits*, SC-31(1):24–29, 1996.

77 H.-M. Rein and M. Möller. Design considerations for very-high-speed Si-bipolar IC's operating up to 50 Gb/s. *IEEE J. Solid-State Circuits*, SC-31(8):1076–1090, 1996.

78 J. J. Morikuni and S.-M. Kang. An analysis of inductive peaking in photoreceiver design. *J. Lightwave Technol.*, LT-10(10):1426–1437, 1992.

79 T. H. Lee. *The Design of CMOS Radio-Frequency Integrated Circuits.* Cambridge University Press, Cambridge, U.K., 2nd edition, 2003.

80 S. Shekhar, J. S. Walling, and D. J. Allstot. Bandwidth extension techniques for CMOS amplifiers. *IEEE J. Solid-State Circuits*, SC-41(11):2424–2439, 2006.

81 P. Starič and E. Margan. *Wideband Amplifiers.* Springer, Dordrecht, The Netherlands, 2006.

82 A. Meaamar, C. C. Boon, K. S. Yeo, and M. A. Do. A wideband low power low-noise amplifier in CMOS technology. *IEEE Trans. Circuits Syst. I*, CASI-57(4):773–782, 2010.

83 J. Han, B. Choi, M. Seo, J. Yun, D. Lee, T. Kim, Y. Eo, and S. M. Park. A 20-Gb/s transformer-based current-mode optical receiver in 0.13-μm CMOS. *IEEE Trans. Circuits Syst. II*, CASII-57(5):348–352, 2010.

84 A. P. Freundorfer and T. L. Nguyen. Noise in distributed MESFET preamplifiers. *IEEE J. Solid-State Circuits*, SC-31(8):1100–1111, 1996.

85 S. Kimura, Y. Imai, and Y. Miyamoto. Direct-coupled distributed baseband amplifier IC's for 40-Gb/s optical communication. *IEEE J. Solid-State Circuits*, SC-31(10):1374–1379, 1996.

86 S. van Waasen, A. Umbach, U. Auer, H.-G. Bach, R. M. Bertenburg, G. Janssen, G. G. Mekonnen, W. Passenberg, R. Reuter, W. Schlaak, C. Schramm, G. Unterbörsch, P. Wolfram, and F.-J. Tegude. 27-GHz bandwidth high-speed monolithic integrated optoelectronic photoreceiver consisting of a waveguide fed photodiode and an InAlAs/InGaAs-HFET traveling wave amplifier. *IEEE J. Solid-State Circuits*, SC-32(9):1394–1401, 1997.

87 T. T. Y. Wong. *Fundamentals of Distributed Amplification.* Artech House, Boston, MA, 1993.

88 T. Shibata, S. Kimura, H. Kimura, Y. Imai, Y. Umeda, and Y. Akazawa. A design technique for a 60 GHz-bandwidth distributed baseband amplifier IC module. *IEEE J. Solid-State Circuits*, SC-29(12):1537–1544, 1994.

89 H.-T. Ahn and D. J. Allstot. A 0.58.5-GHz fully differential CMOS distributed amplifier. *IEEE J. Solid-State Circuits*, SC-37(8):985–993, 2002.

90 B. M. Ballweber, R. Gupta, and D. J. Allstot. A fully-integrated 0.55.5-GHz CMOS distributed amplifier. *IEEE J. Solid-State Circuits*, SC-35(2):231–239, 2000.

91 D. M. Pozar. *Microwave Engineering.* John Wiley & Sons, Inc., Hoboken, NJ, 4th edition, 2012.

8

Advanced Transimpedance Amplifier Design II

In the following, we explore ways to reduce the noise by replacing the noisy feedback resistor of the shunt-feedback TIA with a capacitive, optical, or active-feedback device. Then, we discuss the current-mode TIA in which the voltage amplifier is replace by a current amplifier. After that, we look at photodetector bootstrapping, a technique to reduce the detector's effective capacitance. Finally, we turn to TIAs for specialized applications, such as burst-mode receivers and analog receivers.

8.1 TIA with Nonresistive Feedback

Capacitive-Feedback TIA with Differentiator. In medium- and low-speed applications, a large portion of the TIA's input-referred noise current originates from the feedback resistor(s) (cf. Section 6.3). In principle, we can get rid of this noise by replacing the feedback resistor with a noise-free feedback capacitor. But unfortunately, this replacement turns the TIA into an integrator, which severely distorts the signal waveforms.

One solution to this problem is to follow the integrator with a differentiator, as shown in Fig. 8.1(a), thus equalizing the frequency response [1, 2]. Assuming large voltage gains for A_1 and A_2, the low-frequency transimpedance of this circuit is [1]

$$R_T = \frac{C_2}{C_1} R_F \tag{8.1}$$

and its input-referred noise current PSD due to R_F is [1]

$$I_{n,TIA}^2 = \left(\frac{C_1}{C_2}\right)^2 \frac{4kT}{R_F} + \ldots \tag{8.2}$$

Equation (8.1) suggests that the integrator–differentiator circuit can be re-interpreted as a current amplifier consisting of A_1, C_1, and C_2 with current

Analysis and Design of Transimpedance Amplifiers for Optical Receivers, First Edition. Eduard Säckinger.

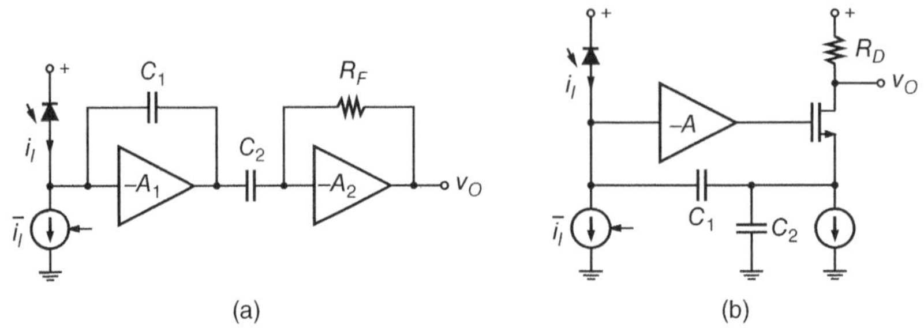

Figure 8.1 TIAs with capacitive feedback: (a) integrator–differentiator and (b) current-divider configuration.

gain C_2/C_1 followed by a resistive shunt-feedback TIA consisting of A_2 and R_F. Comparing Eqs (8.1) and (8.2), we find that for $C_2 > C_1$ the noise due to R_F is lower than that of the equivalent resistive-feedback TIA. For example, for $C_2/C_1 = 2$, the noise contribution of the capacitive-feedback TIA is $I^2_{n,TIA} = kT/R_F$, whereas the noise contribution of the equivalent resistive-feedback TIA with $2R_F$ is $I^2_{n,TIA} = 4kT/(2R_F)$, which is twice as large.

The main drawback of this arrangement is that the integrator quickly saturates (overloads) for input signals that contain a DC component. This issue can be avoided by adding a DC input current control circuit that removes the average input current, $\bar{i}_I$, from the input signal, as indicated with the current source in Fig. 8.1(a). (A feedback circuit similar to that in Fig. 7.11 can be used to control the current source.) Alternatively, a large resistor in parallel to C_1 [1] can provide some tolerance to DC currents. Unfortunately, both of these methods add noise.

Another drawback, familiar from the high-impedance front-end (cf. Section 6.1), is that the integrator saturates for long runs of zeros or ones, making it impossible for the differentiator to recover the original signal waveform.

The main application of the integrator–differentiator circuit in Fig. 8.1(a) has been for high-sensitivity current probes [1, 2].

Capacitive-Feedback TIA with Current Divider. Another capacitive-feedback circuit, which has been proposed with optical receivers in mind [3, 4] and also has found application in microelectromechanical systems (MEMS) [5], is shown in Fig. 8.1(b). This circuit produces an AC drain current that is $(C_1 + C_2)/C_1$ times larger than the AC input current i_i. To understand the origin of this gain, consider that the currents through C_1 and C_2 must sum up to the AC drain current and that, given a virtual ground at the input of the voltage amplifier, the ratio of these two currents must be C_1 to C_2. Thus, the AC drain current is divided into two parts, with the part through C_1 being equal to i_i and the part through C_2 being equal to $C_2/C_1 \cdot i_I$. Finally, the load resistor R_D converts the drain current into a voltage, making this circuit a TIA.

Assuming a large voltage gain A, the low-frequency transimpedance of this circuit is [3, 5]

$$R_T = \left(\frac{C_1 + C_2}{C_1}\right) R_D \tag{8.3}$$

and its input-referred noise current PSD due to R_D is [5]

$$I^2_{n,TIA} = \left(\frac{C_1}{C_1 + C_2}\right)^2 \frac{4kT}{R_D} + \ldots \tag{8.4}$$

Comparing Eqs. (8.3) and (8.4), we find that for $C_2 > 0$, the noise due to R_D is lower than that of a simple common-gate TIA with the same transimpedance. But like the integrator–differentiator circuit, this circuit also suffers from intolerance to DC input currents, necessitating a DC input current control circuit [3]. Although this TIA does not contain an explicit integrator, the voltage amplifier output follows the integral of the input current and thus is susceptible to saturation when long runs of zeros or ones occur.

Note that the integrate-and-dump receiver that we discussed in Section 4.9 can be realized as a TIA with capacitive feedback and a reset switch [6]. In this case, the saturation problem is resolved by periodically discharging the feedback capacitor.

Optical-Feedback TIA. Another way to eliminate the noise of the feedback resistor is to replace it with an optical-feedback scheme [7, 8]. This approach is illustrated in Fig. 8.2. The voltage amplifier drives a light-emitting diode (LED) through the series resistor R_S, producing an optical feedback signal. This optical signal illuminates a p–i–n photodetector, generating an electrical current that is approximately proportional to v_O. Assuming a large voltage gain A, the current from the feedback photodetector tracks the current i_I from the input photodetector.

The entire optical feedback path can be described by the coupling factor K, defined as the ratio of the small-signal p–i–n photodetector current to the small-signal LED current (thus K is the product of the quantum efficiency

Figure 8.2 TIA with optical feedback.

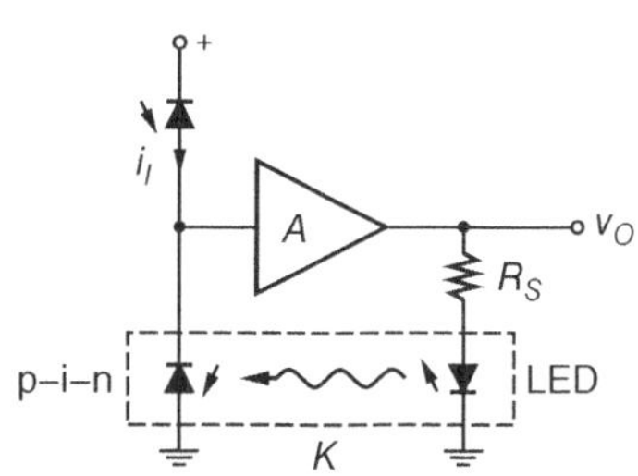

of the p–i–n detector, the differential quantum efficiency of the LED, and the coupling loss from the LED to the p–i–n detector). Typical values for K are in the range 10^{-4} to 10^{-3}. The low-frequency transimpedance of the optical-feedback TIA is [8]

$$R_T = \frac{R_S}{K} \tag{8.5}$$

and its input-referred noise current PSD due to R_S is [8]

$$I^2_{n,TIA} = K^2 \frac{4kT}{R_S} + \ldots \tag{8.6}$$

For $K < 0$, which is easy to realize, the noise due to R_S is lower than that of a resistive shunt-feedback TIA with the same transimpedance.

A detailed analysis [8] shows that optical feedback offers an improved dynamic range over resistive feedback for bit rates up to a few hundred megabits per second, with the exact crossover point depending on the value of K.

The optical-feedback TIA also has the advantage of a very low parasitic feedback capacitance C_F [8]. The bandwidth of TIAs with large feedback resistors and high amplifier gains is often limited by the parasitic feedback capacitance. For the optical-feedback TIA, this capacitance can be made arbitrarily small by increasing the spacing between the LED and the p–i–n photodiode (15 cm of fiber are used in [8]).

The main drawbacks of optical feedback are the cost and board space taken up by the optocoupler, which cannot be integrated in a standard technology.

The aforementioned optical-feedback scheme takes advantage of the fact that a resistor R connected in series with an ideal current attenuator K has the effective resistance R/K for signals but the $1/K$ times larger resistance R/K^2 for noise. Current attenuators realized in the electrical rather than the optical domain can achieve a similar effect [2]. Note that the current attenuator requires a third terminal and thus it is not possible to build a two-terminal resistor with less noise than required by physics.

Active-Feedback TIA. In yet another variation of the shunt-feedback TIA, the feedback resistor R_F is replaced by a transistor. The feedback transistor can be a FET or a BJT. In either case there are three different feedback configurations: common source, common drain, or common gate for the FET case and common emitter, common collector, or common base for the BJT case.

In the following we look at active-feedback TIAs based on a common-source and a common-drain feedback device in more detail. TIAs based on a common-gate (common-base) feedback device [9] are less common.

Common-Source Feedback Device. The basic circuit with a common-source feedback device is shown in Fig. 8.3(a) [10, 11]. Note that the FET here does *not* act as a variable resistor like in Fig. 7.14(a), but as a voltage-controlled

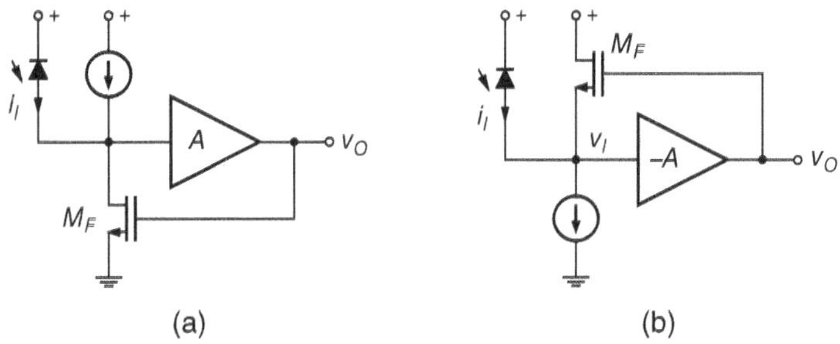

Figure 8.3 TIA with active feedback through an n-MOS device: (a) common-source and (b) common-drain configuration.

current source (transconductor) driven by the TIA output voltage and injecting a current into the TIA input. This type of feedback is known as *active feedback* [11, 12]. To obtain negative feedback through the inverting M_F, the voltage amplifier must be of the noninverting type. It can be implemented, for example, with two cascaded common-source (common-emitter) stages [11] or a combination of common-gate (common-base) and common-drain (common-collector) stages [12–14].

The low-frequency transimpedance of the common-source active-feedback TIA is [10, 11] (cf. Eq. (I.145))

$$R_T = \frac{1}{g_{mF}} \tag{8.7}$$

and its input-referred noise current PSD due to M_F is (cf. Eq. (I.147))

$$I^2_{n,TIA} = 4kT\tilde{\Gamma}_F g_{mF} + \dots, \tag{8.8}$$

where g_{mF} is the transconductance of M_F, $\tilde{\Gamma}_F = \gamma_F g_{d0F}/g_{mF}$ (cf. Section 6.3), and we neglected the output conductance of M_F. The low-frequency input resistance is $R_I = 1/(A_0 g_{mF})$. If we take $A(s)$ as frequency independent, the TIA's bandwidth becomes $BW_{3dB} = A_0 g_{mF}/(2\pi C_T)$, where C_T is the total capacitance at the input node [10, 11]. If we model $A(s)$ as a first-order low-pass and place the pole such that the TIA assumes a second-order Butterworth response, the bandwidth increases by $\sqrt{2}\times$ over the frequency-independent case (cf. Eq. (I.146)). [→ Problem 8.1.]

If we replace the noninverting amplifier in Fig. 8.3(a) with a simple wire ($A = 1$), we end up with a gate-drain connected FET. This degenerate form of the active-feedback TIA can serve as a receiver front-end in the same way a simple resistor can serve as a low- or high-impedance front-end [15].

Common-Drain Feedback Device. Active feedback through a common-drain device is shown in Fig. 8.3(b) [16–18]. This arrangement requires an inverting voltage amplifier, just like in the case of resistive feedback. Here, M_F acts as a voltage-controlled current source producing a feedback current that depends on the voltage *difference* $v_O - v_I$. Note the close analogy to resistive feedback for which the feedback current is given by $(v_O - v_I)/R_F$.

The low-frequency transimpedance of the common-drain active-feedback TIA is (cf. Eq. (I.150))

$$R_T = \frac{A_0}{(A_0 + 1)g_{mF}} \tag{8.9}$$

and its input-referred noise current PSD due to M_F is (cf. Eq. (I.153))

$$I^2_{n,TIA} = 4kT\tilde{\Gamma}_F g_{mF} + \dots, \tag{8.10}$$

where we neglected the bulk transconductance, output conductance, and induced gate noise of M_F. The low-frequency input resistance is $R_I = 1/[(A_0 + 1)g_{mF}]$ [16], which happens to be the same expression as for the regulated-cascode TIA. Note the similarity between this active-feedback TIA and the regulated-cascode TIA, the main difference being that the output signal is taken from the booster amplifier's output rather than the input FET's drain. The frequency response and bandwidth of this active-feedback TIA are identical to those of the resistive feedback TIA, if we substitute R_F by $1/g_{mF}$ (and neglect g_{mb}, g_o, and C_{gs} of M_F). [→ Problem 8.2.]

Active Feedback versus Resistive Feedback. The active-feedback TIA potentially is less noisy at low frequencies than the resistive-feedback TIA [19]. To realize this noise advantage, $\tilde{\Gamma}_F$ must be less than one, a condition that occurs for long-channel devices where theoretically $\tilde{\Gamma}_F \approx \frac{2}{3}$. If we replace M_F with a BJT, the noise due to the feedback device is lower than that of the corresponding feedback resistor if $R_{bF} < 1/(2g_{mF})$, where R_{bF} is the base resistance of the feedback BJT [20] (cf. Eq. (I.148)). Another advantage of the active-feedback TIA is that the voltage amplifier needs to drive only the small capacitive load presented by M_F.

On the downside, active feedback tends to result in a higher total capacitance C_T at the input and more high-frequency (f^2) noise than resistive feedback.

Furthermore, active feedback is less linear than resistive feedback which may be a disadvantage (because of signal distortions) or an advantage (because of desirable signal compression) depending on the application. If linearity is important, the active-feedback TIA can be followed by a nonlinearity that cancels the feedback nonlinearity. In the case of common-source (or common-emitter) active feedback this is easily implemented by driving a matched common-source (or common-emitter) transistor from the voltage amplifier output [19, 20].

Besides the optical receiver application shown in Fig. 8.3, active-feedback TIAs also find application in wideband LNAs for wireless applications [16–18], MEMS [13], and as load elements in broadband amplifier stages [21–23].

8.2 Current-Mode TIA

What happens if we replace the voltage amplifier in the shunt-feedback TIA with a current amplifier, as shown in Fig. 8.4? The current amplifier senses the input current, i, with a low-impedance input ($R_I = R_S$) and outputs the amplified current, $A\,i$, at a high-impedance output ($R_O \to \infty$). Circuits like this, where signals are represented by currents rather than voltages, are called *current-mode circuits.*

Simple Analysis. Let us assume for now that the current amplifier has a high gain, $A(s) \to \infty$. Then, for the output current to remain finite, the input current must stay close to zero. Like a voltage amplifier has a small input voltage (virtual ground), a current amplifier has a small input current ("virtual open") when negative feedback is applied and the loop gain is high. Thus, almost all of the photodetector current i_I flows into R_F. With the input voltage of the current amplifier being close to zero (grounded through R_S), the output voltage becomes $v_O = -R_F i_I$. Interestingly, no matter whether a (high-gain) voltage amplifier or a (high-gain) current amplifier is used, the overall transimpedance is the same and is determined only by the feedback element R_F.

There are however two interesting differences: First, whereas the voltage-mode TIA features a low input impedance only as a result of the feedback action, the current-mode TIA has a low input impedance *even before* feedback is applied [24, 25] (feedback reduces it further). This feature promises a strong suppression of the total capacitance at the input ($C_T = C_D + C_I$), resulting in a low sensitivity to the photodetector capacitance C_D.

Second, whereas the AC feedback current in the voltage-mode TIA decreases with increasing R_F, the current source output of the current-mode TIA *forces* a feedback current that is independent of R_F (neglecting C_L). As a result, the TIA's bandwidth becomes independent of R_F and R_T. This property is called *gain-bandwidth independence* [24, 25]. We return to this topic later.

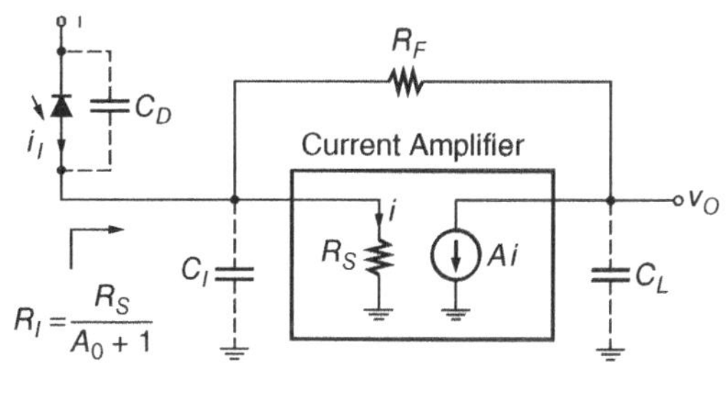

Figure 8.4 Shunt-feedback TIA with current amplifier.

Current Amplifier with Finite Gain and Bandwidth. Now dropping the assumption of a high current gain A, we find the low-frequency transimpedance of the circuit in Fig. 8.4 as [24, 25] (cf. Eq. (I.155))

$$R_T = \frac{A_0}{A_0 + 1}\left(R_F - \frac{R_S}{A_0}\right), \tag{8.11}$$

where A_0 is the low-frequency current gain. Usually, R_S/A_0 is much smaller than R_F and thus the result is essentially identical to that of the voltage-mode TIA in Eq. (6.9). The input resistance of the current-mode TIA is $R_I = R_S/(A_0 + 1)$ at low frequencies and $R_I = R_S$ at high frequencies. [→ Problem 8.3.]

Assuming a current amplifier with a single pole at f_A and $C_L = 0$, we find the bandwidth of the current-mode TIA as $BW_{3\text{dB}} \approx \sqrt{2}A_0 f_A$, which is essentially the gain-bandwidth product of the current amplifier (cf. Eq. (I.157)). As expected, this expression does *not* depend on R_F or R_T, that is, there is no trade-off between bandwidth and transimpedance. This is in stark contrast to the voltage-mode TIA, Eqs (6.4) and (6.13), where the bandwidth depends strongly on the transimpedance.

Transimpedance Limit. Does this mean that the current-mode TIA somehow gets around the transimpedance limit? No, the load capacitance C_L, which we neglected thus far, is the spoiler. Repeating the bandwidth calculation with $C_L \neq 0$ (but $R_S = 0$ to avoid third-order expressions), we find $BW_{3\text{dB}} \approx \sqrt{2A_0(A_0 + 1)/(2\pi R_F C_L)}$ (cf. Eq. (I.161)). Now, the bandwidth does depend on the transimpedance and the transimpedance limit becomes [26] (cf. Eq. (I.162))

$$R_T \leq \frac{A_0 f_A}{2\pi C_L \cdot BW_{3\text{dB}}^2}. \tag{8.12}$$

Interestingly, this limit has the same form as that of the voltage-mode TIA, Eq. (6.14), but the role of C_T is now played by C_L. Considering the adjoint network [27] for the voltage-mode TIA in Fig. 6.3(a) explains this switch: The adjoint network, which has the same transfer function as the original network, matches the network of the current-mode TIA in Fig. 8.4 with C_T moved from the input to the output. In conclusion, we can expect the current-mode TIA to have a higher transimpedance only if $C_L < C_T$.

Implementation Examples. In the following, we illustrate the implementation of current-mode TIAs with two transistor-level examples.

Figure 8.5(a) shows a simplified version of the HBT circuit in [25]. A common-base input stage, Q_1, provides the low input resistance of the current amplifier ($R_S \approx 1/g_{m1}$) and R_C converts the input current into a voltage. This voltage is level shifted with a coupling capacitor and converted back into a current with Q_2 (biasing network for Q_2 is not shown). The current amplifier

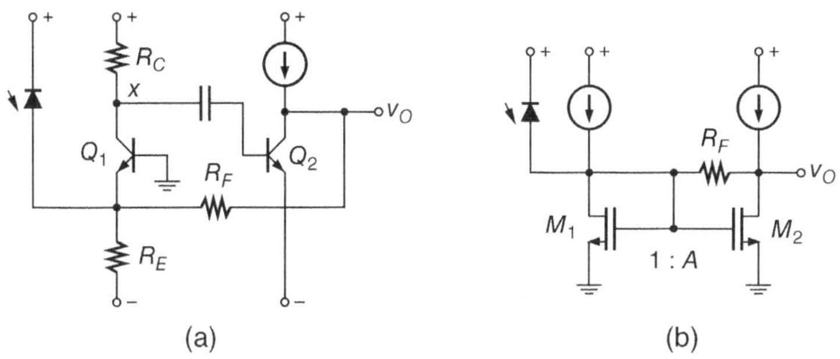

Figure 8.5 Current-mode TIA: simplified implementation examples.

thus has the low-frequency gain $A_0 = g_{m2}R_C$ and the dominant pole of its response $A(s)$ is determined by node x.

Figure 8.5(b) shows a simplified version of the MOSFET circuit in [28]. Here, a current mirror with a small input transistor M_1 and a large output transistor M_2 amplifies the input current. This current amplifier has the low-frequency gain $A_0 = g_{m2}/g_{m1}$.

Some current-mode TIAs can operate from very low supply voltages, making them attractive for low-power applications [28]. However, current-mode TIAs tend to be noisier than voltage-mode TIAs.

8.3 TIA with Bootstrapped Photodetector

The effective photodetector capacitance C_D seen by the TIA can be reduced by means of a circuit technique known as bootstrapping. Figure 8.6(a) shows the block diagram of a TIA with a *bootstrapped photodetector* [29–32]. Instead of biasing the photodetector at a fixed voltage above ground, it is biased at a fixed voltage above the TIA's input node. In the figure, the voltage buffer B senses the TIA's input voltage, while loading this node as little as possible. Then, the reverse bias voltage V_{BI} is added to the buffer output voltage and the sum drives the photodetector. For an ideal bootstrap buffer with gain $B_0 = 1$, the AC voltage across C_D becomes zero, suppressing any charging or discharging currents, effectively making $C_D = 0$. For a bootstrap buffer with $B_0 \leq 1$, the residual photodetector capacitance becomes $(1 - B_0)C_D$. Without bootstrapping, $B_0 = 0$, all of the detector capacitance is visible, as expected.

The bootstrap buffer is typically implemented with a source follower [33, 34] or an emitter follower [35]. An example circuit with a source follower is shown in Fig. 8.6(b). The source follower has the low-frequency gain $B_0 = (g_{mB} - 1/R_I)/(g_{mB} + 1/R_S)$, where R_I is the input resistance of the TIA ($R_I = R_F/[A_0 + 1]$) and R_S is the total load resistance at the source of M_B ($R_S = 1/g_{oB}$, if the bias current source is ideal). This gain is always less than one, but approaches one

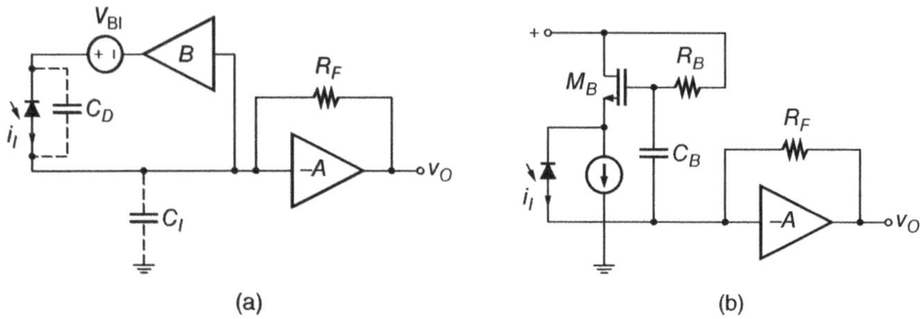

Figure 8.6 TIA with bootstrapped photodetector: (a) block diagram and (b) a typical transistor-level implementation.

for large values of g_{mB}. (There are two mechanisms responsible for keeping the gain below one: the finite loop gain of the follower, $g_{mB}R_S$, and the loading of the follower by the AC current i_i.)

As shown in Fig. 8.6(b), a coupling capacitor C_B together with a large biasing resistor R_B can be used to add the reverse bias voltage $V_{\rm BI}$ to the buffer output voltage [34, 35]. Unfortunately, the gate–source voltage drop of the source follower reduces the reverse bias compared with the nonbootstrapped case and a higher supply voltage may be necessary. Alternatively, the follower can be DC coupled to the TIA input and the AC coupling capacitor is then added at the output of the buffer [33]. Noise injection from the bootstrap buffer is an important concern. A large bias resistor R_B is needed to keep this noise low [34, 35].

The bootstrap TIA in Fig. 8.6 comes in a number of variations. A separate bootstrap buffer B in parallel with the voltage amplifier A can be avoided, if the first stage of the voltage amplifier is a buffer stage. In this case, the bootstrap signal can simply be tapped off that buffer [29, 30]. In another variation, the drain of the source follower in Fig. 8.6(b) (or the collector of the emitter follower) is connected to the second input of a differential TIA, thus using the currents from both terminals of the photodetector [36, 37].

Another way of reducing the photodetector capacitance C_D is to shunt it with a negative capacitance. A noninverting amplifier with gain $B_0 > 1$ and feedback capacitor C_F synthesizes the negative capacitance $-(B_0 - 1)C_F$ at its input, which can be used for this purpose. This approach, which is sometimes known as a grounded-source bootstrap [31], results in the residual input capacitance $C_T - (B_0 - 1)C_F$.

8.4 Burst-Mode TIA

The Burst-Mode Problem. How does a burst-mode TIA differ from the continuous-mode TIAs that we have discussed so far? A burst-mode TIA

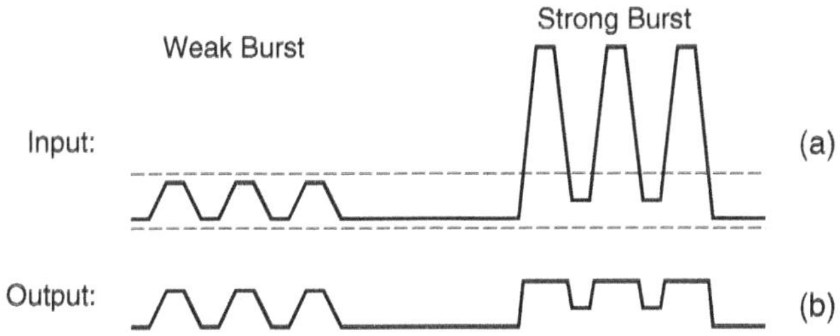

Figure 8.7 Pulse-width distortions resulting from a TIA with fixed transimpedance: (a) input current and (b) output voltage.

must be able to respond quickly to an input signal whose amplitude varies significantly from burst to burst. In passive optical networks (PON), the most common application for burst-mode TIAs, the amplitude variation (loud/soft ratio) at the central office (CO) can be up to 30 dB (cf. Chapter 1). The burst-mode TIA adjusts to the amplitude of the individual bursts by means of a preamble. The preamble typically contains a 111...000... or a 101010... pattern or a combination of both [38] and its duration is between 10 and 1,000 bit periods, depending on the standard (e.g., BPON, GPON, EPON). The TIA must complete its adjustments in a fraction of this duration, because the main amplifier, the equalizer (if present), and the CDR require a settled TIA signal to make their adjustments.

Figure 8.7(a) illustrates a weak burst followed by a strong burst at the input of the TIA. (For clarity, the bursts are shown schematically as 5-bit 10101 patterns; in practice, the bursts are much longer.) Such a situation arises, for example, at the CO side of a PON system when a distant subscriber and a nearby subscriber transmit bursts in adjacent time slots (cf. Chapter 1). When received by a TIA with a fixed transimpedance, the distorted output signal shown in Fig. 8.7(b) is obtained. The horizontal dashed lines in Fig. 8.7(a) show the (fixed) input range of the TIA; outside of these lines, the TIA clips the signal. Clearly, such a TIA introduces severe pulse-width distortions that depend on the burst's strength and extinction ratio. (TIA overload effects, which are not considered in Fig. 8.7(b), make the situation worse.) In the case of a strong burst with a low extinction ratio (high zero-level signal), the burst may even be lost altogether.

These problems can be avoided with an adaptive TIA that adjusts the transimpedance from burst to burst such that the output signal does not get clipped or distorted. Visualize the dashed lines in Fig. 8.7(a) changing their vertical positions for every burst such that they are just outside of the burst's envelope. The correct decision threshold voltage, which minimizes pulse-width distortions and maximizes sensitivity, can then be determined based on the zero and one levels at the output of the TIA.

The length of the preamble and the coding scheme used for the payload impact the design choices for a burst-mode receiver. If the preamble is very long and the code has short runs (e.g., if the 8B/10B code is used), it is possible to use the same or similar circuits as for continuous-mode receivers [39]. The time constants for the gain control, offset control, and AC coupling simply must be chosen short enough such that the receiver settles within the preamble. Unfortunately, a PON system with such a long preamble has a poor bandwidth efficiency because a substantial fraction of the transmitted bits consist of preamble bits rather than payload bits. In contrast, if the preamble is short and the code has long runs (e.g., if the 64B/66B code is used), it is not possible to find a (single) time constant that simultaneously allows the receiver to settle within the preamble *and* that prevents the gain and offset control from drifting during long runs of zeros or ones. In this case, specialized burst-mode circuits are required.

To simultaneously achieve fast gain and offset control and high tolerance to long runs, burst-mode receivers make use of a *reset signal* that becomes active in between bursts. The reset signal erases the receiver's memory of the previous burst and prepares it for the next burst. The reset signal can be generated, for example, by tracing the burst's envelope or by counting the bits in the burst [40, 41]. We discuss this reset signal in more detail shortly.

Fast Gain Control. In Section 7.4, we discussed several ways of controlling the transimpedance in response to the received signal strength. The TIA with a nonlinear feedback or shunt network (cf. Figs 7.18 and 7.19), which compresses the dynamic range like a logarithmic amplifier, is a possible solution for burst-mode receivers [33]. It has the advantage of responding instantaneously and avoiding the need for a fast control loop. However, the nonlinear transfer function causes some pulse-width distortion [42] and strong bursts with low extinction ratio result in small output signals [43].

A linear TIA with a fast burst-by-burst gain control loop is shown in Fig. 8.8(a). The main difference to the continuous-mode AGC scheme shown in Fig. 7.13(a) is that the low-pass filter for measuring the signal swing has been replaced with a resetable peak detector. This peak detector responds to negative peaks and hence is also known as a bottom-level hold. The waveforms in Fig. 8.8(b) illustrate the operation for a strong burst followed by a weak burst. Before the first burst arrives, the peak detector is reset to the dark level of the TIA's output voltage, $v_B = v_O$. Because this voltage is above V_{REF}, the gain-control switch opens and the higher of two transimpedance values is selected. When the strong burst arrives, the peak detector traces the lower edge of the envelope and as soon as it crosses below V_{REF}, the gain-control switch closes, thus reducing the transimpedance. For comparison, the response without gain control is also plotted (dashed line). At the end of the burst, the peak detector is reset and the gain-control switch opens in preparation for the next burst. When the weak burst arrives, the peak detector voltage never crosses below V_{REF} and the transimpedance remains high. The relationship

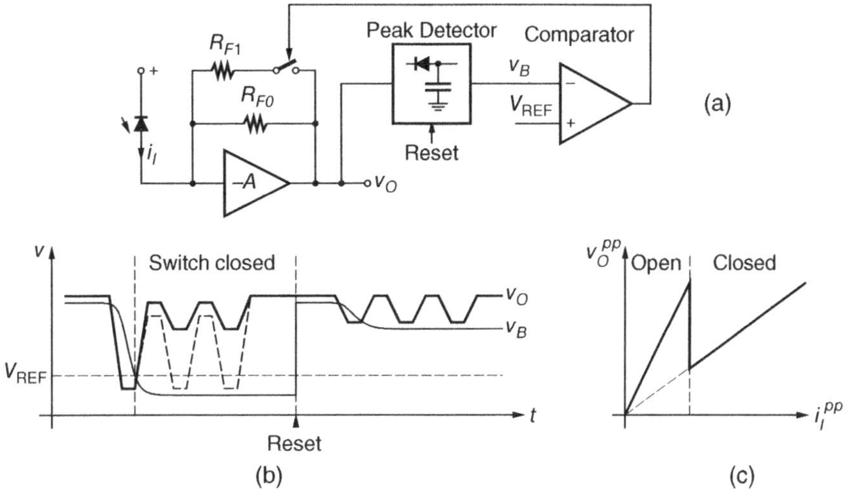

Figure 8.8 Burst-mode TIA with fast gain control: (a) block diagram, (b) waveforms, and (c) output swing versus input swing.

between the input swing and output swing for this burst-mode TIA is plotted in Fig. 8.8(c).

The circuit in Fig. 8.8(a) comes in a number of variations: More than two gain settings can be implemented with additional feedback resistors, switches, and comparators, resulting in a circuit similar to that in Fig. 7.14(b). (To keep Fig. 8.8(a) simple, the compensation capacitor C_{F1} is not shown.) Burst-mode TIAs with two gain settings [40, 44–47] and three gain settings [42] have been reported.

Instead of discrete gain control (a.k.a. gain switching), continuous gain control, similar to that in Fig. 7.14(a), can be used [41, 43, 48]. Continuous gain control is of interest in applications that require good linearity and constant amplitude, such as in receivers with electronic dispersion compensation [48].

Instead of a variable feedback resistor, a variable shunt element at the input, similar to that in Fig. 7.16, can be used [49].

In a variation of the fast AGC loop in Fig. 8.8(a), the peak detector can be omitted if the subsequent comparator is designed with a large hysteresis [42, 47]. The idea is that once the signal crosses the threshold of the hysteresis comparator its output remains high (due to the hysteresis) and the switch remains closed, even when the signal returns to the dark level. The reset pulse must now be applied to the hysteresis comparator instead of the peak detector. (See Fig. 9.9 in Chapter 9 for a circuit example.)

In all burst-mode AGC schemes, gain switching *after* the preamble must be avoided to prevent corrupting the payload data. This issue may occur when the signal strength is just below the switching threshold during the preamble and a fluctuation in amplitude (or a noise peak) pushes it over the threshold during the payload. Freezing the gain at the end of the preamble resolves this issue [50].

Fast Offset and Threshold Control. Many burst-mode TIAs feature differential outputs that make the output signal more immune to coupled noise and that facilitate DC coupling to the subsequent burst-mode main amplifier. Burst-mode TIAs with differential outputs often include an offset control circuit that removes the output offset voltage [40, 42, 49]. Alternatively, the offset voltage can be removed at the input of the subsequent main amplifier [51]. Like gain control, offset control must be completed in a small fraction of the preamble's duration.

Adjusting the offset of a signal relative to a fixed decision threshold has the same effect as adjusting the decision threshold. Thus, in this context the terms automatic offset control (AOC) and *automatic threshold control* (ATC) are often used interchangeably.

A burst-mode TIA with differential outputs and fast threshold control is shown in Fig. 8.9(a) [41, 45, 46]. A single-ended burst-mode TIA is followed by a threshold voltage generator similar to the arrangement in Fig. 7.9, except that the slow *RC* low-pass filter has been replaced by two fast peak detectors and a voltage divider. One peak detector operates as a bottom-level hold and outputs v_B; the other peak detector operates as a top-level hold and outputs v_T. The voltage divider produces the average voltage $v_M = \frac{1}{2}(v_T + v_B)$.

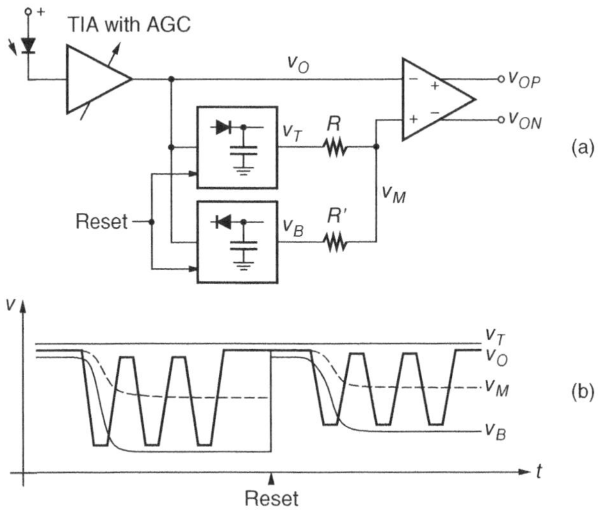

Figure 8.9 Burst-mode TIA with differential outputs and fast threshold control: (a) block diagram and (b) waveforms (simplified).

Figure 8.9(b) illustrates the output signal waveforms from the peak detectors for two sequential bursts with unequal amplitude. In between bursts, both detectors are reset to the dark level, $v_T = v_B = v_O$. When the burst arrives, the peak detectors trace the upper and the lower edge of the envelope and the midpoint between the two edges represents the threshold voltage v_M. A subsequent differential stage amplifies the difference $v_O - v_M$ and produces the complementary output signals v_{OP} and $_{ON}$. The differential output voltage $v_{OP} - v_{ON}$ swings symmetrically around the zero level and thus is offset free.

The timing of the reset signal is critical. Asserted in between bursts, it must be released only *after* the AGC process in the preceding TIA has settled (not shown in Fig. 8.9(b)) [40]. Otherwise, the bottom-level hold may pick up a transient, such as the first large negative peak in Fig. 8.8(b), resulting in the incorrect threshold voltage. Even in the absence of an AGC transient, the reset signal must be released only *after* the burst has started to correctly acquire the zero level [51]. If released too early, the top-level hold acquires the dark level, which is different from the zero level in the case of finite extinction.

The design of the peak detectors is critical. Their rise and fall times should be fast enough to track the envelope but not too fast to avoid tracking the noise in the received signal. Their droop (capacitor discharging) must be small enough to prevent the threshold voltage from drifting too much during long runs of zeros or ones [52]. Because the peak detectors in the AGC and the AOC/ATC circuits are both driven by the output of the single-ended TIA, they can be shared between these two functions [40, 41].

The circuit in Fig. 8.9(a) comes in a number of variations: To save chip area, the top-level hold can be replaced by a replica TIA with an open input [40] (cf. Fig. 7.8). However, the threshold is then based on the dark level instead of the zero level and the drift of the bottom-level hold due to the accumulation of noise peaks [51, 53] is no longer counterbalanced by a similar drift of the top-level hold.

In another variation, the threshold voltage v_M is derived with an *RC* low-pass filter, just like in Fig. 7.9, thus getting rid of both peak detectors. But for this approach to work, the system must switch between two time constants: During the preamble, a fast time constant is selected to quickly acquire the average level of the burst, then during the payload, a slow time constant is selected to minimize drift of the threshold voltage during long runs of zeros or ones [45, 46, 54].

The circuit in Fig. 8.9(a) derives the offset-compensation voltage v_M from the input signal, a method known as *feedforward* offset control. Alternatively, v_M can be derived from the output signals v_{OP} and v_{ON} [54], a method known as *feedback* offset control. Feedback offset control has the advantage of compensating offset errors in the whole receiver chain, but it is typically slower than feedforward offset control.

Burst-Mode Penalty. In Section 4.2, we discussed the problem of acquiring the decision threshold based on a small number of preamble bits. The limited amount of averaging over the noisy zeros and ones results in a decision threshold that is necessarily corrupted by noise [52, 53, 55, 56]. The degree of corruption depends on the number of preamble bits that are averaged.

The *burst-mode penalty* resulting from this effect can be derived from Eq. (4.11) as [53]

$$PP = \sqrt{1 + \frac{1}{n}}, \tag{8.13}$$

where n is the number of preamble bits used for threshold estimation and an equal number of zeros and ones is assumed. The expression in Eq. (8.13) is valid for a p–i–n receiver that acquires the zero and one levels by means of averaging. Similar expressions can be derived for an APD receiver (with unequal and non-Gaussian noise on the zeros and ones) [56] and for level acquisition with peak detectors (affected by drift due to the accumulation of noise peaks and discharge currents) [52, 53].

If the threshold is estimated from a single zero and a single one bit ($n = 2$) the power penalty according to Eq. (8.13) is 0.88 dB, which is rather large. Thus, a larger number of preamble bits is commonly used. In practice, burst-mode penalties of 0.7 dB [47] and 0.3 dB [51] (and about 3 dB in older burst-mode receivers [57, 58]) have been measured.

For a continuous-mode receiver, the number of bits that are averaged to establish the decision threshold is very large ($n \to \infty$). In this case, Eq. (8.13) goes to one and the "burst-mode" penalty disappears.

Chatter. There is one more peculiarity about burst-mode receivers. In between bursts there are extended periods of time when no optical signal is received. During these dead periods the transimpedance is usually set to its highest value and the decision threshold may be set close to zero in anticipation of a burst. Unfortunately, with these settings, the amplified TIA noise crosses the decision threshold randomly, generating a random bit sequence called *chatter* at the output of the receiver.

A brute force solution to this problem is to intentionally offset the decision threshold [57]. To suppress the chatter, the applied offset voltage must be larger than the expected peak noise voltage: $V_{OS} = Q v_{n,TIA}^{rms}$ (cf. Section 4.2). Moreover, the offset voltage must track the temperature dependence of the noise voltage [57]. However, an offset voltage of that magnitude degrades the sensitivity of the receiver by around 3 dB.

A better approach is to monitor the received signal strength to detect the presence or absence of a burst (activity detector) [51]. In the absence of a burst, the decision threshold is forced well above zero to avoid chatter. When

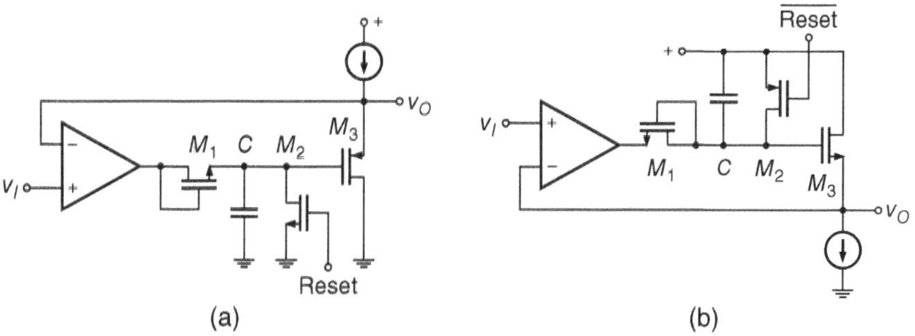

Figure 8.10 CMOS implementations of (a) top-level hold and (b) bottom-level hold.

a burst is detected, the ATC circuit acquires the optimum threshold voltage, thus avoiding the penalty associated with an intentional offset.

Peak Detectors. Given the importance of peak detectors in burst-mode circuits, a short discussion of their implementation is in order. Popular CMOS implementations of the top-level hold and the bottom-level hold are shown in Fig. 8.10 [41, 43, 45, 46, 51, 59].

The top-level hold in Fig. 8.10(a) operates as follows: If the input voltage v_I exceeds the output voltage v_O (the peak value acquired thus far), the output of the amplifier goes high. In response, the gate-drain connected MOSFET M_1, which acts as a diode, turns on and charges capacitor C. The voltage of the capacitor is buffered and level-shifted by the source follower M_3 and fed back to the amplifier input. As soon as the capacitor is charged enough for v_O to match v_I, the amplifier turns diode M_1 off and the charging stops. Thanks to the feedback loop through the amplifier, the voltage drops across the diode M_1 and the source follower M_3 are suppressed. To reset the peak detector, capacitor C is discharged with M_2.

The bottom-level hold in Fig. 8.10(b) operates in the same way, except that the polarity of diode M_1 is reversed to acquire the bottom level.

Some peak detector circuits merge the diode function with the amplifier function. In fact, an amplifier that can only source (for a top-level hold) or only sink (for a bottom-level hold) current may be faster than a regular amplifier followed by a diode. Amplifiers with an open collector [50], open drain [60], or an unbiased emitter follower [61] can be used for this purpose.

A potential issue with peak detectors is that they may overshoot (for a top-level hold) or undershoot (for a bottom-level hold) their target value [50, 60]. This overshoot (or undershoot) is caused by the delay through the amplifier, M_1, and M_3, which results in a late turn off of the current that charges C. One way to compensate for overshoot (undershoot) is to artificially increase (decrease) the voltage that is fed back to the amplifier [50].

8.5 Analog Receiver TIA

Optical fiber communication also has a significant impact on analog RF and microwave applications that traditionally have relied on coaxial cable. The low loss of optical fiber permits the elimination of many RF boosting amplifiers. The elimination of these amplifiers, in turn, results in a link with lower noise and distortion.

A typical application of this type is the hybrid fiber-coax (HFC) network used to distribute CATV signals from the head end to the neighborhood through optical fiber (cf. Chapter 1). Another example is the distribution of CATV signals over a PON by means of a dedicated wavelength (video overlay) [62]. Analog optical links are also used to connect wireless base stations with remote antennas. This application is referred to as *microwave photonics* [63] or *radio over fiber* (RoF) [64].

In contrast to digital receivers, analog receivers must be highly linear to keep distortions in the fragile analog signals (e.g., QAM or AM-VSB signals) to a minimum. The linearity of CATV receivers is specified in terms of the composite second order (CSO) distortion and the composite triple beat (CTB) distortion. For these and other nonlinearity measures, see Appendix D.

In addition, analog receivers require a much higher ratio of signal to noise than digital receivers. The noise performance of CATV receivers is specified in terms of the carrier-to-noise ratio (CNR), the passband equivalent of the signal-to-noise ratio (SNR) (cf. Section 4.3). The noise performance of microwave photonic links is specified in terms of the noise figure [63].

Low-Impedance Front-End. A simple implementation of an analog receiver is shown in Fig. 8.11(a). It consists of a low-impedance front-end followed by a linear low-noise amplifier [65]. Typically, the front-end impedance and the amplifier input impedance are 50 Ω (or 75 Ω in CATV systems) such that standard cables and connectors can be used.

To achieve high linearity, the p–i–n photodetector must be operated below its saturation current (cf. Section 3.1). A beam splitter, multiple photodetectors,

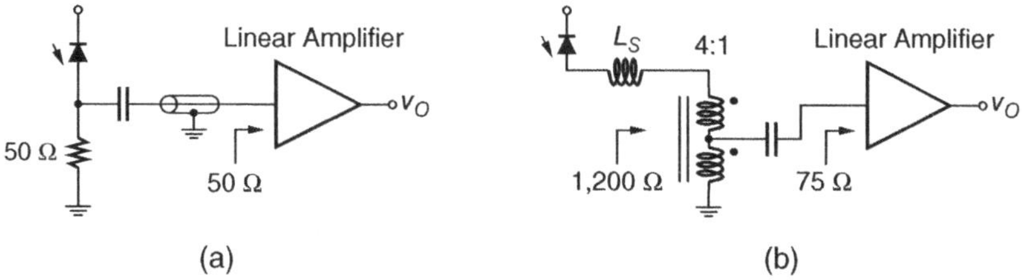

Figure 8.11 Receivers for analog signals: (a) low-impedance front-end and (b) front-end with matching network.

and a power combiner can be used to increase the effective saturation current of the photodetector, if necessary [66].

Amplifiers for CATV applications typically consist of a balun, that is, a transformer that converts the single-ended input signal to a balanced differential signal, followed by a differential amplifier with good symmetry (a push–pull amplifier), and another balun for converting the differential signal back to a single-ended output signal [67, 68]. Owing to its symmetry, this topology exhibits low even-order (in particular, low second-order) distortions.

Noise Matching. As we know, the low-impedance front-end in Fig. 8.11(a) is rather noisy. A transformer that matches the low input impedance of the amplifier to the higher impedance of the photodetector, as shown in Fig. 8.11(b), can be used to improve the noise performance [69]. The example shown in the figure uses an autotransformer with a turns ratio of 4:1 to transform the 75-Ω input impedance of the CATV amplifier up to 1.2 kΩ (16:1 impedance ratio). Such a transformer can be constructed, for example, by winding six turns from the input to the center tap (output) and two turns from the center tap to the ground on a ferrite core. This matching technique eliminates the load resistor to ground and the noise associated with it. Furthermore, because the transformer has a current gain of 4×, the input-referred noise current of the amplifier is attenuated by the same factor 4× (corresponding to 12 dB) when referred back to the photodetector.

While the transformer helps with the noise, it also reduces the bandwidth. For example, a photodetector with a 0.2-pF capacitance loaded by the 1.2-kΩ resistance seen through the transformer limits the bandwidth to about 660 MHz, which is not sufficient for most CATV applications (CATV signals occupy the frequency band from 50 MHz up to 1 GHz). However, adding a small series inductor L_S (see Fig. 8.11(b)) can increase the bandwidth to the desired value (cf. Section 7.7) and even provide some uptilt in the frequency response, if desired [69].

More generally, the input-referred noise of analog receivers can be shaped and reduced by placing a noise matching network between the photodetector and the TIA [69–71]. We discussed this technique in Section 6.5. Compared to digital receivers, the higher low-frequency cutoff typical for analog receivers (e.g., 50 MHz) permits more flexibility in the design of the noise matching network.

Balanced Photodetector. In Section 4.4, we discussed the noise of analog receivers and found that it is composed of circuit noise, shot noise, and relative intensity noise (RIN). While noise matching can reduce the circuit noise, it cannot help with the shot and RIN noise.

To suppress the RIN noise, which originates from the laser in the transmitter, a receiver with a balanced detector can be used [63]. For examples

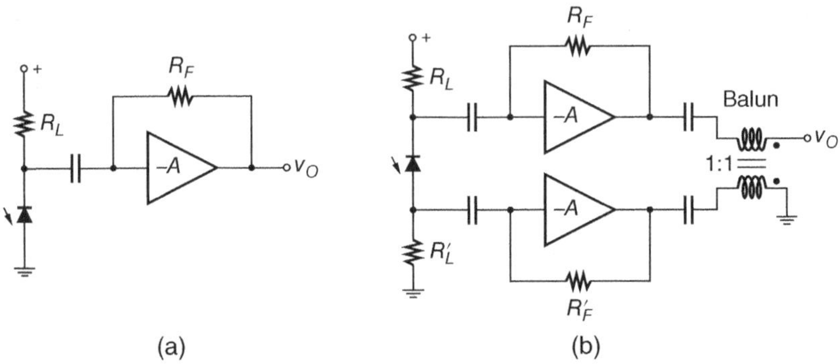

Figure 8.12 Shunt-feedback TIAs for analog receivers: (a) single ended topology and (b) pseudo-differential topology with differentially coupled photodetector.

of receivers with balanced detectors, see Section 3.5. In this scheme, the transmitter generates complementary optical signals with a Mach–Zehnder modulator, which are transmitted through two separate fibers. At the receiver, the balanced photodetector subtracts the two optical inputs, recovering the modulated signal but rejecting the common-mode RIN noise. With this method photonic microwave links with noise figures below 6.9 dB up to 12 GHz have been realized [63].

Shunt-Feedback TIA. Shunt-feedback TIAs can achieve better noise performance than low-impedance front-ends (cf. Chapter 6). Figure 8.12(a) shows an analog receiver with a single-ended shunt-feedback TIA [72]. The photodetector is AC coupled to the TIA input. The bias resistor R_L sets the reverse voltage of the photodetector to $V_{DD} - R_L \mathcal{R} \overline{P}_S$, where V_{DD} is the supply voltage, $\mathcal{R}$ is the responsivity of the photodetector, and $\overline{P}_S$ is the average received optical power, usually between −3 and 0 dBm for CATV applications. Alternatively, an RF choke or a combination of RF choke and resistor can be used to reduce the voltage drop and noise of the bias network. In any case, the impedance of the bias network must be high compared with the input resistance of the TIA in the frequency band of interest, such that almost all AC current from the detector flows through the coupling capacitor into the TIA. Similar to the DC input current control circuits discussed in Section 7.3, the bias network removes the DC component of the photodetector current, leaving only the AC current for the TIA. For CATV applications, in which the minimum frequency is 50 MHz, a relatively small coupling capacitor (1 to 10 nF) is sufficient.

To achieve the necessary high linearity with a single-ended TIA, the voltage amplifier must operate at a high supply voltage and a large quiescent current (e.g., 15 V and 100 mA for CATV applications) such that the signal excursions are only a small percentage of the operating point [72, 73]. Moreover, to keep the output signal within the linear operating region of the amplifier, the TIA must have an adaptive transimpedance (variable R_F) [72] (cf. Section 7.4).

The large power dissipation that results from the high supply voltage and quiescent current makes copackaging an analog TIA with a photodetector challenging [73]. Therefore, analog TIA circuits that dissipate less power while maintaining high linearity are needed.

Differential Shunt-Feedback TIA. TIAs with differential inputs and outputs depend less on transistor linearity because the symmetry of the differential topology ensures low even-order distortions. The resulting lower operating-point voltages and currents lead to a lower power dissipation.

The pseudo-differential TIA shown in Fig. 8.12(b) relies on good matching between its two halves to minimize the even-order distortions. (A slight imbalance may be introduced on purpose to compensate for the photodetector nonlinearity [69].) A balun can be used to connect the single-ended photodetector to the differential TIA inputs, but the preferred method is to take complementary current signals from the two terminals of the photodetector [67, 69, 71], as shown in Fig. 8.12(b). This arrangement avoids the bandwidth limitation inherent with a balun. The bias resistors R_L and R'_L, together with the average photodetector current, set the reverse voltage of the photodetector. Again, RF chokes can be used as part of the bias network to reduce voltage drops and noise [71].

To obtain well balanced input signals, the parasitic capacitances of the photodetector and the interconnects must be similarly well balanced. Interestingly, the pseudo-differential TIA is more resilient to an input-capacitance imbalance than the regular differential TIA. Whereas both topologies have the same differential input resistances, $R_{I,d} = 2R_F/(A_0 + 1)$, the pseudo-differential TIA has a lower common-mode input resistance, $R_{I,c} = \frac{1}{2}R_F/(A_0 + 1)$, that the regular differential TIA, $R_{I,c} = \frac{1}{2}R_F$. This lower common-mode input resistance helps to suppress the AC common-mode voltage that otherwise would throw the voltage amplifier off balance.

Differential coupling of the photodetector to the TIA eliminates the balun at the input, but another balun is still needed to convert the differential output signals back to a single-ended output signal (cf. Fig. 8.12(b)). In a current-mode implementation, where the amplifier output signals are in the form of currents, the current subtraction can be done by means of Kirchhoff's current law, avoiding the balun necessary for a voltage subtraction [74].

Distortion Cancellation. Another technique for simultaneously obtaining high linearity and low power dissipation is *distortion cancellation*. The idea is to follow a first stage, which is slightly nonlinear, by a second stage that attempts to undo the nonlinearity of the first one.

For example, in [73] a cascoded common-source stage is followed by a common-drain stage that approximately cancels the distortions of the first one. This linearized voltage amplifier forms the core of a single-ended shunt-feedback TIA. Careful tuning of the bias currents in the two stages results in inversely related nonlinearities and an overall linear response without the need for large operating-point voltages and currents.

Another example of distortion cancellation is the nonlinear active-feedback TIA followed by an inversely related nonlinear post amplifier that we mentioned in Section 8.1 [19, 20].

8.6 Summary

Capacitive feedback eliminates the noise generated by the feedback resistor in a conventional shunt-feedback TIA. However, simply replacing the feedback resistor with a capacitor turns the TIA into an integrator. One way to resolve this issue is to follow the integrator with a differentiator; another way is to use a capacitive current divider as the feedback element.

Optical feedback also has the potential of lowering the noise, but unlike resistive feedback it cannot be integrated in a standard circuit technology.

Active feedback (feedback through a transistor) comes in several forms: common source, common drain, or common gate for a FET and common emitter, common collector, or common base for a BJT. Active feedback also has the potential of lowering the noise, but its linearity is inferior to that of resistive feedback.

Current-mode TIAs promise gain-bandwidth independence, but when taking all parasitic capacitances into account, their performance is often not much different from that of voltage-mode TIAs.

Bootstrapping makes the effective photodetector capacitance appear smaller. In one implementation, the photodetector is driven such that the AC voltage drop across the detector becomes small. In another implementation, a negative capacitance is synthesized and applied in parallel to the detector.

Burst-mode TIAs, used for example in receivers for passive optical networks (PON), feature fast gain control and fast offset (or threshold) control to deal with bursty signals that have large burst-to-burst amplitude variations. To quickly acquire the burst amplitude, peak detectors, which are reset in between bursts, are commonly used. The gain and offset of burst-mode TIAs are set on a burst-by-burst basis.

Analog TIAs, used for example in receivers for hybrid fiber-coax (HFC) networks and microwave photonic links, feature high linearity and low noise to minimize signal distortions. Noise matching with transformers and inductors, symmetric (push–pull) designs to minimize even-order distortions, and distortion cancellation with inversely related nonlinearities are some of the techniques used.

Problems

8.1 Active-Feedback TIA with Common-Source Feedback Device. (a) Calculate the transimpedance $Z_T(s)$ of the active-feedback TIA shown in Fig. 8.3(a). Assume a single-pole model (with time constant T_A) for the voltage amplifier, define the total input capacitance as C_T, and neglect g_o and C_{gd} of M_F. (b) What time constant T_A is needed to obtain a Butterworth response? (c) What is the 3-dB bandwidth of the TIA given a Butterworth response? (d) Does the transimpedance limit Eq. (6.14) apply to this TIA? (e) Calculate the input-referred noise current PSD of this active-feedback TIA, taking the noise contributions from the feedback FET and the front-end FET into account. (f) Repeat the noise calculation for a TIA with a BJT front-end and feedback through a common-emitter BJT.

8.2 Active-Feedback TIA with Common-Drain Feedback Device. (a) Calculate the transimpedance $Z_T(s)$ of the active-feedback TIA shown in Fig. 8.3(b). Assume a single-pole model (with time constant T_A) for the voltage amplifier, define the total input capacitance as C_T, and neglect g_o and C_{gs} of M_F. (b) Does the transimpedance limit Eq. (6.14) apply to this TIA? (c) Calculate the input-referred noise current PSD of this active-feedback TIA, taking the noise contributions from the feedback FET and the front-end FET into account.

8.3 Current-Mode TIA. (a) Calculate the transimpedance $Z_T(s)$ of the current-mode TIA shown in Fig. 8.4. Assume a single-pole model (with time constant T_A) for the current amplifier, $R_S \neq 0$, and $C_L = 0$. What is the 3-dB bandwidth of the TIA given a Butterworth response and $R_S C_T < T_A$? (b) Repeat (a), but now assume $R_S = 0$, and $C_L \neq 0$. What is the 3-dB bandwidth of the TIA given a Butterworth response and $R_F C_L > T_A$? Derive the transimpedance limit.

References

1 C. Ciofi, F. Crupi, C. Pace, G. Scandurra, and M. Patanè. A new circuit topology for the realization of very low-noise wide-bandwidth transimpedance amplifier. *IEEE Trans. Instrum. Meas.*, IM-56(5):1626–1631, 2007.

2 G. Ferrari, F. Gozzini, A. Molari, and M. Sampietro. Transimpedance amplifier for high sensitivity current measurements on nanodevices. *IEEE J. Solid-State Circuits*, SC-44(5):1609–1616, 2009.

3 B. Razavi. A 622Mb/s, 4.5pA/$\sqrt{Hz}$ CMOS transimpedance amplifier. In *ISSCC Digest of Technical Papers*, pages 162–163, February 2000.

4 S. Shahdoost, B. Bozorgzadeh, A. Medi, and N. Saniei. Low-noise transimpedance amplifier design procedure for optical communications. In *IEEE Austrian Workshop on Microelectronics (Austrochip)*, pages 1–5, October 2014.

5 J. Salvia, P. Lajevardi, M. Hekmat, and B. Murmann. A 56MΩ CMOS TIA for MEMS applications. In *Proceedings of IEEE Custom Integrated Circuits Conference*, pages 199–202, September 2009.

6 A. E. Stevens. *An Integrate-and-Dump Receiver for Fiber Optic Networks.* PhD thesis, Columbia University, New York, 1995.

7 B. L. Kasper. Receiver using optical feedback. U.S. Patent No. 4,744,105, May 1988.

8 B. L. Kasper, A. R. McCormick, C. A. Burrus Jr., and J. R. Talman. An optical-feedback transimpedance receiver for high sensitivity and wide dynamic range at low bit rates. *J. Lightwave Technol.*, LT-6(2):329–338, 1988.

9 G. F. Williams. Nonintegrating receiver. U.S. Patent No. 4,540,952, September 1985.

10 E. Braß, U. Hilleringmann, and K. Schumacher. System integration of optical devices and analog CMOS amplifiers. *IEEE J. Solid-State Circuits*, SC-29(8):1006–1010, 1994.

11 C. Rooman, D. Coppée, and M. Kuijk. Asynchronous 250-Mb/s optical receivers with integrated detector in standard CMOS technology for optocoupler applications. *IEEE J. Solid-State Circuits*, SC-35(7):953–958, 2000.

12 S. B. Baker and C. Toumazou. Low noise CMOS common gate optical preamplifier using active feedback. *Electron. Lett.*, 34(23):2235–2237, 1998.

13 H. M. Lavasani, W. Pan, B. Harrington, R. Abdolvand, and F. Ayazi. A 76 dBΩ 1.7 GHz 0.18 μm CMOS tunable TIA using broadband current pre-amplifier for high frequency lateral MEMS oscillators. *IEEE J. Solid-State Circuits*, SC-46(1):224–235, 2011.

14 Z. Lu, K. S. Yeo, W. M. Lim, M. A. Do, and C. C. Boon. Design of a CMOS broadband transimpedance amplifier with active feedback. *IEEE Trans. Very Large Scale Integr. Syst.*, VLSI-18(3):461–472, 2010.

15 J. S. Yun, M. Seo, B. Choi, J. Han, Y. Eo, and S. M. Park. A 4Gb/s current-mode optical transceiver in 0.18 μm CMOS. In *ISSCC Digest of Technical Papers*, pages 102–103, February 2009.

16 S. Andersson, C. Svensson, and O. Drugge. Wideband LNA for a multistandard wireless receiver in 0.18 μm CMOS. In *Proceedings of European Solid-State Circuits Conference*, pages 655–658, September 2003.

17 J. Borremans, P. Wambacq, C. Soens, Y. Rolain, and M. Kuijk. Low-area active-feedback low-noise amplifier design in scaled digital CMOS. *IEEE J. Solid-State Circuits*, SC-43(11):2422–2433, 2008.

18 B. Razavi. *RF Microelectronics*. Prentice-Hall, Inc., Upper Saddle River, NJ, 1998.

19 A. Kopa and A. B. Apsel. 124dB $Hz^{2/3}$ dynamic range transimpedance amplifier for electronic-photonic channelizer. In *Proceedings of IEEE International Symposium on Circuits and Systems*, pages 189–192, Seattle, WA, May 2008.

20 A. Kopa and A. B. Apsel. Common-emitter feedback transimpedance amplifier for analog optical receivers. In *Proceedings of IEEE International Symposium on Circuits and Systems*, pages 5479–5482, Island of Kos, May 2006.

21 S. Galal and B. Razavi. 10-Gb/s limiting amplifier and laser/modulator driver in 0.18-μm CMOS technology. *IEEE J. Solid-State Circuits*, SC-38(12):2138–2146, 2003.

22 E. Säckinger. *Broadband Circuits for Optical Fiber Communication*. John Wiley &Sons, Inc., Hoboken, NJ, 2005.

23 K.-Y. Toh, R. G. Meyer, D. C. Soo, G. M. Chin, and A. M. Voshchenkov. Wide-band, low-noise, matched-impedance amplifiers in submicrometer MOS technology. *IEEE J. Solid-State Circuits*, SC-22(6):1031–1040, 1987.

24 B. Wilson and J. D. Drew. Novel transimpedance amplifier formulation exhibiting gain-bandwidth independence. In *IEEE International Symposium on Circuits and Systems Proceedings*, pages 169–172, 1997.

25 B. Wilson and J. D. Drew. High-performance transimpedance formulation for MESFET- and HBT-based monolithic microwave integrated circuits. *IEE Proc. Circuits Devices Syst.*, 145(6):429–436, 1998.

26 E. Säckinger. The transimpedance limit. *IEEE Trans. Circuits Syst. I*, CASI-57(8):1848–1856, 2010.

27 A. S. Sedra and G. W. Roberts. Current conveyor theory and practice. In C. Toumazou, F. J. Lidgey, and D. G. Haigh, editors, *Analogue IC Design: The Current-Mode Approach*, pages 93–126. Institution of Engineering and Technology (IET), London, UK, 1990.

28 K. Phang and D. A. Johns. A 1V 1mW CMOS front-end with on-chip dynamic gate biasing for 75Mb/s optical receiver. In *ISSCC Digest of Technical Papers*, pages 218–219, February 2001.

29 M. Abraham. Design of Butterworth-type transimpedance and bootstrap-transimpedance preamplifiers for fiber-optic receivers. *IEEE Trans. Circuits Syst.*, CAS-29(6):375–382, 1982.

30 M. Abraham. Bootstrap-transimpedance preamplifier for a fiber optic receiver. U.S. Patent No. 4,535,233, August 1985.

31 C. Hoyle and A. Peyton. Bootstrapping techniques to improve the bandwidth of transimpedance amplifiers. In *IEE Colloquium on Analog Signal Processing*, pages 7/1–7/6, Oxford, UK, October 1998.

32 C. Hoyle and A. Peyton. Shunt bootstrapping technique to improve bandwidth of transimpedance amplifiers. *Electron. Lett.*, 35(5):369–370, 1999.

33 S. Brigati, P. Colombara, L. D'Ascoli, U. Gatti, T. Kerekes, and P. Malcovati. A SiGe BiCMOS burst-mode 155-Mb/s receiver for PON. *IEEE J. Solid-State Circuits*, SC-37(7):887–894, 2002.

34 C.-M. Tsai. A 20mW 85dBΩ 1.25Gb/s CMOS transimpedance amplifier with photodiode capacitance cancellation. In *Symposium on VLSI Circuits Digest of Technical Papers*, pages 408–409, June 2004.

35 A. Maxim. A 54 dBΩ + 42 dB 10 Gb/s SiGe transimpedance-limiting amplifier using bootstrap photodiode capacitance neutralization and vertical threshold adjustment. *IEEE J. Solid-State Circuits*, SC-42(9):1851–1864, September 2007. This paper may contain falsified information, see the June 2008 issue.

36 C.-M. Tsai. A 40 mW 3 Gb/s self-compensated differential transimpedance amplifier with enlarged input capacitance tolerance in 0.18 μm CMOS technology. *IEEE J. Solid-State Circuits*, SC-44(10):2671–2677, 2009.

37 C.-M. Tsai and L.-R. Huang. A 21mW 2.5Gb/s 15kΩ self-compensated differential transimpedance amplifier. In *ISSCC Digest of Technical Papers*, pages 234–235, February 2005.

38 X.-Z. Qiu, P. Ossieur, J. Bauwelinck, Y. Yi, D. Verhulst, J. Vandewege, B. De Vos, and P. Solina. Development of GPON upstream physical-media-dependent prototypes. *J. Lightwave Technol.*, LT-22(11):2498–2508, 2004.

39 K. Hara, S. Kimura, H. Nakamura, N. Yoshimoto, and H. Hadama. New AC-coupled burst-mode optical receiver using transient-phenomena cancellation techniques for 10 Gbit/s-class high-speed TDM-PON systems. *J. Lightwave Technol.*, LT-28(19):2775–2782, 2010.

40 Q. Le, S.-G. Lee, H.-Y. Kang, and S.-H. Chai. A CMOS burst-mode TIA with step AGC and selective internally created reset for 1.25Gb/s EPON. In *ISSCC Digest of Technical Papers*, pages 50–51, February 2007.

41 Q. Le, S.-G. Lee, Y.-H. Oh, H.-Y. Kang, and T.-H. Yoo. A burst-mode receiver for 1.25-Gb/s Ethernet PON with AGC and internally created reset signal. *IEEE J. Solid-State Circuits*, SC-39(12):2379–2388, 2004.

42 M. Nakamura, Y. Imai, Y. Umeda, J. Endo, and Y. Akatsu. 1.25-Gb/s burst-mode receiver ICs with quick response for PON systems. *IEEE J. Solid-State Circuits*, SC-40(12):2680–2688, 2005.

43 S. Yamashita, S. Ide, K. Mori, A. Hayakawa, N. Ueno, and K. Tanaka. Novel cell-AGC technique for burst-mode CMOS preamplifier with wide dynamic range and high sensitivity for ATM-PON system. *IEEE J. Solid-State Circuits*, SC-37(7):881–886, 2002.

44 W.-Z. Chen and R.-M. Gan. 1.8V variable gain transimpedance amplifiers with constant damping factor for burst-mode optical receiver. In *IEEE Radio Frequency Integrated Circuits (RFIC) Symposium*, pages 691–694, Long Beach, CA, June 2005.

45 W.-Z. Chen and R.-M. Gan. A single chip 2.5 Gbps CMOS burst mode optical receiver. In *Symposium on VLSI Circuits Digest of Technical Papers*, pages 120–121, June 2006.

46 W.-Z. Chen, R.-M. Gan, and S.-H. Huang. A single-chip 2.5-Gb/s CMOS burst-mode optical receiver. *IEEE Trans.Circuits Syst. I*, CASI-56(10):2325–2331, 2009.

47 S. Nishihara, S. Kimura, T. Yoshida, M. Nakamura, J. Terada, K. Nishimura, K. Kishine, K. Kato, Y. Ohtomo, N. Yoshimoto, T. Imai, and M. Tsubokawa. A burst-mode 3R receiver for 10-Gbit/s PON systems with high sensitivity, wide dynamic range, and fast response. *J. Lightwave Technol.*, LT-26(1):99–107, 2008.

48 P. Ossieur, N. A. Quadir, S. Porto, C. Antony, W. Han, M. Rensing, P. O'Brien, and P. D. Townsend. A 10 Gb/s linear burst-mode receiver in 0.25 μm SiGe:C BiCMOS. *IEEE J. Solid-State Circuits*, SC-48(2):381–390, 2013.

49 K. Nishimura, H. Kimura, M. Watanabe, T. Nagai, K. Nojima, K. Gomyo, M. Takata, M. Iwamoto, and H. Asano. A 1.25-Gb/s CMOS burst-mode optical transceiver for Ethernet PON system. *IEEE J. Solid-State Circuits*, SC-40(4):1027–1034, 2005.

50 T. De Ridder, P. Ossieur, B. Baekelandt, C. Mélange, J. Bauwelinck, C. Ford, X.-Z. Qui, and J. Vandewege. A 2.7V 9.8Gb/s burst-mode TIA with fast automatic gain locking and coarse threshold extraction. In *ISSCC Digest of Technical Papers*, pages 220–221, February 2008.

51 P. Ossieur, D. Verhulst, Y. Martens, W. Chen, J. Bauwelinck, X.-Z. Qui, and J. Vandewege. A 1.25-Gb/s burst-mode receiver for GPON applications. *IEEE J. Solid-State Circuits*, SC-40(5):1180–1189, 2005.

52 C. Su, L.-K. Chen, and K.-W. Cheung. Theory of burst-mode receiver and its applications in optical multiaccess networks. *J. Lightwave Technol.*, LT-15(4):590–606, 1997.

53 P. Menéndez-Valdés. Performance of optical direct receivers using noise corrupted decision threshold. *J. Lightwave Technol.*, LT-13(11):2202–2214, 1995.

54 R. Weber, M. Levy, and R. Dvir. Transimpedance (TIA) circuit usable for burst mode communications. U.S. Patent No. 7,583,904 B2, September 2009.

55 C. A. Eldering. Theoretical determination of sensitivity penalty for burst mode fiber optic receivers. *J. Lightwave Technol.*, LT-11(12):2145–2149, 1993.

56 P. Ossieur, X.-Z. Qiu, J. Bauwelinck, and J. Vandewege. Sensitivity penalty calculation for burst-mode receivers using avalanche photodiodes. *J. Lightwave Technol.*, LT-21(11):2565–2575, 2003.

57 Y. Ota and R. G. Swartz. Burst-mode compatible optical receiver with a large dynamic range. *J. Lightwave Technol.*, LT-8(12):1897–1903, 1990.

58 Y. Ota, R. G. Swartz, and V. D. Archer III. DC-1Gb/s burst-mode compatible receiver for optical bus applications. *J. Lightwave Technol.*, LT-10(2):244–249, 1992.

59 M. Nakamura, N. Ishihara, and Y. Akazawa. A 156-Mb/s CMOS optical receiver for burst-mode transmission. *IEEE J. Solid-State Circuits*, SC-33(8):1179–1187, 1998.

60 H.-L. Chu and S.-I. Liu. A 10Gb/s burst-mode transimpedance amplifier in 0.13μm CMOS. In *IEEE Asian Solid-State Circuits Conference (ASSCC)*, pages 400–403, November 2007.

61 Y. Ota, R. G. Swartz, V. D. Archer III, S. K. Korotky, M. Banu, and A. E. Dunlop. High-speed, burst-mode, packet-capable optical receiver and instantaneous clock recovery for optical bus operation. *J. Lightwave Technol.*, LT-12(2):325–331, 1994.

62 F. Effenberger. PONs: state of the art and standardized. In I. P. Kaminow, T. Li, and A. E. Willner, editors, *Optical Fiber Telecommunications VIB*, pages 891–926. Academic Press, 2013.

63 C. H. Cox III and E. I. Ackerman. Recent advances in high-frequency (>10GHz) microwave photonic links. In I. P. Kaminow, T. Li, and A. E. Willner, editors, *Optical Fiber Telecommunications VIB*, pages 853–872. Academic Press, 2013.

64 G.-K. Chang, Y.-T. Hsueh, and S.-H. Fan. Advances in 1–100GHz microwave photonics: all-band optical wireless access networks using radio over fiber technologies. In I. P. Kaminow, T. Li, and A. E. Willner, editors, *Optical Fiber Telecommunications VIB*, pages 873–890. Academic Press, 2013.

65 J. Chen, F. Saibi, E. Säckinger, J. Othmer, M.-L. (Mark) Yu, F. Yang, J. Lin, T. Huang, T.-P. Liu, and K. Azadet. A multi-carrier QAM transceiver for ultra-wideband optical communication. *IEEE J. Solid-State Circuits*, SC-41(8):1876–1893, 2006.

66 S. Itakura, K. Sakai, T. Nagatsuka, E. Ishimura, M. Nakaji, H. Otsuka, K. Mori, and Y. Hirano. High-current backside-illuminated photodiode array module for optical analog links. *J. Lightwave Technol.*, LT-28(6):965–971, 2010.

67 B. L. Kasper. Lightwave receiver having differential input. U.S. Patent No. 5,013,903, May 1991.

68 Philips Semiconductors. A hybrid wideband amplifier module for digital CATV networks with the BGD902. Philips Semiconductors, Application Note AN98109, February 1999.

69 H. A. Blauvelt, I. Ury, D. B. Huff, and H. L. Loboda. Broadband optical receiver with passive tuning network. U.S. Patent No. 5,179,461, January 1993.

70 F. R. Little Jr., H. A. Kruse, and J. Megna. Push-pull optical receiver. U.S. Patent No. 5,239,402, August 1993.

71 J. Skrobko. Push-pull optical receiver with cascode amplifiers. U.S. Patent No. 5,347,389, September 1994.

72 H. A. Kruse. Fiber optic transimpedance receiver. U.S. Patent No. 4,998,012, March 1991.

73 B. L. Kasper and E. M. Peral. Distortion cancellation in a transimpedance amplifier circuit. U.S. Patent Application No. 2007/0003291 A1, January 2007.

74 R. A. Gosser and E. P. Jordan. Low-distortion transimpedance amplifier structures and methods. U.S. Patent No. 6,529,078 B1, March 2003.

9

Transimpedance Amplifier Circuit Examples

In the following, we examine some representative transistor-level TIA circuits taken from the literature. These circuits illustrate how the design principles discussed in the previous chapters can be applied and combined. Circuits in a broad range of technologies are discussed. This includes the bipolar junction transistor (BJT), the heterojunction bipolar transistor (HBT), the complementary metal–oxide–semiconductor (CMOS), the metal–semiconductor field-effect transistor (MESFET), and the heterostructure field-effect transistor (HFET) technologies. For a discussion of these technologies and its devices see Refs [1–3]. We conclude with a table summarizing the performance, topology, and technology of some recently published high-speed TIAs.

9.1 BJT, HBT, and BiCMOS Circuits

High-Speed TIA. Figure 9.1 shows the TIA reported in [4]. This shunt-feedback TIA, which is optimized for high speed and low noise, is implemented in SiGe BiCMOS HBT technology.

The voltage amplifier of this shunt-feedback TIA is implemented with a single transistor Q_1 and the load resistor R_C. The emitter degeneration resistor R_E improves the amplifier's linearity and the feedback resistor R_F closes the loop around the voltage amplifier.

The on-chip spiral inductors L_B and L_C boost the TIA's bandwidth and reduce its noise. The shunt capacitor C_E further improves the high-frequency response and suppresses the noise of R_E.

The single-transistor shunt-feedback TIA is followed by the common-emitter stage Q_2 to drive 50-Ω off-chip loads.

TIA with Adaptive Transimpedance. Figures 9.2–9.4 show a simplified version of the TIA reported in [5]. This shunt-feedback TIA with adaptive transimpedance is implemented in BiCMOS technology.

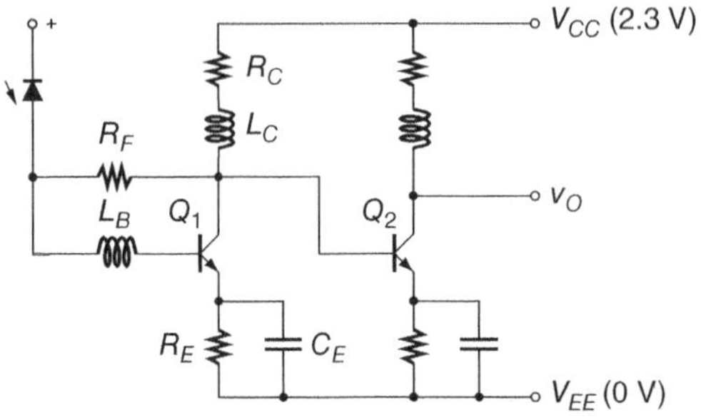

Figure 9.1 SiGe HBT implementation of a high-speed TIA based on [4].

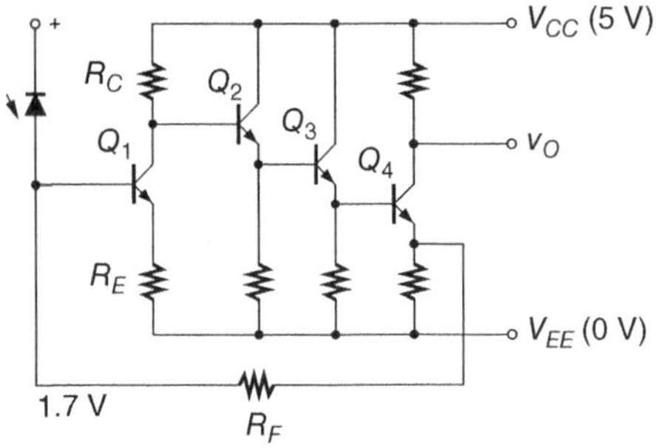

Figure 9.2 BJT implementation of a TIA based on [5].

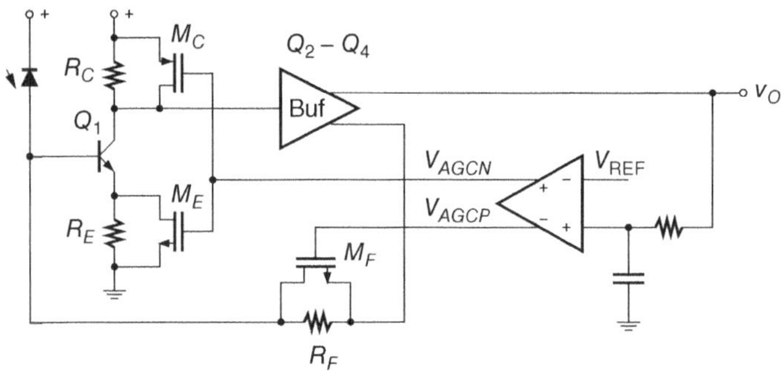

Figure 9.3 The TIA from Fig. 9.2, enhanced with a control loop to adapt the transimpedance to the signal strength (based on [5]).

The core of the TIA, excluding the AGC circuit, is shown in Fig. 9.2. The voltage amplifier is implemented with transistors Q_1 through Q_4. The gain is provided by the common-emitter stage consisting of Q_1, R_E, and R_C. The ratio of the collector resistor, R_C, to the emitter resistor, R_E, sets the DC gain. This gain stage is followed by a cascade of three emitter followers (Q_2 through Q_4), which buffer and level-shift the output signal. The level shift ensures that the collector of Q_1 is biased well above the base. The last stage has two outputs, one for the

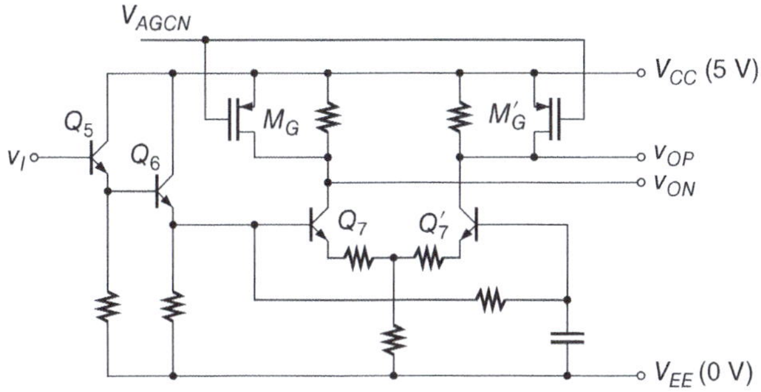

Figure 9.4 Single-ended to differential converter and post amplifier following the TIA core in Figs 9.2 and 9.3 (based on [5]).

feedback signal at the emitter and one for the output signal at the collector. The feedback resistor R_F closes the loop around this inverting amplifier.

The AGC circuit, shown in Fig. 9.3, extends the dynamic range of the TIA core in Fig. 9.2. For large input signals, the MOSFET M_F lowers the transimpedance by gradually shunting the feedback resistor. To maintain a flat frequency response, MOSFETs M_C and M_E simultaneously reduce the gain of the voltage amplifier. The maximum transimpedance (and the maximum voltage amplifier gain) is reached when the control voltage V_{AGCP} is low and V_{AGCN} is high. Under this condition, M_F and M_C are turned off and M_E is fully turned on. A differential control amplifier, implemented with mostly MOSFETs, generates the necessary complementary control voltages V_{AGCP} and V_{AGCN} from the difference between the time-averaged TIA output voltage and the reference voltage V_{REF}.

The single-ended TIA from Figs 9.2 and 9.3 is followed by the single-ended to differential converter shown in Fig. 9.4. The signal from the core is buffered and level-shifted by the emitter followers Q_5 and Q_6. The differential amplifier around Q_7 and Q'_7 amplifies the difference between the single-ended signal and its time-averaged value and outputs the balanced differential signals v_{OP} and v_{ON}. To prevent the differential amplifier from overloading, its gain is controlled by the MOSFETs M_G and M'_G, which gradually shunt the load resistors for large signals. The gates of M_G and M'_G are controlled by the same AGC circuit that we discussed in Fig. 9.3. The differential amplifier is followed by an output buffer to drive 50-Ω off-chip loads (not shown).

A single-ended TIA similar to that in Fig. 9.2, but without the emitter degeneration resistor R_E and a fixed transimpedance, has been reported in [6]. Eliminating R_E not only boosts the gain of the voltage amplifier, but also gets rid of the noise associated with this resistor. On the downside, eliminating R_E makes the voltage amplifier less linear and its gain more dependent on the supply voltage and temperature. For the adaptive TIA in Fig. 9.3, R_E is effectively shorted by M_E when the TIA operates at its maximum transimpedance. This is good because the noise of R_E is eliminated when it is most important, namely when receiving a weak input signal. For stronger input signals, when the AGC loop increases R_E, the extra noise can be tolerated.

A similar circuit as in Fig. 9.2, but with an additional shunt capacitor C_E in parallel to R_E, is proposed [7]. The zero introduced by this capacitor is used to optimize the TIA's jitter performance.

A number of authors have suggested to add a second feedback path to the shunt-feedback circuit in Fig. 9.2 to flatten the frequency response [8–10]. The proposed feedback path leads from the output, v_O, back to the emitter of Q_1 and may consist of an RC series circuit [9], an emitter follower with an emitter resistor [8], or an emitter follower with a parallel RC circuit [10].

Differential TIA with Inductive Input Coupling. Figure 9.5 shows the core of the TIA reported in [11]. This differential shunt-feedback TIA is implemented in BiCMOS technology.

The signal path of this TIA is implemented with BJTs and the bias network is implemented with MOS transistors. This partitioning is typical for a design in BiCMOS technology: It takes advantage of the BJT's superior high-frequency performance and accurate matching as well as the MOSFET's high output impedance, virtually zero gate current, and low noise. (For a comparison of the noise in MOSFET and BJT current sources, see Refs [12, 13].)

The differential voltage amplifier of this shunt-feedback TIA is implemented with the BJTs Q_1, Q_1', Q_2 and Q_2', and the MOSFETs M_1, M_2, and M_2'. The gain is provided by the differential stage consisting of the differential pair Q_1, Q_1' and the poly-resistor loads R, R'. The constant tail current provided by M_1 in conjunction with the linear load resistors guarantees balanced output signals, that is, a constant common-mode output voltage regardless of the input voltages. The outputs of the differential stage are buffered by emitter followers Q_2,

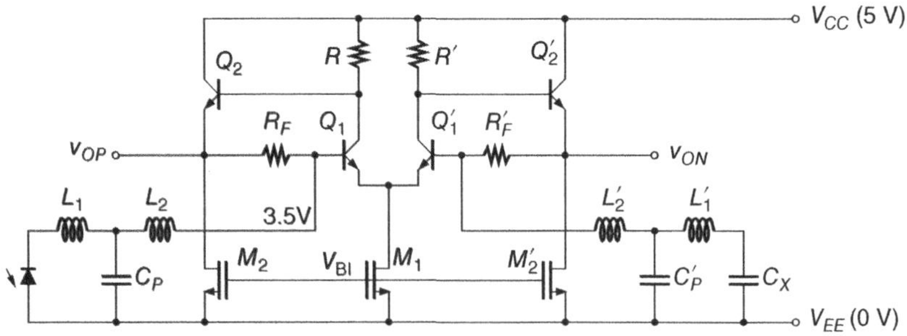

Figure 9.5 BiCMOS implementation of a differential TIA with inductive input coupling based on [11].

M_2 and Q'_2, M'_2. Two feedback resistors, R_F and R'_F, close the loop around the differential amplifier. A bias-voltage generator (not shown) biases the gates of M_1, M_2, and M'_2 at V_{BI}.

A point of interest in this circuit is the LC coupling network between the photodetector and the input transistor, which improves the TIA's noise performance and bandwidth (cf. Section 7.7). The coupling network consists of the bond-wire inductor L_1, the bond-pad capacitance C_P, and the on-chip spiral inductor L_2. To obtain a balanced TIA configuration, the coupling network and the capacitance of the photodetector, C_X, are replicated at the unused input (cf. Section 7.2).

In addition to the circuits shown in Fig. 9.5, the design in [11] includes another pair of emitter followers and a 50-Ω output buffer.

Differential TIA for Coherent Receiver. Figures 9.6 and 9.7 show a simplified version of the TIA reported in [14]. This differential shunt-feedback TIA for a coherent receiver is implemented in SiGe:C BiCMOS technology.

The input section of the TIA, excluding the overload/offset control circuits, is shown in Fig. 9.6. The subtraction of the two photocurrents from the balanced detector is done by a pseudo-differential TIA and a subsequent differential VGA rather than directly at the photodetector (cf. Section 3.5). This configuration permits the cathodes of both photodetectors to be tied to the positive supply (3.3 V). A pseudo-differential TIA consisting of two matched single-ended TIAs (Q_1, Q_2 and Q'_1, Q'_2) was chosen for the input section rather than a differential TIA with a tail current source because this topology has a lower DC input voltage. The lower input voltage results in a larger reverse bias voltage for the photodetectors.

To ensure a supply-voltage independent bias current, a replica biasing scheme is used. A (scaled) replica of the single-ended TIA (Q''_1, Q''_2) is biased

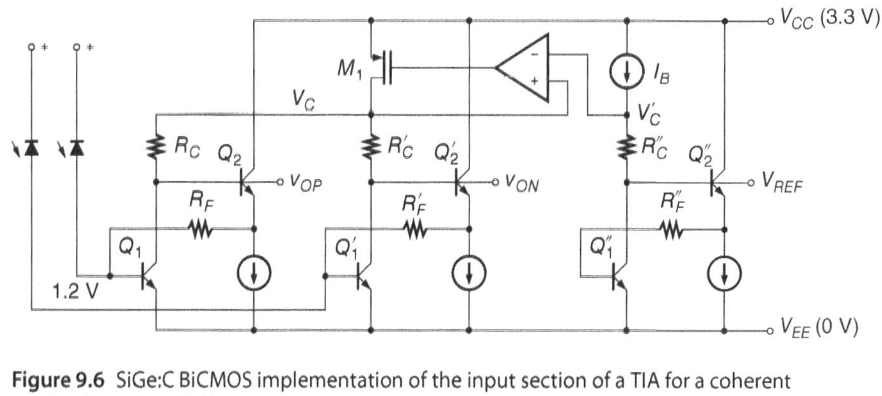

Figure 9.6 SiGe:C BiCMOS implementation of the input section of a TIA for a coherent receiver based on [14].

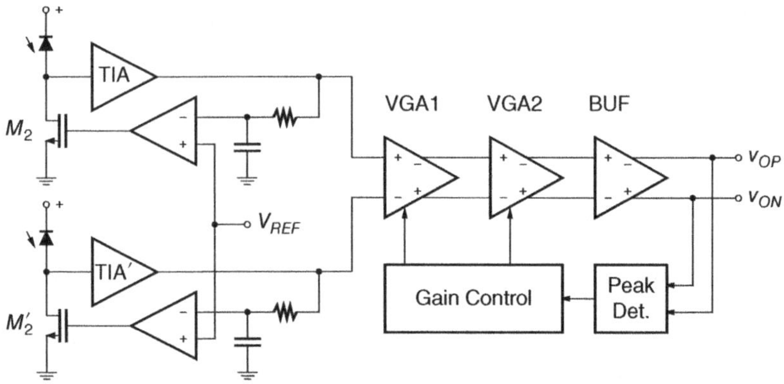

Figure 9.7 The TIA circuit from Fig. 9.6 plus overload/offset control loops and post amplifier with gain control.

at the constant current I_B. The voltage at the replica node V_C' is buffered with a low-dropout regulator (op amp and M_1) and supplied to node V_C of the two main TIAs.

The strong local-oscillator laser used in this coherent receiver leads to large DC currents in both photodetectors. To avoid overloading the TIA circuits in Fig. 9.6, the control loops with M_2 and M_2' shown in Fig. 9.7 are added. The replica TIA in Fig. 9.6 generates the reference voltage V_{REF}, which corresponds to the TIA output voltage under the condition of zero DC input current. Two independent loops control the current sources M_2 and M_2' until the low-pass filtered output voltage of both TIAs match V_{REF}. The use of two loops not only cancels the average DC input current but also the mismatch between the two DC currents (offset control).

The output signals of the two single-ended TIAs in Fig. 9.6 are subtracted by VGA1. The transimpedance of the TIA input section is fixed ($\approx R_F$), but the post amplifier (VGA1 and VGA2) has a variable voltage gain, making the overall transimpedance variable. A gain-control loop maintains a constant output swing.

In addition to the circuits shown in Fig. 9.7, the design in [14] includes negative capacitance circuits to extend the bandwidth and a second offset-voltage control loop around VGA2 and the output buffer.

Burst-Mode TIA. Figures 9.8 and 9.9 show a simplified version of the burst-mode TIA reported in [15]. This shunt-feedback TIA is implemented in SiGe BiCMOS technology.

The core of this TIA is shown in Fig. 9.8. Compared to the similar circuit in Fig. 9.2, cascode transistor Q_2 was added to suppress the Miller multiplication of C_{bc1}, thus increasing the frequency of the open-loop input pole. Instead of cascaded emitter followers, diode D_1 is used to shift the bias voltage by the

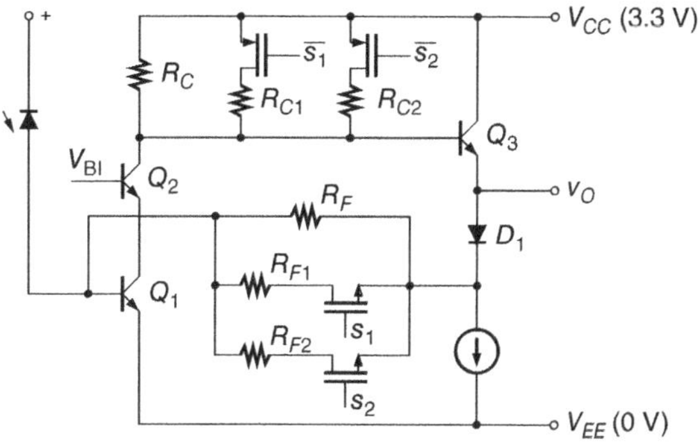

Figure 9.8 SiGe BiCMOS implementation of a burst-mode TIA based on [15].

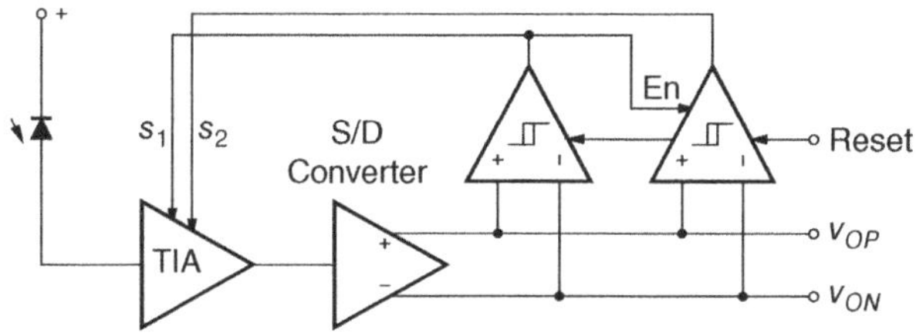

Figure 9.9 The TIA from Fig. 9.8 plus the control loop to adapt the transimpedance to the amplitude of the incoming bursts (based on [15]).

desired amount. Unlike the TIA with continuous AGC in Fig. 9.3, this TIA makes use of discrete transimpedance control. One of three possible transimpedance values is selected with the digital control signals s_1 and s_2 ($\overline{s_1}$ and $\overline{s_2}$ are their complements). When both control signals are low, $s_1 = 0$ and $s_2 = 0$, the total feedback resistance is just R_F; when $s_1 = 1$ and $s_2 = 0$, the feedback resistance is reduced to $R_F || R_{F1}$; and when $s_1 = 1$ and $s_2 = 1$, the feedback resistance is reduced further to $R_F || R_{F1} || R_{F2}$. To maintain a flat frequency response for all settings, the collector load resistance is controlled with a corresponding array of resistors (R_C, R_{C1}, R_{C2}) and switches.

The AGC circuit shown in Fig. 9.9 instantaneously adapts the transimpedance of the TIA in Fig. 9.8 to the amplitude of the incoming bursts. The single-ended signal from the TIA core is converted to differential signals that feed two comparators with hysteresis. If the signal amplitude exceeds the upper threshold of the hysteresis, the output s_1 of the first comparator goes high, instantaneously reducing the transimpedance and thus the TIA's output signal. In addition, the control signal s_1 enables the second comparator. If the reduced output signal still exceeds the upper threshold of the hysteresis, the output s_2 of the second comparator goes high, further reducing the transimpedance. At the end of the burst, both hysteresis comparators are reset to select the highest transimpedance value in anticipation of the next burst.

In addition to the circuits shown in Figs 9.8 and 9.9, the design in [15] includes a limiting amplifier and an output buffer.

A similar burst-mode TIA in SiGe BiCMOS technology has been reported in [16]. In contrast to Figs 9.8 and 9.9, this TIA has only *two* possible transimpedance values and hence requires only one hysteresis comparator.

9.2 CMOS Circuits

Differential Low-Voltage TIA. Figures 9.10 and 9.11 show a simplified version of the TIA reported in [17, 18]. This differential shunt-feedback TIA is part of a single-chip receiver that includes a clock and data recovery (CDR) circuit and a demultiplexer (DMUX). A differential rather that a single-ended topology was chosen for this TIA to better suppress the supply and substrate noise from the CDR and DMUX.

The core of the TIA is shown in Fig. 9.10. The differential voltage amplifier of this shunt-feedback TIA consists of the differential pair M_1 and M_1', the tail current source M_3, and the p-MOS load transistors M_2 and M_2'. The latter have grounded gates and operate in the linear regime. The output signals of the

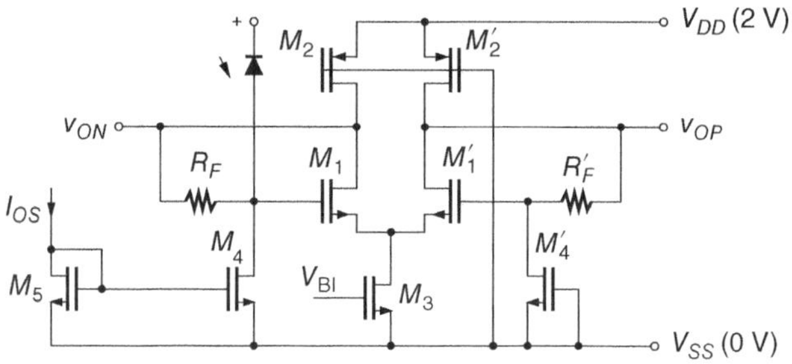

Figure 9.10 CMOS implementation of a differential low-voltage TIA based on [17].

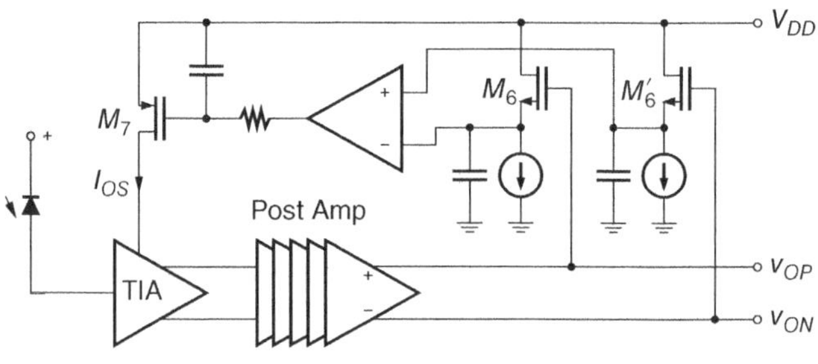

Figure 9.11 The TIA from Fig. 9.10 plus the post amplifier and offset control loop (based on [17, 18]).

differential stage are fed back to the inputs with the resistors R_F and R'_F, which are realized with n-MOS transistors operating in the linear regime. Headroom considerations led to the decision not to use source-follower buffers in this 2-V design. If they were used, the input transistors M_1 and M'_1 would be forced to operate at a very low gate–source voltage leading to low values for f_T and g_m. A low value of g_m relative to the substrate coupling admittance sC_{3dB}, where C_{3dB} is the drain-to-substrate junction capacitance, results in a high sensitivity to substrate noise.

M_5 and M_4 form a current mirror, which subtracts (a scaled version of) the offset control current I_{OS} from the photodetector current. An offset control circuit improves the output dynamic range, which is particularly important in this low-voltage design. Dummy transistor M'_4 at the unused input balances out the capacitance of M_4.

Figure 9.11 shows the offset control circuit that generates the current I_{OS} for the TIA circuit in Fig. 9.10. The TIA core is followed by a 5-stage post amplifier with the outputs v_{OP} and v_{ON}. (One of the gain stages has a variable gain and is controlled by an AGC loop, which is not shown in Fig. 9.11.) Two leaky peak detectors, realized with M_6 and M'_6, determine the top values of v_{OP} and v_{ON}. The difference between these two values is the output offset voltage. An error amplifier followed by an RC low-pass filter amplifies this offset and controls the current source M_7, which, in turn, controls the input offset of the TIA core. The loop slowly adjusts the offset control current I_{OS} until the output offset voltage is essentially zero.

TIA with Push–Pull (CMOS-Style) Voltage Amplifier. Figures 9.12 and 9.13 show a simplified version of the differential shunt-feedback TIA reported in [19].

The core of the TIA is shown in Fig. 9.12. In nanoscale CMOS technologies, an n-MOS transistor with a passive p-MOS load, like M_1 and M_2 in Fig. 9.10, may not provide enough voltage gain. One way to boost the gain is to drive the gates of the n-MOS *and* the p-MOS device resulting in a push–pull

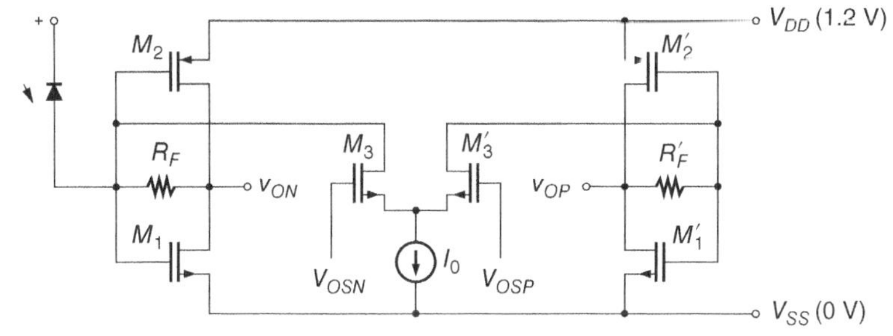

Figure 9.12 CMOS implementation of a pseudo-differential TIA with push–pull voltage amplifier based on [19].

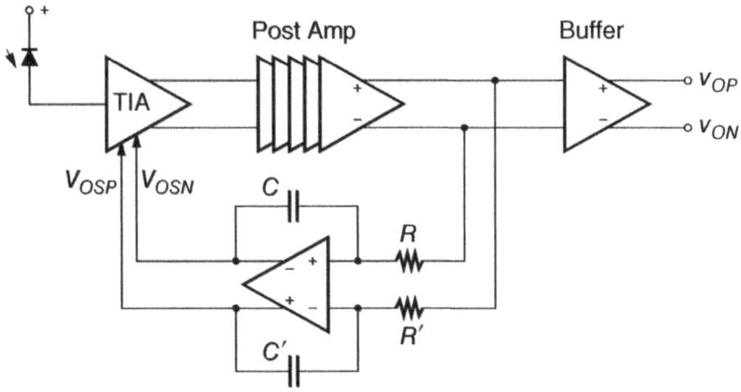

Figure 9.13 The TIA from Fig. 9.12 plus the post amplifier and offset control loop (based on [19]).

topology similar to that of a digital CMOS inverter. The DC gain of the push–pull amplifier in Fig. 9.12 is $A_0 = (g_{m1} + g_{m2})/(g_{o1} + g_{o2})$ as opposed to only $A_0 = g_{m1}/(g_{o1} + g_{o2})$ for the circuit in Fig. 9.10, where we neglected the loading of R_F in both cases. The push–pull amplifier, which stacks only two transistors, also has the advantage that it can operate from a low supply voltage. Finally, the push–pull amplifier is power efficient because the p-MOS transistor reuses the bias current from the n-MOS transistor.

The pseudo-differential TIA with a single photodetector shown in Fig. 9.12 outputs an active signal at v_{ON} and a DC reference voltage at v_{OP}. The differential output voltage $v_{OP} - v_{ON}$ is more immune to supply and substrate noise than v_{ON} alone. The differential pair M_3 and M_3', which is driven by the control voltage $V_{OSP} - V_{OSN}$, steers a fraction of the bias current I_0 into the input of the main TIA and the remainder into the dummy TIA thus controlling the TIA's output offset voltage.

Figure 9.13 shows the offset-control circuit that generates the voltages V_{OSP} and V_{OSN} for the TIA circuit in Fig. 9.12. The TIA core is followed by a 5-stage post amplifier and a buffer with the outputs v_{OP} and v_{ON}. An active low-pass filter, consisting of an amplifier, R, R', C, and C', senses the offset voltage at the output of the post amplifier and slowly adjusts the DC control voltages in response to it.

In [20] a single-ended TIA with push–pull voltage amplifier, similar to the left half of the circuit in Fig. 9.12 (M_1, M_2, and R_F), followed by a single-ended push–pull post amplifier (another CMOS inverter) has been reported. Other examples of CMOS-inverter-based TIAs can be found in [21–25].

TIA with Multistage Voltage Amplifier. Figure 9.14 shows the three-stage shunt-feedback TIA reported in [26].

While most shunt-feedback TIAs are build around a voltage amplifier with a single gain stage, this design uses three gain stages. Each stage consists of three

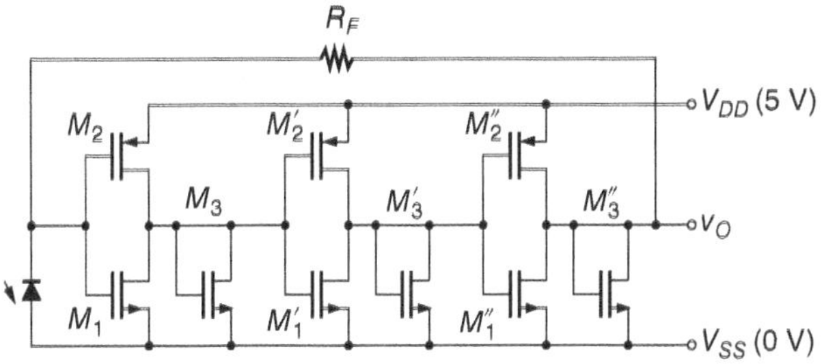

Figure 9.14 CMOS implementation of a multistage TIA based on [26].

MOSFETs: M_1 through M_3 in the first, M_1' through M_3' in the second, and M_1'' through M_3'' in the third stage. M_1 and M_2 act as a push–pull transconductor and M_3 acts as a load. Thus, the stage gain is approximately $(g_{m1} + g_{m2})/g_{m3}$, a quantity that is well defined by the device geometries. Note that this single-ended amplifier must have an odd number of stages to ensure negative feedback through R_F.

The use of three gain stages results in a higher DC gain, but it also produces three poles at similar frequencies in the open-loop frequency response of this TIA. These poles must be placed at sufficiently high frequencies to ensure a good phase margin and a flat closed-loop frequency response (cf. Section 6.2).

TIA with Common-Gate Input Stage. Figure 9.15 shows the core of the differential TIA reported in [27]. This design combines two common-gate input stages with a conventional shunt-feedback section.

The differential voltage amplifier of the shunt-feedback section is implemented with MOSFETs M_1 through M_3 and M_1' through M_3'. The gain is provided by the differential stage consisting of M_1 and M_1' and the load

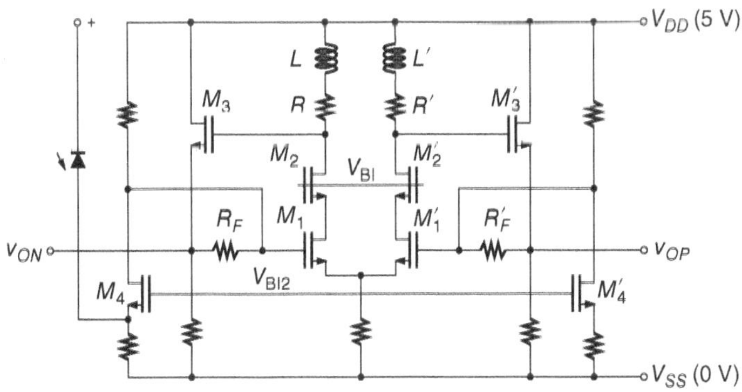

Figure 9.15 CMOS implementation of a differential TIA with common-gate input stages based on [27].

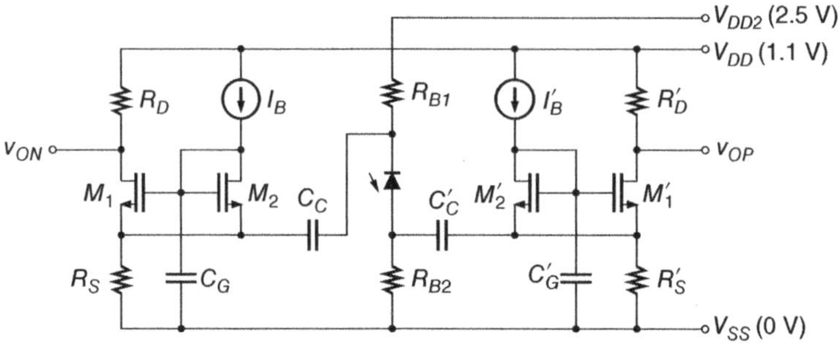

Figure 9.16 CMOS implementation of a pseudo-differential common-gate TIA based on [31].

resistors R and R'. The cascode transistors M_2 and M_2', biased by the DC voltage V_{BI}, as well as the shunt-peaking inductors L and L' broaden the bandwidth of the stage. Source followers M_3 and M_3' buffer the outputs of the differential stage and drive the feedback resistors R_F and R_F'.

The common-gate input stage is implemented with transistor M_4 biased by the DC voltage V_{BI2}. This stage decouples the photodetector capacitance from the critical node at the gate of M_1 and thus reduces the dependence of the TIA's bandwidth on the detector capacitance (cf. Section 7.5). The presence of M_4 also increases the reverse-bias of the photodetector by V_{DS4}, which improves the speed of the detector. To obtain a balanced TIA configuration, the common-gate input stage is replicated at the unused input. In [27], this TIA core is followed by an output buffer to drive off-chip loads (not shown).

A similar TIA circuit, but using regulated-cascode input stages instead of common-gate input stages, has been reported in [28, 29]. The regulated-cascode stage can provide a lower input impedance or reduce the power consumption (cf. Section 7.6). Another example of a (modified) regulated-cascode stage followed by a shunt-feedback TIA can be found in [30].

Common-Gate Feedforward TIA. Figure 9.16 shows the simplified core of the feedforward TIA reported in [31]. An array of sixteen such TIAs together with sixteen laser drivers are integrated on a single chip for an optical interconnect application.

Two resistors, R_{B1} and R_{B2}, bias the photodetector such that the photocurrent can be taken from both ends. Each side of the photodetector connects to a single-ended TIA. Assuming matched parasitic capacitances at the TIA inputs, a balanced differential signal is produced at the output. To overcome the DC voltage difference between the photodetector terminals and the TIA inputs, coupling capacitors C_C and C_C' are used. The size constraints for these on-chip capacitors lead to a relatively high low-frequency cutoff.

Since this pseudo-differential TIA consists of two identical single-ended TIAs, we can restrict ourselves to discuss only one of them. The TIA is of the feedforward type and is implemented around the common-gate transistor M_1. Its gate is AC grounded by means of capacitor C_G. Neglecting the source resistor R_S, the transimpedance of this TIA is approximately equal to the load resistance R_D. The DC bias current is established with the "floating current mirror" consisting of M_2 and M_1. With M_1 being sixteen times larger than M_2, the bias current in M_1 is set to $16I_B$.

In addition to the circuits shown in Fig. 9.16, the design in [31] features inductors in series to R_D and R'_D, similar to those in Fig. 9.15, and inductors in series to C_C and C'_C, similar to those in Fig. 9.5, to broaden the bandwidth. Moreover, the TIA core is followed by a five-stage limiting amplifier and a source-follower buffer for driving 50-Ω off-chip loads.

The circuit in Fig. 9.16 can easily be converted to a single-ended version by omitting one half of it [32, 33]. The photodetector can then be DC coupled to the TIA input, getting rid of the coupling capacitor C_C and the associated low-frequency cutoff. The biasing scheme with the floating current mirror ensures that the bias current through the common-gate MOSFET is not disturbed by the DC current from the photodetector.

Low-Voltage Regulated-Cascode TIA. Figure 9.17 shows the regulated-cascode TIA reported in [34]. This TIA is used for a short-distance optical interconnect application.

In the conventional regulated-cascode TIA, see Fig. 7.26(b), the gate of the cascode transistor M_1 is biased two gate–source voltages above ground. One gate–source voltage, V_{GS1}, is due to the cascode transistor and the other one is due to the input voltage of the booster amplifier (V_{GSB} in Fig. 7.26(b)). As a result, the gate bias voltage must be larger than two threshold voltages, which may present a problem in a low-voltage design.

The circuit in Fig. 9.17 alleviates this problem by adding stage M_2, R_2 to the conventional booster stage M_3, R_3. The common-gate transistor M_2 reduces the

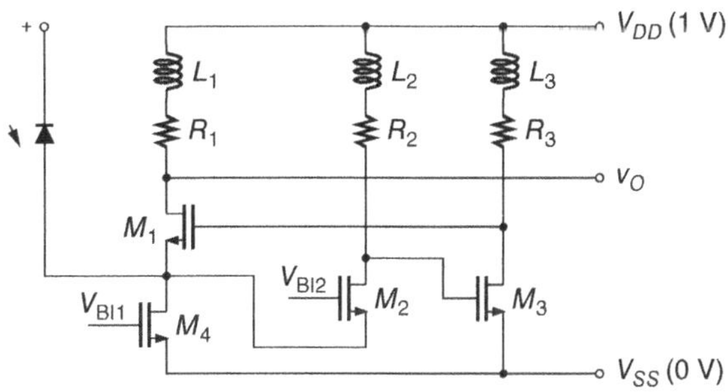

Figure 9.17 CMOS implementation of a low-voltage regulated-cascode TIA based on [34].

input voltage of the booster amplifier to $V_{\mathrm{BI2}} - V_{GS2}$, which can be well below one threshold voltage. The bias voltage at the gate of M_1 is now only $V_{\mathrm{BI2}} - V_{GS2} + V_{GS1}$ instead of $V_{GSB} + V_{GS1}$.

Moreover, the two-stage booster amplifier provides more gain resulting in a lower TIA input resistance. Stability of the feedback loop must be checked carefully to make sure that the added stage does not contribute too much phase shift. Shunt-peaking inductors L_2 and L_3 increase the bandwidth of the booster amplifier.

Burst-Mode TIA. Figures 9.18 and 9.19 show a simplified version of the shunt-feedback TIA for burst-mode applications reported in [35].

The core of this TIA, shown in Fig. 9.18, is similar to its bipolar counterpart depicted in Fig. 9.8. Both TIAs have a discrete gain-control mechanism. The digital control signal s ($\bar{s}$ is the complement of s) selects one of two possible

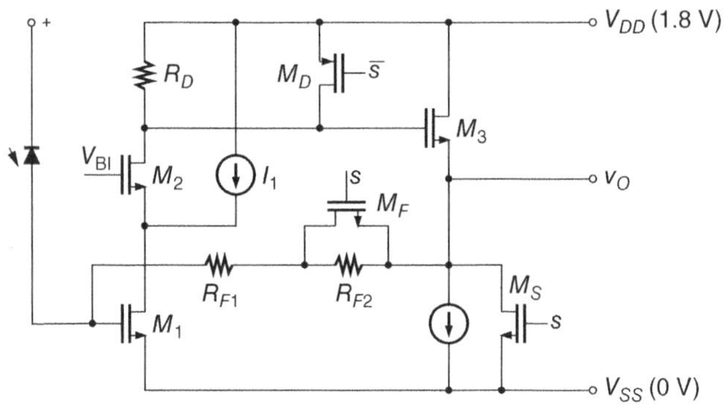

Figure 9.18 CMOS implementation of a burst-mode TIA based on [35].

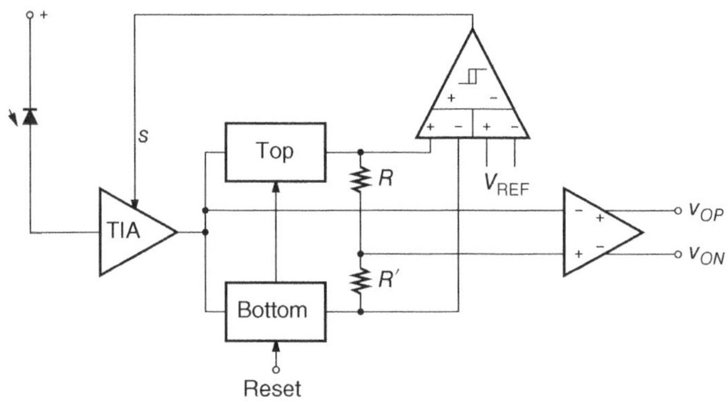

Figure 9.19 The TIA from Fig. 9.18 plus the control loop to adapt the transimpedance to the amplitude of the incoming bursts (based on [35]).

transimpedance values. In the high-gain mode, the control signal is low and the total feedback resistance is $R_{F1} + R_{F2}$; in the low-gain mode, M_F turns on, reducing the feedback resistance to R_{F1}. To maintain a flat frequency response, M_D reduces the load resistance when the TIA is in the low-gain mode. To maintain a constant DC output voltage, M_S increases the bias current in the source follower M_3 when the low-gain mode is selected. The resulting increase in gate-to-source voltage compensates for the increased voltage at the gate of M_3. Current source I_1 injects extra current into M_1 to boost its transconductance, g_{m1}, and thus the voltage gain, $g_{m1}R_D$, without increasing the DC voltage drop across R_D.

The AGC circuit, shown in Fig. 9.19, instantaneously adapts the transimpedance of the TIA in Fig. 9.18 to the amplitude of the incoming bursts. The top and the bottom levels of the signal from the TIA core are determined with two peak detectors. Their difference, which corresponds to the signal swing, is compared with the reference voltage V_{REF}. If the swing exceeds V_{REF}, s goes high and the low-gain mode is selected. A small hysteresis in the comparator ensures that the gain mode remains stable. Resistors R and R' produce a voltage halfway between the top and bottom levels, which is used as a reference to convert the single-ended signal from the TIA core to a differential output signal.

9.3 MESFET and HFET Circuits

Single-Ended TIA. Figure 9.20 shows a simplified schematic of the TIA reported in [36, 37]. This single-ended shunt-feedback TIA is implemented in GaAs-MESFET technology. The transistors in this circuit are depletion-mode FETs, which means that they conduct current when the gate–source voltage is zero. (In the schematics, we use a thin line from drain to source to distinguish depletion-mode from enhancement-mode devices.)

FETs M_1 through M_4 constitute the voltage amplifier of the shunt-feedback TIA. The common-source stage consisting of M_1 and M_2 provides the voltage gain. FET M_2 (like M_4 and M_6) has the gate tied to the source and thus

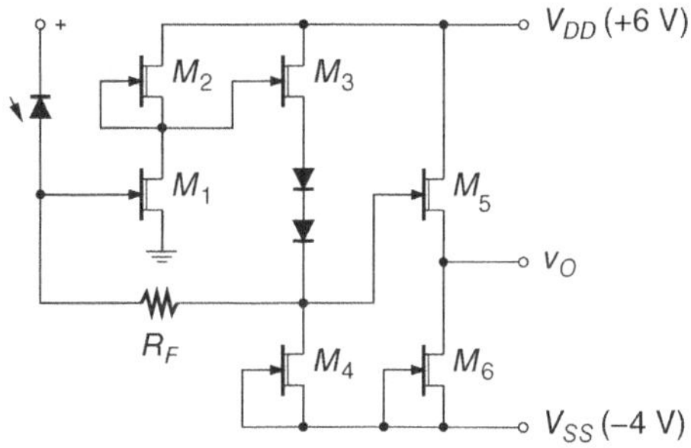

Figure 9.20 GaAs-MESFET implementation of a single-ended TIA based on [37].

acts as a constant current source. The source-follower (common-drain) stage with M_3 and M_4 buffers the output signal and, with a stack of two Schottky diodes, shifts the DC voltage to a lower value. The feedback resistor R_F closes the loop around this inverting amplifier. The level shifter implemented with the source-follower stage and the diode stack ensures that M_1 operates in saturation. Another source-follower stage with M_5 and M_6 serves as an output buffer.

TIA with Two-Stage Voltage Amplifier. Figure 9.21 shows a simplified schematic of the TIA reported in [36, 38]. This two-stage shunt-feedback TIA is also implemented in GaAs-MESFET technology.

The first gain stage, M_1 to M_5, is similar to the voltage amplifier in Fig. 9.20. The cascode transistor M_2 was added to suppress the Miller multiplication of C_{gd1}, thus increasing the frequency of the open-loop input pole. The inductor L was inserted at the source of the load transistor M_3 to improve the bandwidth of the stage and to reduce the noise. Interestingly, the load impedance synthesized by the combination of M_3 and L is $Z_L = r_{o3} + j\omega(g_{m3}r_{o3} + 1)L$, that is, the inductance L of the spiral inductor is multiplied by the intrinsic gain of M_3 plus one [38, 39].

The second stage, M_6 to M_8, is implemented with a common-gate FET to keep the overall amplifier polarity inverting. (A three-stage amplifier with three inverting stages was also considered by the designers, but the two-stage approach was chosen because of its better stability.) The drain current of the common-gate FET M_6 is "folded" with the current source M_7 into the diode stack and the load FET M_8 producing an output voltage that is suitable for biasing the gate of M_1 through the feedback resistor R_F. An interesting detail: the gate of M_6 is not grounded as in a regular common-gate stage, but is connected to the TIA's input. This trick generates a zero in the open-loop transfer function that enhances the stability. The output buffer, M_9 and M_{10}, is the same as in Fig. 9.20.

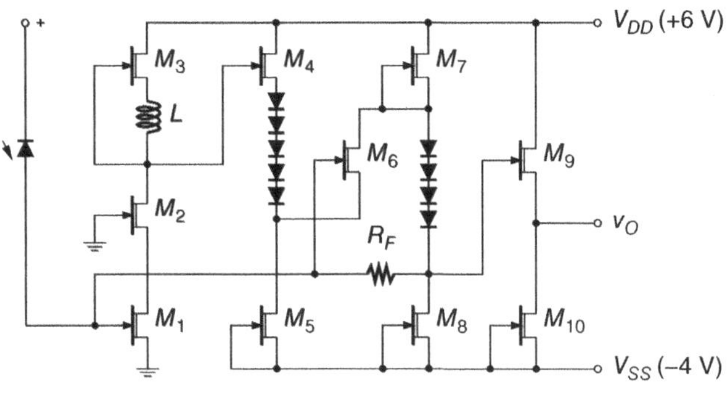

Figure 9.21 GaAs-MESFET implementation of a two-stage TIA based on [38].

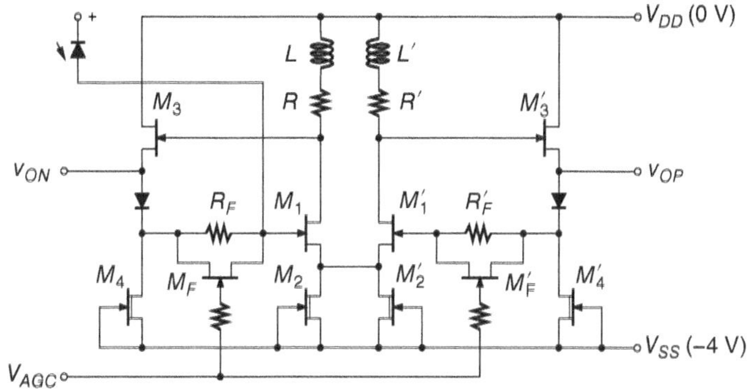

Figure 9.22 GaAs-HFET implementation of a differential TIA with AGC based on [40].

In addition to the circuits shown in Fig. 9.21, the TIA in [38] includes a FET in parallel to R_F, which is controlled by an AGC loop.

Differential TIA with AGC. Figure 9.22 shows a simplified schematic of the TIA reported in [40]. This differential shunt-feedback TIA is implemented in GaAs-HFET technology.

The differential voltage amplifier of the shunt-feedback TIA is implemented with both enhancement-mode and depletion-mode HFETs. Enhancement-mode HFETs M_1, M_1', M_3, M_3', M_F, and M_F' are used in the signal path, whereas depletion-mode HFETs M_2, M_2', M_4, and M_4' are used as bias elements. The gain is provided by the differential pair, M_1 and M_1', with tail current source, M_2 and M_2', and load resistors R and R'. The shunt peaking inductors L and L' reduce the noise and broaden the bandwidth of the stage. The outputs of the differential stage are buffered by the source followers M_3 and M_3'. Schottky diodes lower the DC output voltage and the feedback resistors R_F and R_F' close the loop around this differential amplifier.

The effective value of the feedback resistance is determined by the parallel connection of R_F and M_F (R_F' and M_F'). HFETs M_F and M_F', which operate in the linear regime, are controlled by an AGC loop.

In addition to the circuits shown in Fig. 9.22, the design in [40] includes a four-stage post amplifier and a buffer to drive a 50-Ω off-chip load.

9.4 Summary

Tables 9.1 and 9.2 list the performance, topology, and technology of high-speed (continuous mode) TIAs taken from journal and conference papers. Unlike commercial data sheets, these publications usually give detailed information about the circuit topology and the technology the chip is fabricated in.

Table 9.1 High-speed TIA publications (part 1).

Authors, Year	Speed (Gb/s)	R_T (kΩ)	BW_{3dB} (GHz)	$i^{rms}_{n,TIA}$ (μA)	Topology	Technology
Kim et al., 2001 [11]	10.0	0.56	9.0		SFB (DCE/EF)	0.25-μm BiCMOS
Ikeda et al., 2001 [41]	10.0	1.46	8.0	0.58[a)]	SFB (CS/CG/SF)	0.1-μm HFET
Analui and Hajimiri, 2004 [42]	10.0	0.50	9.2	1.60	SFB (CS/CG/SF)	0.18-μm CMOS
Wu et al., 2005 [25]	10.0	1.12	7.2	0.70[a)]	SFB (INV)	0.18-μm CMOS
Schow et al., 2009 [31]	10.0				CG	0.13-μm CMOS
Morita et al., 2014 [23]	10.0	1.45		0.95	SFB (INV)	65-nm CMOS
Proesel et al., 2012 [19]	25.0	8.20	20.0	2.40	SFB (INV)	90-nm CMOS
Buckwalter et al., 2012 [43]	25.0	2.20	25.0	6.32[a)]	AFB (DCS/DCS)	0.13-μm SOI CMOS
Kalogerakis et al., 2013 [44]	25.0	6.70	23.0	2.40	CB/CE	0.13-μm SiGe BiCMOS
Li and Palermo, 2013 [30]	25.0	0.89	26.0	3.45	RGC + SFB (CE)	0.25-μm SiGe BiCMOS (f_T = 137 GHz)
Li et al., 2014 [21]	25.0	14.10	13.6	1.79	SFB (INV)	65-nm CMOS
Takemoto et al., 2014 [45]	28.0	6.90	21.4		RGC	65-nm CMOS
Chiang et al., 2015 [46]	25.0	4.20	21.0	4.20	SFB (CS)	65-nm CMOS

a) Estimated from the reported noise spectral density and 3-dB bandwidth.

Table 9.2 High-speed TIA publications (part 2).

Authors, Year	Speed (Gb/s)	R_T (kΩ)	BW_{3dB} (GHz)	$i_{n,TIA}^{rms}$ (μA)	Topology	Technology
Weiner et al., 2004 [47]	40.0	0.63	47.0	6.30[a)]	SFB (DCE/EF)	InP HBT (f_T = 160 GHz)
Tran et al., 2004 [48]	43.0	6.00	35.0		SFB (DCE/EF)	InP HBT (f_T = 160 GHz)
Liao and Liu, 2008 [49]	40.0	2.00	22.0	3.30	SFB (CG/DCS/SF)	90-nm CMOS
Jin and Hsu, 2008 [50]	40.0	0.35	30.5	9.70	LZ	0.18-μm CMOS
Knochenhauer et al., 2010 [7]	40.0	6.00	37.6	<3.90[a)]	SFB (CE/CB/EF)	0.25-μm SiGe BiCMOS (f_T = 180 GHz)
Kim and Buckwalter, 2012 [20]	40.0	0.54	30.0	3.54	SFB (INV)	45-nm SOI CMOS
Kim et al., 2014 [51]	50.0	0.40	50.0	5.01	CG/CS	65-nm CMOS
Mohammadnezhad et al., 2015 [52]	50.0	0.11	50.0	7.92[a)]	RGC	0.13-μm SiGe BiCMOS (f_T = 180 GHz)
Dupuy et al., 2015 [53]	64.0	0.90	68.0	5.74[a)]	SFB (DCE/EF)	0.7-μm InP HBT (f_T = 400 GHz)
Awny et al., 2016 [14]	64.0	10.00	53.0	5.72[a)]	SFB (CE/EF)	0.13-μm SiGe BiCMOS (f_T = 250 GHz)
Vasilakopoulos et al., 2015 [4]	120.0	0.22	92.0	5.60	SFB (CE)	55-nm SiGe BiCMOS HBT (f_T = 330 GHz)

a) Estimated from the reported noise spectral density and 3-dB bandwidth.

The table thus provides some insight into which topologies are popular and what performance can be achieved in which technology.

The abbreviations in the Topology column have the following meaning: SFB = (resistive) shunt feedback, AFB = active feedback, CB = common base, CG = common gate, RGC = regulated cascode, LZ low-impedance front-end. In the case of SFB or AFB the abbreviations in parenthesis describe the topology of the voltage amplifier: CE = common-emitter stage, DCE = differential common-emitter stage (a bipolar differential pair), CB = common-base stage (a cascode stage, if preceded by CE or DCE), EF = emitter-follower stage, CS = common-source stage, DCS = differential common-source stage (a FET differential pair), CG = common-gate stage (a cascode stage, if preceded by CS or DCS), SF = source-follower stage, and INV = CMOS-style inverter.

References

1 S. P. Voinigescu, D. S. McPherson, F. Pera, S. Szilagyi, M. Tazlauanu, and H. Tran. A comparison of silicon and III-V technology performance and building block implementations for 10 and 40 Gb/s optical networking ICs. *Int. J. High Speed Electron. Syst.*, 13(1), 2003.

2 S. Voinigescu. *High-Frequency Integrated Circuits*. Cambridge University Press, 2013.

3 S. P. Voinigescu, T. O. Dickson, R. Beerkens, and P. Westergaard. A comparison of Si CMOS, SiGe BiCMOS, and InP HBT technologies for high-speed and millimeter-wave ICs. In *Silicon Monolithic Integrated Circuits in RF Systems, Atlanta, GA*, September 2004.

4 K. Vasilakopoulos, S. P. Voinigescu, P. Schvan, P. Chevalier, and A. Cathelin. A 92GHz bandwidth SiGe BiCMOS HBT TIA with less than 6dB noise figure. In *Proceedings of IEEE Bipolar/BiCMOS Circuits and Technology Meeting*, pages 168–171, October 2015.

5 H. Khorramabadi, L. D. Tzeng, and M. J. Tarsia. A 1.06Gb/s, −31dBm to 0dBm BiCMOS optical preamplifier featuring adaptive transimpedance. In *ISSCC Digest of Technical Papers*, pages 54–55, February 1995.

6 R. G. Meyer and R. A. Blauschild. A wide-band low-noise monolithic transimpedance amplifier. *IEEE J. Solid-State Circuits*, SC-21(4):530–533, 1986.

7 C. Knochenhauer, S. Hauptmann, J. C. Scheytt, and F. Ellinger. A jitter-optimized differential 40-Gbit/s transimpedance amplifier in SiGe BiCMOS. *IEEE Trans. Microwave Theory Tech.*, MTT-58(10):2538–2548, 2010.

8 Y. Kuriyama, J. Akagi, T. Sugiyama, S. Hongo, K. Tsuda, N. Iizuka, and M. Obara. DC to 40-GHz broad-band amplifiers using AlGaAs/GaAs HBT's. *IEEE J. Solid-State Circuits*, SC-30(10):1051–1054, 1995.

9 M. Soda, H. Tezuka, F. Sato, T. Hashimoto, S. Nakamura, T. Tatsumi, T. Suzaki, and T. Tashiro. Si-analog IC's for 20 Gb/s optical receiver. *IEEE J. Solid-State Circuits*, SC-29(12):1577–1582, 1994.

10 T. Suzaki, M. Soda, T. Morikawa, H. Tezuka, C. Ogawa, S. Fujita, H. Takemura, and T. Tashiro. Si bipolar chip set for 10-Gb/s optical receiver. *IEEE J. Solid-State Circuits*, SC-27(12):1781–1786, 1992.

11 H. H. Kim, S. Chandrasekhar, C. A. Burrus, and J. Bauman. A Si BiCMOS transimpedance amplifier for 10-Gb/s SONET receiver. *IEEE J. Solid-State Circuits*, SC-36(5):769–776, 2001.

12 B. Razavi. A 1.5V 900MHz downconversion mixer. In *ISSCC Digest of Technical Papers*, pages 48–49, February 1996.

13 B. Razavi. A 2.5-Gb/s 15-mW clock recovery circuit. *IEEE J. Solid-State Circuits*, SC-31(4):472–480, 1996.

14 A. Awny, R. Nagulapalli, D. Micusik, J. Hoffmann, G. Fischer, D. Kissinger, and A. C. Ulusoy. A dual 64Gbaud 10kΩ 5% THD linear differential transimpedance amplifier with automatic gain control in 0.13μm BiCMOS technology for optical fiber coherent receivers. In *ISSCC Digest of Technical Papers*, pages 406–407, February 2016.

15 M. Nakamura, Y. Imai, Y. Umeda, J. Endo, and Y. Akatsu. 1.25-Gb/s burst-mode receiver ICs with quick response for PON systems. *IEEE J. Solid-State Circuits*, SC-40(12):2680–2688, 2005.

16 S. Nishihara, M. Nakamura, T. Ito, T. Kurosaki, Y. Ohtomo, and A. Okada. A SiGe BiCMOS burst-mode transimpedance amplifier using fast and accurate automatic offset compensation technique for 1G/10G dual-rate transceiver. In *Proceedings of IEEE Bipolar/BiCMOS Circuits and Technology Meeting*, pages 158–161, October 2009.

17 A. Tanabe, M. Soda, Y. Nakahara, T. Tamura, K. Yoshida, and A. Furukawa. A single-chip 2.4-Gb/s CMOS optical receiver IC with low substrate cross-talk preamplifier. *IEEE J. Solid-State Circuits*, SC-33(12):2148–2153, 1998.

18 A. Tanabe, M. Soda, Y. Nakahara, A. Furukawa, T. Tamura, and K. Yoshida. A single chip 2.4Gb/s CMOS optical receiver IC with low substrate crosstalk preamplifier. In *ISSCC Digest of Technical Papers*, pages 304–305, February 1998.

19 J. Proesel, C. Schow, and A. Rylyakov. 25Gb/s 3.6pJ/b and 15Gb/s 1.37pJ/b VCSEL-based optical links in 90nm CMOS. In *ISSCC Digest of Technical Papers*, pages 418–419, February 2012.

20 J. Kim and J. F. Buckwalter. A 40-Gb/s optical transceiver front-end in 45-nm SOI CMOS. *IEEE J. Solid-State Circuits*, SC-47(3):615–626, 2012.

21 D. Li, G. Minoia, M. Repossi, D. Baldi, E. Temporiti, A. Mazzanti, and F. Svelto. A low-noise design technique for high-speed CMOS optical receivers. *IEEE J. Solid-State Circuits*, SC-49(6):1437–1447, 2014.

22 F. Y. Liu, D. Patil, J. Lexau, P. Amberg, M. Dayringer, J. Gainsley, H. F. Moghadam, X. Zheng, J. E. Cunningham, A. V. Krishnamoorthy, E. Alon, and R. Ho. 10-Gbps, 5.3-mW optical transmitter and receiver circuits in 40-nm CMOS. *IEEE J. Solid-State Circuits*, SC-47(9):2049–2067, 2012.
23 H. Morita, K. Uchino, E. Otani, H. Ohtorii, T. Ogura, K. Oniki, S. Oka, S. Yanagawa, and H. Suzuki. A 12 × 5 two-dimensional optical I/O array for 600Gb/s chip-to-chip interconnect in 65nm CMOS. In *ISSCC Digest of Technical Papers*, pages 140–141, February 2014.
24 T. K. Woodward and A. V. Krishnamoorthy. 1 Gb/s CMOS photoreceiver with integrated detector operating at 850nm. *Electron. Lett.*, 34(12):1252–1253, 1998.
25 C.-H. Wu, C.-H. Lee, W.-S. Chen, and S.-I. Liu. CMOS wideband amplifiers using multiple inductive-series peaking technique. *IEEE J. Solid-State Circuits*, SC-40(2):548–552, 2005.
26 M. Ingels, G. Van der Plas, J. Crols, and M. Steyaert. A CMOS 18 THzΩ 240 Mb/s transimpedance amplifier and 155 Mb/s LED-driver for low cost optical fiber links. *IEEE J. Solid-State Circuits*, SC-29(12):1552–1559, 1994.
27 S. S. Mohan, M. D. M. Hershenson, S. P. Boyd, and T. H. Lee. Bandwidth extension in CMOS with optimized on-chip inductors. *IEEE J. Solid-State Circuits*, SC-35(3):346–355, 2000.
28 S. M. Park, J. Lee, and H.-J. Yoo. 1-Gb/s 80-dBΩ fully differential CMOS transimpedance amplifier in multichip on oxide technology for optical interconnects. *IEEE J. Solid-State Circuits*, SC-39(6):971–974, 2004.
29 S. M. Park and H.-J. Yoo. 1.25-Gb/s regulated cascode CMOS transimpedance amplifier for Gigabit Ethernet applications. *IEEE J. Solid-State Circuits*, SC-39(1):112–121, 2004.
30 C. Li and S. Palermo. A low-power 26-GHz transformer-based regulated cascode SiGe BiCMOS transimpedance amplifier. *IEEE J. Solid-State Circuits*, SC-48(5):1264–1275, 2013.
31 C. L. Schow, F. E. Doany, C. Chen, A. V. Rylyakov, C. W. Baks, D. M. Kuchta, R. A. John, and J. A. Kash. Low-power 16 × 10 Gb/s bi-directional single chip CMOS optical transceivers operating at < 5 mW/Gb/s/link. *IEEE J. Solid-State Circuits*, SC-44(1):301–313, 2009.
32 D. Guckenberger, J. D. Schaub, and K. T. Kornegay. A DC-coupled low-power transimpedance amplifier architecture for Gb/s communication system applications. In *IEEE Radio Frequency Integrated Circuits (RFIC) Symposium*, pages 515–518, June 2004.
33 C. L. Schow, L. Schares, R. A. John, L. S. Fischer, and D. Guckenberger. 25 Gbit/s transimpedance amplifier in 0.13 μm CMOS. *Electron. Lett.*, 42(21):1240–1241, 2006.
34 C. Kromer, G. Sialm, T. Morf, M. L. Schmatz, F. Ellinger, D. Erni, and H. Jäckel. A low-power 20-GHz 52-dBΩ transimpedance amplifier in 80-nm CMOS. *IEEE J. Solid-State Circuits*, SC-39(6):885–894, 2004.

35 W.-Z. Chen, R.-M. Gan, and S.-H. Huang. A single-chip 2.5-Gb/s CMOS burst-mode optical receiver. *IEEE Trans. Circuits Syst. I*, CASI-56(10):2325–2331, 2009.

36 D. Estreich. Wideband amplifiers. In R. Goyal, editor, *High-Frequency Analog Integrated Circuit Design*, pages 170–240. John Wiley & Sons, Inc., New York, 1995.

37 O. Wada, T. Hamaguchi, S. Miura, M. Makiuchi, K. Nakai, H. Horimatsu, and T. Sakurai. AlGaAs/GaAs p-i-n photodiode/preamplifier monolithic photoreceiver integrated on semi-insulating GaAs substrate. *Appl. Phys. Lett.*, 46(10):981–983, 1985.

38 N. Scheinberg, R. J. Bayruns, and T. M. Laverick. Monolithic GaAs transimpedance amplifiers for fiber-optic receivers. *IEEE J. Solid-State Circuits*, SC-26(12):1834–1839, 1991.

39 Y.-H. Oh, S.-G. Lee, and H. H. Park. A 2.5Gb/s CMOS transimpedance amplifier using novel active inductor load. In *Proceedings of European Solid-State Circuits Conference*, Villach, Austria, September 2001.

40 Z. Lao, M. Berroth, V. Hurm, A. Thiede, R. Bosch, P. Hofman, A. Hülsmann, C. Moglestue, and K. Köhler. 25Gb/s AGC amplifier, 22GHz transimpedance amplifier and 27.7GHz limiting amplifier ICs using AlGaAs/GaAs-HEMTs. In *ISSCC Digest of Technical Papers*, pages 356–357, February 1997.

41 H. Ikeda, T. Ohshima, M. Tsunotani, T. Ichioka, and T. Kimura. An auto-gain control transimpedance amplifier with low noise and wide input dynamic range for 10-Gb/s optical communication systems. *IEEE J. Solid-State Circuits*, SC-36(9):1303–1308, 2001.

42 B. Analui and A. Hajimiri. Bandwidth enhancement for transimpedance amplifiers. *IEEE J. Solid-State Circuits*, SC-39(8):1263–1270, 2004.

43 J. F. Buckwalter, X. Zheng, G. Li, K. Raj, and A. V. Krishnamoorthy. A monolithic 25-Gb/s transceiver with photonic ring modulators and Ge detectors in a 130-nm CMOS SOI process. *IEEE J. Solid-State Circuits*, SC-47(6):1309–1322, 2012.

44 G. Kalogerakis, T. Moran, T. Nguyen, and G. Denoyer. A quad 25Gb/s 270mW TIA in 0.13μm BiCMOS with <0.15dB crosstalk penalty. In *ISSCC Digest of Technical Papers*, pages 116–117, February 2013.

45 T. Takemoto, H. Yamashita, T. Yazaki, N. Chujo, Y. Lee, and Y. Matsuoka. A 25-to-28 Gb/s high-sensitivity (−9.7 dBm) 65 nm CMOS optical receiver for board-to-board interconnects. *IEEE J. Solid-State Circuits*, SC-49(10):2259–2276, 2014.

46 P.-C. Chiang, J.-Y. Jiang, H. W. Hung, C.-Y. Wu, G.-S. Chen, and J. Lee. 4×25-Gb/s transceiver with optical front-end for 100 GbE system in 65 nm CMOS technology. *IEEE J. Solid-State Circuits*, SC-50(2):573–585, 2015.

47 J. S. Weiner, J. S. Lee, A. Leven, Y. Baeyens, V. Houtsma, G. Georgiou, Y. Yang, J. Frackoviak, A. Tate, R. Reyes, R. F. Kopf, W.-J. Sung, N. G. Weimann, and Y.-K. Chen. An InGaAs–InP HBT differential transimpedance amplifier with 47-GHz bandwidth. *IEEE J. Solid-State Circuits*, SC-39(10):1720–1723, 2004.

48 H. Tran, F. Pera, D. S. McPherson, D. Viorel, and S. P. Voinigescu. 6-kΩ 43-Gb/s differential transimpedance-limiting amplifier with auto-zero feedback and high dynamic range. *IEEE J. Solid-State Circuits*, SC-39(10):1680–1696, October 2004. Also see the Correction in the December 2004 issue.

49 C.-F. Liao and S.-I. Liu. 40 Gb/s transimpedance-AGC amplifier and CDR circuit for broadband data receivers in 90 nm CMOS. *IEEE J. Solid-State Circuits*, SC-43(3):642–655, 2008.

50 J.-D. Jin and S. S. H. Hsu. A 40-Gb/s transimpedance amplifier in 0.18-¯ CMOS technology. *IEEE J. Solid-State Circuits*, SC-43(6):1449–1457, 2008.

51 S. G. Kim, S. H. Jung, Y. S. Eo, S. H. Kim, X. Ying, H. Choi, C. Hong, K. Lee, and S. M. Park. A 50-Gb/s differential transimpedance amplifier in 65nm CMOS technology. In *IEEE Asian Solid-State Circuits Conference (ASSCC)*, pages 357–360, November 2014.

52 H. Mohammadnezhad, A. K. Bidhendi, M. M. Green, and P. Heydari. A low-power BiCMOS 50 Gbps G_m-boosted dual-feedback transimpedance amplifier. In *Proceedings of IEEE Bipolar/BiCMOS Circuits and Technology Meeting*, pages 161–164, October 2015.

53 J.-Y. Dupuy, F. Jorge, M. Riet, V. Nodjiadjim, H. Aubry, and A. Konczykowska. 59-dBΩ 68-GHz variable gain-bandwidth differential linear TIA in 0.7-μm InP DHBT for 400-Gb/s optical communication systems. In *Compound Semiconductor Integrated Circuit Symposium (CSICS)*, pages 1–4, New Orleans, LA, October 2015.

A

Communication Signals

In this appendix, we review the non-return-to-zero (NRZ) signal, the return-to-zero (RZ) signal, the pulse amplitude-modulated (PAM) signal, the amplitude-modulated signal with vestigial sideband (AM-VSB) for analog television, and the quadrature amplitude-modulated (QAM) signal for digital television. For each signal we discuss the signal power, the power spectral density (PSD), the bandwidth, and the required signal-to-noise ratio (SNR). A partial summary of the results is given in Table A.1. Furthermore, we show how the notation and expressions typically found in the literature on optical communication relate to those found in the literature on communication systems such as [1–5].

A.1 Non-Return-to-Zero Signal

The NRZ signal is illustrated in Fig. A.1 with a voltage waveform. The signal is characterized by the swing v_S^{pp}, the mean value (or DC component) $\overline{v}_S$, and the bit rate B. It is assumed that on average the signal contains the same number of zeros and ones such that the mean value is centered halfway in between the zero and one levels (DC balanced signal). It is further assumed that the NRZ signal has a rectangular shape with steep edges (zero rise and fall times), as shown in the figure.

Power. The mean-square value of a signal, $v_S(t)$, is obtained by first squaring it and then averaging it over time, $\overline{v_S^2}$. (For most stationary random signals, the same value is obtained by squaring and averaging over different *realizations* of the random signal, that is, by taking the ensemble average instead of the time average.) This mean-square value is often referred to as the signal power because it corresponds to the physical power that would be dissipated, if the signal voltage $v_S(t)$ were dropped across a 1-Ω resistor. For the NRZ signal in Fig. A.1, the signal voltage alternates between $\overline{v}_S - \frac{1}{2}v_S^{pp}$ and $\overline{v}_S + \frac{1}{2}v_S^{pp}$ and on average spends equal amounts of time at each level. (Similarly, both levels are

Table A.1 Bandwidth and SNR requirement for some digital communication signals.

Signal	Bandwidth (DC to null)	SNR requirement for $BER = 10^{-12}$
NRZ	B	16.9 dB
50%-RZ	$2B$	15.7 dB
4-PAM	$\frac{1}{2}B$	23.9 dB

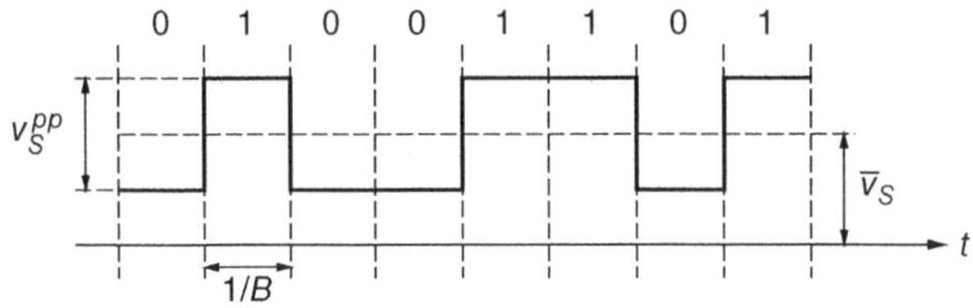

Figure A.1 NRZ-signal waveform.

realized with the same probability.) Thus, the signal power is

$$\overline{v_S^2} = \frac{1}{2}\left(\overline{v}_S - \frac{v_S^{pp}}{2}\right)^2 + \frac{1}{2}\left(\overline{v}_S + \frac{v_S^{pp}}{2}\right)^2 = \overline{v}_S^2 + \frac{1}{4}\,(v_S^{pp})^2. \tag{A.1}$$

The first part, $\overline{v}_S^2$, is the DC power and the second part, $\frac{1}{4}(v_S^{pp})^2$, is the AC power of the NRZ signal.

Power Spectral Density (PSD). It can be shown that an NRZ signal that encodes a random bit sequence has the following one-sided PSD [1, 3]:

$$\frac{\Delta\overline{v_S^2}}{\Delta f} = \overline{v}_S^2\ \delta(f) + \frac{(v_S^{pp})^2}{2B} \cdot \frac{\sin^2(\pi f/B)}{(\pi f/B)^2}, \tag{A.2}$$

where $\delta(\cdot)$ is Dirac's delta function. This PSD is illustrated in Fig. A.2 with the solid line.

When integrating this PSD over all positive frequencies, the signal power calculated in Eq. (A.1) is recovered. Note that the DC power is contained in the delta function at $f = 0$ and the AC power is contained in the sinc-shaped part of the spectrum. It may seem counterintuitive that a signal without DC component, $\overline{v}_S = 0$, has a PSD that extends down to DC. However, because the PSD is a *density*, its contribution to the signal power in an infinitesimal interval around $f = 0$ is zero as long as the PSD is finite (not a delta function).

The sinc-shaped PSD shown in Fig. A.2 holds for an NRZ signal that encodes a truly random bit sequence. In contrast, the PSD of an NRZ signal encoding

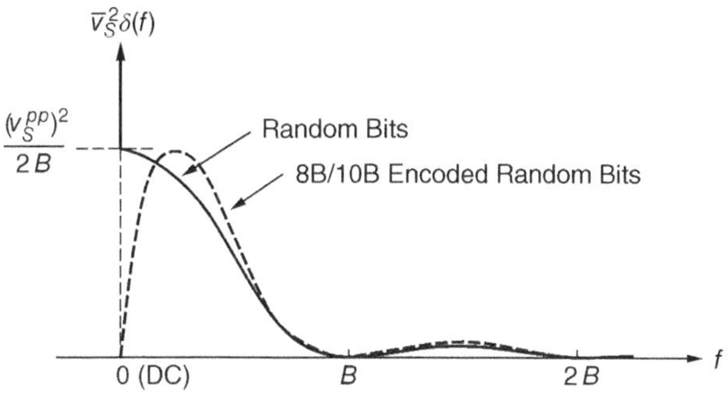

Figure A.2 One-sided power spectral density of the NRZ signal (linear scale).

a pseudorandom bit sequence (PRBS) consists of many finely spaced discrete lines weighted by an envelope that has the sinc shape shown in Fig. A.2. The spacing of the lines is given by the bit rate divided by the length of the PRBS, that is, the repetition rate. For example, the PSD of a $2^{31} - 1$ PRBS at 10 Gb/s has a line spacing of 4.7 Hz. If the random data is first encoded with a modulation code, such as the 8B/10B code, before the NRZ signal is formed (cf. Chapter 1), the PSD also deviates from the sinc form. The run-length limiting effect of the 8B/10B code reduces the low-frequency content of the PSD and introduces a null at DC, as indicated with the dashed line in Fig. A.2. Because the coding does not change the total power of the NRZ signal, the PSD becomes emphasized at higher frequencies.

The sinc-shaped PSD depicted in Fig. A.2 describes a signal with zero rise and fall times (infinitely fast edges). For a more realistic waveform with nonzero rise and fall times, the PSD rolls off faster toward high frequencies. Given the 10-to-90% rise/fall time t_R, the high-frequency cutoff where the actual PSD is about 6 dB below the ideal sinc spectrum occurs at $f_{\text{cutoff}} = 0.5/t_R$ [6].

Bandwidth. The spectrum of the NRZ signal is not band limited and thus has an infinite absolute bandwidth. However, the bandwidth measured from DC to the first null is finite and equals the bit rate, as shown in Fig. A.2:

$$BW_{\text{null}} = B. \tag{A.3}$$

The bandwidth at which the PSD drops to half of its DC value (excluding the delta function at DC), that is the 3-dB bandwidth, is

$$BW_{\text{3dB}} = 0.44B. \tag{A.4}$$

Signal-to-Noise Requirement. It was shown in Section 4.2, Eq. (4.8) that the signal swing required to meet a certain bit-error rate (BER) is given by $v_S^{pp} = 2Q\, v_n^{rms}$, where Q is the Personick Q and v_n^{rms} is the rms-value of the noise. The assumptions were additive Gaussian noise (equal amounts of noise on the zeros and ones) and an optimally set decision threshold. The Personick Q determines the BER by means of the function $BER = \frac{1}{2}\text{erfc}(Q/\sqrt{2})$, which is tabulated in Table 4.1.

With Eq. (A.1), we can convert the required signal swing into the required AC signal power: $\overline{v_S^2} = \frac{1}{4}(v_S^{pp})^2 = Q^2\ \overline{v_n^2}$. Dividing this expression by the noise power, we find the SNR requirement for a given BER:

$$SNR = Q^2 \qquad \text{for} \qquad BER = \frac{1}{2}\,\text{erfc}\left(\frac{Q}{\sqrt{2}}\right). \tag{A.5}$$

This equation corresponds to Eq. (4.13) in the main text. A generalization of this result for a signal-dependent noise (i.e., *non*additive noise) can be found in the solution to Problem 4.4 on p. 503. Note that in the aforementioned SNR calculation we excluded the DC power of the signal. This practice has the advantage of making the SNR requirement independent of biasing conditions.

Evaluating the aforementioned equations for $Q = 7.03$, we find the SNR requirement

$$SNR = 16.9\ \text{dB} \qquad \text{for} \qquad BER = 10^{-12}. \tag{A.6}$$

Remarks about the Literature on Communication Systems. The literature on communication systems usually makes a distinction between *polar NRZ* and *unipolar NRZ*. In our notation, setting $\overline{v}_S = 0$, such that the signal swings symmetrically around zero, corresponds to polar NRZ signaling, whereas setting $\overline{v}_S = \frac{1}{2}v_S^{pp}$, such that the lower signal level becomes zero, corresponds to unipolar NRZ signaling.

To obtain the error probability expressions typically found in the literature on communication systems, we need to make additional assumptions and change the notation. For the aforementioned SNR analysis, we have assumed that the noise is additive and Gaussian. Now, we additionally assume that it is white and that we are using a matched-filter receiver. These assumptions may *not* be valid for an optical receiver and we make them here only for the sake of showing the connection between the results given here and those found in the literature on communication systems. Given the one-sided noise PSD N_0 and the noise bandwidth BW_n of the receiver, we can write $\overline{v_n^2} = N_0 \cdot BW_n$. The matched filter for an NRZ signal has the noise bandwidth $BW_n = \frac{1}{2}B$ (cf. Section 4.8) and thus the noise power becomes $\overline{v_n^2} = \frac{1}{2}N_0B$. Expressing the signal power, $\overline{v_S^2}$, in terms of the average energy per bit, E_b, and the bit rate, B, (the information and channel bit rates are assumed to be equal) results in $\overline{v_S^2} = E_bB$. With these expressions, we find $E_b/N_0 = \frac{1}{2}SNR$ and we can rewrite Eq. (A.5) as

$$\frac{E_b}{N_0} = \frac{Q^2}{2} \qquad \text{for} \qquad BER = \frac{1}{2}\,\text{erfc}\left(\frac{Q}{\sqrt{2}}\right). \tag{A.7}$$

This equation corresponds to Eq. (4.16) in the main text. Next, we make a change in notation. First, to eliminate the Personick Q, which is typical for the optical literature, we solve the equation on the left-hand side for Q and insert the result into the equation on the right-hand side. Second, to state the result more succinctly, we introduce the Q function defined as $Q(x) = \frac{1}{2}\text{erfc}(x/\sqrt{2})$:

$$BER = \frac{1}{2}\,\text{erfc}\left(\sqrt{\frac{E_b}{N_0}}\right) = Q\left(\sqrt{\frac{2E_b}{N_0}}\right). \tag{A.8}$$

Equation (A.8) gives the bit-error probability of a receiver for a polar NRZ signal as found in the literature on communication systems [1–5].

A.2 Return-to-Zero Signal

The 50%-RZ signal is illustrated in Fig. A.3 with a voltage waveform. The signal is characterized by the swing v_S^{pp}, the mean value (or DC component) $\overline{v}_S$, and the bit rate B. Each pulse fills 50% of the available bit period, $1/B$. It is assumed that on average the signal contains the same number of zeros and ones such that the mean value is at the 25% level. It is further assumed that the RZ signal has a rectangular shape with steep edges (zero rise and fall times), as shown in the figure.

Power. The voltage of the 50%-RZ signal in Fig. A.3 alternates between $\overline{v}_S - \frac{1}{4}v_S^{pp}$ and $\overline{v}_S + \frac{3}{4}v_S^{pp}$ and on average spends $\frac{3}{4}$ of the time at the low level and $\frac{1}{4}$ of the time at the high level. Thus, the signal power is

$$\overline{v_S^2} = \frac{3}{4}\left(\overline{v}_S - \frac{1}{4}v_S^{pp}\right)^2 + \frac{1}{4}\left(\overline{v}_S + \frac{3}{4}v_S^{pp}\right)^2 = \overline{v}_S^2 + \frac{3}{16}\,(v_S^{pp})^2. \tag{A.9}$$

The first part, $\overline{v}_S^2$, is the DC power and the second part, $\frac{3}{16}(v_S^{pp})^2$, is the AC power of the RZ signal. Note that for the same voltage swing, the AC power of the 50%-RZ signal is only $\frac{3}{4}$ of the AC power of the NRZ signal. Vice versa, for the same AC power, the 50%-RZ signal has a $\frac{4}{3}$ larger swing than the NRZ signal.

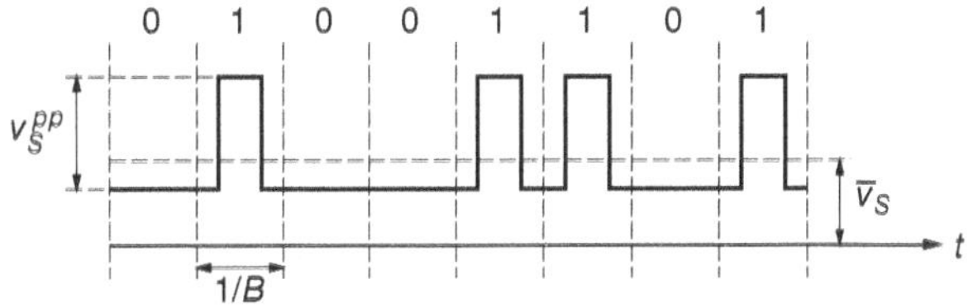

Figure A.3 50%-RZ signal waveform.

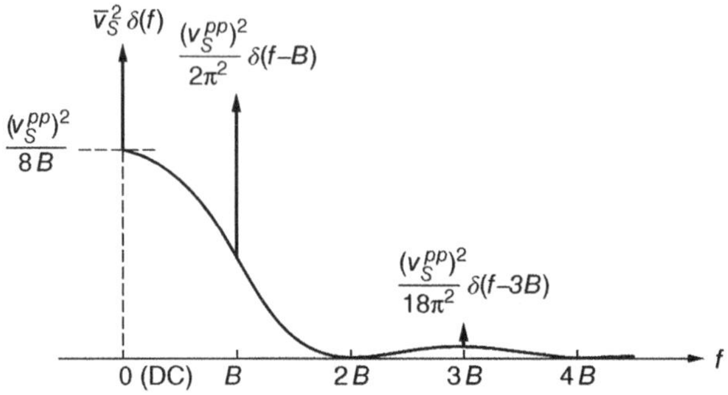

Figure A.4 One-sided power spectral density of the 50%-RZ signal (linear scale).

Power Spectral Density (PSD). It can be shown that a 50%-RZ signal that encodes a random bit sequence has the following one-sided PSD [1, 3]:

$$\frac{\Delta\overline{v_S^2}}{\Delta f} = \overline{v_S}^2\,\delta(f) + \frac{(v_S^{pp})^2}{8B}\cdot\frac{\sin^2(\pi f/2B)}{(\pi f/2B)^2} + \frac{(v_S^{pp})^2}{2\pi^2}\sum_{\text{odd }n}\frac{\delta(f-nB)}{n^2}. \tag{A.10}$$

This PSD is illustrated in Fig. A.4.

When integrating this PSD over all positive frequencies, the signal power calculated in Eq. (A.9) is recovered. Note that the DC power is contained in the delta function at $f = 0$, as in the case of the NRZ signal, while the AC power is contained partly in the continuous sinc-shaped part of the PSD and partly in the discrete delta functions at $f = B, 3B, 5B$, and so forth. In fact, an integration shows that $\frac{2}{3}$ of the total AC power is in the continuous part of the PSD and $\frac{1}{3}$ of the AC power is in the discrete part of the PSD.

In contrast to the NRZ signal, which had no power at $f = B$, the RZ signal has substantial power at that frequency. This feature of the RZ signal is exploited by some clock and data recovery (CDR) schemes.

Bandwidth. The spectrum of the RZ signal is not band limited and thus has an infinite absolute bandwidth. However, the bandwidth measured from DC to the first null is finite and for a 50%-RZ signal equals twice the bit rate, as shown in Fig. A.4:

$$BW_{\text{null}} = 2B. \tag{A.11}$$

The bandwidth at which the PSD drops to half of its DC value (excluding the delta functions), that is the 3-dB bandwidth, is

$$BW_{\text{3dB}} = 0.89B. \tag{A.12}$$

Signal-to-Noise Requirement. The signal swing required to meet a given BER is the same for NRZ and RZ signals, namely $v_S^{pp} = 2Q\,v_n^{rms}$, where v_n^{rms} is the rms-value of the additive Gaussian noise. However, as we have seen, the AC power of the 50%-RZ signal is only $\frac{3}{4}$ of the AC power of the NRZ signal, given the same swing. Thus, the SNR required for a 50%-RZ signal is only $\frac{3}{4}$ of the SNR required for an NRZ signal, given the same BER:

$$SNR = \frac{3}{4}Q^2 \qquad \text{for} \qquad BER = \frac{1}{2}\,\text{erfc}\left(\frac{Q}{\sqrt{2}}\right). \tag{A.13}$$

Evaluating this equation for $Q = 7.03$, we find the SNR requirement

$$SNR = 15.7\ \text{dB} \qquad \text{for} \qquad BER = 10^{-12}. \tag{A.14}$$

A.3 Pulse Amplitude-Modulated Signal

The 4-level PAM (4-PAM) signal is illustrated in Fig. A.5 with a voltage waveform. Successive bits are grouped in pairs, as indicated by the parenthesis, and mapped to one of four voltage levels. A Gray code is normally used for the mapping, such that a decision error mistaking the correct level for one of the nearest levels causes only a single-bit error. The signal is characterized by the swing v_S^{pp}, the mean value (or DC component) $\overline{v}_S$, and the symbol rate, which is half of the bit rate, $\frac{1}{2}B$. It is assumed that all four levels are evenly spaced and occur equally likely, such that the mean value is centered halfway between the minimum and maximum levels. It is further assumed that the 4-PAM signal has steep edges (zero rise and fall times), as shown in the figure.

Power. The voltage of the 4-PAM signal in Fig. A.5 switches between the discrete values $\overline{v}_S - \frac{1}{2}v_S^{pp}$, $\overline{v}_S - \frac{1}{6}v_S^{pp}$, $\overline{v}_S + \frac{1}{6}v_S^{pp}$, and $\overline{v}_S + \frac{1}{2}v_S^{pp}$ and on average spends equal amounts of time at each level. Thus, the signal power is

$$\begin{aligned}\overline{v_S^2} &= \frac{1}{4}\left(\overline{v}_S - \frac{v_S^{pp}}{2}\right)^2 + \frac{1}{4}\left(\overline{v}_S - \frac{v_S^{pp}}{6}\right)^2 \\ &\quad + \frac{1}{4}\left(\overline{v}_S + \frac{v_S^{pp}}{6}\right)^2 + \frac{1}{4}\left(\overline{v}_S + \frac{v_S^{pp}}{2}\right)^2 = \overline{v}_S^2 + \frac{5}{36}\,(v_S^{pp})^2.\end{aligned} \tag{A.15}$$

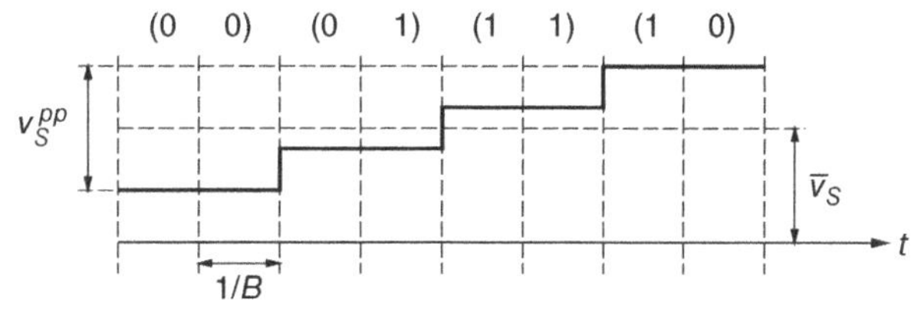

Figure A.5 4-PAM signal waveform.

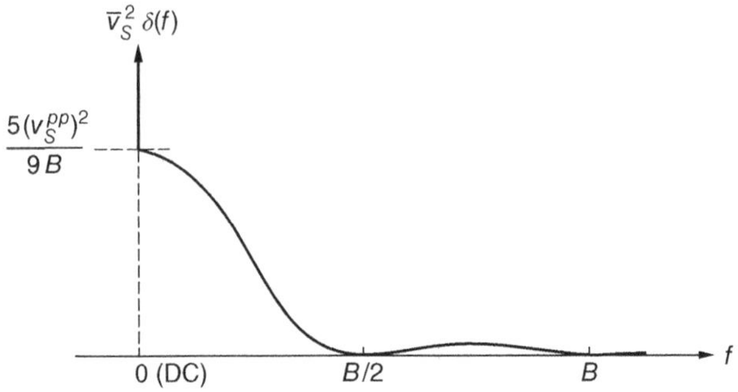

Figure A.6 One-sided power spectral density of the 4-PAM signal (linear scale).

The first part, $\overline{v}_S^2$, is the DC power and the second part, $\frac{5}{36}(v_S^{pp})^2$, is the AC power of the 4-PAM signal.

Power Spectral Density (PSD). It can be shown that a 4-PAM signal that encodes a random bit sequence has the following one-sided PSD [7]:

$$\frac{\Delta \overline{v_S^2}}{\Delta f} = \overline{v}_S^2\,\delta(f) + \frac{5(v_S^{pp})^2}{9B} \cdot \frac{\sin^2(2\pi f/B)}{(2\pi f/B)^2}. \tag{A.16}$$

This PSD is illustrated in Fig. A.6. When integrating this PSD over all positive frequencies, the signal power calculated in Eq. (A.15) is recovered.

Bandwidth. The spectrum of the 4-PAM signal is not band limited and thus has an infinite absolute bandwidth. However, the bandwidth measured from DC to the first null is finite and equals the symbol rate, that is, half the bit rate, as shown in Fig. A.6:

$$BW_{\text{null}} = \frac{1}{2}B. \tag{A.17}$$

The bandwidth at which the PSD drops to half of its DC value (excluding the delta function at DC), that is the 3-dB bandwidth, is

$$BW_{\text{3dB}} = 0.22B. \tag{A.18}$$

Signal-to-Noise Requirement. The precise signal swing required to meet a given BER is a little bit harder to calculate for the 4-PAM signal than it was for the NRZ or RZ signals. As a first approximation we can argue that each step in the 4-PAM signal should be equal to the swing of the NRZ signal necessary for the same BER. Since there are three steps in a 4-PAM signal (see Fig. A.5), the swing becomes $v_S^{pp} = 6Q\, v_n^{rms}$, where v_n^{rms} is the rms-value of the additive Gaussian noise. With Eq. (A.15), we can convert the required signal swing into the required AC signal power: $\overline{v_S^2} = \frac{5}{36}(v_S^{pp})^2 = 5Q^2\ \overline{v_n^2}$. Dividing this expression by the noise power, we find the approximate SNR requirement for a given BER:

$$SNR \approx 5Q^2 \qquad \text{for} \qquad BER \approx \frac{1}{2}\,\mathrm{erfc}\left(\frac{Q}{\sqrt{2}}\right). \tag{A.19}$$

The aforementioned approximation suffers from two small inaccuracies. First, the calculated error rate is the symbol-error rate rather than the bit-error rate, that is, the probability that a level (not a bit) is detected incorrectly. An incorrect symbol (level) could mean that only one of the two bits or both bits are in error. When Gray coding is used, it is very likely that only one of the two bits is in error (for both bits to be in error, a level had to be confused with the next nearest level). Thus, the BER is really only half of our initial estimate in Eq. (A.19). Second, while the outer two levels can result in an error by deviating toward the center only, just like in the case of the NRZ signal, the inner two levels can cause an error by deviating in either direction. Thus, the symbol-error rate for the inner two levels is twice that of an NRZ signal and the overall symbol-error rate for the 4-PAM signal is 1.5 times that of our initial estimate in Eq. (A.19). Updating the error rate expression with these two correction factors, $\frac{1}{2}$ and $\frac{3}{2}$, which partially cancel each other, we obtain the accurate result

$$SNR = 5\tilde{Q}^2 \qquad \text{for} \qquad BER = \frac{3}{8}\,\mathrm{erfc}\left(\frac{\tilde{Q}}{\sqrt{2}}\right), \tag{A.20}$$

where $\tilde{Q}$ is no longer the original Personick Q. Evaluating the aforementioned equation for $\tilde{Q} = 6.99$, we find the SNR requirement

$$SNR = 23.9\ \text{dB} \qquad \text{for} \qquad BER = 10^{-12}. \tag{A.21}$$

Whereas the approximation in Eq. (A.19) yields 23.93 dB, the accurate expression in Eq. (A.20) yields 23.88 dB, both referring to $BER = 10^{-12}$. Because of the exponential growth of erfc(·), a small error in the BER expression translates into a very small error in the SNR requirement.

Remarks about the Literature on Communication Systems. As we did for the NRZ signal, we now show the connection between Eq. (A.20) and the error probability results for 4-PAM signaling found in the literature on communication systems. As before, we need to make the additional assumptions of white noise and a matched-filter receiver to convert SNR to E_b/N_0. Recognizing that the noise bandwidth of the matched filter for a 4-PAM signal is $BW_n = \frac{1}{4}B$, we find $E_b/N_0 = \frac{1}{4}SNR$ and we can rewrite Eq. (A.20) as

$$\frac{E_b}{N_0} = \frac{5\tilde{Q}^2}{4} \qquad \text{for} \qquad BER = \frac{3}{8}\,\mathrm{erfc}\left(\frac{\tilde{Q}}{\sqrt{2}}\right). \tag{A.22}$$

Now, eliminating $\tilde{Q}$ and introducing $Q(x) = \frac{1}{2}\text{erfc}(x/\sqrt{2})$ yields

$$BER = \frac{3}{8}\,\text{erfc}\left(\sqrt{\frac{2E_b}{5N_0}}\right) = \frac{3}{4}Q\left(\sqrt{\frac{4E_b}{5N_0}}\right). \tag{A.23}$$

Equation (A.23) gives the bit-error probability of a receiver for a 4-PAM signal as found in the literature on communication systems [4, 5]. The corresponding symbol-error probability is twice that expression.

A.4 Analog Television Signal

The North American analog TV signal is defined by the National Television Systems Committee (NTSC) [8]. The generation of the RF-modulated analog TV signal from the video and audio baseband signals is illustrated in Fig. A.7. The baseband video signal, which has a bandwidth of about 4.2 MHz, modulates a carrier using amplitude modulation (AM). Mathematically, AM can be described as the addition of a constant to the baseband signal, such that it becomes strictly positive, followed by a multiplication with the carrier. The resulting AM signal has two sidebands and occupies a frequency band of 2×4.2 MHz $= 8.4$ MHz with the carrier located at the center. Since both sidebands contain the same information, one of them can be removed. The vestigial sideband (VSB) filter partially removes the lower sideband shrinking the occupied frequency band to about 5.5 MHz. This modulation scheme is known as ***amplitude modulation with vestigial sideband*** or AM-VSB. (Compared to single-sideband modulation, which removes one sideband entirely, AM-VSB permits a simpler receiver implementation.) The baseband audio signal, which has a bandwidth of about 15 kHz, is frequency modulated (FM) on a separate carrier that lies 4.5 MHz above the video carrier. The modulated video and audio signals are then combined to produce a 6-MHz wide analog TV signal.

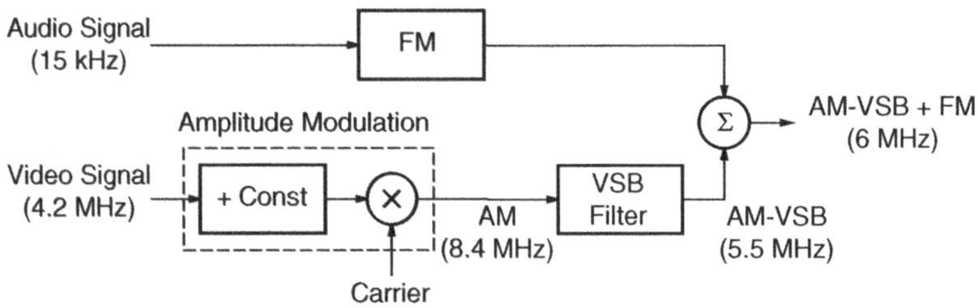

Figure A.7 Generation of the AM-VSB analog TV signal.

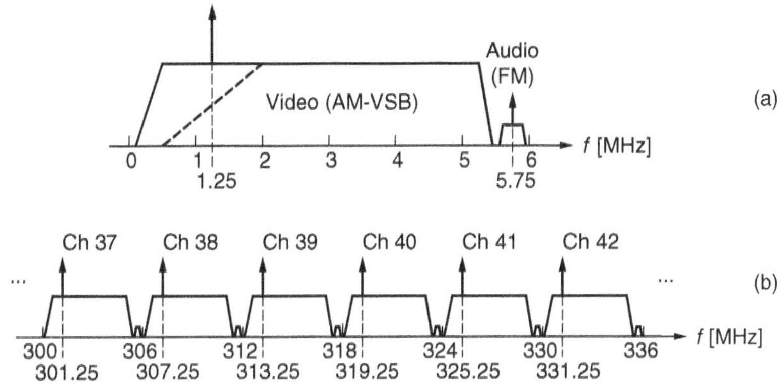

Figure A.8 Power spectral density of (a) the analog TV signal and (b) subcarrier multiplexed analog TV signals.

Power Spectral Density and Bandwidth. The PSD of the NTSC analog TV signal is shown in Fig. A.8(a) [8, 9]. The video and the audio signals fit both into the channel bandwidth of 6 MHz. Because the lower sideband is truncated to less than 1.25 MHz, the video carrier is placed 1.25 MHz above the lower band edge. Before demodulating the AM-VSB signal in the receiver, the lower sideband is truncated further, as indicated by the dashed line in Fig. A.8(a). This step is necessary to prevent a doubling of the demodulated signal in the frequency range where both sidebands are present. The audio carrier is located above the upper sideband of the video spectrum, more precisely, it is 5.75 MHz above the lower band edge. In addition to the spectral components shown in Fig. A.8(a), the analog TV signal also contains a color carrier and its sidebands, which overlap the AM-VSB spectrum, as well as a stereo pilot and the associated audio signals which reside at the upper end of the 6-MHz spectrum.

In a community-antenna television (CATV) or hybrid fiber-coax (HFC) system, the signals of many TV channels (e.g., 80 or 120 TV channels) are combined into a single broadband signal by means of *subcarrier multiplexing* (SCM). This is done by assigning a different carrier frequency to each TV channel and summing them all up (see Fig. 1.8 on p. 9). The carriers are spaced 6 MHz apart such that the channel spectra do not overlap. A small section of the resulting spectrum for the North American standard cable channel plan is illustrated in Fig. A.8(b).[1] Note that SCM keeps a small guard band

1 Basically, the North American standard cable channel plan (ANSI/EIA-542) defines the channel edges at 6 MHz $\cdot$ n and the carriers at 6 MHz $\cdot$ n + 1.25 MHz, where n is a whole number. However, to avoid interference with existing radio communication channels, several small deviations from the 6-MHz raster exist; furthermore, for historical reasons, the channel numbering contains several discontinuities [8].

between adjacent channels that permits their separation by means of filters. Other multicarrier formats, such as *orthogonal frequency division multiplexing* (OFDM) or *discrete multitone* (DMT) modulation, eliminate this guard band and permit the channel spectra to overlap by using orthogonal carriers.

An important consideration of the SCM signal is its *peak-to-average ratio* (PAR). Most of the time, the carriers in the SCM signal have a random phase relationship and add up to a noise-like signal. Occasionally, however, the phases of multiple carriers drift into alignment producing a large signal peak (more precisely, repetitive peaks at a rate of 6 MHz because of the 6-MHz channel spacing). As a result, the peak power of the SCM signal is much larger than its average power. Peak-to-average ratios from 3.5:1 to greater than 7:1 are typical [8]. The problem with these large peaks is that they can saturate amplifiers and optical transmission equipment, resulting in clipping distortions.

Carrier-to-Noise Requirement. The Federal Communications Commission (FCC) regulates that the carrier-to-noise ratio (CNR) of the AM-VSB analog TV signal must be ≥ 43 dB (at the TV input). However, good engineering practice dictates a higher CNR number. While 45 dB results in a picture that is judged as "slighly annoying," only the range of 48 to 51 dB is judged as "not annoying" [8].

Note that cable-television engineers use the term CNR for RF-modulated signals such as the TV signals in an SCM system and reserve the term SNR for baseband signals. Communication engineers do not always make this distinction [8].

A.5 Digital Television Signal

The North American digital TV signal for transmission over cable is defined by the Society of Cable Telecommunications Engineers (SCTE) [10]. The modulation format is *quadrature amplitude modulation* (QAM) with either a 64-point signal constellation (64-QAM) or a 256-point signal constellation (256-QAM), as shown in Fig. A.9.

Figure A.10 illustrates the generation of a 256-QAM signal starting from the digital data stream. Data is fed into the modulator at a symbol rate of about 5.36 MS/s where each symbol consists of 8 bits (=1 byte). Each byte is split into two 4-bit nibbles and run through two 4-bit digital-to-analog converters (DACs) producing two 16-PAM signals. Similar to our discussion of the 4-PAM signal in Section A.3, the DC-to-null bandwidth of the 16-PAM signal equals the symbol rate, which is 5.36 MHz. Next, the two 16-PAM signals are low-pass filtered to soften their steep edges. The filter is designed such that the output signal has a square-root raised-cosine spectrum with an excess bandwidth of 12% (cf. Section 4.9). The bandwidth of the resulting smoothed 16-PAM

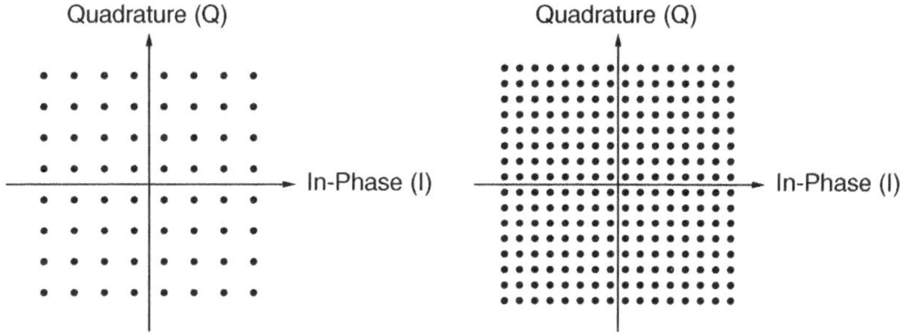

Figure A.9 Constellations for the 64-QAM and 256-QAM signal.

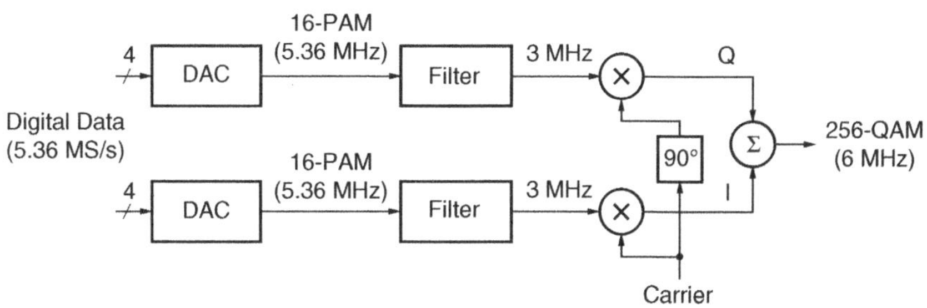

Figure A.10 Generation of the 256-QAM digital TV signal.

signals is now limited to 3 MHz (= 1.12 × 5.36 MHz/2). One 16-PAM signal is multiplied with the carrier (in-phase carrier) and the other 16-PAM signal is multiplied with the 90° phase-shifted carrier (quadrature carrier). The resulting in-phase and quadrature products (I and Q) are summed up to form the 6-MHz wide 256-QAM digital TV signal.

The channel bit rate of the 256-QAM digital TV signal is 5.36 Mb/s × 8 = 42.88 Mb/s. To protect the digital information from transmission errors, forward error correction (FEC) is applied prior to the modulation (cf. Appendix G). A Reed–Solomon code, RS(128,122), followed by a trellis code are used for this purpose. The code rate of this concatenated code including the framing overhead is $r = 0.905$. Thus, each 6-MHz channel can carry an information bit rate of 0.905 × 42.88 Mb/s = 38.81 Mb/s [10]. This means that about ten digital standard-definition TV channels can be broadcast in the bandwidth of a single analog TV channel. (After MPEG-2 compression, standard-definition movies require a bit rate of about 3 Mb/s and high-definition movies require a bit rate of about 12 Mb/s [8].)

For the 64-QAM digital TV signal, only 6 instead of 8 bits are transmitted per symbol. Moreover, the symbol rate and the code rate are chosen slighly lower. The resulting information bit rate is about 26.97 Mb/s [10].

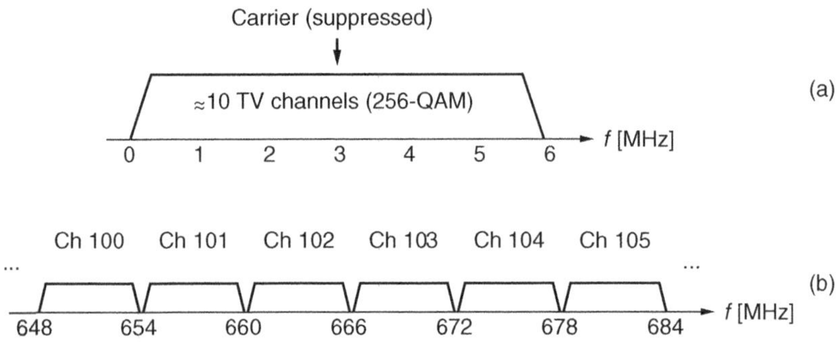

Figure A.11 Power spectral density of (a) the digital TV signal and (b) subcarrier multiplexed digital TV signals.

Power Spectral Density and Bandwidth. The PSD of the SCTE digital TV signal is shown in Fig. A.11(a). To be compatible with the analog TV channel plan, the signal bandwidth is made to be 6 MHz. But recall that this digital signal can carry the video and audio information of about ten standard-definition TV channels. Compared to the AM-VSB PSD shown in Fig. A.8(a), the QAM spectrum is flatter and has no discrete carrier components. The carrier frequency is located at the mid-point of the band, that is 3 MHz above the lower band edge, but because the carrier is suppressed it does not appear in the spectrum. As we have seen, the 3-MHz baseband signals are upconverted to the carrier frequency, resulting in two 3-MHz sidebands, one above and one below the suppressed carrier. In contrast to AM, the sidebands of QAM are *not* redundant and both bands must be transmitted.

In a CATV/HFC system, analog and digital TV signals can be multiplexed into a single signal by means of SCM (cf. Fig. 1.8). Typically, the channels below 550 MHz are used for analog TV and the channels above 550 MHz are used for digital TV. Figure A.11(b) shows a small section of the spectrum above 550 MHz.

Carrier-to-Noise Requirement. For a good quality digital TV experience, error events should occur less often than once per 15 minutes after FEC [10]. To achieve this performance, a CNR of ≥ 33 dB is needed for the 256-QAM digital TV signal (at the TV input). For the more robust, but lower rate, 64-QAM signal, a CNR of ≥ 27 dB is sufficient [8].

In an SCM system with mixed analog and digital TV channels, the digital channels typically are transmitted at a lower power level than the analog channels. This is possible because the CNR requirement for digital channels is at least 10 dB lower than that for analog channels.

References

1 L. W. Couch II. *Digital and Analog Communication Systems*. Prentice-Hall, Upper Saddle River, NJ, 7th edition, 2006.

2 R. D. Gitlin, J. F. Hayes, and S. B. Weinstein. *Data Communications Principles*. Plenum Press, New York, 1992.

3 S. Haykin and M. Moher. *Communication Systems*. John Wiley & Sons, Inc., Hoboken, NJ, 5th edition, 2009.

4 J. G. Proakis and M. Salehi. *Digital Communications*. McGraw Hill, New York, 5th edition, 2008.

5 B. Sklar. *Digital Communications: Fundamentals and Applications*. Prentice-Hall, Englewood Cliffs, NJ, 2nd edition, 2001.

6 H. Johnson. High-speed digital design. *IEEE Microwave Mag.*, 12(5):42–50, 2011.

7 J. Conradi. Bandwidth-efficient modulation formats for digital fiber transmission systems. In I. P. Kaminow and T. Li, editors, *Optical Fiber Telecommunications IVB*, pages 862–901. Academic Press, San Diego, CA, 2002.

8 W. Ciciora, J. Farmer, D. Large, and M. Adams. *Modern Cable Television Technology: Video, Voice, and Data Communications*. Morgan Kaufmann, San Francisco, CA, 2nd edition, 2004.

9 M. R. Phillips and T. E. Darcie. Lightwave analog video transmission. In I. P. Kaminow and T. L. Koch, editors, *Optical Fiber Telecommunications IIIA*, pages 523–559. Academic Press, San Diego, CA, 1997.

10 SCTE. Digital video transmission standard for cable television, ANSI/SCTE 07 2000 (formerly DVS 031). Society of Cable Telecommunications Engineers (SCTE), 2000. http://www.scte.org.

B

Eye Diagrams

Eye diagrams provide an intuitive graphical representation of electrical or optical digital communication signals. The quality of the signal, that is, the rise and fall times, the amount of intersymbol interference (ISI), noise, and jitter, can be judged from the appearance of the eye. In the following, we discuss how to measure and simulate eye diagrams and how to determine the eye openings and eye margins. In Appendix C, we discuss the related subject of jitter and its measurement.

Cutting and Overlaying Waveforms. The waveform of a communication signal, such as a non-return-to-zero (NRZ), a return-to-zero (RZ), or a 4-level pulse amplitude modulation (4-PAM) signal, can be turned into an *eye diagram* (a.k.a. *eye pattern*) by cutting it up into segments that are two bit (or symbol) intervals long and overlaying them. For example, in Fig. B.1, the waveform of an NRZ signal with mild ISI is cut up into segments with half a bit interval on the left, a full bit interval at the center, and half a bit interval on the right. Then, all segments are overlaid to form the eye diagram, as shown at the center of Fig. B.1. Because the ISI in our example is limited to just one bit to the right and left, there are only eight distinct segments corresponding to the three-bit binary words: 000, 001, 010, 011, 100, 101, 110, and 111 (see the left- and right-hand side of Fig. B.1). In the case of a signal with stronger ISI, there are more distinct segments.

An important advantage of the eye diagram over the linear signal representation is that it shows all possible bit transitions in a compact way. Deviations of these transitions from their ideal locations are clearly visible and represent the jitter in the signal. In the eye diagram, the periodic bit (or symbol) sampling instants overlay into a *single* sampling instant, making it easy to see the ISI and noise at the time of sampling.

The eye diagrams for a 50%-RZ and a 4-PAM signal are sketched in Fig. B.2 as further examples.

Analysis and Design of Transimpedance Amplifiers for Optical Receivers, First Edition. Eduard Säckinger.

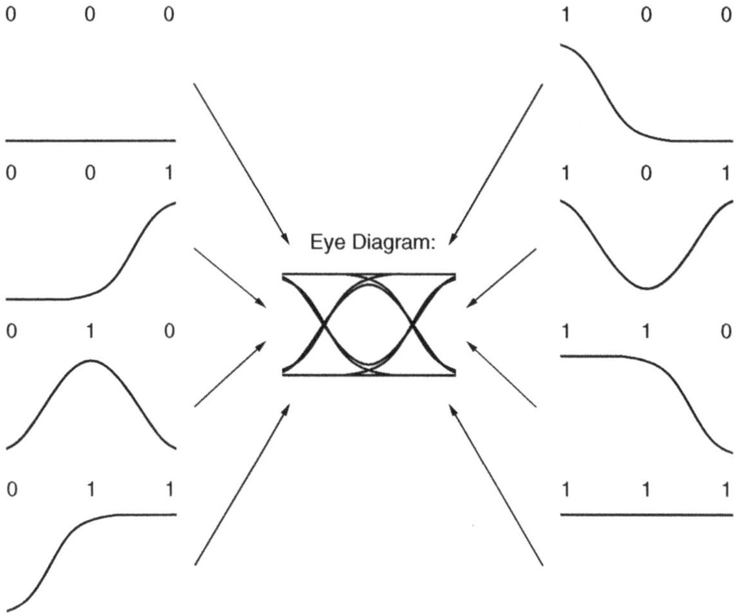

Figure B.1 Construction of an eye diagram by overlaying waveform segments.

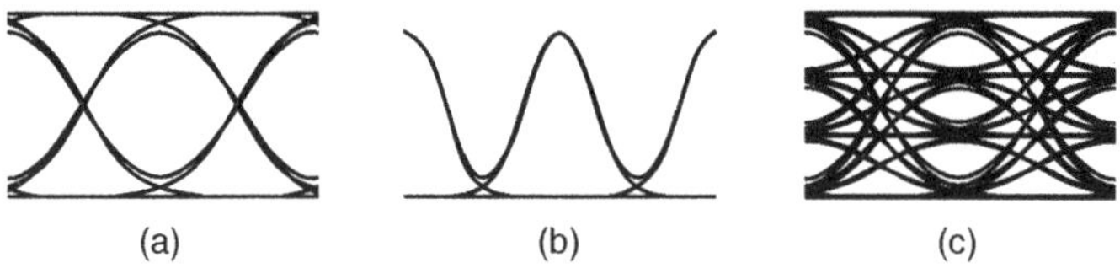

Figure B.2 Eye diagrams for (a) an NRZ signal, (b) a 50%-RZ signal, and (c) a 4-PAM signal.

Measurement. Figure B.3 shows a setup for measuring the eye diagram with an oscilloscope. A pulse pattern generator produces a clean NRZ data signal. A pseudorandom bit sequence (PRBS), which unlike a true random sequence repeats itself after a certain number of bits, is often used for this test. The NRZ data signal passes through the device under test (DUT), which adds ISI, noise, and jitter. The output of the DUT is connected to the vertical input of the oscilloscope. To display the eye diagram, the oscilloscope must be triggered on the clock signal, not the data signal. The bit-clock signal from the pattern generator, as shown in Fig. B.3, or a clock recovered from the data signal can be used for this purpose.

Many scopes offer a built in (golden) phase-locked loop (PLL) that can be used to recover a periodic trigger signal from the output signal of the DUT (cf. Fig. C.9(b)). This PLL may be a piece of hardware (in the case of

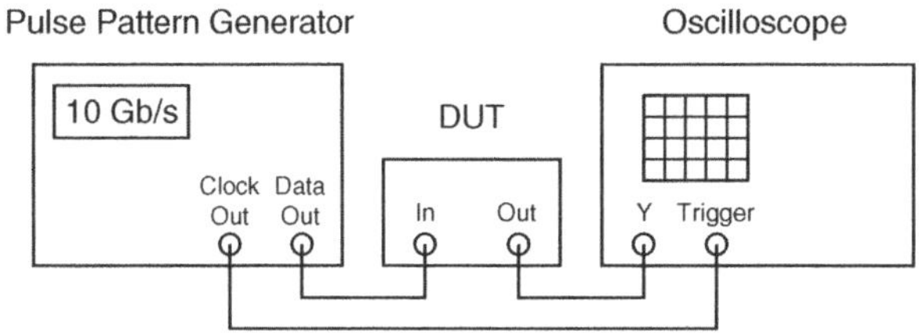

Figure B.3 Measurement of an eye diagram with an oscilloscope.

equivalent-time sampling oscilloscopes) or a software function (in the case of real-time sampling oscilloscopes). The detailed appearance of the eye diagram depends on the PLL's bandwidth: Low-frequency jitter from the DUT that falls into the PLL's bandwidth is tracked and thus does not appear in the eye diagram. A typical clock-recovery PLL has a first-order response with a bandwidth of $B/1667$, where B is the bit rate.

Many scopes have a feature called *color grading*, which uses a color code to indicate how often a certain point in the eye is reached. Figures B.4 and B.5 show measured eye diagrams of an NRZ and a 4-PAM signal, respectively.

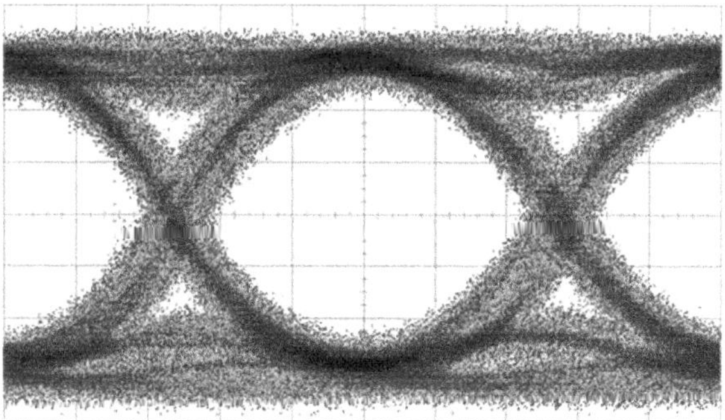

Figure B.4 Eye diagram of an NRZ signal measured with a sampling oscilloscope. Darker regions are sampled more often.

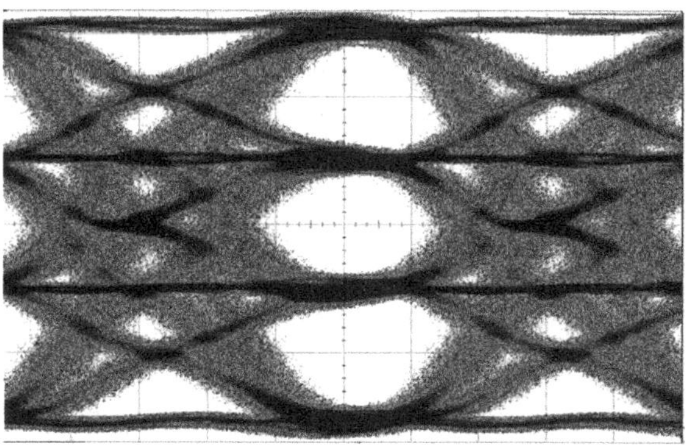

Figure B.5 Eye diagram of an optical 4-PAM signal measured with a sampling oscilloscope. *Source*: Reprinted by permission from Finisar Corporation.

Simulation. After running a transient simulation with a circuit simulator, the linear waveform can be turned into an eye diagram with a post processor. For example, the SKILL function `eyeDiagram()` can take the transient result from a Spectre simulation and convert it into an eye diagram.[1] See Fig. B.6 for an example of a simulated eye diagram.

With older simulation tools that do not provide a post processor for eye diagrams, but permit plotting of one variable against another, the following method can be used to obtain an eye diagram. First, run a transient simulation of the DUT together with a little helper circuit that generates a linearly rising sawtooth voltage with a period of two-bit intervals and a rapid fall time. Then, instead of plotting the data signal as a function of time, as usual, plot it against this sawtooth voltage. This trick turns the waveform into an eye. A drawback of this method is that the sawtooth has a finite fall time, creating spurious trace-back lines across the eye diagram. This problem can be solved by generating a pulse voltage that is always zero except for the trace-back period, where it assumes a large value. When this voltage is added to the data signal,

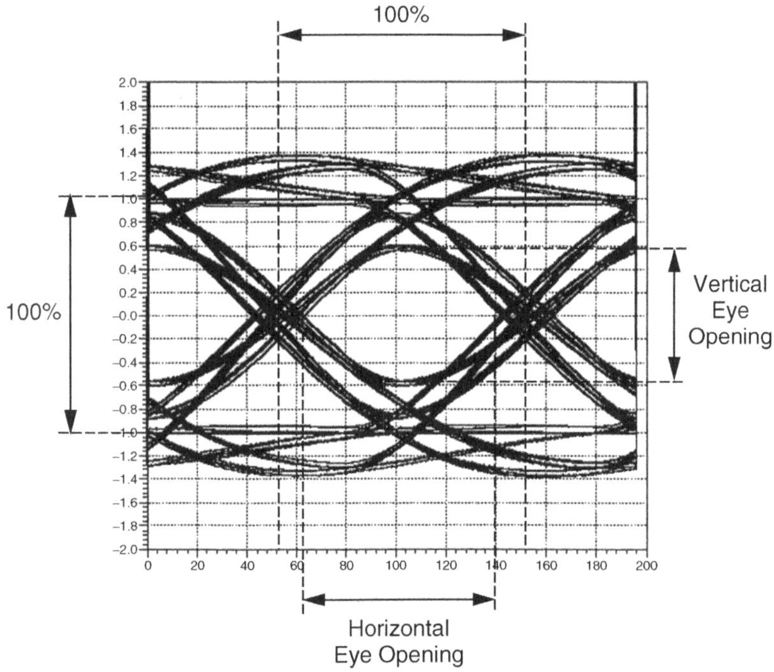

Figure B.6 Eye openings in the simulated eye diagram without noise and random jitter.

1 SKILL is a LISP-like programming language and Spectre is a SPICE-like circuit simulators, both from Cadence Design Systems, Inc.

the trace-back lines are pushed outside of the eye diagram and can be "clipped away" by choosing the appropriate plotting window.

A conventional transient simulation does not include random noise and thus eye diagrams obtained in this way do not show noise or random jitter (cf. Fig. B.6). To produce eye diagrams with noise and random jitter, a special "transient noise" simulation must be run.

Eye Openings. The *vertical eye opening* and the *horizontal eye opening* are important characteristics of the eye diagram that aid in quantifying the signal quality. The vertical eye opening is measured at a particular sampling instant, often the one that maximizes the vertical opening, and is expressed in millivolts (or milliwatts or microamps) or as a percentage of the full eye height (not including over- or undershoots). The horizontal eye opening is measured at a particular decision threshold, often the one that maximizes the horizontal opening, and is expressed in picoseconds or as a percentage of the bit/symbol interval. Sometimes, the complementary terms *vertical eye closure* and *horizontal eye closure* are used instead. Eye closure and eye opening add up to 100%.

In the case of an eye diagram *without* noise and random jitter, the openings can be found in a straightforward way, as illustrated in Fig. B.6. The vertical eye opening is determined by ISI, and the horizontal eye opening is determined by deterministic jitter (including pulse-width distortion). The eye openings may depend on the sequence length of the PRBS used for testing. Typical PRBS sequence lengths are between $2^7 - 1$ and $2^{31} - 1$. The longer the sequence, the longer the runs of zeros or ones contained in the PRBS. Now, if the DUT has a low-frequency cutoff (e.g., due to AC coupling), the eye openings reduce with increasing sequence length. Therefore, it is important to state the test pattern (e.g., PRBS-7 or PRBS-31) along with the eye diagram and the eye openings.

Eye Margins. Real eye diagrams do contain noise and random jitter. As a result, the measured eye openings change over time. The longer we accumulate eye samples, the smaller the openings become. To get around this problem, we need to define the eye openings in a statistical sense.

Each point in the eye diagram corresponds to a decision point (defined by the sampling instant and the decision threshold) and thus has a bit-error rate (BER) associated with it. With this picture in mind, we can define contours of constant BER inside the eye (see Fig. B.7). Clearly, contours for lower BERs are nested inside contours for larger BERs. If we make decisions inside a contour for a given BER, the resulting BER is less than that of the contour. Now, let's define the horizontal and vertical eye openings of the noisy eye as the horizontal and vertical dimensions of the contour for a given BER (reference BER). Note that these eye openings are well defined in the presence of noise and random jitter.

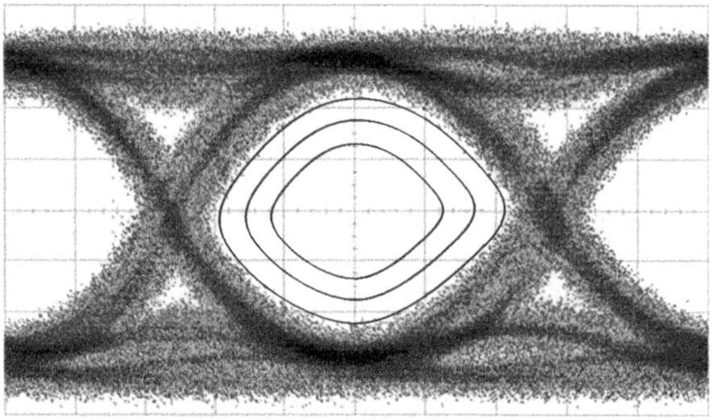

Figure B.7 Contours of constant BER in the eye diagram with noise and random jitter.

To be more precise, we should call the statistically defined eye openings *eye margins* [1]. Instead of vertical eye opening and horizontal eye opening, we should say *vertical eye margin* and *horizontal eye margin*. (Sometimes, the horizontal eye margin is also referred to as the *phase margin* [1], not to be confused with the identical term from stability theory.) In an eye diagram without noise and random jitter, eye openings and eye margins are the same. But in an eye diagram with noise and random jitter, eye openings are not well defined, and eye margins must be used instead. It is important to state the reference BER and the test pattern along with the eye margins.

If the eye margins for a given BER are close to zero, only a near-perfect decision circuit can recover the data at the desired BER. However, if the eye margins are substantially larger than zero, then the decision circuit is permitted to have some decision-threshold and sampling-time error while still meeting the desired BER, hence the term *margin*.

Closely related to the horizontal eye margin is the concept of total jitter. Specifically, the horizontal eye margin and the total jitter, measured peak to peak, add up to 100% or 1 unit interval when determined at the same BER. Similarly, in the noise-free eye, the horizontal eye opening and the deterministic jitter, measured peak to peak, add up to 100%. For a discussion of jitter and its measurement, see Appendix C.

Measurement of Eye Margins. Eye margins can be estimated from the eye on the scope, but a more accurate measurement is possible with a so-called *BERT scan*. For this procedure, a *bit-error rate test set* (BERT), consisting of a pulse pattern generator and an error detector, is needed. The DUT is connected to the BERT as shown in Fig. B.8. The bit-clock signal for the error detector may come from the pattern generator, as shown, or from a (golden) PLL that recovers the clock from the output signal of the DUT. The error detector slices the data signal at the decision threshold V_{DTH} and samples it at the instant t_S. The recovered bit sequence is compared with the transmitted bit sequence and the resulting BER is displayed. Both the decision threshold V_{DTH} and the sampling instant t_S are adjustable.

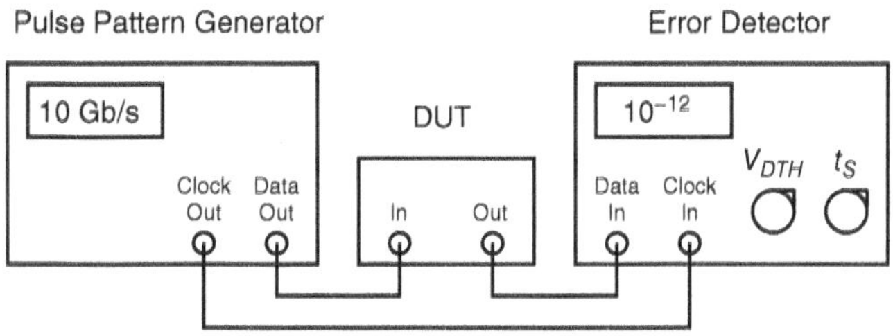

Figure B.8 Measurement setup for a BERT scan.

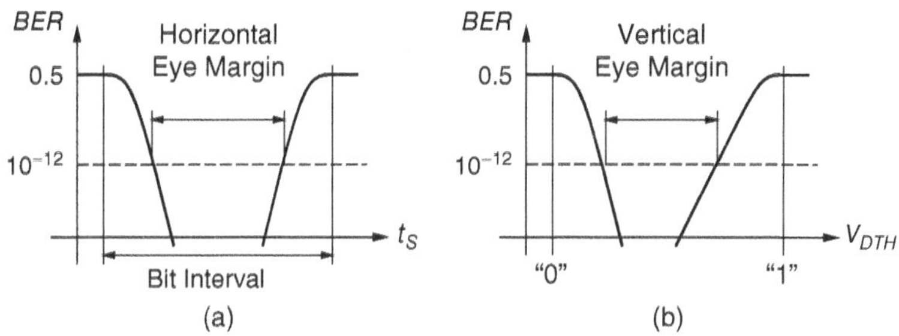

Figure B.9 Bathtub curves resulting from (a) a horizontal BERT scan and (b) a vertical BERT scan.

A horizontal BERT scan is performed by setting V_{DTH} to the center of the eye and scanning t_S horizontally across the eye. The resulting BER curve is shown schematically in Fig. B.9(a). The BER is low when sampling at the center of the eye and goes up when approaching the eye crossings to the left and right. This curve is known as the *bathtub curve* because of its characteristic shape. The horizontal eye margin is the separation of the two sides of the bathtub curve at a specified BER level (see Fig. B.9(a)). For example, in the 10-GbE standard, the horizontal eye margin is specified for a BER of 10^{-12}.

Similarly, a vertical BERT scan is performed by setting the sampling instant t_S to the center of the eye and scanning the decision threshold V_{DTH} vertically across the eye. The resulting bathtub curve is shown in Fig. B.9(b), where it was assumed that the ones are noisier than the zeros. Like before, the vertical eye margin is obtained by comparing this curve with a reference BER.

In practice, measuring a complete bathtub curve down to a BER of 10^{-12} can be very time consuming. For a 10 Gb/s system such a measurement may take more than 40 hours [2]. However, if we are interested in the eye margin or the total jitter only, we can limit our measurement to the sections of the bathtub curve that are close to the desired reference BER. With this and other optimizations it is possible to perform a margin measurement for a 10 Gb/s system in about 20 minutes [2].

References

1 K. Ogawa, L. D. Tzeng, Y. K. Park, and E. Sano. Advances in high bit-rate transmission systems. In I. P. Kaminow and T. L. Koch, editors, *Optical Fiber Telecommunications IIIA*, pages 336–372. Academic Press, San Diego, CA, 1997.

2 M. Müller, R. Stephens, and R. McHugh. Total jitter measurement at low probability levels, using optimized BERT scan method. In *DesignCon East*, Worcester, MA, September 2005.

C

Timing Jitter

Timing jitter is important when dealing with systems operating at high data rates or systems requiring precise clocks (e.g., sampling clocks for data converters). We start by examining data jitter and its components, such as deterministic and random jitter. We discuss the measurement and decomposition of data jitter. Then, we turn to clock jitter in its various forms, such as absolute jitter, period jitter, and cycle-to-cycle jitter. Finally, we show how jitter, phase noise, and bit-error rate are related.

C.1 Data Jitter

Noise, ISI, and Jitter. In Chapter 4, we discussed data signals with noise and ISI. Noise and ISI not only affect the signal voltage at the sampling instants, they also affect the time points where the signal crosses the decision threshold (often referred to as zero crossings). The effect of noise on the zero crossings results in *random jitter* (RJ), and the effect of ISI on the zero crossings results in *data-dependent jitter* (DDJ).

We know that we can characterize noise and ISI in the eye diagram with a histogram of the voltage values at the sampling instant, as shown in Fig. C.1(right). Similarly, we can characterize jitter with a histogram of the time points where the signal crosses the decision threshold, as shown in Fig. C.1(bottom).

In Chapter 4, we discussed how the location of the decision threshold affects the BER. The decision process, however, is not only controlled by the decision threshold voltage V_{DTH}, it is also controlled by the sampling instant t_S. The decision threshold voltage slices the eye diagram horizontally, whereas the sampling instant slices the eye diagram vertically. The two slicing lines intersect in the so-called *decision point*, as shown in Fig. C.1.

We know that bit errors occur when the instantaneous value of the noise and ISI is so large that the signal is pushed to the "wrong" side of the decision threshold. Similarly, bit errors occur if the instantaneous value of the jitter is so

Analysis and Design of Transimpedance Amplifiers for Optical Receivers, First Edition. Eduard Säckinger.

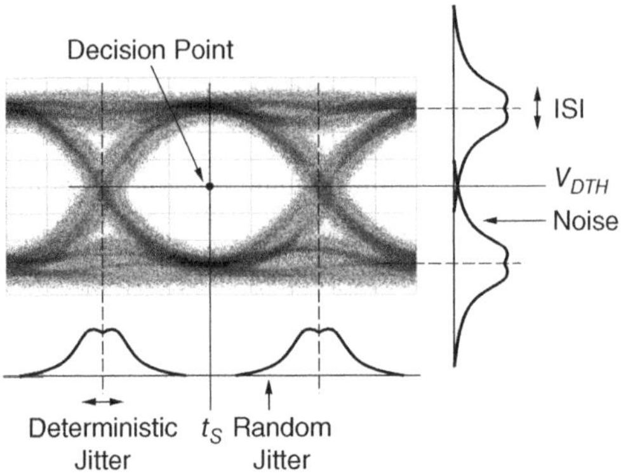

Figure C.1 Eye diagram and histograms of a signal with noise, ISI, random jitter, and deterministic jitter.

large that a bit slides past the sampling instant and is sampled on the "wrong" side of the edge.

Not surprisingly, the subsequent discussion of jitter has many parallels to our discussion of noise and ISI in Chapter 4.

Data-Dependent Jitter. A data signal suffers from data-dependent jitter if the signal's edges are slightly early or late (relative to their ideal time points) depending on the values of the preceding and succeeding bits. In the eye diagram, this condition manifests itself as a breaking up (splitting) of the rising and falling edges into multiple closely spaced edges.

Data-dependent jitter occurs, for example, if the data stream is transmitted through a channel with insufficient bandwidth. In the example shown in Fig. C.2(a), the sequence "00100" has a falling edge that is a little bit earlier than that in the sequence "01100" because the full voltage level was not reached for the isolated one bit in the middle. Data-dependent jitter also occurs if the channel has an insufficient low-frequency cutoff. As illustrated in Fig. C.2(b), the signal drifts up and down depending on the number of consecutive zeros or ones, an effect known as baseline wander. This amplitude error translates into a zero-crossing time error (jitter) when the signal edges have a finite slew rate. Other causes for data-dependent jitter are an insufficient phase linearity, reflections on transmission lines and cables due to an impedance mismatch, and circuits operated beyond their overload limit.

In the examples given earlier (Fig. C.2), the signals suffer not only from jitter but also from ISI. In fact, it is the ISI on the data edges in conjunction with the finite slew rate of these edges that causes the data-dependent jitter. Thus, fast edges help to suppress the jitter. Note that whereas the ISI in the aforementioned signals can be removed by passing the signals through a limiting amplifier, the jitter cannot be removed in this way.

The histogram of data-dependent jitter consists of discrete lines and is bounded. If the edge timing depends mostly on a *single* bit relative to this

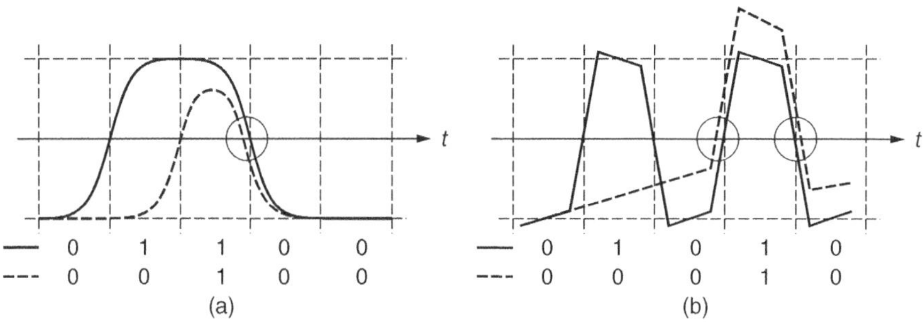

Figure C.2 Data-dependent jitter caused by (a) insufficient bandwidth and (b) baseline wander.

edge, then the time error (jitter) assumes one of two discrete values, one for when the bit is zero and one for when it is one, resulting in a histogram with two lines. Similarly, if the edge timing depends mostly on *two* bits, which can assume four values (00, 01, 10, and 11), then the jitter histogram has four lines. In general, if n bits are involved in the generation of data-dependent jitter, the histogram has 2^n lines, some of which may fall on top of each other [1]. For example, in Fig. C.2(a) the edge timing depends mostly on the values of the preceding two bits (11 and 01 are shown). Because data-dependent jitter is bounded, it is usually specified by its peak-to-peak value t_{DDJ}^{pp}.

Duty-Cycle Distortion Jitter. The simplest case of data-dependent jitter occurs when the edge timing depends on only a single bit, usually the bit immediately preceding the edge. Interestingly, this case cannot occur when an ideal NRZ signal is distorted by a *linear* transfer function, because every data pattern and its complementary data pattern produce the same edge timing. However, in the presence of nonlinearities, threshold errors, or asymmetric rise and fall times this type of jitter can occur and is called *duty-cycle distortion jitter* (DCD). Figure C.3 shows an example of DCD caused by asymmetric rise and fall times illustrated for a "01010" sequence. For such a clock-like data pattern, the widening (or narrowing) of the ones relative to the zeros can be seen clearly. Data-dependent jitter other than the special case of duty-cycle distortion jitter is sometimes called *intersymbol-interference induced jitter*.

Figure C.3 Duty-cycle distortion jitter due to asymmetric rise and fall times.

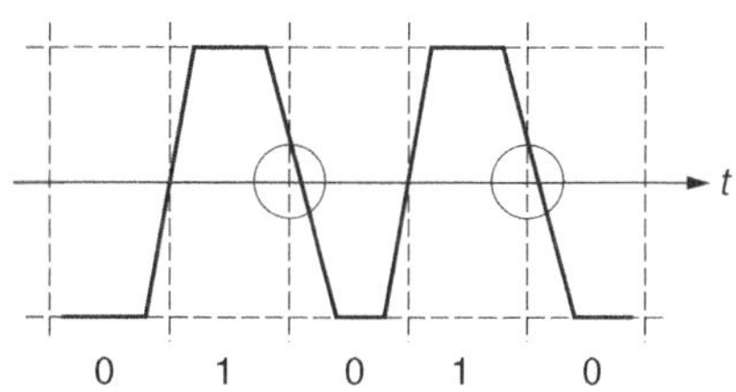

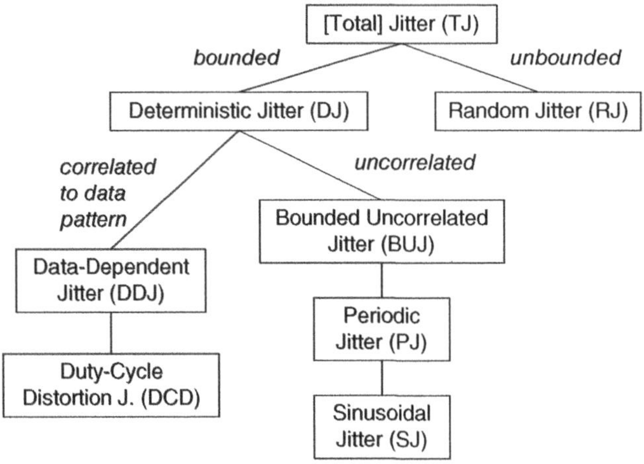

Figure C.4 A nomenclature for jitter types.

The histogram of duty-cycle distortion jitter has two discrete lines, one for edges preceded by a zero, that is rising edges, and one for edges preceded by a one, that is falling edges. In the eye diagram, this jitter manifests itself as an upward or downward shift of the eye crossing.

Deterministic Jitter. As illustrated in Fig. C.4, data-dependent jitter belongs to a larger class of jitter known as *deterministic jitter* (DJ) [2, 3]. Deterministic jitter is systematic and has a bounded histogram. Data-dependent jitter and the special case of duty-cycle distortion jitter are both correlated to the data pattern. Deterministic jitter types that are *not* correlated to the data pattern are summarized by the term *bounded uncorrelated jitter* (BUJ). Examples for BUJ are jitter due to crosstalk from adjacent signal lines [4] or disturbances from the power and ground lines. A switching power supply may cause *periodic jitter* (PJ), which is a subcategory of BUJ. Finally, *sinusoidal jitter* (SJ), a subcategory of periodic jitter, usually is artificially generated and used for compliance testing [2, 5].

Random Jitter. Random jitter is not correlated to the data pattern or any other signal, but, as the name implies, is random. The histogram of random jitter is approximately Gaussian and is unbounded. In the eye diagram, random jitter manifests itself as a blurring of the edges. Because of its unbounded nature, random jitter is usually specified by its rms value t_{RJ}^{rms}, which corresponds to the standard deviation of the associated Gaussian distribution. Alternatively, random jitter can be specified by its peak-to-peak value, but then it must be tied to a sample size or a reference BER. We discuss how to do that shortly.

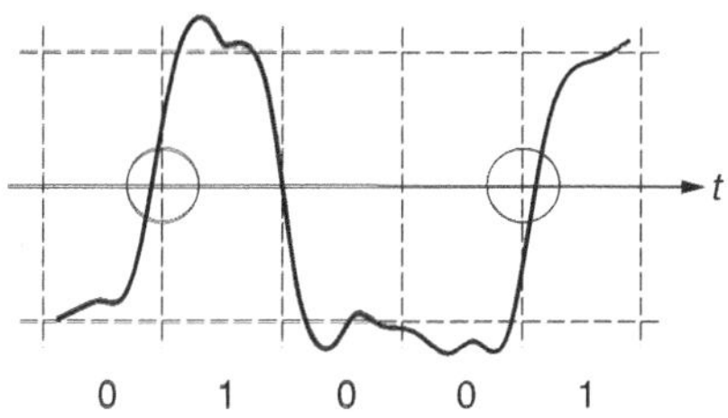

Figure C.5 Random jitter as a result of noise and a finite edge rate.

Any signal with a finite edge rate and noise on it exhibits random jitter. Figure C.5 illustrates how the finite edge rate converts the amplitude uncertainty into a zero-crossing time uncertainty. Again, fast edges help to suppress the jitter. Note that the noise-to-jitter conversion does not exactly preserve the statistical distribution of the noise because the edges have a finite duration and are nonlinear. Although the noise in the amplitude domain may be Gaussian, the resulting random jitter is non-Gaussian in its far-out tails.

Total Jitter. A typical jitter histogram contains deterministic and random jitter. Mathematically, the combined histogram is the convolution of the histogram for the deterministic jitter with the histogram for the random jitter. Figure C.6 illustrates this operation with a simple example. The deterministic jitter is modeled by a histogram that has two discrete lines, shown in Fig. C.6(a). This jitter model is known as the *dual Dirac model*. It corresponds to an eye diagram with two sharp and equally likely zero crossings and can describe, for example, pure duty-cycle distortion jitter or weak data-dependent jitter. Convolving the dual Dirac histogram with the Gaussian histogram in Fig. C.6(b) results in the composite jitter histogram shown in Fig. C.6(c). Note that the inner part of the composite histogram can be identified with the deterministic jitter and that the Gaussian tails can be identified with the random jitter.

If we have multiple deterministic jitter components, t_{DJi}^{pp} and multiple random jitter components, t_{RJi}^{rms}, the combined jitter is again found by convolving all the individual histograms. In particular, multiple bounded jitter components combine into a larger bounded jitter with the peak-to-peak value $t_{DJ}^{pp} = \sum_i t_{DJi}^{pp}$ and

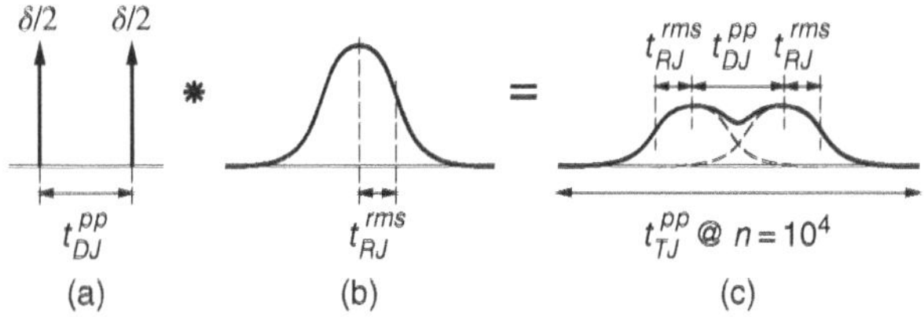

Figure C.6 Dual Dirac model: histogram of (a) deterministic jitter, (b) random jitter, and (c) the combination of both jitters.

multiple Gaussian jitter components combine into a larger Gaussian jitter with the variance $(t_{RJ}^{rms})^2 = \sum_i (t_{RJi}^{rms})^2$.

The combination of all deterministic and random jitter components is known as *total jitter* (TJ). It is usually specified by its peak-to-peak value t_{TJ}^{pp} and because of its unbounded nature must be tied to a sample size or a reference BER. Here are two methods to determine total jitter:

- Acquire a histogram of the composite jitter with a scope. The total jitter, t_{TJ}^{pp}, equals the width of the histogram. The number of samples n taken for the histogram must be specified along with the jitter value because t_{TJ}^{pp} grows with n (cf. Fig. C.6(c)).
- Perform a BERT scan, that is, scan the sampling instant t_S horizontally across the eye while recording the BER. The BER is low when sampling near the center of the eye and goes up when approaching the eye crossings to the left and right. This curve is known as the *horizontal bathtub curve* (cf. Appendix B). The total jitter, t_{TJ}^{pp}, equals one bit period, $T = 1/B$, minus the width of the bathtub curve for a given reference BER, such as 10^{-12}.

The two methods of determining total jitter are related, as demonstrated by Fig. C.7. Imagine that we move the sampling instant to the right end of the left histogram such that $n - 1$ samples are to the left and one sample is to the right of the sampling instant ($n = 10^4$ in our example). In this case, $n - 1$ samples are taken from the correct bits while one sample is taken from a slipped bit. Because the latter sample is incorrect with probability $\frac{1}{2}$, the BER for that sampling instant is $1/(2n)$. This value must match the BER value of the bathtub curve. Thus, the total jitter from a histogram with n samples is roughly the same as the total jitter from a bathtub curve for $BER = 1/n$ ($BER = 10^{-4}$ in our example).

Total Jitter Estimation. Measuring total jitter from the histogram or the bathtub curve can be very time consuming if n is large or BER is small. For example,

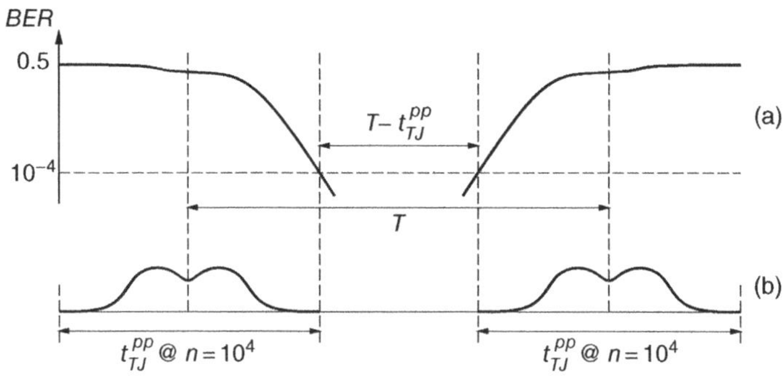

Figure C.7 Total jitter determined (a) from the horizontal bathtub curve and (b) from the histogram.

it takes 10,000 s (2 hours and 47 minutes) to collect 100 errors at 10 Gb/s and $BER = 10^{-12}$. To get around this issue, we can separate composite jitter into a random part, t_{RJ}^{rms}, and a deterministic part, t_{DJ}^{pp}, and then use modeling and extrapolation techniques to estimate the total jitter at the desired reference BER.

Let us assume that we successfully separated the jitter into a random and a deterministic component (we come to that in a moment), that the deterministic jitter can be described by the dual Dirac model, that the random jitter is Gaussian, and that $t_{DJ}^{pp} \gg t_{RJ}^{rms}$. Then, the peak-to-peak value of the total jitter for a given BER can be related to its deterministic and random jitter components as (cf. solution to Problem C.1(a) on p. 531)

$$t_{TJ}^{pp} = t_{DJ}^{pp} + 2Qt_{RJ}^{rms}, \tag{C.1}$$

where Q is the familiar Personick Q (cf. Section 4.2) evaluated for the reference BER of the total jitter. For example, given a deterministic jitter of 0.3 UI peak-to-peak and a random jitter of 0.02 UI rms, the total jitter is $0.3\text{ UI} + 2 \cdot 7.03 \cdot 0.02\text{ UI} = 0.58$ UI peak-to-peak when referred to $BER = 10^{-12}$.

In practice, Eq. (C.1) is often used to estimate total jitter even in situations where the deterministic jitter does not have a dual Dirac distribution. It is important to be aware of the limitations coming with this approximation. For deterministic jitter that is more evenly distributed than the dual Dirac model (more than two Dirac pulses), large deterministic and random jitter fluctuations are less likely do coincide and Eq. (C.1) overestimates the total jitter. For a deterministic jitter value (peak-to-peak) that is similar or smaller than the random jitter value (rms), both tails of the Gaussian distribution matter and Eq. (C.1) underestimates the total jitter. Moreover, the accuracy of the relationship depends on how well the random jitter follows a Gaussian distribution, in particular, in its far-out tails. [→ Problem C.1.]

RJ/DJ Decomposition. Accurately decomposing jitter into a random and a deterministic component is not easy. Here is a choice of three methods [2]:

- *Two-Tests Method*: First, test the system with a clock-like data pattern ("01010101…"), which does not produce data-dependent jitter (other than DCD), to determine the random jitter. Then, test the same system with a repetitive data pattern (e.g., the 20-bit long 8B/10B comma-character sequence) to determine the data-dependent jitter. Triggering on the pattern clock and using the averaging feature of the oscilloscope suppresses the random jitter in the latter measurement.

  The weakness of this method is that deterministic jitter other than data-dependent jitter (e.g., periodic jitter or bounded uncorrelated jitter) is erroneously included in the first (random jitter) measurement and it is not accounted for in the second (deterministic jitter) measurement.

- *Tail-Fit Method*: Measure the total jitter histogram and fit two Gaussians to its tails [6]. The standard deviations of the Gaussians provide an estimate for the rms value of the random jitter. The separation of the centers of the two Gaussians provides an estimate for the peak-to-peak value of the deterministic jitter (cf. Fig. C.6(c)).
  The weakness of the tail-fit method is that low-probability deterministic jitter is counted as random jitter. For the sake of accuracy, the jitter components found by the aforementioned method are sometimes called *Gaussian jitter* (instead of random jitter) and *high-probability jitter* (instead of deterministic jitter).
- *BERT-Scan Method*: Measure the horizontal bathtub curve and fit a parametrized mathematical model for deterministic and random jitter to the measurement (e.g., see Ref. [7]). Roughly speaking, the horizontal portions of the bathtub curve within the unit interval correspond to the deterministic jitter, whereas the sloped portions relate to the random jitter (cf. Fig. C.7(a)).
  The weakness of this method is similar to that listed for the tail-fit method, and therefore the terms Gaussian jitter and high-probability jitter are also preferred in this case.

Having found the random and deterministic components by one of the aforementioned methods, we can use Eq. (C.1) to estimate the total jitter for the desired reference BER. Moreover, knowing the composition of jitter can be helpful in debugging jitter issues. [→ Problem C.2.]

Jitter Trend and Jitter Spectrum. Sometimes, we are interested in the jitter in a particular bandwidth. For example, we may only be interested in the high-frequency jitter of a transmitter, because the low-frequency jitter can be tracked by the receiver and does not cause bit errors. How do we measure jitter in a given bandwidth? Should we pass the data signal through a filter with the specified bandwidth and then measure the jitter? No, we are not supposed to filter the signal itself, but the *jitter* of the signal!

Figure C.8 illustrates the difference between the data signal (v_O) and its jitter (t_J) with an example. The jitter t_J of a given data edge is defined as the deviation

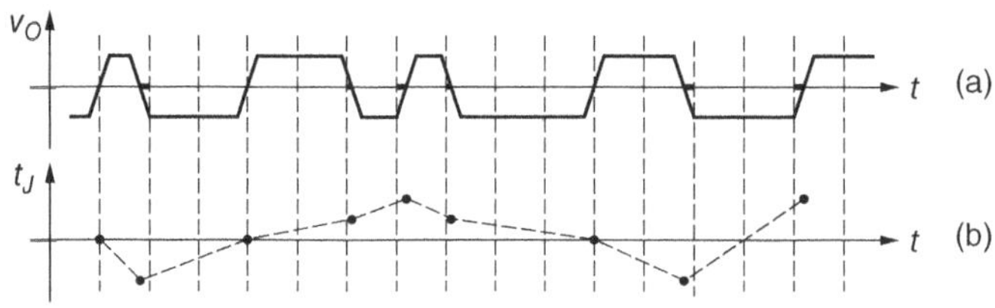

Figure C.8 (a) Data signal with jitter and (b) evolution of the edge jitter with time (jitter trend).

from its ideal location. In Fig. C.8(a), these deviations are indicated with bold horizontal lines; in (b), these same deviations are plotted as a function of time together with a linear interpolation (dashed line). The time evolution of t_J is known as the *jitter trend*. It is the frequency content of this jitter trend that we are interested in.

Mathematically, the data signal, $v_O(t)$, can be written in terms of its ideal, jitter-free counterpart, $v_I(t)$, and the jitter trend, $t_J(t)$:

$$v_I(t) \rightarrow v_O(t) = v_I(t - t_J(t)). \tag{C.2}$$

The jitter trend, $t_J(t)$, thus can be interpreted as the function that distorts or warps the time axis of the ideal data signal to produce the actual data signal. If $t_J(t)$ is constant, there is no jitter; if $t_J(t)$ is a linear function, there is a data-rate offset, and so forth. Now, given the waveforms $v_I(t)$ and $v_O(t)$, we can use Eq. (C.2) to estimate the jitter trend $t_J(t)$. Clearly, we can find a definite value for $t_J(t)$ only at time points t where the waveform $v_O(t)$ has a rising or falling edge. At these time points, the jitter is given by the (signed) time difference between the edges of v_O and v_I, in agreement with Fig. C.8 (the vertical dashed lines indicate the edge locations of the ideal signal v_I). [→ Problem C.3.]

Different types of jitter have different power spectral densities (PSDs). For example, random jitter and data-dependent jitter typically have a wide spectrum extending up to half of the bit rate, $B/2$, duty-cycle distortion jitter has a spectrum that peaks around $B/2$, and periodic jitter tends to have spectral components well below $B/2$. For example, periodic jitter due to a switching power supply has components in the 100 kHz to MHz range.

Jitter Filters. Jitter can be filtered with a clock-recovery circuit that has a well-defined jitter transfer characteristics and adds very little jitter of its own, a so-called *golden phase-locked loop* (PLL) [2].

To measure low-frequency jitter only, we run the data signal from the device under test (DUT) through a golden PLL, as shown in Fig. C.9(a). The PLL's jitter transfer function is one (0 dB) up to the desired jitter bandwidth BW_J, after which it rolls off. Now, the recovered clock from the PLL contains only the low-frequency jitter ($< BW_J$) and can be measured, for example, by taking the histogram with a scope.

To measure high-frequency jitter only, we use a golden PLL with a bandwidth equal to the lower corner of the desired high-pass function, as shown in Fig. C.9(b). This PLL again is fed with the data signal, but now its output is used to trigger the scope. Because the scope Y input and the trigger input both get the same amount of low-frequency jitter ($< BW_J$), it is suppressed (a common-mode signal in the time domain) and only the desired high-frequency jitter ($> BW_J$) appears in the jitter histogram.

To measure jitter in the frequency band from BW_{J0} to BW_{J1}, the two set-ups shown in Fig. C.9(a) and (b) can be combined using one golden PLL with the

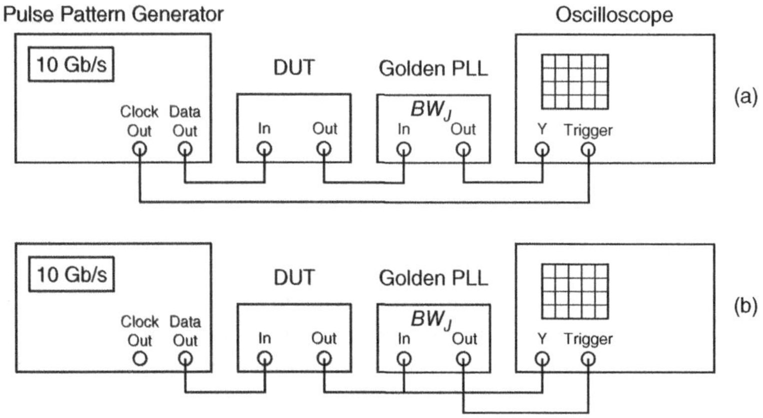

Figure C.9 Measuring (a) low-pass filtered jitter with $f < BW_J$ and (b) high-pass filtered jitter with $f > BW_J$.

bandwidth BW_{J0} in the trigger path and another golden PLL with the bandwidth BW_{J1} in the signal path.

In addition to its bandwidth, the frequency response of the jitter filter (golden PLL) matters as well. Usually, a first- or second-order PLL is used. In the case of a second-order PLL, the damping factor, which controls the jitter peaking, must be given as well. A typical jitter filter for characterizing SerDes transmitters has a first-order high-pass response with the 3-dB bandwidth $BW_J = B/1667$, where B is the bit rate [2].

Jitter can also be filtered by means of software. Many real-time sampling oscilloscopes come with software that can calculate jitter in the desired bandwidth from the digitized and stored waveform. Similarly, the jitter of a simulated waveform can be determined in the desired bandwidth by means of a post processor implementing a sofware jitter filter.

To filter jitter in software, we do the following. First, the zero crossings are compared against the edges of an ideal software-generated constant-frequency clock, resulting in the jitter samples $t_J(n)$ at the nth clock edge. (We simplified our earlier notation of $t_J(nT)$ to just $t_J(n)$.) Next, a suitable interpolation fills in the missing jitter samples (where the data signal has no transitions), such that $t_J(n)$ is defined for all n. Finally, the jitter samples $t_J(n)$ are passed through a discrete-time filter, producing the output jitter samples $t_J'(n)$. For a discretized first-order high-pass filter, we have

$$t_J'(n) = \frac{t_J'(n-1) + t_J(n) - t_J(n-1)}{1 + 2\pi BW_J/B}, \tag{C.3}$$

where B is the bit rate (equal to the jitter sample rate) and BW_J is the 3-dB bandwidth.

Jitter Measurement. A variety of test equipments are available to measure jitter, each one with its own advantages and disadvantages [2, 8].

- *Real-time sampling oscilloscopes* with the necessary jitter analysis software are easy to use and can perform a variety of jitter measurements at moderate to high bit rates. Such measurements include peak-to-peak and rms value of the composite jitter, separated random and deterministic jitter, estimated total jitter at the desired BER, jitter trend, and jitter PSD. One limitation of real-time sampling scopes is that the RJ/DJ separation becomes inaccurate for long data patterns, such as PRBS-31, that do not fit entirely into the scope's memory.
- *Equivalent-time sampling oscilloscopes* can perform similar measurements at higher bit rates and with very low residual jitter (e.g., < 100 fs), but require a periodic data signal and a trigger signal, which makes them harder to use.
- *Time interval analyzers* can be used to measure the time intervals between zero crossings of the data signal with very high accuracy (e.g., 200 fs). From the statistics of all time intervals corresponding to one bit period, two bit periods, three bit periods, and so forth, it is possible to calculate, among other things, the jitter PSD.
- *Bit-error rate test sets* (BERT) can be used to measure total jitter at the desired BER directly and accurately (with a horizontal BERT scan). However, at low reference BERs the measurements become very time consuming.
- Dedicated *jitter analyzers* generally are used to test very high-speed circuits for compliance with specific standards.
- *Spectrum analyzers* can be used to determine the phase noise PSD of a clock signal (see Fig. C.13). From this PSD it is possible to calculate the rms jitter in the desired bandwidth (cf. Eq. (C.5)). Whereas spectrum analyzers can be used to determine clock jitter, they cannot be used for data jitter.

C.2 Clock Jitter

Let us turn from jitter in data signals to jitter in clock signals. Jitter in the clock signal of a data converter degrades the SNR of the converted signal [9]. Jitter in the clock of a digital system may cause setup time violations. Jitter in the clock of a decision circuit may degrade the BER of the receiver, if the clock jitter is uncorrelated to the received data jitter. However, if the clock jitter tracks the received data jitter, it may improve the BER. The latter situation occurs when the sampling clock is recovered from the data signal with a PLL.

Absolute Clock Jitter (Time Interval Error). There are several kinds of clock jitter. The most fundamental clock jitter is *absolute jitter* also known as *time interval error* (TIE) [3]. It is defined as the deviation, t_J, of the actual clock edge (either rising or falling) from its ideal location. This definition corresponds directly to

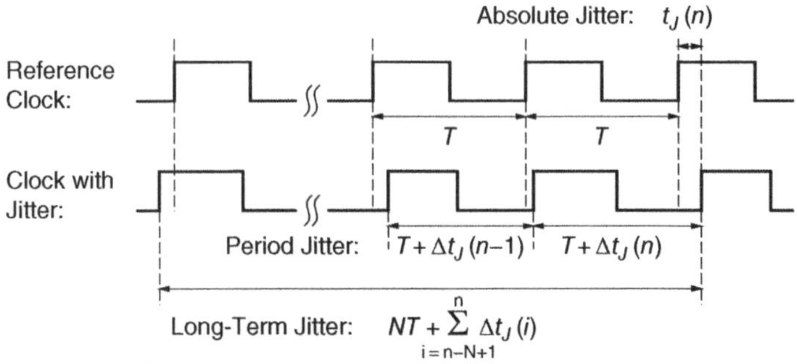

Figure C.10 Absolute, period, and long-term clock jitter.

that of data jitter. Absolute clock jitter, like data jitter, can be characterized by a histogram, can be separated into random and deterministic components, and can be analyzed in the time domain by a trend of in the frequency domain by a PSD.

To measure absolute jitter we need a reference clock that defines the ideal edge location (see Fig. C.10). There are two basic ways to obtain the reference clock. First, the reference clock may come from a clock source in the test system. For example, the sampling jitter of a clock-recovery circuit can be measured relative to the clock of the pattern generator that produced the test signal [10]. Second, the reference clock may be recovered from the clock to be analyzed. A real-time scope, for example, digitizes and stores the clock waveform to be analyzed, determines the ideal edge locations in software (clock recovery), and then calculates the edge deviations (jitter).

To measure the absolute jitter of a free-running oscillator, such as a crystal oscillator, only the second option (recovered clock) is available. When measuring a free-running oscillator over longer and longer periods of time, the absolute rms jitter grows larger and larger. Thus, for a meaningful measurement, it is necessary to remove the low-frequency jitter. A typical jitter filter for characterizing reference clocks has a passband from 12 kHz to 20 MHz.

Period Jitter (Cycle Jitter). *Period jitter* (a.k.a. *cycle jitter*) is defined as the deviation, Δt_J, of the clock period from its ideal value T (in practice, often the average period value). This jitter measures is self-referenced and thus does not rely on a reference clock. Period jitter can be expressed as the difference of the absolute jitter of two adjacent edges. Figure C.10 illustrates the period jitter, Δt_J and its relationship to absolute jitter, t_J, with an example. Period jitter can be measured by triggering the oscilloscope on one clock edge and taking the histogram of the edge a period later.

Cycle-to-Cycle Jitter. *Cycle-to-cycle jitter* is defined as the deviation, $\Delta\Delta t_J$, of the clock period relative to the previous clock period. In Fig. C.10, cycle-to-cycle jitter appears as the difference $\Delta t_J(n) - \Delta t_J(n-1)$. When expressed in terms of absolute jitter, cycle-to-cycle jitter is a doubly differential measure. (Caution: Not all authors define *cycle-to-cycle jitter* in the same way. Sometimes this term refers to what we previously defined as *period jitter* [9].)

Clock Jitter Comparison and Relationships. Table C.1 summarizes the three clock jitter measures and Fig. C.11 illustrates them with an example. The upper half of the table expresses all jitter measures in terms of the absolute jitter measured at the *n*th clock edge, $t_J(n)$, clearly showing the differential nature of period and cycle-to-cycle jitter. The lower half of the table re-expresses all jitter measures in terms of the period jitter measured for the *n*th clock cycle, $\Delta t_J(n)$.

From the jitter definitions in the upper half of Table C.1 it is evident that period jitter and cycle-to-cycle jitter can be regarded as filtered versions of the absolute jitter. The discrete filter function for the period jitter is $t_J'(n) = t_J(n) -$

Table C.1 Clock jitter measures expressed in terms of absolute jitter and period jitter.

Jitter measure	Instantaneous value	RMS value
Absolute jitter	$t_J(n)$	
Period jitter	$t_J(n) - t_J(n-1)$	
Cycle-to-cycle jitter	$[t_J(n) - t_J(n-1)] - [t_J(n-1) - t_J(n-2)]$	
Absolute jitter	$\sum_{i=1}^{n} \Delta t_J(i) + t_J(0)$	$\propto \sqrt{t}$
Period jitter	$\Delta t_J(n)$	Δt_J^{rms}
Cycle-to-cycle jitter	$\Delta t_J(n) - \Delta t_J(n-1)$	$\sqrt{2} \cdot \Delta t_J^{rms}$

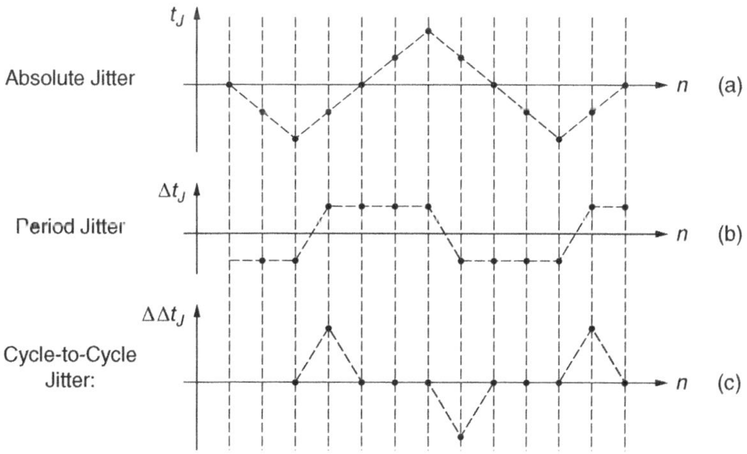

Figure C.11 Example for the time evolution (trend) of (a) absolute, (b) period, and (c) cycle-to-cycle jitter.

$t_J(n-1)$ or $H(z) = 1 - z^{-1}$ in the z-domain. Similarly, the filter function for the cycle-to-cycle jitter is $t_J'(n) = t_J(n) - 2t_J(n-1) + t_J(n-2)$ or $H(z) = 1 - 2z^{-1} + z^{-2}$ in the z-domain. Both filters emphasize high-frequency jitter.

If the period jitter samples, $\Delta t_J(n)$, are uncorrelated and stationary, we obtain the rms jitter expressions in the lower half of Table C.1 (last column). Under these conditions, the absolute jitter performs a one-dimensional random walk and its rms value diverges proportional to $\sqrt{t}$ [11]. This is the situation we encountered for the free-running oscillator without jitter filtering. The rms values of the period and cycle-to-cycle jitter of a free-running oscillator are well defined because these jitter measures naturally suppress low-frequency jitter.

What clock jitter measure should we use in which situation?

- The *absolute* clock jitter is relevant in clocked data converters, where it determines an upper limit for the SNR [9]. In clocked receivers, it determines the variation of the sampling instant in the data eye.
- The *period* clock jitter is relevant when closing timing in digital systems. The minimum clock period from the clock edge that launches the data to the next clock edge that latches the result equals the nominal clock period minus the peak period jitter.
- The *cycle-to-cycle* clock jitter can be used to characterize the quality of clocks with a large intentional jitter, such as spread-spectrum clocks (SSC). More commonly, however, spread-spectrum clocks are characterized by their absolute jitter after passing them through a second-order high-pass jitter filter to suppress the low-frequency spreading jitter.

N-Period Jitter and PLLs. The idea of period jitter can be generalized to that of *N-period jitter*, which is defined as the deviation of N consecutive periods from their ideal value. N-period jitter can be measured by triggering the oscilloscope on one edge and taking the histogram of the edge N periods later. This jitter, illustrated in Fig. C.10, is also known as *long-term jitter* or *accumulated jitter* if N is a large number. (Correspondingly, 1-period jitter is sometimes known as *short-term jitter*.)

Mathematically, the N-period jitter can be written in terms of period jitter as $\sum_{i=n-N+1}^{n} \Delta t_J(i)$, which looks very similar to the absolute-jitter expression in Table C.1 (but it is still a self-referenced measure). For a free-running oscillator, the rms value of the N-period jitter diverges with N. For a phase-locked oscillator in a PLL, the rms value of the N-period jitter initially increases proportional to $\sqrt{N}$, but then saturates at a finite long-term jitter value equal to $\sqrt{2}\times$ the absolute rms jitter [10]. Saturation occurs when NT exceeds the time constant of the PLL (see Fig. C.12). [→ Problem C.4.]

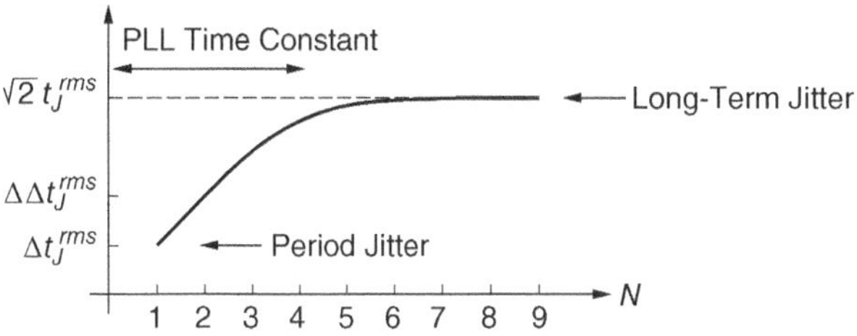

Figure C.12 *N*-period rms jitter (log scale) of a phase-locked oscillator.

C.3 Jitter, Phase Noise, and Bit-Error Rate

Jitter and Phase Noise. An interesting relationship exists between the absolute jitter and the *phase noise* of a clock signal [9, 11–13]. The output signal of a sine-wave oscillator with (constant) amplitude V_O, frequency f_0 $(= 1/T)$, and phase noise $\phi_n(t)$ can be written as

$$v_O = V_O \sin[2\pi f_0 t + \phi_n(t)]. \tag{C.4}$$

Comparing this expression with the definition of absolute jitter given in Eq. (C.2), we find the relationship between jitter trend and phase noise: $t_J(t) = -\phi_n(t)/(2\pi f_0)$.

If the phase noise is small, $\phi_n \ll 1$, Eq. (C.4) can be rewritten in the form $v_O \approx V_O \sin(2\pi f_0 t) + \phi_n(t) \cdot V_O \cos(2\pi f_0 t)$. This form shows that *phase* noise in a sine-wave oscillator also implies *voltage* noise. The voltage PSD, which can be measured with a spectrum analyzer, shows the upconverted phase-noise in the form of a “skirt” around the spectral line at f_0 (see Fig. C.13). Normalizing the skirt to the carrier strength, frequency translating it down to DC, and consolidating positive and negative frequencies results in the (approximate) one-sided phase-noise PSD $\Phi_n^2(f)$. Now, converting phase noise to jitter, using the relationship found earlier, yields the one-sided jitter PSD $\Phi_n^2(f)/(2\pi f_0)^2$. Integrating the latter over all frequencies and taking the square root results in the absolute rms clock jitter [9, 13]

$$t_J^{rms} = \frac{1}{2\pi f_0}\sqrt{\int_0^{\infty} \Phi_n^2(f)\, df}. \tag{C.5}$$

This relationship permits us to estimate the absolute rms jitter of a sine-wave oscillator from the noise PSD measured with a spectrum analyzer.

For an ideal free-running oscillator the phase-noise PSD has the form $\Phi_n^2(f) \propto 1/f^2$ [14], which causes the integral in Eq. (C.5) to diverge. This is consistent with our earlier observation that absolute jitter of a free-running

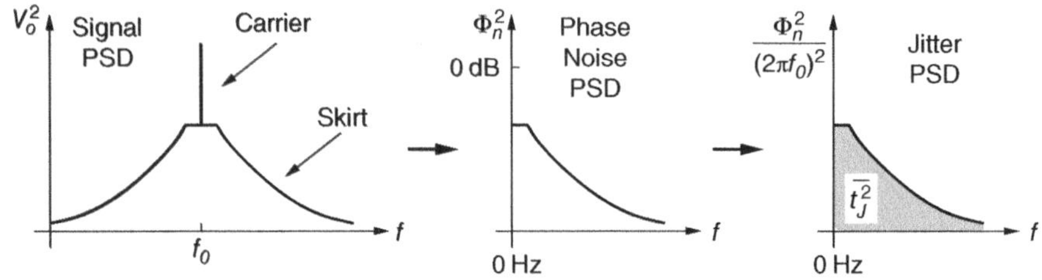

Figure C.13 Determination of the absolute rms jitter from the signal PSD by means of the phase noise PSD and jitter PSD.

oscillator diverges. When the oscillator is phase locked to a clean reference clock, the phase noise at low frequencies (the signal noise close to the carrier) gets suppressed, making the integral and the absolute rms jitter finite.

Jitter and BER. As we pointed out at the beginning of this appendix, excessive jitter in the data signal relative to the sampling clock can cause bit errors. Just like noise may cause the sampled voltage to be on the "wrong side" of the decision threshold, random jitter may cause the data edge to slip past the sampling instant such that one bit is sampled twice while the adjacent bit is not sampled. In general, we have to consider noise, ISI, and the different types of jitter jointly to calculate the BER accurately. Fortunately, the BER in many optical communication system, is *mostly* determined by noise and ISI and our discussion in Chapter 4 remains valid. But with increasing bit rates, jitter does play a more prominent role. In any case, it is useful to know how low the jitter has to be, such that its effect on BER is negligible.

To answer the aforementioned question, we derive the BER for a data signal that contains jitter (DJ and RJ) but is free of noise and ISI. The clock and data recovery circuit (CDR) in our receiver can tolerate a maximum amount of peak-to-peak jitter without making errors, known as the *jitter tolerance*, t_{JTOL}^{pp}. The CDR's jitter tolerance is tested with a sinusoidal jitter (SJ) with frequency f and peak-to-peak magnitude t_{SJ}^{pp}. For $t_{SJ}^{pp} < t_{JTOL}^{pp}(f)$ the CDR operates error free.[1] Figure C.14(a) shows a typical jitter tolerance curve. At low frequencies, the CDR tracks the sinusoidal jitter and thus can tolerate a fair amount of jitter. When the jitter frequency, f, exceeds the CDR's tracking bandwidth, the jitter tolerance curve bottoms out. In this regime, an ideal CDR would be able to tolerate one bit period of jitter, $t_{JTOL}^{pp} = 1/B$, but in practice the setup and hold time of the decision circuit, the sampling time offset, sampling jitter, and so forth limit the jitter tolerance to a lower value.

1 This is a simplified definition of jitter tolerance. In [2, 5], several jitter components (SJ, DJ, and RJ) are applied simultaneously and the receiver is tested for compliance with either a BER (e.g., $BER < 10^{-12}$) or a power penalty (e.g., power penalty due to SJ < 1 dB).

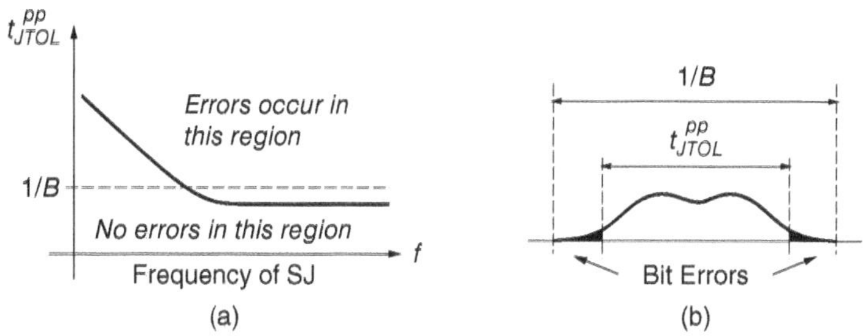

Figure C.14 (a) Jitter tolerance curve and (b) histogram of data jitter.

We are now ready to calculate the BER resulting from jitter. Let us make the worst-case assumption that the jitter in the received signal is untrackable high-frequency jitter. Then, the BER is given by the (normalized) area under the tails of the data jitter distribution that extend beyond the high-frequency jitter tolerance, as illustrated in Fig. C.14(b). (If the jitter contains trackable components, the BER is lower.) Now, if the total jitter of the distribution equals the jitter tolerance of the CDR, then the BER to which the total jitter is referenced to is the BER that we are looking for (cf. Problem C.1(a)). Thus, setting $t_{TJ}^{pp} = t_{JTOL}^{pp}$ and solving Eq. (C.1) for Q reveals the BER as a function of the deterministic and random jitter

$$BER = \int_{Q}^{\infty} \text{Gauss}(x)\, dx \qquad \text{with} \qquad Q = \frac{t_{JTOL}^{pp} - t_{DJ}^{pp}}{2t_{RJ}^{rms}}. \tag{C.6}$$

This equation comes with the same caveats as Eq. (C.1), which we have discussed earlier (dual Dirac model for DJ, Gaussian model for RJ, and $t_{DJ}^{pp} \gg t_{RJ}^{rms}$), but it can still serve as a useful approximation in many practical situations. With Eq. (C.6) we find, for example, that for a CDR with a jitter tolerance of 0.7 UI and a data signal with a deterministic jitter of 0.3 UI peak-to peak and a random jitter of 0.02 UI rms, the expected BER is about 10^{-23} ($Q = 10$). For this scenario, jitter has a negligible impact and we are permitted to calculate the BER based on noise and ISI alone.

Questions to Ask. When reporting jitter numbers, it is important to be clear about what was measured (or simulated). The following check list may be helpful:

- Is it rms or peak-to-peak jitter? If it is peak-to-peak jitter, what BER does it refer to?
- Is it total jitter or were only certain components, such as data-dependent jitter or random jitter, measured? If it is total jitter referenced to a low BER, was it actually measured or is it an estimated based on an RJ/DJ decomposition?

- Is it unfiltered or filtered jitter? In the first case, what is the lowest jitter frequency the measurement equipment can capture? In the second case, what are the upper and lower 3-dB corner frequencies of the filter? What is the shape of the jitter filter: first order, second order, or brick wall?
- In the case of data jitter: What data pattern is used? Is the reference clock recovered or external? If it is recovered, what are the clock-recovery (PLL) parameters?
- In the case of clock jitter: Is it absolute, period, or cycle-to-cycle jitter? If it is absolute jitter, is the reference clock recovered or external? If it is recovered, what are the clock-recovery (PLL) parameters? If it is long-term jitter, over how many periods was it measured? If it is cycle-to-cycle jitter, clarify its definition.

Problems

C.1 Total Jitter. (a) Show analytically that Eq. (C.1) holds for deterministic jitter with a dual Dirac distribution, random jitter with a Gaussian distribution, and $t_{DJ}^{pp} \gg t_{RJ}^{rms}$. (b) Calculate numerically (e.g., using Matlab) the total jitter referred to $BER = 10^{-12}$ and compare the result with the prediction from Eq. (C.1) for the following cases: $t_{DJ}^{pp} = 0.3$ UI with a dual Dirac distribution and $t_{RJ}^{rms} = 0.02$ UI with a Gaussian distribution, $t_{DJ}^{pp} = 0.3$ UI with an even distribution and $t_{RJ}^{rms} = 0.02$ UI with a Gaussian distribution, and $t_{DJ}^{pp} = 0$ UI and $t_{RJ}^{rms} = 0.0413$ UI with a Gaussian distribution.

C.2 Jitter Decomposition. The horizontal eye margin is determined with a BERT scan for three different BERs as 29.8% for $BER = 10^{-12}$, 35.0% for $BER = 10^{-9}$, and 41.2% for $BER = 10^{-6}$. Using Eq. (C.1) as a model, how large are the deterministic and random jitter components?

C.3 Jitter Frequency. (a) How does the sinusoidal jitter $t_J(t) = T_J \sin(2\pi Bt)$ affect an NRZ data signal with bit rate B? (b) How does the sinusoidal jitter $t_J(t) = T_J \sin(\pi Bt)$ affect the clock-like NRZ data signal "01010101 …" with bit rate B?

C.4 Cycle-to-Cycle Jitter. Under which circumstances does the cycle-to-cycle rms jitter equal the 2-period rms jitter?

References

1 J. Buckwalter, B. Analui, and A. Hajimiri. Predicting data-dependent jitter. *IEEE Trans. Circuits Syst. II*, CASII-51(9):453–457, 2004.

2 NCITS. Fibre channel: methodologies for jitter specification 2, T11.2 / Project 1316-DT / Rev 0.0. National Committee for Information Technology Standardization, April 2000. http://www.t11.org.
3 Tektronix. Understanding and characterizing timing jitter: Primer, October 2003.
4 J. F. Buckwalter and A. Hajimiri. Cancellation of crosstalk-induced jitter. *IEEE J. Solid-State Circuits*, SC-41(3):621–632, 2006.
5 Telcordia Technologies. Synchronous optical network (SONET) transport systems: common generic criteria, GR-253-CORE, Issue 3. Telcordia Technologies (formerly Bellcore), Piscataway, NJ, September 2000.
6 M. P. Li, J. Wilstrup, R. Jessen, and D. Petrich. A new method for jitter decomposition through its distribution tail fitting. In *International Test Conference (ITC)*, September 1999.
7 IEEE. 802.3ah. Part 3: Carrier sense multiple access with collision detection (CSMA/CD) access method and physical layer specifications. Amendmend: Media access control parameters, physical layers, and management parameters for subscriber access networks. IEEE Computer Society, New York, NY, September 2004. http://www.ieee802.org/3/efm/.
8 Tektronix. Jitter analysis: a brief guide to jitter, June 2005.
9 C. Azeredo-Leme. Clock jitter effects on sampling: a tutorial. *IEEE Circuits and Systems Magazine*, 11(3):26–37, 2011.
10 J. A. McNeill. Jitter in ring oscillators. *IEEE J. Solid-State Circuits*, SC-32(6):870–879, 1997.
11 F. Herzel and B. Razavi. A study of oscillator jitter due to supply and substrate noise. *IEEE Trans. Circuits Syst. II*, CASII-46(1):56–62, 1999.
12 A. Hajimiri, S. Limotyrakis, and T. H. Lee. Jitter and phase noise in ring oscillators. *IEEE J. Solid-State Circuits*, SC-34(6):790–804, 1999.
13 B. Razavi. *Design of Integrated Circuits for Optical Communications.* McGraw-Hill, New York, 2003.
14 T. H. Lee. *The Design of CMOS Radio-Frequency Integrated Circuits.* Cambridge University Press, Cambridge, UK, 2nd edition, 2003.

D

Nonlinearity

Nonlinearity and the resulting signal distortions are important in systems that perform linear signal processing (equalization, data conversion, etc.) and in applications that use higher-order modulation (4-PAM, QPSK, QAM, etc.) or multicarrier modulation (SCM, OFDM, DMT, etc.). In the following, we review nonlinear effects, such as gain compression, harmonic distortions, intermodulation distortions, and composite distortions. We show how these effects are related to the power-series coefficients of the nonlinearity and to the input signal strength.

Importance of Linearity. In Section 4.1, we introduced the *linear channel* as an abstraction for the chain formed by the photodetector, transimpedance amplifier, and main amplifier. How linear does this channel have to be?

If the linear channel is followed directly by a binary decision circuit, as shown in Fig. 4.1, linearity is of little concern. In this case, even a very nonlinear amplifier such as a limiting amplifier is fine, as long as it does not create undue pulse-width distortion and jitter. However, if the linear channel is followed by an analog-to-digital converter, an equalizer, or some other type of signal processing circuit, linearity becomes important.

If the linear channel is part of a receiver for 4-PAM, QPSK, or another higher-order modulation format, a certain amount of linearity becomes mandatory. For multicarrier signals, such as a SCM, OFDM, or DMT signals, linearity becomes essential. This is particularly so, if the subcarriers are modulated with analog signals (e.g., AM-VSB) or high-order QAM signals.

Nonlinearity Model. A straightforward way to describe a nonlinear DC transfer function $y = f(x)$ is to expand it into a power series:

$$y = A(a_0 + x + a_2x^2 + a_3x^3 + a_4x^4 + a_5x^5 + \dots), \tag{D.1}$$

where A is the small-signal gain and a_i are the power-series coefficients characterizing the nonlinearity. Note that $A\ a_0$ represents the output-referred offset and that $a_1 = 1$ because, for convenience, all coefficients have been normalized to the small-signal gain.

Analysis and Design of Transimpedance Amplifiers for Optical Receivers, First Edition. Eduard Säckinger.

Equation (D.1) is also valid for AC signals, if the nonlinearity is memoryless, that is, if it responds instantaneously. To model nonlinearities with a memory effect (finite bandwidth), Eq. (D.1) must be reformulated as a Volterra series [1]. In addition to the convolution of the input signal with the impulse response, known from linear systems, the Volterra series includes convolutions with higher-order impulse responses, known as Volterra kernels, that model the nonlinearity. For the following simplified analysis, we assume that the bandwidth of the system under consideration is wide enough such that we can use the power-series representation in Eq. (D.1).

D.1 Gain Compression

A simple measure of nonlinearity is the loss of gain experienced by large signals relative to the small-signal gain. As illustrated in Fig. D.1(a), we apply an input signal swinging from $-X$ to X to the nonlinear device. Dividing the output swing by the input swing, $[y(X) - y(-X)]/[X - (-X)]$, we obtain the wideband large-signal gain. Using Eq. (D.1), we find this gain to be $A_{LS} = A(1 + a_3X^2 + a_5X^4 + \ldots)$. After normalizing it to the small-signal gain A, we obtain the ratio

$$\frac{A_{LS}}{A} = 1 + a_3X^2 + a_5X^4 + \ldots, \tag{D.2}$$

which describes how the large-signal gain varies with signal strength X. For practical amplifiers, a_3 and a_5 are negative. Thus, A_{LS}/A, which starts out at 1 for small signals, reduced with increasing signal strength, that is, we have *gain compression.*

The term *gain compression* originated in the RF world where its meaning is slightly different from its wideband cousin introduced earlier. RF

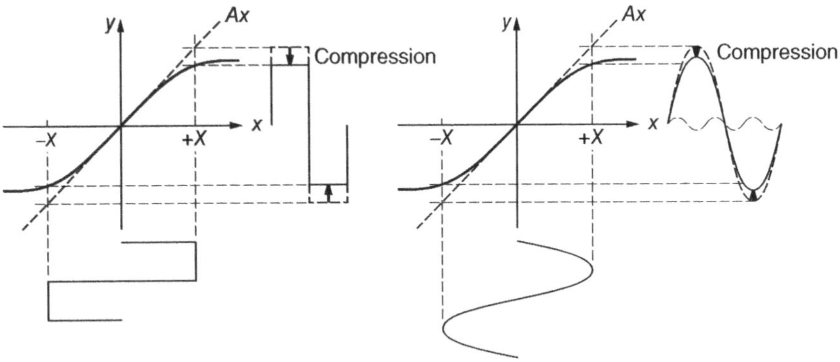

Figure D.1 Gain compression resulting from a nonlinearity for (a) wideband and (b) narrowband signals.

amplifiers and mixers are circuits that are tuned to a particular frequency and hence the gain compression for *narrowband* signals is of interest. To discuss the narrowband large-signal gain, also known as the *describing function* [2], we take the input signal as a sine wave $x(t) = X\sin(2\pi ft)$ with amplitude X and frequency f, as shown in Fig. D.1(b). The distorted output signal contains harmonic components at f, $2f$, $3f$, and so forth. Now, for the narrowband large-signal gain we ignore all harmonics at $2f$, $3f$, and so forth, because they are filtered out by the tuned circuit, and focus on the fundamental component at f only. Using Eq. (D.1), we find the output-signal component at the fundamental frequency to be $y(t) = A(X + \frac{3}{4}a_3X^3 + \frac{5}{8}a_5X^5 + \ldots)\sin(2\pi ft)$. Dividing this expression by the input signal yields the narrowband large-signal gain $\tilde{A}_{LS} = A(1 + \frac{3}{4}a_3X^2 + \frac{5}{8}a_5X^4 + \ldots)$. After normalizing this gain to the small-signal gain A, we obtain the compression factor for narrowband signals as a function of the input amplitude X:

$$\frac{\tilde{A}_{LS}}{A} = 1 + \frac{3}{4}a_3X^2 + \frac{5}{8}a_5X^4 + \ldots. \tag{D.3}$$

Note that for negative values of a_3 and a_5, the narrowband large-signal gain is somewhat larger than the wideband gain, that is, filtering out the harmonic distortion products increased the swing of the output signal. See Fig. D.1(b) for an example where the third harmonic (dashed line) reduces the swing from the fundamental component.

The maximum input signal of a nonlinear device can be specified, for example, as the input amplitude X for which the gain is compressed by ≤ 1 dB ($A_{LS}/A = 0.89$). This amplitude is known as the 1-dB gain compression point.

D.2 Harmonic Distortions

Another measure of nonlinearity is provided by the *harmonic distortions*. As shown in Fig. D.2, we take the input signal as a sine wave $x(t) = X\sin(2\pi ft)$ with amplitude X and frequency f. For the illustration we assumed an odd-order, compressive nonlinearity that results in a soft clipped sine wave at the output (fat solid line). This output signal can be decomposed in a fundamental component at f (dashed line) and a harmonic distortion product at $3f$ (thin solid line). The ratio of the 3rd-order distortion product to the fundamental component is the 3rd-order harmonic distortion HD_3.

More generally, the nth-order harmonic distortion HD_n is defined as the ratio of the output-signal component (distortion product) at the frequency nf to the fundamental component at f. With Eq. (D.1), we find the most significant output-signal components at the frequencies $2f$ and $3f$ to be $y(t) = A[-\frac{1}{2}a_2X^2\cos(4\pi ft) - \frac{1}{4}a_3X^3\sin(6\pi f) + \ldots]$, assuming a relatively small

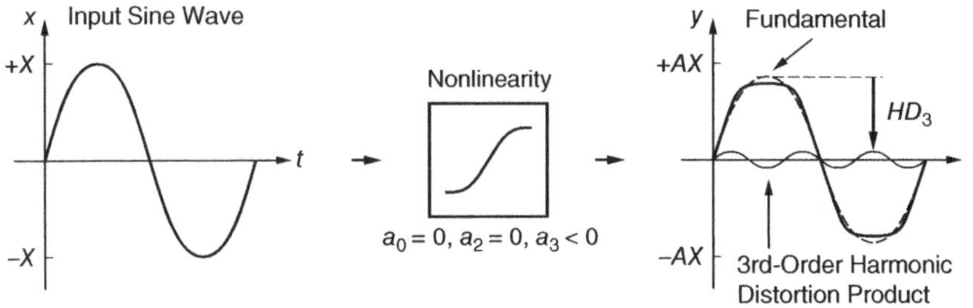

Figure D.2 3rd-order harmonic distortion HD_3.

input signal X (cf. Eq. (I.163) on p. 532). Thus, using AX as an approximation for the fundamental component, we find the following harmonic distortion expressions [2, 3]:

$$HD_2 \approx \frac{1}{2}|a_2|X, \tag{D.4}$$

$$HD_3 \approx \frac{1}{4}|a_3|X^2. \tag{D.5}$$

From these expressions we see that a 1-dB increase in the input signal X causes a 1-dB increase in HD_2 and a 2-dB increase in HD_3. In general, higher-order harmonics depend more strongly on the input signal amplitude: the nth-order harmonic distortion product is proportional to X^n, or equivalently, the nth-order harmonic distortion, HD_n, is proportional to X^{n-1}. In practice, HD_2 and HD_3 are often the only harmonic distortions considered because the higher-order harmonics drop off very rapidly for small signals. Also note that the even-order (odd-order) harmonic distortions originate from the even-order (odd-order) coefficients in the power series. This means that for a differential circuit, which has small even-order coefficients, HD_2 is small compared with HD_3.

The *total harmonic distortion* (THD) describes the nonlinearity with a single number:

$$THD = \sqrt{HD_2{}^2 + HD_3{}^2 + \ldots}. \tag{D.6}$$

The THD can be expressed as a percentage (distortion products as a fraction of the fundamental amplitude) or in dBs using the conversion rule 20 log THD. The maximum input signal of a nonlinear device can be specified, for example, as the input amplitude X for which $THD \leq 1\%$.

D.3 Intermodulation Distortions

In most applications, the input signal to the nonlinear device is not a pure sine wave, but a superposition of multiple sine waves. This means that in addition to the harmonic distortions, we also have to be concerned about *intermodulation distortions*.

Let us start with the simple case of only two tones. We apply a superposition of two equally strong sine waves with frequencies f_1 and f_2 to the input of the nonlinear device: $x(t) = X[\sin(2\pi f_1 t) + \sin(2\pi f_2 t)]$. With Eq. (D.1), we find that the output signal contains two second-order intermodulation products at $f_1 + f_2$ and $|f_1 - f_2|$ and four third-order intermodulation products at $2f_1 + f_2$, $2f_1 - f_2$, $2f_2 + f_1$, and $2f_2 - f_1$. Interestingly, the two second-order products have the same amplitude, and the four third-order products have the same amplitudes. In analogy to the harmonic distortion, we define the intermodulation distortion IMD_n as one of the (equally strong) nth-order distortion products in the output signal normalized to one of the two (equally strong) fundamental tones. We can derive the second- and third-order intermodulation distortions for the two-tone case as [2] (cf. Eq. (I.163))

$$IMD_2 \approx |a_2|X, \tag{D.7}$$

$$IMD_3 \approx \frac{3}{4}|a_3|X^2, \tag{D.8}$$

where X is the amplitude of one of the two (equally strong) tones at the input. As before, the approximations hold for relatively small values of X. Compared with the harmonic distortions in Eqs. (D.4) and (D.5), we find the same dependence on the amplitude X and the power-series coefficients a_i. However, the IMD_2 is twice as strong (+6 dB) as HD_2, and the IMD_3 is three times as strong (+9.5 dB) as HD_3. Besides the intermodulation products, we still have the harmonic distortion products resulting from each tone. Figure D.3 summarizes all the second- and third-order distortion products for the two-tone case. [→ Problem D.1.]

RF engineers who design narrowband systems typically care about the third-order intermodulation products $2f_1 - f_2$ and $2f_2 - f_1$, which fall back into the band of interest (see Fig. D.3). In this situation, the value X for which $IMD_3 = 1$ (extrapolated from $IMD_3(X)$ where $IMD_3 \ll 1$) is a common measure for the linear input range and is known as the *input-referred 3rd-order intercept point* (IIP3).

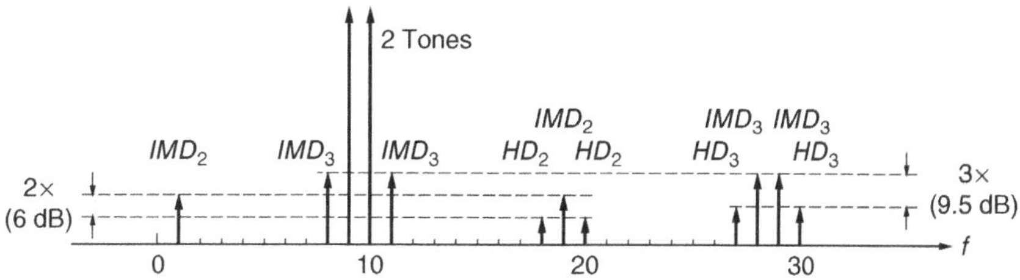

Figure D.3 Second- and third-order distortion products for a two-tone signal with frequencies $f_1 = 9$ and $f_2 = 10$ resulting from a nonlinearity.

Let us add a third tone with the frequency f_3. Again, we get harmonic distortion products for each one of the three tones at the frequencies nf_1, nf_2, and nf_3 for $n = 2, 3, \ldots$. Then, we get second-order intermodulation products at all permutations of $|f_i \pm f_j|$ (6 products in total). Then, we get third-order intermodulation products at all permutations of $2f_i \pm f_j$ (12 products in total). But now we also get third-order intermodulation products at all combinations of $|f_1 \pm f_2 \pm f_3|$ (4 products in total). These products are the so-called *triple beats*. With Eq. (D.1) and sufficient perseverance we can derive the triple-beat distortion for small signals as

$$IMD_{3T} \approx \frac{3}{2}|a_3|X^2. \tag{D.9}$$

This triple-beat distortion is twice as strong (+6.0 dB) as the (double-beat) IMD_3 distortion and six times as strong (+15.6 dB) as the third-order harmonic distortion, HD_3, making it a dominant factor in multicarrier systems.

D.4 Composite Distortions

In multicarrier systems using subcarrier multiplexing (SCM), orthogonal frequency division multiplexing (OFDM), or discrete multitones (DMT), each carrier plays the role of a tone in the aforementioned analysis. All these carriers produce a large number of harmonic and intermodulation products in the presence of a nonlinearity. These distortions are known as *composite distortions*.

Composite Distortions in SCM CATV Systems. Figure D.4 shows part of the spectrum of an SCM CATV signal with 80 TV channels. To measure the impact of nonlinearity on a particular channel, we turn this channel off while keeping all the other channels on. Then, we measure the composite distortion products falling into the bandwidth of the turned-off channel. All channels are tested in this way to find the worst-case channel with the most distortion products.

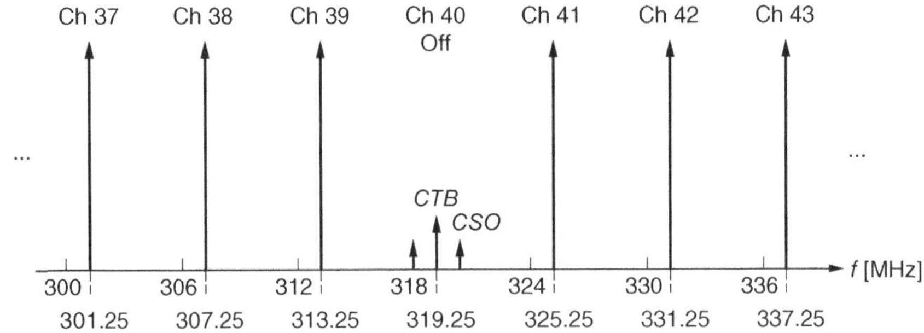

Figure D.4 Composite distortion products falling into channel 40 for an 80-channel SCM CATV system.

In the North American standard cable channel plan, the carriers are spaced 6 MHz apart and are offset by +1.25 MHz from whole multiples of 6 MHz (with some exceptions; cf. Appendix A.4). As a result of this offset, the majority of even-order products fall 1.25 MHz above or 1.25 MHz below the carrier frequencies, whereas the majority of odd-order products fall on the carrier frequencies themselves [4, 5]. Thus, the composite even- and odd-order products have different effects on the picture quality and can be measured separately with the appropriate bandpass filters.

The *composite second-order* (CSO) distortion is defined as the rms value of the distortion products falling into a 30-kHz bandwidth located 1.25 MHz above the carrier frequency, normalized to the carrier amplitude [4]. Most of these distortion products are *second-order* intermodulation products, hence the name composite second order. The second-order products falling 1.25 MHz *below* the carrier frequency usually are ignored because they have little impact on the picture quality (they fall into the gap between channels). The *composite triple-beat* (CTB) distortion is defined as the rms value of the distortion products falling into a 30-kHz bandwidth centered on the carrier frequency, normalized to the carrier amplitude [4]. Most of these distortion products are *triple-beat* intermodulation products, hence the name composite triple beat. The CSO and CTB distortions are expressed in dBc, that is, dB relative to the carrier amplitude, using the 20 log CSO and 20 log CTB conversion rules, respectively.[1]

The CSO and CTB distortions can be calculated from the second-order and triple-beat intermodulation distortions if the number of contributing intermodulation products is known. Assuming carriers with random phases, we can add the power of the individual intermodulation products. Further assuming equal-power carriers, we can write

$$CSO = \sqrt{N_{CSO}} \cdot IMD_2 \approx \sqrt{N_{CSO}} \cdot |a_2|X, \tag{D.10}$$

$$CTB = \sqrt{N_{CTB}} \cdot IMD_{3T} \approx \sqrt{N_{CTB}} \cdot \frac{3}{2}|a_3|X^2, \tag{D.11}$$

where N_{CSO} is the number of second-order intermodulation products falling on the CSO frequency (carrier frequency + 1.25 MHz) of the turned-off channel and N_{CTB} is the number of triple-beat products falling on the carrier frequency of the turned-off channel. These beat counts can be fairly high; for example, in an 80-channel SCM CATV system, the maximum N_{CSO} is about 30 and occurs for the channel at the upper end of the spectrum, whereas the maximum N_{CTB} is about 2,000 and occurs for the channel in the middle of the spectrum [4]. The maximum beat counts for an N-channel system can be estimated as $N_{CSO}(\max) \approx \frac{1}{2}N$ and $N_{CTB}(\max) \approx \frac{3}{8}N^2$.

1 Alternatively, CSO and CTB can be defined as *power* ratios (CSO^2, CTB^2) instead of amplitude ratios (CSO, CTB); the dBc values are then calculated as $10 \log CSO^2$ and $10 \log CTB^2$, which are identical to $20 \log CSO$ and $20 \log CTB$.

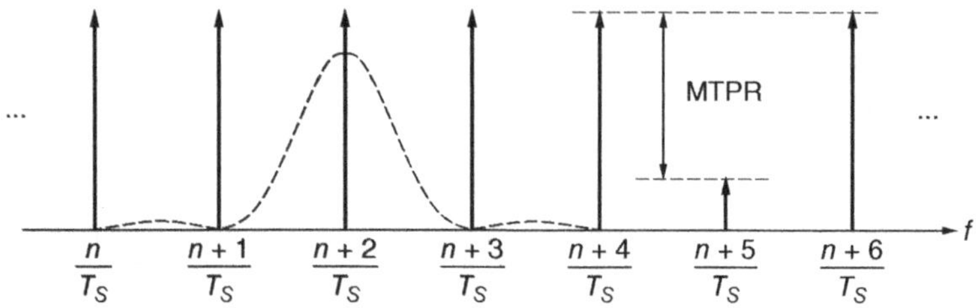

Figure D.5 Composite distortion products falling onto carrier $n + 5$ of an OFDM or DMT signal.

The Federal Communications Commission (FCC) regulates that both CSO and CTB must be ≤ -51 dBc for analog TV signals; however, it is common industry practice to design broadband networks for a CSO and CTB of about -53 dBc [4].

Composite Distortions in OFDM or DMT Systems. Figure D.5 shows part of the spectrum of an OFDM or DMT signal. The main difference to the SCM CATV signal in Fig. A.8(b) is that there is no guard band between adjacent channels. In fact, the spectra of the QAM modulated carriers do overlap. To make this possible without causing interference, the carriers must be chosen to be *orthogonal*. This condition is satisfied when the carrier spacing is a multiple of the QAM symbol rate $1/T_S$ [6, 7]. In Fig. D.5, the modulation side bands of carrier $n + 2$ are sketched to illustrate the orthogonality condition.

To measure the impact of nonlinearity on a particular carrier, this carrier is turned off while all the other carriers are kept on (see Fig. D.5). The carriers are unmodulated and their phases are chosen such that the overall signal has the desired peak-to-average ratio (PAR). Then, the ratio between the rms value of the carrier and the rms value of all distortion products falling into the $1/T_S$ frequency band centered on the turned-off carrier is measured. This ratio is known as the *multitone power ratio* (MTPR) [8, 9]. No distinction between even and odd-order products is made. The MTPR is expressed in dB using the conversion rule 20 log *MTPR*. All carriers are tested in this way to find the carrier with the worst MTPR.

The composite distortions can be estimated from the individual intermodulation products and their count, as discussed for the SCM signal. Alternatively, if the multicarrier signal can be described by a Gaussian amplitude distribution, statistical methods can be used to estimate the variance of the composite distortions [10].

The MTPR requirement depends, among other factors, on the constellation size of the QAM signal modulated on the individual carriers. For every doubling of the constellation size (addition of one bit per symbol), the required MTPR increases by about 3 dB. The ADSL standard [8] requires that the MTPR at the output of the transmitter must be better (larger) than 20 dB + 3 dB $\cdot$ N, where N is the number of bits per symbol. For example, 64-QAM requires an MTPR of 38 dB, whereas 32,768-QAM requires 65 dB.

In another approach to measure the impact of nonlinearity on an OFDM or DMT signal, a vector signal generator (VSG) generates a clean signal (all carriers are on and modulated), which is applied to the input of the nonlinear device. A vector signal analyzer (VSA) demodulates the output signal and measures the ratio between the rms value of the error in the constellation points and the rms value of the ideal constellation points. This ratio is known as the *error vector magnitude* (EVM) [11]. The rms error value is determined over many symbols and all carriers. The EVM can be expressed as a percentage, 100% $\cdot$ EVM, or in dB using the conversion rule 20 log EVM.

The EVM requirement, like the MTPR requirement, depends on the QAM constellation size. For example, the Wi-Fi standard [12] requires that the EVM at the output of the transmitter must be better (smaller) than −13 dB for QPSK (4-QAM), −19 dB for 16-QAM, and −25 dB for 64-QAM (code rate $r = 3/4$). Note that nonlinearity is only one of several mechanisms degrading the constellation.

Problems

D.1 **Second-Order Distortions.** Given the nonlinear transfer function in Eq. (D.1) with $a_i = 0$ for $i \geq 3$ and the two-tone input signal $x(t) = X[\sin(\omega_1 t) + \sin(\omega_2 t)]$, calculate the output signal. How large are the distortion products and the output offset?

References

1. R. G. Meyer, M. J. Shensa, and R. Eschenbach. Cross modulation and intermodulation in amplifiers at high frequencies. *IEEE J. Solid-State Circuits*, SC-7(1):16–23, 1972.
2. T. H. Lee. *The Design of CMOS Radio-Frequency Integrated Circuits*. Cambridge University Press, Cambridge, UK, 2nd edition, 2003.
3. P. R. Gray, P. J. Hurst, S. H. Lewis, and R. G. Meyer. *Analysis and Design of Analog Integrated Circuits*. John Wiley &Sons, Inc., New York, 5th edition, 2009.
4. W. Ciciora, J. Farmer, D. Large, and M. Adams. *Modern Cable Television Technology: Video, Voice, and Data Communications*. Morgan Kaufmann, San Francisco, CA, 2nd edition, 2004.
5. M. M.-K. Liu. *Principles and Applications of Optical Communications*. Irwin, McGraw-Hill, Chicago, 1996.

6 G. P. Agrawal. *Fiber-Optic Communication Systems*. John Wiley &Sons, Inc., Hoboken, NJ, 4th edition, 2010.
7 X. Chen, A. A. Amin, A. Li, and W. Shieh. Multicarrier optical transmission. In I. P. Kaminow, T. Li, and A. E. Willner, editors, *Optical Fiber Telecommunications VIB*, pages 337–380. Academic Press, 2013.
8 ANSI. Network and customer installation interfaces – asymmetric digital subscriber line (ADSL) metallic interface ANSI T1.413-1998. American National Standards Institute, Washington, DC, June 1998.
9 Intersil Corporation. Analog amp linearity characterization via probability weighted multitone power ratio testing. Intersil Application AN9718, April 1997.
10 D. Tsonev, S. Sinanovic, and H. Haas. Complete modeling of nonlinear distortion in OFDM-based optical wireless communication. *J. Lightwave Technol.*, LT-31(18):3064–3076, 2013.
11 M. D. McKinley, K. A. Remley, M. Myslinski, J. S. Kenney, D. Schreurs, and B. Nauwelaers. EVM calculation for broadband modulated signals. In *ARFTG Microwave Measurement Conference*, December 2004.
12 IEEE. 802.11a. Part 11: Wireless LAN medium access control (MAC) and physical layer (PHY) specifications – high-speed physical layer in the 5 GHz band. IEEE Standards Board, Piscataway, NJ, 1999.

E

Adaptive Equalizers

The adaptive equalizer has become an important building block, which is used in almost every communication system, be it for optical, wire-line, or wireless communication. In the following, we discuss the basics of the feedforward and decision-feedback equalizers. How do they cancel intersymbol interference? How do they adapt to the incoming signal? What are the implementation challenges?

Intersymbol Interference Canceler. The signal at the output of the receiver's linear channel invariably contains some intersymbol interference (ISI). This ISI is caused, among other things, by dispersion in the optical fiber (modal, chromatic, and polarization-mode dispersion) as well as the frequency response of the linear channel itself.

ISI can be mitigated with an *ISI canceler*. The effectiveness of the ISI canceler depends on whether the ISI is caused by a linear or a nonlinear distortion. In the linear case, the ISI canceler can apply the inverse transfer function and cancel the ISI almost perfectly. In the nonlinear case, cancellation is much harder. For this reason, a coherent transmission system, which is characterized by a mostly linear transfer function, can be compensated much better than a direct-detection system, which contains a square-law detector (cf. Chapter 2).

In many cases, the transfer function that causes the ISI is not known ahead of time and may even change over time (e.g., in the case of polarization-mode dispersion). Thus, an ISI canceler that automatically *adapts* to the ISI in the incoming signal is very desirable.

The most powerful ISI canceler is the *maximum likelihood sequence estimator* (MLSE). This ISI canceler makes use of a parameterized channel model. The most likely transmitted bit sequence is estimated based on the received signal and the channel model. The sequence estimation can be implemented efficiently with a *Viterbi decoder*. An adaptive MLSE additionally estimates and updates the model parameters based on the incoming signal. The implementation of an MLSE at high bit rates, however, is challenging due to the complexity of the algorithm [1–3].

Analysis and Design of Transimpedance Amplifiers for Optical Receivers, First Edition. Eduard Säckinger.

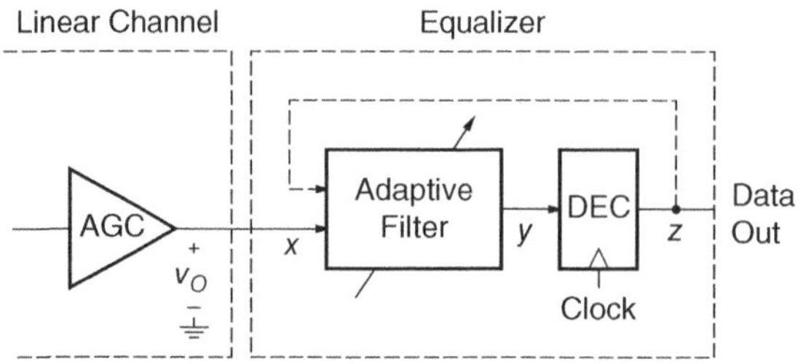

Figure E.1 The linear channel of Fig. 4.1 followed by an adaptive equalizer.

A less demanding ISI canceler is the *equalizer*. Figure E.1 shows the linear channel of Fig. 4.1 followed by an adaptive equalizer. The adaptive filter attempts to undo the distortions introduced by the channel (channel-response inversion) and the decision circuit makes bit-by-bit decisions. The filter may have a second input to receive *decision feedback* from the decision circuit (dashed line). We discuss equalizers with and without this feedback path shortly. Equalizers have been applied successfully to extend the reach of optical links ranging from short multimode-fiber data links to ultra-long-haul transmission systems [4–8].

In the following we provide a couple of simple examples to illustrate how equalizers cancel ISI and how they adapt to varying signal conditions.

E.1 Feedforward and Decision-Feedback Equalizers

Precursor and Postcursor ISI. In Section 4.8, we introduced the concept of precursor and postcursor ISI. Let us review this idea before discussing the equalizers. Figure E.2(top) shows the eye diagram of the input signal (node x in Fig. E.1)

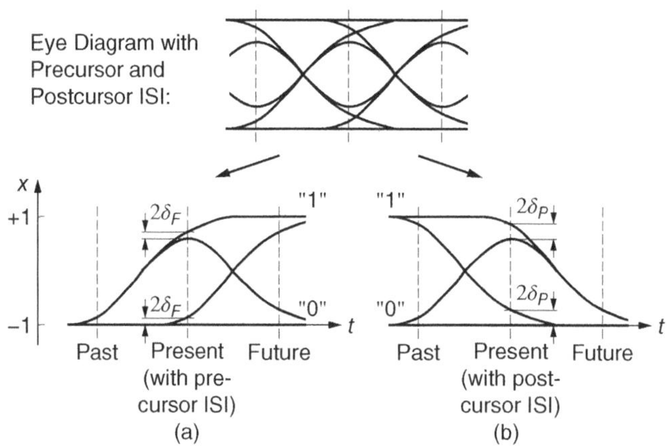

Figure E.2 (a) Precursor and (b) postcursor ISI at the input of the equalizer.

with a moderate amount of ISI. For clarity, the eye traces that are relevant to the analysis of precursor and postcursor ISI are redrawn below. From the traces in Fig. E.2(b) on the right, we see how the past bit influences the signal levels of the present bit. If the past bit is a one, the signal levels of the present bit are slightly shifted *upward* compared with when the past bit is a zero. This shift represents the postcursor ISI and is marked in Fig. E.2(b) with $2\delta_P$. For the subsequent analysis we assume that zeros are represented by $x = -1$ and the ones by $x = +1$ resulting in a signal swing of 2; thus, the postcursor ISI normalized to the swing is simply δ_P. Similarly, from the traces in Fig. E.2(a) on the left, we see how the future bit influences the signal levels of the present bit. If the future bit is a one, the signal levels of the present bit are slightly shifted upward (by $2\delta_F$) compared with when the future bit is a zero. This shift represents the precursor ISI. At first, this may sound like a violation of causality, but because a typical transmission system has a latency of many bits, precursor ISI is possible.

The complete eye diagram in Fig. E.2(top) exhibits precursor ISI, postcursor ISI, and the combination of the two. More generally, precursor ISI may originate from any bit in the future, not just the first one. Similarly, postcursor ISI may originate from any bit in the past. Typically, the stronger the ISI is, the more past and future bits are involved in its generation.

Feedforward Equalizer. Let us start with the *feedforward equalizer* (FFE), which, as its name implies, does not have a feedback path. This equalizer is also known as a *transversal filter* or a *finite impulse response* (FIR) *filter*. Figure E.3 shows the block diagram of an FFE with three taps followed by a decision circuit. The input signal x is fed into a delay line with three taps. Each delay corresponds to one bit interval. The signals at the taps are multiplied by the coefficients c_1, c_2, and c_3, also known as tap weights, and summed up. The resulting equalized signal y is then sliced and retimed in the decision circuit yielding the digital output signal z.

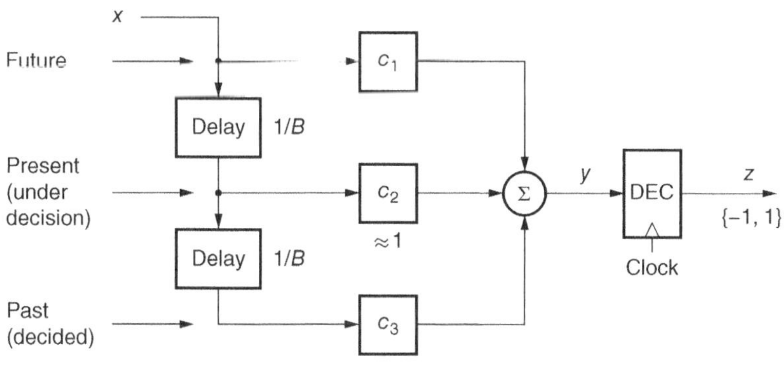

Figure E.3 Feedforward equalizer with three taps.

How does this FFE cancel ISI? Let us assume that initially the filter coefficients are set to $c_1 = 0$, $c_2 = 1$, and $c_3 = 0$. In this case, the output signal y is just a time-delayed copy of the input signal x. But now we can see an opportunity to cancel the ISI: at the first tap we have an estimate of the future bit and at the third tap we have an estimate of the past bit. If we know the value of δ_F we can remove the precursor ISI (at the instant of sampling) by adding the correction term $-\delta_F \times$ ***future bit*** to the signal (remember, the bits are represented by the values $\{-1, +1\}$ rather than the usual $\{0, 1\}$). Thus, setting $c_1 = -\delta_F$ will approximately cancel the precursor ISI. Similarly, if we know the value of δ_P we can remove the postcursor ISI (at the instant of sampling) by adding the correction term $-\delta_P \times$ ***past bit*** to the signal. Thus, setting $c_3 = -\delta_P$ will approximately cancel the postcursor ISI. Note that the signals from the delay line contain ISI and noise and thus are only estimates of the true future and past bit values. For this reason the ISI cancellation with the aforementioned tap weights is only approximate, but it could be improved by fine-tuning the three weights with an adaptation algorithm. [→ Problem E.1.]

Most practical FFEs have more than three taps. The length of the delay line that is necessary for good ISI cancellation depends on the duration of the channel's single-bit response (cf. Section 4.8): channels with high dispersion require long equalizers.

An equalizer with a tap spacing of one bit period ($T = 1/B$), as shown in Fig. E.3, is known as a *synchronous equalizer*. Such an equalizer only cancels the ISI at the sampling instant. In other words, this equalizer improves the vertical eye opening, but does not help with the horizontal eye opening, in fact, it may even make it worse. Although a large vertical eye opening is our primary goal, a reduced horizontal eye opening poses a problem if the sampling clock is offset from its ideal location or contains jitter. In order to enlarge the vertical and horizontal eye openings simultaneously, we can reduce the tap spacing from T to $T/2$ and double the number of taps, thus keeping the equalizer's time span the same. The ISI at the center of the bit is canceled in the same way as before, but now the newly inserted half-bit taps cancel the ISI at the bit edges as well [9]. An equalizer with a tap spacing of less than one bit period is known as a *fractionally spaced equalizer* (FSE).

A disadvantage of the FFE is the so-called *noise enhancement* [10]. For example, in trying to compensate for a channel with a low-pass response, the equalizer adapts to a high-pass response, which has the undesirable side effect of amplifying (or enhancing) high-frequency noise. The result is a reduced signal-to-noise ratio (SNR).

Multipath channels (e.g., fiber with first-order PMD) are hard to equalize with an FFE [11, 12]. In the frequency domain, these channels are characterized by deep notches in the amplitude response. When inverted, these notches lead to noise enhancement. The decision-feedback equalizer, which we discuss next, does a much better job on such channels [11, 13].

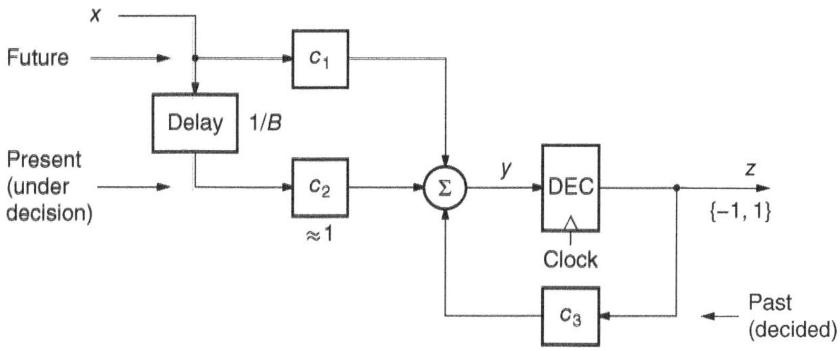

Figure E.4 Decision-feedback equalizer with two feedforward and one feedback tap.

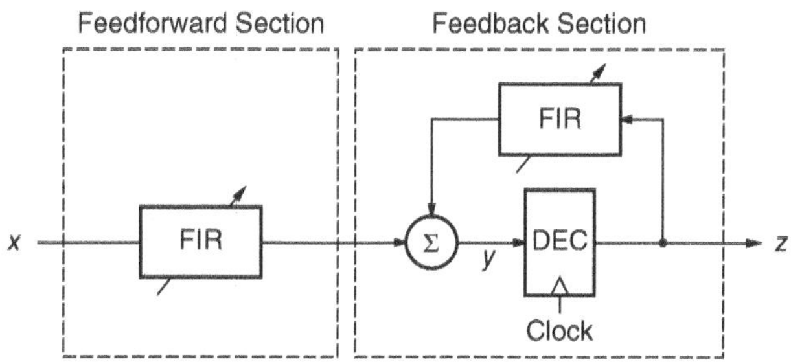

Figure E.5 General architecture of a decision-feedback equalizer.

Decision-Feedback Equalizer. Let us turn to the more powerful and often used *decision-feedback equalizer* (DFE). Figure E.4 shows the block diagram of a DFE with three taps. Of the three taps the two connecting to the input delay line are feedforward taps, and the one connecting to the decision-circuit output is a feedback tap. The feedforward taps work just like in the FFE, but the feedback tap is new. The general DFE consists of two FIR filters, a feedforward filter, which is fed by the input signal x, and a feedback filter, which is fed by the output of the decision circuit z, as shown in Fig. E.5.[1] The output signal of the equalizer, node y, is the sum of the output signals from the two FIR filters.

How does the DFE cancel ISI in our three-tap example? Comparing Fig. E.4 with Fig. E.3, we see that the precursor ISI is removed in the same way as in the FFE. However, the postcursor ISI is now removed using the output from the decision circuit instead of the output from the delay line. This way of estimating the past bit has the advantage that it is free of ISI and noise and, in fact, represents the precise bit value, assuming that the past decision was correct. Thus, if we set $c_3 = -\delta_p$, the postcursor ISI is canceled exactly and no noise is

1 Here we use the term *DFE* for the combination of the feedforward and feedback sections. Some authors use the term *DFE* for the feedback section *only*.

injected through c_3. But there is also a weakness: If a decision happens to be wrong, the equalizer adds *more* ISI to the output signal y, possibly causing further decision errors. This effect is known as *error propagation*. Nevertheless, in most cases the advantages of the DFE (noise and ISI free past bit estimates) outweigh this disadvantage [13].

There is another way of looking at the DFE, namely as an FFE followed by a decision circuit that dynamically adjusts its decision threshold in response to the "bit history." From Fig. E.2(b), we see that the optimum decision threshold is slightly above or below the mid point, depending on whether the past bit was a one or a zero. This suggests that we should use feedback from the decided bit to control the threshold level. Tap weight c_3 together with the adder at the input of the decision circuit performs precisely this function. Of course, this "dynamic threshold" view is equivalent to the "postcursor cancellation" view presented earlier.

As in the case of the FFE, most practical DFEs have more than three taps so that they can equalize channels with long single-bit responses.

Because the feedforward section of the DFE removes the precursor ISI and the feedback section removes the postcursor ISI, these two sections are also known as *precursor equalizer* and *postcursor equalizer*, respectively.

In contrast to the FFE, the DFE is a *nonlinear* equalizer because the decision circuit is part of the equalizer structure.

E.2 Adaptation Algorithms

How can we (or rather the adaptive equalizer) find the filter coefficients $c_1, c_2, \ldots$ that result in the least ISI at node y, or more generally, result in the best receiver performance? And how can we make these coefficients adapt to an input signal with time-varying ISI? To answer these questions we first need to introduce the idea of a *cost function*.

Cost Function. The cost function defines how well the receiver performs for every possible combination of coefficient values. Good performance corresponds to a low cost; bad performance corresponds to a high cost. This function tells us if we are "hot" (low cost) or "cold" (high cost) while searching the coefficient space for the optimum performance. Here are some examples of costs functions and how to measure them:

- Measure the vertical or horizontal eye opening of the equalized signal (node y) with a so-called *eye monitor*. The complement of the eye opening, the eye closure, can serve as a cost function. The vertical eye closure is a measure of ISI and noise and the horizontal eye closure is a measure of pulse-width distortion and jitter (cf. Appendix B). For the implementation of eye monitors see [14–16] and Appendix F.

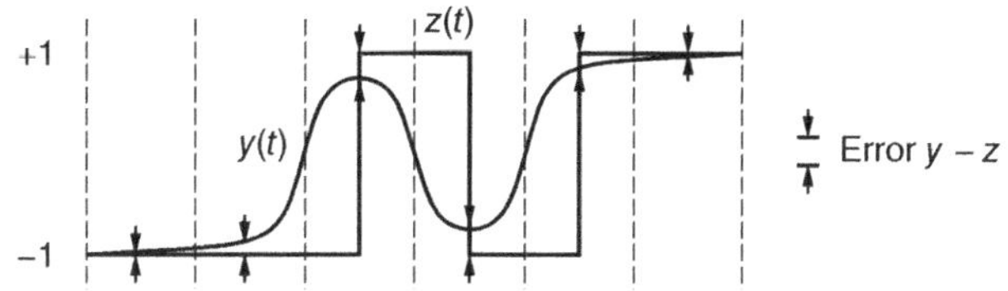

Figure E.6 Error between the equalized signal, y, and the decision-circuit signal, z.

- Measure the difference between the equalized signal (node y) and the (appropriately scaled) signal from the decision circuit (node z). The difference, $y - z$, at the sampling instant is illustrated in Fig. E.6 and is known as the *error*. In the absence of ISI and noise, the error is zero. Squaring and averaging the error, $\overline{(y - z)^2}$, yields the *mean-square error* (MSE), a popular cost function that measures ISI and noise jointly. We discuss an adaptation algorithm based on this cost function shortly.
  Cost functions based on the error, $y - z$, are only meaningful if the eye is sufficiently open such that most decisions, z, are correct. If this is not the case, we can use the modified error $y - z'$, where z' represents bits of an *a priori* known training sequence. Whereas an equalizer using z for adaptation is said to operate in the *decision-directed mode*, an equalizer using z' for adaptation is said to operate in the *training mode*.
- Measure the PSD of the equalized signal (node y) and compare it with the PSD of the desired signal (node z). The absolute difference between the two PSDs can serve as a cost function [17–19].
- Measure the BER at the receiver. In a system with checksums or forward error correction (cf. Appendix G), bit errors can easily be detected and counted. The BER can serve directly as a cost function.
  The BER-based cost function has the advantage of being the most relevant performance measure; the other cost functions (eye closure, MSE, etc.) are only proxies for the BER [20]. In many systems, however, bit errors occur very infrequently, resulting in a slow adaptation when this cost function is used.

Exhaustive Search. Given the cost function, we now have to find a way to optimize the coefficients, c_i, of the equalizer. In the case of only two coefficients, c_1 and c_2, we can visualize the cost function as a warped surface over the two-dimensional coefficient space, as shown in Fig. E.7. Our job is to find the location (c_1, c_2) in the bottom plane for which the surface is lowest.

The simplest approach is to conduct an *exhaustive search*. If there are only a few coarsely quantized coefficients, say 2 coefficients with 4 settings each, the cost function can be evaluated for all 4×4 settings to find the optimum. If there are too many coefficient combinations, however, this method becomes impractical.

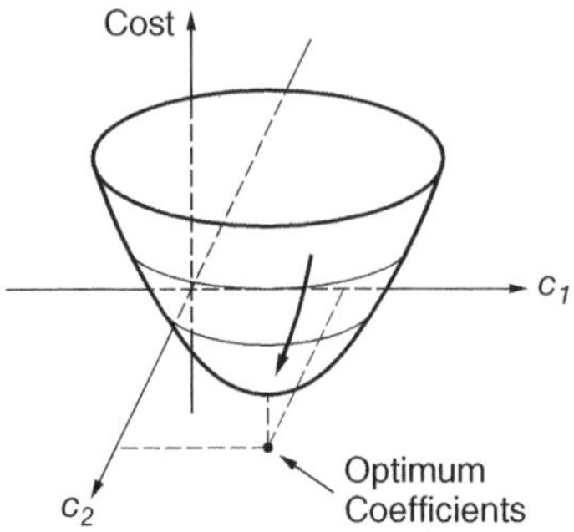

Figure E.7 Visualization of a cost function for the case of two coefficients.

Gradient Descent. Another approach to find the minimum is to start at a convenient point on the cost surface and follow the gradient, or an approximation thereof, in the downhill direction (arrow in Fig. E.7). If the cost function is well behaved and we take small steps, we will eventually arrive at the optimum coefficients (c_1, c_2). This method is known as *gradient descent* or *steepest descent*.

The gradient of the cost function can be estimated in a number of ways. In one procedure, called *weight perturbation* (or *dithering*), each coefficient (weight) is perturbed slightly (in our two-coefficient example by Δc_1 and Δc_2), one by one, and its effect on the cost is recorded. At the end of this somewhat tedious procedure, we have an estimate of the gradient (in our example, the gradient is $[\Delta\text{cost}_1/\Delta c_1, \Delta\text{cost}_2/\Delta c_2]$). Now, we can update the coefficients by taking a small step in the direction of the negative gradient:

$$c_i \leftarrow c_i - \mu_s \frac{\Delta\text{cost}_i}{\Delta c_i}, \tag{E.1}$$

where μ_s is the adaptation step size of the equalizer, typically a small number such as 10^{-6}. This method can find the gradient of any cost function but, unfortunately, is rather slow.

There are two other well-known adaptation algorithms that are faster than weight perturbation because they estimate the gradient in a more efficient way: the *least-mean-square (LMS) algorithm* and the *zero-forcing (ZF) algorithm*. These algorithms, which we outline in the following, were pioneered by Widrow and Hoff [21] and Lucky [22] in the early 1960s. For an entertaining story of how Bob Lucky invented the zero-forcing algorithm on his way home from work while waiting at a traffic light in Red Bank, New Jersey, see [23].

Least-Mean-Square (LMS) Algorithm. The LMS algorithm is an elegant and efficient procedure to minimize the MSE. This algorithm updates the tap weights of an FFE as follows [10]:

$$c_i \leftarrow c_i - \mu_s (y - z) x_i, \tag{E.2}$$

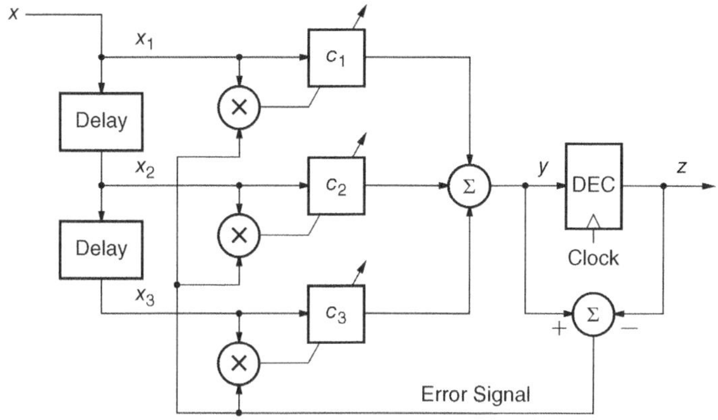

Figure E.8 An adaptive three-tap FFE controlled by the LMS algorithm.

where $y - z$ is the error introduced earlier and x_i is the signal from the ith delay-line tap feeding weight c_i. Figure E.8 shows our three-tap FFE example from Fig. E.3 extended with the blocks needed for LMS adaptation.

The adaptation rule Eq. (E.2) can be understood as follows. The square error is $(y - z)^2$ and its gradient with respect to the weights, c_i, follows as $2(y - z)x_i$ (remember that $y = \sum_i c_i x_i$). Now we see that the rule in Eq. (E.2) follows this (negative) gradient in every adaptation step, the size of which is controlled by μ_s. Taking many small steps down the gradient of the (instantaneous) square error, on average, is equivalent to taking one larger step down the gradient of the *mean*-square error. In other words, the LMS rule performs a *stochastic* gradient descent on the MSE surface.

Zero-Forcing (ZF) Algorithm. The ZF algorithm is similar to the LMS algorithm, but whereas the LMS algorithm minimizes ISI and noise jointly, the ZF algorithm minimizes ISI by itself. Thus, the ZF algorithm finds a solution similar to the raised-cosine filtering approach (cf. Section 4.9). The weight update rule for the ZF algorithm is similar to Eq. (E.2); however, instead of correlating the error with the analog signals from the input delay line, the error is now correlated with digital signals from after the decision circuit [10]:

$$c_i \leftarrow c_i - \mu_s(y - z)z_i. \tag{E.3}$$

The corresponding block diagram is shown in Fig. E.9 [11]. The delayed digital signals, z_i, can be obtained easily with a shift register. [→ Problem E.2.]

The main attraction of the ZF algorithm is that the error signal is multiplied by binary (digital) signals rather than continuous-valued (analog) signals. This feature simplifies the hardware implementation of the ZF algorithm over the LMS algorithm, especially in analog high-speed equalizers. However, because

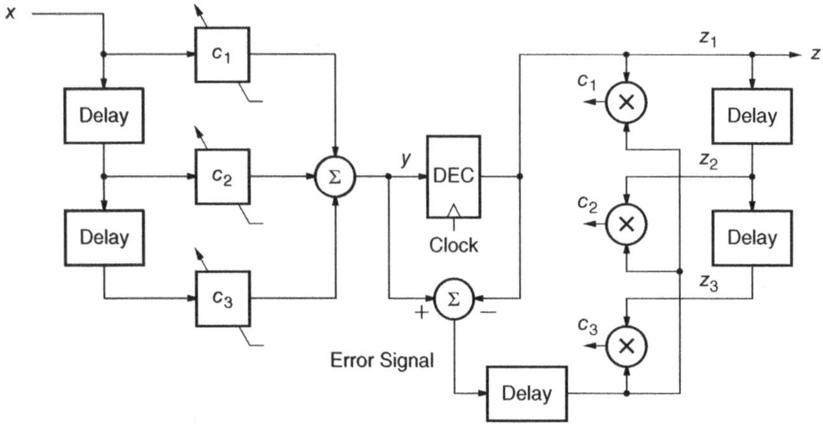

Figure E.9 An adaptive three-tap FFE controlled by the ZF algorithm.

noise is ignored by the underlying cost function, the frequency response found by the ZF algorithm may result in undesirable noise enhancement.

E.3 Hardware Implementations

ISI cancelers can be implemented in the digital or the analog domain. Digital implementations are more amenable to complex algorithms, such as MLSE and equalizers with many taps. Analog implementations can run at higher speeds or consume less power, but generally are limited to a small number of taps (for time-domain equalizers such as FFEs and DFEs) or a small number of poles and zeros (for frequency-domain equalizers).

Digital Implementations. An MLSE for compensating dispersion in multimode fibers (10-Gigabit Ethernet LRM) [1], an MLSE for compensating chromatic dispersion in single-mode fibers [2], and an equalizer for compensating chromatic dispersion and PMD in a coherent DP-QPSK receiver [5] are some of the digital implementation examples found in the literature. The latter design consists of a 256-tap equalizer implemented in the frequency domain (using FFT and inverse FFT) for bulk chromatic dispersion compensation and a 16-tap $T/2$ FFE with multiple inputs and multiple outputs (MIMO) for PMD compensation. For more information on digital equalizers for coherent receivers, see [7, 24].

The biggest challenge for a digital implementation is the analog-to-digital converter (ADC), which samples and digitizes the signal from the linear channel. For a 10-Gb/s NRZ receiver, a converter with about 5 bits of resolution sampling at 10 GHz is needed [25, 26]. For a 100-Gb/s DP-QPSK receiver, four converters with 6 to 8 bits of resolution sampling at 56 to 65 GHz are needed [27]. To alleviate the speed requirements of the ADC, multiple lower-speed converters can be operated in parallel using time interleaving. Nevertheless, the track-and-hold circuit at the input still needs to operate at the full speed. For examples of high-speed data converters and track-and-hold circuits, see [26–31].

Analog Implementations. The challenge in designing analog time-domain equalizers is to come up with delay elements, multipliers, and summers that have enough bandwidth and precision over process, supply voltage, and temperature. For examples of analog equalizer circuits, see [3, 8, 32–34].

The delay element for a continuous-time FFE can be implemented, for example, with a unity-gain stage [35] or a cascade of unity-gain stages [25, 36] for larger delays. A delay-locked loop (DLL) can be used to tune the delay to a precise value [25]. Alternatively, the delay element can be implemented with on-chip artificial transmission lines whose delay is fairly well controlled over process, voltage, and temperature [8, 34, 37, 38]. Finally, the delay line can be implemented with an array of sample-and-hold circuits [39, 40].

The tap weight can be implemented with an analog multiplier or a variable-gain amplifier (VGA) [8]. If the coefficients are digital, a programmable-gain amplifier (PGA) or a multiplying digital-to-analog converter (MDAC) can be used [35, 36]. In either case, a tap weight with a current output has the advantage of permitting easy summation by means of Kirchhoff's current law.

DFE Loop Unrolling. The maximum speed at which a DFE can operate usually is limited by the delay through the feedback loop consisting of the (feedback) FIR filter, the summation node, and the decision circuit (cf. Fig. E.5). Fortunately, this loop can be eliminated by exploiting parallel processing, as shown in Fig. E.10 for the example of a single feedback tap. Two parallel decision circuits are used: one is slicing for the case that the past bit was a zero and the other one

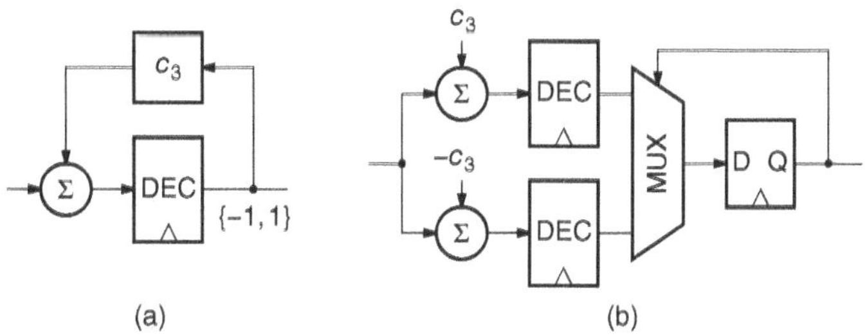

Figure E.10 DFE with a single feedback tap: (a) conceptual representation and (b) high-speed implementation.

is slicing for the case that the past bit was a one. Then, a multiplexer in the digital domain selects which result to use based on the previous decision. This neat trick goes by several names: *speculation, look ahead,* and *loop unrolling*. Note that the equalized signal, which appeared at the input of the decision circuit in Fig. E.10(a), no longer appears at any single node in Fig. E.10(b).

Speculation can be extended to additional feedback taps. Speculating on n taps, requires 2^n parallel decision circuits, each one with its particular threshold level. When speculating on all feedback taps, the DFE turns into an ADC with nonuniform quantization levels (the array of decision circuits) followed by a digital datapath (muxes and flip-flops) [41].

A DFE with four feedback taps, where the first tap is unrolled and the remaining three taps are implemented directly, is described in [42]. For more on high-speed DFE techniques, see [43–45].

Adaptation. As we know, the implementation of the ZF algorithm is simpler than the LMS algorithm. The ZF implementation described in [46], for example, digitizes the analog output signal of the equalizer, y, with a 6-bit ADC permitting the correlation of the error with the recovered bits, z_i, to be done in the digital domain. In [47] a switched-capacitor circuit is used to compute the error and its correlation with the recovered bits.

The implementation of the LMS or ZF algorithms can be simplified further, albeit at a loss of performance, by replacing continuous-valued signals with just their sign values (+1 or −1), thus making them binary signals. The resulting algorithms are called the *sign–sign least-mean-square algorithm* and the *modified zero-forcing algorithm,* and their weight update rules are $c_i \leftarrow c_i - \mu_s \text{sgn}(y-z)\text{sgn}(x_i)$ and $c_i \leftarrow c_i - \mu_s \text{sgn}(y-z) z_i$, respectively [10]. With these modifications all multiplications reduce to simple XOR operations. The 4-tap DFE in [48], for example, uses this approach together with a mixed-signal integrator to implement adaptation. The frequency-domain equalizer with a 1-tap DFE in [49] uses a sign-based ZF algorithm.

Clock Recovery. A related issue is the implementation of the clock-recovery circuit for receivers with an equalizer. Ideally, we would like to extract the clock signal from the equalized data signal, but for a sampled-data equalizer, the equalized signal *depends* on the clock and its phase relationship to the incoming signal. If the sampling clock is offset from the center of the eye, the equalizer cannot do a good job.

A simple solution is to perform clock recovery on the unequalized signal, but this approach is limited to signals with little ISI. A more sophisticated solution is given by the following two-step start-up process [42]. At first, the equalizer is disabled such that the clock-recovery circuit sees the unequalized signal. The clock-recovery circuit acquires lock on this raw signal. Then, the equalizer is enabled and coefficient adaptation takes place. While the equalizer adapts, the clock-recovery circuit fine adjusts the clock phase, eventually arriving at a joint optimum. To suppress unwanted interactions between the phase adjustment of the clock-recovery circuit and the coefficient adjustments of the equalizer, the latter process should be made much slower than the former.

Problems

E.1 **Feedforward Equalizer.** Calculate the frequency response of the three-tap feedforward equalizer shown in Fig. E.3, assuming $c_2 = 1$. How large is the magnitude of this frequency response at DC and $f = B/2$?

E.2 **Equalizer Tap Weights.** A distorted NRZ signal at the input of an equalizer has the following sampled values: +1.0, +1.0, +0.9, −0.8, −1.0, −0.9, +0.7, −0.7, +0.8, +0.9, −0.8, −1.0. (a) How much pre- and postcursor ISI does this sequence contain? (b) Given the three-tap FFE of Fig. E.3, what weight values equalize this sequence optimally? (c) Given the three-tap DFE of Fig. E.4, what weight values equalize this sequence optimally? Tip: Use a spreadsheet to simulate the equalizers and use gradient descent to find the optimum weights.

References

1 O. E. Agazzi, M. R. Hueda, D. E. Crivelli, H. S. Carrer, A. Nazemi, G. Luna, F. Ramos, R. López, C. Grace, B. Kobeissy, C. Abidin, M. Kazemi, M. Kargar, C. Marquez, S. Ramprasad, F. Bollo, V. Posse, S. Wang, G. Asmanis, G. Eaton, N. Swenson, T. Lindsay, and P. Voois. A 90 nm CMOS DSP MLSD transceiver with integrated AFE for electronic dispersion compensation of multimode optical fibers at 10 Gb/s. *IEEE J. Solid-State Circuits*, SC-43(12):2939–2957, 2008.

2 H.-M. Bae, J. B. Ashbrook, J. Park, N. R. Shanbhag, A. C. Singer, and S. Chopra. An MLSE receiver for electronic dispersion compensation of OC-192 fiber links. *IEEE J. Solid-State Circuits*, SC-41(11):2541–2554, 2006.

3 S. Elahmadi, M. Bussmann, D. Baranauskas, D. Zelenin, J. Edwards, K. Tran, L. F. Linder, C. Gill, H. Tan, D. Ng, and S. El-Ahmadi. An 11.1 Gbps analog PRML receiver for electronic dispersion compensation of fiber optic communications. *IEEE J. Solid-State Circuits*, SC-45(7):1330–1344, 2010.

4 F. Buchali and H. Bülow. Adaptive PMD compensation by electrical and optical techniques. *J. Lightwave Technol.*, LT-22(4):1116–1126, 2004.

5 D. E. Crivelli, M. R. Hueda, H. S. Carrer, M. del Barco, R. R. López, P. Gianni, J. Finochietto, N. Swenson, P. Voois, and O. E. Agazzi. Architecture of a single-chip 50 Gb/s DP-QPSK/BPSK transceiver with electronic

dispersion compensation for coherent optical channels. *IEEE Trans. Circuits Syst. I*, CASI-61(4):1012–1025, 2014.

6 T. Nielsen and S. Chandrasekhar. OFC 2004 workshop on optical and electronic mitigation of impairments. *J. Lightwave Technol.*, LT-23(1):131–142, 2005.

7 S. J. Savory, G. Gavioli, R. I. Killey, and P. Bayvel. Electronic compensation of chromatic dispersion using a digital coherent receiver. *OSA Opt. Express*, 15(5):2120–2126, 2007.

8 H. Wu, J. A. Tierno, P. Pepeljugoski, J. Schaub, S. Gowda, J. A. Kash, and A. Hajimiri. Integrated transversal equalizers in high-speed fiber-optic systems. *IEEE J. Solid-State Circuits*, SC-38(12):2131–2137, 2003.

9 K.-L. J. Wong and C.-K. K. Yang. A serial-link transceiver with transition equalization. In *ISSCC Digest of Technical Papers*, pages 82–83, February 2006.

10 J. H. Winters and R. D. Gitlin. Electrical signal processing techniques in long-haul fiber-optic systems. *IEEE Trans. Commun.*, COM-38(9):1439–1453, 1990.

11 G. Sebald, B. Lankl, and J. A. Nossek. Advanced time- and frequency-domain adaptive equalization in multilevel QAM digital radio systems. *IEEE J. Sel. Areas Commun.*, SAC-5(3):448–456, 1987.

12 J. Sewter and A. C. Carusone. Equalizer architectures for 40-Gb/s optical systems limited by polarization-mode dispersion. In Y. Liu and H. Yang, editors, *High-Speed Optical Transceivers, Integrated Circuits Designs and Optical Devices Techniques*, pages 73–90. World Scientific, Singapore, 2006.

13 J. G. Proakis and M. Salehi. *Digital Communications*. McGraw Hill, New York, 5th edition, 2008.

14 B. Analui, A. Rylyakov, S. Rylov, M. Meghelli, and A. Hajimiri. A 10-Gb/s two-dimensional eye-opening monitor in 0.13-μm standard CMOS. *IEEE J. Solid-State Circuits*, SC-40(12):2689–2699, 2005.

15 T. Ellermeyer, U. Langmann, B. Wedding, and W. Pöhlmann. A 10-Gb/s eye-opening monitor IC for decision-guided adaption of the frequency response of an optical receiver. *IEEE J. Solid-State Circuits*, SC-35(12):1958–1963, 2000.

16 F. Gerfers, G. W. den Besten, P. V. Petkov, J. E. Conder, and A. J. Koellmann. A 0.2–2 Gb/s 6x OSR receiver using a digitally self-adaptive equalizer. *IEEE J. Solid-State Circuits*, SC-43(6):1436–1448, 2008.

17 J.-S. Choi, M.-S. Hwang, and D.-K. Jeong. A 0.18-μm CMOS 3.5-Gb/s continuous-time adaptive cable equalizer using enhanced low-frequency gain control method. *IEEE J. Solid-State Circuits*, SC-39(3):419–425, 2004.

18 S. Gondi, J. Lee, D. Takeuchi, and B. Razavi. A 10Gb/s CMOS adaptive equalizer for backplane applications. In *ISSCC Digest of Technical Papers*, pages 328–329, February 2005.

19 J. Lee. A 20-Gb/s adaptive equalizer in 0.13-μm CMOS technology. *IEEE J. Solid-State Circuits*, SC-41(9):2058–2066, 2006.

20 E.-H. Chen, J. Ren, B. Leibowitz, H.-C. Lee, Q. Lin, K. S. (Dan) Oh, F. Lambrecht, V. Stojanović, J. Zerbe, and C.-K. K. Yang. Near-optimal equalizer and timing adaptation for I/O links using a BER-based metric. *IEEE J. Solid-State Circuits*, SC-43(9):2144–2156, 2008.

21 B. Widrow and M. E. Hoff Jr. Adaptive switching circuits. In *1960 IRE Western Electric Show and Convention Record, Part 4*, pages 96–104, August 1960.

22 R. W. Lucky. Automatic equalization for digital communications. *Bell Syst. Tech. J.*, 44(4):547–588, 1965.

23 R. W. Lucky. The adaptive equalizer. *IEEE Signal Process. Mag.*, 23(3):104–107, 2006.

24 P. Bayvel, C. Behrens, and D. S. Millar. Digital signal processing (DSP) and its application in optical communication systems. In I. P. Kaminow, T. Li, and A. E. Willner, editors, *Optical Fiber Telecommunications VIB*, pages 163–212. Academic Press, 2013.

25 K. Azadet, E. F. Haratsch, H. Kim, F. Saibi, J. H. Saunders, M. Shaffer, L. Song, and M.-L. Yu. Equalization and FEC techniques for optical transceivers. *IEEE J. Solid-State Circuits*, SC-37(3):317–327, 2002.

26 J. Lee, P. Roux, U.-V. Koc, T. Link, Y. Baeyens, and Y.-K. Chen. A 5-b 10-GSample/s A/D converter for 10-Gb/s optical receivers. *IEEE J. Solid-State Circuits*, SC-39(10):1671–1679, 2004.

27 Y. M. Greshishchev. CMOS ADCs for optical communications. In A. H. M. van Roermund, A. Baschirotto, and M. Steyaert, editors, *Nyquist AD Converters, Sensor Interfaces, and Robustness — Advances in Analog Circuit Design*, pages 97–114. Springer, 2012.

28 J. Lee, A. Leven, J. S. Weiner, Y. Baeyens, Y. Yang, W.-J. Sung, J. Frackoviak, R. F. Kopf, and Y.-K. Chen. A 6-b 12-GSamples/s track-and-hold amplifier in InP DHBT technology. *IEEE J. Solid-State Circuits*, SC-38(9):1533–1539, 2003.

29 L. Y. Nathawad, R. Urata, B. A. Wooley, and D. A. B. Miller. A 40-GHz-bandwidth, 4-bit, time-interleaved A/D converter using photoconductive sampling. *IEEE J. Solid-State Circuits*, SC-38(12):2021–2030, 2003.

30 P. Schvan, J. Bach, C. Falt, P. Flemke, R. Gibbins, Y. Greshishchev, N. Ben-Hamida, D. Pollex, J. Sitch, S.-C. Wang, and J. Wolczanski. A 24GS/s 6b ADC in 90nm CMOS. In *ISSCC Digest of Technical Papers*, pages 544–545, February 2008.

31 C.-K. K. Yang, V. Stojanovic, S. Modjtahedi, M. A. Horowitz, and W. F. Ellersick. A serial-link transceiver based on 8-GSamples/s A/D and D/A converters in 0.25-μm CMOS. *IEEE J. Solid-State Circuits*, SC-36(11):1684–1692, 2001.

32 H. Bülow, F. Buchali, W. Baumert, R. Ballentin, and T. Wehren. PMD mitigation at 10 Gb/s using linear and nonlinear integrated electronic equaliser circuits. *Electron. Lett.*, 36(2):163–164, 2000.

33 A. Momtaz, D. Chung, N. Kocaman, J. Cao, M. Caresosa, B. Zhang, and I. Fujimori. A fully integrated 10-Gb/s receiver with adaptive optical dispersion equalizer in 0.13-μm CMOS. *IEEE J. Solid-State Circuits*, SC-42(4):872–880, 2007.

34 J. Sewter and A. C. Carusone. A CMOS finite impulse response filter with a crossover traveling wave topology for equalization up to 30 Gb/s. *IEEE J. Solid-State Circuits*, SC-41(4):909–917, 2006.

35 X. Lin, S. Saw, and J. Liu. A CMOS 0.25-μm continuous-time FIR filter with 125 ps per tap delay as a fractionally spaced receiver equalizer for 1-Gb/s data transmission. *IEEE J. Solid-State Circuits*, SC-40(3):593–602, 2005.

36 X. Lin, J. Liu, H. Lee, and H. Liu. A 2.5- to 3.5-Gb/s adaptive FIR equalizer with continuous-time wide-bandwidth delay line in 0.25-μm CMOS. *IEEE J. Solid-State Circuits*, SC-40(8):1908–1918, 2006.

37 C. Pelard, E. Gebara, A. J. Kim, M. G. Vrazel, F. Bien, Y. Hur, M. Maeng, S. Chandramouli, C. Chun, S. Bajekal, S. E. Ralph, B. Schmukler, V. M. Hietala, and J. Laskar. Realization of multigigabit channel equalization and crosstalk cancellation integrated circuits. *IEEE J. Solid-State Circuits*, SC-39(10):1659–1670, 2004.

38 J. Sewter and A. C. Carusone. A 3-tap FIR filter with cascaded distributed tap amplifiers for equalization up to 40 Gb/s in 0.18-μm CMOS. *IEEE J. Solid-State Circuits*, SC-41(8):1919–1929, 2006.

39 T.-C. Lee and B. Razavi. A 125-MHz CMOS mixed-signal equalizer for gigabit Ethernet on copper wire. In *Proceedings of IEEE Custom Integrated Circuits Conference*, pages 131–134, September 2001.

40 N. Sitthimahachaikul, L. P. Rao, and P. J. Hurst. Overcoming the effect of the summation-node parasitic pole in an analog equalizer. *IEEE Trans. Circuits Syst. I*, CASI-59(3):652–663, 2012.

41 J. Kim, E.-H. Chen, J. Ren, B. S. Leibowitz, P. Satarzadeh, J. L. Zerbe, and C.-K. Ken Yang. Equalizer design and performance trad-offs in ADC-based serial links. *IEEE Trans. Circuits Syst. I*, CASI-58(9):2096–2107, 2011.

42 V. Balan, J. Caroselli, J.-G. Chern, C. Chow, R. Dadi, C. Desai, L. Fang, D. Hsu, P. Joshi, H. Kimura, C. Y. Liu, T.-W. Pan, R. Park, C. You, Y. Zeng, E. Zhang, and F. Zhong. A 4.8–6.4-Gb/s serial link for backplane applications using decision feedback equalization. *IEEE J. Solid-State Circuits*, SC-40(9):1957–1967, 2005.

43 S. Ibrahim and B. Razavi. Low-power CMOS equalizer design for 20-Gb/s systems. *IEEE J. Solid-State Circuits*, SC-46(6):1321–1336, 2011.

44 S. Kasturia and J. H. Winters. Techniques for high-speed implementation of nonlinear cancellation. *IEEE J. Sel. Areas Commun.*, SAC-9(5):711–717, 1991.

45 K.-L. J. Wong, A. Rylyakov, and C.-K. K. Yang. A 5-mW 6-Gb/s quarter-rate sampling receiver with a 2-tap DFE using soft decisions. *IEEE J. Solid-State Circuits*, SC-42(4):881–888, 2007.

46 H. Higashi, S. Masaki, M. Kibune, S. Matsubara, T. Chiba, Y. Doi, H. Yamaguchi, H. Takauchi, H. Ishida, K. Gotoh, and H. Tamura. A 5–6.4-Gb/s 12-channel transceiver with pre-emphasis and equalization. *IEEE J. Solid-State Circuits*, SC-40(4):978–985, 2005.

47 Y. Tomita, M. Kibune, J. Ogawa, W. W. Walker, H. Tamura, and T. Kuroda. A 10-Gb/s receiver with series equalizer and on-chip ISI monitor in 0.11-μm CMOS. *IEEE J. Solid-State Circuits*, SC-40(4):986–993, 2005.

48 M. Q. Le, P. J. Hurst, and K. C. Dyer. An analog DFE for disk drives using a mixed-signal integrator. *IEEE J. Solid-State Circuits*, SC-34(5):592–598, 1999.

49 Y. Hidaka, W. Gai, T. Horie, J. H. Jiang, Y. Koyanagi, and H. Osone. A 4-channel 1.25–10.3 Gb/s backplane transceiver macro with 35 dB equalizer and sign-based zero-forcing adaptive control. *IEEE J. Solid-State Circuits*, SC-44(12):3547–3559, 2009.

F

Decision-Point Control

The location of the decision point (the intersection of the decision threshold with the sampling instant) in the eye diagram has a significant impact on the bit-error rate (BER) of the receiver. In the following, we discuss a method to control the decision point based on pseudo-error measurements.

Decision-Threshold Control. The optimum decision threshold (a.k.a. slice level) is at the point where the probability distributions of the zero and the one bits intersect (cf. Figs 4.4 and 4.5). If the noise distributions have equal widths (same rms value), the optimum decision threshold is centered halfway between the zero and one levels. An AC-coupled receiver automatically slices at that threshold level, given a DC-balanced NRZ signal and no circuit offsets. However, if there is more noise on the ones than the zeros, for example, because optical amplifiers or an APD is used, the optimum decision threshold is below the center (see Fig. F.1) and a simple AC-coupled receiver makes more errors than necessary. [→ Problem F.1.]

Making the decision threshold adjustable and setting it for the lowest BER avoids this performance penalty [1]. An intentional offset voltage in the decision circuit or the preceding MA can be used for this purpose [2]. A more sophisticated approach is to make the adjustment automatic and *adaptive* such that the decision threshold tracks variations in the signal swing and noise statistics over time. The latter method is known as *adaptive threshold control* (ATC).

Pseudo BER. To find the optimum decision threshold we need a way to determine the BER, which is what we want to minimize. In a system with checksums or forward error correction (cf. Appendix G), the number of detected errors can be used to estimate the BER. (Moreover, the number of errored zeros vs the number of errored ones can be used to steer the decision threshold.) But how can we determine the BER in the absence of data-integrity checks?

We can use an eye monitor to measure the so-called *pseudo bit-error rate.* This method was first proposed by Gooding in 1968 [3]. The idea is to use two

Analysis and Design of Transimpedance Amplifiers for Optical Receivers, First Edition. Eduard Säckinger.

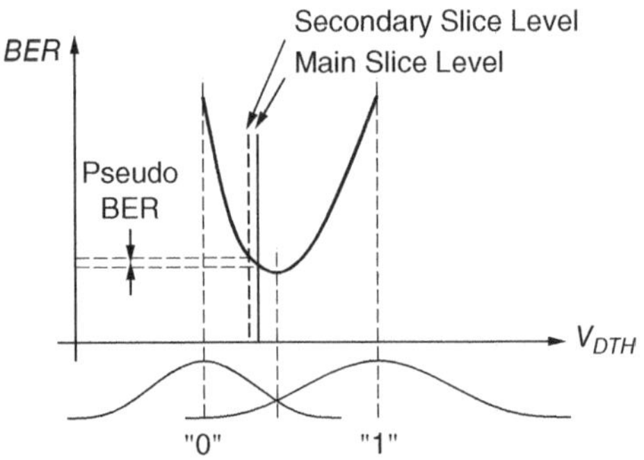

Figure F.1 Optimum decision threshold for unequal noise distributions.

parallel decision circuits: a main decision circuit and a secondary decision circuit. The secondary decision circuit's performance is intentionally degraded by offsetting its decision threshold such that it makes more bit errors. This situation is illustrated in Fig. F.1. Although both BERs are unknown, the disagreements between the main and secondary decision circuit, the so-called *pseudo errors*, are measurable.

Because the pseudo BER and the actual BER have almost identical minima, minimizing the pseudo BER also minimizes the actual BER to a good approximation [4, 5]. The smaller the spacing between the main and secondary decision thresholds is, the better the approximation becomes. [→ Problem F.2.]

Implementation Issues. Figure F.2 shows an implementation of this concept using three decision circuits [4]. The controller generates the main slice level and two secondary slice levels, a lower and an upper slice level, that are spaced

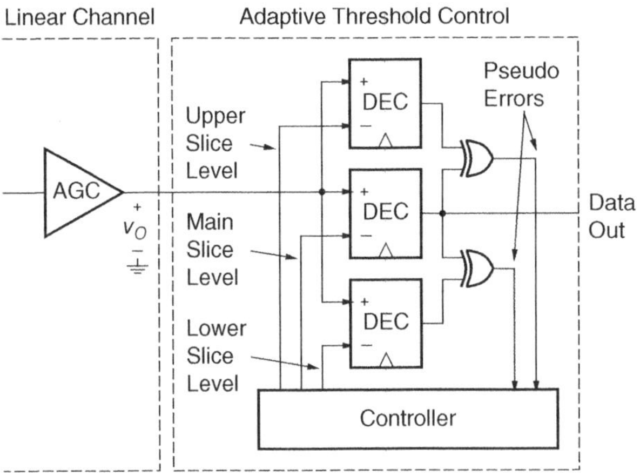

Figure F.2 The linear channel of Fig. 4.1 followed by a circuit for adaptive threshold control.

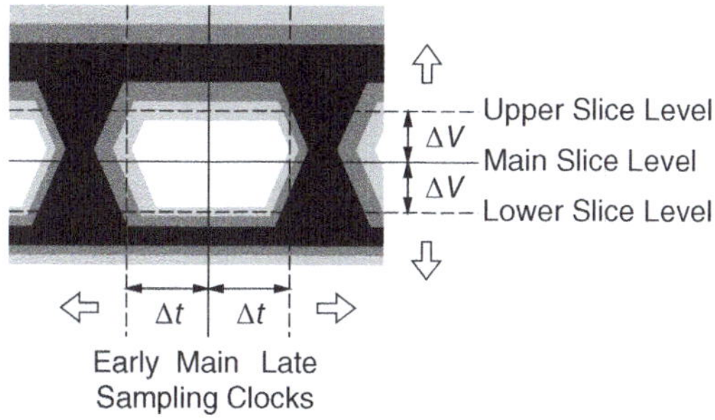

Figure F.3 Optimum placement of the slice levels (and sampling clocks) in the eye diagram.

equidistantly above and below the main slice level. Two XOR gates detect the pseudo errors. The controller measures and compares the two pseudo BERs and if the upper pseudo BER is smaller, it moves all slice levels up; if the lower pseudo BER is smaller, it moves them all down. This function can easily be implemented with an up/down counter [4]. When both pseudo BERs are equal, a close approximation to the optimum slice level has been found.

An important consideration is the spacing between the main and the secondary slice levels [4]. A large spacing results in a high pseudo BER that can be measured quickly and permits a rapid adaptation. However, an overly large spacing results in a significant offset between the pseudo BER minimum and the actual BER minimum, making the slice-level optimization inaccurate. In contrast, an overly small spacing results in a slow adaptation and possibly instability if the spacing becomes smaller than the discrimination sensitivity of the decision circuit. Clearly, an adaptive mechanism to find the optimum slice-level spacing, ΔV, is desirable [6, 7]. When the optimum spacing is found, the secondary slice levels are located right at the edges of the open part of the eye, as illustrated in Fig. F.3.

In a variation of the implementation shown in Fig. F.2, one of the two secondary decision circuits and the associated XOR gate is eliminated at the expense of a more complex controller [8]. The remaining secondary decision circuit, called the monitoring or roaming decision circuit, is used to measure both pseudo BERs *sequentially*. First, the controller places the roaming slice level below the main slice level and measures the pseudo BER; then, the controller places the roaming slice level above the main slice level and measures the pseudo BER again; depending on how the two pseudo BERs compare, the controller moves the main slice level up or down.

Sampling-Instant Control. What we have said about controlling the decision threshold can also be applied to control the sampling instant. A standard clock recovery circuit finds the (average) timing of the data transitions and then samples the data half a bit period later. Although this works fine for an ideal NRZ signal, it is suboptimal in the presence of duty-cycle distortions or asymmetrically distributed jitter. In these cases a better BER performance

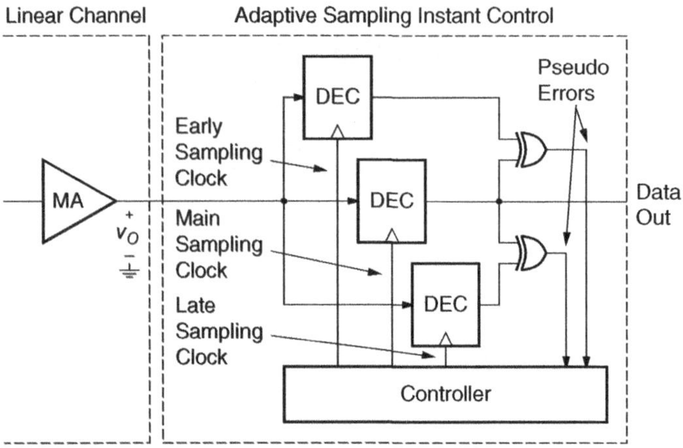

Figure F.4 The linear channel of Fig. 4.1 followed by a circuit for adaptive sampling-instant control.

can be achieved when sampling slightly early or late relative to the centered sampling instant. Again, we can do that by either adjusting the sampling instant manually or using an automatic control mechanism. Figure F.4 shows a receiver with adaptive sampling-instant control, which is analogous to the adaptive threshold control discussed before. Besides the main clock, an early and a late clock sample the input data. The early and late samples are used to generate two pseudo-error streams. A controller adjusts the sampling clocks until the two pseudo BERs become equal.

Just like with adaptive threshold control, choosing the delay between the early and late clocks is critical and preferably is made adaptive to track the edges of the eye [9]. Figure F.3 illustrates the optimum placement of the sampling clocks in the eye.

In a variation of the scheme in Fig. F.4, a single roaming sampler is used to measure pseudo BERs sequentially for early and late sampling [8]. In another variation, the input data rather than the sampling clock is delayed [10]. In this scheme, the input data signal is fed into a tapped delay line that generates three temporally staggered signals. These signals then drive three decision circuits, which are all connected to the same (main) sampling clock.

Adaptive Decision-Point Control. Adaptive threshold control and adaptive sampling-instant control can be combined to form *adaptive decision-point control*. This combined method is also known as *decision-point steering* [5].

A 40-Gb/s CDR with adaptive decision-point control is described in [8].

Adaptive decision-point control and adaptive equalization (cf. Appendix E) have some features in common and are often used together. For example, an adaptive decision threshold is the same as an adaptive signal offset introduced by the preceding equalizer. Moreover, adaptation of the decision point is based on a cost function, just like adaptation of an equalizer. In the implementation discussed here, the cost function is the pseudo BER and its gradient is determined by taking the difference of two pseudo BER measurements.

Problems

F.1 **Optimum Decision Threshold.** Assume that the sampled voltage values v_O follow the general probability distributions Zero(v_O) (for the zeros) and One(v_O) (for the ones) and that zeros and ones are equally likely. What decision threshold V_{DTH} leads to the lowest BER?

F.2 **Pseudo Bit-Error Rate.** Assume that the sampled voltage values v_O follow the general probability distributions Zero(v_O) (for the zeros) and One(v_O) (for the ones) and that zeros and ones are equally likely. (a) Calculate the pseudo BER for two closely spaced decision threshold levels, V_{DTH1} and V_{DTH2}. (b) Calculate the BER for each decision threshold individually. Can the pseudo BER be expressed as the absolute difference of the two actual BERs?

References

1 K. Ogawa, L. D. Tzeng, Y. K. Park, and E. Sano. Advances in high bit-rate transmission systems. In I. P. Kaminow and T. L. Koch, editors, *Optical Fiber Telecommunications IIIA*, pages 336–372. Academic Press, San Diego, CA, 1997.

2 A. Momtaz, D. Chung, N. Kocaman, J. Cao, M. Caresosa, B. Zhang, and I. Fujimori. A fully integrated 10-Gb/s receiver with adaptive optical dispersion equalizer in 0.13-μm CMOS. *IEEE J. Solid-State Circuits*, SC-42(4):872–880, 2007.

3 D. J. Gooding. Performance monitor techniques for digital receivers based on extrapolation of error rate. *IEEE Trans. Commun.*, COM-16(3):380–387, 1968.

4 M. Kawai, H. Watanabe, T. Ohtsuka, and K. Yamaguchi. Smart optical receiver with automatic decision threshold setting and retiming phase alignment. *J. Lightwave Technol.*, LT-7(11):1634–1640, 1989.

5 M. Sherif and P. A. Davies. Decision-point steering in optical fibre communication systems: theory. *IEE Proceedings, Pt. J*, 136(3):169–176, 1989.

6 Y. Tremblay and D. J. Nicholson. Binary data regenerator with adaptive threshold level. U.S. Patent No. 4,823,360, April 1989.

7 Y. Matsumoto, T. Kuriyama, D. Inami, and M. Ohta. An adaptive decision threshold control of the optical receiver for multi-gigabit terrestrial DWDM transmission systems. In *Optical Fiber Communication Conference (OFC)*, Anaheim, CA, 2001.

8 H. Noguchi, N. Yoshida, H. Uchida, M. Ozaki, S. Kanemitsu, and S. Wada. A 40-Gb/s CDR circuit with adaptive decision-point control based on eye-opening monitor feedback. *IEEE J. Solid-State Circuits*, SC-43(12):2929–2938, 2008.

9 S.-H. Lee, M.-S. Hwang, Y. Choi, S. Kim, Y. Moon, B.-J. Lee, D.-K. Jeong, W. Kim, Y.-J. Park, and G. Ahn. A 5-Gb/s 0.25-μm CMOS jitter-tolerant variable-interval oversampling clock/data recovery circuit. *IEEE J. Solid-State Circuits*, SC-37(12):1822–1830, 2002.

10 Y. Miki, T. Saito, H. Yamashita, F. Yuki, T. Baba, A. Koyama, and M. Sonehara. A 50-mW/ch 2.5-Gb/s/ch data recovery circuit for the SFI-5 interface with digital eye-tracking. *IEEE J. Solid-State Circuits*, SC-39(4):613–621, 2004.

G

Forward Error Correction

Forward error correction has become ubiquitous in modern communication systems of all kinds. In the following, we discuss the basics of forward error correction. How does it work? How much performance do we gain?

SNR Requirements. In Section 4.2, we found that we need an SNR of about 17 dB to receive an NRZ signal at a BER of 10^{-12}. Can we do better than that? Yes we can, if we use error-correcting codes! A simple (but impractical) example of such a code is to send each bit three times. At the receiver, we can analyze the (corrupted) 3-bit codewords and correct single-bit errors. For example, if we receive the code word "101," we know that an error occurred and that the correct code word most likely is "111." This method of adding redundancy at the transmitter and correcting errors at the receiver is known as FEC. In practice, sophisticated codes such as the *Reed–Solomon* (RS) codes and the *Bose–Chaudhuri–Hocquenghem* (BCH) codes are used. (The alternative to FEC is to add a checksum at the transmitter and upon detection of an error at the receiver request the retransmission of the errored data block. This method is known as *automatic repeat request* [ARQ]. However, the high latency of long-haul optical communication links makes this method impractical.)

By how much can we lower the SNR requirement if we use FEC? The *channel capacity theorem* (a.k.a. *information capacity theorem*) asserts that with sufficient coding, error-free transmission over a channel with additive white Gaussian noise is possible if

$$B \leq BW \cdot \log_2(SNR + 1), \tag{G.1}$$

where B is the information bit rate and BW is the channel bandwidth. This landmark theorem was published by Shannon in 1948 [1].

To get a rough idea of the potential of FEC, let us assume that we use a code with the code rate r and NRZ modulation. With Eq. (4.66), we can estimate the required channel bandwidth as $BW \approx \frac{2}{3}B/r$, where B/r is the channel bit rate. Solving Eq. (G.1) for SNR and inserting our bandwidth estimate results

Analysis and Design of Transimpedance Amplifiers for Optical Receivers, First Edition. Eduard Säckinger.

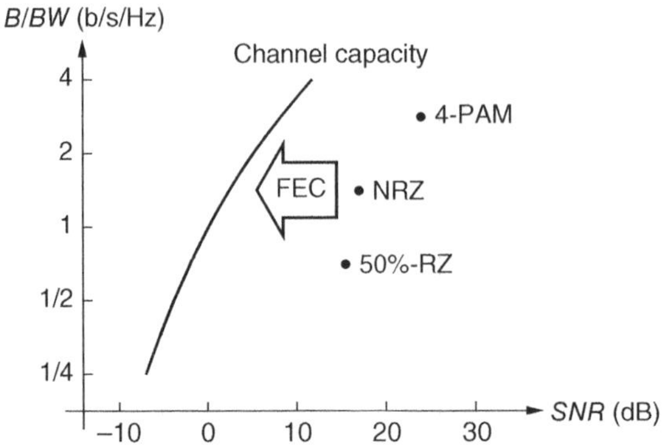

Figure G.1 FEC reduces the required SNR, ideally approaching the channel capacity limit.

in $SNR \geq 2^{B/BW} - 1 \approx 2^{3r/2} - 1$. Thus, for a code with 5% overhead ($r = 0.95$), the necessary SNR is just 2.3 dB. That is about 15 dB less than what we need without coding! Figure G.1 illustrates the difference between the SNR requirements of uncoded NRZ, 50%-RZ and 4-PAM modulation (cf. Appendix A) on one hand and the channel capacity limit given by Eq. (G.1) on the other hand. In all cases, there is plenty of room for FEC to reduce the required SNR.

The lower SNR requirement resulting from FEC can be used to either tolerate more noise and ISI for a given transmit power or to lower the transmit power for a given amount of noise and ISI (while maintaining the same BER in both cases). The first use of FEC helps to extend the reach of optical links, whereas the second use helps to suppress nonlinear fiber effects. Moreover, the addition of FEC in low-cost systems permits the used of lower quality optoelectronic components by correcting for potentially harmful BER floors [2].

FEC Based on Reed–Solomon Code. The Reed–Solomon codes are a popular choice for FEC systems. These codes make efficient use of the redundancy, have an adjustable block length and code rate, and permit efficient hardware implementations. The code is named after its inventors Irving Reed and Gustave Solomon who published their work in 1960 [3].

Figure G.2 illustrates the operation of an FEC encoder based on the RS(255,239) Reed–Solomon code used, for example, in undersea lightwave systems [4] and the OTN standard [5] (cf. Chapter 1). The data stream into the encoder, the so-called payload, is arranged into blocks of 238 data bytes. A framing byte is appended to each data block, making it a 239-byte block. This block is then encoded with the RS(255,239) code, which adds 16 bytes of redundancy, producing a 255-byte block known as an FEC frame. Before transmitting the encoded block, it is run through a so-called 16× *interleaver*. This means that rather than transmitting a complete 255-byte block at a time, one byte is transmitted from the first block, then one byte from the second block, and so forth, until block 16 is reached; then the process continues with

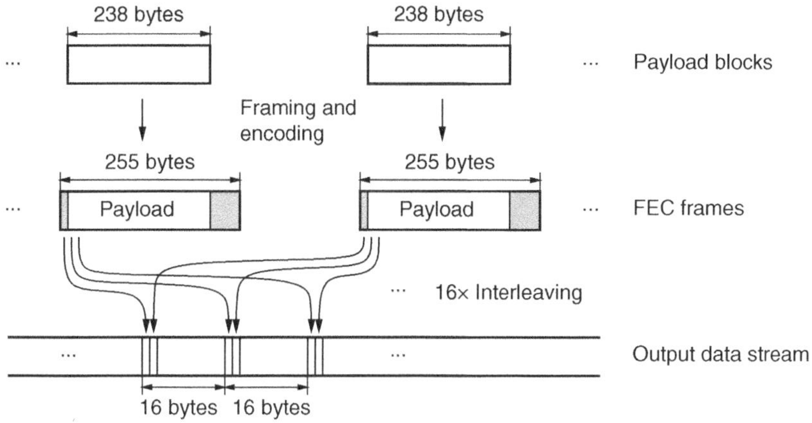

Figure G.2 Data manipulations in a typical RS(255,239) FEC encoder.

the next byte from the first block, and so forth. Interleaving spreads burst errors, which may occur during transmission, into multiple received blocks, thus increasing the error-correcting capacity for bursts.

The RS(255,239) encoder increases the transmitted bit rate by 7% $(255/238 = 15/14 = 1.071)$, which is equivalent to saying that the code rate r is $14/15 = 0.933$. Thus, slightly faster hardware is needed in the transceiver front-end than without coding. In return, the benefit of the code is that up to 8 byte errors can be corrected per block,[1] thus significantly lowering the BER. Furthermore, thanks to the 16× interleaver, burst errors up to 16×8 bytes (1,024 bits) in length can be corrected. The precise improvement in BER depends on the incoming BER and the distribution of the bit errors in the received signal. In a typical transmission system with FEC based on RS(255,239), an incoming BER of 10^{-4} (i.e., BER at the output of the decision circuit) can be boosted to 10^{-12} after error correction. This is an improvement of eight orders of magnitude! [→ Problem G.1.]

The FEC described in the aforementioned example is known as a *first-generation* FEC [6]. FEC based on more powerful codes, such as *turbo codes, turbo product codes* (TPC), and *low-density parity-check codes* (LDPC), have been developed since [7]. For example, the LDPC in [8] can correct a BER of 10^{-3} to 10^{-12} with an overhead of 7% and the BCH-based turbo code in [6] can correct a BER of 10^{-2} to 10^{-12} with an overhead of 21.5%.

Coding Gain. The performance of an FEC code can be described graphically with a plot of BER versus SNR, sometimes called a *waterfall curve* because of

1 In general, an RS(n,k) code with a block length of n symbols and a message length of k symbols can correct $\lfloor (n-k)/2 \rfloor$ symbol errors.

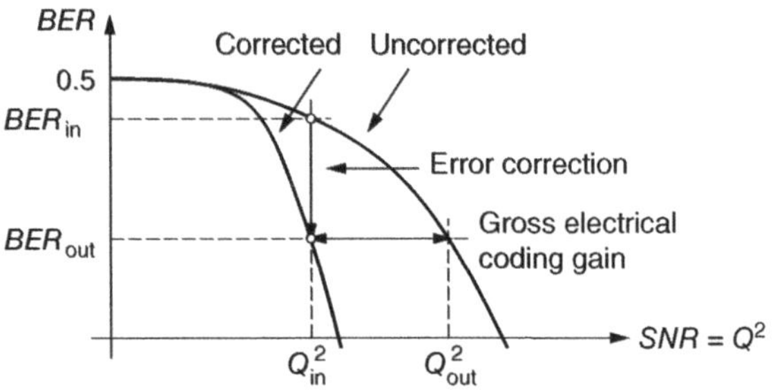

Figure G.3 BER versus SNR on a log–log scale before and after FEC.

its characteristic shape. See the example in Fig. G.3. The x-axis represents the SNR at the input of the decision circuit and the y-axis represents the BER. One curve, labeled "Uncorrected," shows the BER at the input of the FEC decoder and the other curve, labeled "Corrected," shows the BER at the output of the FEC decoder. As expected, the BER at the output is lower than the BER at the input. The farther down the corrected curve is relative to the uncorrected curve, the more powerful the FEC code is.

In the case of an ideal NRZ signal with additive Gaussian noise, the SNR on the x-axis equals $\mathcal{Q}^2$ (cf. Eq. (4.13)) and the uncorrected curve corresponds to the BER expression in Eq. (4.8), which is also tabulated in Table 4.1. Let us assume, for example, that the incoming BER (BER_{in}) is 10^{-4} and that the outgoing BER (BER_{out}) is 10^{-12}. The incoming BER indicates that we have $\mathcal{Q}_{\text{in}} = 3.72$ at the decision circuit (cf. Table 4.1). To obtain the outgoing BER *without* coding, we would need to have $\mathcal{Q}_{\text{out}} = 7.03$ at the decision circuit. Thus, the addition of FEC to the transmission system relaxed the SNR requirement by $\mathcal{Q}^2_{\text{out}}/\mathcal{Q}^2_{\text{in}}$, which in our example is 5.5 dB. This quantity is known as the *electrical coding gain* or simply the *coding gain*. As can be seen from Fig. G.3, the coding gain depends on the desired BER_{out}.

It can be argued that the coding gain as calculated earlier is not quite fair. Without coding, the channel bit rate would be somewhat lower, which would permit us to reduce the bandwidth of the receiver resulting in less noise and thus a somewhat larger $\mathcal{Q}_{\text{in}}$. To be more specific, the receiver bandwidth could be reduced by the code rate r. Then, assuming white noise, the rms noise would reduce by $\sqrt{r}$ and $\mathcal{Q}_{\text{in}}$ would improve to $\mathcal{Q}_{\text{in}}/\sqrt{r}$. For this reason, it is common to distinguish between the *gross electrical coding gain* as defined earlier ($\mathcal{Q}^2_{\text{out}}/\mathcal{Q}^2_{\text{in}}$) and the *net electrical coding gain* (NECG) that corrects for the difference in channel bit rate [9]

$$NECG = r \cdot \frac{\mathcal{Q}^2_{\text{out}}}{\mathcal{Q}^2_{\text{in}}}. \tag{G.2}$$

In the aforementioned example, the gross electrical coding gain is 5.5 dB and the NECG is 5.2 dB assuming $r = 14/15$. For advanced codes, NECGs up to 10 dB have been reported [6, 10].

In Section 4.2, we have introduced E_b/N_0, an SNR-type measure that includes a normalization with respect to the code rate: $E_b/N_0 = Q^2/(2r)$ (cf. Eq. (4.16)). Now, we can express NECG conveniently as the ratio of two 'ebno's [9]

$$NECG = \frac{(E_b/N_0)_{\text{out}}}{(E_b/N_0)_{\text{in}}}, \tag{G.3}$$

where $(E_b/N_0)_{\text{in}} = Q_{\text{in}}^2/(2r)$ is needed to achieve BER_{out} *with* FEC and $(E_b/N_0)_{\text{out}} = Q_{\text{out}}^2/2$ is needed to achieve BER_{out} *without* FEC. Thus, if we plot the BER as a function of E_b/N_0 rather than Q^2 or SNR, the horizontal displacement between the coded and uncoded curves directly represents the NECG.

Unamplified versus Amplified Transmission Systems. It is instructive to compare the benefits of FEC when applied to an unamplified and an amplified transmission system. In an unamplified system, the optical sensitivity of a p–i–n receiver is proportional to Q (cf. Eq. (4.27)). Thus, the improvement in sensitivity is given by the *square root* of the NECG. In our example, the optical sensitivity increases by 5.2 dB/2 = 2.6 dB. The system reach increases proportional to the improvement in sensitivity measured in dBs. Thus, given a system with a 16-dB fiber loss budget, the reach increases by 2.6 dB/16 dB = 16%.

The FEC advantage is much more dramatic in an amplified transmission system. There, the required OSNR is proportional to Q^2 (cf. Eq. (4.53)) and thus the improvement in required OSNR is equal to the NECG. Moreover, the OSNR at the receiver measured in decibels degrades proportional to the logarithm of the system reach, if we keep the amplifier spacing fixed and increase the reach by adding more amplifiers (cf. Eq. (4.55)). In other words, the system reach increases *exponentially* with the improvement in required OSNR measured in decibels. In our example, the required OSNR is reduced by 5.2 dB resulting in the remarkable reach increase of $10^{5.2/10} = 3.3\times$. For every 3 dB of NECG, the reach of an amplified system is doubled [2]. For this reason, FEC is particularly attractive for ultra-long-haul amplified transmission systems.

Soft-Decision Decoding. An FEC system that corrects errors based on the binary values from the decision circuit is known as a *hard-decision decoder*. Although many transmission errors can be corrected in this way, more errors can be corrected if the *analog* values of the received samples are known. These values are a measure of the confidence of the received bits and can be taken into account when correcting errors. The latter system is known as a *soft-decision decoder* [11–13].

A theoretical analysis shows that the coding gain of soft-decision decoders is about 2 dB better than that of hard-decision decoders [13, 14]. Soft-decision decoding is of particular interest with turbo codes and LDPC codes, which are naturally based on soft information [15]. For example, a soft-decision decoder for a BCH-based turbo code is described in [16]. This code has an overhead of 24.6% ($r = 0.8$) and achieves a NECG of 10.1 dB at $BER = 10^{-13}$, which is 1.2 dB better than the performance of the corresponding hard-decision decoder. A soft-decision decoder for an LDPC code is described in [17].

Implementation Issues. Although FEC is implemented in the digital domain, its presence has consequences for the analog front-end. Because the channel bit rate increases by an amount ranging from a few percents up to about 25% for high-performance codes, correspondingly faster front-end circuits are required.

The fact that the receiver front-end has to work with very high BERs ranging from 10^{-4} up to 10^{-2} for high-performance codes, puts strain on the clock-and-data recovery (CDR) circuit and the decision-feedback equalizer (DFE). The CDR must be able to recover a clean clock from a very noisy signal. If the DFE is fed with uncorrected decisions (from the decision circuit before the FEC decoder) many of them are wrong resulting in harmful error propagation (cf. Appendix E). Unfortunately, the corrected decisions, which would avoid error propagation, arrive only after passing through the FEC decoder, which is too late to be useful for the DFE. If the decoder latency is not too long, it is possible to use multiple parallel DFEs, each one equalizing for another speculative bit sequence, and once the correct sequence is known from the FEC decoder to select the corresponding DFE output [18].

To implement soft-decision decoding, a multilevel slicer is needed in the front-end. Figure G.4 shows a receiver with a slicer that has four different output states (similar to a 2-bit flash analog-to-digital converter) allowing a coarse form of soft-decision decoding. The four states correspond to a "hard zero," a "soft zero," a "soft one," and a "hard one." They can be encoded into two bits such that one bit represents the likely bit value and the other bit represents the

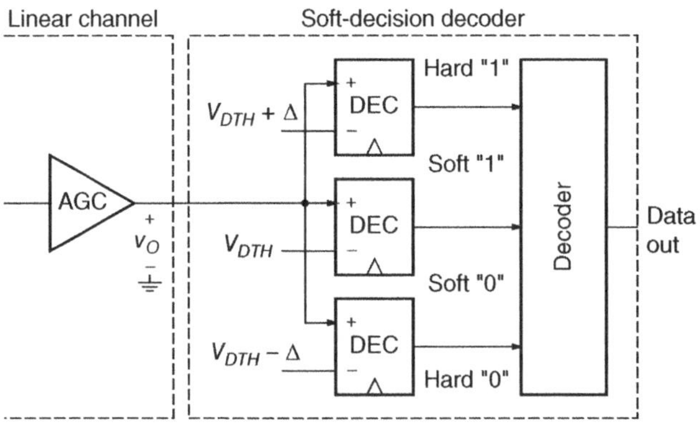

Figure G.4 The linear channel of Fig. 4.1 followed by a 2-bit soft-decision decoder.

confidence level. The slicer is followed by the decoder logic that detects and corrects errors.

The simple 2-bit soft-decision decoder of Fig. G.4 can be extended to an N-bit decoder with $2^N - 1$ slicers. Typically, most of the coding gain is realized with a small number of bits (e.g., a 3-bit decoder is used in [16] and a 4-bit decoder is used in [17]). For optimum error-correcting performance, the slice levels can be spaced nonuniformly to better capture the unequal noise distributions of the zeros and ones in optically amplified systems [16].

Problems

G.1 **Forward Error Correction.** Assume that the bit errors at the input of the RS(255,239) FEC decoder are random and independent (no bursts) and have $BER = 10^{-4}$. For simplicity, assume that the probability for two or more errors in the same byte is very small. What is the BER at the output of the decoder?

References

1 C. E. Shannon. A mathematical theory of communication. *Bell Syst. Tech. J.*, 27:379–423 and 623–656, 1948.
2 V. W. S. Chan. Coding and error correction in optical fiber communications systems. In I. P. Kaminow and T. L. Koch, editors, *Optical Fiber Telecommunications IIIA*, pages 42–62. Academic Press, San Diego, CA, 1997.
3 I. S. Reed and G. Solomon. Polynomial codes over certain finite fields. *SIAM J. Appl. Math.*, 8:300–304, 1960.
4 ITU-T. Forward error correction for submarine systems, recommendation G.975. International Telecommunication Union, Geneva, Switzerland, October 2000.
5 ITU-T. Interfaces for the optical transport network (OTN), recommendation G.709/Y.1331. International Telecommunication Union, Geneva, Switzerland, February 2001.
6 T. Mizuochi, K. Kubo, H. Yoshida, H. Fujita, H. Tagami, M. Akita, and K. Motoshima. Next generation FEC for optical transmission systems. In *Optical Fiber Communication Conference (OFC), Anaheim, CA*, 2003.
7 I. B. Djordjevic. Advanced coding for optical communications. In I. P. Kaminow, T. Li, and A. E. Willner, editors, *Optical Fiber Telecommunications VIB*, pages 221–296. Academic Press, 2013.
8 J. H. Sinsky, A. Adamiecki, A. Gnauck, C. A. Burrus Jr., J. Leuthold, O. Wohlgemuth, S. Chandrasekhar, and A. Umbach. RZ-DPSK transmission

using a 42.7-Gb/s integrated balanced optical front end with record sensitivity. *J. Lightwave Technol.*, LT-22(1):180–185, 2004.
9 F. Kerfoot and H. Kidorf. Forward error correction for optical transmission systems. In *Optical Fiber Communication Conference (OFC) Tutorials, Anaheim, CA*, 2002.
10 H. Haunstein and R. Urbansky. Application of electronic equalization and error correction in lightwave systems. In *European Conference on Optical Communication (ECOC), Stockholm, Sweden*, 2004.
11 R. D. Gitlin, J. F. Hayes, and S. B. Weinstein. *Data Communications Principles.* Plenum Press, New York, 1992.
12 E. A. Lee and D. G. Messerschmitt. *Digital Communication.* Kluwer Academic Publishers, Boston, MA, 2nd edition, 1994.
13 J. G. Proakis and M. Salehi. *Digital Communications.* McGraw Hill, New York, 5th edition, 2008.
14 B. Sklar. *Digital Communications: Fundamentals and Applications.* Prentice Hall, Englewood Cliffs, NJ, 2nd edition, 2001.
15 D. J. Costello Jr. and G. D. Forney Jr. Channel coding: the road to channel capacity. *Proc. IEEE*, 95(6):1150–1177, 2007.
16 H. Tagami, T. Kobayashi, Y. Miyata, K. Ouchi, K. Sawada, K. Kubo, K. Kuno, H. Yoshida, K. Shimizu, T. Mizuochi, and K. Motoshima. A 3-bit soft-decision IC for powerful forward error correction in 10-Gb/s optical communication systems. *IEEE J. Solid-State Circuits*, SC-40(8):1695–1705, 2005.
17 A. J. Blanksby and C. J. Howland. A 690-mW 1-Gb/s 1024-b, rate-1/2 low-density parity-check code decoder. *IEEE J. Solid-State Circuits*, SC-37(3):404–412, 2002.
18 M. Hatamian, O. E. Agazzi, J. Creigh, H. Samueli, A. J. Castellano, D. Kruse, A. Madisetti, N. Yousefi, K. Bult, P. Pai, M. Wakayama, M. M. McConnell, and M. Colombatto. Design considerations for gigabit Ethernet 1000Base-T twisted pair transceivers. In *Proceedings of IEEE Custom Integrated Circuits Conference*, pages 335–342, May 1998.

H

Second-Order Low-Pass Transfer Functions

Let us define the second-order low-pass transfer function

$$H(s) = \frac{1}{1 + s/(\omega_0 Q) + s^2/\omega_0^2}, \tag{H.1}$$

which plays an important role in the analysis of TIAs (cf. Chapter 6). In the following, we calculate the 3-dB bandwidth, noise bandwidths, and Personick integrals of this transfer function. Then, we specialize the results for the case when the poles are real, relevant for common-gate and common-base TIAs. After that, we discuss four special cases of the transfer function (critically damped, Bessel–Thomson, Butterworth, and $Q = 1$) in the frequency and time domain.

3-dB Bandwidth. The transfer function in Eq. (H.1) has two poles, no zeros, and a low-frequency gain of one. Its 3-dB bandwidth is found by solving $|H(j2\pi BW_{3dB})|^2 = \frac{1}{2}$, which results in

$$BW_{3dB} = \rho(Q)\,\frac{\omega_0}{2\pi}, \tag{H.2}$$

where

$$\rho(Q) = \sqrt{\sqrt{\left(1 - \frac{1}{2Q^2}\right)^2 + 1} + \left(1 - \frac{1}{2Q^2}\right)}. \tag{H.3}$$

As shown in Fig. H.1, this somewhat complicated expression can be bounded by two linear terms:

$$Q\,\frac{\omega_0}{2\pi} \leq BW_{3dB} \leq \sqrt{2}Q\,\frac{\omega_0}{2\pi}, \tag{H.4}$$

where the lower bound holds for $Q \leq \sqrt{2}$ (none to moderate amplitude peaking), and the upper bound holds for any Q. The equality of the lower bound is reached asymptotically for $Q \to 0$ and for $Q = \sqrt{2}$; the equality of the upper

Analysis and Design of Transimpedance Amplifiers for Optical Receivers, First Edition. Eduard Säckinger.

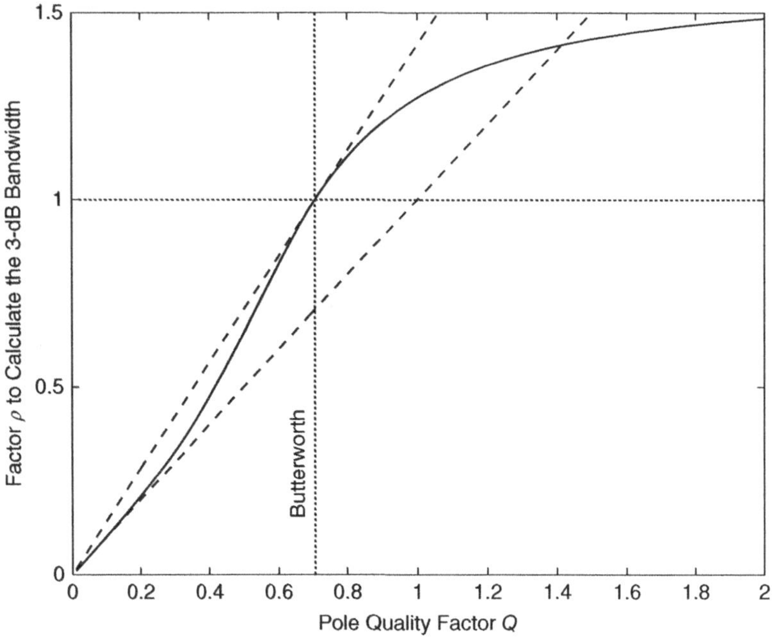

Figure H.1 Factor ρ ($= 2\pi BW_{3dB}/\omega_0$) as a function of the pole quality factor Q (solid line, Eq. (H.2)) and three bounds (dashed lines, Eq. (H.4); dotted horizontal line, Eq. (H.5)).

bound is reached for $Q = 1/\sqrt{2}$. Another bound, frequently used in Chapter 6, follows directly from the upper bound in Eq. (H.4) if we require $Q \leq 1/\sqrt{2}$ (no amplitude peaking):

$$BW_{3dB} \leq \frac{\omega_0}{2\pi}. \tag{H.5}$$

This bound is shown with the dotted horizontal line in Fig. H.1.

Noise Bandwidths and Personick Integrals. The noise bandwidths BW_n and BW_{n2} of the transfer function in Eq. (H.1) are found by evaluating the integrals in Eqs. (4.44) and (4.45), which leads to the following surprisingly simple analytical expressions [1]:

$$\begin{aligned} BW_n &= \int_0^\infty |H(f)|^2 \, df \\ &= \int_0^\infty \frac{1}{1 + \left(\frac{1}{Q^2} - 2\right)\left(\frac{2\pi}{\omega_0}\right)^2 f^2 + \left(\frac{2\pi}{\omega_0}\right)^4 f^4} \, df \\ &= \frac{\pi Q}{2} \cdot \frac{\omega_0}{2\pi}, \end{aligned} \tag{H.6}$$

$$\begin{aligned} BW_{n2}^3 &= 3\int_0^{\infty} |H(f)|^2 f^2 \, df \\ &= 3\int_0^{\infty} \frac{f^2}{1+\left(\frac{1}{Q^2}-2\right)\left(\frac{2\pi}{\omega_0}\right)^2 f^2+\left(\frac{2\pi}{\omega_0}\right)^4 f^4} \, df \\ &= \frac{3\pi Q}{2}\cdot\left(\frac{\omega_0}{2\pi}\right)^3. \end{aligned} \tag{H.7}$$

With Eq. (H.2), the noise bandwidths can be expressed in terms of the 3-dB bandwidth:

$$BW_n = \frac{\pi Q}{2\rho}\cdot BW_{3\text{dB}}, \tag{H.8}$$

$$BW_{n2}^3 = \frac{3\pi Q}{2\rho^3}\cdot BW_{3\text{dB}}^3. \tag{H.9}$$

These noise bandwidths are plotted in Fig. H.2.

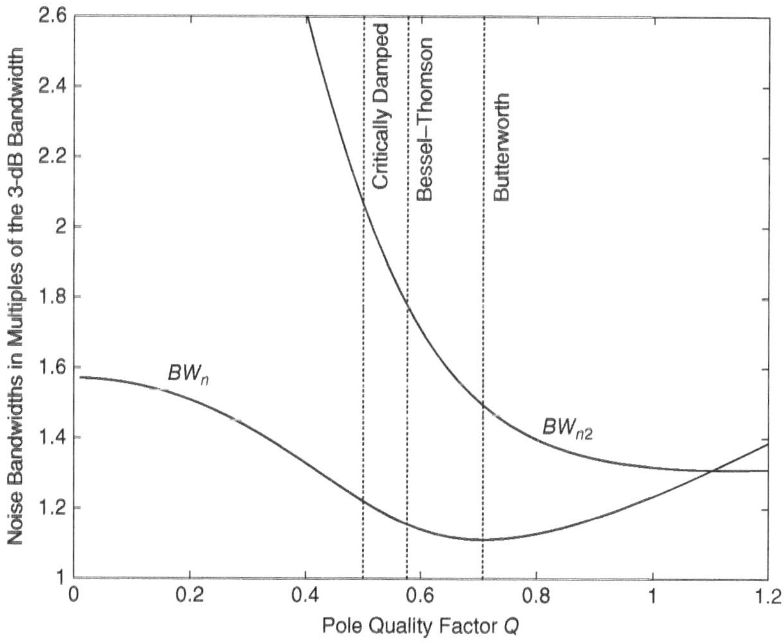

Figure H.2 Noise bandwidths BW_n and BW_{n2}, normalized to the 3-dB bandwidth $BW_{3\text{dB}}$, as a function of the pole quality factor Q (Eqs. (H.8) and (H.9)).

With Eqs. (4.48) and (4.49), the values of the second and third Personick integral are found as

$$I_2 = \frac{\pi Q}{2\rho} \cdot \frac{BW_{3dB}}{B}, \tag{H.10}$$

$$I_3 = \frac{\pi Q}{2\rho^3} \cdot \left(\frac{BW_{3dB}}{B}\right)^3. \tag{H.11}$$

Special Case: Real Poles. When $Q \leq 0.5$, the two poles become real and Eq. (H.1) can be rewritten in the form

$$H(s) = \frac{1}{(1 + s/\omega_{p1})(1 + s/\omega_{p2})}. \tag{H.12}$$

The poles are located at the (angular) frequencies $-\omega_{p1}$ and $-\omega_{p2}$. Let us define the pole spacing as the ratio

$$\chi = \frac{\omega_{p1}}{\omega_{p2}}. \tag{H.13}$$

By comparing Eq. (H.1) with Eq. (H.12), we find

$$\omega_0 = \sqrt{\omega_{p1}\omega_{p2}} = \frac{\omega_{p1}}{\sqrt{\chi}} = \sqrt{\chi}\ \omega_{p2} \tag{H.14}$$

and

$$Q = \frac{\sqrt{\omega_{p1}\omega_{p2}}}{\omega_{p1} + \omega_{p2}} = \frac{\sqrt{\chi}}{\chi + 1}. \tag{H.15}$$

Thus, the pole spacing χ and the pole quality factor Q contain the same information, but in the case of real poles, χ is the more meaningful parameter.

3-dB Bandwidth (Real Poles). The 3-dB bandwidth of the transfer function in Eq. (H.12), expressed in terms of ω_{p1} and χ or ω_{p2} and χ, can be found from Eq. (H.2) with Eqs. (H.14) and (H.15):

$$BW_{3dB} = \rho\left(\frac{\sqrt{\chi}}{\chi + 1}\right)\frac{1}{\sqrt{\chi}} \cdot \frac{\omega_{p1}}{2\pi} = \rho\left(\frac{\sqrt{\chi}}{\chi + 1}\right)\sqrt{\chi} \cdot \frac{\omega_{p2}}{2\pi}, \tag{H.16}$$

where

$$\rho\left(\frac{\sqrt{\chi}}{\chi + 1}\right) = \sqrt{\frac{\sqrt{\chi^4 + 6\chi^2 + 1} - \chi^2 - 1}{2\chi}}. \tag{H.17}$$

For large pole spacings $\rho \to 1/\sqrt{\chi}$.

Noise Bandwidths and Personick Integrals (Real Poles). The noise bandwidths BW_n and BW_{n2} of the transfer function in Eq. (H.12), expressed in terms of ω_{p1} and χ or ω_{p2} and χ, can be found from Eqs. (H.6) and (H.7) with Eqs. (H.14) and (H.15):

$$BW_n = \frac{\pi}{2(\chi+1)} \cdot \frac{\omega_{p1}}{2\pi} = \frac{\pi\chi}{2(\chi+1)} \cdot \frac{\omega_{p2}}{2\pi}, \tag{H.18}$$

$$BW_{n2}^3 = \frac{3\pi}{2(\chi+1)\chi} \cdot \left(\frac{\omega_{p1}}{2\pi}\right)^3 = \frac{3\pi\chi^2}{2(\chi+1)} \cdot \left(\frac{\omega_{p2}}{2\pi}\right)^3. \tag{H.19}$$

With Eq. (H.16) the noise bandwidths can be expressed in terms of the 3-dB bandwidth:

$$BW_n = \frac{\pi\sqrt{\chi}}{2(\chi+1)\rho} \cdot BW_{3dB}, \tag{H.20}$$

$$BW_{n2}^3 = \frac{3\pi\sqrt{\chi}}{2(\chi+1)\rho^3} \cdot BW_{3dB}^3. \tag{H.21}$$

These noise bandwidths are plotted in Fig. H.3.

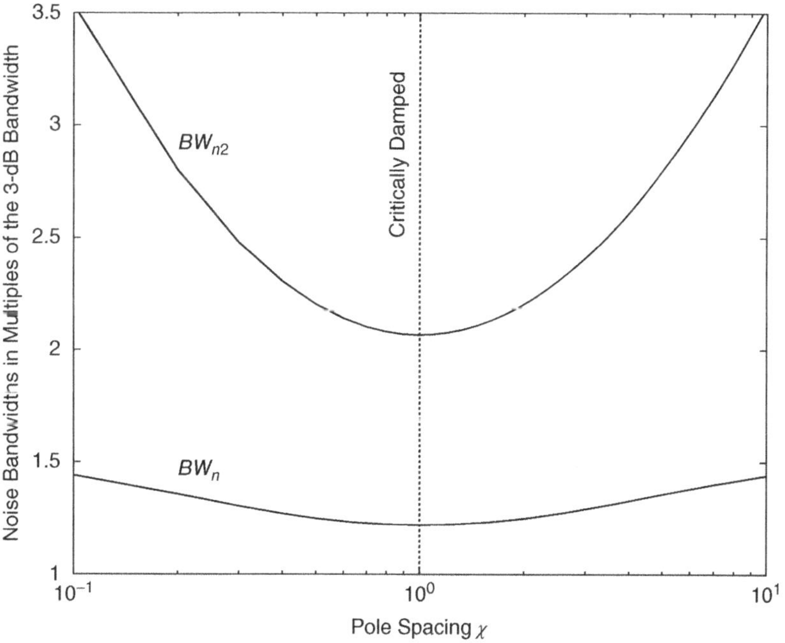

Figure H.3 Noise bandwidths BW_n and BW_{n2}, normalized to the 3-dB bandwidth BW_{3dB}, as a function of the pole spacing χ (Eqs. (H.20) and (H.21)).

The values of the second and third Personick integral can be found from Eqs. (H.10) and (H.11) with Eq. (H.15):

$$I_2 = \frac{\pi\sqrt{\chi}}{2(\chi+1)\rho} \cdot \frac{BW_{3dB}}{B}, \tag{H.22}$$

$$I_3 = \frac{\pi\sqrt{\chi}}{2(\chi+1)\rho^3} \cdot \left(\frac{BW_{3dB}}{B}\right)^3. \tag{H.23}$$

Critically Damped, Bessel–Thomson, and Butterworth. Next, we choose four representative values for the pole quality factor Q in the range from 0.5 to 1.0 and set the pole (angular) frequency ω_0 for each Q such that the 3-dB bandwidth becomes 1 Hz. The values for Q and ω_0 are listed in Table H.1.

- For $Q = 1/2$, we obtain the critically damped response, characterized by identical and real poles.
- For $Q = 1/\sqrt{3}$, we obtain the Bessel–Thomson response, characterized by a maximally flat group delay. The Bessel–Thomson response can be regarded as a rational approximation to the delay response e^{-sT} [2].
- For $Q = 1/\sqrt{2}$, we obtain the Butterworth response, characterized by a maximally flat amplitude. For the nth-order Butterworth response, the first $n-1$ derivatives of $|H(j\omega)|$ are all zero at $\omega = 0$ [2].
- For $Q = 1$, we obtain a response with significant peaking (ripple) in the frequency domain and overshoot (ringing) in the time domain.

Pole Locations. Figure H.4 shows the locations of the poles for our four responses in the s plane (root-locus plot). Both poles lie in the left half plane. As Q increases from 0.5 to 1.0, the poles move closer to the imaginary axis. The parameters ω_0 and Q determine the geometric pole locations as follows:

- For $Q \geq 0.5$, the poles of Eq. (H.1) are conjugate complex. Parameter ω_0 equals the distance from the poles to the origin and parameter Q equals $1/(2\cos\varphi)$, where φ represents the angle between the lines connecting the

Table H.1 Parameters of some important second-order low-pass responses with a 3-dB bandwidth of 1 Hz.

Response	Q	ω_0 (rad/s)
Critically damped	0.5	$2\pi/\sqrt{\sqrt{2}-1} = 9.76$
Bessel–Thomson	$1/\sqrt{3} = 0.577$	$2\pi/\sqrt{\sqrt{1.25}-0.5} = 7.99$
Butterworth	$1/\sqrt{2} = 0.707$	$2\pi = 6.28$
$Q = 1$	1.0	$2\pi/\sqrt{\sqrt{1.25}+0.5} = 4.94$

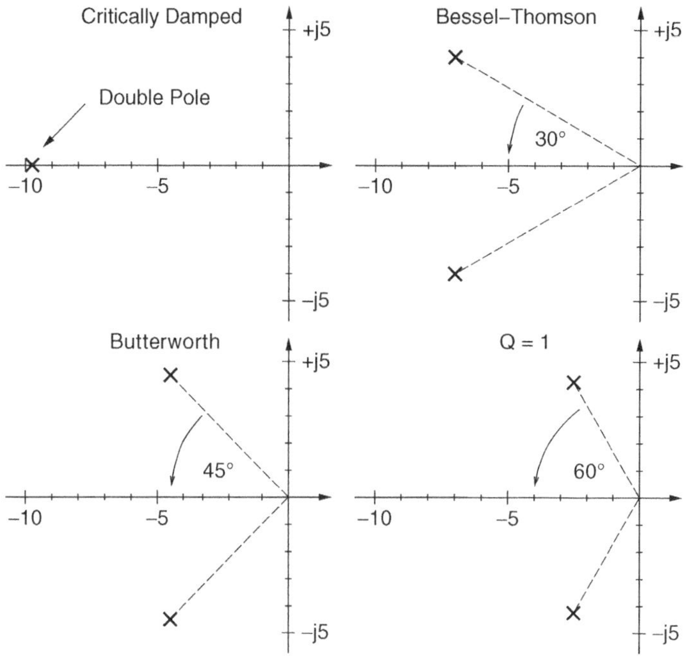

Figure H.4 Pole locations of some important second-order low-pass responses with a 3-dB bandwidth of 1 Hz (s is in radian per second).

poles to the origin and the real axis. Hence, the distance from the poles to the imaginary axis is $\omega_0/(2Q)$.

- For $Q \leq 0.5$, the poles become real and Eq. (H.1) can be rewritten as Eq. (H.12). The poles are located at $-\omega_{p1}$ and $-\omega_{p2}$. Their geometrical mean $\sqrt{\omega_{p1}\omega_{p2}}$ equals ω_0. The pole spacing $\chi = \omega_{p1}/\omega_{p2}$ is related to the pole quality factor as $Q = \sqrt{\chi}/(\chi + 1)$.

The amplitude response resulting from a particular pole configuration can be pictured by imagining a circus tent set up on top of the s-plane. Two tall vertical poles at the pole locations support the tent's canvas. The intersection of the canvas with the vertical plane rising up from the imaginary axis represents the amplitude response. Intuitively, we can understand that poles located closer to the imaginary axis result in an amplitude response with stronger peaking.

Frequency-Domain Response. The amplitude responses of our four transfer functions are shown in Fig. H.5. Only the response with the highest pole quality factor, $Q = 1$, exhibits peaking. The amount of peaking can be calculated as $Q^2/\sqrt{Q^2 - 0.25}$ for $Q \geq 1/\sqrt{2}$ [3] and is listed in Table H.2 for each response.

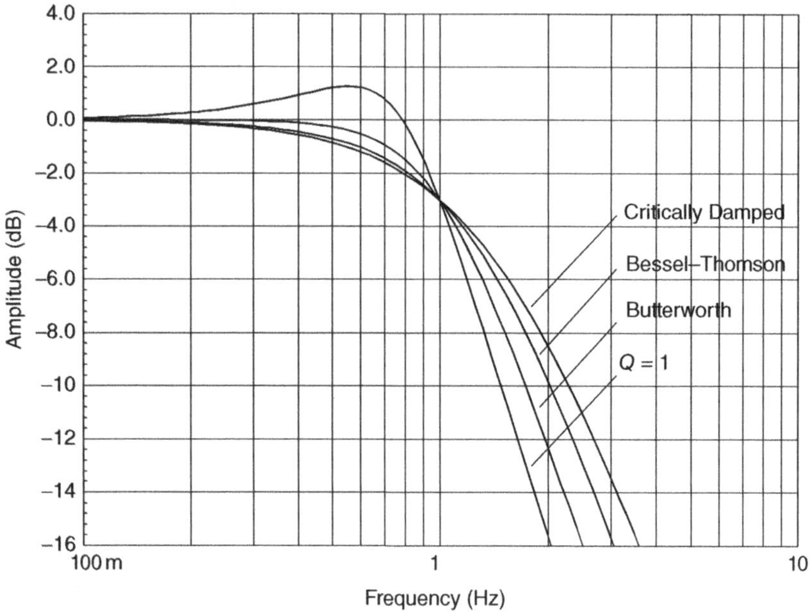

Figure H.5 Amplitude response of some important second-order low-pass transfer functions.

Table H.2 Frequency-domain characteristics of some important second-order low-pass responses

Response	Peaking	Phase lin. $\lvert\Delta\phi\rvert$	Group-delay var. $\Delta\tau$
Critically damped	0 dB	8.2°	0.058 s (0.087 UI)
Bessel–Thomson	0 dB	3.7°	0.039 s (0.058 UI)
Butterworth	0 dB	9.0°	0.047 s (0.070 UI)
$Q = 1$	1.25 dB (15.5%)	43.0°	0.234 s (0.350 UI)

The phase responses, $\phi(f)$, of our four transfer functions are shown in Fig. H.6. Linear frequency and phase scales are used for this graph such that a linear phase response appears as a straight line. The deviation of the phase from the linear phase, $\Delta\phi(f) = \phi(f) - [\partial\phi(f)/\partial f]_{f=0} \cdot f$, is shown in Fig. H.7. The phase linearity (or, more accurately, the phase *non*linearity), that is, the maximum of $|\Delta\phi(f)|$ for $f < BW_{3dB}$, is listed in Table H.2 for each response. As expected, the Bessel response has the best phase linearity.

The group-delay responses, $\tau(f) = -\partial\phi(f)/\partial(2\pi f)$ (with ϕ in radians), of our four transfer functions are shown in Fig. H.8. The group delay in the

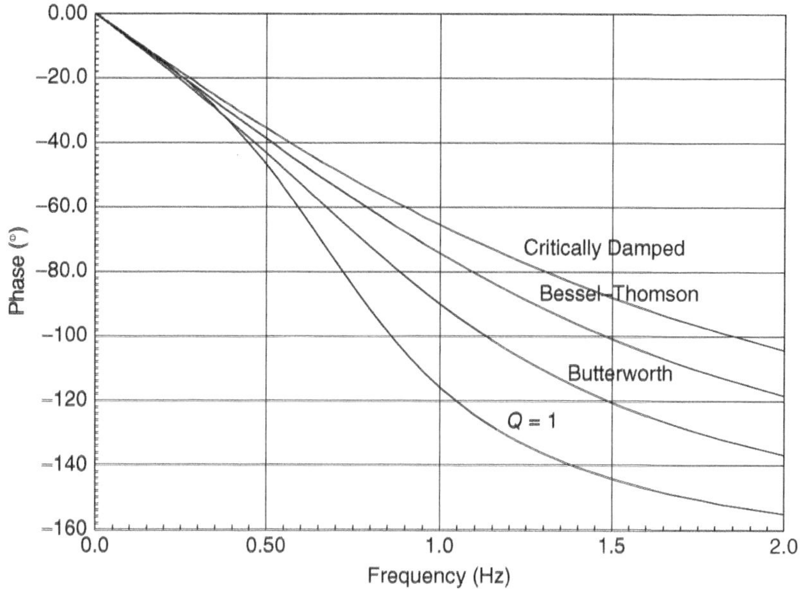

Figure H.6 Phase response of some important second-order low-pass transfer functions.

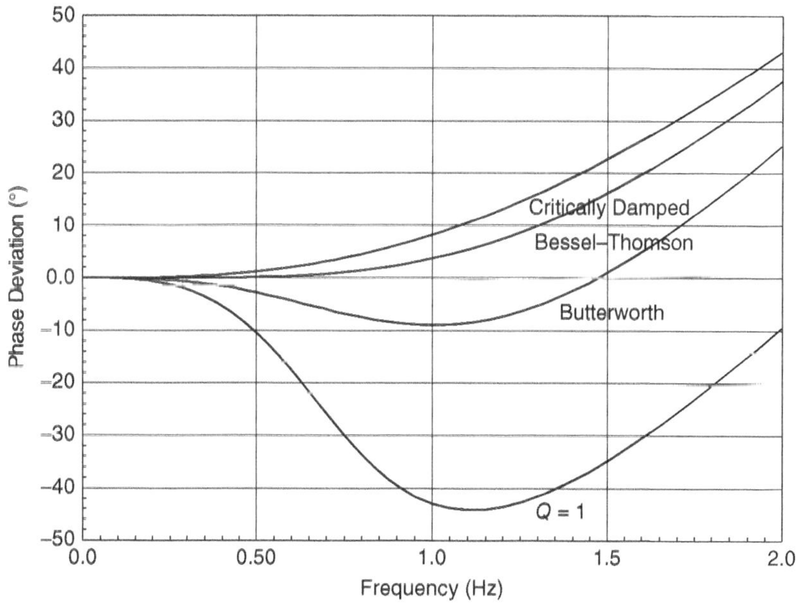

Figure H.7 Phase-deviation response (from linear phase) of some important second-order low-pass transfer functions.

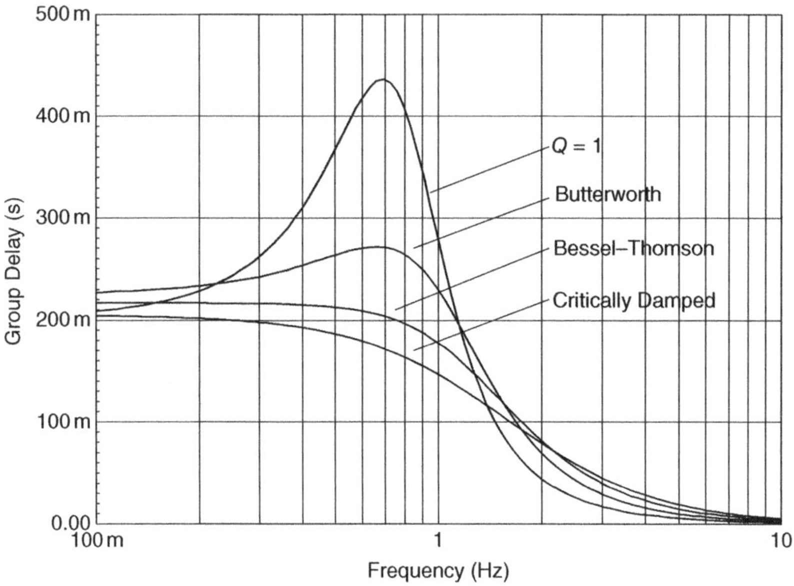

Figure H.8 Group-delay response of some important second-order low-pass transfer functions.

passband is approximately 0.25 s (or more generally, $0.25/BW_{3dB}$) because a second-order response turns the phase by about 90° (0.25 cycles) when the frequency is swept from zero to its 3-dB bandwidth (see Fig. H.6). The group-delay variation, that is, the maximum of $\Delta\tau(f) = |\tau(f) - \tau(0)|$ for $f < BW_{3dB}$, is listed in Table H.2 for each response. The group-delay variations are given in seconds (for $BW_{3dB} = 1$ Hz) as well as in unit intervals (assuming $1\ \text{UI} = \frac{2}{3}$ s, corresponding to $BW_{3dB} = \frac{2}{3}B$). Again, the Bessel response has the smallest group-delay variation.

Interestingly, the phase deviations from the linear phase in Fig. H.7 at first remain close to zero and start to build up only in the last 25 to 50% of the bandwidth. Based on this observation, we can estimate the group-delay variation from the phase linearity as $\Delta\tau \approx |\Delta\phi|/(2\pi BW_{3dB}\alpha)$, where ϕ is in radians and $\alpha = 0.25 \ldots 0.5$ for our responses.

Time-Domain Response. To discuss the time-domain response of our four transfer functions, we apply an ideal non-return-to-zero (NRZ) signal to the input and observe the eye diagram at the output. The 3-dB bandwidth of all transfer functions is now set to $\frac{2}{3}$ Hz and the bit rate is chosen 1 b/s (corresponding to $BW_{3dB} = \frac{2}{3}B$). The resulting eye diagrams are shown in Figs H.9–H.12.

All responses, except the critically damped one, show some overshoot. The amount of overshoot is listed in Table H.3 for each response. The overshoot in

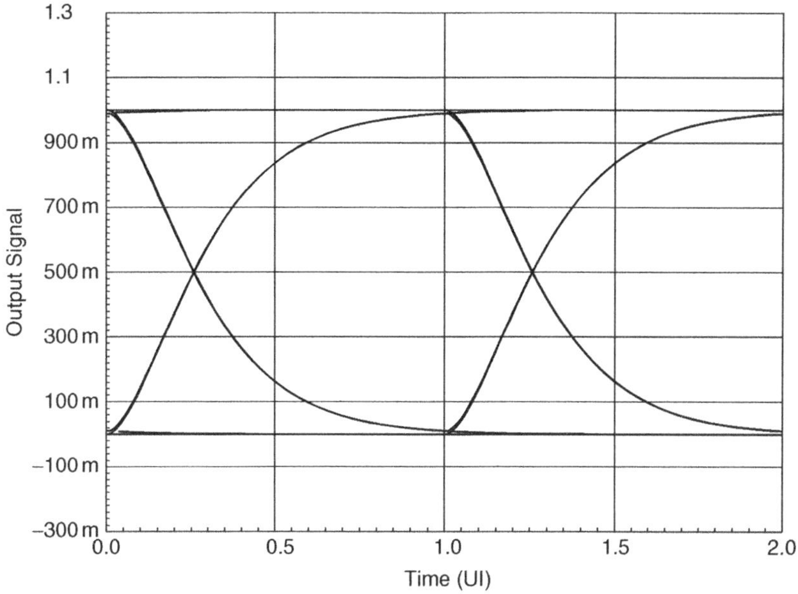

Figure H.9 Eye diagram for a critically damped second-order low-pass response.

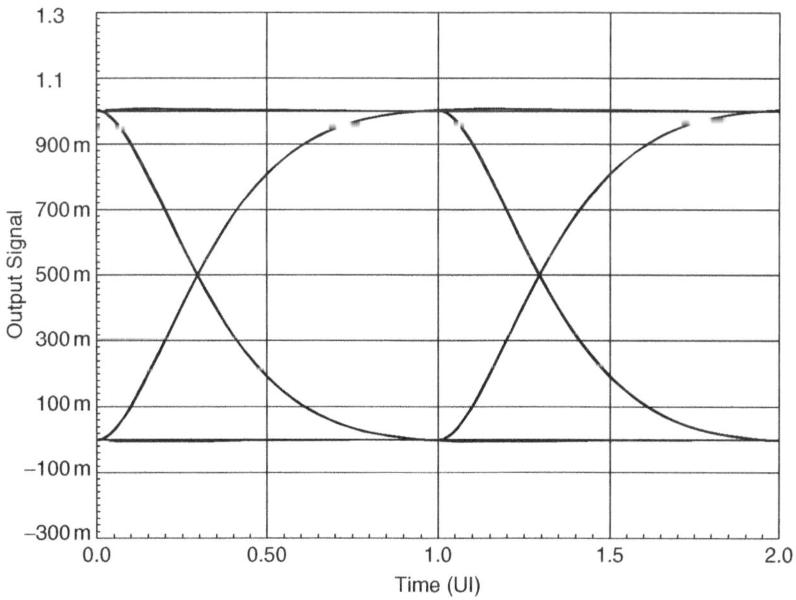

Figure H.10 Eye diagram for a Bessel–Thomson second-order low-pass response.

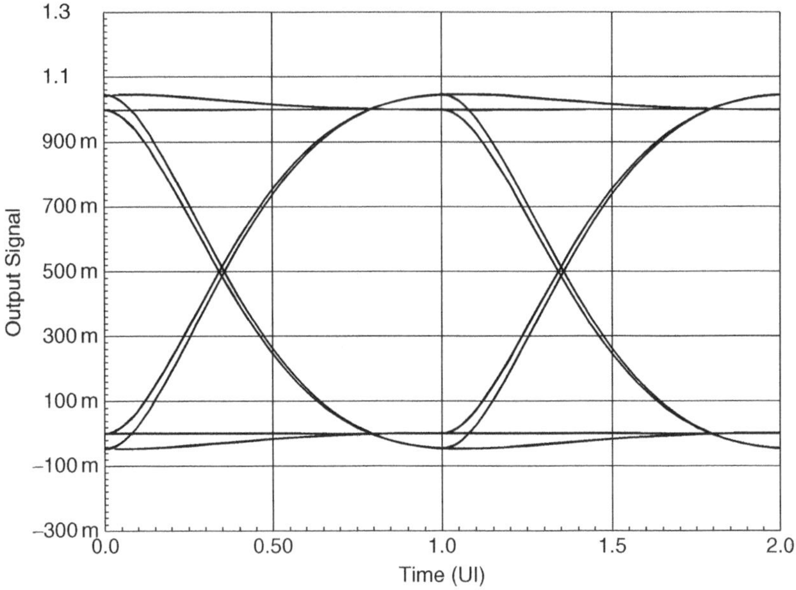

Figure H.11 Eye diagram for a Butterworth second-order low-pass response.

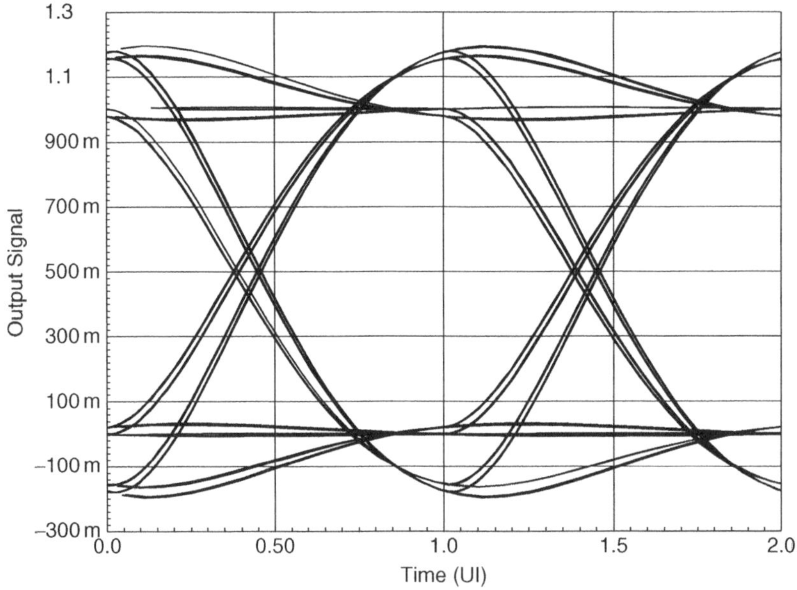

Figure H.12 Eye diagram for a second-order low-pass response with $Q = 1$.

Table H.3 Eye-diagram characteristics of some important second-order low-pass responses

Response	Overshoot (%)	Deterministic jitter t_{DJ}^{pp} (UI)
Critically damped	0	0.001
Bessel–Thomson	0.54	0.002
Butterworth	4.8	0.016
$Q = 1$	19.6	0.084

the eye diagram is slightly larger than that in the step response because of intersymbol interference (ISI). (The overshoot in the step response can be calculated as $\exp(-\pi/(2\sqrt{Q^2 - 0.25}))$ for $Q \geq 1/2$ [3] yielding 0%, 0.43%, 4.3%, and 16.3% for our four responses.)

The deterministic jitter measured peak-to-peak, t_{DJ}^{pp}, is also listed in Table H.3 for each response (cf. Appendix C). This jitter is significantly smaller than the corresponding group-delay variation, $\Delta\tau$ and the two measures are not necessarily monotonically related (cf. Table H.2). For example, the smallest jitter occurs for the critically damped response, whereas the smallest group-delay variation occurs for the Bessel response.

References

1. J. J. Morikuni, A. Dharchoudhury, Y. Leblebici, and S. M. Kang. Improvements to the standard theory for photoreceiver noise. *J. Lightwave Technol.*, LT-12(7):1174–1184, 1994.
2. P. Starič and E. Margan. *Wideband Amplifiers.* Springer, Dordrecht, The Netherlands, 2006.
3. P. E. Allen and D. R. Holberg. *CMOS Analog Circuit Design.* Holt, Rinehart and Winston, New York, 1987.

I

Answers to the Problems

Chapter 2

2.1(a) $f = c/\lambda = (299.8\ \text{Mm/s})/(1.55\ \mu\text{m}) = 193.4\ \text{THz}$.

2.1(b) Differentiating $f = c/\lambda$ with respect to λ, we find $\Delta f = c/\lambda^2 \cdot \Delta\lambda = (299.8\ \text{Mm/s})/(1.55\ \mu\text{m})^2 \cdot 0.1\ \text{nm} = 12.48\ \text{GHz}$.

2.2 The linear expression for $D(\lambda)$ is

$$D(\lambda) = 17\ \text{ps/(nm} \cdot \text{km)} \cdot \frac{\lambda - 1{,}300\ \text{nm}}{1{,}550\ \text{nm} - 1{,}300\ \text{nm}}.$$

Integrating $\partial\tau/\partial\lambda = D(\lambda) \cdot L$ (Eq. (2.2)) results in

$$\tau(\lambda) = 17\ \text{ps/(nm} \cdot \text{km)} \cdot \left(\frac{\lambda^2}{500\ \text{nm}} - 5.2 \cdot \lambda\right) \cdot L + \xi, \tag{I.1}$$

where ξ is an arbitrary constant. This is the quadratic relationship plotted in Fig. 2.4.

2.3 Convolving the Gaussian input pulse $x(t)$ with the impulse response $h(t)$ results in

$$\begin{aligned} y(t) &= \int_{-\infty}^{\infty} h(t - t') \cdot x(t')\, dt' \\ &= h(0) \cdot x(0) \cdot \int_{-\infty}^{\infty} \exp\left(-\frac{1}{2} \cdot \frac{(t - t')^2}{\sigma_T^2}\right) \cdot \exp\left(-\frac{1}{2} \cdot \frac{t'^2}{\sigma_{\text{in}}^2}\right) dt' \\ &= y(0) \cdot \exp\left(-\frac{1}{2} \cdot \frac{t^2}{\sigma_{\text{in}}^2 + \sigma_T^2}\right). \end{aligned} \tag{I.2}$$

It thus follows that $\sigma_{\mathrm{out}} = \sqrt{\sigma_{\mathrm{in}}^2 + \sigma_T^2}$, which, when multiplied by two, is equivalent to Eq. (2.7).

2.4 Calculating the Fourier transform of the impulse response $h(t)$ results in

$$H(f) = \int_{-\infty}^{\infty} h(t) \cdot \exp(-j\,2\pi f\,t)\,dt$$
$$= h(0) \cdot \int_{-\infty}^{\infty} \exp\left(-\frac{1}{2} \cdot \frac{t^2}{\sigma_T^2}\right) \cdot \exp(-j\,2\pi f\,t)\,dt$$
$$= H(0) \cdot \exp\left(-\frac{1}{2} \cdot \frac{(2\pi f)^2}{1/\sigma_T^2}\right), \tag{I.3}$$

which is equivalent to Eq. (2.9).

2.5 Inserting $f = B/2$ into Eq. (2.9) and comparing it with $0.794 \cdot H(0)$ for 1 dB of attenuation yields

$$\exp\left(-\frac{(\pi B)^2(\Delta T/2)^2}{2}\right) \geq 0.794.$$

Solving for B gives $B \leq \sqrt{-8\,\ln(0.794)}/(\pi \Delta T)$, or approximately $B \leq 1/(2 \cdot \Delta T)$. This, in fact, is how the spreading limit given in Eq. (2.8) was derived in [1].

2.6 From Fig. 2.4, we see that for $D > 0$, shorter wavelengths propagate faster than longer wavelengths. A pulse with negative chirp has a longer wavelength during the leading edge (red shift) and a shorter wavelength (blue shift) during the trailing edge. Thus, the trailing edge will "catch up" with the leading edge, effectively compressing the pulse.

2.7(a) $L = 21.5\ \mathrm{dB}/(0.35\ \mathrm{dB/km}) = 61.4\ \mathrm{km}$.

2.7(b) With $|D| \cdot \Delta\lambda \cdot L = 1/(2B)$ we find $L = 1/(2 \cdot 0.3\ \mathrm{ps/(nm \cdot km)} \cdot 2\ \mathrm{nm} \cdot 10\ \mathrm{Gb/s}) = 83.3\ \mathrm{km}$. The maximum transmission distance is 61.4 km, limited by attenuation.

2.8(a) $L = 21.5\ \mathrm{dB}/(0.2\ \mathrm{dB/km}) = 107.5\ \mathrm{km}$.

2.8(b) $L = 1/(2 \cdot 17\ \mathrm{ps/(nm \cdot km)} \cdot 2\ \mathrm{nm} \cdot 10\ \mathrm{Gb/s}) = 1.47\ \mathrm{km}$. The maximum transmission distance is 1.47 km, limited by chromatic dispersion.

2.9(a) The dispersion-limited system of Problem 2.8.

2.9(b) The dispersion limit increases to 147 km. The system is now limited by attenuation to a distance of 107.5 km.

2.10 We do not have to worry about PMD. Even for 107.5 km of fiber, we have $\overline{\Delta T} = 0.1\ \text{ps}/\sqrt{\text{km}} \cdot \sqrt{107.5\ \text{km}} = 1.04$ ps, which is significantly lower than $0.1/(10\ \text{Gb/s}) = 10$ ps, thus the outage probability is extremely small.

Chapter 3

3.1 The optical attenuation is $40\ \text{km} \cdot 0.2\ \text{dB/km} = 8$ dB; hence the electrical attenuation is $2 \cdot 8\ \text{dB} = 16$ dB.

3.2 The missing power is supplied by the voltage source used to reverse bias the photodiode (V_{PIN}). The photodiode itself generates electrical power only in the fourth quadrant ($V > 0, I < 0$, see Fig. 3.2) and energy conservation requires that $-I \cdot V < P$ (a hyperbola limiting how far the I/V curve can go to the lower-right side), which is always met.

3.3(a) The rms signal current is $\sqrt{P/R_{\text{ANT}}}$ under matched conditions and the rms noise current is $\sqrt{kT/R_{\text{ANT}} \cdot BW_n}$ under matched conditions. They become equal for $P = kT \cdot BW_n$.

3.3(b) With $T = 300$ K and $BW_n = 100$ MHz, the sensitivity is -94 dBm.

3.3(c) For $f = 1$ GHz and $T = 300$ K, we have $hf \ll kT$, and Eq. (3.35) can be approximated as $P \approx (kT + hf/2) \cdot BW_n \approx kT \cdot BW_n$, which agrees with 3.3(a).

3.4 Conventional answer: The shot-noise current in the battery/resistor circuit is strongly "suppressed" and usually is not measurable. Shot noise in its full strength, $\overline{i_n^2} = 2qI \cdot BW_n$, only occurs if the carriers cross from one electrode to another electrode, without "obstacles." This is the case to a good approximation in p–n junctions and vacuum tubes but not in resistors. However, the resistor R produces a thermal noise current, $\overline{i_n^2} = 4kT/R \cdot BW_n$, which is independent of the DC current.

Unconventional answer [2]: The instantaneous current i in the battery/resistor circuit can be decomposed into three (additive) currents. A drift current i_D, a forward diffusion current i_F, and a backward diffusion current i_R. Whereas the current pulses of i_D are periodic (noise free), the current pulses of i_D and i_F are uncorrelated and Poisson distributed. The total average current is $I = \overline{i_D}$ because the average diffusion currents $\overline{i_F} = -\overline{i_R} = kT/(qR)$ cancel each other. The total noise current results from the shot noise of the two diffusion currents and is calculated as $\overline{i_n^2} = 2q|\overline{i_F}| \cdot BW_n + 2q|\overline{i_R}| \cdot BW_n$. This expression evaluates to the thermal noise expression $\overline{i_n^2} = 4kT/R \cdot BW_n$!

3.5 The shot noise equation $\overline{i_n^2} = 2qI \cdot BW_n$ applies only to *randomly* arriving carriers. However, in the deterministic APD, each photon generates a group of M carriers with highly correlated arrival times. In fact, we could say that the current in this APD ($I_{APD} = MI_{PIN}$) consists of "coarse" carriers with the charge Mq. Substituting these quantities into the shot noise equation yields the correct result: $\overline{i_{n,APD}^2} = 2 \cdot (Mq) \cdot (MI_{PIN}) \cdot BW_n$.

3.6 The noise current of an optically preamplified p–i–n detector, including the shot noise terms, is

$$\begin{aligned}\overline{i_{n,OA}^2} = {} & 2R^2 P_S S_{ASE} \cdot BW_n + R^2 S_{ASE}^2 \cdot BW_O \cdot BW_n \\ & +2qRP_S \cdot BW_n + 2qRS_{ASE} \cdot BW_O \cdot BW_n \\ & +2qI_{DK} \cdot BW_n.\end{aligned} \tag{I.4}$$

The 5 terms are (i) signal–spontaneous beat noise, (ii) spontaneous–spontaneous beat noise, (iii) shot noise due to the signal power, (iv) shot noise due to the ASE power, and (v) shot noise due to the detector dark current (cf. [3] for terms (i)–(iv)).

3.7 The average electrical signal power is $\overline{(i_{OA}(t) - I_{OA})^2} = R^2 \cdot \overline{(P_S(t) - \overline{P_S})^2}$, which is equal to $R^2 \overline{P_S}^2$ for a DC-balanced ideal NRZ signal with high extinction. The average electrical noise power is $\overline{i_{n,OA}^2} = R^2(2\overline{P_S} S_{ASE} + S_{ASE}^2 \cdot BW_O) BW_n$. Thus, Eqs. (3.20) and (3.21) remain valid for a DC-balanced ideal NRZ signal after the substitution $P_S \rightarrow \overline{P_S}$.

3.8(a) The noise figure according to the definition Eq. (3.22) is $F = \overline{i_{n,OA}^2}/(2R^2 GP_S \cdot hc/\lambda \cdot BW_n)$, where we used $I_{PIN} = RP$ and $R = \lambda q/(hc)$ for the ideal p–i–n detector current and $P = P_S/G$

for the input signal power. Inserting $\overline{i_{n,OA}^2}$ from Eq. (I.4) yields

$$F = \frac{S_{ASE}}{G} \cdot \frac{\lambda}{hc} \cdot \left(1 + \frac{S_{ASE} \cdot BW_O}{2P_S}\right) + \frac{1}{G} \cdot \left(1 + \frac{S_{ASE} \cdot BW_O}{P_S} + \frac{I_{DK}}{\mathcal{R}P_S}\right). \tag{I.5}$$

The 5 terms are due to (i) signal–spontaneous beat noise, (ii) spontaneous–spontaneous beat noise, (iii) shot noise due to the signal power, (iv) shot noise due to the ASE power, and (v) shot noise due to the detector dark current.

3.8(b) For $S_{ASE} = 0$, we have $F = 1/G$ from the third term. This is less than one for $G > 1$ (corresponding to a negative noise figure when expressed in dBs). Note that a noiseless amplifier, as in Fig. 3.22(b), needs $S_{ASE} > 0$ to account for the quantum (shot) noise of the source. However, for a noiseless amplifier with $G < 1$ (a passive attenuator), the equation $F = 1/G$ predicts the correct noise figure.

3.8(c) Neglecting the second, fourth, and fifth term, we find $F = S_{ASE}/G \cdot \lambda/(hc) + 1/G$. Thus for $F = 1$ (noiseless amplifier), we need $S_{ASE} = (G-1) \cdot hc/\lambda$.

3.9(a) Following the definition for the noise figure. The total output noise is the shot noise $\overline{i_n^2} = 2q\mathcal{R}G_1P \cdot BW_n$ where P is the optical input power. The output noise due to the source is $\overline{i_{n,S}^2} = G_1^2 \cdot 2q\mathcal{R}P \cdot BW_n$, as in Fig. 3.22(b) with $I_{PIN} = \mathcal{R}P$. The ratio $\overline{i_n^2}/\overline{i_{n,S}^2}$ is the noise figure $F = 1/G_1$.

3.9(b) Following the definition for the noise figure. The total output noise is $\overline{i_n^2} = F_2G_2^2 \cdot 2q\mathcal{R}G_1P \cdot BW_n$, as in Fig. 3.22(c) with $I_{PIN} = \mathcal{R}G_1P$. The output noise due to the source is $\overline{i_{n,S}^2} = (G_2G_1)^2 \cdot 2q\mathcal{R}P \cdot BW_n$. The ratio $\overline{i_n^2}/\overline{i_{n,S}^2}$ is the noise figure

$$F = \frac{1}{G_1} \cdot F_2. \tag{I.6}$$

In dBs, the total noise figure is the sum of the fiber loss and the amplifier noise figure.

3.9(c) Following the definition for the noise figure. The total output noise is $\overline{i_n^2} = nF_2G^2 \cdot 2q\mathcal{R}P/G \cdot BW_n$ (note that the input power of each segment is P and that the gain from each segment output to the system output is one; thus, the n (uncorrelated) noise contributions are equal and add up directly). The output noise due to the source is $\overline{i_{n,S}^2} = 2qRP \cdot BW_n$. The ratio $\overline{i_n^2}/\overline{i_{n,S}^2}$ is the noise figure

$$F = nGF_2. \tag{I.7}$$

In dBs, the total noise figure is the sum of 10 log n, the amplifier gain, and the amplifier noise figure.

Chapter 4

4.1 If the input noise PSD is time independent on the interval $[t - \xi \ldots t]$ for a given t, that is, $I_{n,PD}^2(f, [t - \xi \ldots t]) = I_{n,PD}^2(f, t)$, then we can rewrite Eq. (4.4) as $V_{n,PD}^2(f, t) = H(f) \cdot I_{n,PD}^2(f, t) \cdot \int_{-\infty}^{\infty} h(t') \cdot e^{j\,2\pi f\,t'}\,dt' = H(f) \cdot I_{n,PD}^2(f, t) \cdot H^*(f) = |H(f)|^2 \cdot I_{n,PD}^2(f, t)$. Thus, the approximation is valid if we sample the output signal ξ time units (the duration of the impulse response) after a change in the input noise PSD (i.e., after a bit transition).

4.2(a) Integrating the one-sided output PSD Eq. (4.4) results in $\int_0^{\infty} \int_{-\infty}^{\infty} H(f) \cdot I_{n,PD}^2(t - t') \cdot h(t') \cdot \exp(j2\pi f t')\,dt'\,df$. Swapping the integrals and moving frequency independent terms out of the frequency integral yields $\int_{-\infty}^{\infty} I_{n,PD}^2(t - t') \cdot h(t') \int_0^{\infty} H(f) \cdot \exp(j2\pi f t')\,df\,dt'$. The frequency integral evaluates to $h(t')/2$ and thus

$$\overline{v_{n,PD}^2}(t) = \frac{1}{2} \cdot \int_{-\infty}^{\infty} I_{n,PD}^2(t - t') \cdot h^2(t')\,dt', \tag{I.8}$$

where $I_{n,PD}^2(t)$ is the *one-sided* detector noise PSD.

4.2(b) The first-order low-pass frequency response $H(f) = H_0/(1 + jf/BW_{3dB})$ has the impulse response $h(t) = H_0 \cdot 2\pi BW_{3dB} \cdot \exp(-2\pi BW_{3dB} t)$ for $t > 0$ (and zero for $t < 0$). Evaluating Eq. (I.8) at $t = T$ $(T = 1/B)$ with $I_{n,PD}^2(t) = 0$ if $t \in [0 \ldots T]$ and $I_{n,PD}^2(t) = I_{n,PD}^2$ elsewhere (…11011 … bit pattern) yields

$$\overline{v_{n,PD,0}^2}(T) = \frac{H_0^2 \pi BW_{3dB}}{2} \cdot I_{n,PD}^2 \cdot \exp\left(-4\pi \frac{BW_{3dB}}{B}\right). \tag{I.9}$$

Evaluating Eq. (I.8) at $t = T$ with $I_{n,PD}^2(t) = I_{n,PD}^2$ for all t (…11111 … bit pattern) yields

$$\overline{v_{n,PD,1}^2}(T) = \frac{H_0^2 \pi BW_{3dB}}{2} \cdot I_{n,PD}^2 \cdot \tag{I.10}$$

Note that $H_0^2 \pi BW_{3\text{dB}}/2$ is equal to the integral $\int_0^\infty |H(f)|^2\,df$ and thus the noise for the ones is predicted correctly by Eq. (4.5). The spillover term $\exp(-4\pi BW_{3\text{dB}}/B)$ affecting the noise for the zeros is a puny 0.023% for $BW_{3\text{dB}}/B = \frac{2}{3}$.

4.3 The two unequal Gaussian distributions are $(1/v_{n,0}^{rms})\,\text{Gauss}[v_O/v_{n,0}^{rms}]$ and $(1/v_{n,1}^{rms})\,\text{Gauss}[(v_O - v_S^{pp})/v_{n,1}^{rms}]$, where we assumed, without loss of generality, that the zero level is at 0. Equating these two distributions and solving for v_O yields the optimum decision threshold voltage V_{DTH}. Neglecting the different heights of the distributions, we find $V_{DTH} = v_S^{pp} \cdot v_{n,0}^{rms}/(v_{n,0}^{rms} + v_{n,1}^{rms})$. Integrating the two tails results in $BER = \frac{1}{2}(1/v_{n,0}^{rms}) \int_{V_{DTH}}^{\infty} \text{Gauss}[v_O/v_{n,0}^{rms}]\,dv_O + \frac{1}{2}(1/v_{n,1}^{rms}) \int_{-\infty}^{V_{DTH}} \text{Gauss}[(v_O - v_S^{pp})/v_{n,1}^{rms}]\,dv_O$. Substituting $x_0 = v_O/v_{n,0}^{rms}$ and $x_1 = (v_O - v_S^{pp})/v_{n,1}^{rms}$ yields $BER = \frac{1}{2}\int_Q^\infty \text{Gauss}(x_0)\,dx_0 + \frac{1}{2}\int_Q^\infty \text{Gauss}(x_1)\,dx_1 = \int_Q^\infty \text{Gauss}(x)\,dx$, where $Q = v_S^{pp}/(v_{n,0}^{rms} + v_{n,1}^{rms})$.

4.4 With $\xi = v_{n,0}^{rms}/v_{n,1}^{rms}$, we can rewrite Eq. (4.10) as $Q^2 = (v_S^{pp})^2/[(\xi+1)^2 \cdot \overline{v_{n,1}^2}]$ and Eq. (4.12) as $SNR = (v_S^{pp})^2/[2(\xi^2+1) \cdot \overline{v_{n,1}^2}]$. Thus, Eq. (4.13) can be generalized to

$$SNR = \frac{(\xi+1)^2}{2(\xi^2+1)} \cdot Q^2. \tag{I.11}$$

4.5(a) If we normalize the noise power to 1, the swing of the finite slope NRZ signal must be $2Q$ to achieve the specified BER. The time-averaged mean-free signal power, and thus the SNR, of that signal is $SNR = [\xi/6 + (1-\xi) + \xi/6]Q^2 = (1 - \frac{2}{3}\xi)Q^2$.

4.5(b) For a 0.3-UI rise/fall-time NRZ signal and $BER = 10^{-12}$, $SNR = 0.8 \cdot 7.034^2 = 39.6$ (16.0 dB).

4.5(c) When sampling in the flat parts of the signal, the sampled SNR does not depend on ξ and is always $SNR = Q^2$.

4.6 They are all right, but they use different SNR definitions. The student following the SNR convention used in this book comes up with the result (c).
$10\log Q = 10\log(1/[0.14 + 0.02]) = 7.96$ dB (a)
$10\log Q^2 = 20\log(1/[0.14 + 0.02]) = 15.92$ dB (d)
$10\log$(peak signal power/average continuous-time noise power) =
$10\log([1^2]/[0.25 \cdot 0.14^2 + 0.75 \cdot 0.02^2]) = 22.84$ dB (g)

10 log (average c.-t. signal power/average c.-t. noise power) = 10 log ($[0.25 \cdot 1^2]/[0.25 \cdot 0.14^2 + 0.75 \cdot 0.02^2]$) = 16.82 dB (e)
10 log (av. c.-t. signal power w/o DC power/av. c.-t. noise power) = 10 log ($[0.25 \cdot 1^2 - 0.25^2]/[0.25 \cdot 0.14^2 + 0.75 \cdot 0.02^2]$) = 15.57 dB (c)
10 log (average sampled signal power/average sampled noise power) = 10 log ($[0.5 \cdot 1^2]/[0.5 \cdot 0.14^2 + 0.5 \cdot 0.02^2]$) = 16.99 dB (f)
10 log (av. sampled signal power w/o DC power/av. sampled noise power) = 10 log ($[0.5 \cdot 1^2 - 0.5^2]/[0.5 \cdot 0.14^2 + 0.5 \cdot 0.02^2]$) = 13.98 dB (b)

4.7(a) E_b is the *signal power* times the information bit period $1/(rB)$. N_0 is the *noise power* divided by the noise bandwidth BW_n of the linear channel. Hence

$$\frac{E_b}{N_0} = SNR \cdot \frac{BW_n}{rB}. \tag{I.12}$$

4.7(b) When the noise bandwidth BW_n is equal to the information bit rate (a.k.a. the system bit rate), rB.

4.8 Given the average optical power $\overline{P}_S$ and the extinction ratio ER, the power for zeros and ones are $P_0 = 2\overline{P}_S/(ER+1)$ and $P_1 = 2\overline{P}_S \cdot ER/(ER+1)$, respectively; thus, $i_S^{pp} = \mathcal{R}(P_1 - P_0) = 2\mathcal{R}\overline{P}_S \cdot (ER-1)/(ER+1)$. Solving for $\overline{P}_S$ and inserting Eq. (4.20) for i_S^{pp} yields

$$\overline{P}_{\text{sens}} = \frac{ER+1}{ER-1} \cdot \frac{\mathcal{Q}(i_{n,0}^{rms} + i_{n,1}^{rms})}{2\mathcal{R}}. \tag{I.13}$$

4.9 The rule is equivalent to Eq. (4.27) when written in the log domain and specialized for $BER = 10^{-12}$. The value $10 \log \mathcal{Q}(BER = 10^{-12}) - 30 = -21.53$ [dBm] is the sensitivity of a p–i–n receiver with $\mathcal{R} = 1$ A/W and $i_{n,\text{ckt}}^{rms} = 1$ μA.

4.10(a) With Eq. (3.19), we find the noise power for the zeros as $\overline{i_{n,0}^2} = \overline{i_{n,\text{ckt}}^2} + \mathcal{R}^2 S_{ASE}^2 \cdot BW_O \cdot BW_n$ and the noise power for the ones as $\overline{i_{n,1}^2} = \overline{i_{n,0}^2} + 4\mathcal{R}^2 G\overline{P}_S S_{ASE} \cdot BW_n$. With Eq. (4.22), the sensitivity is $\overline{P}_{\text{sens},OA} = \mathcal{Q}(i_{n,0}^{rms} + i_{n,1}^{rms})/(2G\mathcal{R})$. Setting $\overline{P}_S = \overline{P}_{\text{sens},OA}$ and solving for $\overline{P}_{\text{sens},OA}$ yields

$$\overline{P}_{\text{sens},OA} = \frac{\mathcal{Q}}{G\mathcal{R}} \cdot \sqrt{\overline{i_{n,\text{ckt}}^2} + (\mathcal{R}S_{ASE})^2 \cdot BW_O \cdot BW_n} + \frac{\mathcal{Q}^2 S_{ASE} \cdot BW_n}{G}. \tag{I.14}$$

4.10(b) Replacing S_{ASE} with $\eta \tilde{F} q G / \mathcal{R}$ (from Eqs. (3.27) and (3.3)) yields

$$\overline{P}_{\text{sens},OA} = \frac{\mathcal{Q}}{G\mathcal{R}} \cdot \sqrt{\overline{i_{n,\text{ckt}}^2} + (\eta \tilde{F} q G)^2 \cdot BW_O \cdot BW_n} + \eta \tilde{F} \cdot \frac{\mathcal{Q}^2 q \cdot BW_n}{\mathcal{R}}, \tag{I.15}$$

which agrees with the result given in [4] under the condition that $2BW_O \gg BW_n$.

4.11 According to Eq. (4.52), the necessary transmit power is $\overline{P}_{\text{out}} = \eta n G F \cdot \mathcal{Q}^2 q \cdot BW_n / \mathcal{R}$ when neglecting the circuit noise. With Eq. (3.3), this power can be rewritten as $\overline{P}_{\text{out}} = nGF \cdot \mathcal{Q}^2 \cdot hc/\lambda \cdot BW_n$. Solving Eq. (4.54) for $\overline{P}_{\text{out}}$ and inserting Eq. (4.53) for $OSNR$ also leads to $\overline{P}_{\text{out}} = nGF \cdot \mathcal{Q}^2 \cdot hc/\lambda \cdot BW_n$.

4.12(a) The Q-factor budget is shown in the following table. The worst-case Q factor corresponds to $BER = 10^{-4}$.

Parameter	**Symbol**	**Value**
Transmitter power	$\overline{P}_{\text{out}}$	4.0 dBm
Constant in Eq. (4.55)		+58.0 dB
Amplifier gain (or span loss)	G	−10.0 dB
Amplifier noise figure	F	−5.0 dB
Number of amplifiers (10 log n in dB)	n	−23.0 dB
OSNR according to Eq. (4.55)	$OSNR$	24.0 dB
Ideal $\mathcal{Q}$ from OSNR value (20 log $\mathcal{Q}$ in dB)	$\mathcal{Q}$	19.4 dB
Penalty for disp., nonlin., crosstalk, etc.		−3.0 dB
Penalty for manuf. and time variations		−3.0 dB
Margin for aging and repair		−1.0 dB
System margin		−1.0 dB
Worst-case $\mathcal{Q}$ factor (20 log $\mathcal{Q}$ in dB)	$\mathcal{Q}$	11.4 dB

where the ideal $\mathcal{Q}$ is calculated as $3.72/\sqrt{39.8} \cdot \sqrt{OSNR}$ for a receiver that requires $OSNR = 39.8$ (16 dB) at $\mathcal{Q} = 3.72$ ($BER = 10^{-4}$).

4.12(b) At the beginning of life, 20 log $\mathcal{Q} = 13.4$ dB should be observed.

4.13 The noise power for the zeros is $\overline{i_{n,0}^2} = \overline{i_{n,\text{ckt}}^2}$, and with Eq. (3.19), the noise power for the ones is $\overline{i_{n,1}^2} = \overline{i_{n,\text{ckt}}^2} + 4\mathcal{R}^2\overline{P}_S S_{ASE} \cdot BW_n$. With $OSNR = \overline{P}_S/(S_{ASE} \cdot BW_O)$, the noise for the ones can be rewritten in terms of OSNR as $\overline{i_{n,1}^2} = \overline{i_{n,\text{ckt}}^2} + 4\mathcal{R}^2\overline{P}_S^2 \cdot BW_n/(OSNR \cdot BW_O)$. Inserting into Eq. (4.22), setting $\overline{P}_S = \overline{P}_{\text{sens}}$, and solving for $\overline{P}_{\text{sens}}$ yields

$$\overline{P}_{\text{sens}} = \frac{1}{1 - Q^2/OSNR \cdot BW_n/BW_O} \cdot \frac{Q\, i_{n,\text{ckt}}^{rms}}{\mathcal{R}}. \tag{I.16}$$

4.13(a) For $OSNR \to \infty$, the sensitivity becomes identical to $\overline{P}_{\text{sens},PIN}$ in Eq. (4.27).

4.13(b) With decreasing OSNR, more received power is needed. In particular, for $OSNR = Q^2 \cdot BW_n/BW_O$, $\overline{P}_{\text{sens}} \to \infty$. Thus for a high received power, we need at least $OSNR = Q^2 \cdot BW_n/BW_O$ to meet the specified BER, in agreement with Eq. (4.53).

4.14 For a finite extinction ratio (and neglecting spontaneous–spontaneous beat noise), the noise for the zeros is $i_{n,0}^{rms} = \mathcal{R}\sqrt{4/(ER+1)\overline{P}P_{ASE}}\sqrt{BW_n/BW_O}$ and the noise for the ones is $i_{n,1}^{rms} = \mathcal{R}\sqrt{4ER/(ER+1)\overline{P}P_{ASE}}\sqrt{BW_n/BW_O}$. The signal swing is $i_S^{pp} = 2\mathcal{R}(ER-1)/(ER+1)\overline{P}$. With $Q = i_S^{pp}/(i_{n,0}^{rms} + i_{n,1}^{rms})$, we can generalize Eq. (4.53) to [5]

$$OSNR_{\text{req}} = \frac{\sqrt{ER}+1}{\sqrt{ER}-1} \cdot \frac{ER+1}{ER-1} \cdot Q^2 \cdot \frac{BW_n}{BW_O}. \tag{I.17}$$

4.15(a) Including the spontaneous–spontaneous beat noise (and assuming high extinction), we find the noise for the zeros as $i_{n,0}^{rms} = \mathcal{R}P_{ASE}\sqrt{BW_n/BW_O}$ and the noise for the ones as $i_{n,1}^{rms} = \mathcal{R}\sqrt{4\overline{P}P_{ASE} + P_{ASE}^2}\sqrt{BW_n/BW_O}$. The signal swing is $i_S^{pp} = 2\mathcal{R}\overline{P}$. With $Q = i_S^{pp}/(i_{n,0}^{rms} + i_{n,1}^{rms})$, we can generalize Eq. (4.53) to [6, 7]

$$Q = \frac{2OSNR_{\text{req}}}{\sqrt{4OSNR_{\text{req}}+1}+1} \cdot \sqrt{\frac{BW_O}{BW_n}}. \tag{I.18}$$

4.15(b) For a finite extinction ratio (and including spontaneous–spontaneous beat noise), the noise for the zeros is $i_{n,0}^{rms} = \mathcal{R}\sqrt{4/(ER+1)\overline{P}P_{ASE} + P_{ASE}^2}\sqrt{BW_n/BW_O}$ and the noise for the ones is $i_{n,1}^{rms} = \mathcal{R}\sqrt{4ER/(ER+1)\overline{P}P_{ASE} + P_{ASE}^2}\sqrt{BW_n/BW_O}$. The signal swing is $i_S^{pp} = 2\mathcal{R}(ER-1)/(ER+1)\overline{P}$. With $Q = i_S^{pp}/(i_{n,0}^{rms} + i_{n,1}^{rms})$, we can generalize Eq. (4.53) to [8]

$$Q = \frac{2\frac{ER-1}{ER+1}OSNR_{\text{req}}}{\sqrt{\frac{4ER}{ER+1}OSNR_{\text{req}} + 1} + \sqrt{\frac{4}{ER+1}OSNR_{\text{req}} + 1}} \cdot \sqrt{\frac{BW_O}{BW_n}}. \tag{I.19}$$

4.16 At the sensitivity limit, the optical preamplifier produces the following optical signal and noise powers at the output: $\overline{P}_S = G\overline{P}_{\text{sens},OA}$ and $P_{ASE} = S_{ASE} \cdot BW_O = \tilde{F}G \cdot hc/\lambda \cdot BW_O$ (using Eq. (3.27)). Thus, the required OSNR is

$$OSNR_{\text{req}} \approx \frac{\overline{P}_{\text{sens},OA}}{F \cdot hc/\lambda \cdot BW_O}. \tag{I.20}$$

Transforming Eq. (I.20) into the log domain and specializing for $\lambda = 1.55$ µm and a BW_O corresponding to 0.1 nm, we obtain [5]

$$OSNR_{\text{req}}\ [\text{dB}] \approx 58\ \text{dB} + \overline{P}_{\text{sens},OA}\ [\text{dBm}] - F\ [\text{dB}]. \tag{I.21}$$

4.17(a) The noise due to the APD dark current is $\overline{i_{n,DK}^2} = FM^2 \cdot 2qI_{DK} \cdot BW_n$. With $F = k_A M$ and $M^2 = i_{n,\text{ckt}}^{rms}/(Qk_A qBW_n)$ for the optimum gain, the noise becomes $\overline{i_{n,DK}^2} = 2I_{DK} i_{n,\text{ckt}}^{rms}/Q \cdot \sqrt{i_{n,\text{ckt}}^{rms}/(Qk_A qBW_n)}$. Assuming that the avalanche noise equals the circuit noise, the (squared) power penalty becomes $PP^2 = (2\overline{i_{n,\text{ckt}}^2} + \overline{i_{n,DK}^2})/(2\overline{i_{n,\text{ckt}}^2})$. Inserting $\overline{i_{n,DK}^2}$ results in

$$PP = \sqrt{1 + \frac{I_{DK}}{Q\sqrt{Q\, i_{n,\text{ckt}}^{rms} k_A q \cdot BW_n}}}. \tag{I.22}$$

4.17(b) Solving this equation for I_{DK} results in the following requirement for the APD dark current:

$$I_{DK} < (PP^2 - 1)Q\sqrt{Q\, i_{n,\text{ckt}}^{rms} k_A q \cdot BW_n}. \tag{I.23}$$

This expression is an approximation to the solution given in [9].

4.18(a) The smallest output value for a one occurs if it is preceded by a long sequence of zeros. This value is given by the step response ($0 \rightarrow 1$ at $t = 0$) of the filter evaluated at $t = 1/B$. Similarly, the largest output value for a zero occurs if it is preceded by a long sequence of ones, which is given by the inverse step response ($1 \rightarrow 0$ at $t = 0$) evaluated at $t = 1/B$. The difference between these two values is the worst-case output swing, and its reciprocal value is the power penalty (assuming the full swing is normalized to one).

For a first-order low-pass filter, the step response is $1 - \exp(-2\pi BW_{3dB}t)$; the inverse step response is $\exp(-2\pi BW_{3dB}t)$. Thus, the power penalty is

$$PP = \frac{1}{1 - 2\exp(-2\pi BW_{3dB}/B)}. \tag{I.24}$$

4.18(b) For a second-order Butterworth low-pass filter, the step response is [10] $1 - \sqrt{2}\exp(-\sqrt{2}\pi BW_{3dB}t) \cdot \sin(\sqrt{2}\pi BW_{3dB}t + \pi/4)$. Following the same procedure as in Problem 4.18(a), the power penalty is

$$PP = \frac{1}{1 - 2\sqrt{2}\exp(-\sqrt{2}\pi BW_{3dB}/B) \cdot \sin(\sqrt{2}\pi BW_{3dB}/B + \pi/4)}. \tag{I.25}$$

If this equation yields $PP < 1$ as a result of over/undershoot, we set $PP = 1$.

4.18(c) The power penalty values are

Bandwidth BW_{3dB}/B	1st-Order filter PP (dB)	2nd-Order Butterworth PP (dB)
1/3	1.23	2.97
2/3	0.13	0.00
4/3	0.00	0.01

4.19 Calculating the Fourier transform of the impulse response $h(t)$ results in $H(f) = \int_{-\infty}^{\infty} h(t)\exp(-j\,2\pi f\,t)\,dt = \int_0^{1/B} \exp(-j\,2\pi f\,t)\,dt = j/(2\pi f) \cdot [\exp(-j\,2\pi f/B) - 1]$. The magnitude is

$$|H(f)| = \frac{1}{2\pi f}\sqrt{\left[\cos\left(\frac{2\pi f}{B}\right) - 1\right]^2 + \left[\sin\left(\frac{2\pi f}{B}\right)\right]^2} = \frac{\sin(\pi f/B)}{\pi f}. \tag{I.26}$$

The tangent of the phase is given by

$$\frac{\text{Im}[H(f)]}{\text{Re}[H(f)]} = \frac{\cos(2\pi f/B) - 1}{\sin(2\pi f/B)} = \tan\left(-\frac{\pi f}{B}\right). \tag{I.27}$$

Thus, $H(f)$ can be written as

$$H(f) = \frac{\sin(\pi f/B)}{\pi f} \cdot e^{-j\ \pi f/B}. \tag{I.28}$$

When normalized such that $|H(0)| = 1$, this expression is identical to Eq. (4.70).

Chapter 5

5.1(a) The ABCD parameters are defined by $V_1 = \mathcal{A}V_2 - \mathcal{B}I_2$ and $I_1 = \mathcal{C}V_2 - \mathcal{D}I_2$. With the load admittance Y_L at the output of the two-port, the second equation yields $V_2 = I_1/(\mathcal{C} + \mathcal{D}Y_L)$. The load admittance presented by the S-parameter two-port to the ABCD-parameter two-port is $Y_L = (1 - S_{11})/[(1 + S_{11})R_0]$ [11, p. 331]. The loaded voltage gain from V_2 to the TIA output V_O is $S_{21}/(1 + S_{11})$ [11, p. 332]. Putting all these equations together, we find

$$V_o = \frac{S_{21}}{1 + S_{11}} \cdot \frac{I_1}{\mathcal{C} + \mathcal{D}\frac{1-S_{11}}{(1+S_{11})R_0}}. \tag{I.29}$$

Solving for Z_T, we find

$$Z_T = \frac{V_o}{I_1} = \frac{S_{21}R_0}{\mathcal{C}(1 + S_{11})R_0 + \mathcal{D}(1 - S_{11})}. \tag{I.30}$$

This result agrees with Eq. (32) in [12].

5.1(b) The transimpedance is the product of the following three factors: (i) the current referral function from the intrinsic photodetector to the TIA input node, H_{PD}; (ii) the total impedance at the TIA input node converting the current into a voltage, $1/(Y_D + Y_L)$, where $Y_L = (1 - S_{11})/[(1 + S_{11})R_0]$ [11, p. 331]; and (iii) the loaded voltage gain from the TIA input to the TIA output, $S_{21}/(1 + S_{11})$ [11, p. 332]. Multiplying these factors, we find

$$Z_T = H_{PD} \cdot \frac{1}{Y_D + \frac{1-S_{11}}{(1+S_{11})R_0}} \cdot \frac{S_{21}}{1 + S_{11}} = \frac{H_{PD}S_{21}R_0}{Y_D(1 + S_{11})R_0 + (1 - S_{11})}. \tag{I.31}$$

This result agrees with Eq. (5.3). Comparing Eq. (I.30) with Eq. (I.31), we can identify

$$Y_D = \frac{C}{D} \quad \text{and} \quad H_{PD} = \frac{1}{D}. \tag{I.32}$$

5.1(c) The ABCD matrix for the photodetector/package model in Fig. 5.6 is

$$\begin{bmatrix} A & B \\ C & D \end{bmatrix} = \begin{bmatrix} 1 & sL_B \\ sC_D & s^2L_BC_D + 1 \end{bmatrix}. \tag{I.33}$$

Thus, $Y_D = C/D = sC_D/(s^2L_BC_D + 1)$ and $H_{PD} = 1/D = 1/(s^2L_BC_D + 1)$.

5.2(a) The rms noise in the differential mode is $\sqrt{2}$ mV $\approx$ 1.41 mV, and the rms noise in the common mode is $\sqrt{2}/2$ mV $\approx$ 0.71 mV.

5.2(b) When reproducing the single-ended output noise, $i_{n,TIA}^{rms} = 1$ mV/0.5 kΩ = 2 μA. When reproducing the differential output noise, $i_{n,TIA}^{rms} = 1.41$ mV/1.0 kΩ = 1.41 μA.

5.3(a) The optical dynamic range is $3 - (-19) = 22$ dB; the electrical dynamic range is twice that: 44 dB.

5.3(b) The input overload current must be $i_{ovl}^{pp} > 2\mathcal{R}\overline{P}_{ovl} = 3.2$ mA, and the input-referred rms noise current must be $i_{n,TIA}^{rms} < \mathcal{R}\overline{P}_{\text{sens}}/\mathcal{Q} = 1.43$ μA.

5.3(c) The averaged input-referred noise current density must be $I_{n,TIA}^{\text{avg}} < 1.43\ \mu\text{A}/\sqrt{7.5\ \text{GHz}} = 16.5\ \text{pA}/\sqrt{\text{Hz}}$.

5.4(a) The output admittance of an ABCD two-port with an open input port is $Y_D = C/D$. The current referral function is $H_{PD} = 1/D$ (cf. Eq. (I.32)). Thus,

$$G_{Di} = \frac{\text{Re}\{Y_D\}}{|H_{PD}|^2} = \text{Re}\left\{\frac{C}{D}\right\}|D|^2 = \frac{\text{Re}\{CD^*\}}{DD^*}DD^* = \text{Re}\{C^*D\}. \tag{I.34}$$

5.4(b) The ABCD matrix for the two-port network in Fig. 5.16 is

$$\begin{bmatrix} A & B \\ C & D \end{bmatrix} = \begin{bmatrix} 1 & R_{PD} \\ sC_{PD} & sR_{PD}C_{PD} + 1 \end{bmatrix}. \tag{I.35}$$

Thus, $G_{Di} = \text{Re}\{C^*D\} = \omega^2C_{PD}^2R_{PD}$.

Chapter 6

6.1(a) The transimpedance is $R_T = 25\ \Omega \cdot A = 2.5\ \text{k}\Omega$.

6.1(b) An amplifier with a 2-dB noise figure ($F = 1.58$) connected to a 50-Ω source ($R_0 = 50\ \Omega$) has an input-referred noise power that is 1.58 × larger than that of the 50-Ω source alone. Thus, the input-referred rms noise current is [13]

$$i_n^{rms} = \sqrt{F\,\frac{4kT}{R_0}\cdot BW_n}, \tag{I.36}$$

which evaluates to 2.29 μA. (This result is slightly inaccurate, because the source impedance presented to the amplifier is not exactly 50 Ω, but 50 Ω in parallel with Z_D, and thus the noise figure at this impedance is slightly different from 2 dB.)

6.1(c) The optical sensitivity of the TIA receiver is better by $10\ \log(2.29/1.0) = 3.6$ dB.

6.2(a) Let the open-loop pole spacing be $\chi = R_F C_T / T_A$. From Eq. (6.11), we find that $Q = \sqrt{(A_0+1)\ \chi/(\chi+1)}$. For $\chi^2 \gg 1$, we can simplify this expression to $Q \approx \sqrt{(A_0+1)/(\chi+2)}$; thus, the required pole spacing is

$$\chi \approx \frac{A_0+1}{Q^2} - 2. \tag{I.37}$$

For a Butterworth response, we have $Q = 1/\sqrt{2}$, and thus $\chi \approx 2A_0$.

6.2(b) For a Bessel response, we have $Q = 1/\sqrt{3}$, and thus $\chi \approx 3A_0 + 1$.

6.2(c) For a critically damped response, we have $Q = 1/2$, and thus $\chi \approx 4A_0 + 2$.

6.3 By simply plugging the numbers into Eq. (6.14), we find that at 2.5 Gb/s, $R_T \leq 7.62\ \text{k}\Omega$, at 10 Gb/s, $R_T \leq 476\ \Omega$, and at 40 Gb/s, $R_T \leq 29.8\ \Omega$.

6.4(a) Comparing Eqs. (6.24)–(6.26) with Eqs. (6.9)–(6.11), we find the mapping

$$A_0' = A_0\left(1+\frac{R_L}{R_F}\right)+\frac{R_L}{R_F} \quad \text{and} \quad T_A' = T_A\left(1+\frac{R_L}{R_F}\right). \tag{I.38}$$

6.4(b) The input impedance of the single-transistor shunt-feedback TIA is

$$Z_I(s) = R_I \cdot \frac{1 + s/\omega_z}{1 + s/(\omega_0 Q) + s^2/\omega_0^2}, \tag{I.39}$$

where

$$R_I = \frac{R_F + R_L}{A_0' + 1}, \quad \omega_z = \frac{R_F + R_L}{R_F T_A'}, \tag{I.40}$$

and the expressions for ω_0 and Q are the same as in Eqs. (6.25) and (6.26), respectively.

6.4(c) Inserting A_0' and T_A' from Eq. (I.38) into Eq. (I.39) yields Eq. (6.21).

6.5(a) The transimpedance of the single-transistor shunt-feedback TIA with feedback capacitor C_F is

$$Z_T(s) = -R_T \cdot \frac{1 - s/\omega_z}{1 + s/(\omega_0 Q) + s^2/\omega_0^2}, \tag{I.41}$$

where

$$R_T = \frac{A_0' R_F - R_L}{A_0' + 1}, \quad \omega_z = \frac{g_m - 1/R_F}{C_F}, \tag{I.42}$$

$$\omega_0 = \sqrt{\frac{A_0' + 1}{R_F R_L (C_T C_F + C_T C_L + C_F C_L)}}, \tag{I.43}$$

$$Q = \frac{\sqrt{(A_0' + 1) R_F R_L (C_T C_F + C_T C_L + C_F C_L)}}{R_F [C_T + (A_0' + 1) C_F] + R_L (C_T + C_L)}, \tag{I.44}$$

and $A_0' = g_m R_L$. This result is consistent with the expressions in [14].

6.5(b) Ignoring the zero, which is located at a very high frequency, the transimpedance limit can be found by combining the expressions for ω_0 and R_T:

$$R_T \leq \frac{A_0' f_A'}{2\pi C_T \cdot BW_{3dB}^2} \cdot \frac{1 - 1/(g_m R_F)}{1 + C_F/C_T + C_F/C_L}, \tag{I.45}$$

where the factor on the left-hand side is the transimpedance limit in Eq. (6.14) for $R_L \to 0$. Alternatively, the limit can be written as

$$R_T \leq \frac{A_0' f_A'}{2\pi (C_T + C_F) \cdot BW_{3dB}^2} \cdot \frac{1 - 1/(g_m R_F)}{1 + C_T C_F/(C_T + C_F)/C_L}, \tag{I.46}$$

where the factor on the left-hand side is the transimpedance limit in Eq. (I.59) for $R_L \to 0$.

6.6(a) The transimpedance of the shunt-feedback TIA with a series resistance R_L at the output of the voltage amplifier can be obtained easily from Eqs. (6.8)–(6.11) by substituting $R_F \to R_F + R_L$ and by multiplying the result with the transfer function from the ideal output *before* R_L to the actual output after R_L, $[R_F - R_L/A(s)]/(R_F + R_L)$:

$$Z_T(s) = -R_T \cdot \frac{1 - s/\omega_z}{1 + s/(\omega_0 Q) + s^2/\omega_0^2}, \tag{I.47}$$

where

$$R_T = \frac{A_0 R_F - R_L}{A_0 + 1}, \quad \omega_z = \frac{A_0 R_F - R_L}{R_L T_A}, \tag{I.48}$$

$$\omega_0 = \sqrt{\frac{A_0 + 1}{(R_F + R_L) C_T T_A}}, \quad Q = \frac{\sqrt{(A_0 + 1)(R_F + R_L) C_T T_A}}{(R_F + R_L) C_T + T_A}. \tag{I.49}$$

6.6(b) Ignoring the zero, which is located at a very high frequency, the transimpedance limit can be found by combining the expressions for ω_0 and R_T:

$$R_T \leq \frac{A_0 f_A}{2\pi C_T \cdot BW_{3dB}^2} \cdot \frac{1 - R_L/(A_0 R_F)}{1 + R_L/R_F}. \tag{I.50}$$

6.6(c) The output impedance of the TIA is

$$Z_O(s) = R_O \cdot \frac{(1 + s/\omega_{z1})(1 + s/\omega_{z2})}{1 + s/(\omega_0 Q) + s^2/\omega_0^2}, \tag{I.51}$$

where

$$R_O = \frac{R_L}{A_0 + 1}, \quad \omega_{z1} = \frac{1}{R_F C_T}, \quad \omega_{z2} = \frac{1}{T_A}, \tag{I.52}$$

and the expressions for ω_0 and Q are the same as in Eq. (I.49).

6.7(a) The transimpedance expression including the feedback capacitor C_F can be found easily from Eq. (6.8) by substituting $R_F \to R_F/(1 + sR_F C_F)$

$$Z_T(s) = -R_T \cdot \frac{1}{1 + s/(\omega_0 Q) + s^2/\omega_0^2}, \tag{I.53}$$

where

$$R_T = \frac{A_0}{A_0 + 1} \cdot R_F, \tag{I.54}$$

$$\omega_0 = \sqrt{\frac{A_0 + 1}{R_F(C_T + C_F)T_A}}, \tag{I.55}$$

$$Q = \frac{\sqrt{(A_0 + 1)R_F(C_T + C_F)T_A}}{R_F[C_T + (A_0 + 1)C_F] + T_A}, \tag{I.56}$$

which is consistent with the expressions in [15].

6.7(b) Setting $Q = 1/\sqrt{2}$ and solving for C_F (with $C_F \ll C_T$) yields

$$C_F \approx \sqrt{\frac{2C_T T_A}{(A_0 + 1)R_F}} - \frac{T_A}{(A_0 + 1)R_F} - \frac{C_T}{A_0 + 1}, \tag{I.57}$$

which is consistent with the expression in [16].

6.7(c) For large values of C_F, the TIA is so strongly damped that a Butterworth response cannot be achieved for any value of T_A, that is, $Q(T_A) = 1/\sqrt{2}$ with Eq. (I.56) does not have a solution. Thus, we cannot generally calculate the bandwidth as $BW_{3dB} = \omega_0/(2\pi)$, which presumes $Q = 1/\sqrt{2}$.
Instead, we use the bandwidth estimate $BW_{3dB} \approx \omega_0 Q/(2\pi)$, which works well for small Q values (cf. Appendix H). From Eqs. (I.55) and (I.56) and assuming $T_A \ll R_F[C_T + (A_0 + 1)C_F]$ we obtain

$$BW_{3dB} \approx \frac{(A_0 + 1)}{2\pi R_F[C_T + (A_0 + 1)C_F]}. \tag{I.58}$$

If we further assume that $C_T \ll (A_0 + 1)C_F$, the expression reduces to $BW_{3dB} \approx 1/(2\pi R_F C_F)$.

6.7(d) For $Q \leq 1/\sqrt{2}$, $\omega_0/(2\pi)$ is an upper bound for the bandwidth. Combining Eqs. (I.54) and (I.55) and solving for R_T thus yields the transimpedance limit [17]

$$R_T \leq \frac{A_0 f_A}{2\pi(C_T + C_F) \cdot BW_{3dB}^2}, \tag{I.59}$$

which is lower than the limit in Eq. (6.14). However because $Q \ll 1/\sqrt{2}$ for large values of C_F, this bound is very loose. In fact, for $C_F > C_T$ it cannot be reached.

For any Q, $\sqrt{2}\omega_0 Q/(2\pi)$ is another upper bound for the bandwidth (cf. Appendix H). Combining Eqs. (I.54)–(I.56) we find

$$BW_{3dB} \leq \frac{\sqrt{2}A_0}{2\pi\{R_T[C_T + (A_0+1)C_F] + T_A A_0/(A_0+1)\}} < \frac{\sqrt{2}}{2\pi R_T C_F}, \tag{I.60}$$

where the second bound results from removing positive terms from the denominator of the first bound. Solving for R_T yields another transimpedance limit [17]:

$$R_T < \frac{\sqrt{2}}{2\pi C_F \cdot BW_{3dB}}. \tag{I.61}$$

For large values of C_F and small values of BW_{3dB}, this bound is tighter than Eq. (I.59).

6.8(a) The equivalent drain noise current source has a direct contribution from the channel noise and an indirect contribution from the induced gate noise:

$$\underline{I}_{n,D} = \underline{I}_{n,\text{chan}} + H_D \underline{I}_{n,\text{ind}}, \tag{I.62}$$

where H_D is the referral function from the gate current source to the drain current source and underlined quantities represent phasors. The mean-square-magnitude of the drain noise current phasor is

$$\overline{|\underline{I}_{n,D}|^2} = \overline{|\underline{I}_{n,\text{chan}} + H_D \underline{I}_{n,\text{ind}}|^2}. \tag{I.63}$$

Using the complex identity $|z_1 + z_2|^2 = |z_1|^2 + 2\,\text{Re}\{z_1^* z_2\} + |z_2|^2$, we can expand

$$\overline{|\underline{I}_{n,D}|^2} = \overline{|\underline{I}_{n,\text{chan}}|^2} + 2\,\text{Re}\{H_D \overline{\underline{I}^*_{n,\text{chan}} \underline{I}_{n,\text{ind}}}\} + |H_D|^2 \overline{|\underline{I}_{n,\text{ind}}|^2}. \tag{I.64}$$

Using definition Eq. (6.107) to introduce the correlation coefficient c in the middle term yields

$$\overline{|\underline{I}_{n,D}|^2} = \overline{|\underline{I}_{n,\text{chan}}|^2} - 2\,\text{Re}\{c^* H_D\}\sqrt{\overline{|\underline{I}_{n,\text{chan}}|^2}\;\overline{|\underline{I}_{n,\text{ind}}|^2}} + |H_D|^2 \overline{|\underline{I}_{n,\text{ind}}|^2}. \tag{I.65}$$

Making use of the fact that mean-square-magnitude noise phasors are proportional to their PSDs, we can write

$$I^2_{n,D} = I^2_{n,\text{chan}} - 2\,\text{Re}\{c^* H_D\} I_{n,\text{chan}} I_{n,\text{ind}} + |H_D|^2 I^2_{n,\text{ind}}. \tag{I.66}$$

Inserting $H_D = -g_m/(j\omega\tilde{C}_T)$, where $\tilde{C}_T = C_D + C_{gs} + C_{gd}$ (feedforward through C_{gd} is neglected in H_D), inserting Eqs. (6.39) and (6.40), and simplifying finally yields the white noise:

$$I_{n,D}^2 = 4kT\left[\gamma g_{d0} - 2\ \mathrm{Im}\{c\} g_m \sqrt{\frac{\gamma\delta}{5}\frac{C_{gs}}{\tilde{C}_T}} + \frac{\delta g_m^2}{5 g_{d0}}\left(\frac{C_{gs}}{\tilde{C}_T}\right)^2\right]. \tag{I.67}$$

6.8(b) Comparing Eq. (I.67) with Eq. (6.41) reveals the full expression for the Ogawa Γ factor:

$$\Gamma = \gamma\frac{g_{d0}}{g_m} - 2\ \mathrm{Im}\{c\}\sqrt{\frac{\gamma\delta}{5}\frac{C_{gs}}{\tilde{C}_T}} + \frac{\delta g_m}{5 g_{d0}}\left(\frac{C_{gs}}{\tilde{C}_T}\right)^2. \tag{I.68}$$

6.8(c) Inserting the numerical values, we find

$$\Gamma = \gamma\left[1 - 0.5\frac{C_{gs}}{\tilde{C}_T} + 0.4\left(\frac{C_{gs}}{\tilde{C}_T}\right)^2\right]. \tag{I.69}$$

For $C_{gs}/\tilde{C}_T = \frac{1}{2}$, the induced gate noise *reduces* the channel noise by 15%; for $C_{gs}/\tilde{C}_T = \frac{1}{3}$, by 12%; and for $C_{gs}/\tilde{C}_T = \frac{1}{4}$, by 10%.

6.9(a) The output voltage phasor $\underline{V}_o$ of a general linear TIA model can be written in terms of the input noise current phasor $\underline{I}_{n,TIA}$ and the drain noise current phasor $\underline{I}_{n,D}$ as $\underline{V}_o = R_1(s)\underline{I}_{n,TIA} + R_2(s)\underline{I}_{n,D}$, where $R_1(s)$ and $R_2(s)$ are the transfer functions from the input current to the output and the drain current to the output, respectively.
According to the first definition, the input-referral function $H_D^{-1}(s)$ is $R_2(s)/R_1(s)$. According to the second definition, we set $R_1(s)\underline{I}_{n,TIA} = R_2(s)\underline{I}_{n,D}$ and solve for $\underline{I}_{n,TIA}/\underline{I}_{n,D}$, which is again $R_2(s)/R_1(s)$. According to the third definition, we set $R_1(s)\underline{I}_{n,TIA} + R_2(s)\underline{I}_{n,D} = 0$ and solve for $-(\underline{I}_{n,TIA}/\underline{I}_{n,D})$, which is again $R_2(s)/R_1(s)$.

6.9(b) The transfer functions are

$$R_1(s) = \frac{R_D R_F (g_m - sC_{gd})}{D(s)} \tag{I.70}$$

and

$$R_2(s) = \frac{-R_D[sR_F(C_D + C_{gs} + C_{gd}) + 1]}{D(s)}, \tag{I.71}$$

where the denominator is

$$D(s) = (g_m R_D + 1) + sR_F[C_D + C_{gs} + C_{gd}(g_m R_D + 1)] + s^2 R_D R_F (C_D + C_{gs}) C_{gd}. \tag{I.72}$$

Both transfer functions containing the Miller factor $(g_m R_D + 1)$ in the denominator. Dividing the two transfer functions results in the much simpler input-referral function

$$H_D^{-1}(s) = \frac{R_2(s)}{R_1(s)} = -\frac{sR_F(C_D + C_{gs} + C_{gd}) + 1}{g_m R_F} \cdot \frac{1}{1 - sC_{gd}/g_m}, \tag{I.73}$$

which corresponds to Eq. (6.43) if the RHP pole at g_m/C_{gd} is neglected.

6.10(a) Equating the f^2 noise of Eq. (6.44) to the white noise of Eqs. (6.37) and (6.44), ignoring the gate shot noise, and solving for f, we find the f^2-noise corner frequency

$$f_{c2} = \frac{1}{2\pi R_F \tilde{C}_T} \cdot \sqrt{\frac{g_m R_F}{\Gamma} + 1}. \tag{I.74}$$

6.10(b) Generally, higher bit-rate TIAs have a higher f^2-noise corner frequency (lower R_F and lower $\tilde{C}_T$).

6.11 With Eqs. (6.37) and (6.43), ignoring the gate shot noise, we find the noise PSD

$$I_{n,TIA}^2(f) = \frac{4kT}{R_F} + 4kT\Gamma \left[\frac{1}{g_m R_F^2} + \frac{f_c}{g_m R_F^2} f^{-1} + \frac{(2\pi \tilde{C}_T)^2}{g_m} f^2 + \frac{(2\pi \tilde{C}_T)^2 f_c}{g_m} f \right]. \tag{I.75}$$

6.12(a) Dividing Eq. (6.10) by Eq. (6.11) to eliminate A_0 and solving for $f_A = 1/(2\pi T_A)$, we find

$$f_A = \frac{\omega_0}{2\pi Q} - \frac{1}{2\pi R_F C_T}. \tag{I.76}$$

Inserting this solution into Eq. (6.10) and solving for A_0, we find

$$A_0 = \frac{\omega_0^2 R_F C_T}{\omega_0/Q - 1/(R_F C_T)} - 1. \tag{I.77}$$

C_T is given by $C_D + C_I$ and ω_0 is given by $2\pi BW_{3dB}/\rho$ (cf. Eq. (H.2)) where ρ is a function of Q (cf. Eq. (H.3)).

6.12(b) For f_A to remain positive in Eq. (I.76), the TIA bandwidth must be larger than

$$BW_{3\text{dB}}(\text{min}) = \frac{\rho Q}{2\pi R_F C_T}. \quad (\text{I.78})$$

Multiplying Eqs. (I.76) and (I.77) yields the gain-bandwidth product of the voltage amplifier

$$A_0 f_A = 2\pi R_F C_T \left(\frac{BW_{3\text{dB}}}{\rho} \right)^2 - \frac{BW_{3\text{dB}}}{\rho Q} + \frac{1}{2\pi R_F C_T}. \quad (\text{I.79})$$

In order for this product to remain below $A_0 f_A(\text{max})$, the TIA bandwidth must be smaller than

$$BW_{3\text{dB}}(\text{max}) = \frac{\rho(\sqrt{4Q^2[2\pi R_F C_T A_0 f_A(\text{max}) - 1] + 1} + 1)}{4\pi R_F C_T Q}. \quad (\text{I.80})$$

6.13(a) The noise fraction due to the FET is

$$\frac{\overline{i^2_{n,TIA,3}}}{\overline{i^2_{n,TIA}}} = \frac{\frac{(C_D+\tilde{C}_I)^2}{\tilde{C}_I}\xi_2}{(C_D + \tilde{C}_I)\xi_1 + \frac{(C_D+\tilde{C}_I)^2}{\tilde{C}_I}\xi_2}. \quad (\text{I.81})$$

For the optimum input capacitance $\tilde{C}_I = \psi C_D$, we have

$$\frac{\overline{i^2_{n,TIA,3}}}{\overline{i^2_{n,TIA}}} = \frac{1 + \psi}{\psi \xi_1/\xi_2 + 1 + \psi}. \quad (\text{I.82})$$

Finally, using $\xi_1/\xi_2 = 1/\psi^2 - 1$ from Eq. (6.62), the whole expression simplifies to ψ.

6.13(b) Substituting $2\pi R_F C_T = 1/f_I$ into Eq. (I.79), the following quadratic equation for f_I is found:

$$f_I^2 - \left(A_0 f_A + \frac{BW_{3\text{dB}}}{\rho Q} \right) f_I + \left(\frac{BW_{3\text{dB}}}{\rho} \right)^2 = 0. \quad (\text{I.83})$$

Assuming $f_I \ll A_0 f_A$ and $BW_{3\text{dB}} \ll \rho Q A_0 f_A$, the solution of Eq. (I.83) can be approximated as

$$f_I \approx \frac{BW^2_{3\text{dB}}}{\rho^2 A_0 f_A}. \quad (\text{I.84})$$

This result can also be found from the transimpedance limit with $R_F \approx R_T$. The solution for f_A is found by substituting $2\pi R_F C_T = 1/f_I$ into

Eq. (I.76)

$$f_A = \frac{BW_{3\text{dB}}}{\rho Q} - f_I. \tag{I.85}$$

6.13(c) The minimum bandwidth is reached when R_F exceeds $R_F(\text{max})$. From Eq. (I.79) we find

$$BW_{3\text{dB}}(\text{min}) = \frac{\rho\left(\sqrt{4Q^2[2\pi R_F(\text{max})C_T A_0 f_A - 1] + 1} + 1\right)}{4\pi R_F(\text{max})C_T Q}. \tag{I.86}$$

The maximum bandwidth is reached when Eq. (I.83) no longer has a real solution for f_I:

$$BW_{3\text{dB}}(\text{max}) = \frac{\rho A_0 f_A}{2 - 1/Q}. \tag{I.87}$$

6.13(d) From Eq. (6.61) we have

$$\frac{\xi_1}{\xi_2} = \frac{3 f_T f_I \cdot BW_n}{\Gamma \cdot BW_{n2}^3}. \tag{I.88}$$

Expressing the noise bandwidths of a second-order TIA in terms of $BW_{3\text{dB}}$ and Q (cf. Eqs. (H.8) and (H.9)) and using Eq. (6.62) to calculate ψ, we find

$$\psi = \sqrt{\frac{\Gamma \cdot BW_{3\text{dB}}^2}{\Gamma \cdot BW_{3\text{dB}}^2 + \rho^2 f_T f_I}}. \tag{I.89}$$

Inserting the approximation for f_I from Eq. (I.84) leads us to Eq. (6.63).

6.14(a) The input-referred noise consists of (i) the source noise, (ii) the feedback resistor noise, (iii) the gate shot noise, (iv) the induced gate noise, and (v) the channel noise referred to the input:

$$\underline{I}_{n,TIA} = \underline{I}_{n,S} + \underline{I}_{n,\text{res}} + \underline{I}_{n,G} + \underline{I}_{n,\text{ind}} + H_D^{-1}\underline{I}_{n,\text{chan}}, \tag{I.90}$$

where H_D^{-1} is the input-referral function from the drain current source to the gate current source and underlined quantities represent phasors. The mean-square-magnitude of the input-referred noise current phasor is

$$\overline{|\underline{I}_{n,TIA}|^2} = \overline{|\underline{I}_{n,S}|^2} + \overline{|\underline{I}_{n,\text{res}}|^2} + \overline{|\underline{I}_{n,G}|^2} + \overline{|\underline{I}_{n,\text{ind}} + H_D^{-1}\underline{I}_{n,\text{chan}}|^2}. \tag{I.91}$$

Using the complex identity $|z_1 + z_2|^2 = |z_1|^2 + 2\,\mathrm{Re}\{z_1^* z_2\} + |z_2|^2$, we can expand the last term to

$$\overline{|\underline{I}_{n,\mathrm{ind}}|^2} + 2\,\mathrm{Re}\left\{H_D^{-1}\overline{\underline{I}^*_{n,\mathrm{ind}}\underline{I}_{n,\mathrm{chan}}}\right\} + |H_D^{-1}|^2\overline{|\underline{I}_{n,\mathrm{chan}}|^2}. \quad \text{(I.92)}$$

Using definition Eq. (6.107) to introduce the correlation coefficient c yields

$$\overline{|\underline{I}_{n,\mathrm{ind}}|^2} - 2\,\mathrm{Re}\{cH_D^{-1}\}\sqrt{\overline{|\underline{I}_{n,\mathrm{ind}}|^2}\;\overline{|\underline{I}_{n,\mathrm{chan}}|^2}} + |H_D^{-1}|^2\overline{|\underline{I}_{n,\mathrm{chan}}|^2}. \quad \text{(I.93)}$$

Splitting the induced gate noise term into $\overline{|\underline{I}_{n,\mathrm{ind}}|^2}(1-|c|^2) + \overline{|\underline{I}_{n,\mathrm{ind}}|^2}|c|^2$ and using $|z_1|^2 + 2\,\mathrm{Re}\{z_1^* z_2\} + |z_2|^2 = |z_1 + z_2|^2$, we can simplify to

$$\overline{|\underline{I}_{n,\mathrm{ind}}|^2}(1-|c|^2) + \left|c^*\sqrt{\overline{|\underline{I}_{n,\mathrm{ind}}|^2}} - H_D^{-1}\sqrt{\overline{|\underline{I}_{n,\mathrm{chan}}|^2}}\right|^2. \quad \text{(I.94)}$$

Making use of the fact that mean-square-magnitude noise phasors are proportional to their PSDs, we can write

$$I^2_{n,TIA} = I^2_{n,S} + I^2_{n,\mathrm{res}} + I^2_{n,G} + I^2_{n,\mathrm{ind}}(1-|c|^2) + |c^* I_{n,\mathrm{ind}} - H_D^{-1} I_{n,\mathrm{chan}}|^2. \quad \text{(I.95)}$$

Inserting $H_D^{-1} = -(Y_S + 1/R_F + j\omega\tilde{C}_I)/g_m$, where $\tilde{C}_I = C_{gs} + C_{gd}$ (feedforward through C_{gd} is neglected in H_D^{-1}) yields

$$\begin{aligned} I^2_{n,TIA} &= I^2_{n,S} + I^2_{n,\mathrm{res}} + I^2_{n,G} + I^2_{n,\mathrm{ind}}(1-|c|^2) \\ &+ \frac{I^2_{n,\mathrm{chan}}}{g_m^2}\left|Y_S + \frac{1}{R_F} + j\omega\tilde{C}_I + c^* g_m \frac{I_{n,\mathrm{ind}}}{I_{n,\mathrm{chan}}}\right|^2. \end{aligned} \quad \text{(I.96)}$$

Finally, inserting Eqs. (6.37)–(6.40) results in

$$\begin{aligned} I^2_{n,TIA}(\omega) &= 4kT|G_S(\omega)| + \frac{4kT}{R_F} + 2qI_G + 4kT\delta\frac{\omega^2 C_{gs}^2}{5g_{d0}}(1-|c|^2) \\ &+ 4kT\gamma\frac{g_{d0}}{g_m^2}\left(G_S + \frac{1}{R_F} + \mathrm{Re}\{c\}\;\omega C_{gs}\sqrt{\frac{\delta}{5\gamma}\frac{g_m}{g_{d0}}}\right)^2 \\ &+ 4kT\gamma\frac{g_{d0}}{g_m^2}\left(B_S + \omega\tilde{C}_I - \mathrm{Im}\{c\}\;\omega C_{gs}\sqrt{\frac{\delta}{5\gamma}\frac{g_m}{g_{d0}}}\right)^2. \end{aligned} \quad \text{(I.97)}$$

6.14(b) The optimum source admittance is (cf. [18])

$$\tilde{Y}_{\mathrm{opt}}(\omega) = -j\omega\left(\tilde{C}_I - \mathrm{Im}\{c\}\;C_{gs}\sqrt{\frac{\delta}{5\gamma}\frac{g_m}{g_{d0}}}\right), \quad \text{(I.98)}$$

under the conditions that $g_m^2 R_F > 2\gamma g_{d0}$ and $\mathrm{Re}\{c\} = 0$.

6.14(c) Equating $Y_M(\omega) = j[\omega C_D - 1/(\omega L_P)]$ to $\tilde{Y}_{\rm opt}(\omega)$ and solving for L_P results in (cf. [18])

$$L_P = \frac{1}{\omega^2 \left(C_D + \tilde{C}_I - \mathrm{Im}\{c\}\; C_{gs} \sqrt{\frac{\delta}{5\gamma} \frac{g_m}{g_{d0}}} \right)}. \tag{I.99}$$

6.14(d) The residual noise for $Y_M = \tilde{Y}_{\rm opt}$ (and $\mathrm{Re}\{c\} = 0$) is

$$I^2_{n,TIA}(\omega) = \frac{4kT}{R_F} + 2qI_G + 4kT\delta \frac{\omega^2 C_{gs}^2}{5g_{d0}}(1 - |c|^2) + 4kT\gamma \frac{g_{d0}}{g_m^2 R_F^2}. \tag{I.100}$$

For $I_G R_F \ll 2V_T$ and $g_m^2 R_F \gg \gamma g_{d0}$ only feedback resistor noise and the uncorrelated induced gate noise remains (cf. [18]).

6.15 Comparing Eq. (I.96) with Eq. (6.92) reveals the noise parameters of the TIA with FET front-end:

$$V_{n4}^2 = \frac{I^2_{n,\rm chan}}{g_m^2} = 4kT\gamma \frac{g_{d0}}{g_m^2}, \tag{I.101}$$

$$\begin{aligned} I^2_{n4u} &= I^2_{n,\rm res} + I^2_{n,G} + I^2_{n,\rm ind}(1 - |c|^2) \\ &= \frac{4kT}{R_F} + 2qI_G + 4kT\delta \frac{\omega^2 C_{gs}^2}{5g_{d0}}(1 - |c|^2), \end{aligned} \tag{I.102}$$

$$\begin{aligned} Y_c &= \frac{1}{R_F} + j\omega \tilde{C}_I + c^* g_m \frac{I_{n,\rm ind}}{I_{n,\rm chan}} \\ &= \frac{1}{R_F} + j\omega \tilde{C}_I + c^* \omega C_{gs} \sqrt{\frac{\delta}{5\gamma} \frac{g_m}{g_{d0}}}. \end{aligned} \tag{I.103}$$

6.16(a) Substituting $Y_S = G_S + jB_S$ and $Y_c = G_c + jB_c$ into Eq. (6.92), taking the derivatives with respect to G_S and B_S, and setting them to zero yields

$$4kT\mathrm{sgn}(G_S) + 2(G_c + G_S)V_{n4}^2 = 0, \quad 2(B_c + B_S)V_{n4}^2 = 0. \tag{I.104}$$

Thus, the optimum source admittance is

$$\tilde{G}_{\rm opt} = \begin{cases} -G_c + \frac{2kT}{V_{n4}^2}, & \text{if } G_c > +\frac{2kT}{V_{n4}^2}, \\ 0, & \text{if } |G_c| \leq \frac{2kT}{V_{n4}^2}, \\ -G_c - \frac{2kT}{V_{n4}^2}, & \text{if } G_c < -\frac{2kT}{V_{n4}^2}, \end{cases} \tag{I.105}$$

and

$$\tilde{B}_{\text{opt}} = -B_c. \tag{I.106}$$

6.16(b) Dividing Eq. (6.92) by the noise current of the source admittance yields the noise figure

$$F = 1 + \frac{I_{n4u}^2 + |Y_c + Y_S|^2 V_{n4}^2}{4kT|\text{Re}\{Y_S\}|}. \tag{I.107}$$

Taking the derivatives with respect to G_S and B_S and setting them to zero yields the optimum source admittance (cf. [19, p. 259])

$$G_{\text{opt}} = \sqrt{\frac{I_{n4u}^2}{V_{n4}^2} + G_c^2} \quad \text{and} \quad B_{\text{opt}} = -B_c. \tag{I.108}$$

6.16(c) For the FET front-end we have $G_c = 1/R_F$, $B_c = \omega\tilde{C}_I$ (neglecting induced gate noise), $I_{n4u}^2 = 4kT/R_F$ (neglecting I_G), and $V_{n4}^2 = 4kT\gamma g_{d0}/g_m^2$. Thus, the optimum source admittances are

$$\tilde{Y}_{\text{opt}} = -j\omega\tilde{C}_I \quad \text{for} \quad g_m^2 R_F \geq 2\gamma g_{d0} \tag{I.109}$$

and

$$Y_{\text{opt}} = \sqrt{\frac{g_m^2}{\gamma g_{d0} R_F} + \frac{1}{R_F^2}} - j\omega\tilde{C}_I. \tag{I.110}$$

6.17 The ABCD parameters are defined by $V_1 = AV_2 - BI_2$ and $I_1 = CV_2 - DI_2$. To find the output admittance Y_M, we solve the equations for $Y_M = I_2/V_2$ under the condition that the input is terminated with Y_D, that is, $I_1/V_1 = -Y_D$:

$$Y_M = \frac{AY_D + C}{BY_D + D}. \tag{I.111}$$

To find the current referral function H_M, we apply a test current I_1' at the input and a test current I_2' at the output of the two port and solve the ABCD equations for $H_M = I_2'/I_1'$ under the condition that both test currents result in the same output voltage $V_2(I_1') = V_2(I_2')$. The output voltage $V_2(I_1')$ follows from setting $I_1 = -Y_D V_1 + I_1'$ and $I_2 = 0$ as $V_2 = I_1'/(AY_D + C)$. The output voltage $V_2(I_2')$ follows from setting $I_1 = -Y_D V_1$ and $I_2 = I_2'$ as $V_2 = I_2'(BY_D + D)/(AY_D + C)$. Equating the two output voltages and solving for I_2'/I_1' results in

$$H_M = \frac{1}{BY_D + D}. \tag{I.112}$$

Note that the solutions in Eq. (I.32) are a special case of Eqs. (I.111) and (I.112) for when the input termination admittance is zero.

6.18 Substituting the ABCD parameter expansions for Y_M and $|H_M|^2$ from Eq. (6.99) into Eq. (6.105) yields

$$\frac{1}{|BY_D + D|^2} = \frac{\text{Re}\left\{\frac{AY_D+C}{BY_D+D}\right\}}{\text{Re}\{Y_D\}}. \quad \text{(I.113)}$$

Using $\text{Re}\{x/y\} = \text{Re}\{xy^*\}/(yy^*) = \text{Re}\{xy^*\}/|y|^2$ results in

$$\text{Re}\{(AY_D + C)(BY_D + D)^*\} = \text{Re}\{Y_D\}, \quad \text{(I.114)}$$

which for a lossless network becomes

$$(AD + \text{Im}\{B\}\text{Im}\{C\})\text{Re}\{Y_D\} = \text{Re}\{Y_D\}, \quad \text{(I.115)}$$

which for a reciprocal network is an identity.

Chapter 7

7.1 Given the shunt-feedback TIA bandwidth BW_S, its maximum transimpedance follows from Eq. (6.14) as $A_0 f_A/(2\pi C_T \cdot BW_S^2)$. Given the post-amplifier stage bandwidth BW_S and the gain-bandwidth product $A_0 f_A$, the post-amplifier voltage gain is $(A_0 f_A/BW_S)^n$. Multiplying the two expressions results in the overall transimpedance $R_T \leq (A_0 f_A)^{n+1}/(2\pi C_T \cdot BW_S^{n+2})$. The overall bandwidth is $BW_{3dB} \leq (2^{1/(n+1)} - 1)^{1/4} \cdot BW_S$, where equality occurs for $Q = 1/\sqrt{2}$, that is, the case of $n + 1$ cascaded second-order Butterworth stages. Solving for BW_S and inserting into the former equation results in the transimpedance limit [17, 20]

$$R_T \leq \sqrt[4]{\left(\sqrt[n+1]{2} - 1\right)^{n+2}} \cdot \frac{(A_0 f_A)^{n+1}}{2\pi C_T \cdot BW_{3dB}^{n+2}}. \quad \text{(I.116)}$$

7.2 From the transimpedance and the stage gains, we can estimate the value of the feedback resistor in the first stage: $R_F = 1/A_1 \cdot (A_0 + 1)/A_0 \cdot R_T = 1\ \text{k}\Omega$ (cf. Eq. (7.1)). The white-noise contribution from this $1\ \text{k}\Omega$ resistor alone is about $4\ \text{pA}/\sqrt{\text{Hz}}$. The measured noise density is suspiciously low!

7.3(a) The (low-frequency) relationship between the detector current, i_I, and the single-ended TIA input voltage, v_{IP}, is $v_{IP} =$

$V_{OCM} + (A_0 + 2)/(2A_0 + 2) \cdot R_F i_I$, where V_{OCM} is the output common-mode voltage and A_0 is the differential gain of the feedback amplifier. Thus, the single-ended input resistance of the balanced TIA is

$$R_{I,se} = \frac{A_0 + 2}{2A_0 + 2} R_F \approx \frac{1}{2} R_F. \tag{I.117}$$

7.3(b) For $i_{IP} = i_I$ and $i_{IN} = 0$, we find that the single-ended input resistance $R_{I,se} = \Delta v_{IP}/\Delta i_I$ can be written in terms of the differential and common-mode resistances as $R_{I,se} = R_{I,c} + \frac{1}{4} R_{I,d}$ [11, Appendix B], which leads to the same result as in Problem 7.3(a).

7.4(a) With the given simplifying assumptions, Eq. (6.11) becomes $Q \approx \sqrt{A_0 T_A/(R_F C_T)}$ and the bandwidth becomes $BW_{3dB} \approx A_0/(2\pi R_F C_T)$ (cf. Eq. (6.4)). Thus, varying R_F and keeping A_0, T_A fixed results in

$$Q \propto 1/\sqrt{R_F} \qquad \text{and} \qquad BW_{3dB} \propto 1/R_F. \tag{I.118}$$

7.4(b) Varying $A_0 \propto R_F$ and keeping A_0/T_A fixed results in

$$Q \propto \sqrt{R_F} \qquad \text{and} \qquad BW_{3dB} \propto 1. \tag{I.119}$$

7.4(c) Varying $A_0 \propto \sqrt{R_F}$ and keeping A_0/T_A fixed results in

$$Q \propto 1 \qquad \text{and} \qquad BW_{3dB} \propto 1/\sqrt{R_F}. \tag{I.120}$$

7.5(a) The transimpedance expression including the shunt resistor R_S can be found easily from Eq. (6.8) by substituting $sC_T \rightarrow sC_T + 1/R_S$:

$$Z_T(s) = -R_T \cdot \frac{1}{1 + s/(\omega_0 Q) + s^2/\omega_0^2}, \tag{I.121}$$

where

$$R_T = \frac{A_0}{A_0 + 1 + R_F/R_S} \cdot R_F, \tag{I.122}$$

$$\omega_0 = \sqrt{\frac{A_0 + 1 + R_F/R_S}{R_F C_T T_A}}, \tag{I.123}$$

$$Q = \frac{\sqrt{(A_0 + 1 + R_F/R_S) R_F C_T T_A}}{R_F C_T + T_A (1 + R_F/R_S)}. \tag{I.124}$$

7.5(b) Combining the expressions for ω_0 and R_T, we find the transimpedance limit (for $Q \leq 1/\sqrt{2}$) to be

$$R_T \leq \frac{A_0 f_A}{2\pi C_T \cdot BW_{3dB}^2}, \tag{I.125}$$

which is identical to the limit in Eq. (6.14).

7.6 The transimpedance expression for the common-base TIA including the base resistance R_b is

$$Z_T(s) = \frac{R_T}{[1 + s/(\omega_0 Q) + s^2/\omega_0^2][1 + s/\omega_{p2}]}, \tag{I.126}$$

where

$$R_T = R_C, \quad \omega_{p2} = \frac{1}{R_C C_L}, \tag{I.127}$$

$$\omega_0 = \sqrt{\frac{g_m}{R_b C_D C_I}}, \quad Q = \frac{\sqrt{g_m R_b C_D C_I}}{C_T}, \tag{I.128}$$

$C_I = C_{be}$, C_D is the photodetector capacitance plus all capacitances in parallel to it, and $C_T = C_D + C_I$ is the total input capacitance at low frequencies. To obtain $Q \leq 1/\sqrt{2}$, the base resistance must be limited to $R_b \leq C_T^2/(2g_m C_D C_I)$. For $Q = 1/\sqrt{2}$, the bandwidth due to the input poles is given by $BW_{3dB} = \sqrt{2}g_m/(2\pi C_T)$.

7.7 The precise 3-dB bandwidth as a function of the pole spacing χ can be calculated from Eq. (7.12) by setting $|Z_T(j2\pi BW_{3dB})|^2 = \frac{1}{2}R_T^2$ or by using $BW_{3dB} = \rho\sqrt{\chi} \cdot \omega_{p2}/(2\pi)$ from Eq. (H.16) with ρ from Eq. (H.17):

$$BW_{3dB} = \sqrt{\frac{\sqrt{\chi^4 + 6\chi^2 + 1} - \chi^2 - 1}{2}} \cdot \frac{\omega_{p2}}{2\pi}. \tag{I.129}$$

With $\omega_{p2} = \sqrt{2\pi A_0 f_A/(\chi R_C C_T)}$ (see main text) and $R_T = R_C$, the transimpedance follows as [17]

$$R_T = \frac{\sqrt{\chi^4 + 6\chi^2 + 1} - \chi^2 - 1}{2\chi} \cdot \frac{A_0 f_A}{2\pi C_T \cdot BW_{3dB}^2}. \tag{I.130}$$

For $\chi = 1$, the first factor reaches its maximum value of $\sqrt{2} - 1$.

7.8 The input-referred noise current PSD of the common-gate TIA is

$$\overline{I_{n,TIA}^2}(f) = \frac{4kT}{R_D} + \frac{4kT}{R_S} + 2qI_G + 4kT\tilde{\Gamma} g_m \frac{(2\pi C_T)^2}{(g_m + g_{mb})^2} f^2 + \ldots, \tag{I.131}$$

where $\tilde{\Gamma} = \gamma g_{d0}/g_m$ (the induced gate noise is neglected) and we assumed that $(g_m + g_{mb})^2 R_S/g_m \gg \tilde{\Gamma}$ (the white noise due to R_S dominates that due to the FET) and $g_m R_D \gg 1/\tilde{\Gamma}$ (the f^2 noise due to the FET dominates that due to R_D).

7.9 With the asymptotic noise bandwidths ($\chi \to \infty$) $BW_n \approx \pi/2 \cdot BW_{3dB}$ and $BW_{n2}^3 \approx 3\pi\chi/2 \cdot BW_{3dB}^3$ and Eq. (7.15), we find the total input-referred noise current PSD of the common-base TIA as

$$\overline{i_{n,TIA}^2} \approx \left(\frac{4kT}{R_C} + \frac{4kT}{R_E} + \frac{2qI_C}{\beta} + \frac{2qI_C}{\chi} \right) \frac{\pi}{2} \cdot BW_{3dB} + 2kTR_b(2\pi C_D)^2 \pi\chi \cdot BW_{3dB}^3 + \dots, \tag{I.132}$$

where we have used $g_m/(2\pi C_T) \approx \chi \cdot BW_{3dB}$.
With the same asymptotic noise bandwidths and Eq. (I.131), we find the total input-referred noise current PSD of the common-gate TIA as

$$\overline{i_{n,TIA}^2} \approx \left(\frac{4kT}{R_D} + \frac{4kT}{R_S} + 2qI_G + \frac{4kT\tilde{\Gamma} g_m}{\chi} \right) \frac{\pi}{2} \cdot BW_{3dB} + \dots, \tag{I.133}$$

where we have used $(g_m + g_{mb})/(2\pi C_T) \approx \chi \cdot BW_{3dB}$.

7.10 The transimpedance of the transistor-level RGC TIA is

$$Z_T(s) = R_T \frac{1 + s/\omega_z}{[1 + s/(\omega_0 Q) + s^2/\omega_0^2][1 + s/\omega_{p2}]}, \tag{I.134}$$

where

$$R_T = R_D, \quad \omega_z = \frac{g_{mB}R_B + 1}{R_B C_B}, \quad \omega_{p2} = \frac{1}{R_D C_L}, \tag{I.135}$$

$$\omega_0 = \sqrt{\frac{(g_{mB}R_B + 1)g_{m1}}{R_B[(C_D + C_I)C_M + C_T C_B]}}, \tag{I.136}$$

$$Q = \frac{\sqrt{(g_{mB}R_B + 1)g_{m1}R_B[(C_D + C_I)C_M + C_T C_B]}}{C_T + g_{mB}R_B C_M + g_{m1}R_B C_B}, \tag{I.137}$$

$C_T = C_D + C_I + C_M$ is the total input capacitance at low frequencies, C_D is the photodetector capacitance plus all parasitic capacitances in parallel to it, $C_I = C_{gsB} + C_{sb1}$, $C_M = C_{gs1} + C_{gdB}$, $C_B = C_{dbB}$, and C_L is the total load capacitance at the output node.

7.11 The (simplified) input resistance of the common-gate stage with V_{GS}-mirror biasing in Fig. 7.28(a) is

$$R_I = \frac{g_{m2}g_{m3} - g_{m1}g_{m4}}{g_{m1}g_{m2}g_{m3}}, \tag{I.138}$$

which is zero for $g_{m1} = g_{m2}$ and $g_{m3} = g_{m4}$.
The (simplified) input resistance of the translinear cross-quad in Fig. 7.28(b) is

$$R_I = \frac{g_{m3}g_{m4} - g_{m1}g_{m2}}{g_{m1}g_{m3}g_{m4}}, \tag{I.139}$$

which is zero for $g_{m1} = g_{m3}$ and $g_{m2} = g_{m4}$.

7.12(a) A third-order Butterworth transfer function is of the form [21, p. 272]

$$H(s) = \frac{1}{1 + 2sT + 2s^2T^2 + s^3T^3}, \tag{I.140}$$

where the 3-dB bandwidth is given by $1/(2\pi T)$. Identifying the polynomial coefficients with those of Eq. (7.27), $2T = R_I C_T$, $2T^2 = L_S C_D$, $T^3 = R_I C_I L_S C_D$, and solving for L_S, C_D, and C_I yields

$$L_S = \frac{2}{3}R_I^2 C_T, \quad C_D = \frac{3}{4}C_T, \quad C_I = \frac{1}{4}C_T. \tag{I.141}$$

The original bandwidth is $BW_{3dB} = 1/(2\pi R_I C_T)$ and the new bandwidth is $BW_{3dB} = 1/(2\pi T) = 1/(\pi R_I C_T)$, that is, the bandwidth doubles.

7.12(b) A third-order Bessel transfer function is of the form [21, p. 279]

$$H(s) = \frac{1}{1 + sT + \frac{2}{5}s^2T^2 + \frac{1}{15}s^3T^3}, \tag{I.142}$$

where the 3-dB bandwidth is given by $1.756/(2\pi T)$. Identifying the polynomial coefficients with those of Eq. (7.27), $T = R_I C_T$, $\frac{2}{5}T^2 = L_S C_D$, $\frac{1}{15}T^3 = R_I C_I L_S C_D$, and solving for L_S, C_D, and C_I yields

$$L_S = \frac{12}{25}R_I^2 C_T, \quad C_D = \frac{5}{6}C_T, \quad C_I = \frac{1}{6}C_T. \tag{I.143}$$

The original bandwidth is $BW_{3dB} = 1/(2\pi R_I C_T)$ and the new bandwidth is $BW_{3dB} = 1.756/(2\pi T) = 1.756/(2\pi R_I C_T)$, that is, the bandwidth increased by 75.6%.

Chapter 8

8.1(a) The transimpedance of the active-feedback TIA with common-source feedback device is

$$Z_T(s) = R_T \cdot \frac{1}{1 + s/(\omega_0 Q) + s^2/\omega_0^2}, \tag{I.144}$$

where

$$R_T = \frac{1}{g_{mF}}, \quad \omega_0 = \sqrt{\frac{A_0 g_{mF}}{C_T T_A}}, \quad Q = \sqrt{\frac{A_0 g_{mF} T_A}{C_T}}. \tag{I.145}$$

8.1(b) Setting $Q = 1/\sqrt{2}$ and solving for T_A yields $T_A = C_T/(2A_0 g_{mF})$.

8.1(c) Inserting this time constant into the expression for ω_0 yields

$$BW_{3\text{dB}} = \frac{\sqrt{2}A_0 g_{mF}}{2\pi C_T}. \tag{I.146}$$

This is $\sqrt{2}\times$ larger than the approximation for $T_A = 0$ given in [22, 23].

8.1(d) Combining the expressions for ω_0 and R_T, we find the same transimpedance limit as in Eq. (6.14).

8.1(e) For the MOS common-source active-feedback TIA, the input-referral functions for the noise sources are as follows. From $i_{n,G}$ to $i_{n,TIA}$ and from $i_{n,DF}$ to $i_{n,TIA}$: $H^{-1}(s) = 1$; from $i_{n,D}$ to $i_{n,TIA}$: $H_D^{-1}(s) = s\tilde{C}_T/g_m$ (assuming $g_{oF} = 0$). Thus the input-referred noise current PSD is

$$I_{n,TIA}^2(f) = 4kT\tilde{\Gamma}_F g_{mF} + 2qI_G + 4kT\Gamma\frac{(2\pi\tilde{C}_T)^2}{g_m}f^2 + \ldots, \tag{I.147}$$

where $\tilde{\Gamma}_F = \gamma_F g_{d0F}/g_{mF}$ and the first term of Eq. (I.147) is due to M_F.

8.1(f) For a BJT common-emitter active-feedback TIA, the input-referral functions for the noise sources are as follows. From $i_{n,B}$ to $i_{n,TIA}$ and from $i_{n,CF}$ to $i_{n,TIA}$: $H^{-1}(s) = 1$; from $i_{n,C}$ to $i_{n,TIA}$: $H_C^{-1}(s) = s\tilde{C}_T/g_m$ (neglecting R_b); from $i_{n,Rb}$ to $i_{n,TIA}$: $H_{Rb}^{-1}(s) = sR_bC_D$; from $i_{n,RbF}$ to $i_{n,TIA}$: $H_{RbF}^{-1}(s) = g_{mF}R_{bF}/(1 + sR_{bF}C_{beF})$. Thus the input-referred noise current PSD is

$$I_{n,TIA}^2(f) = 2kTg_{mF} + \frac{4kTg_{mF}^2 R_{bF}}{1 + (2\pi R_{bF}C_{beF})^2 f^2} + \frac{2qI_C}{\beta} + 2qI_C\frac{(2\pi\tilde{C}_T)^2}{g_m^2}f^2 + 4kTR_b(2\pi C_D)^2 f^2 + \ldots, \tag{I.148}$$

where the relationship $g_{mF} = q/(kT) \cdot I_{CF}$ was used in the first term and R_{bF} and C_{beF} are the base resistance and the base–emitter capacitance of the feedback device, respectively. The first two terms of Eq. (I.148) are due to the feedback BJT.

8.2(a) The transimpedance of the active-feedback TIA with common-drain feedback device is

$$Z_T(s) = -R_T \cdot \frac{1}{1 + s/(\omega_0 Q) + s^2/\omega_0^2}, \tag{I.149}$$

where

$$R_T = \frac{A_0}{(A_0 + 1)g_{mF} + g_{mbF}}, \tag{I.150}$$

$$\omega_0 = \sqrt{\frac{(A_0 + 1)g_{mF} + g_{mbF}}{C_T T_A}}, \tag{I.151}$$

$$Q = \frac{\sqrt{[(A_0 + 1)g_{mF} + g_{mbF}]C_T T_A}}{C_T + T_A(g_{mF} + g_{mbF})}. \tag{I.152}$$

With $g_{mbF} = 0$ and $g_{mF} \to 1/R_F$, these equations become identical to Eqs. (6.8)–(6.11).

8.2(b) Combining the expressions for ω_0 and R_T, we find the same transimpedance limit as in Eq. (6.14).

8.2(c) The input-referral functions for the noise sources are as follows. From $i_{n,G}$ to $i_{n,TIA}$ and from $i_{n,DF}$ to $i_{n,TIA}$: $H^{-1}(s) = 1$; from $i_{n,D}$ to $i_{n,TIA}$: $H_D^{-1}(s) = (s\tilde{C}_T + g_{mF} + g_{mbF})/g_m$ (assuming $g_{oF} = 0$). Thus the input-referred noise current PSD is

$$I^2_{n,TIA}(f) = 4kT\tilde{\Gamma}_F g_{mF} + 2qI_G + 4kT\Gamma\frac{(2\pi\tilde{C}_T)^2}{g_m}f^2 + \ldots, \tag{I.153}$$

where we assumed $g_m/g_{mF} \gg \Gamma/\tilde{\Gamma}_F \cdot (1 + g_{mbF}/g_{mF})^2$ (the white noise due to M_F dominates that due to the input FET) and neglected the shot and induced gate noise of M_F ($\tilde{\Gamma}_F = \gamma_F g_{d0F}/g_{mF}$). The first term of Eq. (I.153) is due to M_F.

8.3(a) The transimpedance of the current-mode TIA for $R_S \neq 0$, $C_L = 0$ is

$$Z_T(s) = -R_T \cdot \frac{1 - s/\omega_z}{1 + s/(\omega_0 Q) + s^2/\omega_0^2}, \tag{I.154}$$

where

$$R_T = \frac{A_0 R_F - R_S}{A_0 + 1}, \quad \omega_z = \frac{A_0 R_F - R_S}{R_S T_A}, \tag{I.155}$$

$$\omega_0 = \sqrt{\frac{A_0 + 1}{R_S C_T T_A}}, \quad Q = \frac{\sqrt{(A_0 + 1) R_S C_T T_A}}{R_S C_T + T_A}. \tag{I.156}$$

Neglecting the zero, which usually is located at a very high frequency, setting $Q = 1/\sqrt{2}$ for a Butterworth response, and solving for T_A, we find $T_A \approx 2A_0 R_S C_T$ (solution for $T_A > R_S C_T$). Inserting this solution into ω_0, we find the bandwidth

$$BW_{3\text{dB}} = \frac{\omega_0}{2\pi} = \frac{\sqrt{2A_0(A_0 + 1)}}{2\pi T_A} \approx \sqrt{2} A_0 f_A. \tag{I.157}$$

Note: The bandwidth is independent of R_F no matter which solution we pick ($T_A > R_S C_T$ or $T_A < R_S C_T$).

8.3(b) The transimpedance of the current-mode TIA for $R_S = 0$, $C_L \neq 0$ is

$$Z_T(s) = -R_T \cdot \frac{1}{1 + s/(\omega_0 Q) + s^2/\omega_0^2}, \tag{I.158}$$

where

$$R_T = \frac{A_0 R_F}{A_0 + 1}, \tag{I.159}$$

$$\omega_0 = \sqrt{\frac{A_0 + 1}{R_F C_L T_A}}, \quad Q = \frac{\sqrt{(A_0 + 1) R_F C_L T_A}}{R_F C_L + T_A}. \tag{I.160}$$

Setting $Q = 1/\sqrt{2}$ for a Butterworth response, and solving for T_A, we find $T_A \approx R_F C_L/(2A_0)$ (solution for $T_A < R_F C_L$). Inserting T_A into ω_0, we find the bandwidth

$$BW_{3\text{dB}} = \frac{\omega_0}{2\pi} = \frac{\sqrt{2A_0(A_0 + 1)}}{2\pi R_F C_L}. \tag{I.161}$$

Combining the expressions for ω_0 and R_T, we find the transimpedance limit [17]

$$R_T \leq \frac{A_0 f_A}{2\pi C_L \cdot BW_{3\text{dB}}^2}. \tag{I.162}$$

Appendices

C.1(a) The composite jitter distribution is $\frac{1}{2}(1/t_{RJ}^{rms})$ Gauss$[(t_J - t_{DJ}^{pp}/2)/t_{RJ}^{rms}]$ $+ \frac{1}{2}(1/t_{RJ}^{rms})$ Gauss$[(t_J + t_{DJ}^{pp}/2)/t_{RJ}^{rms}]$. Integrating the two (equal) tails outside of $\pm t_{TJ}^{pp}/2$ results in $BER = (1/t_{RJ}^{rms}) \int_{t_{TJ}^{pp}/2}^{\infty}$ Gauss$[(t_J - t_{DJ}^{pp}/2)/t_{RJ}^{rms}]$ + Gauss$[(t_J + t_{DJ}^{pp}/2)/t_{RJ}^{rms}]\, dt_J$. For $t_{DJ}^{pp} \gg t_{RJ}^{rms}$, the second term under the integral can be neglected. Substituting $x = (t_J - t_{DJ}^{pp}/2)/t_{RJ}^{rms}$ yields $BER = \int_Q^{\infty}$ Gauss$(x)\, dx$, where $Q = (t_{TJ}^{pp} - t_{DJ}^{pp})/(2t_{RJ}^{rms})$. Thus, $t_{TJ}^{pp} = t_{DJ}^{pp} + 2Qt_{RJ}^{rms}$.

C.1(b) For the dual Dirac distribution, $t_{TJ}^{pp} = 0.581$ UI and Eq. (C.1) correctly predicts $t_{TJ}^{pp} = 0.581$ UI. For the even distribution, $t_{TJ}^{pp} = 0.559$ UI and Eq. (C.1) overestimates $t_{TJ}^{pp} = 0.581$ UI (+4%). Without deterministic jitter, $t_{TJ}^{pp} = 0.589$ UI and Eq. (C.1) underestimates $t_{TJ}^{pp} = 0.581$ UI (−1.4%).

C.2 Based on the model equations

$$t_{DJ}^{pp} + 2 \times 7.034\ t_{RJ}^{rms} = (1 - 0.298)\ \text{UI},$$
$$t_{DJ}^{pp} + 2 \times 5.998\ t_{RJ}^{rms} = (1 - 0.350)\ \text{UI},$$
$$t_{DJ}^{pp} + 2 \times 4.753\ t_{RJ}^{rms} = (1 - 0.412)\ \text{UI},$$

we find the best fit for the parameters: $t_{DJ}^{pp} = 0.35$ UI and $t_{RJ}^{rms} =$ 0.025 UI.

C.3(a) The data signal is not affected (all edges are translated by the same amount).

C.3(b) The clock-like data signal undergoes a duty-cycle distortion.

C.4 Cycle-to-cycle jitter is defined as $\Delta t_J(n) - \Delta t_J(n-1)$ and the 2-period jitter is defined as $\Delta t_J(n) + \Delta t_J(n-1)$. Their rms values are equal if the period jitter $\Delta t_J(n)$ has a symmetric distribution, is uncorrelated (from cycle to cycle), and stationary.

D.1 The output signal is $y(t) = A\{a_0 + X[\sin(\omega_1 t) + \sin(\omega_2 t)] + a_2 X^2[\sin(\omega_1 t) + \sin(\omega_2 t)]^2\}$. After expanding and sorting with respect to frequencies, we find

$$y(t) = A[a_0 + a_2 X^2]$$
$$+ AX[\sin(\omega_1 t) + \sin(\omega_2 t)]$$

$$- A\frac{1}{2}a_2X^2[\cos(2\omega_1 t) + \cos(2\omega_2 t)] \qquad (I.163)$$
$$+ Aa_2X^2[\cos((\omega_1 - \omega_2)t) - \cos((\omega_1 + \omega_2)t)].$$

The first line represents the output offset, the second line represents the fundamental tones, the third line represents the second-order harmonic products, and the fourth line represents the second-order intermodulation products.

E.1 The output signal equals c_1 times the input signal plus the input signal delayed by $1/B$ plus c_3 times the input signal delayed by $2/B$; thus, the frequency response is $H(f) = c_1 + \exp(-j\,2\pi f/B) + c_3\exp(-j\,4\pi f/B)$. Its squared magnitude is

$$|H(f)|^2 = \left[c_1 + \cos\left(\frac{2\pi f}{B}\right) + c_3\cos\left(\frac{4\pi f}{B}\right)\right]^2 + \left[\sin\left(\frac{2\pi f}{B}\right) + c_3\sin\left(\frac{4\pi f}{B}\right)\right]^2 \qquad (I.164)$$

and thus $|H(0)| = |1 + c_1 + c_3|$ and $|H(B/2)| = |1 - c_1 - c_3|$, corresponding to a high-pass response for $c_1, c_3 < 0$.

E.2(a) The pre- and postcursor ISI are $\delta_F = 0.05$ and $\delta_P = 0.1$, respectively.

E.2(b) The FFE weights yielding the smallest MSE for the middle 10 bits are $c_1 = -0.073$, $c_2 = +1.192$, and $c_3 = -0.138$. The MSE is $0.16 \cdot 10^{-3}$.

E.2(c) The DFE weights yielding the smallest MSE for the middle 10 bits are $c_1 = -0.070$, $c_2 = +1.184$, and $c_3 = -0.120$. The MSE is $0.01 \cdot 10^{-3}$.

F.1 The error probability as a function of the decision threshold is $BER = \frac{1}{2}[\int_{V_{DTH}}^{\infty} \text{Zero}(v_O)\,dv_O + \int_{-\infty}^{V_{DTH}} \text{One}(v_O)\,dv_O]$, which can be rewritten as $\frac{1}{2}[1 - \int_{-\infty}^{V_{DTH}} \text{Zero}(v_O)\,dv_O + \int_{-\infty}^{V_{DTH}} \text{One}(v_O)\,dv_O]$. Taking the derivative and setting it to zero leads to $\partial BER/\partial V_{DTH} = \frac{1}{2}[-\text{Zero}(V_{DTH}) + \text{One}(V_{DTH})] = 0$; hence the optimum decision threshold is at the intersection point $\text{Zero}(V_{DTH}) = \text{One}(V_{DTH})$.

F.2(a) The probability for a pseudo error is $PBER = \frac{1}{2}\int_{V_{DTH1}}^{V_{DTH2}} [\text{Zero}(v_O) + \text{One}(v_O)]\,dv_O$.

F.2(b) The probability for an actual error given the decision threshold V_{DTH1} is $BER = \frac{1}{2}[\int_{V_{DTH1}}^{\infty} \text{Zero}(v_O)\,dv_O + \int_{-\infty}^{V_{DTH1}} \text{One}(v_O)\,dv_O]$ and the error probability given the decision threshold V_{DTH2} is $BER =$

$\frac{1}{2}[\int_{V_{DTH2}}^{\infty} \text{Zero}(v_O)\, dv_O + \int_{-\infty}^{V_{DTH2}} \text{One}(v_O)\, dv_O]$ (cf. Problem F.1). The absolute difference of these two BERs is $|\Delta BER| = \frac{1}{2} \int_{V_{DTH1}}^{V_{DTH2}} |\text{Zero}(v_O) - \text{One}(v_O)|\, dv_O$.

Thus, the pseudo bit-error rate is *not* equal to the difference of the two actual bit-error rates. However, they are approximately equal, if one distribution dominates the other within the two decision thresholds (integral bounds).

G.1 A frame size of 255 bytes and $BER = 10^{-4}$ results in an average of $M = 255 \times 8 \times 10^{-4} = 0.204$ bit errors per frame. The probability for 9 bit errors per frame (which is not correctable with RS(255,239), assuming each error is in a different byte) can be found with the Poisson distribution as $\exp(-M) \cdot M^9/9! = 1.38 \cdot 10^{-12}$. Neglecting the small possibility of more than 9 bit errors per frame, this number is the frame error rate at the output of the decoder. Converting the frame error rate back to the (payload) bit error rate yields $BER = 9/(239 \times 8) \times 1.38 \cdot 10^{-12} = 6.47 \cdot 10^{-15}$. (A more precise analysis yields $BER = 5 \cdot 10^{-15}$ [24].)

References

1 P. S. Henry, R. A. Linke, and A. H. Gnauck. Introduction to lightwave systems. In S. E. Miller and I. P. Kaminow, editors, *Optical Fiber Telecommunications II*, pages 781–831. Academic Press, San Diego, CA, 1988.

2 R. Sarpeshkar, T. Delbrück, and C. A. Mead. White noise in MOS transistors and resistors. *IEEE Circuits Devices Mag.*, 9(6):23–29, 1993.

3 R. C. Steele, G. R. Walker, and N. G. Walker. Sensitivity of optically preamplified receivers with optical filtering. *IEEE Photonics Technol. Lett.*, 3(6):545–547, 1991.

4 B. L. Kasper, O. Mizuhara, and Y.-K. Chen. High bit-rate receivers, transmitters, and electronics. In I. P. Kaminow and T. Li, editors, *Optical Fiber Telecommunications IVA*, pages 784–851. Academic Press, San Diego, CA, 2002.

5 R.-J. Essiambre, G. Raybon, and B. Mikkelsen. Pseudo-linear transmission of high-speed TDM signals: 40 and 160 Gb/s. In I. P. Kaminow and T. Li, editors, *Optical Fiber Telecommunications IVB*, pages 232–304. Academic Press, San Diego, CA, 2002.

6 N. S. Bergano. Undersea amplified lightwave systems design. In I. P. Kaminow and T. L. Koch, editors, *Optical Fiber Telecommunications IIIA*, pages 302–335. Academic Press, San Diego, CA, 1997.

7 R. Ramaswami, K. N. Sivarajan, and G. Sasaki. *Optical Networks: A Practical Perspective*. Morgan Kaufmann Publishers, San Francisco, CA, 3rd edition, 2010.

8 N. S. Bergano. Undersea communication systems. In I. P. Kaminow and T. Li, editors, *Optical Fiber Telecommunications IVB*, pages 154–197. Academic Press, San Diego, CA, 2002.

9 T. V. Muoi. Receiver design for high-speed optical-fiber systems. *J. Lightwave Technol.*, LT-2(3):243–267, 1984.

10 P. E. Allen and D. R. Holberg. *CMOS Analog Circuit Design*. Holt, Rinehart and Winston, New York, 1987.

11 E. Säckinger. *Broadband Circuits for Optical Fiber Communication*. John Wiley &Sons, Inc., Hoboken, NJ, 2005.

12 A. Leven, R. Reuter, and Y. Baeyens. Unified analytical expressions for transimpedance and equivalent input noise current of optical receivers. *IEEE Trans. Microwave Theory Tech.*, MTT-48(10):1701–1706, 2000.

13 B. L. Kasper. Receiver design. In S. E. Miller and I. P. Kaminow, editors, *Optical Fiber Telecommunications II*, pages 689–722. Academic Press, San Diego, CA, 1988.

14 J. J. Morikuni, A. Dharchoudhury, Y. Leblebici, and S. M. Kang. Improvements to the standard theory for photoreceiver noise. *J. Lightwave Technol.*, LT-12(7):1174–1184, 1994.

15 M. Abraham. Design of Butterworth-type transimpedance and bootstrap-transimpedance preamplifiers for fiber-optic receivers. *IEEE Trans. Circuits Syst.*, CAS-29(6):375–382, 1982.

16 K. Kieschnick and H. Zimmermann. High-sensitivity BiCMOS OEIC for optical storage systems. *IEEE J. Solid-State Circuits*, SC-38(4):579–584, 2003.

17 E. Säckinger. The transimpedance limit. *IEEE Trans. Circuits Syst. I*, CASI-57(8):1848–1856, 2010.

18 S. D. Greaves and R. T. Unwin. The design of tuned front-end GaAs MMIC optical receivers. *IEEE Trans. Microwave Theory Tech.*, MTT-44(4):591–597, 1996.

19 T. H. Lee. *The Design of CMOS Radio-Frequency Integrated Circuits*. Cambridge University Press, Cambridge, UK, 1998.

20 J. Kim and J. F. Buckwalter. Bandwidth enhancement with low group-delay variation for a 40-Gb/s transimpedance amplifier. *IEEE Trans. Circuits Syst. I*, CASI-57(8):1964–1972, 2010.

21 U. Tietze and C. Schenk. *Halbleiter-Schaltungstechnik*. Springer-Verlag, 5th edition, 1980.

22 E. Braß, U. Hilleringmann, and K. Schumacher. System integration of optical devices and analog CMOS amplifiers. *IEEE J. Solid-State Circuits*, SC-29(8):1006–1010, 1994.

23 C. Rooman, D. Coppée, and M. Kuijk. Asynchronous 250-Mb/s optical receivers with integrated detector in standard CMOS technology for optocoupler applications. *IEEE J. Solid-State Circuits*, SC-35(7):953–958, 2000.

24 ITU-T. Forward error correction for submarine systems, recommendation G.975. International Telecommunication Union, Geneva, Switzerland, October 2000.

J

Notation

Voltages and Currents

- Constant voltages and currents (i.e., DC voltages and currents) are designated with uppercase letters and uppercase subscripts (e.g., V_{GS} or I_D).
- Total instantaneous voltages and currents are designated with lowercase letters and uppercase subscripts (e.g., v_{GS} or i_D).
- Small-signal voltages and currents in the time domain are designated with lowercase letters and lowercase subscripts (e.g., v_{gs} or i_d). These small-signal quantities represent a small change in the total instantaneous values (e.g., $v_{gs} = \Delta v_{GS}$ or $i_d = \Delta i_D$).
- Phasors for voltages and currents (complex quantities that describe the amplitude and phase of a small-signal sinusoid) are designated with uppercase letters and lowercase subscripts (e.g., V_{gs} or I_d). To avoid confusion with the notation for root spectral densities (see below), phasors may be underlined (e.g., $\underline{V}_{gs}$ or $\underline{I}_d$).
- The average value of a voltage or current signal is designated with a lowercase letter and a horizontal bar (e.g., $\overline{v}$ or $\overline{i}$).
- The mean-square value of a voltage or current signal is designated with a lowercase letter and a horizontal bar over the square (e.g., $\overline{v^2}$ or $\overline{i^2}$).
- The root-mean-square (rms) value of a voltage or current signal is designated with a lowercase letter and an *rms* superscript (e.g., v^{rms} or i^{rms}).
- The peak-to-peak value (swing) of a voltage or current signal is designated with a lowercase letter and a *pp* superscript (e.g., v^{pp} or i^{pp}).

Noise Quantities

- Instantaneous noise voltages and currents in the time domain are designated with lowercase letters and the subscript n (e.g., v_n or i_n).
- Mean-square noise voltages and currents, measured in a given bandwidth, are designated with $\overline{v_n^2}$ and $\overline{i_n^2}$, respectively. These quantities are also referred to as *noise powers* for short. The mean indicates either the time average

Analysis and Design of Transimpedance Amplifiers for Optical Receivers, First Edition. Eduard Säckinger.

(for stationary noise) or the ensemble average (for nonstationary noise), as appropriate.

The mean-square noise quantities are the integral of the (output-referred) power spectral densities over the given bandwidth. The integral of input-referred power spectral densities, however, does not lead to meaningful input-referred mean-square noise quantities.

- Root-mean-square (rms) noise voltages and currents, measured in a given bandwidth, are designated with v_n^{rms} and i_n^{rms}, respectively.

  The rms noise quantities are the square root of the mean-square noise quantities.
- Power spectral densities (PSD) of voltages and currents are designated with $V_n^2(f)$ and $I_n^2(f)$, respectively. These quantities are also referred to as *power spectra* for short. The PSDs used in this book are all *one sided*, and thus directly represent the noise power in a 1-Hz bandwidth at frequency f.

  The PSDs are the Fourier transform of the autocorrelation function of the time domain signals.
- Root spectral densities (or amplitude spectral densities) of voltages and currents are designated with $V_n(f)$ and $I_n(f)$, respectively. These quantities are also referred to as *noise densities* for short.

  The root spectral densities are the square root of the PSDs and represent the rms noise voltage or current in a 1-Hz bandwidth at frequency f.

K

Symbols

Latin Symbols

A	(i) voltage gain; (ii) current gain
$\mathcal{A}$	element of ABCD (chain) matrix
A_0	voltage gain at low frequencies
a_ν	normalized power-series coefficient
A_E	emitter area
A_{LS}	large-signal gain
B	(i) bit rate; (ii) susceptance
$\mathcal{B}$	element of ABCD (chain) matrix
B_0	voltage gain at low frequencies of booster amplifier or bootstrap buffer
B_c	correlation susceptance ($Y_c = G_c + jB_c$)
B_{opt}	source susceptance resulting in the minimum noise figure ($Y_{\text{opt}} = G_{\text{opt}} + jB_{\text{opt}}$)
$\tilde{B}_{\text{opt}}$	source susceptance resulting in the minimum input-referred noise current ($\tilde{Y}_{\text{opt}} = \tilde{G}_{\text{opt}} + j\tilde{B}_{\text{opt}}$)
B_S	source susceptance ($Y_S = G_S + jB_S$)
BER	bit-error rate (actually, bit-error probability)
BER_{in}	bit-error rate before error correction
BER_{out}	bit-error rate after error correction
BW	bandwidth
$BW_{3\text{dB}}$	3-dB bandwidth
BW_{APD}	bandwidth of avalanche photodetector
BW_D	bandwidth of decision circuit
BW_n	noise bandwidth (for white noise)
$BW_{n\nu}$	noise bandwidth for f^ν noise ($BW_{n0} = BW_n$)
BW_{null}	DC-to-null bandwidth

Analysis and Design of Transimpedance Amplifiers for Optical Receivers, First Edition. Eduard Säckinger.

BW_O	optical bandwidth
BW_{PIN}	bandwidth of p–i–n photodetector
BW_S	bandwidth of a stage
C	capacitor or capacitance
$\mathcal{C}$	element of ABCD (chain) matrix
c	(i) speed of light in vacuum, $c = 299.8 \cdot 10^6\ m/s$; (ii) correlation coefficient
c_ν	filter coefficient number ν
C_B	load capacitance of booster amplifier
C_{bc}	base–collector capacitance
C_{be}	base–emitter capacitance
C_D	photodetector capacitance including parallel parasitic capacitances
C_{db}	drain–bulk capacitance
C_E	emitter capacitor
C_F	feedback capacitor
C_{gd}	gate–drain capacitance
C_{gs}	gate–source capacitance
C_I	input capacitance
$\tilde{C}_I$	input capacitance for shorted output
C_L	load capacitance
C_O	output capacitance
C_P	bond-pad capacitance
C_{PD}	internal capacitance of photodiode
C_R	replica capacitor
C_{sb}	source–bulk capacitance
C_T	total capacitance at TIA input ($= C_D + C_I$)
$\tilde{C}_T$	total capacitance at TIA input for shorted output ($= C_D + \tilde{C}_I$)
$CMRR$	common-mode rejection ratio
CNR	carrier-to-noise ratio
CSO	composite second-order distortion
CTB	composite triple-beat distortion
D	chromatic dispersion parameter
$\mathcal{D}$	element of ABCD (chain) matrix
D_{PMD}	polarization-mode dispersion parameter
E	(i) energy; (ii) electric field
E_b	energy per information bit
E_g	bandgap energy
ER	extinction ratio
F	(i) noise figure; (ii) excess noise factor
$F_{\min}$	minimum noise figure (F at $Y_S = Y_{\text{opt}}$)
$\tilde{F}$	optical noise figure
f	frequency

Δf	bandwidth
f_A	pole frequency of amplifier ($= 1/(2\pi T_A)$)
f_B	pole frequency of booster amplifier ($= 1/(2\pi T_B)$)
f_b	high-frequency cutoff (bandwidth)
f_c	$1/f$-noise corner frequency of noise spectrum
f_{c2}	f^2-noise corner frequency of noise spectrum
f_{cutoff}	cutoff frequency of transmission line
$f_{\text{cutoff},I}$	cutoff frequency of an input transmission line
$f_{\text{cutoff},O}$	cutoff frequency of an output transmission line
f_I	open-loop input pole frequency
F_M	noise figure of the matching network
f_p	frequency of pole
f_{Rb}	pole frequency due to the base resistance of a BJT ($= 1/(2\pi R_b \tilde{C}_I)$)
f_T	unity current-gain frequency of a transistor (transition frequency)
F_{tot}	total noise figure
f_u	upper frequency bound in integral
G	(i) power gain; (ii) conductance
G_c	correlation conductance ($Y_c = G_c + jB_c$)
G_D	photodetector conductance ($Y_D = G_D + jB_D$)
G_{Di}	photodetector intrinsic conductance
g_{d0}	drain–source conductance at zero drain–source voltage
G_M	matching network output conductance ($Y_M = G_M + jB_M$)
g_m	transconductance of a transistor
g_{mb}	bulk-input transconductance of a transistor
g_o	output conductance of a transistor
G_{opt}	source conductance resulting in the minimum noise figure ($Y_{\text{opt}} = G_{\text{opt}} + jB_{\text{opt}}$)
$\tilde{G}_{\text{opt}}$	source conductance resulting in the minimum input-referred noise current ($\tilde{Y}_{\text{opt}} = \tilde{G}_{\text{opt}} + j\tilde{B}_{\text{opt}}$)
G_S	source conductance ($Y_S = G_S + jB_S$)
$\text{Gauss}(x)$	normalized Gaussian distribution: $1/\sqrt{2\pi} \cdot \exp(-x^2/2)$
$H(f)$	transfer function of a linear system
h	Planck constant, $h = 6.626 \cdot 10^{-34}$ Js
$h(t)$	impulse response of a linear system
H_0	passband value of the transfer function $H(f)$
$H_C(f)$	current referral function from the TIA input to the collector
$H_D(f)$	current referral function from TIA input to the drain
$H_{FRC}(f)$	full raised-cosine spectrum
$H_M(f)$	current referral function through the matching network
$H_{NRZ}(f)$	spectrum of the ideal NRZ signal
$H_{PD}(f)$	current referral function through the extrinsic photodetector

$H_{Rb}(f)$	current referral function from the TIA input to R_b
$H_{RC}(f)$	current referral function from the TIA input to R_C
HD_ν	νth-order harmonic distortion
i	current
I_ν	Personick integral ($\nu = 1, 2, 3$)
i_{ANT}	antenna current
i_{APD}	avalanche photodetector current
I_B	base current
i_{BD}	balanced p–i–n photodetector current
$i_{BD,\text{het}}$	balanced p–i–n photodetector current with optical heterodyne front-end
I_C	collector current
I_D	drain current
I_{DK}	dark current
I_f	Personick integral for f noise
I_G	gate current
i_I	input current
i_{lin}^{pp}	maximum input current for linear operation
i_n	noise current
$i_{n,0}$	noise current when receiving a zero
$i_{n,1}$	noise current when receiving a one
i_{n4}	noise current in the 4-parameter noise model
i_{n4c}	fully correlated noise current in the 4-parameter noise model
i_{n4u}	uncorrelated noise current in the 4-parameter noise model
$i_{n,APD}$	noise current of an avalanche photodetector
$i_{n,APD,0}$	noise current of an avalanche photodetector when receiving a zero
$i_{n,APD,1}$	noise current of an avalanche photodetector when receiving a one
$i_{n,B}$	noise current at the base of a BJT
$i_{n,BD}$	noise current of a balanced p–i–n photodetector
$i_{n,BD,\text{het}}$	noise current of a balanced p–i–n photodetector with optical heterodyne front-end
$i_{n,C}$	noise current at the collector of a BJT
$i_{n,\text{chan}}$	channel noise of an FET
$i_{n,\text{ckt}}$	input-referred noise current of the linear channel (circuit noise)
$i_{n,D}$	noise current at the drain of an FET
$i_{n,DK}$	noise current due to dark current
$i_{n,\text{front}}$	input-referred noise current of the amplifier front-end
$i_{n,\text{in},C}$	input-referred collector noise current
$i_{n,\text{in},D}$	input-referred drain noise current
$i_{n,\text{in},Rb}$	input-referred base-resistance noise current

$i_{n,G}$	noise current at the gate of an FET
$i_{n,\text{ind}}$	induced gate noise current of an FET
$i_{n,OA}$	noise current of an optically preamplified p–i–n detector
$i_{n,OA,0}$	noise current of an optically preamplified p–i–n detector when receiving a zero
$i_{n,OA,1}$	noise current of an optically preamplified p–i–n detector when receiving a one
$i_{n,OA,S}$	noise current of an optically preamplified p–i–n detector that is due to the quantum noise of the optical source
$i_{n,PD}$	noise current of a photodetector
$i_{n,PD,0}$	noise current of a photodetector when receiving a zero
$i_{n,PD,1}$	noise current of a photodetector when receiving a one
$i_{n,PIN}$	noise current of a p–i–n photodetector
$i_{n,PIN,0}$	noise current of a p–i–n photodetector when receiving a zero
$i_{n,PIN,1}$	noise current of a p–i–n photodetector when receiving a one
$i_{n,PIN,\text{het}}$	noise current of a p–i–n photodetector with optical heterodyne front-end
$i_{n,Rb}$	noise current of the base resistance of a BJT
$i_{n,\text{res}}$	noise current of a resistor
$i_{n,RIN}$	noise current of a p–i–n photodetector due to laser RIN noise
$i_{n,S}$	noise current of the (signal) source
$i_{n,TIA}$	input-referred noise current of a TIA
i_{OA}	current of an optically preamplifier p–i–n detector
I_{OS}	offset current
i_{ovl}^{pp}	input overload current
i_{PD}	photodetector current
i_{PIN}	p–i–n photodetector current
$i_{PIN,\text{het}}$	p–i–n photodetector current with optical heterodyne front-end
i_S	signal current
i_{sens}^{pp}	electrical sensitivity
IMD_ν	νth-order intermodulation distortion
IMD_{3T}	triple-beat intermodulation distortion
j	imaginary unit, $j = \sqrt{-1}$
K	coupling factor of optical feedback
k	(i) Boltzmann constant, $k = 1.381 \cdot 10^{-23}\ J/K = 86.18\ \mu eV/K$; (ii) magnetic coupling factor
k_A	ionization-coefficient ratio
K_M	available power gain of matching network (G_M is taken)
L	(i) length; (ii) inductor or inductance
L_B	bond-wire inductance
L_E	emitter length
L_I	input inductor

L_O	output inductor
L_P	parallel inductor
L_S	series inductor
M	(i) avalanche gain; (ii) mean of Poisson distribution
m	modulation index
M_ν	FET number ν
M_{opt}	optimum avalanche gain
$MTPR$	multitone power ratio
N	integer number
n	(i) integer number; (ii) refractive index
N_0	noise power spectral density
N_1	atomic population for the ground state
N_2	atomic population for the excited state
n_{clad}	refractive index of the fiber cladding
n_{cor}	refractive index of the fiber core
N_{CSO}	second-order beat count
N_{CTB}	triple-beat count
$NECG$	net electrical coding gain
$OSNR$	optical signal-to-noise ratio
$OSNR_{req}$	optical signal-to-noise ratio required to meet a given BER
P	power
$\overline{P}$	average power
P_0	optical power when transmitting a zero
P_1	optical power when transmitting a one
P_{ASE}	optical noise power due to amplified spontaneous emission
P_{LO}	optical power of local oscillator
P_{out}	optical output power (from transmitter)
$\overline{P}_{ovl}$	optical overload power
P_S	optical signal power
$\overline{P}_{sens}$	optical receiver sensitivity
$\overline{P}_{sens,APD}$	sensitivity of receiver with an avalanche photodetector
$\overline{P}_{sens,het}$	sensitivity of coherent heterodyne receiver
$\overline{P}_{sens,OA}$	sensitivity of receiver with an optically preamplified p–i–n detector
$\overline{P}_{sens,OAC}$	sensitivity of receiver with a cascade of optical amplifiers
$\overline{P}_{sens,PIN}$	sensitivity of receiver with a p–i–n photodetector
P_{sens}^{OMA}	optical modulation amplitude sensitivity
Poisson(n)	Poisson distribution: $\exp(-M) \cdot M^n/n!$
PP	power penalty (in unamplified systems)
$PSD(f)$	power spectral density
Q	quality factor of a pole pair $(= 1/(2\zeta))$
$Q(x)$	Q function: $\frac{1}{2}\text{erfc}(x/\sqrt{2})$

Q	Personick Q(Q parameter)
q	electron charge, $q = 1.602 \cdot 10^{-19}\ C$
Q_ν	bipolar transistor number ν
$\mathcal{Q}_{\text{in}}$	Personick Q needed for BER_{out} with error correction
$\mathcal{Q}_{\text{out}}$	Personick Q needed for BER_{out} without error correction
R	resistor or resistance
$\mathcal{R}$	responsivity
r	(i) run length; (ii) code rate
R_0	characteristic impedance (if real)
R_{ANT}	antenna resistance
$\mathcal{R}_{APD}$	responsivity of an avalanche photodetector
R_B	(i) base resistor; (ii) load resistor of booster amplifier
R_b	base resistance
R_C	collector resistor
R_D	(i) drain resistor; (ii) resistor in photodiode simulator
R_E	emitter resistor
R_F	feedback resistor
R_g	gate resistance
R_I	input resistance (resistor)
R_L	load resistor
R_n	noise resistance $(= V_{n4}^2/(4kT))$
R_O	output resistance (resistor)
R_{PD}	internal series resistance of photodiode
R_S	(i) shunt resistor; (ii) source resistor
R_T	(i) transresistance ($= Z_T$, if Z_T is real); (ii) thermistor
RIN	relative intensity noise spectrum
S	(i) optical power spectral density; (ii) S parameter
s	complex frequency variable
$S_{\mu\nu}$	S parameter from port ν to port μ
S_{ASE}	power spectral density due to amplified spontaneous emission in both polarization modes $(= 2S'_{ASE})$
S'_{ASE}	power spectral density due to amplified spontaneous emission in a single polarization mode
SNR	signal-to-noise ratio
T	(i) temperature; (ii) bit interval $(= 1/B)$
ΔT	pulse spreading $(= 2\sigma_T)$
t	time
T_A	time constant of amplifier $(= 1/(2\pi f_A))$
T_B	time constant of booster amplifier $(= 1/(2\pi f_B))$
t_{DDJ}	data-dependent jitter
t_{DJ}	deterministic jitter
T_{in}	input pulse width $(= 2\sigma_{\text{in}})$
t_J	timing jitter (absolute jitter or time interval error)

Δt_J	period jitter (cycle jitter)
t_{JTOL}	jitter tolerance
T_{out}	output pulse width ($= 2\sigma_{\text{out}}$)
t_{RJ}	random jitter
T_S	symbol interval
t_S	sampling instant
t_{SJ}	sinusoidal jitter
t_{TJ}	total jitter
THD	total harmonic distortion
v	voltage
V_{AGC}	automatic gain control voltage
V_{APD}	avalanche photodetector bias voltage
v_{AS}	aggressor signal voltage
v_B	voltage from bottom-level hold
V_{BE}	base–emitter voltage
V_{BI}	bias voltage
v_c	carrier velocity
V_{CC}	positive power-supply voltage of a bipolar circuit
V_{DD}	positive power-supply voltage of an FET circuit
V_{DS}	drain–source voltage
V_{DTH}	decision threshold voltage (slice level)
V_E	voltage of vertical eye opening
V_{EE}	negative power-supply voltage of a bipolar circuit
V_{GS}	gate–source voltage
v_I	input voltage
$v_{I,ovl}^{pp}$	maximum permissible input voltage swing
v_{IN}	voltage at inverting input
v_{IP}	voltage at noninverting input
V_M	voltage at midpoint of voltage divider
v_n	noise voltage
$v_{n,0}$	noise voltage when receiving a zero
$v_{n,1}$	noise voltage when receiving a one
v_{n4}	noise voltage in the 4-parameter noise model
$v_{n,TIA}$	noise voltage at the output of the TIA
v_O	output voltage
$v_{O,ovl}^{pp}$	maximum permissible output voltage swing
V_{OCM}	common-mode output voltage
v_{ON}	voltage at inverting output
v_{OP}	voltage at noninverting output
V_{OS}	offset voltage
V_{OSN}	offset-control voltage at the inverting input
V_{OSP}	offset-control voltage at the noninverting input
V_{PIN}	p–i–n photodetector bias voltage

V_R	voltage drop across resistor
V_{REF}	reference voltage
v_S	(i) signal voltage; (ii) source voltage
V_{SS}	negative power-supply voltage of an FET circuit
V_T	thermal voltage, $V_T = kT/q = 25.256$ mV @ 20°C
v_T	voltage from top-level hold
V_{TH}	threshold voltage
v_X	crosstalk voltage
W	width
W_A	absorption layer width
W_E	emitter width
W_M	multiplication layer width
X	amplitude of the input signal x
x	(i) input signal to filter or amplifier; (ii) space coordinate
XT_ν	crosstalk from aggressor number ν
Y	admittance
y	output signal from filter or amplifier
Y_c	correlation admittance $(= G_c + jB_c)$
Y_D	photodetector admittance $(= G_D + jB_D)$
Y_M	matching network output admittance $(Y_M = G_M + jB_M)$
Y_{opt}	source admittance resulting in the minimum noise figure $(= G_{\text{opt}} + jB_{\text{opt}})$
$\tilde{Y}_{\text{opt}}$	source admittance resulting in the minimum input-referred noise current $(= \tilde{G}_{\text{opt}} + j\tilde{B}_{\text{opt}})$
Y_S	source admittance $(= G_S + jB_S)$
Z	impedance
z	output signal from decision circuit
Z_D	photodetector impedance $(= 1/Y_D)$
Z_I	input impedance
Z_O	output impedance
Z_T	transimpedance
$Z_{TL,I}$	characteristic impedance of an input transmission line
$Z_{TL,O}$	characteristic impedance of an output transmission line

Greek Symbols

α	absorption coefficient
α_ν	coefficient for f^ν noise
β	current gain of a bipolar transistor
Γ	Ogawa's noise factor (for channel and induced gate noise of a FET)

$\tilde{\Gamma}$	Ogawa's noise factor without the induced gate noise ($= \gamma g_{d0}/g_m$)
γ	channel-noise factor of FET
δ	(i) decision threshold offset error normalized to the signal swing; (ii) Dirac's delta function; (iii) gate-noise factor of FET
δ_F	precursor ISI normalized to the signal swing
δ_P	postcursor ISI normalized to the signal swing
ζ	damping factor ($= 1/(2Q)$)
η	quantum efficiency
λ	wavelength
$\Delta\lambda$	spectral linewidth
$\Delta\lambda_S$	spectral linewidth of the unmodulated source
μ_s	adaptation step size
ρ	factor controlling the 3-dB bandwidth of a second-order low pass ($= 2\pi BW_{3dB}/\omega_0$)
σ	standard deviation or rms value
σ_{in}	input pulse rms width
σ_{out}	output pulse rms width
σ_T	rms impulse spread
τ	(i) group delay; (ii) transit time; (iii) delay time
$\Delta\tau$	group delay variation
τ_I	section delay of an input transmission line
τ_O	section delay of an output transmission line
ϕ	phase
ϕ_m	phase margin
ϕ_n	phase noise
χ	ratio of two pole frequencies or two cutoff frequencies
ψ	optimum input capacitance normalized to the detector capacitance ($= \tilde{C}_I/C_D$)
ω	angular frequency
ω_0	angular frequency of a pole pair
ω_b	angular frequency of a high-frequency cutoff (bandwidth)
ω_p	angular frequency of a pole
ω_z	angular frequency of a zero

Special Symbols

$\approx$	approximately equal to
$\propto$	proportional to ($y \propto x$ means $y = \text{const} \cdot x$)
$*$	complex conjugate

L

Acronyms

A/D	analog-to-digital (converter)
AC	alternating current
ADC	analog-to-digital converter
AGC	automatic gain control
AM	amplitude modulation
AOC	automatic offset control (or cancellation)
APD	avalanche photodetector (or photodiode)
APON	ATM passive optical network
ARQ	automatic repeat request
ASD	amplitude spectral density
ASE	amplified spontaneous emission
ASK	amplitude-shift keying
ATC	automatic (or adaptive) threshold control
ATM	asynchronous transfer mode
AWG	arrayed waveguide grating
BCH	Bose–Chaudhuri–Hocquenghem (code)
BER	bit-error rate
BERT	bit-error rate test set
BGA	ball grid array
BiCMOS	BJT + CMOS
BJT	bipolar junction transistor
BPON	broadband passive optical network
BPSK	binary phase-shift keying
BUJ	bounded uncorrelated jitter
BW	bandwidth
CATV	community-antenna television
CDR	clock and data recovery
CID	consecutive identical digits
CML	current-mode logic
CMOS	complementary MOS

Analysis and Design of Transimpedance Amplifiers for Optical Receivers, First Edition. Eduard Säckinger.

CMRR	common-mode rejection ratio
CMU	clock multiplication unit
CNR	carrier-to-noise ratio
CO	central office
CPM	cross-phase modulation
CRZ	chirped return-to-zero (modulation)
CS-RZ	carrier-suppressed return-to-zero (modulation)
CSO	composite second-order (distortion)
CTB	composite triple-beat (distortion)
CW	continuous wave
D/A	digital-to-analog (converter)
DA	distributed amplifier
DAC	digital-to-analog converter
DBPSK	differential binary phase-shift keying
DC	direct current
DCD	duty-cycle distortion (jitter)
DCF	dispersion compensating fiber
DDA	differential difference amplifier
DDJ	data-dependent jitter
DEC	decision circuit
DFB	distributed feedback (laser)
DFE	decision-feedback equalizer
DGD	differential group delay
DHBT	double heterojunction bipolar transistor
DJ	deterministic jitter
DLL	delay-locked loop
DMT	discrete multitone (modulation)
DMUX	demultiplexer
DOP	degree of polarization
DP-QPSK	dual-polarization quadrature phase-shift keying
DPSK	differential phase-shift keying
DQPSK	differential quadrature phase-shift keying
DSF	dispersion-shifted fiber
DSL	digital subscriber line
DSP	digital signal processor
DUT	device under test
DWDM	dense wavelength division multiplexing
E/O	electrical to optical (converter)
ECL	emitter-coupled logic
EDC	electronic dispersion compensation
EDFA	erbium-doped fiber amplifier
EFEC	enhanced forward error correction
EFM	Ethernet in the first mile

EM	electromagnetic
EPON	Ethernet passive optical network
Er	erbium
ER	extinction ratio
ESCON	enterprise systems connection
ESD	electrostatic discharge
EVM	error vector magnitude
FDDI	fiber distributed data interface
FEC	forward error correction
FET	field-effect transistor
FFE	feedforward equalizer
FIR	finite impulse response (filter)
FITL	fiber in the loop
FM	frequency modulation
FOPA	fiber optical parametric amplifier
FP	Fabry–Perot (laser)
FSAN	full service access network
FSE	fractionally spaced equalizer
FSK	frequency-shift keying
FTTC	fiber to the curb
FTTdp	fiber to the distribution point
FTTH	fiber to the home
FTTP	fiber to the premise
FWHM	full width at half maximum
FWM	four-wave mixing
GaAs	gallium-arsenide
GbE	gigabit Ethernet
GBW	gain-bandwidth (product)
Ge	germanium
GPON	gigabit-capable passive optical network
GRIN	graded index
GVD	group-velocity dispersion
HBT	heterojunction bipolar transistor
HD	harmonic distortion
HEMT	high electron-mobility transistor
HFC	hybrid fiber-coax (network)
HFET	heterostructure field-effect transistor
HPF	high-pass filter
I/V	current versus voltage
IC	integrated circuit
IEEE	institute of electrical and electronics engineers
IF	intermediate frequency
IFWM	intrachannel four-wave mixing

IIN	interferometric intensity noise
IIP3	input-referred 3rd-order intercept point
IM/DD	intensity modulation with direct detection
IMD	intermodulation distortion
InGaAs	indium–gallium-arsenide
InGaAsP	indium–gallium-arsenide–phosphide
InP	indium-phosphide
IP	Internet protocol
	intellectual property
ISDN	integrated services digital network
ISI	intersymbol interference
IXPM	intrachannel cross-phase modulation
JFET	junction field-effect transistor
JTOL	jitter tolerance
LA	limiting amplifier
LAN	local-area network
LD	laser diode
LDPC	low-density parity-check (code)
LED	light-emitting diode
LF	low frequency
LMS	least-mean-square (algorithm)
LO	local oscillator
LOP	loss of power
LOS	loss of signal
LPF	low-pass filter
MA	main amplifier
MAC	medium access control
MAN	metropolitan-area network
MCM	multi-chip module
MES	metal–semiconductor
MESFET	MES + FET
MLSE	maximum likelihood sequence estimation
MMF	multimode fiber
MMIC	monolithic microwave IC
MODFET	modulation-doped field-effect transistor
MOS	metal–oxide–semiconductor
MOSFET	MOS + FET
MPEG	moving picture experts group
MPI	multipath interference
MQW	multiple quantum well
MSE	mean-square error
MSM	metal–semiconductor–metal (photodetector)
MTPR	multitone power ratio

MUX	multiplexer
NA	numerical aperture
Nd	neodymium
NECG	net electrical coding gain
NF	noise figure
NIC	negative impedance converter
NRZ	non-return-to-zero (modulation)
NRZ1	non-return-to-zero change-on-ones (modulation)
NTSC	national television systems committee
NWA	network analyzer
NZ-DSF	nonzero dispersion-shifted fiber
O/E	optical to electrical (converter)
OA	optical amplifier
OADM	optical add-drop multiplexer
OC	optical carrier
OEIC	optoelectronic integrated circuit
OFDM	orthogonal frequency division multiplexing
OLT	optical line termination
OMA	optical modulation amplitude
ONU	optical network unit
OOK	on–off keying
OPA	optical parametric amplifier
OSI	open systems interconnection
OSNR	optical signal-to-noise ratio
OTDM	optical time-division multiplexing
OTN	optical transport network
OTU	optical transport unit
OXC	optical cross connect
P2MP	point-to-multipoint (network)
P2P	point-to-point (connection)
PAM	pulse amplitude modulation
PAR	peak-to-average ratio
PC	polarization controller
PCS	physical coding sublayer (Ethernet sublayer)
PD	photodetector (or photodiode)
PDF	probability density function
PDM	polarization division multiplexing
PHEMT	pseudomorphic high electron-mobility transistor
PHFET	pseudomorphic heterostructure field-effect transistor
PHY	physical layer (OSI model)
PIC	photonic integrated circuit
PJ	periodic jitter
PLL	phase-locked loop

PM	phase modulation
PMA	physical medium attachment (Ethernet sublayer)
PMD	polarization-mode dispersion
	physical medium dependent (Ethernet sublayer)
PMF	polarization-maintaining fiber
POF	plastic optical fiber
PON	passive optical network
POTS	plain old telephone service
PP	power penalty
	peak-to-peak (value)
PRBS	pseudorandom bit sequence
PSD	power spectral density
PSK	phase-shift keying
PSP	principal state of polarization
PSRR	power-supply rejection ratio
PSTN	public switched telephone network
QAM	quadrature amplitude modulation
QPSK	quadrature phase-shift keying
RCE	resonant-cavity enhanced (photodetector)
RF	radio frequency
RFC	radio-frequency choke
RGC	regulated cascode
RIN	relative intensity noise
RJ	random jitter
RMS	root-mean-square (value)
RN	remote node
ROSA	receiver optical subassembly
RS	Reed–Solomon code
RSSI	received signal strength indicator
RX	receiver
RXEQ	receive equalizer
RZ	return-to-zero (modulation)
SAM	separate absorption and multiplication (APD)
SBR	single-bit response
SBS	stimulated Brillouin scattering
SCFL	source-coupled FET logic
SCM	subcarrier multiplexing
SCTE	society of cable telecommunications engineers
SDH	synchronous digital hierarchy
SDM	space division multiplexing
SDV	switched digital video
SerDes	serializer/deserializer
SFF	small form-factor (module)

SFI	SerDes framer interface
SFP	small form-factor pluggable (module)
Si	silicon
SiGe	silicon–germanium
SiO_2	silicon oxide
SJ	sinusoidal jitter
SMF	single-mode fiber
SML	spatially modulated light (detector)
SNR	signal-to-noise ratio
SOA	semiconductor optical amplifier
SOI	silicon on insulator
SONET	synchronous optical network
SOP	state of polarization
SPICE	simulation program with integrated circuit emphasis
SPM	self-phase modulation
SRS	stimulated Raman scattering
SSC	spread spectrum clock
STM	synchronous transport module
STS	synchronous transport signal
TAS	transadmittance stage
TCM	time compression multiplexing
TCP	transmission control protocol
TDD	time-division duplexing
TDM	time-division multiplexing
TDMA	time-division multiple access
TEGFET	two-dimensional electron-gas field-effect transistor
THD	total harmonic distortion
Ti	titanium
TIA	transimpedance amplifier
	time interval analyzer
TIE	time interval error
TIS	transimpedance stage
TJ	total jitter
TOSA	transmitter optical subassembly
TPC	turbo product code
TQFP	thin quad flat pack
TTL	transistor–transistor logic
TV	television
TWA	traveling-wave amplifier
TWPD	traveling-wave photodetector
TX	transmitter
TXEQ	transmit equalizer
TZA	transimpedance amplifier

UI	unit interval
UTC	uni-traveling carrier (photodetector)
VCSEL	vertical-cavity surface-emitting laser
VGA	variable-gain amplifier
VLSI	very large-scale integration
VNA	vector network analyzer
VPD	vertically illuminated photodetector
VSB	vestigial sideband
WAN	wide-area network
WDM	wavelength division multiplexing
WGPD	waveguide photodetector
XOR	exclusive-or (gate)
XPM	cross-phase modulation
XT	crosstalk
ZF	zero-forcing (algorithm)

Index

$1/f$ noise 236, 238
16-PAM 408
256-QAM 408
4-PAM 6, 30, 403
4B/5B code 11
64-QAM 408
64B/66B code 12
8B/10B code 12, 119, 399

a

ABCD parameter 185, 197, 265
absolute jitter 431
absorption layer 43, 45, 46, 48, 49, 58, 59, 61, 63, 82, 83
absorption length 44, 45, 82
accumulated jitter 434
active feedback 347
active star network 15
active-feedback TIA 252, 346
adaptive decision-point control 472
adaptive equalizer 451
adaptive sampling-instant control 472
adaptive threshold control 469
adaptive transimpedance 188, 292, 297
  of burst-mode TIA 354
adjoint network 350
aggressor channel 197
amplified spontaneous emission 67, 69, 74, 75
amplitude modulation 7, 406
amplitude modulation with vestigial sideband 9, 397, 406, 441
amplitude spectral density 538
amplitude-shift keying 7
analog receiver 9
analog transmitter 9
analog-to-digital converter 4, 460, 480
antireflection coating 44
artificial transmission lines 326, 327
asynchronous transfer mode 18
asynchronous transfer mode PON 18
AT&T Bell laboratories 66
automatic gain control 60, 69, 292
automatic gain control amplifier 1
automatic offset control 290, 356
automatic repeat request 475
automatic threshold control 356
autotransformer 361
available power gain 266
avalanche build-up time 61, 62
avalanche gain 59, 60, 62, 64, 65, 125, 128
avalanche noise 63, 65
avalanche photodetector 43, 57–60, 63–65, 69, 74, 96, 108, 125–129, 133, 151, 153, 253, 469

Analysis and Design of Transimpedance Amplifiers for Optical Receivers, First Edition. Eduard Säckinger.

avalanche region 61
average value 186, 537
averaged input-referred noise current density 194, 202, 241

b

balanced detection 222
balanced detector 87, 88, 90, 128, 284
balun 361, 363
bandgap 45
bandwidth
 noise 137, 239
 of 4-PAM signal 404
 of analog TV signal 407
 of avalanche photodetector 61
 of CMOS photodiode 83–85
 of communication signals 397
 of decision circuit 136
 of digital TV signal 410
 of EDFA 68
 of fiber 25, 26, 37
 of high-impedance front-end 209
 of low-impedance front-end 207
 of MSM photodetector 82
 of NRZ receivers 161
 of NRZ signal 399
 of optical preamplifier 68
 of p–i–n photodetector 46
 of Raman amplifier 76
 of receiver 153
 of RZ receivers 163
 of RZ signal 402
 of transimpedance amplifier 188, 211
 of voltage amplifier 214, 280
 optical 66
bandwidth allocation 164
bandwidth headroom 229, 280, 311, 315
bandwidth-efficient modulation 6, 8
base resistance 237
baseband signals 7
baseline wander 11, 156, 422
bathtub curve 419, 426
BERT scan 418, 426
Bessel–Thomson response 137, 214, 488
bias voltage
 of avalanche photodetector 59, 60
 of p–i–n photodetector 46
BiCMOS technology 376
binary phase-shift keying 8, 90
bipolar junction transistor 212, 373
birefringence 30
bit error 109
bit-error rate 113, 417
 and E_b/N_0 119
 and dynamic range 123
 and forward error correction 478
 and jitter 186, 437
 and OSNR 141
 and Personick Q 113, 399, 402, 404
 and sensitivity 120
 and SNR 117, 399, 402, 404
bit-error rate floor 133, 144
bit-error rate plot 132, 144
bit-error rate test set 134, 418
bond-wire inductor 184, 323
booster amplifier 312, 314, 316, 386
bootstrap buffer 351, 352
bootstrapped photodetector 351
Bose–Chaudhuri–Hocquenghem code 475
bottom-level hold 354, 356, 359
bounded uncorrelated jitter 424
broadband PON 18
buried-layer shield 283
burst 13, 17, 353
burst-mode main amplifier 356
burst-mode penalty 358
burst-mode receiver 115, 116, 129, 352, 353, 358, 378, 386
burst-mode transimpedance amplifier 352, 378, 386

burst-mode transmission 13, 17
Butterworth response 137, 156, 214, 321, 488

C

capacitive-feedback TIA 252, 343
carrier-suppressed return-to-zero 7, 35
carrier-to-noise ratio 118, 360, 408, 410
central office 14, 17, 353
chain parameter 185, 197, 265
channel bits 10
channel capacity theorem 475
channel noise 232, 233
channel noise factor 232
characteristic impedance 326
chatter 358
chi-square distribution 71, 88, 128
chirp 38
chirped return-to-zero 7
chromatic dispersion 26, 28–30, 35, 37, 38
chromatic dispersion compensation 29
circuit noise 108, 131
clad fiber 26
clock and data recovery circuit 2, 208, 402, 436
clock multiplication unit 1
clock-recovery circuit 109, 168, 429, 462
closed-loop frequency response 213, 227, 297, 383
code rate 119, 477
coding gain 478, 479
coherent detection 4, 35
coherent detection with phase and polarization diversity 90
coherent detector 90, 94, 108
coherent receiver 30, 32, 53, 90, 377
color grading 415
common-base feedforward TIA 301
common-base input stage 301
common-emitter input stage 305
common-gate feedforward TIA 302, 384
common-gate input stage 301, 383
common-source input stage 305
community-antenna television 8, 58, 360, 407, 410
complementary metal–oxide–semiconductor 82, 84, 373
composite distortions 446
composite second order 360, 447
composite triple beat 360, 447
confocal waveguide 24
continuous wave 1
continuous-mode receiver 115, 116
continuous-mode transmission 12
copackaging of receiver 77, 323, 363
coplanar waveguides 283
Corning glass works 24
correlated noise current 262
correlation admittance 262
correlation coefficient 233, 261
cost function 456
coupling factor of optocoupler 345
critically damped response 488
cross-phase modulation 32
crosstalk 32–34, 156, 197–199
current amplifier 343, 349
current attenuator 298
current buffer 301, 320
current conveyor 320
current conveyor TIAs 320
current follower 301
current-mode circuit 349
current-mode transimpedance amplifier 349
cutoff-frequency spacing 310, 311, 312, 315
cycle jitter 432
cycle-to-cycle jitter 432

d

dark current 45, 56, 57, 64, 65, 74, 151
data-dependent jitter 189, 190, 421, 422
DC balance 10, 12, 117
DC input current control 289, 292, 344, 345, 362
DC overload control 291, 377
DC wander 11
decision circuit 109, 110, 112, 119, 136, 165, 167, 431, 455, 469, 478, 479
decision feedback 452
decision point 421, 469
decision threshold control 469
decision threshold offset 123, 149
decision threshold voltage 109, 112–114, 134, 149, 353, 421, 469
decision-directed mode 457
decision-feedback equalizer 455
decision-point control 469, 472
decision-point steering 472
decoupling capacitors 283
delay interferometer 87, 88
demultiplexer 2
dense wavelength division multiplexing 26
depletion layer 82–85
describing function 443
detected optical power 122
detector model 108
detector noise 108
deterministic jitter 191, 417, 418, 424
dielectric stack 83
differential binary phase-shift keying 88
differential group delay 30
differential phase-shift keying 8, 86, 147
differential quadrature phase-shift keying 8, 88, 90, 285
differential TIA 85, 87, 93, 182, 192, 238, 284, 285, 308, 356, 376, 380, 381, 383, 389
diffusion current 46, 47, 61
digital signal processor 4, 10, 30, 91, 108, 460
digital-to-analog converters 4
direct bandgap 82
direct detection 35
discrete multitone 408, 441, 446
dispersion
  chromatic 28, 35
  group-velocity 28
  modal 27
  polarization-mode 30
dispersion compensating fiber 29
dispersion compensation 29, 30
dispersion parameter 28
dispersion-shifted fiber 29
distortion cancellation 363
distributed amplifier 50, 325
distributed amplifier front-end 326
distributed gain 76, 77
dithering 458
double-core waveguide photodetector 49
double-pass scheme 44
downstream direction 17
drift field 43, 46, 57, 82
drift of baseline 11
driving point impedance 181
dual Dirac model 425
dual-polarization quadrature phase-shift keying 90
dummy TIA 287, 382
duobinary 7
duty-cycle distortion jitter 423, 471
dynamic range
  electrical 124
  of p–i–n photodetector 57
  of receiver 61, 123, 133
  of transimpedance amplifier 186, 294, 297
  optical 124

e

E_b/N_0 118, 120, 400, 405, 479
edge-coupled APD 63
edge-coupled photodetector 46, 49, 58
effective noise figure 77
electrical dB 118
electrical receiver sensitivity 120
electron–hole pair 82
electronic dispersion compensation 28, 30, 31
envelope delay 189
equalizer 30, 31, 83, 86, 170, 209, 452
  adaptive 451
  decision-feedback 455
  feedforward 453
  fractionally spaced 454
  postcursor 456
  precursor 456
  synchronous 454
equivalent input noise current 191
equivalent-time sampling oscilloscope 415, 431
erbium ion 67–69, 75
erbium-doped fiber 66, 67, 69
erbium-doped fiber amplifier 25, 66, 69, 72
error propagation 456, 480
error vector magnitude 449
Ethernet 5, 6, 13, 14
Ethernet PON 18
evanescently coupled waveguide photodetector 49
excess bandwidth 169
excess noise factor 64, 233
exhaustive search 457
external quantum efficiency 45
extinction ratio 56, 123, 146, 152, 173, 174
eye diagram 157, 166, 413, 421, 422, 492
eye margin 418
eye monitor 456, 469
eye opening 162
eye pattern 413

f

fast Ethernet 5, 11
federal communications commission 408, 448
feeder fiber 17
feedforward equalizer 453
feedforward TIA 301, 384
fiber attenuation 23
fiber bandwidth 25, 26, 37
fiber distributed data interface 5, 12, 15
fiber heater 88
fiber loss 23
fiber optical parametric amplifier 25, 75
fiber-to-the-curb system 17
fiber-to-the-home system 17, 26
fibre channel 12, 15
finite impulse response filter 453
flicker noise 139, 236
flip-chip bonding 78
forward error correction 14, 123, 143, 409, 457, 469, 475
four-parameter noise model 261
four-wave mixing 32
fractionally spaced equalizer 454
frequency modulation 7, 406
Friis noise formula 266

g

gain
  of avalanche photodetector 59, 125
  of coding 478
  of EDFA 69
  of optical amplifier 68, 125
  of Raman amplifier 76
  of voltage amplifier 211, 279
gain compression 57, 442, 443
gain switching 355
gain-bandwidth independence 349
gain-bandwidth product 62

gallium-arsenide technology 387, 389
gas lens 24
gate noise factor 233
Gaussian distribution 55, 71, 112, 113, 128, 133, 424, 427
Gaussian jitter 428
Gaussian noise 112, 120, 162, 475
Gaussian pulse 36
Gaussian spectrum 36
germanium 78
gigabit Ethernet 12, 14, 15
gigabit-capable PON 18
golden PLL 429
graded-index multimode fiber 27
gradient descent 458, 459
Gray code 6
gross electrical coding gain 478
group delay 490
group-delay variation 156, 189, 492
group-velocity dispersion 28

h

half sections 330
hard-decision decoder 479
harmonic distortion 443, 445, 446
heterodyne receiver 53, 54, 91
heterojunction bipolar transistor 373
heterostructure field-effect transistor 373
high-impedance front-end 207, 252, 344
high-probability jitter 428
higher-order modulation format 6, 8
homodyne receiver 90
horizontal eye closure 417, 456
horizontal eye margin 418, 419
horizontal eye opening 417, 454, 456
hybrid fiber-coax 8, 58, 360, 407, 410
hysteresis comparator 355

i

impact ionization 59, 60
induced gate noise 232, 233
information bits 10
information capacity theorem 475
input overload current 123, 186, 212, 292, 294, 298, 300
input-referral function 234, 237, 238, 256, 259, 265, 266
input-referred 3rd-order intercept point 445
input-referred mean-square noise current 193, 239, 243
input-referred noise current 191, 192, 230, 261, 305, 311
input-referred noise current density 193
input-referred noise current PSD 192, 195, 196, 231, 235, 238, 239, 241
input-referred rms noise current 193, 195, 238–240, 242, 252, 305
integrate and dump 167, 210, 345
integrated photodetector 77, 253
integrating front-end 210, 343
intensity modulation 7, 9, 35
interferometer 86
interleaver 476
intermediate frequency 91
intermodulation distortion 445, 447
internal quantum efficiency 45
Internet 14, 18
intersymbol interference 153, 154, 166, 169, 413, 451
intersymbol-interference induced jitter 423
intrachannel cross-phase modulation 33
intrachannel four-wave mixing 33
intradyne receiver 91
intrinsic photodetector 181, 185, 192, 197
ionization-coefficient ratio 62, 64, 80
ISI canceler 451

j

jitter 182, 186, 413, 495
- absolute 431
- accumulated 434
- bounded-uncorrelated 424
- cycle 432
- cycle-to-cycle 432
- data-dependent 189, 421, 422
- deterministic 417, 424
- duty-cycle distortion 423
- Gaussian 428
- high-probability 428
- intersymbol-interference-induced 423
- long-term 434
- narrowband 428
- period 432
- periodic 424
- random 417, 421, 424
- sampling 165, 167, 436
- short-term 434
- sinusoidal 424
- total 426, 427, 437

jitter analyzer 431
jitter generation 191
jitter histogram 421
jitter PSD 429, 431
jitter tolerance 436, 437
jitter transfer 429
jitter trend 429

k

Kerr effect 32
Kirchhoff's current law 87, 93, 363, 461

l

large-signal gain 442, 443
laser diode 1
laser driver 1
lattice mismatch 78
least-mean-square algorithm 458
level separation mismatch ratio 188
light-emitting diode 345
limiting amplifier 1, 109, 188, 422
limiting transimpedance amplifier 279
line code 11, 353
linear channel 108–111, 119, 441, 451
linear regime 294, 297
link budget 124
local area network 14
local oscillator 53, 90
logarithmic amplifier 299, 354
long-haul transmission 6, 14, 23, 25, 27, 35, 77, 139
long-term jitter 434
look ahead 462
loop unrolling 462
loss
- of fiber 23
- of transmission line 330

loss of signal 291
low-density parity-check code 477, 479
low-frequency cutoff 156, 183, 417, 422
low-impedance front-end 207, 208, 242, 325, 360
low-noise amplifier 222

m

m-derived half sections 330
main amplifier 2, 107, 183, 279
mark density 11, 12
matched filter 119, 147, 165, 166, 400, 405
maximum input current for linear operation 187, 199, 292
maximum likelihood sequence estimator 166, 451
mean-square error 457, 458
mean-square value 537
metal–semiconductor field-effect transistor 80, 373

metal–semiconductor–metal photodetector 79
microstrip lines 283
microwave photonics 360
minimum noise figure 196, 263
minority carrier 82, 83
modal dispersion 27
modified matched filter 165
modified zero-forcing algorithm 462
modulation code 11
modulation index 131
modulator 1
modulator driver 1
molecular vibration 34, 76
multi-chip module 78
multiband OFDM 10
multichannel TIA 197
multimode fiber 27, 452
multipath channel 454
multipath interference noise 70
multiple quantum well 62
multiplexer 1, 462
multiplication factor 59
multiplication region 58, 59, 63
multiplying digital-to-analog converter 461
multistage amplifier 228, 382
multitone power ratio 448

n

national television systems committee 406
negative capacitance 352
net electrical coding gain 478
network analyzer 183, 184
noise
  and bit-error rate 111
  non-Gaussian 128
  of analog receiver 131
  of avalanche photodetector 63
  of balanced detector 88, 93
  of coherent detector 93
  of distributed amplifier front-end 329
  of optical amplifier 69
  of p–i–n photodetector 54
  of receiver 110
  of transimpedance amplifier 191, 230
noise bandwidth 119, 120, 126, 127, 137, 139, 167, 170, 194, 239, 305
noise canceling 252, 309
noise corrupted decision threshold 115, 129, 358
noise enhancement 454
noise equivalent bandwidth 137
noise figure 72–76, 97, 126, 142, 196, 257, 263, 266, 267
noise matching 252, 254, 257, 322, 361
noise optimization 242
noise parameter 196, 260
noise resistance 196, 263
noise-loaded BER 145
non-Gaussian noise 55, 65, 71, 88, 128, 129
non-return-to-zero 5, 25, 37, 112, 117, 119, 153, 161, 168, 186, 397, 413
non-return-to-zero change-on-ones 5
nondominant pole 227
nonlinear TIA 299, 354
nonlinearity 441
  and composite second-order distortion 447
  and composite triple-beat distortion 447
  and error vector magnitude 449
  and gain compression 442, 443
  and harmonic distortion 443
  and intermodulation distortion 445
  and multitone power ratio 448
  of decision circuit 109
  of decision-feedback equalizer 456

of fiber 32
of intensity modulation with direct detection 35
nonzero dispersion-shifted fiber 35
Nyquist bandwidth 169

o

offset control 289, 356, 381
offset voltage 289, 290
Ogawa's noise factor 233, 256
on–off keying 7
open-loop frequency response 213, 215, 227, 383
open-loop pole 214, 269, 295, 297, 301
open-loop TIA 301
optical amplifier 68, 72, 140, 208, 469, 295
erbium-doped fiber amplifier 25, 66
fiber optical parametric amplifier 25, 75
Raman amplifier 25, 75
semiconductor optical amplifier 66
optical carrier 14, 123
optical dB 118
optical duobinary 7, 30
optical fiber 23, 25, 26
optical field 32
optical filter 66, 68, 70
optical in-line amplifier 25, 77, 139, 141
optical isolator 66, 67
optical modulation amplitude 122
optical noise figure 74
optical overload power 123, 133, 294
optical receiver sensitivity 121, 125, 126, 140, 294
optical signal-to-noise ratio 71, 120, 141, 142
optical signal-to-noise ratio budget 142
optical signal-to-noise ratio limit 144
optical signal-to-noise ratio penalty 143
optical transport network 14
optical-feedback TIA 252, 345
optically preamplified p–i–n detector 43, 63, 66, 69, 73, 74, 108, 125–127, 129, 253
optimum APD gain 65, 128
optimum decision threshold 469
optimum receiver bandwidth 153
optimum receiver response 165, 166
optimum slice level 471
optimum source admittance 196, 257, 264
optoelectronic integrated circuit 78
optoelectronic repeater 24
orthogonal frequency division multiplexing 10, 408, 441, 446
oscilloscope 195, 414, 427, 431, 432, 434
output buffer 279, 384, 388
overhead section 13
overload limit 123, 133, 143, 294, 422
overshoot 359, 492
overshoot velocity 58

p

p–i–n FET 78
p–i–n photodetector 43, 47, 51, 54, 56, 73, 78, 87, 108, 125, 126, 253, 345, 360
package parasitics 46, 48, 61, 192, 323
parallel optical interconnect 197
passband signals 7
passive optical network 17, 18, 353, 360
passive star network 15
peak detector 290, 293, 354, 357–359
peak-to-average ratio 408, 448
peak-to-peak value 186, 537
peaking 489

penetration depth 44
period jitter 432, 434
periodic jitter 424
Personick integral 138, 139
Personick Q 113, 143, 399, 401, 405, 427
phantom zero 227
phase distortion 189
phase linearity 156, 189, 422, 490, 492
phase margin 418
phase modulation 7, 86
phase noise 431, 435
phase synchronization 327, 330
phase-insensitive amplifier 75
phase-locked loop 90, 414, 418, 429, 431, 434
phase-sensitive amplifier 75
phasor 537
phonon 32, 34, 76
photodetector 1, 43, 183, 188, 191, 195
photodetector capacitance 184, 188, 192, 211, 243, 244, 249, 256, 301
photodetector intrinsic conductance 197
photodetector simulator 183
photonic integrated circuit 78, 91
photonic microwave link 362
physical coding sublayer 4
physical layer 4
physical medium attachment 4
physical medium dependent 4
piezoelectric fiber stretcher 88
planarity 78
plastic optical fiber 24
point-to-multipoint network 15, 17
point-to-point connection 14
Poisson distribution 55, 135
polarization controller 31
polarization diversity 32
polarization division multiplexing 8
polarization filter 70
polarization maintaining fiber 31
polarization-mode dispersion 30, 451
polarization-mode dispersion compensation 31
polarization-mode dispersion parameter 31
pole quality factor 213, 488
pole spacing 303, 306, 310, 312
population inversion 69, 75
post amplifier 2, 248, 279
postcursor equalizer 456
postcursor ISI 155, 453, 454
power budget 124
power penalty 148, 153
  and crosstalk 198
  and decision threshold offset 149
  and detector dark current 151
  and extinction ratio 152
  and fiber alignment 27
  and finite receiver bandwidth 158
  and intersymbol interference 158
  and polarization-mode dispersion 31
  and pulse spreading 37
  and signal distortion 158
power spectral density 397, 538
  of 4-PAM signal 404
  of analog TV signal 407
  of digital TV signal 410
  of NRZ signal 398
  of RZ signal 402
power splitter/combiner 17
preamble 13, 115, 353
preamble bit 116
precursor equalizer 456
precursor ISI 155, 453, 454
primary dark current 65
principle state of polarization 31
programmable-gain amplifier 461
pseudo bit-error rate 469, 471, 472
pseudo error 470, 472
pseudo-differential amplifier 287

pseudo-differential TIA 287, 288, 308, 363, 382
pseudorandom bit sequence 11, 134, 399, 414, 417
pulse amplitude modulation 6, 397, 403
pulse spreading 29, 35–38
pulse-width distortion 182, 186, 353, 354
pump laser 66, 69, 76
pump power 69, 75, 76
push–pull amplifier 361

q

quadrature amplitude modulation 9, 397, 408, 441
quadrature phase-shift keying 8, 30, 90, 285
quantum efficiency 44–46, 51, 59, 63, 73, 81–83, 122, 253

r

radio frequency 9
radio over fiber 360
raised-cosine filtering 139, 166, 168, 170, 408
Raman amplifier 25, 75
Raman effect 75
Raman pump 76, 77
random jitter 191, 417, 421, 424
Rayleigh distribution 72
real-time sampling oscilloscope 415, 431
receive equalizer 2, 451
received signal strength indicator 291, 358
receiver bandwidth 153, 157, 161, 164, 189
receiver model 107, 109
receiver optical subassembly 3, 78
rectangular filter 165–167
Reed–Solomon code 409, 475, 476
reflection 156, 422
regenerator 139
regulated cascode 312, 324, 348, 384, 385
relative intensity noise 92, 131, 361
remote node 9, 17
repeater 25
replica TIA 287, 288
required OSNR 141, 143, 144
reset signal 354, 355
resonant front-end 257
responsivity 51, 52, 57, 59, 69, 81, 83–85, 121, 125, 187, 253
return-to-zero 5, 121, 147, 163, 397, 401
return-to-zero differential phase-shift keying 8, 35, 86
return-to-zero differential quadrature phase-shift keying 8
Rician distribution 72
ring network 15
root spectral density 538
root-mean-square value 537
RS(128, 122) 409
RS(255, 239) 476
run length 10, 11, 417

s

S parameter 184
sampling instant 109, 134, 413, 421, 436, 454, 469
sampling-instant control 471
sampling jitter 165, 167, 436
sampling offset 167, 436
sampling oscilloscope 415, 430, 431
saturation current (of photodiode) 57
saturation of bipolar junction transistor 212
Schottky diode 79–81
scrambler 11
self-coherent 90
self-coherent detector 94
self-phase modulation 32, 38
semiconductor optical amplifier 66

sensitivity 120
and bit rate 252
and detector capacitance 254
and dynamic range 123
and noise 120, 239
and optical signal-to-noise ratio 141
and receiver bandwidth 153
and reference bit-error rate 123
electrical 120
of analog receiver 131
of APD receiver 125, 126
of coherent receiver 129
of p–i–n receiver 125, 126
of receiver with optical preamplifier 125, 126
of transimpedance amplifier 194
optical 121
sensitivity limit 123, 133, 143, 294
serializer/deserializer 3
short-term jitter 434
shot noise 54, 64, 70, 72, 97, 126, 127, 131, 151, 231, 237
shunt feedback 210, 287
shunt peaking 324
shunt-feedback TIA 210, 212, 217, 218, 230, 279, 284, 301, 310
sign–sign least-mean-square algorithm 462
signal–spontaneous beat noise 70, 74, 141, 146
signal–spontaneous beat noise limited noise figure 74
signal-to-noise ratio 71, 113, 117, 118, 120, 360, 397, 475
and 4-PAM signal 404
and NRZ signal 399
and RZ signal 402
silica glass 23, 24, 68, 76
silicon on insulator 82
single-bit response 155
single-mode fiber 27–29, 36, 38
sinusoidal jitter 424, 436
slice-level adjustment 469
SNR per bit 118
society of cable telecommunications engineers 408
soft-decision decoder 479, 481
soliton 38
space division multiplexing 17
spatially modulated light detector 85, 285
spectral efficiency 10, 25
spectrum analyzer 195, 431, 435
speculation 462
spiral inductor 323, 324, 377, 388
spontaneous emission 67
spontaneous–spontaneous beat noise 70, 71, 74, 146
stagger charging 322, 323
steepest descent 458
step response 495
stimulated Brillouin scattering 32
stimulated emission 67
stimulated Raman scattering 32, 33, 76
subcarrier multiplexing 9, 257, 407, 410, 441, 446
substrate coupling 283
superheterodyne radio receiver 53
synchronous digital hierarchy 11, 14, 15
synchronous equalizer 454
synchronous optical network 11, 14, 15, 123

t

T-coil network 325, 330
temperature dependence of APD 60
thermal noise 108
threshold control 356
time compression multiplexing 15
time division duplexing 15
time division multiple access 17

time division multiplexing 18
time interval analyzer 431
time interval error 431
top-level hold 356, 359
total dark current 65
total harmonic distortion 187, 444
total input-referred noise current 193
total jitter 418, 426, 427, 437
total responsivity 59
training mode 457
transceiver 2, 5, 15, 477
transfer impedance xi, 181
transformer 361
transimpedance 164, 181–184, 188, 193, 207–209, 211, 213
transimpedance amplifier 1, 107
  active-feedback 252, 346
  adaptive 292, 373
  analog 360
  and bootstrapped photodetector 351
  and DC input current control 289
  and DC overload control 291, 377
  and gain control 292, 354, 373, 378, 386, 389
  and inductive broadbanding 320, 376
  and noise canceling 252, 309
  and noise matching 252, 254
  and noise optimization 242
  and offset control 289, 380, 381
  and post amplifier 279
  burst-mode 352, 378, 386
  capacitive-feedback 252, 343
  common-base 301
  common-gate 301, 383, 384
  current-mode 349
  differential 284, 308, 376, 377, 380, 389
  distributed-amplifier 325
  feedforward 301
  optical-feedback 252, 345
  regulated-cascode 312, 385
  shunt-feedback 210, 373, 376, 377, 380–382, 387, 388
  specifications 181
transimpedance limit 216–218, 224, 228–230, 243, 247, 280, 304, 310, 350
transit time 46, 48, 61, 62, 81
transition density 10, 11
translinear cross-quad 319
transmission line 49, 282
transmit equalizer 1
transmitter optical subassembly 3
transponder 3
transresistance 183
transversal filter 453
traveling-wave amplifier 330
traveling-wave photodetector 48–50
triple beat 446
tuned front-end 257
turbo code 477, 479
turbo product code 477
typical values 18

u

uncertainty principle 75
uncorrelated noise current 262
undershoot 359
uni-traveling-carrier photodiode 58
upstream direction 17

v

variable-gain amplifier 461
vector signal analyzer 449
vector signal generator 449
vertical bathtub curve 161
vertical eye closure 417, 456
vertical eye margin 162, 418, 419
vertical eye opening 154, 159, 198, 417, 454, 456
vertically illuminated photodetector 44, 46, 48, 49, 58, 63

vestigial sideband filter 406
victim channel 197
video overlay 360
Viterbi decoder 451
voltage swing
  at TIA input 212, 286
  at TIA output 212
  of differential circuit 284
  of high-impedance front-end 209

W

water peak 23
waterfall curve 477
waveguide photodetector 49
wavelength division multiplexing 17, 143, 163
wavelength division multiplexing PON 18
weight perturbation 458
wideband jitter 428
wire bonding 78

X

XOR gate 471

Z

zero-forcing algorithm 458

Printed and bound by CPI Group (UK) Ltd, Croydon, CR0 4YY
17/09/2026
14964312-0008